Avian breeding cycles

Male Willow Tit *Parus montanus* courtship feeding the female at the entrance to their nest hole. (Photograph taken in Surrey, England.)

Avian breeding cycles

R. K. MURTON

AND

N. J. WESTWOOD

Institute of Terrestrial Ecology
Natural Environment Research Council
Monks Wood Experimental Station
Huntingdon

CLARENDON PRESS   OXFORD

1977

Oxford University Press, Walton Street, Oxford OX2 6DP

OXFORD LONDON GLASGOW NEW YORK
TORONTO MELBOURNE WELLINGTON CAPE TOWN
IBADAN NAIROBI DAR ES SALAAM LUSAKA ADDIS ABABA
KUALA LUMPUR SINGAPORE JAKARTA HONG KONG TOKYO
DELHI BOMBAY CALCUTTA MADRAS KARACHI

British Library Cataloguing in Publication Data

Murton, Ronald Keir
  Avian breeding cycles.
  1. Birds — Reproduction
  I. Title II. Westwood, N J
  598.2'1'66   QL698
  ISBN 0−19−857357−X

Printed in Great Britain
by the Pitman Press, Bath

Preface

We aim in this book to stimulate research and so have addressed ourselves to post-graduate workers and senior undergraduates. Notwithstanding this objective, we hope that the book will appeal to ornithologists in general. Our aims have to some extent delimited the style. It would certainly have been easier to draw general broad conclusions, which may as a result have been oversimplifications, than to summarize other workers' results accurately and we hope that we have adopted the most useful course. We have not neglected the early literature, for many of the experiments and observations made in the past remain valid, even if they have been repeated with the greater sophistication that comes with hindsight. Some topics are particularly difficult to present in a readable manner and we are very conscious of the technical jargon referring to the so-called 'pituitary gland'. There are no simple names for the structural components of the neuro-hypophysial system, yet knowledge of its structure is essential if the problems of understanding avian photoperiodism are to be resolved. We hope that the liberal use of text-figures will help the reader.

When Lofts and Murton (1968) reviewed photoperiodic and physiological adaptations regulating avian breeding cycles they pleaded for a greater synthesis of ecological and physiological research, rather than a multiplicity of photoperiodic manipulations of a few north-temperate species; at the same time they suggested that ecologists had neglected the field consequences that may follow from a primarily physiological adaptation. It is still true, in spite of the increasing use of such terms as eco-physiology in the biological vocabulary, that research on avian biology is too polarized between ecology or physiology. Our thesis is that this represents an out-moded approach and that the exciting advances will be made only by those prepared to adopt a much more integrated and multidisciplinary view-point. The text concentrates on avian breeding cycles but the viewpoint is applicable to other aspects of avian biology. This is particularly true for some of the topics which today are important in an applied context.

Environmental pollution in general and, in particular, the subtle effects of organochlorine insecticides on the reproductive performance of individuals and populations of raptorial birds, has highlighted our lack of knowledge of avian eco-physiology. Often one observes a marked inter-specific variation in response to a given pollutant, presumably partly reflecting differences in biochemical make-up of the species and partly ecological differences in exposure risk. Why are raptorial species so sensitive to organo-chlorine compounds? Dieldrin has been responsible for the direct death of adult birds, DDT and its metabolites cause egg-shell thinning and a reduced breeding success, while PCBs (polychlorinated biphenyls) cause egg infertility as well as other harmful effects. Each organohalide compound interferes with a particular range of physiological mechanisms. Normally population processes operate on wild animals in ways which allow a decrease in adult survival rate to be compensated for by a decrease in the mortality rate of the egg and nestling. However, when dieldrin and related compounds caused adult losses among raptorial birds, other chemicals were also in use which caused reductions in breeding success and productivity. Attacking an animal from different directions in this way is the object of integrated methods of pest control and it is not surprising that populations of many birds of prey declined drastically when the more persistent and toxic insecticides were introduced.

A multidisciplinary approach requires knowledge of a wide spectrum of subjects and is attended by the risk of becoming a 'jack of all trades and master of none'; we were conscious of this pitfall in writing this book. But there can be no real excuse for not trying. For example, much time would have been saved and research effort spared if those of us who were interested in the photoperiodic control of avian breeding seasons had appreciated the importance of a circadian rhythm of light sensitivity in animal photoperiodism. The relevant research was published as early as the late 1950s but it referred to plants and invertebrates. It required the pioneer efforts of Professor W. M. Hamner in the mid-1960s to show that birds measure daylength by essentially the same mechanism as that employed by invertebrates and plants. Now we realize that a multiplicity of physiological functions exhibit an endogenous circadian rhythmicity so that under appropriate experimental conditions a free-running periodicity close to, but not exactly equal to, a 24-hour duration can be demonstrated. These rhythms are entrained by the daily cycle of night and day, thereby assuming an exact 24-hour periodicity. However, entrainment also determines the phasing of rhythmic functions relative to the light–dark cycle and relative to each other. There is a need to measure how phasing effects alter with the seasonal daylength cycle and how species have evolved appropriate adaptations to suit their normal latitude range.

At present interest centres on the manifestation of circannual rhythms of body function, for example, of seasonal body moult, fat deposition, the waxing and waning of the reproductive system. These circannual rhythms appear to be entrained by the amplitude produced by the seasonal variations in daylength throughout the year. It is not yet fully established whether these circannual rhythms are controlled by endogenous circannual oscillators, in the way that circadian rhythms appear to be controlled by circadian oscillators. We incline to the view that they are compounded from a diversity of circadian rhythms which alter in phasing throughout the year, thereby producing new rhythms. However, the essential point is to define the manner in which rhythms are entrained. In the tropics—where twilight effects can be very important—there are many species whose rhythms are entrained by the daylength oscillations of very low amplitude, but entrainment is weak so that different individuals within a population are not well synchronized, one with another; special adaptations are needed to ensure synchrony. In other species entrainment may not occur so that rhythmic functions run free to result in breeding activity intervals of less that 12 months. The assertion of some field workers that the slight change in daylength near the equator cannot affect the timing of breeding cycles therefore results from a misunderstanding of the mechanisms involved.

The above does not deny that other factors are also important in timing seasonal events. Males, whether of tropical or temperate species, are brought into full reproductive condition by appropriate responses to the light cycles under which they have evolved adaptive responses. On the other hand, in females the ovarian follicles can be stimulated only to a point at which limited quantities of yolk are added and some other factor is needed for the final expression of ovarian development. (Lofts and Murton 1968). The final growth phase during which much yolk is added, resulting in a ten-fold increase in weight prior to ovulation, occurs over the 6–12-day period of pre-incubation courtship from the male. It now seems from the work of Jones and Ward (1976) that a labile protein reserve is accumulated in the flight muscle of the female before egg-laying—in *Quelea* the reserve increases by 80% in females prior to breeding but by only 14 per cent in males—and this is rapidly transferred to the developing follicles during the period of courtship; the amount available could have a direct effect in determining clutch size. The achievement of an appropriate nutritional state obviously provides the final proximate regulation for the onset of egg-laying, that is, the critical day-to-day adjustment within the broad time spectrum of reproductive activity which is dependent on the photoperiod and on male courtship. But it would be wrong to conclude with Jones and Ward that no other environmental releasers are required for birds to breed at the appropriate time of year.

The extent to which photoperiodic responses serve to control breeding, or other seasonally occurring events in the bird's calendar, in a permissive or obligatory fashion broadly changes with increase in latitude and the intensity of the selective disadvantages of any imprecision. In short, the seasonal cycle of a wild bird implicates an extremely complex interaction between endogenous, autonomous components, which are expressed as self-sustaining diurnal and annual rhythms, and environmental stimuli. A considerable species-specific variability in adaptive response is discerned, although the degree of adaptiveness is necessarily open to subjective assessment. One of the points we wish to stress is that evolutionary changes are constrained by the complexity of the physiological mechanisms involved so that closely related species can be shown to have photoperiodic responses which conform to a common pattern. The food and feeding ecology of ducks of the genus *Anas* differ quite considerably between species yet all have reproductive cycles which are determined by a photoperiodic mechanism which is relatively constrained within the genus (Chapter 14). To what extent such physiological restraints inhibit the full ecological exploitation of a niche it is impossible to say, when competition from better adapted phylogenetic lines is absent. It is not easy to identify biological compromise. There are many indications that species may evolve physiological adaptations at a particular latitude that effectively pre-adapt them for niches in a totally different range and we provide some examples referring to montane species and those which have adapted to drastically man-altered environments.

Many people have helped in the preparation of this book. Professors Brian Lofts

and Brian Follett are former associates whose work we have quoted extensively
and they have been generous in allowing us to reproduce figures from their own
papers. Many publishers and editors have kindly granted permission for us to
reproduce text-figures, which to a variable extent we may have altered or incorporated into compound drawings. Our thanks in this respect are due to the following:
the Zoological Society of London for Figs. 1.1, 4.8, 5.8, 7.4c, 7.14, 12.1j; the
National Academy of Sciences for Figs. 1.4, 7.2, 11.7, 11.9, 11.10; Academic Press,
Inc. for Figs. 1.5b, 2.3d,e, 3.7, 3.9, 7.17, 9.8b, 12.7a,b,c, 13.4; the Editor, *Nature*
(Macmillan Journals Ltd.), for Figs. 1.7, 8.11; Professor J. G. Koritke, for Fig. 2.1;
Springer-Verlag, Heidelberg, for Figs. 2.3b,c, 5.9, 5.10, 12.6b; S. Karger AG, Basel
for Fig. 2.3c; *Exerpta Medica* for Fig. 3.2; Duke University Press for Figs. 3.4, 8.14,
8.15, 12.1a,b, 12.9a; the American Association for the Advancement of Science for
Figs. 3.11, 7.15b, 11.14, 12.9b (copyright 1971, 1965, 1967, 1970 respectively);
Academic Press, Inc. (London), Ltd. for Figs. 4.2, 7.7, 7.8, 9.13; Centre National de
la Recherche Scientifique for Fig. 4.3; G.T.E. Parsons, Liquidator, British Egg
Marketing Board for Fig. 4.5b; the Editor, *Journal of Endocrinology* for Figs. 4.5c,
6.4; the Editor, *Journal of Reproduction and Fertility* for Figs. 4.5d, 5.4, 11.12,
12.1c,d,e, 12.7a, 12.15; the Editor, *Poultry Science* for Fig. 4.7; the Editor, *Ardea*
for Fig. 4.9; the Society for Experimental Biology and Medicine for Figs. 5.6, 5.11;
E.J. Brill, Leiden for Figs. 5.12, 11.1, 11.3, 11.4, 12.16a; the Editor, *Wilson Bulletin*
for Fig. 7.1; the Editor, *Bird Study* for Figs. 7.5, 7.6, 10.5; Thomas Nelson and Sons
Ltd for Figs. 7.7, 7.8, 7.10, 9.13; the American Ornithologists' Union for Fig. 7.12;
the University of Chicago Press for Figs. 7.13, 9.3, 9.4, 11.8, 12.11; the Wildlife
Society for Fig. 7.15a; The *Ibis*, journal of the British Ornithologists' Union for
Figs. 8.1c, 8.10, 8.16, 10.1, 10.3a; Chapman and Hall Ltd. for Fig. 8.2a, first
published 1968 by Methuen and Co. Ltd.; the Wistar Press for Fig. 9.5; the Company
of Biologists for Fig. 9.7; the Editor, *Biological Bulletin Marine Biological Laboratory*,
Woods Hole for Fig. 9.9; the Regents of the University of California for Fig. 9.12,
copyright © the University of California Press; the Editor, *Condor* for Figs. 10.12,
10.13; North Holland Publishing Company, Amsterdam for Fig. 11.5; Stephen R.
Jarowski, Cold Spring Harbor Laboratory for Fig. 11.6; ASP Biological and Medical
Press B.V. for Fig. 11.13; the Pergamon Press for Fig. 12.1f; the Editor, *Journal of
Interdisciplinary Cycle Research* for Fig. 12.12, the Editor, *Biology of Reproduction*
for Figs. 12.13, 12.14, Deutsche Ornithologen-Gesellschaft for Fig. 12.16b; Masson
et Cie for Fig. 13.3a, the Wildfowl Trust for Fig. 14.10b; Dr. W. Junk B.V., Publishers
for Fig. 14.21; the Royal Society for Fig. 15.5; the Society of Systematic Zoology
for Fig. 15.7; the Oxford University Press for Fig. 16.9, from *Population studies of
birds* by D. Lack; the Royal Astronomical Society for Fig. A.1.

   The following authors have kindly given us permission to reproduce figures:
   Dr. J.R. Baker, Fig. 1.1; Dr. R.E. Ricklefs, Figs. 1.4 1.7, 8.16, Dr. M.L. Cody,
Figs. 1.5b,c; Prof. B.K. Follett, Figs. 2.1, 11.12, 12.1d,e, 12.6b, 12.7a,b,c, 12.10;
Prof. M.H. Stetson, Figs. 2.3b,c,e, 3.9; Prof. H. Kobayashi & M. Wada, Fig. 2.3d;
Prof. B. Lofts, Figs. 3.7, 10.3a,c, 12.1c; Dr. A.H. Meier, Figs. 3.11, 7.17; Dr. A.B.
Gilbert, Fig. 4.2; Dr. F.J. Cunningham, Fig. 4.5c; Dr. P.J. Sharp, Fig. 4.5d;

J.C.D. Hutchinson, Fig. 4.6; Dr. Rosemary E. Hutchison, Figs. 4.8, 5.8; H. Klomp, Fig. 4.9; Dr. J.B. Hutchison, Fig. 5.4; Dr. D.M. Vowles, Fig. 6.4; Dr. S.C. Kendeigh, Fig. 7.2; Dr. P. Ward, Figs. 7.4c, 7.13, 7.14, 10.1; Dr. K.H. Voous, Figs. 7.7, 7.8, 7.10, 9.13; Prof. W.D. Schmid, Fig. 7.15b; Dr. C.M. Perrins, Fig. 8.1c; L. Schifferli, Fig. 8.10; Dr. J. Parsons, Fig. 8.11; Dr. J.T. Rutledge, Figs. 9.5, 13.1; Prof. D.S. Farner, Figs. 9.8b, 9.9; Prof. R.G. Anthony, Fig. 10.12; Dr. H. Tordoff, Fig. 10.13; Prof. J. Aschoff, Fig. 11.5; Prof. C.S. Pittendrigh, Figs. 11.6, 11.7, 11.8, 11.9; Dr. A. Eskin, Fig. 11.10; Prof. M. Menaker, Figs. 11.14, 12.9b; Dr. G.B. Shellswell and Prof. R.A. Hinde, Fig. 12.1j; Prof. E. Haase, Dr. E. Paulke, and Dr. P.J. Sharp, Fig. 12.7a; Dr. E. Gwinner, Fig. 12.15; Prof. I. Assenmacher, Fig. 13.4, Dr. P. Berthold, Fig. 13.5; Prof. R.J. Berry, Fig. 15.5; Dr. D. McKenzie, Fig. A1.

Mrs. Gladys Sanderson, Mrs. Jean Pilcher, and Mrs. Margaret Haas typed the original manuscript and various drafts and we offer them our grateful thanks. Mrs. Haas additionally undertook the compilation of the bibliography and was exclusively responsible for checking the proofs of this section of the book. She also read the proofs of the whole book, checking bibliographical references in the process, and finding many mistakes which we had overlooked. We owe her a special debt of gratitude. Finally Daniel Osborn, Dr. Stuart Dobson, and Dr. Peter Ward read or advised on sections of the book and helped us to make various corrections.

Institute of Terrestrial Ecology   R.K.M.
Monks Wood   N.J.W.
September 1976

Contents

1 Introduction to avian reproductive strategies

Summary

Successful reproduction demands the temporal organization of gamete preparation in relation to the mate and environmental resources of energy. Photoperiodism is the process whereby endogenous circadian and circannual rhythms of body function are synchronized with external daily and seasonal rhythms. Latitudinal gradients in the adaptive phasing of a wide range of these body functions, involving also breeding seasons, can be identified. Since rhythmic processes also control rates of development and attainment of maturity a host of physiological parameters, including body size and metabolic rates, are correlated, although it is difficult to identify causality. For example, body size increases with increase in latitude in closely related species (Bergmann's rule) while the metabolic rate declines. The body weight of *Anser* and *Branta* goose species is inversely correlated with mid-latitude of the breeding range (positively with the mid-latitude of the winter range) but is more closely correlated with the photoperiodic response which itself is related to the latitude. This photoperiodic response is in turn very closely related to the survival rate of various species, suggesting that the 'biological clock' is the fundamental mechanism controlling a wide range of co-adapted physiological and ecological responses.

In very predictable environments, such as equatorial evergreen forests, the survival rate and life span of birds increases, although their reproductive rate declines. The evolution of species-specific survival rates, conversely mortality rates, is mentioned in the context of the increased risks consequent on reproduction, particularly in the tropics. Age of attainment of reproductive maturity and the reproductive rate will depend on a balance of survival probabilities between breeding and increasing the risk of adult death, or of not breeding and leaving no progeny. These species-specific survival rates are not to be confused with the factors which regulate animal numbers (see Chapter 16).

Breeding seasons

In the past, most studies of avian reproduction have been divided between two major disciplines, the domain of ecologists and physiologists respectively. Ecologists have been interested in the evolutionary significance of clutch size, productivity rates, nesting dispersion, and other parameters of reproduction that can be related to the external environment. They have concerned themselves with the so-called *ultimate* factors (see below) which determine breeding adaptations, whereas physiologists have studied the mechanism of reproduction and the internal environment of the individual. Whereas all workers recognize the existence of a co-adapted system, some ecologists have tended to make the tacit assumption that given sufficient environmental pressure physiological adjustment is relatively easy. In attempting to synthesize some of our physiological and ecological understanding of avian reproduction, one object of this book will be to illuminate some of the patterns of ecological adaptation that result from the restraints imposed by complex physiological mechanisms. Another aim is to persuade physiologists to relate their experiments to the natural conditions under which their subjects live. We believe that future advances will involve a multi-disciplinary approach and we hope our book will provide a starting-point and stimulus.

Mention above of *ultimate* factors introduces a clarification brought to the subject many years ago by J. R. Baker (1938*a*). Baker was concerned with the determination of breeding seasons but his ideas apply to clutch size, nest construction, migration, and a host of other topics. He distinguished *ultimate* factors as being those which essentially have survival value and not much causal function, from the *proximate* factors, which provide the actual mechanisms whereby breeding adaptations are achieved; both factors are of course determined by natural selection. Most bird species breed around the time when food supplies for themselves and their young are most readily available (Thomson 1950). This follows because natural selection favours those genotypes whose progeny are reared at the most appropriate times. Attempts to raise young at less advantageous seasons are nullified by a wasteful and disproportionate nestling mortality; indeed, parental survival may even be endangered as well. A differential survival rate in relation to food resources can define the spread and characteristic pattern of the breeding season, thereby constituting an ultimate regulator of breeding periodicity. Similarly, egg size must ultimately be set in relation to the needs of the young chick and resource availability. Ultimate factors affecting reproduction have provided the subject-matter of three masterly books by David Lack (1954, 1966, 1968*a*).

To achieve breeding condition and anticipate an approaching season of good food supplies and the availability of nesting cover, etc., birds have evolved response mechanisms to a host of environmental stimuli which function as signals heralding the approach of suitable times. The most important proximate factor is the seasonal change in length of day which stimulates, via neurohypothalamic pathways, endocrine secretions that allow the gonads to assume a functional state. However, many other proximate factors, including temperature, population density, appropriate behavioural stimuli from the mate, and availability of nesting sites, influence the birds' capacity to reproduce. The study of proximate factors has primarily been the prerogative of the physiologist and experimental studies of photoperiodism have featured prominently in their repertoire (major reviews by Burger 1949; Farner 1959, 1964; Marshall 1959; Wolfson 1959; Benoit 1961; Lofts and Murton 1968; Farner and Lewis 1971).

Reproduction is essentially a matter of temporal organization in relation to the mate and environmental resources. Photoperiodism is the process whereby the animal adaptively synchronizes an internal rhythm of body function with the external chronology of day and night and the seasons. To understand avian reproductive strategy requires knowledge of the way in which the 'biological clock' functions. In reviewing

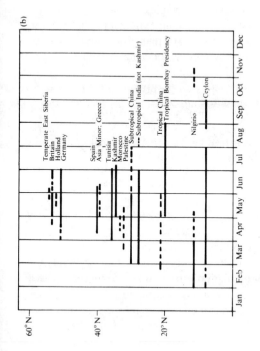

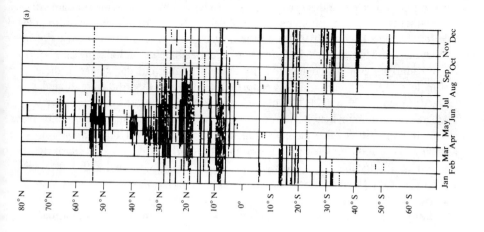

Fig. 1.1. Egg-laying seasons of passeriform birds at different latitudes: (a) passerines in general; (b) Great Tit *Parus major*. In (a) the line representing *Passer domesticus* in the original has been shortened since it was clearly inaccurate in the original and breeding is known to end in late July (see also p. 219). from Baker (1938a).

avian breeding seasons (and also flowering seasons and periodic leaf-fall in plants), Baker long ago appreciated that he was dealing with an internal rhythm interacting with the external environment (Baker and Baker 1936) and he recognized patterns of change in breeding season with latitude (Baker 1938*a, b*). Much of what Baker wrote has been neglected or overlooked by later workers; for example the kind of data reproduced in Fig. 1.1, from which he concluded that 'neither a long day nor yet a rapidly increasing day are necessary concomitants of rapid reproduction by birds'. And

Despite all the intensively interesting experiments on the effects of light on the reproduction of birds, for which we are indebted mostly to the pioneer work of Rowan (1929 and earlier and later papers) and of Bissonnette (1930, etc.), yet clearly length of day stands in no direct and obvious relation to the breeding season of birds under natural conditions. One is forced to the conclusion that light is only one of the factors concerned. Why, otherwise, should birds breed later in the boreal zone than in the temperate? After the March equinox the days are longer the further one goes north.

It has taken nearly 30 years to resolve this topic, partly because the dichotomy of physiology and ecology failed to identify it as a problem. However, it has also been necessary to wait for the realization that circadian clocks are composed of systems of endogenous coupled oscillators entrained by the environmental 24-hour oscillation of day and night and, even more important, that circannual rhythms of animal function also exist, entrained by the yearly oscillation produced as the sun moves back and forth across the equator and the duration of day and night (and twilight) change accordingly (Chapters 11, 12, and 13). The significance of photoperiodism has been sadly misunderstood by many ecologists, Phillips (1971) exemplifying their quandary.

Baker recognized that the breeding seasons of birds varied in a systematic manner between the equator and poles but not in direct agreement with the way length of day alters with latitude. Other workers too have been impressed by latitudinal gradients in reproductive strategies, which appear to have some underlying fundamental basis rather than being a simple reflection of ecological changes from tropics to poles (see for example, Slagsvold 1975*a*). In the tropics clutch size is smaller (Lack 1948, 1968*a*; Lack and Moreau 1965; Cody 1966) the breeding season longer (Baker 1938*a, b*), and adult survival rates higher than in temperate zones but the change in these parameters with latitude is usually proportionately greater than the increase in day-length (but see Hussell 1972). This has led various workers to seek an alternative to a general explanation which is derived from Lack's specific explanation for an increase in clutch with latitude: that increase in the available hours of daylight enables birds to collect more food for their young. Ashmole (1963*a*) modified Lack's thesis by pointing out that stable tropical ecosystems would also be saturated with individuals, so that the real difference in resource availability could be between a saturated tropical habitat, and a temperate habitat, in which days are long and the pre-breeding population is much reduced by the mortality occurring during the contra-nuptial season.

The concept of stability is not easily quantified but it is widely accepted and Cody (1971) has justified it in a valuable review of ecological aspects of avian reproduction. The essential point is the extent to which ecological resources are predictable, and this is eventually dependent on the degree of seasonal fluctuation in climate. The Australian desert is frequently instanced as an unpredictable environment but in reality in many parts the rain arrives at a very regular time, even though the quantity may be very low and variable. Some plants are able to flower and fruit with extreme regularity and temporal precision in the same month each year, while others may be very irregular and bear variable crops. For this reason it is necessary to consider the actual resource which a bird demands in order to make generalizations about the predictability of the habitat. A really stable environment, such as equatorial evergreen forest, offers little stimulus for a change in well-adapted animals and once an individual becomes established the probability will be small that any offspring it produces will be better adapted than itself. In such conditions plants can resort to vegetative reproduction. In birds, the life-span in tropical species is on average higher than those of closely related temperate forms, whereas eggs and nestling success is lower. A relatively high survival rate is also noted in birds which inhabit the more stable habitats in temperate zones as does the Swift *Apus apus* (Lack 1954) and various oceanic birds (Wynne-Edwards 1962). Genetic variability is further reduced by the tendency for successful males and females to remain paired for life. Cody (1971) suggests that the relative stability of tropical ecosystems compared with those of temperate zones facilitates predators becoming specialists of eggs and young thereby reducing breeding success (e.g. Skutch 1966). On the other hand, adult losses are more likely to result from accidents, and these are perhaps more often provoked in fluctuating temperate systems compounded by inclement climates. In addition, competition from young must be increased in temperate areas as a result of the higher productivity rates. In unstable ecosystems there is more premium on realizing as much of the genetic variability that is potentially provided by random gene recombination, for this will offer a greater opportunity for a species to adapt to changing ecological conditions (but this might also result in a bigger wastage of genotypically unfit animals (Murton 1972)). In contrast to the long pair bonds and monomorphism characteristic of tropical birds, temperate species tend to dimorphism and short pair bonds.

Survival rates

As mentioned above, natural selection must favour those individuals which leave the largest number of surviving offspring, yet reproduction is attended by various risks which increase the probability of accidental death, for example, changes in metabolism occur, eggs must be attended, and the presence of young increase the pressure on food searching. This is clearly illustrated later in this book (Fig. 10.5, p. 257): juvenile mortality in the Rook *Corvus frugilegus* occurs during the summer droughts, when food supplies are restricted, whereas adult losses are concentrated in the breeding season (see also Tables 2 and 3 in Vowles and Harwood 1966). In the Starling *Sturnus vulgaris* it has been suggested that the parents are slightly less vigilant when busily collecting food for their chicks and that this enables them to be surprised more easily by cats and other predators (Coulson 1960). In this species the females suffer a higher annual mortality than males in their first year (70 per cent against 39 per cent) and the difference mostly arises with the onset of the breeding season. Since first-year females breed, but first-year males do not, it is likely that the difference arises because the female is exposed to greater predation risks. The annual survival rate of the Shag *Phalacrocorax aristotelis* is on average 5 per cent lower in the female than the male (Potts 1969) and a difference of this order is also apparent in the Yellow-eyed Penguin *Medadyptes antipodes* (Richdale 1957); in both these species the females begin breeding at an earlier age than the males. What then is the optimum strategy that a bird should adopt—increase its reproductive effort yet reduce its own survival prospects or reproduce more slowly but live longer to have the prospect of further breeding efforts? Williams ((1966), quoted by Cody (1971)) considered the problem in terms of a series of choices or commitments such as whether to lay one more egg or not; whether to breed this year or wait till next. The extent of the commitment should as a result of natural selection be a function of two variables, the probability of success of the commitment and the probability of the parent surviving to breed again. If the mortality resulting from reproduction is higher than non-breeding mortality, selection may favour an increase in length of life at the expense of early reproduction (Williams 1966). In quoting Murphy (1968), Cody (1971) has succinctly emphasized that the commitment to reproduction must be related to the non-reproductive and reproductive portions of the survival curve, as discussed in the following paragraphs.

Adult survival rates s (number of birds surviving/total number of birds) or, conversely, mortality rates $(1-s)$ are species-specific and appear to remain more or less constant with increasing age, primarily because deaths result from accidental causes whose risk remains constant. The majority of individual birds do not live long enough to suffer from senescence except in captivity (e.g. Comfort 1962), but there is some evidence that age-dependent mortality may occur and assume increasing importance in the field with advancing age (Botkin and Miller 1974). The survival rates of closely related species are usually very similar in spite of often marked differences in their ecology and productivity. The average annual survival rate for the Wood Pigeon *Columba palumbus* living in rural England is 0·644 ± 0·02 compared with 0·665 ± 0·049 for Feral Pigeons *C. livia* inhabiting the Salford Docks, Manchester (Murton 1966a, b; Murton, Thearle, and Thompson 1972). Moreover, mortality rates seem to be largely independent of population size (see below). Fig. 1.2 in effect plots the probability that any one individual will survive to age x (denoted as l_x). At first, the probability of dying is high and it declines rapidly to the point at which adult status is achieved and thereafter remains approximately constant with age. The age of sexual maturity is denoted by α so that from this point female fecundity with age (m_x) multiplied by the probability of being alive (strictly the probability of a female being alive) is the number of young produced, that is, $l_x m_x$. The end of reproductive life is denoted by ω, but as mentioned this point is rarely reached in wild subjects. Hence, the total number of young produced during the lifetime of the female or the net reproductive rate per generation is:

$$R_0 = \Sigma \, l_x m_x \, . \tag{1.1}$$

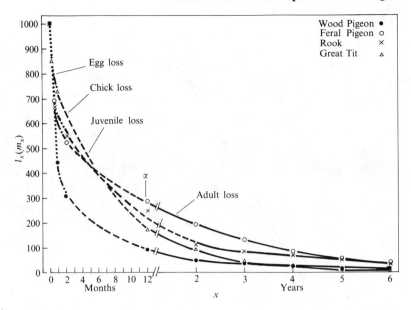

Fig. 1.2. Survivorship curves for four bird species in terms of 1000 eggs initially laid. Note difference in the horizontal scale used for the first and subsequent years. Loss of eggs indicated by dotted lines, loss of nestlings by dot–dash, loss of juveniles by dash, and adult losses by solid lines. l_x is the probability of survival to age x; α is the age of sexual maturity, and ω, not shown, the end of the reproductive span. The survival curve can also be used as a measure of the distribution of female births (m_x) whereupon $l_x m_x$ gives the reproductive output of females in relation to age within the limits α to ω. Compiled from data in Holyoak (1971); Lack (1966); Murton (1965); Murton and Westwood (1974a); Murton et al. (1972).

Changes in the $l_x m_x$ curve alter the generation time T.

The intrinsic rate of natural increase of a population is defined as:

$$\Sigma\, l_x m_x\, e^{-rx} = 1 \,, \tag{1.2}$$

where r is a species-specific constant and e the base of the natural logarithms.

Birds produce more young than are needed for the replacement of adult losses and this reproductive surplus could potentially result in an exponential or Malthusian increase in population size as the young themselves reproduce thus:

$$dN/dt = rN \,, \tag{1.3}$$

where r is the constant representing the exponential growth rate of the population during each time interval and N is the initial population. Integrated with respect to time t the equation becomes:

$$N_t = N_0\, e^{rt} \,, \tag{1.4}$$

with N_0 the initial population; N_t the population after time t.

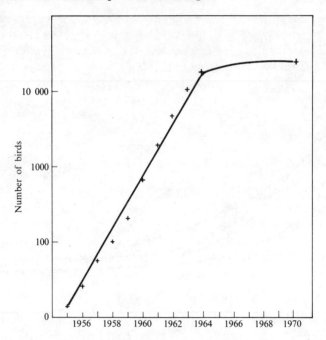

Fig. 1.3. Logarithmic increase of Collared Dove *Streptopelia decaocto* in Britain since original colonization in 1955. Based on data in Hudson (1965, 1972).

The Collared Dove *Streptopelia decaocto* has increased in Britain in this way (Fig. 1.3) and so did the Pheasant *Phasianus colchicus* when introduced to Protection Island (Einarsen 1945). As population size in the Pheasant rose the birth-rate declined in consequence of increased nesting losses but the winter mortality rate remained approximately constant.

If t in eqn. (1.3) is made equal to the generation time T we get

$$e^{rT} = N_T/N_0 = R_0. \tag{1.5}$$

Any increase in R_0, such as an increase in clutch size or hatching success, or a decrease in T will increase r and will confer survival value in competition with species having lower values of r (Fisher 1928). This is true when species are not living at saturation density, as applies in seasonally changing environments in which population size fluctuates between high and low limits following seasonal restrictions on resource availability. But in saturated habitats, in which breeding is difficult and the survival prospects for chicks poor, and where considerable experience is needed to become established in the adult social system, ecological conditions enforce a deferment of reproductive maturity and a low R_0 and thereby encourage an increase in T. So when the probability of young surviving to reproductive age is low more progeny will eventually be raised if reproduction is extended over a long time-span (see more detailed discussions by Murphy (1968), Levins (1968), and Cody (1966, 1971)). Any decrease in R_0 and increase in T must result in a low r, this being a consequence of ecological pressure rather than an adaptation to limit population growth. Fig. 1.4, which is based on Ricklefs (1973), illustrates the positive correlation between the age of first breeding and the annual adult survival rate. It shows that first-year survival varies around an average trend of 50–60 per cent of the adult level, irrespective of the absolute adult survival rate or the occurrence of delayed maturity. Delayed maturity occurs when the adult survival rate exceeds about 60 per cent. In many species it has been demonstrated that experience is gained with increase in age and this becomes reflected in an improved capacity to breed. For example, breeding success increases with age in the White Stork *Ciconia ciconia* (Hornberger 1954; Schüz 1957) and Kittiwake *Rissa tridactyla* (Coulson and White 1960; Coulson 1971).

Changes in mode of development or metabolic function that affect survival curves will influence reproductive strategy. For example, Bergmann's rule relates an increased body size in a polytypic species to cooler parts of the range and a latitudinal and altitudinal gradient is observed. However, Kendeigh (1969a, 1970) emphasizes that any advantages of larger body size in cold climates relative to temperature regulation could be offset by the need to eat more food and so body size will only increase if the physiological advantages outweigh the ecological disadvantages. Since an increase in body size is associated with a decrease in metabolic rate, more efficient use can be made of food resources in proportional, as distinct from absolute, terms. More recently, Kendeigh (1972) has suggested that increase in body size is limited by the thermo-regulatory capacity of the bird at high ambient temperature, while decrease in body size is limited by its capacity to mobilize energy. Because non-passerines

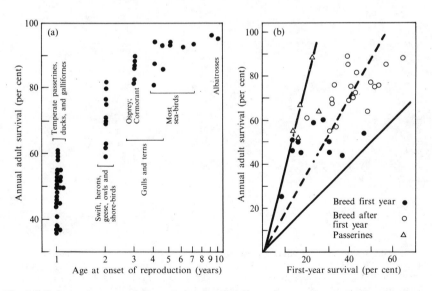

Fig. 1.4. Relation between adult survival rate and: (a) age of onset of reproductive capacity; (b) first-year survival. From Ricklefs (1973).

have lower rates of standard metabolism than passerines, they can achieve both smaller and larger body sizes. Evolution towards small size imposes greater energy demands on the individual but it occurs because otherwise some ecological niches would go vacant. There is a trend for phylogenetically older lines to be composed of larger species than newly radiating groups. If a large body size can be achieved, the animal's habitat is effectively stabilized, because the large body is a buffer against environmental changes and it also serves as a defence against predators. Not surprisingly, large body size in birds is associated with a reduced mortality rate in both young and adults. This is illustrated for waders in Fig. 1.5(a) and for various ducks (Anatidae) in Fig. 1.6(a).

The body weight of the *Anser* and *Branta* geese is fairly well correlated with the photoperiod under which breeding is initiated in each species, when they are all kept under the same seasonal light regime (see Fig. 14.15; p. 400) and it will emerge in Chapter 14 that this photoperiod is strongly dependent on the latitude of origin of

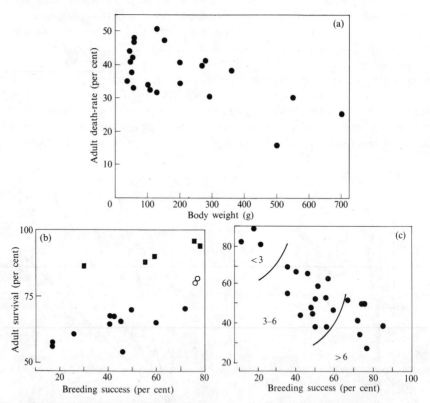

Fig. 1.5. Relationship between adult survival rate of birds and: (a) adult body weight in various wading birds (Charadrii) (data in Boyd 1962*a*); (b) pre-reproductive survival measured as the percentage hatching success of eggs in Charadrii (solid circles), Alcidae and Procellariformes (squares), and Apodidae (open circles). All the species have a fixed clutch size. (From Cody (1971) based partly on Boyd (1962*a*).) (c) Breeding success of temperate and tropical passerines with variable clutch size in the approximate size range indicated. From Cody (1971).

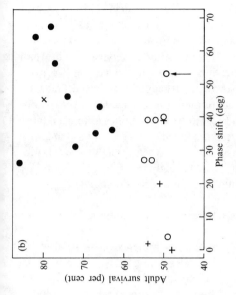

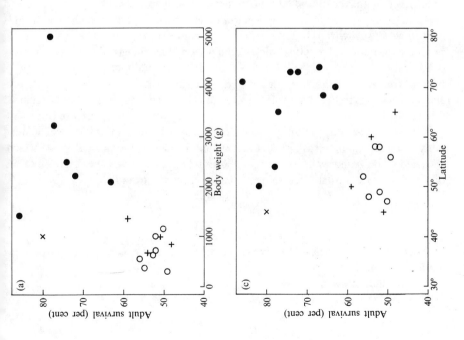

Fig. 1.6. Relationship between adult survival rate of various Anatidae (data in Boyd (1962b) and: (a) adult body weight; (b) phase shift of egg-laying cycle to seasonal photoperiod as in Fig. 14.4 (p.370); (c) mid-latitude of the natural breeding range.

Solid circles refer to *Anser* and *Branta* geese, these being *A. brachyrhynchus*, *A. albifrons* (two races), *A. anser*, *A. caerulescens* (two races, Greater and Lesser), *Branta canadensis* (two races), and *B. bernicla*. The cross refers to *Tadorna tadorna* and open circles to *Anas* ducks, *A. acuta*, *A. crecca*, *A. platyrhynchos*, *A. rubripes*, *A. penelope*, *A. dicors*, *A. clypeata*. Plus signs refer to *Aythya* species, *A. vallisneria*, *A. americana*, *A. fuliga*, *Aythya merila*. *A. platyrhynchos*, arrowed, is probably atypical because stocks of this species in Britain have been much affected by interbreeding with domesticated strains. Correlation coefficients and regression equations are:

(a) $r_{16} = 0.649$; $y = 0.52 + 0.0001x$;
(b) $r_{18} = 0.649$; $y = 0.48 + 0.004x$;
(c) $r_{19} = 0.323$; not significant.

the various species (Fig. 14.2; p. 368). By Chapter 14 the photo-response mechanism will have been considered in terms of the phasing of the circadian–circannual oscillator system which is the basis of the 'biological clock'. Since the 'clock' regulates all metabolic function, including growth and development, it is of interest at this stage to note that 42 per cent of the interspecific variation in survival rate in Anatidae is accounted for by at least one component of the oscillator system (Fig. 1.6(b)). This highlights the fact that we are dealing with a co-adapted system and that ecological and physiological aspects of reproduction must be considered together. For the present it can simply be noted that in Fig. 1.6(b) survival rates have been plotted against the phase difference existing between the breeding season and a standard entraining light cycle (see p. 370). The weak trend for survival rate of Anatidae to be negatively related to latitude (Fig. 1.5(c)), and already mentioned for birds in general, presumably depends on a more fundamental relationship between physiological function and latitude.

The association noted between survival rate and body size could reflect a relationship between metabolic rate, growth, and body size. Ricklefs (1969a) has examined the manner in which the development pattern of birds influences their survival rates. He assumes that the development rate of each species during its evolution has been optimized by natural selection so that the total mortality M occurring throughout the whole developmental period (number of birds dying/total number of birds) is minimized. The value of M also reflects the organizational properties of the organism. For example, an increase in the survival expectation of a young growing bird might be achieved by making use of embryonic tissues that would otherwise be transformed into mature organs or functions; if this was done the overall development rate would have to be slowed. Given Ricklefs' reasonable premise, it becomes evident that the survival expectation of a young chick would not be independent of the attainment of adult function (Fig. 1.5(b) is but one example). The total mortality M from egg to adult can be represented as a decaying exponential whose slope and intercept vary among species (as in Fig. 1.1). It is then found that M remains remarkably constant among birds suggesting, in Ricklefs' words, 'that the outcome of natural selection on strategies of development is constrained by properties of organization which are fairly constant among birds'. Ricklefs calculated the decrease in mortality rate which occurred with age in various species (as in Fig. 1.1) and expressed the results in terms of the percentage mortality occuring per day m using the formula:

$$m = \frac{\ln s_0 - \ln s}{t - t_0},$$

in which s is the proportion of young surviving at any time t; s_0 is the proportion of young surviving at time t_0.

This formula allowed regression lines of mortality rate on age since hatching to be fitted, using either semi-logarithmic or logarithmic expressions thus:

$$m_t = m_0 e^{-kt} \tag{1.6}$$

$$m_t = m_0{}^{t-k} , \tag{1.7}$$

where m_t is the mortality rate at time t, m_0 the initial mortality rate, and k the slope of the regression.

Ricklefs showed that the semi-logarithmic expression was applicable to the Red-billed and Yellow-billed Tropic birds *Phaethon aethereus* and *P. lepturus* and perhaps also to the Glaucous-winged Gull *Larus glaucescens* in which mortality is reduced at a rate proportional to the mortality rate ($dm/dt = -km$). This means that the force of selection for a change in development rate during a given interval is proportional to the overall mortality rate during that interval. A decrease in development rate during the early stages of development, when mortality rates are high, must have a proportionately greater effect on the total survival of offspring than a similar change at a time of lower mortality risk. Ricklefs showed that change k in mortality rate with age was related to the initial mortality rate m_0 (Fig. 1.7(a)). An adjustment of k to reduce the total force of mortality may result in an increase in m_0 while adjustment to minimize M is feasible only while the relative change in k is greater than in m_0. Fig. 1.7(b) shows that k may be related to the rate of body-weight growth, for k increases with a progression from altricial to a precocial mode of development (see also p. 212), while within any category of development k is expected to increase with growth rate. So, as Ricklefs states, change k in mortality rate during development and initial mortality m_0 are both functions of the mode of development and rates of physical growth, while the growth rate of species possessing a given mode of development is a strict function of mature body size (Ricklefs 1968). The trend

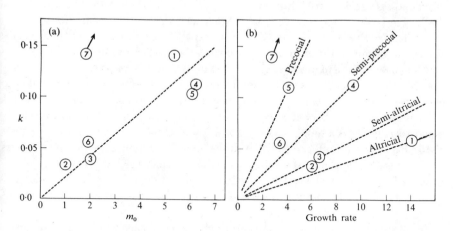

Fig. 1.7. The way in which k (which is the slope of the regression which relates the initial mortality rate to the mortality rate at time t and is, therefore, a measure of the way in which the mortality rate changes with time) is related to: (a) initial mortality rate (m_0); (b) the rate of increase in body weight for various modes of development.

Species depicted are (1) Blackbird *Turdus merula*; (2) Yellow-billed Tropic Bird *Phaethon lepturus*; (3) Red-billed Tropic Bird *P. aethereus*; (4) Glaucous-winged Gull *Larus glaucescens*; (5) California Quail *Lophortyx californicus*; (6) Pheasant *Phasianus colchichus*; (7) Dunlin *Calidris alpina* (off the graph). From Ricklefs (1969a).

of natural selection is for an increase in growth rate, and Ricklefs assumes that the observed rates are the physiological maxima for each mode of development. Thus k is determined by body size and mode of development and alteration of the latter appears to be the best means of adjusting k. Ricklefs concludes that species have evolved varied strategies of development in response to their respective environments. But in spite of this diversity of adaptation, or perhaps, more correctly, because of it the total force of mortality acting on the development period, the 'outcome' of natural selection, is similar for each species. The extent to which birds can reduce the cost of development through evolution must be limited primarily by counter-adaptation of other organisms causing mortality. That the outcome of this interaction is so uniform may be the result of competitive interactions which set limits for evolutionary success.

Reproduction must first involve the actual physical preparation and development of the eggs and sperm (gametogenesis) in the testes and ovary followed by their transport and union in fertilization and eventual oviposition into a prepared nest. The reproductive apparatus of birds basically repeats a vertebrate pattern which can be traced to the Agnatha. A range of environmental stimuli is transmitted through the central nervous system to effect the liberation of neurohormones or releasing factors (RF) in the median eminence of the pituitary gland (hypophysis). These neurohormones are conveyed via portal blood-vessels to the anterior lobe of the pituitary (the pars distalis), where they stimulate specific glandular cells to release the pituitary hormones. These include the gonadotrophins, which have two major functions in reproduction: the initiation and regulation of germ cell development (gametogenesis) and the stimulation of endocrine tissues sited in the testes and ovaries, whereupon gonadal (steroid) hormones are synthesized and secreted. These are involved in facilitating gametogenesis, but they have the additional function of mediating the development of the accessory sexual characters (sperm and egg ducts, cloaca, etc.) and also the secondary sexual characters of the bird (including song, breeding behaviour, and in some cases special nuptial display plumage or colour changes). The endocrine system, therefore, provides a precisely co-ordinated link between the physical breeding apparatus of the bird and the environment, enabling the expression of feed-back relationships which can monitor neural and hormonal functions (Fig. 1.8 is a simplification).

In Chapter 2 the form and function of the pituitary and its secretions are considered, together with the mechanisms by which environmental information is processed, while in Chapters 3 and 4 the gonads and other target organs are considered in more detail. The integration of gametogenesis between the sexes so that effective fertilization can be achieved in a temporally changing, sometimes unpredictable environment, depends on neural and endocrine-mediated behaviours, which are discussed in Chapters 5 and 6. Reproduction also demands energy additional to maintenance requirements and breeding must be timed in relation to moult, migration, winter fat accumulation, and other, often conflicting, demands for extra resources (Chapters 7 and 8). All these factors help mould the diverse breeding responses shown by birds in which physiological and ecological factors are intimately integrated (Chapters 9 and 10). As implied above, all birds are photoperiodic in the sense that the cycle of day and night entrains endogenous periodicities to produce adaptively entrained diurnal rhythms of body functions and feeding and other behaviour. In many cases environmental light cycles are reliably correlated with seasonal environmental changes so that the bird can use them as predictive signals to prepare for reproduction, perhaps first by migrating. In some circumstances endogenous periodicities may be expressed as free-running breeding cycles with periodicities of less

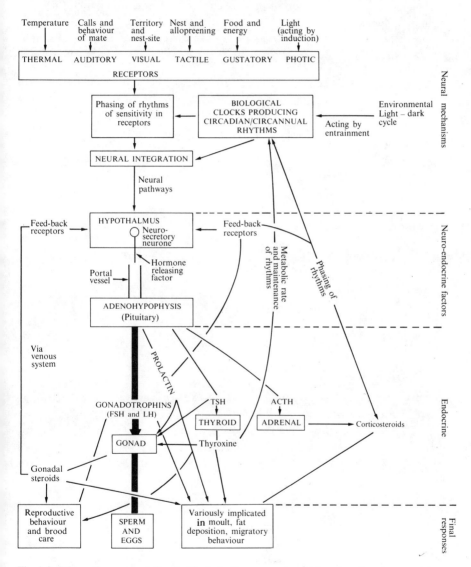

Fig. 1.8. Schematic representation to illustrate the manner in which external environmental stimuli and internal stimuli are integrated in the bird in regulating reproduction. After various schemes by Farner and also Follett (1973*b*).

than 12 months, the classic example being the 9·6-month cycle of the Wideawake Tern *Sterna fuscata* on Ascension Island. Contemporary understanding of these topics is outlined in Chapters 11, 12 and 13 and extended in Chapter 14, which considers the adaptive radiation of photoperiodic phenomena with special reference to wildfowl (Anatidae). Chapter 15 is devoted to special adaptations for reproduction such as polygamy and the reversal of sexual roles, and in Chapter 16 we end by considering what happens to the population surplus which results from reproduction, that is, population dynamics.

2 The neurohypophysial system

Summary

In this chapter the internal control of reproduction and integration with environmental stimuli is shown to depend on the neurohypophysial system, whose complex anatomy is described. This system allows sensory electrical impulses arriving at the brain to be converted into neurohormones in the median eminence of the hypothalamus. The exact neural pathways linking the neurosecretory cells with different thalami in the brain remain to be defined and so it is still not known which and how many centres control the activity of the cells. However, it is reasonably certain that in the main specific cells secrete the specific releasing factors which are carried by portal vessels to the pituitary gland, whose close proximity to the brain makes it the 'master gland' of the body. Specific cells in the pituitary gland, more correctly the adenohypophysis, in response to stimulation by neurohormones, that is, the releasing factors, secrete a range of pituitary hormones: the gonadotrophins, follicle stimulating hormone (FSH) and luteinizing hormone (LH), are produced by beta- and gamma-cells respectively, prolactin by the eta-cells, thyroid stimulating hormone (TSH) by delta-cells, adrenocorticotrophic hormone (ACTH) by epsilon-cells, melanophore stimulating hormone (MSH) by kappa-cells, and somatotropic hormone (STH) by alpha-cells. The mode of action of these hormones, particularly the gonadotrophins and prolactin, has for long been studied by the injection of purified mammalian extracts, usually in non-physiological doses, and studying the resulting histological effects.

Bioassays for endogenous circulating protein hormones remained insufficiently sensitive until the mid to late 1960s, when highly sensitive and specific radioimmunoassays depending on antigen—antibody reactions to specific proteins were developed. These developments heralded the fractionization and preparation of highly purified avian pituitary hormones during the early 1970s leading to the arrival of sensitive and specific avian radioimmunoassay techniques; the next decade will see these being employed to an increasing extent.

The structure of the pineal gland is described in this chapter. It too is a secretory organ and appears to be implicated in synchronizing the biological oscillators that underlie circadian rhythmicity of locomotor and other activity—it is not the biological clock *per se*.

Anatomy of pituitary

The logical starting-point for a consideration of the avian reproductive apparatus is the pituitary gland and its connections to the brain via the hypothalamus. Unfortunately this approach immediately forces us to adopt a highly technical terminology, which is liable to deter all except specialist readers. While there is no alternative to using these technical terms in referring to specific structures it is hoped that frequent reference to the diagrams will clarify the subject for the more general ornithologist. The structure and embryological development of the avian pituitary has been fully elaborated in a monograph by Wingstrand (1951) and his conclusions remain generally valid today. His researches indicated a relative uniformity throughout the class and, while differences in the proportions of some of the component parts of the gland were noted between different avian orders there is still no indication that these are adaptive. More recently Wingstrand has produced a comparative review of the gland in vertebrates in a treatise devoted to the pituitary, edited by Harris and Donovan (1966). Subsequent research has concentrated on identifying the secretory cells of the pituitary using modern histochemical techniques, and in correlating their activity with changes in target tissues and, equally important, in attempts to unravel the neurosecretory fibres and vascular pathways whereby the functioning of the secretory cells is eventually linked to the nervous system. Advances in these fields have recently been well reviewed by Follett (1973a), Tixier-Vidal and Follett (1973), and Kobayashi and Wada (1973) and these sources should be consulted for more detailed information.

The unique status of the pituitary as the master gland depends on its location at the base of the brain, where it sits in a depression of the sphenoid bone just posterior to the optic chiasma. Its autopsy is best achieved by removing the eyes and cranium so that the brain can be pulled back and the optic chiasma located. Hypophysectomy is a delicate surgical procedure usually achieved by drilling through the dorsal floor of the palate. The median part of the hollow fore-brain is the diencephalon its cavity being the third ventricle. It is a complex mass of nerve cells, the thalami, which con- nect with the brain stem and with bilateral expansions which comprise the small cerebral hemispheres. These cerebral hemispheres are divisible into dorsal and ventral parts. In mammals the dorsal part becomes highly convoluted and developed relative to the ventral region of basal ganglia, but the converse is the striking feature of the bird brain. In birds, the basal ganglia are increased in size and complexity by a hyper- striatum thought to be important in breeding behaviour. Indeed, the whole consider- able ventral development of the cerebral hemisphere is thought to be involved with the integration of the complex instinctive behaviour which is so characteristic of the class (see Wright, Caryl, and Vowles 1975).

The floor of the third ventricle of the diencephalon is the hypothalamus which during embryological development becomes linked by a narrow stalk to an outgrowth from the oral epithelium—called Rathke's pouch—to form the pituitary gland or hypophysis. The gland (see Fig. 2.1), therefore, embraces two embryologically dis- tinct components:

1. The adenohypophysis, is derived from oral epithelium, and is subdivisible into two parts:
   (a)   the pars distalis or anterior lobe, and
   (b)   the pars tuberalis, which partly forms the stalk of the pituitary and is partly joined to the tissue derived from the diencephalon.

2. The second component is the neurohypophysis, which comprises neural tissue derived from the diencephalon. This too can be divided into two parts:

(a)   The part extended posteriorly to form the infundibular process is often referred to as the neural lobe. During the evolution of the vertebrates and adaptation to a terrestrial mode of life the caudal region of the neural lobe has become differentiated into the pars nervosa.

(b)   The second part is the infundibulum. This differentiates into:

(i)   an important area lying dorsal to the pars distalis and pituitary stalk known as the median eminence, and

(ii)   some of the tissue of the infundibular stem joins with the pars tuberalis to form the actual stalk of the pituitary—properly called the hypophysial stalk. Actually a sheath of connective tissue separates the two components of the stalk. The lumen of the infundibulum and its stem—the recessus infundibuli—are continuous with the third ventricle.

In fish, reptiles, and birds Rathke's pouch becomes partially constricted during development producing an oral and aboral cavity of which the former contributes significantly to the pars distalis thereby forming a caudal lobe. In higher mammals, but not monotremes and marsupials, the oral cavity contributes relatively little and the pars distalis is not divisible into caudal and cephalic lobes. The consequence of these differences is that in fish, reptiles, and birds there is a more marked spatial segregation of different cell types in the two lobes of the organ.

The median eminence provides the neural—vascular link between the hypothalamus and adenohypophysis. It is not present in primitive vertebrates such as the cyclostomes (Gorbman, Kobayashi, and Uemura 1963), which must rely on the general blood circulation to transport neurohormones, but it becomes a well-developed structure in the Anura (Assenmacher and Tixier-Vidal 1965). In birds neural—endocrine connections exist between the adenohypophysis and the thalami. These in turn receive sensory messages from those parts of the brain concerned with instinctive behaviour, as well as serving as the main ganglia of the autonomic nervous system, integrating metabolic and visceral functions. On the roof of the diencephalon is the pineal gland, although there is no parietal eye as in some reptiles. But while the gland appears to be vestigial it has recently been shown to be involved in the biological-clock mechanism of the bird (see below).

The median eminence is V-shaped in cross-section (Fig. 2.2) and comprises four distinct layers and one other which is less easily identified. Dorsally there is an ependymal layer consisting of the bodies of the ependymal cells and only a few fibres and nerve cells. Below them are a layer or two of hypendymal cells. Processes from both these cell types extend to the ventral part of the median eminence and form a fibre layer where bundles of nerve fibres are concentrated. Ventrally these fibres end in a glandular or palisade layer. Here few nerve fibres occur but axons from the ependymal cells ramify to form a network of neurosecretory fibres close to the gland and the glial cells and the blood-vessels that form the primary capillary net. Axon tracts, the supraoptico-hypophysial and tubero-hypophysial tracts, pass through the fibre layer on their way to the posterior pituitary and the intermixing of axons produces a

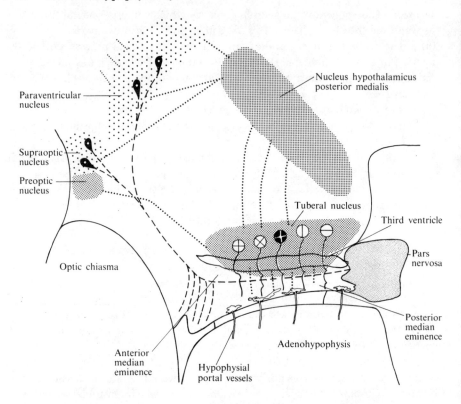

Fig. 2.1. Diagramatic side-view of the hypothalamo-hypophysial system of the bird. Various nuclei are shown by degrees of stippling: the nucleus hypothalamicus posterior medialis; the tuberal nucleus, this being the basal part of the infundibular nucleus; the supraoptic nucleus; the preoptic nucleus behind preoptic recess; the paraventricular nucleus. Cell fibres of the AF-positive or Gomori-positive neurosecretory system are depicted as dashed lines. These arise from cell bodies in the paraventricular and supraoptic nuclei and pass back and down into the median eminence (divided into anterior and posterior parts) and the pars nervosa. A second Gomori-negative neurosecretory system arises in the posterior portions of the hypothalamus from which axons enter the median eminence, particularly the tuberal nucleus, these being indicated by dotted lines; see also below. These axons are associated with a number of different neurosecretory neurones whose cell bodies are situated in the tuberal nucleus and whose axons in turn pass into the ventral regions of the median eminence. The neurosecretory cells are thought to be responsible for the production of different neurohormones, or releasing factors (RF). The releasing factors diffuse into the sinuses of the hypophysial portal vessels from where they pass into the adenohypophysis. Many catecholamine terminals and fibres (flourescent, aminergic system) are also present in the hypothalamus and this system is shown by stippling and dotted lines, areas of most intensive flourescence being given more intensive stippling, notably the tuberal nucleus and nucleus hypothalamicus posterior medialis. Other terminals occur in nuclei associated with the AF-positive system. Based on Follett and Sharp (1968); Sharp and Follett (1970); Oksche *et al.* (1971).

reticulated layer. The essential relationship to be considered is the proximity of ependymal cells in the dorsal median eminence that are capable of producing releasing factors, which the neurosecretory fibres carry to the primary capillary plexus.

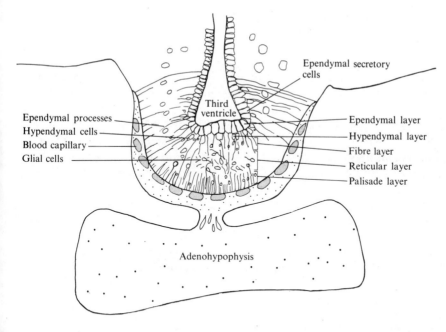

Fig. 2.2. Diagramatic cross-section of the avian median eminence and adenohypophysis. Secretory AF-positive ependymal cells, presumed responsible for the production of releasing factors, send ependymal processes to the blood capillaries which form the primary plexus of the hypophysial portal system. Beneath the ependymal layer are one or more layers of hypendymal cells and the processes from these, and those from the ependymal cells, form a fibre layer, where they are intermixed with glial cells; these last have processes extending in various directions. Below the fibre layer is a reticular layer where neurosecretory axons passing from the supraoptico-hypophyseal tract, also Gomori-negative axons, together with the ependymal and hypendymal processes, are intermixed with glial cells. After leaving the reticular layer all fibres and processes proceed in a pallisade fashion to the basal surface of the median eminence.

Vascular components of the pituitary

The capillary system is supplied by the superior hypophysial artery. Vessels of the hypophysial portal system then collect this blood and carry it via the hypothalamic stalk to the pars distalis where the vessels again ramify throughout the gland. A large number of thin-walled vascular channels, which are continuous with the sinusoids surrounding the gland cells of the anterior lobe, are found where the pars tuberalis and pars distalis join. Sectioning of these vessels in the chicken leads to permanent gonad regression, provided that regeneration is prevented (Shirley and Nalbandov 1956). In the White-crowned Sparrow *Zonotrichia leucophrys* the porto-vascular supply to the adenohypophysis arises from two anatomically separate primary capillary networks sited in the anterior median eminence and posterior median eminence respectively; the latter supplies the caudal lobe of the pars distalis and the former the cephalic (anterior) lobe (Vitums, Mikami, Oksche, and Farner 1964). Electron-microscope studies show that the small arteries supplying the primary capillary plexus are structurally typical arterioles, whereas those within the plexus have many fenestrations and pinocytic vesicles typical of capillaries in other endocrine organs; endothelial cells of the portal vessels protrude into the lumen appearing as, and probably functioning as, valves to regulate blood-flow (Mikami, Oksche, Farner, and Vitums 1970). No capillaries exist in the pars distalis; only sinusoids surrounding the gland cells. A similar system has been found in various other species (Dominic and Singh 1969) but in the Japanese Quail *Coturnix coturnix japonica* a slightly more primitive condition is noted (Sharp and Follett 1969*a*) as also is the case in the Cattle Egret *Bubulcus ibis,* and some coraciiformes. Instead of the marked point-to-point system, several (15–20) portal vessels are seen to enter the pars distalis at a number of points along its length. None the less, a dual portal system probably exists in the quail, for the anterior and posterior divisions of the median eminence are supplied by separate branches of the hypophysial arteries. It will become clear that the neurosecretory cells of the median eminence produce different neurohormones and that these can be distributed to specific target areas in the caudal or cephalic lobes of the pars distalis in consequence of the discrete functional anatomy of the capillary plexus. In addition, the neural lobe of the neurohypophysis also receives a distinct blood-supply from the primary capillary plexus. There are interspecific variations in the proportions of the neurohypophysis, although in all species the tissue is the same, being comprised of the endings of nerve fibres (Oksche, Oehmke, and Farner 1970; Oehmke 1971).

Neurosecretory components of the hypothalamus

Reference has been made to the nuclei and nerve tracks of the hypothalamus, which in the early studies were located by staining with pyridine–silver preparations. The involvement of both adrenergic fibres (motor nerve fibres which secrete the catecholamine, noradrenaline, at their terminals) and cholinergic fibres (secreting acetylcholine) was long ago suspected, since both catecholamines and acetylcholine were found in the hypothalamus of mammals in the 1940s. Furthermore, drugs that interfere with the enzymes regulating neurosecretion in turn alter the secretions of the pituitary hormones (Fuxe and Hökfelt 1967). Modern histological and histochemical methods have enabled regions and neurones of adrenergic or cholinergic activity to be mapped. The past few years have also seen big advances in our knowledge of the formation, storage, release, and metabolism of the catecholamines and the metabolic steps in their biosynthesis elucidated as: tyrosine → dopa → dopamine → noradrenaline → adrenaline, the first stage involving the hydroxylation of tyrosine by tyrosine hydroxylase (Axelrod 1969). The same enzyme is involved in the adrenergic neurotransmitter mechanism via the biosynthetic sequence: tyrosine → dihydroxyphenylalanine → dihydroxyphenylethylamine → norepinephrine (Weiner 1969).

A system which appears to be essentially cholinergic is stainable with Gomori's chrome–alum haematoxylin or with aldehyde–fuschin (AF) (also with Alcian blue or Victoria blue) and is, therefore, termed the Gomori-positive or AF-positive neurosecretory system. Generally, but not invariably, acetylcholinesterase occurs in areas and along nerve tracts which are Gomori-positive. In particular, these cholinesterases have been identified in the paraventricular nucleus and the supraoptic nucleus, from each of which bundles of axons run to end in the anterior median eminence and in the pars nervosa (Fig. 2.1); this has been demonstrated in the chicken (Graber and Nalbandov 1965), White-crowned Sparrow (Kobayashi and Farner 1964), White-eye *Zosterops palpebrosa* (Uemura 1964a), Tree Sparrow *Passer montanus* (Uemura 1965), and Japanese Quail (Sharp and Follett 1970). In addition, acetylcholine-sterases also occur in the posterior division of the median eminence of the White-crowned Sparrow (Follett, Kobayashi, and Farner 1966). That part of the system ending in the pars nervosa is unaffected by photoperiod manipulation (Farner, Kobayashi, Oksche, and Kawashima 1964). Histologically the pars nervosa is very different from the median eminence and it exhibits changes in cholinesterase content and in ultrastructure when experimental birds are subjected to osmotic stress (Matsui 1964; Bern, Nishioka, Mewaldt, and Farner 1966). It appears to be involved in the secretion of vasotocin and oxytocin which are implicated in osmoregulation and oviposition (see p 89, also Jackson and Nalbandov (1969a)).

The acetylcholinesterase (AChE) which can be identified histochemically in the anterior median eminence of White-crowned Sparrows and other species, increases following exposure of the subject to a long photoperiod (Oksche, Laws, Kamemoto, and Farner 1959; Laws 1961; Farner *et al.* 1964; Matsui 1966). A similar concentration occurs in the anterior median eminence of Japanese Quail (Follett and Sharp 1968) while in the Brambling *Fringilla montifringilla* there is a correlation between the number, size, and shape of the AChE-responsive cells in the anterior pituitary and testicular activity of birds under natural spring and summer photoperiods (Haase 1970). In White-crowned Sparrows AChE activity has been shown to increase in the pars distalis following photo-stimulation and it appears to be associated with those cells which are responsible for gonadotrophin secretion (PAS-positive cells, see below) according to Haase and Farner (1969). Additional measurements of AChE in the caudal lobe of this species were made biochemically by Russell and Farner (1968) and again increases occurred following photo-stimulation at critical times of day (see Fig. 12.1, p.310).

It is to be noted that Gomori-positive neurones that arise in the paraventricular and supraoptic nuclei have axons that mostly terminate in the anterior median eminence associated with cholinesterases; some axons from the same sources also pass to the pars nervosa (Matsui 1966). In spite of these observations, Wilson (1967) has shown by lesion experiments (see below) that sectioning the anterior median eminence at the site of accumulation of AF-positive material does not prevent testicular growth in photostimulated White-crowned Sparrows. On the other hand, lesions in the preoptic hypothalamus suppress normal ovulation in domestic hens *Gallus gallus* (Ralph 1959) or ovulation induced by progesterone (Ralph and Fraps 1959). Butyrylcholinesterase (a non-specific cholinesterase) has also been identified in the pars distalis of White-crowned Sparrows in association with amphophilic cells in the cephalic lobe and in the pars tuberalis but, according to Haase and Farner (1971), these cells are involved in production or release of corticotrophic hormone (ACTH). Aldehyde—fuschin will also stain fine bundles of connective tissue and it may have been such material which was regarded by some authors as fine neurosecretory fibres entering the pars distalis. In fact, the only fibres in the adenohypophysis appear to be automatic fibres associated with the blood-vessels and these are not implicated in neurosecretion.

A second series of nerve tracts and nuclei are Gomori-negative and do not fluoresce, and they have been termed parvocellular nuclei (Dodd, Follett, and Sharp 1971) They originate from cell bodies in the basal hypothalamus in the infundibular nucleus (nucleus tuberis) and their axons form the large tubero-infundibular tract which extends into the median eminence (Wingstrand 1951). The cell bodies of these are known to be responsible for secreting various neurohormones that stimulate the cells of the adenohypophysis. Since many hormones are produced in the adenohypophysis several kinds of neurosecretory fibres are to be expected, each kind responsible for a specific releasing factor. So far, differences have not been established and it cannot be shown whether particular kinds of fibre are grouped into discrete tracts originating from discrete groups of cells. More information is available for mammals (Meites 1970

The adrenergic nervous system of the hypothalamus has been much studied since histochemical and pharmacological methods that made it possible to locate cellular monoamines were introduced (Falck, Hillarp, Thieme, and Torp 1962; Falck and Owman 1965). After appropriate histological preparation, 5-hydroxytryptamine (5-HT) and primary catecholamines produce a highly specific yellow or green fluorescence, respectively, when viewed in a microscope incorporating a mercury-vapour light source; the specificity can be checked (Dahlström and Fuxe 1964; Corrodi and Jonsson 1967). Catecholamines have been found in the sub-ependymal pallisade layer of the median eminence of the Feral Pigeon *Columba livia* var. (Fuxe and Ljunggren 1965) Japanese Quail (Sharp and Follett 1968, 1970), House Sparrow *Passer domesti* (Oehmke, Priedkalns, Vaupel-von Harnack, and Oksche 1969), and White-crowned Sparrow (Warren 1968). Previously, monoamine oxidase, which can be demonstrated with tetrazolium salts (Glenner, Burtner, and Brown 1957) and is thought to catabolis monoamines in the central nervous system, had been demonstrated in the median emi nence of the Tree Sparrow (Matsui and Kobayashi 1965) and White-crowned Sparrow (Follett et al. 1966). In both quail and pigeon fluorescing nerve axons extend from the lateral wall of the third ventricle, in the region of the ventro-medial nucleus (nucleus hypothalamicus posterior medialis), and converge medially on the floor of the ventricle in the tuberal nucleus (that is, the ventral part of the infundibular nucleus); dense areas of catecholamine fibres also occur in the preoptic recess and the paraventricular nucleus. Fig. 2.1 summarizes the main areas of fluorescence of this adrenergic system. A few adrenergic fibres can be located in the median eminence but at present it is not clear whether they are neurosecretory or function

[†]Follett (1973a) has clarified the terminology of the infundibular nucleus. The German workers distinguish various subdivisions of this large nucleus, which merges dorsally with the ventro-medial nucleus. Some American workers have combined the dorsal segment of the infundibular with the ventral part of the ventro-medial nucleus as the ventro-medial nucleus. This corresponds to the nucleus hypothalamicus posterior medialis of Sharp and Follett who used the term 'tuberal nucleus' for the remaining ventral part of the infundibular nucleus.

purely in nervous control mechanisms, for example, in regulating capillary dilation. It is probably significant that monoaminergic fibres do form synaptic connections with neurosecretory cells in the House Sparrow (Preidkalns and Oksche 1969).

It is evident that the median eminence is probably the site where neurohormones specifically involved in gonadotrophin secretion are produced. But, as mentioned previously, it seems unlikely that secretions in this area are primarily stimulated via the Gomori-positive cholinergic fibres originating in the supraoptic and paraventricular nuclei. The role of the adrenergic system just described needs further definition.

At present it does seem that the Gomori-negative fibres originating in the tuberal nucleus are the most important in stimulating gonadotrophin secretion. This conclusion is strongly supported by various experiments in which part of the basal hypothalamus is destroyed by electrolytic lesions. Fig. 2.3 summarizes some experiments performed on White-crowned Sparrows by Stetson (1969a). It is seen that lesions which destroyed both anterior and posterior divisions of the median eminence prevented photoperiodically induced testicular growth, a conclusion that was also reached by Wilson (1967). Destruction of the posterior median eminence alone abolished testicular growth in photostimulated subjects whereas loss of the interior median eminence had no inhibiting effect. Interruption of most of the fibres of the supraoptico-hypophysial tract had no effect on testicular growth, although it did result in a decrease in AF-positive material in the anterior median eminence (contrast the results of photostimulation previously discussed: p. 23). Destruction of regions of the ventral infundibular nucleus, without interference to the median eminence, sometimes suppressed the gonad response. The results supported the idea that hypothalamic control of testicular growth in this species is mostly mediated in the region of the parvocellular elements; presumably via the posterior part of the median eminence. However, lesions in the tuberal nucleus still implicate both the Gomori-negative nerve system and aminergic fluorescent elements. Lesions in the tuberal nucleus which do not interfere with the cholinergic AF-positive system also cause testicular regression or prevent growth in the domestic cockerel (Graber, Frankel, and Nalbandov 1967) and Japanese Quail (Sharp and Follett 1969b; Stetson 1969b, 1973). Lesions which do destroy the cholinergic tract of the quail before it enters the median eminence are without effect on the gonad response (see also results for the American Tree Sparrow Spizella arborea: Wilson and Hands (1968)).

These observations certainly suggest that the aldehyde fuchsin-positive neurosecretory system which arises in the magnocellular, supraoptic, and paraventricular nuclei and extends back into the median eminence and pars nervosa is not directly involved in controlling gonadotrophin secretion. Yet, some rôle for this system must exist because lesion of the AF-positive system does appear to prevent testicular development in the Pekin Duck (Gogan, Kordan, and Benoit 1963) and more recently Bouillé and Baylé (1973) found that lesions in the preoptic region caused ovarian and testicular regression in pigeons. Furthermore, infusion of 6-hydroxy-dopamine, which causes degeneration of adrenergic neurones, prevented testicular growth in photostimulated quail and inhibited luteinizing hormone (LH) (Davies and Follett 1974a). But, in spite of these observations, attention has been focused

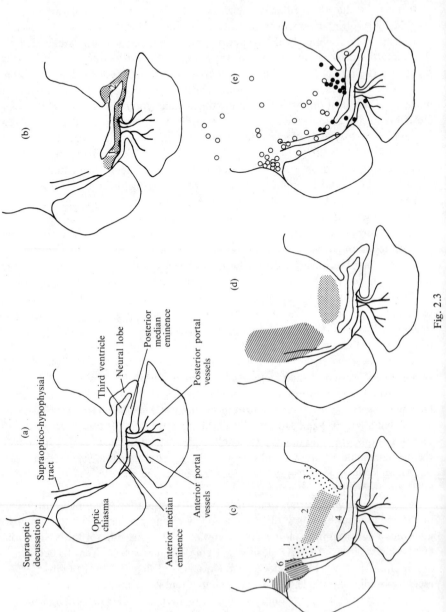

Fig. 2.3

on the posterior (tuberal) region of the hypothalamus, rather than the anterior region, as the source of gonadotrophin releasing sites.

Within the tuberal hypothalamus, two areas have been identified in quail which seem to be implicated in gonadotrophin secretion, and they may be the sites where gonadotrophin releasing neurohormones are synthesized. One is in the ventral part of the infundibular nuclear complex and the other is more dorsally situated in the posterior region of the nucleus. These two areas were regarded as all part of the same general area by Stetson, although they were distinguished by Sharp and Follett (1969b). Accordingly, in more recent studies, Davies and Follett (1975a) have sectioned various areas of the tuberal hypothalamus with micro-knives. Complete deafferentiation blocked testis growth and LH secretion remained low. It also blocked the normal rise in LH level that follows the castration of subjects held on long photoperiods. Severing anterior and antero-lateral afferents blocked gonadotrophin secretion, while posterior cuts were without effect. Electrolytic lesions placed in the postero-dorsal part of the infundibular nuclear complex blocked testis growth, while those in the ventral portion of the nucleus only partly disrupted gonadotrophin secretion. It appears that afferent nerve fibres entering the tuberal nucleus do so in a diffuse manner from both anterior and lateral directions and that only extensive lesions sever all connections.

The source of the fibres entering the tuberal nucleus was investigated further by Davies and Follett (1975b) in quail by making lesions and knife cuts in the anterior ventral hypothalamus. Lesions in the preoptic region of the anterior ventral hypothalamus blocked testis growth as did cuts immediately to the posterior. However, cuts which were slightly more posterior and others made rostral to the preoptic area, had no effect on gonadotrophin secretion. The effective region seems to be located immediately caudal to the septo-mesencephalic tract and extends only for about

---

Fig. 2.3. (see opposite). Sites that lesion experiments indicate are regulatory centres or pathways controlling adeno-hypophysial function. Lesion sites shown by stippling and hatching.

(a) Schematic sagittal section of the median eminance and adenohypophysis showing supraoptic decussation, supraoptico-hypophysial tract, infundibular nucleus, anterior and posterior median eminence, neural lobe, and the anterior and posterior portal vessels entering the caudal and cephalic lobes of the adenohypophysis respectively.

(b) Lesion of both divisions (1 and 2) of the median eminence of White-crowned Sparrows *Z. I. gambelii* prevents testis growth. Lesion of the anterior region only (1) allows gonad growth but lesion of the posterior part does not (Stetson 1969a).

(c) Lesions of the infundibular nucleus at sites (1) and (3) in the White-crowned Sparrow cause moderate suppression of gonad growth (Stetson 1969a) while at site (2) testis growth was virtually prevented. The same applied at site (2) in the Japanese Quail *Coturnix* (Sharp and Follett 1969b). In the Quail lesion of the nucleus tuberis (4) blocked gonad growth (Sharp and Follett 1969b). Site (5) in the preoptic area is apparently involved in ovulation in the domestic hen (Ralph 1959). Lesions of the supraoptico-hypophysial tract (6) do not prevent normal testicular growth in the White-crowned Sparrow.

(d) Possible regulatory sites for TSH (hatched) and ACTH (stipple) release based on Kobayashi and Wada (1973) quoting Kanematsu and Mikami (1970) and Frankel *et al.* (1967).

(e) Sites where micro-implants of testosterone propionate were effective (●) or ineffective (○) in causing testicular regression in Japanese Quail and a very low level of assayable gonadotrophin.

From Stetson (1972).

0·5 mm so that only a few cells appear to be involved in gonadotrophin secretion. The results suggest that the fibres from the preoptic region fan out directly behind this area and then enter the tuberal nucleus as a diffuse network (see also Oksche, Kirschstein, Hartwig, and Oehmke 1974). For this reason there is a large and rapid release of LH when electrical stimulation is applied to the preoptic and postero-dorsal part of the infundibular nucleus but a smaller release occurs if the supraoptic region is stimulated and virtually none if other sites are tested (Davies and Follett 1974b). Thus the preoptic region appears to regulate, by neural or neurosecretory pathways, neurones in the tuberal hypothalamus and Barry, Dubois, and Poulain (1973) have immunofluorescent evidence from the guinea pig that at least some LH-releasing factor cell bodies exist in this region and send processes to the median eminence. Davies and Follett (1975a) point out that LH secretion continues for 7—14 days when quail are transferred to short days (Nicholls, Scanes, and Follett 1973) but that following deafferation of fibres in the tuberal nucleus LH levels fall rapidly. This indicates that the carry-over effect of LH secretion following photoperiodic change resides outside the tuberal nucleus. At some site the rhythmic mechanism whereby the daily photoperiod is measured (see p. 318) is converted by an unknown neural transducer into a continuous stimulus which results in continuous neurohormone and gonadotrophin production.

Hormone-release factors

It has been claimed for the cockerel (Graber *et al.* 1967) that the posterior part of the tuberal nucleus controls the secretion of an LH-releasing factor (LRF) through the posterior portion of the anterior median eminence, while the anterior portion of the anterior median eminence may secrete an FSH-releaser (FRF). A similar claim has been made in the Japanese Quail (Stetson 1969b). In spite of these observations the work of Follett and his colleagues indicates that there is only a single releasing factor for luteinizing hormone and follicle stimulating hormone. A synthetic mammalian gonadotrophin hormone releasing hormone (GHRH but termed synthetic LRF by the makers) injected intravenously into quail caused both LH and FSH release, while plasma levels of FSH and LH appear to rise simultaneously following the exposure of quail to stimulatory photoperiods (Follett, 1975, 1976). A separate release factor seems to exist for thyroid stimulating hormone (TSH) and this also stimulates prolactin secretion (Bolton, Chadwick, and Scanes 1973).

A problem in the identification of neurosecretory pathways arises because it is by no means clear whether the systems operate with excitatory or inhibitory synaptic endings to regulate the production of various releasing factors, nor which synaptic transmitter could be involved in inducing their secretion. In mammals the presence of neurones containing dopamine, norepinephrine, and serotonin have been demonstrated. Furthermore, in rats the ability of stalk median eminence extract to release LH from the adenohypophysis (LH was measured by the ovarian ascorbic acid depletion assay, OAAD method, see below) is potentiated by dopamine, but not by noradrenaline, and it has been suggested that hypothalamic dopaminergic neurones synapse with the neurosecretory neurones, which produce the gonadotrophin-releasing factor (Schneider and McCann 1969, 1970). But, whereas the two primary catecholamines, noradrenaline and dopamine, can be distinguished in mammals using the drugs *m*-tyrosine and α-methyl-*m*-tyrosine, these agents do not differentiate the amines in the pigeon or quail (Fuxe and Ljunggren 1965; Sharp and Follett 1969*c*). It is possible that this is because there is a much greater proportion of noradrenaline in the bird (Juorio and Vogt 1967). Drugs such as reserpine, dibenamine, and dibenzyline interfere with catecholamine metabolism and affect the endocrine glands. It is possible that such adrenergic drugs have direct local effects on, say, the ovary, and it ought not to be inferred that they are influencing central nervous mechanisms (Ferrando and Nalbandov 1969). Indeed, there is good evidence in mammals that the releasing factors are not monoamines but short-chain polypeptides (Cross 1972) and the same appears to hold for the LRF of the domestic hen (Jackson 1971).

Cross (1972) remarks on the potency of these releasing factors in mammals: although 3 kg of pig hypothalamus may be needed to isolate 1 mg of LRF the minimum effective dose for a rat can be as little as 5×10^{-9} g. Follett (1970) showed that the potency of a material isolated from homogenized hypothalami of photostimulated quail, which was capable of releasing pituitary gonadotrophins (measured by the uptake of ^{32}P by the testes of chicks (Breneman, Zeller, and Creek 1962); see below), was not affected by boiling, whereas such treatment destroys the gonadotrophins. In this study, he also showed the vasotocin, oxytocin, vasopressin, noradrenalin, dopamine, putrecine, histamine, and spermidine were ineffective in stimulating gonadotrophin release *in vitro* from chicken adenohypophyses. Jackson and Nalbandov (1969*a,b,c*) found extracts of chicken hypothalamus to contain a factor (LRF) which stimulated the release of LH from the rat adenohypophysis *in vitro*, LH release being measured by the OAAD method. Arginine vasotocin (AVT) did not release LH when tested under these same conditions. Subsequent fractionation by ultrafiltration of the material obtained from the cockerel adenohypophyses gave two components which were each capable of depleting ascorbic acid when separately injected directly into the rat. One component had a molecular weight over, and the other less than, 12 000. Avian LH was retained in the heavy fraction and the active component in the lighter fraction appeared to resemble arginine vasotocin. The function of large quantities of AVT in the adenohypophysis is not yet clear but it does mean that the OAAD method cannot be used to measure LRF in crude pituitary extracts *in vivo*, as Jackson and Nalbandov point out. Smith and Follett (1972) used a technique of superfusing quail pituitaries for up to 12 hours and measuring the LH by radioimmunoassay. Hypothalamic extracts from quail were then added and significantly increased the rate at which LH was released.

Cytology of the adenohypophysis

Just as a lack of unequivocal histological techniques has, until relatively recently, hindered progress towards an understanding of the neurosecretory system of the hypothalamus, so has a full definition of the structure and function of the cellular components of the adenohypophysis been hindered. It has long been realized that different cell types could be classified according to their staining reactions: acidophils have an affinity for acid dyes, basophils for basic ones, and chromophobes are difficult to stain with any agent. And many years ago, Rahn (1939) demonstrated two types of acidophil in the pars distalis of fowl, while Payne (1942) subsequently distinguished the acidophils of the caudal lobe as A_1 cells in contrast to the A_2 cells of the cephalic lobe. But confusion has arisen because there are so many staining techniques—many of them requiring a degree of skill—which have proliferated new terminologies for the cells that are coloured. Unfortunately, the tinctorial affinities of the cells demonstrated by the early workers were not related to functional differences and in some cases simply visualized the same cells at different stages in their morphological and histochemical development. New histophysiological techniques changed the situation so far as chromophile cells were concerned by demonstrating that these comprised two classes: one containing granules of glycoproteins, the other granules of simple proteins. Since the hypophysial hormones also comprise two classes, one having a glycoprotein element and the other containing no carbohydrate, the base for a more sound functional differentiation was being prepared. With these advantages Purves (1966) was able to bring order to the existing chaos so far as mammals were concerned in a valuable review. In the case of the birds, Herlant (1964) was able to classify the cells of the pars distalis of the Pekin Duck using morphological and functional criteria. Evidence emerges that cells of the avian pituitary which have the same staining properties as those of the mammalian gland serve the same function (Gilbert and Amin 1969; Amin and Gilbert quoted by Stockell-Hartree and Cunningham (1971)).

Much of our knowledge of the histology of the avian pars distalis depends on the excellent studies of Dr. Tixier-Vidal and her colleagues in the laboratory of Molecular Biology at the College of France, Paris, and this account relies extensively on her studies. A cell classification using Greek letters, which was first proposed by Romeis (1940), is now generally accepted and it is now also possible to use a functional terminology. In birds and lower vertebrates there is a marked spatial segregation of different cellular forms in the two lobes of the pars distalis, resulting from the early development of Rathke's pouch (p. 18) but these topographical localizations become much less distinct in mammals. Table 2.1 summarizes the cellular composition of the avian adenohypophysis based on cytological studies of the White-crowned Sparrow (Matsuo, Vitums, King, and Farner 1969) and Japanese Quail (Tixier-Vidal, Follett, and Farner 1968) and emphasizes the confusion possible if reliance is placed solely on a tinctorial classification. Tixier-Vidal and her colleagues have made considerable use of the Herlant tetrachrome staining method in studies of the Pekin Duck (Tixier-Vidal 1963), Japanese Quail (Tixier-Vidal *et al.* 1968), and domestic pigeon (Tixier-Vidal and Assenmacher 1966). It is a histological technique first recommended by Herlant (1960), being a combination of Alcian blue, PAS (Periodic acid—Schiff), and Orange G. Its use allows the simultaneous identification of various glycoprotein-containing cells (see Pearse 1968). Matsuo and his collaborators were unsuccessful with this method and instead used their own modification (Matsuo tetrachrome method; see Table 2.1). The old category of acidophil

TABLE 2.1

Staining properties and functional designation of the cells of the avian adenohypophysis

Staining technique	Species	Caudal lobe		Caudal and Cephalic		Cephalic		
		Acidophil (A_1) alpha	Basophil gamma	Basophil delta	Acidophil kappa	Basophil beta	†Acidophil (A_2) epsilon	†Acidophil (A_2) eta
Tetrachrome (Herlant)	JQ	Red–orange	Purple	Pale blue	Very deep blue	Purple–blue	Purple–blue	Rose
Tetrachrome (Matsuo)	WCS	Orange	Purple	Light green	?	Violet	Possibly violet	Possibly violet
Iron–haematoxylin	WCS	Black	Not stained	Not stained		–	Black	Black
Azan (Heidenhain)	WCS	Orange	Not stained	Not stained		Not stained	Deep blue	Deep blue
Modified PAS–methyl blue (Matsuo)	WCS	Very slight	Red	Red		Red	Deep or light purple	Deep or light purple
Brookes	JQ	Yellow–orange	Brown–orange	Pale green	Light green	Blackish	Rose and green	Rose
Alcian blue–PAS–Orange G	JQ	Pale yellow	Brick (yellow + rose)	Pale blue	Not distinguished	Deep rose	Yellow brick	Very pale yellow
PAS–haemalum–Orange G	JQ	Yellow	Brick (yellow + rose)	Pale rose	Brown	Deep rose	Yellow brick	Yellow
Lead haematoxylin–Orange G	JQ	–	Grey (or black in some species)	Not stained	Deep black	Deep rose and grey	Grey	–
Function of cells		Somatotrophic STH serous	LH mucoid	TSH mucoid	MSH	FSH mucoid	Corticotrophic ACTH serous	Prolactin (serous)‡

JQ = Japanese Quail (based on Tixier-Vidal *et al.* 1968).

WCS = White-crowned Sparrow (based on Matsuo *et al.* (1969) modified by Mikami, Hashikawa, and Farner 1973).

†Collectively designated as amphophils in White-crowned Sparrow.

‡Prolactin is not a glycoprotein but the prolactin cells respond histologically as mucoid cells (see text).

cells correspond with those whose granules are simple proteins and they are referred to as serous cells. The mucoid cells with mucoproteinaceous granules represent the old basophil cells. Both kinds are described below.

Chromophobes of several kinds have been described in the pars distalis of birds. They are always small cells with a central nucleus and they usually occur in clusters throughout the gland. Actually they probably represent a single kind of undifferentiated cell at various stages of development into other cell types; differentiation of the chromophobes is most readily demonstrated by experimental manipulation of the subject. These cells arise soon after and in association with the penetration of blood-vessels into the embryonic hypophysis and their subsequent development is consequent on their location. During embryonic development blood-vessels divide the hypophysial primordium into trabecules or cords which anastomose throughout the organ. The cells become arranged in tiers along these cords, with the peripheral elements sometimes forming a columnar layer and with those inside organized as an unbroken lining round small particles or pools of colloid material. These are not real follicles, for they are really only pseudovesicles formed by the bridging cords, but they become surrounded by a capsule of connective tissue containing numerous blood-vessels so that gland cells and vascular elements are closely juxtaposed. The organization of cells into such 'follicles' is more obvious in the case of differentiated cells and these often have cytoplasmic processes extending into the lumen of the glandular trabecule. Electron microscopy reveals an ultrastructure of endothelial cells separating the gland cells from the sinusoids, the cytoplasm of the endothelial cells, at least in the mole and bat, being pierced with pores (Herlant 1964). In some mammals particulate material can be seen to flow from the hypothalamus to the sinusoids of the hypophysis, whereupon it presumably enters the gland cells. It is not clear whether a reciprocal discharge of material from the gland cells occurs into the sinusoid. Instead, spaces filled with amorphous material develop between the endothelial cells and the gland cells while the endothelial cells also contain numerous small vesicles (see Herlant 1974).

The different kinds of gland cells are not homogeneously distributed along the cords and relatively more of a particular kind can be found in different parts of the organ. Thus, although all kinds of cells originate from a common ancestral cell and are mixed together (except in the case of cells confined to a particular lobe which are morphologically distinct from their first appearance), it seems that at different stages of development in the life cycle of the animal specific conditions induce the formation of predominantly one kind of cell. However, the mucoid and serous cells are differentiated early. In the chick embryo Mikami *et al.* (1973) have shown that membrane-bound secretory granules begin to occur in the cytoplasm of cephalic lobe cells by the seventh to eighth day of incubation and by the ninth day two types of glandular cells are distinguishable. Differentiation of acidophils and basophils occurs in the 11-day embryo. The subsequent development of these cells to attain a secretory capacity is delayed until appropriate stimulation from higher centres, as in the case of prolactin cells described below. Experimental photostimulation of a Japanese Quail *Coturnix coturnix* previously reared to maturity on short daylengths causes the glycoprotein-containing beta-cells to increase in abundance and in turn the testes become enlarged. Hence, messenger releasing factors entering the sinusoids of the pars distalis presumably pass through the endothelium and determine the development of the gland cells.

Mucoid cells of the adenohypophysis

The mucoid cells containing glycoproteins (basophiles) can be differentiated into three types according to their response to PAS, aldehyde fuschin, and alcian blue.

Beta- (β-) or FSH cells

These are markedly PAS-positive, remain pink with Alcian blue, do not react with lead haematoxylin, and stain dark green with aldehyde–fuchsin light green. They also exhibit a selective affinity for aniline blue or methyl blue, if stained by a trichrome method such as the Masson–Mallory –Heidenhain. According to Herlant's (1964) nomenclature, these cells are the beta-cells which are confined to the cephalic lobe of the pars distalis. They are small cells with a cytoplasm full of fine granules. In Japanese Quail, the beta-cells are very small and thinly scattered in immature birds but many become differentiated from embryonic cells in subjects held on stimulatory photoperiods. They initially increase in volume and the cytoplasm becomes filled with PAS-positive material which then disappears coincidentally with the period of maximum testis growth and pituitary gonadotrophin content (using the assay method discussed below of Breneman et al. 1962). This strongly suggests an FSH-secreting function, and a general gonadotrophin-producing function for 'basophils' was initially proved in the domestic fowl (Kato 1939; Matsuo and Kato 1961). In both the Pekin Duck (Tixier-Vidal and Benoit 1962) and the ploceid Red Bishop *Euplectis orix* (Gourdji 1965), spermatogenetic development is associated with activity in the beta-cells, while in the duck and quail the beta-cells are noticeably activated in photostimulated castrates, in which there is a marked elevation of assayable gonadotrophin in the adenohypophysis. Even more convincing evidence for an FSH-producing rôle of these cells is noted in the temperate bat *Myotis myotis*. Following copulation in autumn, ovulation and fertilization are delayed until spring. During the female's intervening suboestrous, successive waves of abortive follicular growth occur at which time the pars distalis contains no other gonadotrophic cells but beta-cells (the gamma-cells are completed involuted). In the male, spermatogenesis begins at the beginning of summer and ceases in autumn, at which time the interstitial cells and secondary sexual organs develop: during the summer the hypophysis contains only well developed beta-cells (see Herlant 1964).

Gamma- (γ-) or LH cells

Other basophils are found in the caudal lobe which are only slightly PAS-positive, though they do give a dense purple reaction to Herlant's tetrachrome in the quail. In the fowl these cells have been referred to as amphophils or PAS-purple cells, because of their ability to stain with basic dyes, yet additionally having an affinity for acid ones. In the fowl these cells are confined to the cephalic lobe. In the White-crowned Sparrow similar cells which are confined to the cephalic lobe appear to correspond to epsilon-cells (see below). The use of the term amphophil is best avoided for other acidophilic cells of the cephalic lobe have similar properties. These last are the cells designated as A_2-cells by Payne (1942) and V-cells by Matsuo (1954) and Mikami (1954), and are discussed below as epsilon-cells. To return to the gamma-cells, their appearance associated with the rut of the male bat and mole and with gestation in the female infers an LH activity (Herlant 1964). Confusion in other species has depended on their variable staining affinities and due in the main to a high degree of inter-specific variation in glycoprotein content and to seasonal factors. In the quail they lose their granular inclusions at the start of photostimulation and become very markedly chromophilic in subjects held on non-stimulatory photo-periods.

Delta- (δ-) or TSH cells

These basophils respond to alcian blue but have granular inclusions which are some-times markedly PAS-positive. Generally they are difficult to stain, being nearly chromophobic. They occur in both lobes of the Pekin Duck, Japanese Quail, and Feral Pigeon but, according to Mikami (1958), only in the cephalic lobe of the fowl. These cells are altered by agents which affect thyrotropic function, for example, by the inhibition of secretion with thyroxine or a stimulation of secretion and synthesis with thiourea, and this indicates them as the source of TSH. They are activated by photostimulation in the quail (Tixier-Vidal et al. 1968) which indicates that a rise in thyroid hormone secretion occurs under these conditions. Baylé and Assenmacher (1967) noted a reduction of radioactive iodine fixation by the thyroid of male quail held on LD 18:6 (18 h light : 6 h dark) and suggested that an increase of thyroid hormones occurs at this time. The significance of this is not known at present, although thyroid hormones are involved in moulting and other photoperiodically induced functions (but see p. 322).

It is evident that while the mucoid cells are readily identifiable as a group, they are less readily distinguished individually. Indeed, in White-crowned Sparrows, Matsuo et al. (1969) obtained basophils which stained either pale green (large cells) or dark green (smaller cells and more oval in shape) with their modified tetrachrome and these cells were distributed in both lobes. They judged from the results of photostimulation or castration, which caused these cells to become activated, that they were the source of gonadotrophins. It is possible that the large cells they ident-ified (light basophils) were delta-cells but more studies will be required to resolve this problem. It has been realized since the original studies of Kato and Nishida (1935) that gonadectomy leads to the hypertrophy of certain cells of the pars distalis. Some of the cells assume a very large size and characteristic shape and have been called signet-ring cells (Payne 1942, 1961). It is now clear that castration, which eliminates feed-back inhibition from the gonad (see p. 65), leads to the hypertrophy of delta- (TSH-) cells as well as of gonadotrophs. In fact, in Pekin Ducks (Tixier-Vidal and Benoit 1962) and Quail (Tixier-Vidal et al. 1968) the delta-cells respond more to castration than the beta- or gamma-cells, while in the quail these latter two cell types only hypertrophy in photostimulated castrates. Similarly in White-crowned Sparrows, Matsuo *et al.* (1969) found no castration cells in subjects which were gonadectomized during the refractory period (p. 227) while morphologically and histochemically the castration cells were similar to the large light basophils (thought to be delta-cells).

Eta- (η-) or prolactin cells

The cells appear to correspond with the A_2-cells of Payne and the V-cells of Wingstrand (1951). They vary in their histology. They may contain fine granules in which case the cytoplasm tends to colour purple–blue with tetrachrome, and this makes them very similar to the beta-cells. On the other hand, if the granule inclusions are distinct they are markedly erythrosinophilic and stain rose–red. Prolactin is not a glycoprotein although the prolactin cells respond histochemically as mucoid cells containing glycoproteins (Tixier-Vidal and Picart 1971). They are PAS-negative but after fixation in glutaraldehyde and paraffin embedding, or GMA embedding, become slightly PAS-positive. Eta-cells could be seen in the peripheral regions of the pituitary of the quail but not very easily. Photostimulation of castrates caused the eta-cells to become degranulated and they were judged to be prolactin cells, since they gave many histochemical responses comparable with the prolactin cells of mammals. These cells are much more easily seen in the Pekin Duck, Feral Pigeon, and domestic fowl because they have a very acid cytoplasm, stain bright yellow with Orange G, and possess large erythrosinophilic granules. These luteotrophs are not very visible in the laying hen, which secretes little or no prolactin, but they are readily visible in non-laying birds. During the initial studies of the White-crowned Sparrow by Matsuo *et al.* (1969) and Mikami, Vitums, and Farner (1969), prolactin cells were not distinguished with certainty. More recently, Mikami, Farner, and Lewis (1973*a*) have used subjects which were incubating or brooding to confirm, by electron microscopy, that eosinophilic cells of the cephalic lobe are involved in prolactin secretion. These cells are infrequent, granulated components at the beginning of the breeding cycle. They increase in number to become the most prominent cell type with the onset of incubation and at this stage become degranulated, develop a prominent endoplasmic reticulum and Golgi apparatus. During the later stages of brooding the prolactin cells regress with involution of the endoplasmic reticulum and Golgi apparatus, and granules and lysosomes again accumulate in the cytoplasm.

Serous cells

Originally two kinds of cell which were acidic in staining reaction were recognized, confined either to the caudal or cephalic lobes respectively. Payne (1942) called those of the caudal lobe A_1 acidophils to distinguish them from the lighter staining cephalic A_2 acidophils. The former are synonymous with the alpha-cells.

Alpha- (α-), somatotropic, or STH cells

These are the classic acidophil cell of the caudal lobe, where they comprise the most common cellular element. They are spherical cells with a prominent nucleus and many granules and they stain orange—red with Herlant's tetrachrome, in distinction to the gamma-cells with which they are associated. Cells that appear analogous with the STH cells of the rat can be seen to be numerous in young Japanese Quail but decrease in winter with the approach of maturity (Tixier-Vidal *et al.* 1968). Alpha-cells appear to be restricted to the caudal lobe in the chicken (Amin and Gilbert 1970).

Two acidophilic cells which are confined to the cephalic lobe are difficult to demonstrate.

Epsilon- (ε-) ACTH cells

These form palisade layers in the cephalic lobe of the quail but they become prominent only after experimental treatment with metapirone, which blocks the synthesis of adrenocortical steroids, and other drugs which interfere with ACTH. Tixier-Vidal *et al.* (1962, 1968) have shown the homology between these corticotropic cells in quail and Pekin Duck. They draw attention to the PAS-positive fine granules of these cells and the flocculent appearance of the cytoplasm which stains with Orange G. But, since no cells secreting adrenocortical hormone have been identified in the domestic fowl, it is conceivable that a separate source of ACTH exists in this species (Nalbandov 1966).

Kappa- (κ-) MSH cells

These were initially demonstrated in the cephalic lobe of the Pekin Duck (Tixier-Vidal *et al.* 1962) but could be identified in both lobes of the quail, forming strands of rounded cells. They are medium-sized ellipsoidal cells, with very acidic cytoplasm containing pronounced chromophilic granules which stain deep blue with Herlant's tetrachrome. A distinctive feature of these cells is their intense black colouration with lead haematoxylin. In the Pekin Duck they are located in zones that are rich in melanophorotropic hormone (Tougard 1971) and for various other reasons this function has been attributed to them in both species. It should be mentioned that an MSH activity has also been attributed to the V-cells (Wingstrand 1951; Matsuo 1954). The close association between MSH cells and gonadotrophs is of interest in view of a relationship between melanism and photoperiodicity (see p. 432).

Before terminating this survey of pituitary cytology it ought to be remarked that Matsuo *et al.* (1969) detected three kinds of acidophils in the caudal lobe of the White-crowned Sparrow. With tetrachrome, one kind stained orange and these were probably the alpha-cells but the identity of red acidophils and small acidophils with a slight affinity for basic dyes is still not clear.

Hormones of the adenohypophysis

The pituitary hormones secreted from the cells of the adenohypophysis are: thyrotrophin (TSH), which acts on the thyroid gland; adrenocorticotrophin (ACTH), whose target gland is the adrenal; the somatotrophic or growth hormone (GH);

intermedin or melanophore stimulating hormone (MSH), which regulates melanin deposition; the luteotrophic hormone prolactin (LTH); and the two gonadotrophins, follicle stimulating hormone (FSH) and luteinizing hormone (LH). FSH and LH (together with TSH) are glycoproteins with a molecular weight in the range 25 000– 30 000 (Butt 1967). Mammalian FSH has a peptide chain of 17 amino–acids, which are shared with LH, plus tryptophan (LH additionally has ammonium ions and the carbohydrates galactose, mannose, glucosamine, fucose, and N-acetyl-D-neuraminic acid. Mammalian LH is thought to be a globular protein, with little or no α-helix, having a rigid structure resulting from disulphide cross-linkages (Butt 1967). It has the same carbohydrates as FSH but is relatively low in neuraminic acid and unlike FSH also contains uronic acid. Since they have a similar chemical composition, it is not surprising that mammalian LH and FSH exhibit many similar biological properties in birds. To date most research involving the treatment of birds with exogenous gonadotrophins has relied on these mammalian preparations, although currently avian preparations are being purified and a degree of immunological specificity identified. Until the chemical composition of avian gonadotrophins is known no detailed comparison of avian and mammalian FSH and LH is possible. Indeed some workers, notably Nalbandov (1959a) and van Tienhoven (1959), have claimed that separate FSH and LH fractions do not exist in birds, there being instead a single gonadotrophin complex. This view is refuted by pituitary cytology, and by assays made in conjunction with histological observations (see below).

The first workers to separate two gonadotrophic fractions from chicken pituitaries were Stockell-Hartree and Cunningham (1969) who used methods employed for the extraction of mammalian gonadotrophins; they treated acetone-dried chicken pituitary powder with 6 per cent ammonium acetate (pH 5·1) and 40 per cent ethanol and then separated the glycoprotein fraction by chromatography using a CM-cellulose column with a gradient of ammonium acetate. The first fraction (CM1) was not bound to the column and eluted directly. When tested by mammalian assays it had properties that were essentially equivalent to those of mammalian FSH. The fraction was then purified by chromatography on DEAE-cellulose and hydroxyl-apatite to yield a 100-fold increase in potency over the crude pituitary powder. An increase in the ammonium acetate eluant enabled a second fraction (CM2) to be obtained and this predominantly exhibited the properties of mammalian LH. It could be purified by chromatography on DEAE-cellulose using Amberlite IRC-50. The same extraction method was also followed by Follett, Scanes, and Nicholls (1972b) (see also Scanes and Follett (1972)) who used the LH fraction to develop a radioimmunoassay procedure which enables the corresponding material—designated immuno-reactive LH or IR-LH—to be assayed in other avian species. These workers purified the FSH fraction (CM1) with calcium phosphate gells and chromatography on DEAE-cellulose and Sephadex G^{100} to yield fractions giving less LH activity. A major problem encountered in purifying the CM2 fraction was to remove activity due to TSH, which in mammals has properties which are very similar to those of LH (Pierce, Liao, Howard, Shome, and Cornell 1971). Moreover, it seems that neither of the purified preparations was homogeneous on acrylamide gel electrophoresis or gel-filtration (Stockell-Hartree quoted by Stockell-Hartree and Cunningham (1971)). Since the same extraction procedure can be used for mammalian and avian gonado-trophins it may be inferred that they have chemical properties which are basically the same. Nevertheless, there is no doubt that they differ in biological effectiveness and this has been the primary cause of the uncertainty attached to various assays and studies. Thus, as will be shown in the next chapter, mammalian ovine FSH as supplied by National Institute of Health, Bethesda (N.I.H.FSH) is less effective than mammalian N.I.H.LH in stimulating spermatogenesis in the avian testis, in the Feral Pigeon, House Sparrow *Passer domesticus*, and Japanese Quail.

Knowledge of the assay methods used for detecting and quantifying the gonado-trophins is important for an informed interpretation of results. Until relatively recently the most reliable bioassay for LH has been the rat ovarian ascorbic acid depletion assay (OAAD) due to Parlow (1961), (see also Heald, Furnival, and Rookledge 1967), but as mentioned above it is influenced by arginine vasotocin. The Weaver Finch assay developed by Witschi (1955) is supposedly specific for LH but it has not been tested using purified avian gonadotrophins. The best assay for FSH has depended on the method of Steelman and Pohley (1953). These workers showed that a high dosage of human chorionic gonadotrophin (HCG) makes the ovaries of intact immature rats highly sensitive to exogenous FSH, the relationship between FSH and ovarian weight being linear for certain dose ranges. Reasonable quantities of other pituitary hormones, including LH, do not interfere with this response. Both this ovarian augmentation assay and the OAAD method have been shown to be suitable for measuring avian FSH and LH respectively (Furr and Cunningham 1970), but chicken (avian) hormones are less potent than mammalian hormones in these assays. Even in mammals, where these assays have a high speci-ficity, their sensitivity is low so that pooled pituitary extracts have been required and plasma assay has been impossible.

One of the first studies of seasonal variations in gonadotrophin output in a wild bird involved Pheasants *Phasianus colchicus.* Greeley and Meyer (1953) injected their dried pituitary extracts into day-old domestic chicks and measured the increase in testis weight. Since then it has been discovered that the ability of the chick testis to take up radioactively labelled phosphorous [^{32}P], which can be expressed as counts/min per mg, provides a much more sensitive assay end-point than the small increase in gonad weight which results from stimulation with anterior pituitary extracts (Breneman *et al.* 1962). The uptake of [^{32}P] is markedly increased by admin-istration of N.I.H.LH, for which it has been used as an assay for quantities in the order of 1·0–1·5 μg. A dosage of nearly ten times as much N.I.H.FSH by weight was needed to result in responses comparable with those caused by LH. The potency of pituitary gonadotrophins from the chicken assayed against mammalian LH as stan-dard is 40 times greater with the chick testis than with the OAAD assay using the rat (Follett, unpublished). But although this is a more sensitive assay to chicken gonadotrophins than the mammalian tests it has been found that purified FSH and LH are less active than crude gonadotrophin extract, so it has become apparent that the test is really sensitive only to a synergistic mixture (Furr and Cunningham 1970).

Burns (1969) modified the [^{32}P] method by first treating his extracts with neuraminidase, which destroys FSH activity in chicken pituitary preparations.

The introduction of radioimmunoassay techniques greatly increased the sensitivity of gonadotrophin assay, but at first some controversy attached to their specificity. Only weak cross-reactions, if any, were obtained between chicken and mammalian gonadotrophins as demonstrated by such immunological techniques as gel diffusion (Moudgal and Li 1961), complement fixation (Desjardins and Hafs 1964), or haemagglutination-inhibition (Stockell-Hartree and Cunningham 1969), though Cunningham (quoted in Stockell-Hartree and Cunningham (1971)) has noted a cross-reaction between human hormones and chicken LH by complement fixation. Nevertheless, Bagshawe, Wilde, and Orr (1966) developed a method of assaying antigenically active HCG and LH (these two hormones cross-react in the system employed) in human body fluids and their particular antiserum appeared to cross-react with avian LH. The method depends on allowing an unknown sample of HCG–LH to compete with a measured HCG sample labelled with radioactive iodine, for binding sites on rabbit antibodies to HCG. To assist further aggregation the mixture is incubated with a second antiserum—an antiserum to rabbit γ-globulin raised in guinea-pigs—and the resulting aggregates filtered. The amount of radioactivity in the filtrate is inversely related to the amount of HCG or LH in the samples; 100 per cent recovery of isotopic HCG is equivalent to no unknown LH or HCG in the sample. This particular antiserum has been shown to cross-react with heterologous plasma and pituitary extracts from various species, which compete with [^{125}I] HCG for HCG binding sites on the anti-HCG antibody, including the domestic fowl. In fact, this last species gave dilution curves which approached parallelism with HCG more closely than any of the mammal species examined (Bagshawe, Orr, and Godden 1968). Good dilution curves have since been obtained with plasma from other birds—it can of course be assumed that these contain no HCG—and these include the Mallard *Anas platyrhynchos*, Feral Pigeon *Columba livia*, Wood Pigeon *Columba palumbus*, House Sparrow *Passer domesticus*, and Greenfinch *Carduelis chloris*. Parallelism in this case between the dilution curves of the sample and the reference preparation is not proof that the active factors are identical but it does indicate that antigenic groups of some avian species are closely related to the antigenic groups present in HCG and human LH; it could of course, indicate a complete cross-reaction in the assay system.

Further evidence that the Bagshawe antibody system measured avian LH was obtained from histological studies of the testes of Greenfinches that were exposed to skeleton asymmetric light schedules, in the manner shown in Fig. 2.4 (see also p. 296). Schedules which gave high plasma titres of IR-LH coincided with a marked stimulation of the interstitial Leydig cells without any spermatogenetic development, as discussed in the next chapter (see also Murton, Bagshawe, and Lofts 1969*a*; Murton, Lofts, and Westwood 1970*b*). In contrast, full spermatogenetic development was noted under different light schedules which were associated with an exhaustion of the Leydig cells and only base-line titres of plasma LH. It was assumed that spermatogenetic development depended on FSH activity, which unfortunately could not be measured directly. These same experiments established that there was a tem-

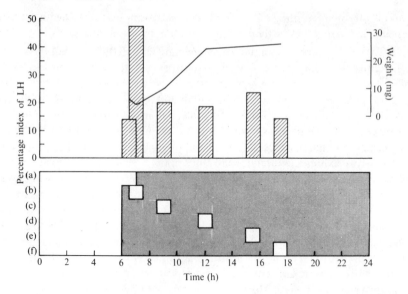

Fig. 2.4. Effect of six different light schedules (a)–(f) for 19 days on plasma IR-LH (Bagshawe) titres in groups of photosensitive Greenfinches *Carduelis chloris* (histograms) and on mean testis weights. All birds had been held on schedule (a) at the start of the experiment and a control group was continued on this regime throughout. From Murton *et al.* (1970*b*).

poral difference in circulating plasma concentrations of IR-LH and a tubule stimulating agent (thought at the time to be FSH, but it could have indicated androgen
activity), consistent with them being controlled by different circadian rhythms.
This topic is discussed in more detail in the next chapter and again in Chapter 11
(in passing, it is also worth noting that Fig. 12.1 (p. 310) suggests a diurnal variation
in acetylcholinesterase (AChE) activity in the pars distalis of the White-crowned
Sparrow). It was fortuitous that an apparently distinct FSH and LH activity could
be partitioned in this way in the Greenfinch and at the time it was argued that this
was evidence that two gonadotrophins were implicated in the control of the avian
gonad.

The prime drawback of the Bagshawe assay for use with bird plasmas was its lack
of sensitivity so that pooled samples were needed to obtain significant results. It was
also a lucky chance that it cross-reacted at all for a large number of anti-HCG sera
were subsequently tested by Follett and his colleagues for their ability to bind
[^{125}I] chicken LH, or for unlabelled avian LH to inhibit the binding of [^{125}I]HCG,
and only exceptionally could a weak cross-reaction be obtained. The ability of some
radioimmunoassays to cross-react may well depend on the uniqueness of the antiserum (Midgley, Niswender, Gay, and Reichert 1971) and certainly the Bagshawe
anti-HCG serum appears to be exceptional in its capacity to cross-react with avian
material. It remains to be resolved exactly what each assay is detecting in the bird,
for it is possible that only part of glycoprotein molecule reacts antigenically. There
is also the problem of homology between avian and mammalian gonadotrophins to

be resolved. Both the 'FSH' and 'LH' fractions obtained from the chicken pituitary stimulate gonadal growth in young chicks, judging by the Breneman assay (Furr and Cunningham 1970; Scanes and Follett 1972). If these fractions are injected into immature Japanese Quail they both stimulate Leydig cell and tubule development. So do both ovine and bovine LH and FSH. Since we must suspect that synergistic actions will prove to be important it is evident that much more work will need doing before this topic is settled. We are only just acquiring an immunoassay for avian FSH. Scanes and Godden (1975) have developed an anti-serum against avian FSH while Follett and his colleagues have employed a heterologous assay for avian FSH using an anti-ovine FSH serum (Follett, 1975, 1976).

To avoid as much as possible problems of specificity, Cunningham, Myres, and McNeilly (1970) used a rabbit antiserum to chicken LH, extracted and purified as above, and an antigen of the same LH labelled with [^{125}I]. Their method gave a linear inhibition curve against the unlabelled LH fraction as standard, but it appeared to be influenced by a TSH cross-reaction, although not by FSH. Subsequently the method was used to develop an LH assay of the same post-precipitation, double-antibody type, using an antiserum to purified LH from the Japanese Quail (Follett et al. 1971, 1972a; Scanes and Follett 1972). This was made possible by the separation and purification of two avian gonadotrophic factors (an LH and FSH from chicken pituitaries), as mentioned above (Stockell-Hartree and Cunningham 1969; Scanes and Follett 1972). The standard chicken LH antiserum cross-reacts well with quail pituitary extracts and also gives parallel dose—response curves with various other birds, but no, or poor, cross-reaction with mammalian material (Scanes, Follett, and Goos 1972a). TSH cross-reaction in this system is thought to be minimal, but the last point remains to be fully resolved (Sharp, personal communication). In quail, the plasma level of IR-LH is reduced in sexually mature birds given testosterone, it is elevated by castration, is unaffected by thyroidectomy or by injections of thyroxine, and is virtually eliminated by hypophysectomy (Follett et al. 1972a). Moreover, injection of purified chicken LH into quail held on short days causes Leydig cell development (Follett, Scanes, and Nicholls 1972b; Brown, Baylé, Scanes, and Follett 1975). IR-LH plasma titres in male quail reared for 28 days on short daily photoperiods (LD 8:16) before transfer on day 0 to long daylengths (LD 20:4) reached a plateau concentration after about 8 days coinciding with growth of the seminiferous tubules; this growth continued for 25—30 days (Scanes, Nicholls, and Follett 1972b).

Prolactin is a protein hormone with a molecular weight of about 23 000 (Lyons and Dixon 1966) which was first discovered by Oscar Riddle, in the early 1930s, through its ability to stimulate the production of pigeon milk from the crop gland of the pigeon (see recent review by Riddle (1963a)). Pigeons are unique among birds in producing a milk having a chemical composition which is close to that of mammals (see Appendix 1). It is remarkable that its secretion should depend on the same pituitary hormone in birds and mammals and that no other avian crop has evolved the same capacity. Of course the uniqueness of the pigeons is in the response of their oesophageal glands not in the possession of prolactin. The capacity of prolactin to stimulate the pigeon crop-sac gland cells provides the basis for its bioassay, details of which are given by Lyons and Dixon (1966) and Nicoll (1967). A more precise but less sensitive method relies on gel electrophoresis to separate prolactin from other pituitary hormones and details are given by Nicoll, Fiorindo, McKenee, and Parsons (1970). In mammals excess amounts of prolactin are secreted if the pituitary is separated from the hypothalamus, whereas this is not the case with TSH, ACTH, and the gonadotrophins. This suggests that prolactin secretion is normally inhibited by hypothalamic factors which must be blocked for secretion to occur (Schally, Arimura, Bowers, Kastin, Sawano, and Redding 1968). The converse seems to apply in birds, at least in chickens, for hypothalamic extracts stimulate prolactin secretion *in vitro* (Meites and Nicoll 1966) while transplantation of the chicken pituitary does not result in an increase in prolactin secretion (Ma and Nalbandov 1963); a prolactin releasing factor has been identified in turkeys (Chen, Bixler, Weber, and Meites 1968). The functional significance of these differences are not clear, although it is known that prolactin secretion in birds can be initiated by appropriate visual stimuli from the eggs (see Chapter 6).

The action of prolactin and the gonadotrophins are considered in more detail in relation to their effect on target organs in the next chapter. Feed-back mechanisms are also conveniently discussed following consideration of the gonadal hormones. The contribution of the other pituitary hormones, particularly TSH, MSH, ACTH, and the neurohypophysial hormone vasotocin, to the avian breeding cycle will be mentioned in appropriate chapters. Scanes and Follett (1972) have commented on chicken TSH and shown it to be chemically similar to mammalian TSH, while chicken ACTH has been assayed (Salem, Norton, and Nalbandov 1970*a*); these same authors (1970*b*) have shown vasotocin to be involved in ACTH release in the chicken. The chemistry and nature of the other avian pituitary hormones is not at present defined and in most cases their existence is presumed from causal effects which parallel the action of equivalent hormones in mammals.

Pineal gland

Until relatively recently it was assumed that the pineal gland in birds was a non-functional rudiment of the reptilian 'third eye'. Its ablation appeared to have no obvious adverse effect on the gonads or on other body components. But it is now clear that the pineal is a secretory gland and that it is implicated in the circadian periodicity of motor activity and in the clock mechanism that underlies photosensitivity. This aspect is dealt with in Chapter 11 and our concern here is to summarize the structure and physiology of the gland. As with so many other aspects of functional anatomy, more is known about the mammalian situation and it is not necessarily justifiable to assume that avian organs are the same. Wurtman, Axelrod, and Kelly (1968) have given an up-to-date comparative synopsis of the pineal while Wight (1971) has specifically reviewed knowledge pertinent to the domestic fowl. The ultrastructure of the pineal of the House Sparrow has been well covered (Quay and Renzoni 1963; Ueck 1970), and so has that of the Feral Pigeon (Oksche, Morita, and Vaupel-von Harnack 1969) and White-crowned Sparrow (Oksche, Kirschstein,

Kobayashi, and Farner 1972) and we draw on these accounts. Menaker and Oksche (1974) have also reviewed the subject.

The pineal gland is sited on the dorsal part of the fore-brain in the triangular space enclosed by the margins of the two cerebral hemispheres and the cerebellum. Here it is protected by meningeal tissue and by the skull and is surrounded by a rich network of blood-vessels, which ultimately originate from the left internal carotid artery and arrive via the posterior meningeal artery. Primitively there are two outgrowths of the diencephalon of the 3–4 day old chick embryo, corresponding in ontogeny to two outgrowths sited one behind the other (they may have arisen as a left and right pair) which in cyclostomes become vestigial eyes. The anterior lobe is the parietal organ, which in some reptiles is an eye-like structure although it becomes vestigial in birds. The posterior lobe develops into the pineal body or epiphysis. It arises as a thickening of the diencephalon from which ependyma-lined vesicles are budded to give a central lumen from which extend numerous diverticulae. The organ is attached to the fore-brain by a stalk extending between the habenular and posterior commissure, but it is not certain whether the stalk has a lumen connecting with the third ventricle. In chickens the gland measures about 3·5 × 2·0 mm and weighs around 5 mg. The vesicles are supported by septae and divided into lobules by trabeculae which arise from a connecting tissue capsule surrounding the organ. These trabeculae contain blood-vessels, nerve fibres, fibroblasts, and also lymphocytes. Characteristic parenchyma cells called pinealocytes form a pallisade round the vesicles and irregular groups of cells in the more distal regions; they are sometimes called hypendymocytes but although distinct during development are not readily distinguished by light microscopy in the mature organ.

The pinealocytes are acidophilic with oval to round nuclei and a prominent nucleolus. Electron microscopy reveals that some of them are supporting cells, broadly sited on the basal lamina surrounding the vesicles. In some species these cells have been termed secretory cells and it is evident that there is some variation in appearance, and perhaps function, of these pinealocytes. Nevertheless, the characteristic pinealocyte is an elongated columnar cell with an ultrastructure which is recognisably similar to the pineal photo-receptor cells of lower vertebrates and reminiscent of the rods and cone cells of the retina. These slender pinealocytes generally have lobulated end-feet, although these have not been identified in all the avian species examined. Near their base is a nucleus, many mitochondria, a prominent Golgi apparatus, numerous small vesicles with smooth endoplasmic reticulum, and free ribosomes; in addition, the cells carry numerous lipid droplets, lysosomes, and small, dense-cored vesicles with dense granules surrounded by membranes. In fact, these are all indicators of a secretory function and the Golgi inclusions are characteristic of monoamines, which can be visualized by the fluorescent formaldehyde-condensation method. According to Wight and MacKenzie (1971), who showed the fowl pineal to possess considerable enzyme activity, they may correspond to serotonin. Histochemical tests for ATP are usually very positive, and, with other signs, probably incidate protein synthesis, especially of the methoxyindole, melatonin. The apex of the pinealocytes protrudes into the lumen as a light, finely granulated club with no organelleles, but a cilium and associated centrioles. It may exhibit membranous lamellations reminiscent of the outer segment of rudimentary retinal rod or cone (Oksche and Vaupel-von Harnack 1966; Bischoff 1969) but it lacks the cone-like outer segment typical of the pineal photo-receptors of lower vertebrates. The pinealocyte of the birds is, therefore, a rudiment of the primitive pineal light receptor of lower vertebrates which has become modified as a secretory cell.

The pallisade pinealocytes are associated with complex fibrils running parallel with their long axes which are visualized by silver preparations. These fibrils often surround the nucleus in looping configurations, while their terminal portions form a reticular zone in the distal parenchyma. They appear not to be secretory inclusions but instead connective tissue performing a rather complex supporting function. They are also stained by aldehyde—fuchsin, a stain developed for elastic fibres, which means that they are easily confused with neural tissue with which they are also associated (see below).

Basophilic or plasma cells are sited near the basal lamina. They are not true parenchymal elements of the pineal and have also been described in the rabbit (Leonhardt 1970). Oksche et al. (1972) were unable to ascribe a function to them.

Also prominent in the pineal are basophilic glial and nerve cells which are associated with a neuropile area containing numerous unmyelinated nerve fibres from other nerve cells and with axon bundles in the pineal stalk. These neuropile areas are also associated with caudally bent basal processes from the pinealocytes, where they form a longitudinal fibre system. In some lower vertebrates studied by Oksche and Vaupel-von Harnack (1965) these appear to be contact zones between the receptors and post-synaptic neurones which pass to the brain, and a similar situation apparently holds for birds. Most of this nervous system is ultrastructurally and histochemica typical of adrenergic, sympathetic fibres. However, Wight and MacKenzie (1970) identified acetylcholine and this may come from cholinergic fibres. Autonomic unmyelinated nerve fibres are numerous in the connective tissue capsule of the organ, and are embedded in Schwann cells.

In the amniotes the pineal cells function to transduce photic stimuli and to transmit them to the dendrites of nerve cells. In birds and mammals, however, the pinealocytes produce and store indole derivatives which are then released into the circulation to affect distant target tissues. An autonomic innervation has developed with this change to a secretory role to replace the cholinergic system typical of lower vertebrates (Wurtman et al. 1968; Kappers 1969). In mammals neural inputs into the pineal result in hormone discharge into the blood-stream, so that any stimulation by light of the organ is indirect, via its innervation from the superior cervical ganglion (Hedlund 1970). However, whether this situation is completely true in birds is still debatable. While ablation of the superior cervical ganglion of the quail reduced the serotonin content of the pineal (Hedlund, Ralph, Chepko, and Lynch 1971), Lauber, Boyd, and Axelrod (1968) claim that the light-dependent synthesis or melatonin in the chick is independent of the retina or the sympathetic innervation. The existence of extra-retinal light receptors in the bird and the rôle of the pineal will be discussed again when the mechanisms of photoperiodism are considered (p. 305).

In mammals melatonin synthesis is initiated by the uptake of circulating tryptophan into the pineal, where it is oxidized to 5-hydroxytryptophan by tryptophan hydroxylase and then decarboxylated to serotonin (5-hydroxytriptamine) (Wurtman 1969). Much of the serotonin is converted via monoamine oxidase to 5-hydroxyindole acetic acid but a small part is methoxylated to melatonin by an enzyme unique to the pineal, hydroxyindole-O-methyl transferase (HIOMT). This enzyme activity is increased in birds kept in the light so that the rate of melatonin synthesis is increased (Lauber et al. 1968). In experiments with Canaries Serinus canaria, Munns (1970) found the highest rate of melatonin synthesis occurred in birds kept in constant light and diurnal light. This led to testicular growth, presumably via hypothalamic pathways. Melatonin synthesis was inhibited in birds kept in the light, but supplied with hoods. Melatonin synthesis showed an inverse relationship with testicular weight in wild male House Sparrows sampled throughout the year (Barfuss and Ellis 1971). Birds photostimulated to testicular activity with red light showed decreased melatonin synthesis, but white light with no red portion of the spectrum increased testes size and also melatonin synthesis. Because darkness also increased melatonin synthesis it appears that a light–dark cycle is followed by melatonin synthesis in the pineal. In both the Japanese Quail (Hedlund et al. 1971) and Feral Pigeon (Quay 1966) there is a diurnal rhythm of serotonin concentration, so that maximum titres occur just after the onset of light in a cycle of LD 14:10, while in the quail and chicken melatonin exhibits a rhythmic increase during the dark part of the cycle (Ralph, Hedlund, and Murphy 1967; Sturkie and Meyer 1972). Quay also found that a diurnal rhythm of 5-hydroxytryptophan (on the biosynthetic pathway to serotonin) could be rephased by altering the light schedule. In the rat, which is of course a nocturnal animal, HIOMT activity rises with the onset of darkness but its 24-h rhythm appears to be exogenous and is extinguished by constant light or dark, although the serotonin rhythm is not lost (Wurtman and Axelrod 1965). In mammals, the sympathetic nerves probably affect the system by releasing noradrenaline for this facilitates melatonin production in vitro (see Wurtman et al. 1968).

Although some conflicting results have been obtained, it is generally the case that pinealectomy results in a hypertrophy of the avian gonad while injected pineal extracts result in hypotrophy. Hypotrophy results when melatonin is injected (Singh and Turner 1967; Homma, McFarland, and Wilson 1967). Oksche et al. (1972) found that pinealectomy had no effect on photoperiodically induced testicular growth in White-crowned Sparrows kept on LD 20:4, but it resulted in a limited but significant increase in testicular weight in birds held on LD 8:16 which these authors discussed in terms of a possible phase-shift effect on the circadian clock (see p. 290). Thus, as also suspected in mammals, it may be that the pineal modifies gonadal function by secreting more or less melatonin. In mammals, the action is thought not to be direct but instead via the hypothalamo-hypophysial axis, particularly the median eminence, to influence the rate of gonadotrophin secretion. In the rat melatonin and 5-hydroxytryptophol reduce pituitary LH whereas 5-methoxytryptophol and serotonin reduce FSH but not LH (Fraschini 1969; Reiter 1969). Evidently rhythmic and photoperiodically induced changes in protein synthesis in the pineal could influence the rhythm of gonadotrophin secretion and enzyme activity in the median eminence but we lack positive information in birds.

Quay (1972) surveyed the size and state of the pineal organ of adult birds of 23 orders covering 121 species. Pineal reduction or atrophy was found in two primitively nocturnal groups, the Procellariformes and Strigiformes. Atrophy was not found in nocturnal species belonging to other diverse genera from *Apteryx* to *Burhinus*.

3   Reproductive apparatus of the male

Summary

During ontogeny primordial germ cells migrate through the blood from the embryonic splanchnopleur to form the germinal epithelium, where they become associated with Sertoli cells in the stromal connective tissue of the intermediate mesoderm. These cells become surrounded in tubules by growths of the coelomic epithelium and between these seminiferous tubules are secretory interstitial Leydig cells derived from fibroblast-like cells in the coelomic connective tissue. Both Sertoli and Leydig cells secrete steroid hormones in response to stimulation from pituitary FSH and LH, respectively. There are indications that these gonadotrophins must arrive at the target secretory cells in the right temporal sequence and there is evidence for a rhythmic secretion of androgen. It is not yet known whether there is a rhythm of sensitivity in the target tissue. The avian testes undergo a seasonal cycle from a fully active to a resting non-functional state; this is described in detail in this chapter. Regression to inactivity can be induced by hypophysectomy or by treatment with drugs and hormones which inhibit pituitary secretion, for example, oestrogen and progesterone have an anti-gonad action in the male. Photoperiodic stimuli, modified by other physiological factors including the plane of nutrition, allow the gonads to attain a functional condition. Following continued exposure to stimulatory photoperiods spontaneous gonad regression occurs as the neurohypophysial–gonad axis is driven into a non-functional phase, the refractory phase, which lasts until the cycle is re-initiated under the influence of 'short' photoperiods (see also Chapter 12).

Spermatogenesis is initiated under the action of FSH, apparently acting on the spermatogonia and also on the Sertoli cell, whereas LH is responsible for the synthesis and secretion of testosterone by the Leydig cells. It seems likely that other steroids on the biosynthetic pathway to testosterone will also be found to have specific rôles in spermatogenesis (and behaviour) while prolactin and ACTH also have influences. Androgens are implicated in the later stages of spermatogenesis being taken into the Sertoli cells by an androgen binding protein whose production is stimulated by FSH. Androgens released from the testes feed-back on target cells in the median eminence to regulate the further secretion of pituitary hormone, particularly LH. In mammals, a substance produced in the testis, apparently from adrogen binding protein and called inhibin, feeds back on the pituitary to inhibit FSH secretion, but the same has not yet been proved in birds.

Development and structure of the testis

Embryologically both the testes and ovary are derived from paired and sexually undifferentiated primordia associated with the intermediate mesoderm (nephrotome) of the early embryo. Within these structures are primordial germ cells (gonocytes) which initially arise in the embryonic splanchnopleur. Thereafter, they migrate, with amoeboid movement, through the blood to reach the intermediate mesoderm, where they sink through the surface into the stromal connective tissue (Dubois 1965). These germ cells have a basophilic cytoplasm and are supported by other cells, which in the male will subsequently develop into Sertoli cells. Interstitial cells develop quite independently from fibroblast-like cells, these being situated in connective tissue which develops as a growth of the coelomic epithelium to surround the primordial germ cells. The left presumptive gonad has been shown to receive more primordial germ cells than the right in the House Sparrow *Passer domesticus,* Red-winged Blackbird *Agelaius phoeniceus,* and domestic hen *Gallus gallus,* and throughout life the left testis of birds is usually the larger of the two (Witschi 1935). The embryonic left germinal ridge also grows more rapidly than the right and this asymmetry persists after hatching.

The initial proliferation of the germinal epithelium in the presumptive gonads of either sex forms a potential testis, but subsequently there occurs a second proliferation of cells in females and this gives rise to a cortex in the left gonad which now becomes the potential ovary. A potential cortex of ovarian tissue also has a transient existence from days 7—11 in the left testis of the domestic chick, American Robin *Turdus migratorius,* sparrow, tern, and duck. Between the seventh and ninth day of development the right gonad of the female atrophies leaving a single left ovary, although functional bilateral ovaries persist in some species (see below). Thus all birds possess paired testes and the majority have a single ovary. However, in some taxa, particularly the Kiwi *Apteryx australis* (Kinsky 1971) and the families Accipitrinae, Falconidae, Buteoninae, and Carthartidae, paired ovaries regularly occur associated with a functional oviduct (Domm 1939; Stanley and Witschi 1940); in a few others the condition may exist exceptionally as in the pigeons (Brambell and Marrian 1929) and Herring Gull *Larus argentatus* (Boss and Witschi 1947).

In many species surgical removal of the functional ovary, or its loss through such a pathological agency as a neoplasm, or a disease such as tuberculosis, allows the left remnant to develop into a functional gonad, usually a testis or ovo-testis. If this happens early in life full germ cell development culminating in the production of spermatozoa is possible, otherwise only ovulations occur. The unique laterality or asymmetry which occurs in the embryological development of the male and female reproductive systems provides conditions which can result in varying degrees of intersexuality (Taber 1964). Females are most likely to be affected in this way and they become masculinized following atrophy of the functional ovary. Cases of such sex reversal are well known in chickens, often in old hens suffering senile changes, and they cause the females to acquire male plumage and the capacity to crow, tread, and actually fertilize other hens. Sex reversal of this kind is also relatively common, and often spectacular, in pheasant species, causing, for example, a sombre-coloured female Golden Pheasant *Chrysolophus pictus* to assume the resplendent plumage of a male (see also Harrison 1932). Sex dimorphism in pheasants is primarily determined by hereditary sex differences but in part is modified by endocrine factors. Both oestrogenic and androgenic hormones have a feminizing effect on the feather pattern (Koch 1939).

Inbred strains of domestic hens can produce a high incidence of intersexuals. Riddle, Hollander and Schooley (1945) encountered the same condition in a strain of domestic pigeons *Columba livia* var. which they kept for eight generations, the hermaphrodites usually having a left ovo-testis and a normal right testis. Lahr and Riddle (1945) in studying the embryological development of this same strain demonstrated that there was a delay in the degeneration of cortical tissue. Presumably then the hermaphrodites were genetic males in which the cortical component of the left testis had persisted and differentiated into ovarian tissue. Hybridization may also yield intersexual genotypes (Riddle 1925; Riddle and Johnson 1939; Bissonnette 1941). Occasionally individuals without gonads are found, and they usually lack both right and left Mullerian ducts. They have a generally masculine disposition and resemble subjects which are experimentally castrated as embryos. Taber (1964) points out that such gonadal agenesis in man (The Turner syndrome) is associated with the loss of one of the sex chromosomes, thereby producing a neuter genotype resembling the female (genetic composition X0 which resembles XX). Unlike man, males are the homogametic sex in birds and it may be significant that individuals lacking gonads resemble the male. In his review of intersexuality Taber also mentioned gynadromorphs and mosaics. Thus cases where the left side of the body has an ovary and externally female plumage and the right side male plumage and a testis have been reported in the Bullfinch *Pyrrhula pyrrhula* (see Witschi 1961), Chaffinch *Fringilla coelebs*, Canary *Serinus canaria*, bantam hen, pheasant, and domestic fowl. Chromosonal aberration is considered to be the cause of these bizarre effects.

In most birds the paired testes undergo marked seasonal changes in size between a regressed 'inactive' condition in the non-breeding season to an expanded functional state in response to changes in pituitary gonadotrophin secretion. They are round to ovoid organs sited in the body cavity ventral to the anterior end of the kidneys. Here they receive spermatic arteries from the dorsal aorta and pass blood via testicular veins to the vena cava. Each testis is surrounded by a thick outer fibrous sheath, the tunica albuginea, which is bounded by a more fragile and thinner coat, the tunica vaginalis. Since the testes expand in volume up to 70-fold during the reproductive season the coat is liable to become stretched and torn and so it is replaced annually by fibroblasts which build a new coat from below (Marshall and Serventy 1957), possibly as a direct response to the collapse of the old tunic (Marshall 1961). For a while the two tunics are discernible and this can provide a means whereby individuals which have not yet bred can be distinguished from those that have (Lofts, Murton and Westwood 1966).

The testes are composed of a mass of convoluted seminiferous tubules which are lined with an epithelium of developing germ cells together with non-germinal, somatic, Sertoli cells. Between the seminiferous tubules are interstices containing areolar connective tissue with blood capillaries, lymph spaces, interstitial Leydig cells, and, in some species, melanoblasts; in the Starling *Sturnis vulgaris* these melanoblasts impart a grey to black colour to the collapsed testes. The testes have both germinal and endocrine functions and these are interdependent.

Endocrine elements of the testis

There appear to be two sources of steroid hormones in the testes, these being the interstitial Leydig cells and the tubuler Sertoli cells. The Sertoli cells were once termed nurse
cells and were thought simply to provide support and nutrients for the germ cells
with which they are associated. They are attached to the basement epithelium of the
tubule by a broad base and extend into the cell lumen as a triangular wedge. Proto-
plasmic extensions from the cell surround a group of germ cells which appear to
comprise a discrete population, analogous with the germinal cysts found in the testes
of anamniotes (Lofts 1968). It is now clear that, like the Leydig cells, the Sertoli
components function as miniature endocrine organs, possibly being capable of producing oestrogens. Both the Sertoli and Leydig cells possess an ultrastructure typical
of steroid-producing tissue. The nucleus has a large nucleolus and several deeply
staining masses of chromatin, while the cytoplasm contains mitochondria with tubu-

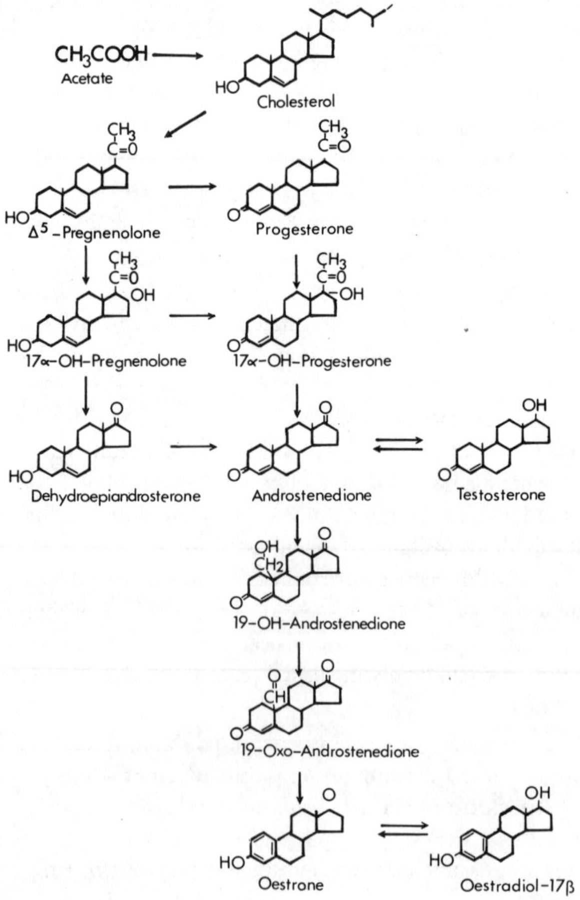

Fig. 3.1. Major routes for the biosynthesis of androgens and oestrogens from precursor acetates
and cholesterol.

lar cristae, a prominent Golgi complex, granular and agranular endoplasmic reticula, and numerous lysosomes; in fact, on structural evidence the Sertoli cells appear to be more active centres of steroidogenic activity than the Leydig cells (Lacy and Pettitt 1969). Both cell types react positively to tests for those enzymes involved in steroid synthesis; Δ^5-3β-hydroxysteroid dehydrogenase (3β-HSDH) is present and it converts Δ^5-3β-hydroxysteroids to Δ^4-3-ketosteroids, a stage in the production of progesterone and androstenedione (Woods and Domm 1966). The main stages involved in the biosynthesis of the gonadal steroids are set out in Fig. 3.1. Human subjects with ectopic testes in which the tubules are lined by Sertoli cells and the germinal epithelium is atrophied, give a marked histochemical reaction for 3α-, 3β-, 11β-, 16β-, 17β-, and 20β-hydroxysteroid dehydrogenases (Baillie and Mack 1966). Using *in vitro* incubation techniques it has been shown that the Sertoli cells of rats are a source of both androgens and various hydroxylated and/or reduced pregnenes and pregnanes (Lacy, Vinson, Collins, Bell, Fyson, Pudney, and Pettitt 1969). Moreover, during these studies heat treatment destroyed the germinal epithelium of rats but left the Sertoli cells intact and the capacity to elaborate steroids was enhanced, 20α-hydroxypregn-4-en-3-one being present in large amounts.

The Leydig and Sertoli cells are subjected to regulation by pituitary hormones and in recent years knowledge has accumulated about the mechanism by which many of these may act (Sutherland, Hardman, Butcher, and Broadus 1969). The general scheme is indicated in Fig. 3.2. LH, which acts as the first messenger on the Leydig

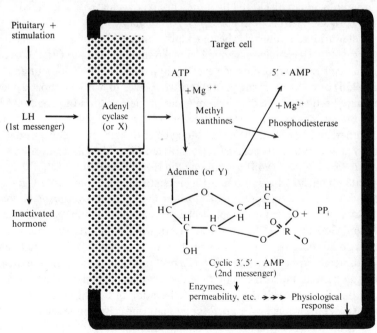

Fig. 3.2. Mode of target action of a hormone such as LH, illustrating the concept of the second messenger system involving adenylatecyclase. Based on Sutherland *et al.* (1969).

cell, is known to activate the enzyme adenylate cyclase in the target cell. This enzyme then catalyzes the conversion of adenosine triphosphate (ATP) to cyclic $3',5'$-adenosine monophosphate (cyclic AMP). This serves as a second messenger within the cell to stimulate further protein synthesis, for example, the conversion of cholesterol to pregnenolone in the mitochondria, leading to the synthesis and release of steroid hormone by the cell. The released hormone then serves as a third messenger to stimulate other target cells. Various steroid hormones (testosterone, progesterone, corticosterone) have been shown to regulate RNA synthesis by direct action on the cell nucleus in the target organ, that is, the hormones act on genes to release information required by the cell cytoplasm for the synthesis of specific proteins (Sekeris 1969). Although LH is known to stimulate the male Leydig cell, as well as various cell types in the female, it is not known whether it can have any direct effect on the Sertoli cell. The enzymes which catalyze the synthesis of steroids are widely distributed. Hence, any specificity of response by a target cell is likely to depend on the precursor material it is able to synthesize, this probably being influenced by the presence of co-factors, rather than on a specific response to gonadotrophin secretion.

Endocrine tissue which can be stimulated by pituitary endocrine secretion occurs inside and outside the testes tubules and its function is best considered in relation to the seasonal cycle of reproductive activity exhibited by the majority of birds. After breeding the testis tubules collapse and any remaining germ cells and their debris become phagocytosed by the Sertoli cells. These now become heavily lipoidal as revealed by colouring with Sudan Black, and their presence is now most easily detected by light microscopy (Fig. 3.3; see between pp. 82-3). With collapse of the tubules to a minimum diameter the interstitium also becomes more noticeable, for at this stage new immature Leydig cells with spindle-shaped nuclei make their appearance (Fig. 3.3). Similar histological manifestations can be achieved by hypophysectomy (Coombs and Marshall 1956) or by transferring photosensitive species to artificial short daylengths, and they have been interpreted as consequences of the cessation of gonadotrophin secretion (Benoit, Assenmacher and Walter 1950a; Lofts and Marshall 1958; King, Follett, Farner, and Morton 1966). It has long been thought from conventional light microscopy that Leydig cells become differentiated from fibroblasts, as suggested by Pfeiffer and Kirschbaum (1943) in studying the House Sparrow, and that they do not originate by mitotic division of existing interstitial cells. There is now good evidence based on electron microscopy of Japanese Quail *Coturnix coturnix japonica* that the Leydigs do arise from fibroblast-like progenitors in the inter-tubular areas (Nicholls and Graham 1972), but it is not known what stimulates their formation. These cells do not develop if immature quail are held on 6-h (LD 6:18) daily photoperiods but metamorphosis into mature cells occurs within three days of exposure to 20-h photoperiods and at the same time there is an increase in plasma androgen. Oestrogen therapy can result in a marked proliferation of Leydig cells (see p. 68) and even using conventional light microscopy maturing cells can be seen arising from the tubule epithelium (Fig. 3.5; see between pp. 82-3). Oestrogen appears to block FSH manifestations and stimulates LH secretion (see below), but whether or not the inhibition of FSH facilitates Leydig formation remains to be determined.

Exogenous purified mammalian FSH has been shown to stimulate the Leydig cell nuclei of House Sparrows causing the nucleolus to be less conspicuous and chromatin to accumulate (Lofts, Murton, and Thearle 1973; Table 3.1). It was also found to stimulate androgen-type behaviour in male Feral Pigeons, posing the question of

TABLE 3.1

Effects of exogenous mammalian hormones on the testes and bill pigmentation of House Sparrows captured from the wild in October

Treatment	Mean weight ± S.D. of testes (mg) (N)	Most advanced germ cells	Nuclear diameter ± S.D. of Leydig cells (μm)	Bill colour scores[†]
Controls kept on natural photoperiods until early December	1·6 ± 0·2 (4)	Spermatogonia	2·9 ± 0·7	1,1,1,1
Birds were kept on LD 8:16 until early December and injected 12 times on alternate days being autopsied the day after injections finished:				
1. Testosterone propionate (0·4 mg)	1·7 ± 0·8 (5)	Occasional prophase primary spermatocytes	1·5 ± 0·5	2,2,2,2,1
2. N.I.H:LH (0·4 mg)	16·2 ± 9·6 (5)	Primary spermatocytes	5·8 ± 1·6	2,2,2,1,1
3. N.I.H.FSH (0·4 mg)	8·3 ± 3·8 (5)	Primary spermatocytes	6·4 ± 1·4	2,2,2,1,1
4. LH (0·4 mg) + FSH (0·4 mg) as mixture	25·6 ± 11·6 (5)	Primary spermatocytes	5·0 ± 1·7	4,4,4,3,3
5. LH (10 × 0·4 mg) followed by FSH (10 × 0·4 mg)	9·2 ± 5·4 (5)	Primary spermatocytes	5·6 ± 1·6	3,3,3,3,2
6. Testosterone (0·4 mg) + FSH (0·4 mg)	9·4 ± 3·8 (5)	Spermatogonia	3·0 ± 0·5	4,3,2,2,2
Held on LD 16:8 July–December (kept refractory)	4·5 ± 2·7 (5)	Spermatogonia		1,1,1,1,1
Held on LD 8:16 until 8 November then LD 16:8 until December (shortened refractory)	343·8 ± 103·4 (5)	Sperm		4,4,4,4,1
Birds kept like controls until late November then on LD 16:8 until late December				
1. Held on LD 16:8 November–December (controls for following groups)	95·5 ± 54·1 (5)	Sperm		4,4,2,1,1
2. Testosterone (12 × 0·4 mg)	178·6 ± 34·8 (2)	Sperm		4,4
3. LH (12 × 0·4 mg)	229·0 ± 63·3 (5)	Sperm		4,4,4,4,3

[†] 1 = non-pigmented breeding condition; 4 = fully pigmented black bill

whether it facilitates the release of steroid material from the Leydig cells (Murton, Thearle, and Lofts 1969*b*; see also p. 108). Recent work with mammals helps answer this question with the discovery of an androgen-binding protein (ABP) having a high affinity for dihydrotestosterone (DHT) and testosterone in homogenates of rat testis, and also of the rete testis and efferent ducts (French and Ritzén 1973*a, b*). It is now well established that LH stimulates androgen production in the interstitial cells of the rat, where tissue receptor sites having a high affinity for LH and HCG have been defined (Catt, Tsuruhara, Mendelson, Ketelslegers, and Dufau 1974). In passing, it may be noted that highly specific binding proteins for 17β-oestradiol have been demonstrated in the testicular interstitial cells of rats but the physiological significance remains unclear (Mulder, Beurden-Lamers, De Boer, Brinkman, and van der Molen 1974). In similar fashion, specific receptor sites to FSH exist in the Sertoli cell (French, McLean, Smith, Tindall, Weddington, Petrusz, Sar, Stumpf, Nayfeh, Hansson, Trygstad, and Ritzen 1974; Hansson, Reusch, Trygstad, Torgerson, Ritzen, and French 1973; Means and Huckins 1974; Desjardins, Zelznik, Midgley, and Reichert 1974). FSH seems first to bind to receptors present in the plasma membranes of the target cells. This stimulates membrane-bound adenylate cyclase which leads to an increase in the intracellular concentration of cyclic AMP. (Evidence has accumulated during the last decade that a large group of hormones act through a mediation of the adenylate cyclase system of the target cell plasma membrane and these systems seem to be hormone selective, many having a specific metal or nucleotide requirement for optimal catalytic activity or hormonal activation. The new cyclic AMP interacts with the regulatory sub-unit of inactive cyclic AMP-dependent protein kinase resulting in increased catalytic activity.) So while it has long been realized that FSH stimulates RNA and protein synthesis it is now appreciated that it stimulates the production of a specific Sertoli cell protein which is the androgen binding protein (Tindall, Schrader, and Means 1974). This ABP can, therefore, increase the binding and accumulation of androgen inside the tubules, against a concentration gradient, in close juxtaposition to androgen dependent cells in the tubule epithelium. FSH stimulates ABP secretion into the testicular fluid and the epididymis where it is taken up by the epididymal cells and activated by 5 α-reduction to form DHT. DHT bound to cytoplasmic receptors is transported into the cell nuclei where the androgen complex binds to chromatin and is involved in sperm maturation.

A substance is produced from the epididymides which feeds back on the pituitary to inhibit FSH secretion. Its composition has not yet been fully established but it appears to originate in the germinal epithelium and has been called inhibin (Franchimont, Chari, Hagelstein and Duraiswami 1975). At the time of writing inhibin has not been identified in the bird but it may be expected that there will soon be developments in this respect.

FSH may also have a direct effect on spermatogonia, for an FSH-sensitive adenylate cyclase is associated with these cells (Braun 1974). However, the evidence is presently conflicting for, although an *in vitro* stimulation of testis cells has been achieved the identity of the cell type remains uncertain; pre-leptotene spermatocytes appeared to be unaffected (Davis and Schuetz 1975). Current thinking is that the effect of

FSH mediated via ABP is to amplify the testosterone stimulus to androgen-sensitive cells in the germinal epithelium. Androgen alone is capable of maintaining Sertoli cell secretory function in immature hypophysectomized rats but it cannot restore secretory activity of the Sertoli cell after post-hypophysectomy regression (Weddington, Hansson, Ritzen, Hagenas, French, and Nayfeh 1975). It appears that the Sertoli cell is responsive to androgens as well as to FSH but that following regression of the cells in the absence of hormone stimulation, FSH is needed for full restoration of secretory activity.

To return to birds, it would, therefore, seem likely that the stimulation of Leydig cells by FSH noted above was an indirect consequence of the stimulation of the Sertoli cell. However, it should be remembered that an androgen binding protein has not yet been demonstrated in birds, although it may be suspected that mechanisms similar to those described for mammals will be discovered. Exogenous *mammalian* LH caused the Leydig nuclei of refractory House Sparrows to become enlarged and vesiculate, with a conspicuous nucleolus (Lofts *et al.* 1973; see also Table 3.1). In these subjects there was not a marked increase in the lipid content of the cells but the same batch of hormone did cause an extensive increase in interstitial lipids when injected into Feral Pigeons that were already in full breeding condition. It also caused an increase in cell numbers (Table 3.2). Purification of avian LH extracts has enabled these to be injected into Japanese Quail (50 μg per day for 10 days) kept on short (6-h non-stimulatory) daily photoperiods. Electron micrographs clearly demonstrate an enlargement and maturation of juvenile Leydig cells but also an apical move-

TABLE 3.2

Effects of exogenous hormones on tubule diameter and Leydig cell count in paired Feral Pigeons

Treatment	Mean tubule diameter ± S.D. (μm)	Mean number Leydig cells ± S.D.[†]
Unpaired controls at start	233 ± 20	53 ± 13
Males pretreated with hormones for week (5 injections) then paired with untreated female for ½-hour per day for two weeks with 10 further injections of:		
1. FSH (14 × 1 mg)	252 ± 22	88 ± 23
2. LH (14 × 1 mg)	232 ± 16	214 ± 150
3. Prolactin (14 × 2 mg)	208 ± 13	72 ± 15
4. Testosterone (14 × 2 mg)	195 ± 23	42 ± 27
5. Oestradiol benzoate (14 × 2 mg)	89 ± 16	170 ± 72
6. Progesterone (14 × 2 mg)	143 ± 58	83 ± 61
Untreated controls paired to female for two weeks	209 ± 21	69 ± 41

[†]The cells visible in 10 interstitial spaces and between adjacent tubules were counted and a correction applied to allow for any increase or decrease in tubule size relative to the untreated controls. Means refer to 6 individuals in each group except for 7 in group 4 and only 4 unpaired controls.

ment of the Sertoli nuclei towards the tubule lumen and the appearance of primary spermatocytes (Follett *et al.* 1972*b*). More recent observations confirm that avian LH stimulated a differentiation of the interstitium of hypophysectomized and intact quail, producing mature Leydig cells with the full complement of organelles typical of steroid-secretion (Brown, Baylé, Scanes, and Follett 1975). These workers found that in addition, LH stimulated some spermatogonial division and partial differentiation of the Sertoli cells, perhaps because it stimulated androgen secretion, but possibly via a direct stimulation. Avian FSH, stimulated testicular growth and development, but had a minimal effect on Leydig cell differentiation. In intact birds the results were similar to those obtained with intact LH treated birds, but in hypophysectomized subjects the testicular weights were greater and the tubules larger than in hypophysectomized LH treated subjects. FSH stimulated the Sertoli cells and pachytene spermatocytes became numerous. Ishii and Furuyu (1975) have examined the effects of purified chicken gonadotropins on the chick testis with comparable results.

If a variety of bird species are held on artificial long days at the end of the breeding season, during the refractory phase which develops at this stage of the cycle (see p. 227), the interstitium becomes expanded with juvenile Leydig cells (Lofts and Murton (1968) give examples; see also Fig. 3.3(f)). These Leydigs do not acquire lipids but their nuclei do become rounded. It is likely that the stimulated interstitium contributes to an increased testis weight in such subjects (Table 3.1) and the observations indicate that LH secretion continues during the photo-refractory phase (see p. 227). The same seems to apply to female Canaries *Serinus canaria* for several photo-refractory subjects were found to have an elevated LH plasma content (Steel, Follett, and Hinde 1975).

Although a new interstitium appears with the post-nuptial collapse of the testis, it does not always at first respond positively to tests for sudanophilic lipids, which denote the presence of unsaturated sterols, even though the tubules may be occluded with dense lipoidal deposits. In both the House Sparrow and House Finch *Carpodacu mexicanus* (Hamner 1968) the interstitium does not become lipoidal until almost 2 weeks after post-nuptial regression and at a time when the tubule lipids are disappearing (see Fig. 3.4). In the Rook *Corvus frugilegus,* lipids appear in both interstitium and tubule simultaneously (Marshall and Coombs 1957) whereas in the Wood Pigeon *Columba palumbus* post-nuptial regression is not accompanied by any deposition of tubule lipid (Lofts, Murton, and Westwood 1967*a*). The variations possibly depend on whether FSH secretion stops completely at the end of the breeding season.[†] Similar histological manifestations can be achieved by agents which differentially suppress or inhibit gonadotrophin secretion. For example, both hypophysectomy and prolacti injections cause tubule steatogenesis in various finches, and these lipids can be cleared by exogenous mammalian FSH injections, which also stimulate tubule expansion and some spermatogenetic development (Coombs and Marshall 1956; Lofts and Marshall 1956, 1958).

[†]Or variation in their temporal phasing relative to each other and to the thyroid hormones, a topic that needs more study.

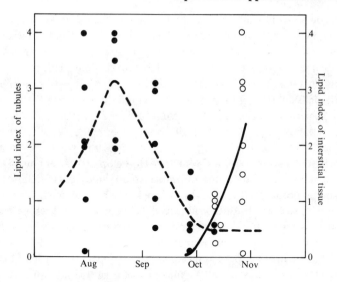

Fig. 3.4. Lipid index of testis tubules (solid circles and dotted lines) and of interstitial tissue (open circles; solid lines) during photo-refractory period of the House Finch *Carpodacus mexicanus.* From Hamner (1968).

There is no doubt that the Leydig cells are the main source of androgen from the testes. Virilizing tumours of these cells in humans lead to increased secretion of 17-ketosteroids and testosterone (Jungck, Thrash, Ohlmacher, Knight, and Dryenforth 1957; Sharma, Racz, Dorfman, and Schoen 1967). In birds, the lipoidal material in the Leydig cell disappears as the tubules expand and spermatogenesis occurs and at the same time manifestations of androgenic behaviour become apparent. These temporal relationships between the seasonal spermatogenetic cycle and changes in the number and activity of the interstitial cells have been conveniently quantified by Jones (1970) using wild-caught California Quail *Lophortyx californicus*; Jones's data have been used to construct Fig. 3.6, which will be referred to again below. We need more details of the precise timing of androgen secretion and whether and under what conditions other steroids may be released. In addition, the factors initiating biosynthetic activity probably need to be distinguished from the secretory mechanisms of the cell. The endocrine role of the Sertoli cell is discussed again in the next section.

Germinal cells and spermatogenesis

The tubules of the fully regressed testis contain only a single layer of type A (or stem) spermatogonia, plus, of course, Sertoli cells. These spermatogonia have an ovoid nucleus with very fine chromatin granules round the nuclear membrane and a prominent central nucleolus. During the first phase of spermatogenesis these cells divide mitotically to produce further type A spermatogonia and these then divide to give both type A and type B spermatogonia. The renewal of the stem cells provides for later generations of cells while the B-type cells give rise to other germ cells. Type

B spermatogonia divide again to form pre-leptotene primary spermatocytes (Fig. 3.5). These now begin DNA replication in readiness for the stages of meiotic reduction division. This marks a second, and distinct stage of spermatogenesis, which probably involves a different hormone balance from the preceding one. This is also suggested by some atypical manifestations which were noted when subjects were exposed to certain artificial skeleton light schedules of the kind shown in Fig. 2.4 (p. 40). In both House Sparrows and Greenfinches *Carduelis chloris* expanded tubules could be produced, containing layer upon layer of primary spermatocytes without any sign of reduction divisions occurring (Fig. 3.5). In the same way when exogenous LH (N.I.H.-bovine) was given to photo-refractory House Sparrows held on LD 8:16 there was an abnormal proliferation of primary spermatocytes (Fig. 3.5(c)). The spermatokinetic property of mammalian LH has long been appreciated and is reflected in the testis weight of the subjects detailed in Table 3.1 whose testes contained numerous prophase primary spermatocytes. It has usually been assumed that such stimulation results from contamination of the hormone with FSH or else depends on mammalian LH possessing some of the properties of avian FSH.[†] However, the result quoted above involving injection of avian LH into quail (p. 55) does make it possible that LH functions in the early stages of spermatogenesis. It is possible that exogenous LH acts synergistically with endogenous FSH rather than by itself. This is suggested by the results in Fig. 2.4, where peak plasma titres of endogenous IR-LH were not correlated with germ cells in advance of spermatogonia. Bovine LH had no obvious influence on the germ cell count of sexually mature and paired Feral Pigeons (Table 3.3). It should be remembered that androgen produced in the Leydig cell under the action of LH is drawn through the tubule wall as a result of FSH stimulating ABP production in the Sertoli cell. Thus, an apparent LH/FSH action might really be an androgen/FSH interaction. There remain many problems to be resolved.

Exogenous androgen can induce the division of spermatogonia in the regressed testis of the Red-billed Quelea *Quelea quelea* (Lofts 1962*a*) and it re-establishes spermatogenesis in pigeons whose gonads have completely regressed following hypophysectomy. Pfeiffer (1947) reported that a similar stimulation could not be achieved in the regressed testis of the House Sparrow, but his results are at variance with those summarized in Table 3.1. It is conceivable, but unlikely, that LH was effective in stimulating spermatogenesis in the above-mentioned experiments because it caused androgen synthesis and release. For one thing, it may be noted from Fig. 3.6 that maximum Leydig cell numbers and glandular activity occur after primary spermatocytes have been formed, so it is unlikely that the experimentally achieved spermatokinetic effect of androgen on the regressed testis has significance in the natural cycle.

It appears that FSH and possibly also LH are involved in the initial divisions of spermatogonia that result in the appearance of primary spermatocytes. FSH appears to be essential for the meiotic division of primary spermatocytes, at least in mammals, exerting its influence via the Sertoli cell (Lacy 1967; Lacy *et al.* 1969) where,

[†]Chick testes are between 2·5 to 9 times more responsive to ovine LH than to ovine FSH (see Breneman *et al.* 1962).

TABLE 3.3

Effects of exogenous hormones on the number of germ cells in paired Feral Pigeons

Treatment (N)		Mean number ± S.D. of:					
			spermatocytes				Total
		spermatogonia	Primary	Secondary	spermatids	spermatozoa	cells
Unpaired controls at start	(4)	38 ± 3	50 ± 5	91 ± 19	85 ± 14	92 ± 20	356
FSH	(6)	21 ± 5	24 ± 6	79 ± 16	73 ± 11	61 ± 17	258
LH	(6)	25 ± 8	21 ± 4	71 ± 12	68 ± 17	40 ± 11	225
Prolactin	(6)	25 ± 7	35 ± 10	52 ± 6	62 ± 30	28 ± 10	202
Testosterone	(7)	25 ± 4	26 ± 7	72 ± 19	47 ± 15	53 ± 26	223
Oestradiol	(6)	48 ± 12	5 ± 5	0			53
Progesterone	(6)	33 ± 12	12 ± 6	19 ± 26	24 ± 39	8 ± 12	96
Untreated controls paired with female	(6)	25 ± 6	38 ± 10	73 ± 18	69 ± 12	44 ± 7	249

These are the same birds referred to in Table 3.2, which gives experimental details. Germ cells were counted across the widest diameters of ten tubule sections per individual bird using the rectangle made by the eyepiece graticule as a guide.

as we have seen, one of its functions is to stimulate the production of androgen binding protein. During spermateleosis (the maturation of spermatids, see below) residual bodies become detached from the spermatids and are subsequently phagocytosed by the Sertoli cells. The residual bodies contain much RNA, mitochondria, and Golgi apparatus which appear to be re-cycled by the Sertoli element (Smith and Lacy 1959). In the rat, oestrogen therapy leads to loss of all germ cells other than types A and B spermatogonia. Spermatocytes and spermatids degenerate, many assuming the appearance of polar bodies, and they are phagocytosed by the Sertoli cells, which become charged with lipids in the process. It seemed that these changes might result from the inhibitory action of oestrogen on the gonadotrophins and so exogenous FSH or LH were given simultaneously with the oestrogen. FSH caused the lipid content of the tubule to be reduced and spermatogenesis to be restored, whereas LH was ineffective in countering the oestrogenic effects (Lacy and Lofts 1965).

Lacy suggested that the residual bodies might act as highly localized chemical messengers and as a source of raw material to be used by the Sertoli cell in the formation of specific steroids. He suggested a convenient reference to be Sertoli cell hormone (SCH) and its elaboration was envisaged to be dependent on FSH. In fact this agent is now seen to be ABP. In Lacy's experiments irradiation destroyed the germ cells but not the availability of FSH and led to accumulation of lipid in the

Sertoli cell. Lacy noted that at first this lipid was not responsive to histochemical tests for unsaturated sterols but gradually it came to give a positive reaction. Thus lipoidal material which was normally used during spermatogenesis could under some circumstances be acquired by the Sertoli cell. This is now to be explained by the passage of steroids from the Leydig cells. In the pigeon *Columba livia* cadmium chloride results in drastic necrotic damage if injected direct into the testis but given intra-muscularly it dislocates the normal radial co-ordination of germ cell development, presumably by interfering with the biosynthetic activity of the Sertoli cell (Lofts and Murton 1967). It has been suggested that cadmium ions interfere with sites of steroid biosynthesis (Parizek 1964). Whether or not oestrogenic steroids are produced in the Sertoli cells remains to be discovered but it is known that a Sertoli cell tumour in a Brown Leghorn capon led to its feminization (Siller 1956).

The stages of meiotic division have been well studied in mammals in relation to DNA and RNA synthesis (see Monesi 1971) but not in birds. The process involves two consecutive cell divisions but only one duplication of chromosomes. As meiosis proceeds the primary spermatocytes derived from type B spermatogonia move towards the tubule lumen and increase in size. At the end of meiosis the first cell division produces two daughter secondary spermatocytes which have a haploid set of chromosomes. These cells divide again and the sister chromatids of each chromosome become separated into the four daughter cells or spermatids. Once spermatids have been formed they do not divide further but undergo a complex metamorphosis into spermatozoa, the process of spermiogenesis or spermateleosis.

Clermont (1958) described the basic changes affecting the avian spermatid some time ago but has dealt more recently and in greater detail with the ultrastructure changes during spermiogenesis in the rat (1967) for which 19 different development stages can be recognized. The basic changes, which apply also to birds, involve a Golgi phase, when proacrosomal granules appear whose glycoprotein reacts strongly to periodic acid–Schiff reagent (PAS reaction). These coalesce to form one granule, the acrosome, which attaches the nucleus at what will become the anterior end. During this stage two centrioles in the presumptive posterior region begin to form a flagellum. This phase is followed by the cap phase during which the acrosome spreads over the end of the nucleus. Next follows an acrosome phase when the nucleus and acrosome becomes positioned at the periphery of the cell. A centriole moves to the base of the nucleus while another remains to form a basal body for the flagellum. Towards the end of this stage the nucleus becomes deflected to one side by rotation of the spermatid and in consequence the acrosome, attached to the Sertoli cell, points inwards towards the seminiferous epithelium. At the same time the nuclear chromatin condenses into dense granules. A fourth and final maturation phase involves the elongation of the cytoplasm behind the nucleus to enclose the proximal part of the flagellum. A chromatoid body forms an annulus round the flagellum and this is also surrounded by mitochondria, which develop into a mitochondrial sheath eventually surrounding the middle section of the sperm. As mentioned, during final spermatid maturation most of the cell cytoplasm, degenerating mitochondria, ribosomes, and other debris are sloughed off as a residual body to be phagocytosed by the Sertoli cells. The spermatozoa as they have now become are released by the Sertoli cell and float free in the tubule lumen. Final maturation occurs in the epididymides.

Spermatogenesis proceeds in an orderly sequence of waves so that at any particular stage of development the various germ cells bear a quantitatively and qualitatively fixed relationship (Clermont 1958). This happens because the spermatogonia undergo mitotic divisions at fixed and regular intervals and this determines the cycle length. In mammals, this takes from 8·6 ± 0·2 days in the mouse (Monesi 1962) to 16 ± 1 days in man (Monesi 1967; Heller, Heller, and Rowley 1969), but unfortunately comparable data are not available for birds. The rate is constant for any particular species and is not affected by such physiological changes as elevation of temperature or hypophysectomy. Spermatogenetic development of the individual cells now proceeds at a fixed rate, taking 34·5 days in the mouse and 74 ± 4 days in man. Thus, before a given generation of germ cells can become mature new generations are initiated and begin their development (four such cycles occurring in the mouse). Species-specific cell associations occur, and Table 3.4, based on Clermont, gives details for the domestic duck in which waves of spermatogenesis occur. It is clear that primary spermatocytes at both zygotene and metaphase will not be encountered at the same time. An orderly progression of germ cell development also occurs with the seasonal recovery of spermatogenesis in those species which do not remain in continuous breeding condition.

It is evident from Table 3.4 that the same environment of plasma hormones must occur during both spermatogonial multiplication and some of the later stages of germ cell development. In other words, there is a problem in understanding what mechanisms could allow mitotic divisions to occur in one cell, while adjacent cells

TABLE 3.4

Germ cell associations in the seminal epithelium of the domestic duck (from Clermont 1958)

Stage of cycle		I	II	III	IV	V	VI	VII	VIII
Resting spermatogonia	(G)	G	G	G	G	G	G	G	G
Spermatogonia in mitosis	(Gm)			Gm		Gm			Gm
Primary spermatocytes at:									
Interphase	(I)				I	L	L	L	L
Leptotene	(L)	L,Z	L,Z	Z					
Zygotene	(Z)								
Pachytene	(P)				P	P	P	P,D	D,M
Diakineses	(D)								
Metaphase	(M)								
Secondary spermatocytes	(S)								S
Secondary spermatocytes in division	(Sm)								Sm
Spermatid stages 1–10	(1–10)	1	2	3	4	5	6	7	8
		9	9	10	10				

undergo reduction divisions, or in appreciating the factors determining whether an A-type or B-type spermatogonium is produced from a stem cell. In part, regulation may be achieved by the Sertoli cell acting selectively on those cells first primed to a specific developmental stage. Also it is clear that diurnal and short-term cyclic variations in gonadotrophin output can initiate successive waves of biochemical activity in the target tissues. This could be deduced from Fig. 2.4 and is a topic which will be discussed in greater detail in Chapters 11 and 12. Exceptional circumstances are created, however, by drug therapy or the artificial manipulation of the photo-regime (Fig. 3.5).

The endocrine state existing up to the time of secondary spermatocyte production has been considered. Subsequent maturation of the germ cells into spermatozoa appears to require androgen. It is not clear whether androgen facilitates the transformation of the secondary spermatocytes into spermatids, as suggested for the House Sparrow by Kumaran and Turner (1949), or is only needed for spermiogenesis, that is, maturation of the actual spermatids. We noted above that injections of testosterone can stimulate spermatogenesis provided they are given to birds with fully regressed gonads. This spermatokinetic effect is thought to depend on large non-physiological doses having a direct effect on the tubule. But these same dosage rates have been shown to retard spermatogenesis in Starlings *Sturnus vulgaris,* House Sparrows, and *Quelea* if they are administered after the onset of spermatogonial divisions and before primary spermatocytes have been formed (Burger 1944; Pfeiffer 1947; Lofts 1962*a*). Yet testosterone injections will accelerate spermatogenesis if given after the production of primary spermatocytes (Pfeiffer 1947; Kumaran and Turner 1949; Lofts 1962*a*). These relationships are summarized in Fig. 3.7, which is

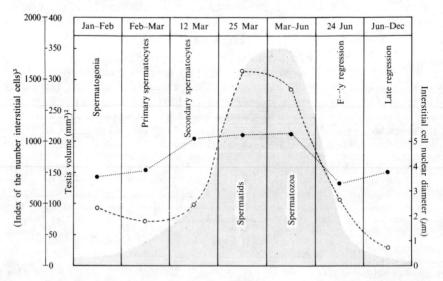

Fig. 3.6. Seasonal variations in testis volume (stippled), index of Leydig cell number (open circles, dashed line) and Leydig cell nuclear diameter (filled circles, dotted line) in the California Quail *Lophortyx californicus.* Based on data in Jones (1970).

based on studies of the Red-billed Quelea by Lofts, who also noted that androgen administration at the end of the breeding cycle prevented testicular regression, presumably because it could allow germ cell maturation to continue. Table 3.1 also showed how testosterone can stimulate an increase in testis weight if given to intact birds with regressed organs or those advanced in spermatogenetic development. It was early reported by Chu (1940) and Chu and You (1946) that testosterone prevented the gonad regression which could be induced by hypophysectomy in pigeons, and that it restored testicular weights if given after regression had occurred. This was not confirmed by Baylé, Kraus, and van Tienhoven (1970) in a study of hypophy-

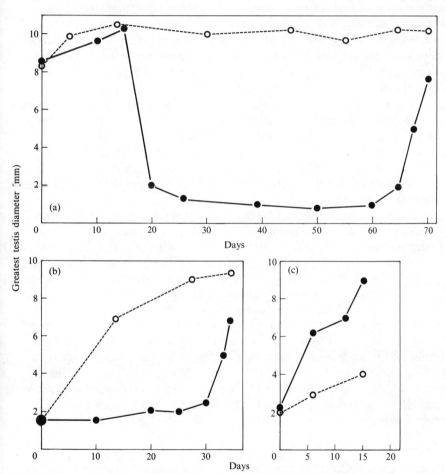

Fig. 3.7. Effect of testosterone injections (1 mg per day) on the testis cycle of *Quelea quelea* (solid lines and solid circles–untreated controls; dotted lines and open circles–experimental subjects: (a) androgen given to birds in full breeding condition maintained spermatogenesis while controls showed spontaneous regression into refractoriness;

(b) androgen administered to birds during the photo-refractory phase stimulated spermatogenetic recovery; (c) androgen given to birds that had begun post-refractory testis recrudescence inhibited gonad growth. From Lofts (1962*a*).

sectomized Japanese Quail, although testosterone did slow down the rate of degeneration of the germinal epithelium.

When testosterone was given to paired Feral Pigeons which were in full breeding condition it produced variable results for in some there was an acceleration of germ cell development; in others there was a retardation of spermatogenetic activity, the number of germ cells was reduced and there was a build-up of lipoidal material within the tubules (Fig. 3.8(a); see between pp. 82–3). Androgen is thought to inhibit FSH activity (see p. 66) so that during the critical stage when FSH is required for the meiotic division of primary spermatocytes, androgen administration retards spermatogenesis. Once secondary spermatocytes have been formed exogenous androgen can facilitate their development into spermatozoa. Variable results would be expected following androgen injection depending on the supply of secondary spermatocytes and spermatids already available, and with prolonged androgen therapy eventual gonad regression would be expected (see also p. 321).

Spermiation, the equivalent of ovulation in the female, involves the release of mature spermatozoa by the Sertoli cell so that they become free in the tubule lumen and can move to the efferent ducts. Ovulation in the female mammal is known to require a surge of LH; a rise in circulating LH and a depletion of LH in the hypophysis occurs in both male and female rats at copulation and ovulation in the female is reflexly achieved by the stimulation of coitus (Taleisnik, Caligaris, and Astrada 1966; Aron, Asch, and Roos 1966). Burgos and Vitale-Calpe (1969) showed that following LH administration to the rat, or 1 h after copulation, a swelling occurs in the apical cytoplasmic matrix of the Sertoli cell together with a dilation of vesicles and cisternae in the smooth endoplasmic reticulum. The cell cytoplasm assumes a watery appearance and swelling causes the apical recesses to unfold. This pushes the spermatozoa towards the tubule lumen and is followed by rupturing of the apical portion of the cell membrane so that the spermatozoa and cellular debris are released. These authors considered that LH achieved these effects by blocking the sodium-activated enzyme ATPase (the sodium pump) located at the cell membrane and microsomal membranes. Fig. 6.2 (p. 133) shows that spermiation was correlated during the pre-incubation behaviour cycle of the Feral Pigeon with high titres of plasma IR-LH and it seems possible that similar mechanisms could be involved.

There are considerable interspecific variations in avian sperm morphology (van Tienhoven 1961b; Bishop 1961) as well as in mammals (Fawcett 1970; Fawcett and Phillips 1970) although the basic structure and components are common to all birds. The adaptive significance of the variations are totally unknown.

The gonad—pituitary axis

It has long been realized that the integration of physiological function involves some kind of self-regulation or homoeostatis. The science of communication and control (cybernetics) has allowed a more explicit mathematical presentation of the properties possessed by feed-back systems and their ability to stabilize biological systems. Yates and Brown-Grant (1969) illustrated the use of cybernetic equations to show how in the absence of negative feed-back a 10-fold increase in the gain of the forward element of a process causes the same increase in output. In contrast, with negative feed-back a 100 per cent change in a parameter leads to only an 80 per cent increase in output, that is, greater stability. The gonadal steroids regulate the rate of pituitary secretion via important feed-back mechanisms which are now being defined in mammals. A similar role has long been suspected in birds from such observations as the increase in size of the pituitary of photostimulated castrated ducks compared with untreated controls (Benoit *et al.* 1950; Benoit 1961). In hemi-castrated domestic ducks the remaining testis developed to about twice its normal size, primarily in consequence of a lengthening of the seminiferous tubules (Benoit 1930*a, b*). No such increase occurred in ducks hemi-castrated during the period of photo-refractoriness (see p. 227), when little or no gonadotrophin would have been released (Benoit *et al.* 1950). Similarly, in photostimulated hemi-castrated White-crowned Sparrows *Zonotrichia leucophrys gambelii* the single testis exhibited a higher logarithmic growth rate than either testis of intact controls; moreover, the final weight of the single organ was nearly equal to the weight of the paired testes of the controls (Farner, Morton, and Follett 1968). The gonadotrophin content (Breneman assay) of the pituitary did not differ between groups suggesting that feed-back mechanisms were of little importance during the period of testicular growth. We should recall that at this stage the Leydig cells have not reached full development and may guess that androgen titres have not reached a peak (see Figs. 3.6 and 3.9). An increased growth rate of the single testis presumably results from it using most of the gonadotrophin previously shared between the two organs.

Gonadectomy has been shown to increase the activity of the enzyme acid phosphatase in the hypothalamo-hypophysial system of the Rustic Bunting *Emberiza rustica* (Uemura 1964*b*), White-eye *Zosterops palpebrosa* (Uemura and Kobayashi 1963), and White-crowned Sparrow (Kobayashi and Farner 1966). Subsequently, Stetson and Erickson (1971) measured the gonadotrophic potency of the adenohypo-

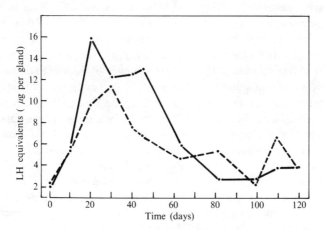

Fig. 3.9. Pituitary gonadotrophin potencies (in μg-equivalents of N.I.H.-LH-S12 per gland) of intact (solid line) and castrated (broken line) White-crowned Sparrows which were photostimulated for 119 days on LD 20:4. From Stetson and Erickson (1971).

physis of this last species in photostimulated castrates and intact controls, again using the Breneman assay. It should be remembered that this assay is not specific for either FSH or LH (see p. 38). As Fig. 3.9 shows, gonadotrophic potency increased towards the end of the experiment in the castrates, whereas in the intact controls it declined, consistent with the idea that the testes were inhibiting gonadotrophin secretion, which in turn resulted in their regression. Gonadotrophin potency did eventually decline in the castrates, indicating that other factors are of prime importance in regulating the release of pituitary hormones (see p. 321 below). These same workers showed that testosterone propionate (200 μg every 2 days for 8 days) given to intact photostimulated birds reduced the level of pituitary gonadotrophins in proportion to dose, and it also resulted in a decrease in testis weight.

In castrated birds the same dose of exogenous testosterone did not immediately depress pituitary gonadotrophins (actually there was a slight but not statistically significant increase). However, in castrates given protracted high doses of testosterone a reduction was eventually noted. These results suggest a priming effect in that intact birds would have produced endogenous androgen at the start of the experiment. Subjects which regenerated their testes because of inadequate castration responded in the same way as intact birds. These feed-back effects are probably also important under non-photostimulatory daylengths; castrated birds kept for 120 days on LD 8:16 had a higher pituitary gonadotrophin content than identically treated intact controls. The adenohypophysis of such castrates acquire B-type gonadotrophin castration cells whereas in photostimulated castrates A-type castration cells occur (Mikami *et al.* 1969). The latter appears to correspond with the δ-(TSH) cell of Tixier-Vidal *et al.* (1968) but the identity of the B-type cell is not yet clear. This seems to indicate that different gonadotrophins are affected and of course testosterone need not be, and is unlikely to be, the only steroid influencing hypothalamic mechanisms. In Japanese Quail, the pituitary cytology is not much modified in non-photostimulated castrates, while the β-(FSH), γ-(LH), and δ-(TSH) cells hypertrophy following photostimulation (see also Tixier-Vidal and Follett 1973).

The feed-back receptors have mostly not been specifically identified for birds although they seem to be located in the median eminence (see p. 20). In mammals, implantation of testosterone in the median eminence of dogs and rats causes testicular atrophy whereas implantation of the anti-androgen cyproterone stimulates the reproductive system (Davidson, Feldman, Smith, and Weick 1969). Similarly, the main site for negative feed-back by both oestrogen and progesterone appears to reside in the median eminence but positive feed-back actions of oestrogen in the female mammal were located in the anterior hypothalamic—medial preoptic area and were ineffective in the basal medial hypothalamus (Smith and Davidson 1968). Wada (1972) implanted testosterone pellets in the third ventricle of the brain of Japanese Quail, which completely inhibited photostimulated testicular growth, as did implants in the nucleus tuberis. Implants in the adenohypophysis were without effect. One hour after injection of [^{3}H] testosterone into Barbary Doves *Streptopelia 'risoria'* (see p. 103) some cell nuclei could be isolated from the cerebrum with radioactivity, but the main site of retention of the hormone was in the hypothalamus

(Zigmond, Stern, and McEwen 1972). Target cells for testosterone, and other steroids on the biosynthetic pathway leading to testosterone, have since been identified using autoradiographic methods (Zigmond 1975) and the relevance to behavioural mechanisms has also been considered (Hutchison 1975). Since the median eminence is the storage site for releasing factors it is likely that neurones involved are also sensitive receptors. The complexity of feed-back systems can be judged from discoveries made in mammals (principally rats) and conveniently summarized by Ramirez (1969). The basic relationship during ovulation is that FSH stimulates oestrogen secretion which in turn inhibits further FSH output and in the presence of progesterone stimulates the LH-RF cells. This results in an ovulatory surge of LH. Negative feed-back of LH then blocks the LH-RF mechanism. Unlike LH, if FSH is stereotaxically implanted in the median eminence it acts as a positive feed-back stimulating further FSH secretion. The inhibitory effect of oestrogen (and perhaps also testosterone) on FSH secretion apparently depends on the fact that both FSH and oestrogen compete for the same receptor neurones in the median eminence; these neurones can be activated by FSH or inhibited by oestrogen. Ramirez has made the intriguing suggestion that control of FSH secretion involves a positive FSH loop and a negative oestrogen one. He has followed Milhorn's (1966) control-theory scheme in which S is an initial stimulus and G the gain in a system to propose that:

$$\text{Total FSH secretion} = S + S(G_{FSH} - G_{oest.}) + S(G_{FSH} - G_{oest.})^2 \dots + S(G_{FSH} - G_{oest.})^\infty$$

If $(G_{FSH} - G_{oest.}) > 1$ there will be positive feed-back that will lead to increased FSH output. If $0 < (G_{FSH} - G_{oest.}) < 1$ feed-back with a finite value occurs. It appears that the secretion of oestrogen prevents the system from deteriorating so that FSH is produced continuously. This example has been given in some detail for other results presented below indicate that some kind of testosterone—oestrogen/FSH interaction occurs in birds also.

Effects of exogenous hormones on testis function
Several references have been made to the rôle of pituitary FSH and LH and of androgens produced by the interstitial cells of the testis, both in stimulating gametogenesis and in regulating gonadotrophin secretion. Brief mention should be made of the effects of other hormones on the testis as this helps in understanding the normal functioning of the organ and to this end some details are given in Tables 3.2 and 3.3.

Oestrogen and progesterone
It can be seen that both oestrogen and progesterone had an anti-gonad action when injected into mature and *paired* male Feral Pigeons which were allowed to undergo a pre-incubation behaviour cycle; details of the experimental technique are given in Chapter 5. In addition, some further experiments were done on Feral Pigeons, proved by laparotomy and brief pairing with a female to be in full breeding condition, which were given oestrogen and progesterone while being kept in *isolation*. Controls received only ten injections of saline during a period of 14 days while the remainder were

either given 10 intra-muscular injections of 2 mg oestradiol benzoate (total dose 20 mg over 14 days) or the same dosing of progesterone. Some of the oestrogenized pigeons were allowed a variable period of recovery following the last injection but all progesterone-treated subjects were killed on day 15. The spermatogenetic condition of the subjects at the beginning and end of the experiments is shown in Fig. 3.10, the means being based on 10 tubule counts per individual and the number of individuals mentioned in the Figure. Progesterone and more so oestrogen caused a marked reduction in testis weight, as Table 3.5 shows.

TABLE 3.5

Combined weight ± S.D. of paired testes of Feral Pigeons after treatment with progesterone or oestradiol benzoate

Controls at start $N = 6$	10 × 2 mg (= 20 mg) progesterone over 14 days $N = 6$	10 × 2 mg (= 20 mg) oestradiol benzoate over 14 days $N = 6$	14 days following end of oestrogen therapy $N = 5$	28 days following end of oestrogen therapy $N = 4$
2467 ± 719	1259 ± 306	273 ± 95	367 ± 168	1691 ± 305

Based on Murton and Westwood 1975.

The histological appearance of the testes from oestrogenized birds was distinctive. Most tubules contained only a resting ring of spermatogonia with a few degenerating primary spermatocytes (Fig. 3.8; see between pp. 82–3). None of the individuals had an intact layer of primary spermatocytes and few of these cells were advanced beyond the meiotic stage of prophase. Moreover, the cells other than spermatogonia were morphologically and topographically abnormal and contained vacuoles where the cell cytoplasm had contracted from the nucleus. As in the case of the domestic rat (Lacy and Lofts 1965), spermatogonia were not harmed, indeed there was an increase in their number, and it was the primary spermatocytes which appeared to be most sensitive to oestrogen treatment and few survived to the pachytene-stage of meiosis. The interstitial areas were enlarged and of unusual appearance. The mature Leydig cell generation could be distinguished because the cell nuclei were large, with slightly crenulated boundary membranes, and they contained little chromatin. These cells were aggregated as nests of cells in which the inner and adjacent cell walls had disintegrated, but not those on the outside, so that between 2 and 12 Leydig nuclei were encapsulated within one membrane (Fig. 3.8). In addition, a new generation of juvenile Leydig cells with small spindle-shaped nuclei had been formed and large numbers of these cells could be seen as a ring surrounding the tubules, apparently being proliferated from fibroblasts in the basement membrane of the testis tubules. Further from the tubules membrane, the cells and their nuclei became bigger and rounder and the nuclei contained numerous dark chromatin granules. Surprisingly, there was virtually no lipid inside the testis tubules as revealed by Sudan Black, but the interstitium was occluded with amorphous deposits. These occurred as a dense mass in the cytoplasm of the old Leydig generation so that the encapsulated aggregations noted in the wax sections above appeared as black spheres in the gelatin-prepared material (Fig. 3.8). In addition, the juvenile Leydig cells contained numerous sudanophilic droplets. Fig. 3.10 depicts the way in which spermatogenesis resumed once oestrogen treatment was terminated and more details are given in Murton and Westwood (1975).

Compared with the oestrogen-treated birds, those given progesterone exhibited a more variable histological appearance, for in some tubules a few spermatozoa could be distinguished, while in others germ cells beyond the secondary spermatocyte stage were not present (Fig. 3.10). The tubules were not much regressed in size and numerous meiotic figures were apparent in the primary spermatocytes ranging from early prophase to metaphase and all seemed normal in appearance. Essentially, development beyond the stage of secondary spermatocyte had been inhibited and these germ cells were accumulating in the tubule lumini. The interstitial cells did not

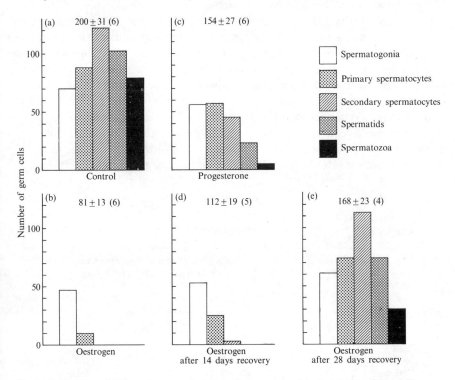

Fig. 3.10. Number of different germ cells in sample segments of testis tubules of Feral Pigeons (number in brackets is number of birds examined) given either (b) 10 injections each of 2 mg of oestradiol benzoate over 14 days, or (c) 10 × 2 mg injections of progesterone over the same period, compared with the control condition at the beginning of the experiment (a). Recovery of spermatogenesis in oestrogen-treated subjects allowed 14 and 28 days recovery from the end of the treatment period is also depicted ((d) and (e)). Based on data in Murton and Westwood (1975).

differ in appearance from the control situation, judging by the wax-embedded material, but gelatin-prepared sections revealed an increased deposition of sudanophilic lipid in the Leydig cytoplasm. Appreciably more lipid was also visible in the Sertoli cytoplasm than was present in control or oestrogen-treated subjects (Fig. 3.8). This lipid was apparent as droplets located near the periphery of the tubule and sometimes the droplets had coalesced to form large globules.

The two gonadal hormones had distinctly different effects on gonadotrophin secretion, for oestrogen stimulated a marked elevation in plasma IR-LH as measured by the Bagshawe radioimmunoassay in all but one of the subjects, whereas progesterone gave somewhat variable results, but did not produce a significant increase in the mean LH titres compared with the control situation (Table 3.6). It is well established that oestrogen stimulates LH secretion in the female mammal, the so-called Hohlweg effect (1934), but it has not so far been certain whether a similar response mechanism exists in birds, particularly in the male. Hinde, Steel, and Follett (1974) have data for the *female* Canary *Serinus canaria* to indicate that exogenous oestrogen inhibited LH release, plasma concentrations of the latter being measured by the Follett radioimmunoassay. It is not clear whether the discrepancies depend on different gonadotrophin factors being measured in the two experiments or whether there is a difference in response to exogenous oestrogen in the two species, or between males and females. In the female mammal oestrogenic hormone reduces the secretion of FSH, as mentioned above, and the same appears to be true in the male rat for FSH treatment immediately restores spermatogenetic development, in particular meiotic divisions of the primary spermatocytes (Lacy and Lofts 1965). It seems reasonable to conjecture that suppression of spermatogenesis in the present studies also depended on an inhibition of FSH synthesis or release (see discussion above on p. 68), the massive steatogenesis of the interstitium, consequent on the accumulation of unsaturated

TABLE 3.6

Plasma concentrations of IR-LH in individual Feral Pigeons following
injection of 2 × 10 mg progesterone or 2 × 10 mg oestradiol benzoate

I.U. per ml plasma in untreated controls	I.U. per ml plasma in subjects treated over a period of 14 days with total of:	
	20 mg progesterone	20 mg oestradiol benzoate
0·0025	0·0050	0·0064
0·0050	0·0050	0·0116
0·0050	0·0066	0·0126
0·0052	0·0085	0·0210
0·0060	0·0140	0·0640
0·0062	0·0260	0·1000
Means 0·0050 ± 0·0013	0·0109 ± 0·0081	0·0359 ± 0·0377

Based on Murton and Westwood (1975). Using Mann Whitney *U*-test the oestrogen treated and controls were significantly different (P = 0·002, two-tailed), but the progesterone and control groups were not. Original measurements expressed as HCG I.U. per ml plasma.

precursor steroidogenic material, being consistent with an increased influence of LH.
Two distinct processes were recorded: (a) the accumulation of lipid in the mature
Leydig cells as if a releasing factor was absent, (b) the proliferation of new Leydig
cells which also accumulated lipid material. An increase in number of Leydig cells
following oestrogen therapy is also apparent in the data in Table 3.2 and is consistent
with an increased involvement of endogenous LH. Table 3.2 also shows that exogenous
mammalian LH results in an increase in the Leydig cell count in the pigeon.

The observations of Erickson, Bruder, Komisaruk, and Lehrman (1967) suggest
that progesterone selectively insulates the central nervous system against the action
of testosterone. It is possible that progesterone more generally inhibits androgenic
function in view of the evidence above (p. 62) that the maturation of germ cells
from the stage of secondary spermatocyte to mature sperm depends on androgen
and not FSH.

Prolactin

Table 3.3 shows that large doses of prolactin caused relatively little inhibition of
spermatogenesis when administered to Feral Pigeons during the pre-incubation
behaviour cycle. This result contrasts with experiments reported in the early literature which also refer to pigeons (Riddle and Bates 1933) as well as to cockerels
(Yamashima 1952). A single injection of 1 mg prolactin into the pectoral muscle
apparently caused gonad regression with massive tubule steatogenesis in House
Sparrows *Passer domesticus* caught wild between June and August (Lofts and
Marshall 1956). In view of other conflicting results reported in the literature this
experiment needs repeating to clarify that the subjects were not entering a period
of post-nuptial refractoriness (p.227) which would have led to similar histological
manifestations. Similarly, when post-refractory finches were photostimulated and
then given a single injection of 1 mg prolactin there appeared to be an inhibition of
spermatogenesis. Only one control Chaffinch *Fringilla coelebs* and one control
Greenfinch *Carduelis chloris*—which was in any case atypical for it did not respond to
photostimulation—were available so the results are somewhat ambiguous. Inhibitory
effects of prolactin on the hypothalamo-gonadal axis have also been reported for
the White-crowned Sparrow *Zonotrichia leucophrys pugetensis* (Bailey 1950), Song
Sparrow *Melospiza melodia,* House Sparrow, and Nuttal Sparrow *Zonotrichia l.
nuttalli* by Meier and Dusseau (1968) and for female White-throated Sparrows
Z. albicollis by Meier (1969). In contrast, no effect of prolactin was detected on any
phase of testicular growth or regression by Laws and Farner (1960), who studied
another race of the White-crowned Sparrow *Z. l. gambelii,* (Gambel's sparrow), by
Hamner (1968), who examined the House Finch *Carpodacus mexicanus,* by Jones
(1969c) studying California Quail *Lophortyx californicus* or in the Junco *Junco
hyemalis* or male White-throated Sparrows (Meier and Dusseau 1968).

An explanation for the conflicting results outlined above is apparently provided
by the more recent studies of Meier, Martin, and MacGregor (1971b). They found
that a temporal synergism was important between corticosterone and prolactin so
that the time of day of injection was important. As Fig. 3.11 shows, when cortico-

sterone and prolactin injections were given to photo-refractory House Sparrows captured from the field in October, they stimulated gonad growth if spaced 8 h apart. In photosensitive White-crowned Sparrows captured in winter, effective synergism was noted when injections were spaced 12 h apart (see Fig. 3.11).

The involvement of prolactin in behaviour cycles is discussed in Chapter 6 (p. 134). Diurnal rhythms in hormone release and the importance of temporal synergisms in allowing responses are considered in Chapter 7 in connection with the involvement of prolactin in migratory fattening, while circadian rhythm mechanisms are discussed in detail in Chapters 11 and 12. It will become apparent that corticosterone probably functions by re-phasing the rhythms of gonadotrophin secretion and that some of the effect depicted in Fig. 3.11 depended on an altered phasing of gonadotrophin secretion. Prolactin probably functions in gonadal growth by virtue of its somatotrophic properties. This is indicated in Table 3.7, where the effect of giving exogenous hormone injections to photo-refractory White-crowned Sparrows for 23 days is summarized. It is of interest to compare Tables 3.7 and 3.1.

When Meier and Dusseau made their 1968 studies they wished to compare the effect of prolactin on the photoperiodically induced gonadal growth of six species. Inhibitory effects were noted in resident or weakly migratory species (Nuttal Sparrow, Song Sparrow, and House Sparrow) but not the strongly migrant ones (Gambel Sparrow, White-throated Sparrow, and Junco) and this correlation appears to be generally true. One possible explanation they suggested was that since prolactin is also important in the physiology of migration (see Chapter 7) a resistance to its effects was an adaptation to allow gonad development during migration. We suggest that the difference reflects a difference in the phasing of the rhythms of hormone release in migrants and non-migrants and that all subjects would respond in the same way if the time of hormone administration were adjusted relative to endogenous rhythms.

The site of prolactin inhibition of gonadotrophic function presumably resides in the hypothalamus, whereas the somatotrophic effect is perhaps mediated in the

TABLE 3.7

Effect of exogenous hormones on testis weight of photorefractory
Zonotrichia leucophrys gambelii

Treatment	Testis weight ± S.D. (mg) (N)		Ovary weight ± S.D. (mg) (N)	
Control	$2 \cdot 2 \pm 0 \cdot 3$	(3)	$7 \cdot 7 \pm 1 \cdot 4$	(5)
Prolactin	$2 \cdot 4 \pm 0 \cdot 6$	(3)	$9 \cdot 5$	(2)
Gonadotrophin	$9 \cdot 1 \pm 6 \cdot 1$	(3)	$11 \cdot 4 \pm 3 \cdot 1$	(5)
Prolactin + gonadotrophin	$56 \cdot 2 \pm 24 \cdot 0$	(6)	$26 \cdot 3 \pm 9 \cdot 9$	(4)
ACTH	$1 \cdot 7$	(1)	$5 \cdot 0 \pm 1 \cdot 3$	(4)
Prolactin + ACTH	$3 \cdot 7$	(1)	$9 \cdot 4 \pm 0 \cdot 7$	(3)

From Meier and Farner (1964). Injections were given for 23 days and each involved 300 μg gonadotrophin, this being a 50:50 mixture of N.I.H.LH and N.I.H.FSH, or 200 μg prolactin or 200 μg ACTH (using Armour Prednisone) or additions of these quantities.

target tissue. Stetson and Erickson (1970) have measured the rate of $^{32}PO_4$ uptake by the testes of chicks (i.e. the Breneman assay) following injection of LH, FSH, and crude pituitary extract alone or each combined with prolactin. No difference was found between the potency of purified gonadotrophin or pituitary extracts when given alone or in combination with prolactin, while large amounts (500 μg) of prolactin enhanced the action of gonadotrophin. These results indicate that if gonadotrophin action is disrupted by prolactin it is via a mechanism different from that involved in the uptake of ^{32}P. Instead, it is thought that prolactin acts by inhibiting the release of LH or FSH or both from the adenohypophysis. Adminis-

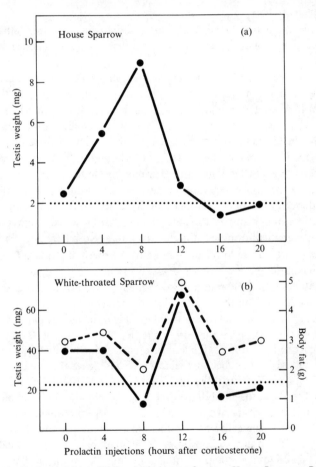

Fig. 3.11. The testicular weight (solid lines) of photo-refractory House Sparrows *Passer domesticus* (a) and photosensitive White-throated Sparrows *Zonotrichia albicollis* (b) given prolactin injections (25 μg per injection over approximately 14 days) at various times following injection of corticosterone (25 μg × 14). Birds were held on continuous light. The effect of the same experimental procedure on the body fat of White-throated Sparrows is shown as a dashed line. Dotted lines indicate the testicular weight of controls killed at the start of the experiment. Based on Meier, Martin, and MacGregor (1971*b*).

tration of testosterone has been recorded as causing an increase in the prolactin content of the adenohypophysis of the domestic duck (Gourdji 1967, 1970) but testosterone given to pigeons does not lead to a development of the crop sac gland (Meites and Turner 1947). As Bern and Nicoll (1969) emphasize, prolactin is a hormone with marked versatility in function throughout the vertebrates: it has a metabolic rôle being protein-anabolic, lipotrophic, and hyperglycaemic, and is effective in regulating sodium retention; it stimulated the pigeon crop sac gland and the mammalian mammary glands; it is also involved in the development of incubation patches and synergistically with other hormones in several other functions.

Accessory sexual organs

These comprise the sperm ducts in the male and the oviducts in the female, together with any special cloacal differentiation into papillae or an intromissive organ as in Anatidae.

Two pairs of ducts, the Mullerian and Wolfian ducts, derived from primitive kidney ducts, develop in both sexes and each pair runs from the gonad to the urogenital sinus. The Wolfian ducts subsequently become the vasa deferentia and epididymides in the male, while they usually remain vestigial in the female. In the female, the left Mullerian duct becomes the oviduct while the right, and both these ducts in the male, remain vestigial. Occasionally the ducts do not atrophy in the normal manner, the occasional persistence of the right Mullerian duct in the female appearing to be genetically controlled since a high proportion of certain inbred strains of fowl exhibit two ducts instead of one (Morgan and Kohlmeyer 1957). Development is controlled by sex hormones, and *in vitro* studies show that the left and right ducts respond differently to the same hormonal stimulus (Iwamura, Koshihara and Noumara 1975).

Spermatozoa are released into the lumina of the spermatic tubules which form the highly convoluted network of the testis. From here they are passed by a small number of tubules called the rete tubules to the highly convoluted vasa efferentia. In the non-breeding season the rete tubules are inconspicuously sited in the tunica albuginea of the testis but they become enlarged and distinct as two-four tubes extending from the testis with the onset of reproductive activity (Bailey 1953). The sperm next pass to thin tubules, the vasa efferentia, which also show some seasonal enlargement and which may become secretory. These leave the tunica albuginea and join to form a long coiled tube, the epididymis. This comprises a compact structure closely bound to the testis and lying adjacent to the kidney. With the onset of sexual activity the epididymis hypertrophies to become a large prominent white knob on the side of the testis. The cells of its epithelium are columnar and ciliated, and secrete a seminal fluid of unknown composition. In the mammal the histological differentiation of the epididymis is complex, indicating that it serves more than one function (Martan 1969) and the same may be true of birds. Sperm are not stored in the epididymis but progress into a deferent duct called the vas deferens; two muscular tubes one from each of the paired epididymides, therefore, lead to the urodaeum of the cloaca. Some kind of maturation change occurs in the sperm after leaving the testis tubules for those taken from the epididymides show a much lower capacity to fertilize than do those from the ductus deferens (Munro 1938). Differences in individual male fertility related to the mass of the testis when mature are common in the domestic goose. This is an inherited characteristic susceptible to alteration by selective breeding (Szumowski and Theret 1965).

The vasa deferentia become enlarged, more convoluted and distinct from the kidney during the breeding season and before entering the cloaca are expanded to form a seminal sac. This increases 30 to 40-fold in weight during the breeding season becoming charged with stored spermatozoa. Middleton (1972) observed in European Goldfinch *Carduelis carduelis,* American Goldfinch *Spinus tristis,* and Canary that sperm cells penetrate the epithelium of the seminal sac. Histochemical tests indicated that this epithelium produces glycogen, suggesting that the sac serves for sperm maturation. A muscular sheath forces the posterior wall of the sperm sacs to expand into the cloaca as erectile papillae and doubtless facilitates the extrusion of sperm during copulation (see Nishiyama 1955; Lake 1957). However, these papillae are minute in most species and during copulation the cloaca of each sex is everted and the papillae are brought into contact with the opening of the oviduct. Low temperature is desirable for sperm storage, as in mammals, and the siting of the sperm sacs near the cloaca is probably important in this respect. In cyclostomes spermatozoa probably mature inside the testis but the evolutionary trend in higher vertebrates is for sperm maturation to occur in the upper coiled portion of the efferent ducts and for storage to be confined to more distal and cooler sites. Thus the cloacal protuberances of birds are reminiscent of the mammalian scrotum (Wolfson 1954). Humphreys (1972) has noted that the semen of the passerine Tree Creeper *Certhia familiaris,* Canary, and Robin *Erithacus rubecula* is thick, with relatively few sperm. These have a spiral configuration and move by a helical forward movement. Non-passerines studied had a more liquid semen containing many spermatozoa, these having an undulating mobility reminiscent of the mammalian condition.

In some species part of the cloaca is modified as a penis-like structure. In ducks the penis is a vascularized sac which can be protruded by a muscle and retracted by a ligament; the ligament is bound spirally round the sac and gives the penis a twisted appearance. Sperm move along a spiral groove extending from the cloacal papillae at the organ's base to its tip. Höhn(1960) has shown that the weight of the penis of the Mallard *Anas platyrhynchos* increases 6-fold with the onset of the breeding season and that castration prevents such development. Benoit (1936) had earlier noted that at puberty the phallus became enlarged under the influence of testicular hormones, but outside the breeding season it did not regress again to its juvenile dimensions and, instead, remained fairly constant in size. Females of those species in which the males possess a 'penis' have a homologous but reduced structure termed a clitoris. Their cloacal opening is slightly modified to receive the intromissive organ of the male.

Penis-like organs occur in tinamous(Tinamidae), curassows (Cracidae), the ratite birds, and the ducks and geese (Anseriformes). Alone among the weavers (Ploceidae) the Black Buffalo Weaver *Bubalornis albirostris* possesses an imperforate penis-like external organ. The groups possessing these modified genitalia are diverse and the species appear to share no common feature which would indicate an adaptive function for their special attributes. However, it may be significant that all those concerned, except the ducks and geese, habitually practice polygamy or polyandry (see p. 435). Conceivably each male must be ready to fertilize the female with a reduced opportunity for pre-copulatory displays which would help synchronize the process, but this view is speculative. The Ostrich *Struthio camelus*, Rhea *Rhea americana*, amd other ratites, and such tinamous as *Nothocercus bonapartei* (Schafer 1954) perform co-operative laying whereby several females lay their eggs in the same nest and this is then attended solely by the male; in some cases the females may repeat the performance for another male. Again selection may have favoured a more effective copulatory mechanism in species which do not have the benefit of a long period of intra-pair courtship. The ducks are also unusual in that pairing and copulation usually occur on the water.

Seasonal changes in the size and degree of development of the vasa deferentia, the epididymis, and penis, if present, parallel the seasonal changes observed in the primary sex organs in those species showing cyclical activity; these structures remain permanently developed in continuous breeding species such as the domestic fowl (Domm 1939). This seasonal development apparently depends on androgenic hormones in the case of the vasa deferentia and epididymides (Witschi 1945) and the seminal sacs (Riddle 1927). Testosterone injections are effective in causing enlargement of the ductus deferens in fringillids and the Starling *Sturnus vulgaris* (Bailey 1953; Hilton 1968). Large daily doses of testosterone (3 mg) for a week will also stimulate the ductus deferens of day-old chicks while inhibiting testicular growth (Lake and Furr 1971). Lake refers to the work of Tveter, Unhjem, Attramadal, Aakvaag and Hansson (1971) to suggest that androgen acts on specific receptor proteins, present only in target tissues, to increase RNA polymerase activity. Some reservation is needed about the rôle of androgens in that some of the pioneer experiments have not been repeated since purified hormones became available and it is possible that early extracts were contaminated with gonadotrophins; this reservation is occasioned by recent experiments into the factors inducing the black bill pigmentation of sparrows (see p. 226). Moreover, in the case of the female Canary it seems that oestrogenic hormones will induce oviduct development only in subjects kept on stimulatory photoperiods and not in birds treated in winter (see p. 116). This suggests that the synergistic action of a gonadotrophin is necessary and it is likely that a parallel situation applies in the case of the male. Critical experiments with hypophysectomized birds have not been performed.

4 Reproductive apparatus of the female

Summary

Like the testis, the vertebrate ovary is a mixture of endocrine and germinal tissues derived from different embryonic sources. Secretory components having a typical steroidogenic structure and containing the enzyme Δ^5-3β-hydroxysteroid dehydrogenase are the interstitial cells (equivalent to Leydig cells) and the granulosa cells which are of epithelial origin (perhaps homologous with Sertoli cells); these last comprise the follicles which surround the ova. Thecal cells become incorporated into the outer layer of the follicles and they are derived from stromal fibroblasts. LH and FSH in varying synergistic mixtures, and probably in specific temporal sequences, control ovary function but we still lack a complete understanding of all the biochemical processes involved. Oogenesis begins in embryonic life but meiosis is arrested at diplotene in the embryo, and is not completed until adult life, an hour or two before ovulation, in response to an ovulatory surge of LH. During the breeding season follicles grow under the influence of FSH and LH, which stimulate the thecal and interstitial cells to secrete oestrogen. Oestrogen feed-back is involved in stimulating the surge of LH which allows a mature follicle to be ovulated. Most follicles do not mature and degenerate at the multi-layer granulosa stage, undergoing a fatty atresia. The granulosa cells, apparently under the influence of LH and prolactin, produce progesterone which by feed-back inhibits further ovulation cycles and allows behaviour associated with incubation and brood care to be expressed.

Ovulation cycles appear to depend on circadian rhythms which are not perfectly entrained by the daily photo-regime but they can be manipulated by changing the amount of light in a 24-h day—night cycle or by varying the period length of the cycle from 24 h. Vitellogenesis and egg formation and oviposition are discussed.

The ovary

The ovary is usually a flattened, pear-shaped organ attached to the body wall by a short mesovarian ligament in close apposition to the cephalic end of the kidneys. It is supplied by the ovarian artery, usually a branch of the left reno-lumbar artery, but sometimes from a branch of the aorta, and is drained by large veins which anastomose and converge into an anterior and posterior ovarian vein, which then discharges into the vena cava. Compared with the testis, the ovarian stalk is relatively well innervated with branches from the ovarian, adrenal, renal, and aortic plexuses. The ovary comprises an inner medulla and an outer cortex at the periphery of which is a germinal epithelium from which are proliferated primary oocytes. By the time the chick hatches the full complement of these have been formed and they must suffice for the rest of the bird's life. Since in the chicken 2000 primary oocytes are reckoned to be visible to the naked eye, 12 000 can be distinguished microscopically, and various estimates of totals of up to many millions have been made (Pearl and Schoppe 1921; Hutt 1949; Krohn 1967) the supply is more than adequate. These oocytes become ensheathed by flattened epithelial cells to form primordial follicles, enormous numbers being present at hatching, and they form the most prominent feature of the female gonad (Fig. 4.1(a) and (b); see between pp. 82–3). Like the testes, the ovary undergoes a seasonal cycle in most avian species. The quiescent post-nuptial stage is characterized by small follicles not exceeding 0·5 mm in diameter, usually too small to differentiate with the unaided eye, but with the onset of breeding a proportion mature and become extruded from the surface of the ovary, to which they remain fixed by a narrow stalk. At this stage the follicles resemble a bunch of variable-sized grapes.

Endocrine cells of the ovary, and oogenesis

The ovarian medulla is highly vascular and comprised of connective tissue from which develop stromal interstitial cells (Fig. 4.1(b)) and these appear to be homologous with the interstitial Leydig cells of the testis. These interstitial cells have the typical steroidogenic structure already detailed and they contain the enzyme Δ^5-3β-hydroxysteroid dehydrogenase, which catalyzes the oxidation of the Δ^5-3β-hydroxy group of steroids and is capable of transforming pregnenolone into progesterone (Chieffi and Botte 1965). It is possible that, like their male counterparts, these cells can produce androgens (Taber 1951; Marshall and Coombs 1957) but their function remains in doubt. They conceivably function during early differentiation of the ovary, for if mammalian FSH is given to chicks younger than 100 days the medulla region hypertrophies and both oestrogen (Nalbandov and Card 1946) and androgen secretion (Das and Nalbandov 1955) are increased. Burns (1961) gives evidence that oestrogen secretion starts during embryogenesis, but whether it is influenced by FSH or not remains to be resolved. Steroid dehydrogenase has been demonstrated in the interstitial cells of the embryonic chick (Boucek, Gyori, and Alvarez 1966; Kannankeril and Domm 1968; Sayler, Dowd, and Wolfson 1970) but evidently the relationship between steroids and inductor substances in the embryo is complex and not well understood, in spite of the importance of the androgenic or gynogenic hormone environment to subsequent development (Gallien 1962; Atz 1964).

Androgens certainly have a function in the adult female and they may additionally or exclusively derive from thecal cells (see below) which also have a stromal origin. Androgens appear to be implicated in the natural behaviour patterns of the female (see p. 117). The female pigeon begins to exhibit aggressive male-type behaviour in the days immediately preceding ovulation (p. 114). In the seasonally breeding Rook *Corvus frugilegus* the stromal interstitial cells become lipoidal in late winter and spring, suggesting that their hormones could be facilitating sexual behaviour (Marshall and Coombs 1957). Possibly androgens generally influence secondary sexual characteristics, for example, they regulate comb growth in the domestic hen. In passing it may be noted that the principal steroid secreted by the ovary at oestrous in the sheep is androstenedione (Short 1972). A gametogenetic function may also exist for androgens in the female, for they can stimulate follicular growth and increase the height of the follicular epithelium (Breneman 1955, 1956). But, as mentioned, androgenic steroids can be synthesized by other gland cells of the ovary, including those of the follicle, and so it is not clear what separate function stromal epithelial cells may have. It is possible that it is functionally advantageous for a source of androgen secretion to exist outside the follicle, allowing differential feedback mechanisms to operate, but this is speculative. The ovarian interstitial cells become renewed during the period of post-nuptial ovarian regression in the Rook. In some species this period of post-nuptial stromal rehabilitation is marked by an upsurge of androgenic-type behaviour, for example, female Robins *Erithacus rubecula* establish autumn territories from which the males and other females are excluded (Lack 1946), while female Starlings *Sturnus vulgaris* sing in the autumn and undergo a change in bill colour which in the male is known to depend on androgenic hormone (Witschi 1961; Wydoski 1964).

As mentioned, oogenesis begins in embryonic life, for after the primordial germ cells have migrated to the genital ridge they divide mitotically to give oogonia and these divide again to produce further oogonia. Each oogonium becomes a primary oocyte when it enters the prophase stage of meiotic division. Eventually haploid secondary oocytes will be formed but meiosis in the female is a remarkably protracted affair compared with what happens in the male. This is because the prophase stage (comprising leptotene, zygotene, pachytene, diplotene, and then diakinesis), which begins in the embryo, becomes arrested at diplotene and is not completed until adult life, 2 hours before the time of ovulation. This happens in response to LH secretion which also controls the extrusion of the first polar body. It is worth recalling that LH is suspected of functioning in early gametogenesis in the male and of also being implicated in spermiation (pp. 64 and 133). Since the female is the heterogametic sex in birds the sex of the future embryo is determined prior to ovulation and not, as in mammals, at fertilization (Olsen and Fraps 1950).

Soon after its embryonic formation the oocyte becomes ensheathed by flattened epithelial cells to form the primordial follicle. These cells become cuboidal and multiply to form a multi-layered zone of granulosa cells (Fig. 4.1(c)). In mammals development to this stage is independent of hormones and it continues in hypophysectomized subjects. As the follicles mature and the granulosa cells multiply thecal cells, which are differentiated from stromal fibroblasts, become incorporated as an outer layer of thecal tissue (Fig. 4.1(d)). This zone becomes highly vascularized and the thecal cells lipoidal, and at the same time liquid becomes incorporated to give a *liquor folliculi*. Unlike the mammalian follicle, a liquid-filled antrum does not form whereby the oocyte is isolated from the walls of the follicle. But, as in mammals, growth and differentiation of the follicle at this stage is much influenced by FSH secretion (Lofts and Murton 1973) and the number of follicles which develop in this way is probably, as in mammals, dependent on the quantity of circulating FSH. Exogenous FSH increases the incidence of polyovular follicles (Taber, Clayton,

Knight, Gambrell, Flowers, and Ayres 1958) comparable with the multiple births in human females consequent on FSH therapy. Hypophysectomy leads to extensive atresia (Opel and Nalbandov 1958; Nalbandov 1959a).

In a seasonally breeding species like the Rook or Wood Pigeon *Columba palumbus* the follicles do not exceed a diameter of about 0·5 mm during the contra-nuptial season. Development to a size range of 1–5 mm occurs with the recovery of reproductive capacity under the influence of gonadotrophin secretion, but not of a mate at this stage. At this point of development the follicles of wild seasonally breeding birds are comparable with the so-called 'resting' follicle of the domestic hen. In the domestic hen development of the post-embryonic oocyte is divisible into three stages (Gilbert 1971e). The first involves a slow deposition of neutral lipid droplets and it lasts for months or even years. The second intermediate phase lasts about 60 days during which time some yolk is deposited in vacuoles in the oocyte; this yolk has a high protein content and is known as the 'white yolk'. This stage corresponds to the period of ovarian recrudescence in seasonally breeding species. The third and final growth stage lasts 7–11 days in the hen and the main yolk mass is now added. This 'final-growth' phase culminates either in ovulation or atresia. In wild species stage 3 normally only occurs when the female is mated and the endocrine mileau necessary is stimulated by courtship and nest-building behaviour (see Chapter 5). In wild species stage 2 clearly depends on gonadotrophin secretion but the balance of LH and FSH which is necessary remains to be elucidated. Both these hormones are apparently essential for stage 3 and a steroid may also be synergistically implicated. Ferrando and Nalbandov (1969) used the drug dibenzyline to block gonadotrophin secretion in hens, thereby causing a follicular atresia. This atresia could not be prevented by FSH or LH alone but a combination was necessary. Amin and Gilbert (1970) showed that both the FSH and LH cells of the adenohypophysis were active during follicle maturation. Imai, Tanaka, and Nakajo (1972) found that plasma LH titres (OAAD method) were the same in moulting and laying hens but that FSH activity (HCG augmentation assay) was about 2-fold higher in the moulting birds. Their results suggested that the failure of follicular growth in moulting hens was due to an imbalance of gonadotrophins, rather than a shortage.

Stetson, Lewis, and Farner (1973) found that subcutaneous injections of an ovine LH + FSH mixture (100 µg LH + 100 µg FSH for 10 consecutive days) augmented ovaria and oviduct growth in photostimulated White-crowned Sparrows *Zonotrichia leucophry* Injections given 21·00h after the subjective dawn were more effective than others given : 09·00 h, presumably because the former coincided with an appropriate endogenous hormone balance, or there was a diurnal variation in target tissue sensitivity. Prolactin given simultaneously with the other gonadotrophins (100 µg/day) did not significantly alter the response of ovary or oviduct. LH and FSH had only a small, but significant, effect on the reproductive organs of photo-refractory subjects, no diurnal variations in sensitivity were found and again prolactin did not modify the response.

The exact site of FSH action appears to be the thecal and interstitial cells but not the granulosa cells. Very little is known about the avian situation and so we quote what is thought to apply to mammals, for there must be similarities. In the rat the

oxygen consumption of isolated groups of granulosa cells could be measured following the *in vitro* addition of LH or FSH to the medium or injection into the animal
prior to micro-dissection. In either circumstance FSH failed to affect the granulosa
cells but it increased the oxygen consumption of the thecal cells, whereas LH stimulated the granulosa cells but was without effect on the thecal component (Ahrén
and Hamberger 1969). These authors also found that FSH stimulated the stromal
interstitial cells of the prepubertal rat whereas LH was without effect. Two distinct
effects of FSH could be demonstrated. First, FSH had an acute influence on amino
acid transport and biosynthesis in the theca and interstitial cells, provided it was
injected into intact animals prior to removal and incubation of the ovaries. The
second effect was a marked stimulation of glucose uptake and lactic acid production
and this could be achieved even if FSH was added to the incubation medium. The
first action could be blocked by the injection of poromycin but not the second. The
biochemical action of FSH on glycolysis is also an important function of LH but in
this case the granulosa cell seems to be the target cell implicated (Hamberger and
Ahrén 1967).

The granulosa cells function in the nutrition of the developing oocyte and in
mammals they have been shown to produce the essential nutrients lactic and pyruvic
acids (Baker 1963, 1971). With the oocyte they form a zona pellucida, in which
microvilli from the oocyte surface, together with processes from the granulosa cells,
occupy canals in the zona and probably serve in the transfer of maternal proteins.
In birds, the granulosa cells become separated into intercellular spaces filled with
perivitelline substance, and fibrillar connections arise on the oocyte cell membrane
to form with the granulosa cells a zona radiata (van Tienhoven 1968). This apparently
functions like its mammalian counterpart (Wyburn, Johnston, and Aitken 1966).
A parallel exists, therefore, between the origin, structure, and function of the granulosa cells and the Sertoli cells of the testis tubules. This raises the question of whether
LH has any effect on the Sertoli cell (p. 54).

The mature follicle comprises a theca externa, a vascular interna, a basement
membrane, the granulosa layer, and an inner vitelline membrane surrounding the
yolk. The zona radiata lies between the granulosa and the oocyte. Gilbert (1971*b*)
gives a good account of its structure and Fig. 4.2 is based on his account.

In the mammal, the hypertrophied appearance of the thecal cells prior to ovulation (see Fig. 4.1(d)) suggests an involvement in steroidogenesis. Indeed, these cells
are thought to be the source of oestrogen of which most is able to escape into the
ovarian vascular supply, although some may enter the follicular fluid and stimulate
the granulosa cells. In turn, it is thought that the granulosa cells become capable of
steroid synthesis only following the gross morphological changes that will accompany
the rupture of the follicle; thereafter these granulosa cells become the source of progesterone (Short 1972). The combined action of FSH and LH appears to be necessary for the secretion of ovarian oestrogen (Lostroh and Johnson 1966). In birds,
Marshall and Coombs (1957) also considered the thecal cells to be the source of
ovarian oestrogen but this view is perhaps contradicted by the studies of Chieffi and
Botte (1965). They have examined histochemically sections from hen's ovaries for

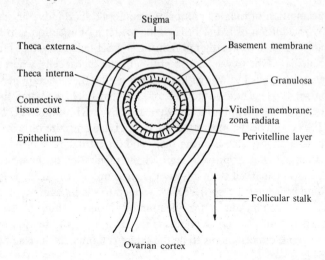

Fig. 4.2. Diagram of the structure of the avian ovarian follicle. From Gilbert (1971*b*).

the enzymes Δ^5-3β-hydroxysteroid dehydrogenase and 17β-hydroxysteroid dehydro-genase which catalyzes the dehydrogenation of 17β-hydroxysteroids to 17-ketosteroid (or 17β-oestradiol to oestrone). The last enzyme could be demonstrated in only the granulosa of the growing ovarian follicle and of the atretic follicle, suggesting that oestrogen metabolism is confined to these cells during follicular growth and the first stages of atresia. In contrast, Δ^5-3β-HSDH, whose presence merely indicates a poten-tiality for the first steps of steroid biosynthesis, was visualized histochemically in the theca interna, the interstitial tissue, and also the granulosa in both pre- and post-ovulation follicles and in the atretic follicle. There seems no doubt that the thecal cells of the fowl produce steroids (Dahl 1970*a*, *b*, 1971*a*) and they respond histologi-cally to exogenous steroids, gonadotrophins, and clomiphene—an oestrogen inhibitor (Dahl 1970*c*, 1971*a*, *b*, *c*). This topic needs to be resolved but it does seem likely that both cells have some ability to synthesize steroids prior to ovulation. In the mammal a rising oestrogen secretion causes the boost in LH secretion necessary for ovulation and this is considered again below.

Follicular atresia

Although the number of follicles which mature is finite and species-specific, the fac-tors responsible for their selection remain unknown (see Gilbert (1971*a*) for more discussion). Of several, identical and juxtaposed follicles only one is seen to develop, and the majority, which underwent a partial development, degenerate at the one to multi-layered granulosa stage. The chromosomes condense, the nucleus of the oocyte and the nuclei of the granulosa cells become pycnotic, while the centre of the follicle undergoes a fatty atresia; during this a lipoidal area first develops round the oocyte and spreads to occlude the whole follicle with a dense sudanophilic mass of choles-terol positive material (Fig. 4.1*b*, *e*, *f*). In the Rook, according to Marshall and Coombs

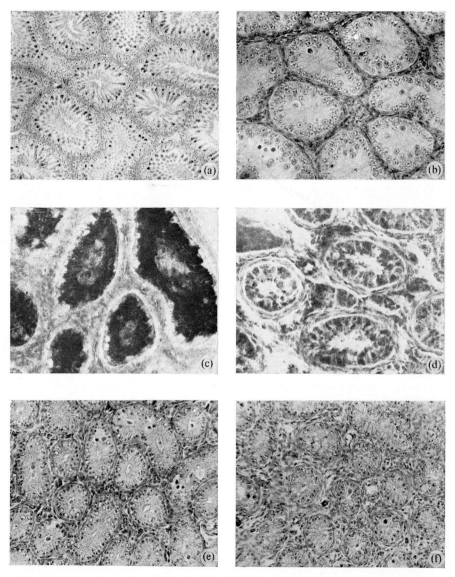

FIG. 3.3 Transverse section testis of various birds to show histological manifestations at end of breeding season. (a) Fully active testis of House Sparrow at height of breeding season showing tubules packed with bunches of spermatozoa. (b) At the end of the breeding season the tubules (larger scale used) and any remaining germ cells become engulfed in vacuoles in the Sertoli cytoplasm. Only a resting ring of spermatogonia remain and juvenile Leydig cells appear in the interstitial spaces. (c) Fully regressed testis of Weaver Finch *Quelea quelea* during the refractory phase with dense mass of sudanophilic lipid (black) enclosed in the Sertoli cell cytoplasm and occluding the lumin of the tubule. (d) Fully regressed testis of Wood Pigeon *Columba palumbus* in late October at a stage comparable with (c) showing virtual absence of tubule lipids. (e) Fully regressed testis of Turtle Dove *Streptopelia turtur* which was kept on an artificial LD 8 : 16 at the end of breeding season (4 August) until 20 October. (f) Turtle Dove captured as at (e) but held on LD 16 : 8 until 20 October. Note that in both (e) and (f) the tubules contain only resting spermatogonia and Sertoli cells but in (f) the interstitium is expanded compared with (e). (a), (b), (e), and (f) Bouin fixation and stained in iron–haematoxylin and Orange G; (c) and (d) fixed in formol–calcium and coloured with Sudan Black B and carmalum.

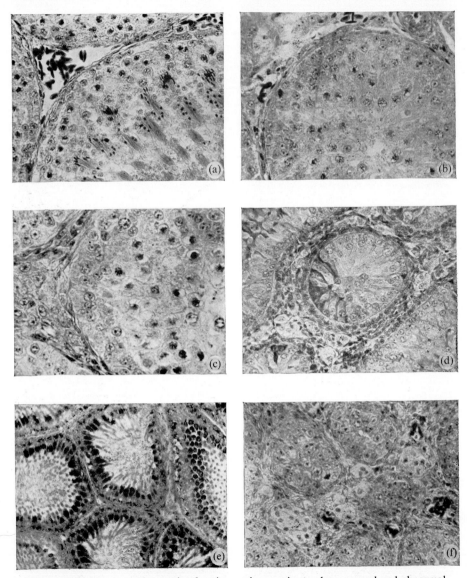

Fig. 3.5. Transverse section testis of various avian species to show normal and abnormal spermatogenesis. (a) Full spermatogenetic development has just been achieved in this House Sparrow *Passer domesticus* and an ordered and radially co-ordinated sequence of germ cell stages from spermatogonia to bunches of spermatozoa can be distinguished. Phagocytes have invaded the exhausted interstitium and removed the spent Leydig cells. (b) Expanded tubule of Greenfinch *Carduelis chloris* in which there has been an inhibition of development beyond the stage of primary spermatocyte and these cells are accumulating in the tubule lumin. Subject began with resting gonad and was held on a skeleton schedule of 6L 5½D 1L 11½D for 19 days. (c) Proliferation of primary spermatocytes in a photo-refractory House Sparrow held on LD 8 : 16 and given 12 × 0·4 mg NIH–LH (bovine) over 28 days. (d) Greenfinch captured in May when in full breeding condition and held on LD7 : 17 until 20 November. Spermatogenesis has stopped and the gonad regressed before normal exhaustion of the Leydig cells. The old generation of unused Leydigs has become encapsulated and a new generation of interstitial cells has appeared. (e) Subject as at (d) above exposed to skeleton 6L 4D 1L 13D for 19 days from 20 November. Spermatogenetic development with the production of bunches of spermatozoa has occurred yet the Sertoli cell cytoplasm contains dense (black) deposits of sudanophilic lipids. (f) Feral Pigeon *Columba livia* var., originally in full breeding condition, given 14 × 2 mg oestradiol benzoate over 3 weeks. The old generation of Leydig cells has become encapsulated and a new generation of interstitial cells has appeared. See also Fig. 3·8(c) and (d). (a), (b), (c), (d), and (f) wax-embedded and stained Haematoxylin and Orange G. (e) Gelatin-embedded and coloured Sudan Black B and carmalum.

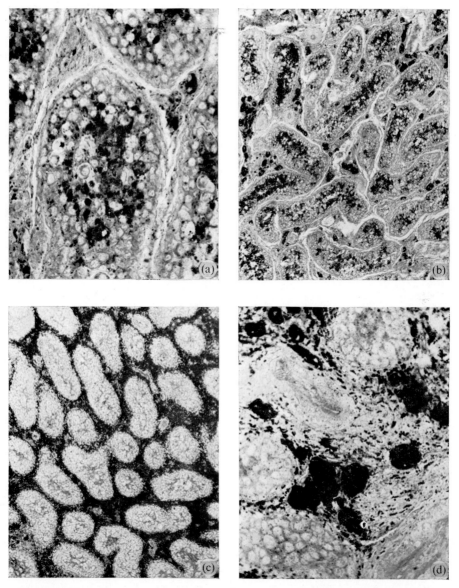

FIG. 3.8. Effect of exogenous hormones on the spermatogenetically mature testis of the Feral Pigeon *Columba livia* var. (a) After administration of 14 × 2 mg testosterone over 3 weeks spermatogenesis has been retarded and there has been a build-up of lipodal material within the seminiferous tubules. (b) After administration of 14 × 2 mg progesterone over 3 weeks spermatogenesis has been inhibited and there has been a build-up of tubule lipids. In addition, the Leydig cells have become heavily lipoidal as if biosynthesis and secretion have been inhibited. (c) The effect of treatment with 14 × 2 mg oestrodiol benzoate over 3 weeks contrasts with the result at (b) for there is an absence of sudanophilic tubule lipid but a very heavy build-up of such lipids in the interstitium. The histological picture is consistent with an enhanced or continued biosynthesis of material without subsequent release. (d) Same treatment as (c) but at higher magnification to show how lipid has accumulated in both the old encapsulated generation of Leydig cells and also in the new cells that have appeared following hormone injection. Compare with Fig. 3.5(f).

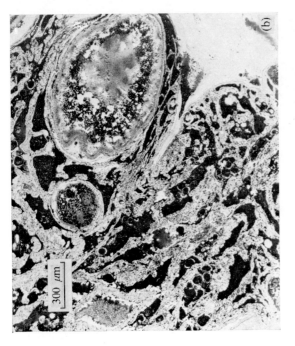

(a)

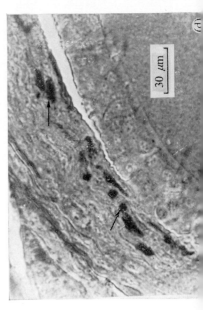

(b)

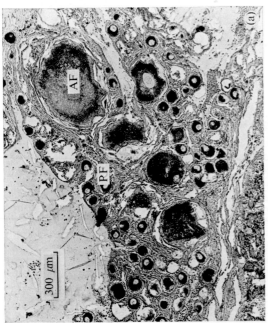

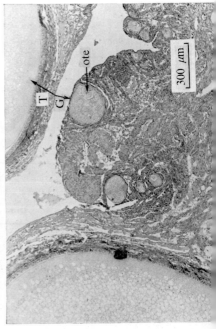

(d)

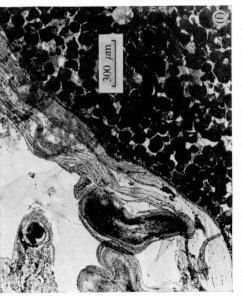

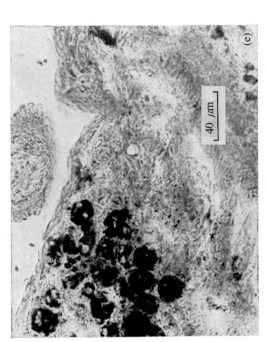

Fig. 4.1. T.S. ovary of Wood Pigeon *Columba palumbus*. (a) Low power gelatin embedded section from specimen killed in early April at the beginning of the breeding season. Numerous primary follicles (PF) can be seen, particularly near the surface of the ovary. Also present are some larger follicles which are becoming atretic (AF) and collapsing and these are distributed throughout the stroma. X40. (b) Gelatin embedded section to same scale as (a) of bird killed in July at the height of the breeding season. Dense lipoidal areas (black stain) are distributed throughout the stroma as a result of numerous atretic follicles disintegrating and becoming incorporated into the stroma (see (e)). In addition, numerous spots of lipids are distributed in the interstitial cells of the stroma. (c) Wax embedded section of another bird killed in April to same scale as (a) and (b). A young primary follicle near the surface of the ovary is composed of an oocyte (Ote) and is surrounded by yolk and a layer of granulosa cells (G). The granulosa layer of an enlarged follicle at top (also far left) is very much more hypertrophied and is surrounded by a thecal layer (T). (d) High power of gelatin section of granulosa and thecal layer of a large follicle in a bird killed in late October. Heavily lipoidal thecal gland cells are arrowed. X400. (e) High power to show glandular cells from an atretic follicle that have become scattered in the ovarian stroma. X250. (f) Low power of part of a large atretic follicle which has been invaded by a mass of heavily lipoidal gland cells. Note the collapsed follicle wall. X40. (c) Bouin fixed, wax embedded and stained with Erhlich's Haematoxylin and eosin. All other sections fixed in formal—calcium and coloured with Sudan Black B and carmalum.

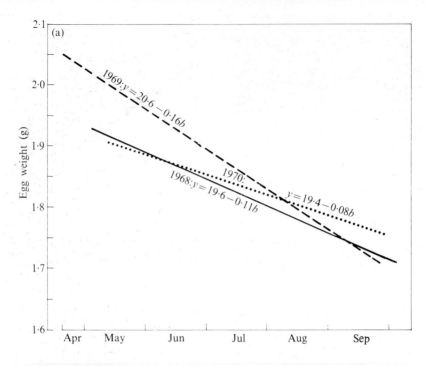

FIG. 8.8. (a) Regression lines showing change in mean weight of Wood Pigeon *Columba palumbus* eggs with date of laying (Murton *et al.*, 1974*b*). (b) Wood Pigeon feeding squab by regurgitating 'milk' from the crop (Cambridgeshire, England)

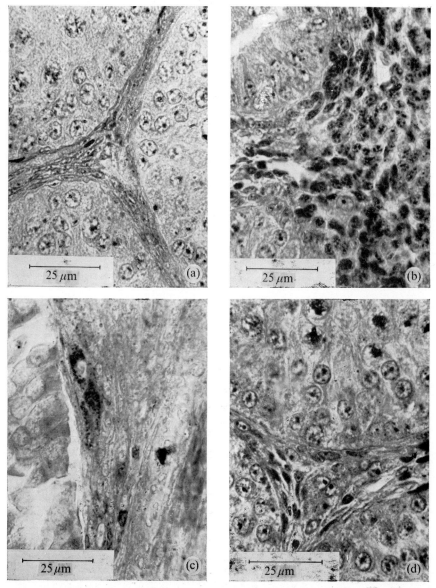

FIG. 12.5. Transverse section of testis of Greenfinch *Carduelis chloris* to show effect of different skeleton light schedules on Leydig cell development. (a) Control held on LD 7 : 17 for 14 days from 24 January. (b) As (a) and then transferred to 6L ½D 1L 16½D for 19 days. Stimulation of the Leydig cells has occurred in known presence of endogenous LH. Note dark-staining nucleolus and other nuclear chromatin. Very little spermatogenetic development has occurred. (c) Bird initially held as at (a) and then given 6L 11D 1L 6D. Showing depletion of lipid from Leydig cell as secretion occurs during spermatogenetic development. (d) Originally as (a) but then exposed to 6L 2½D 1L 14½D for 19 days. Spermatogenesis has occurred and the Leydig cells have been secreting. The nuclei are much swollen but lack the dark chromatin inclusions seen in (b). (a), (b) and (d) haematoxylin and Orange G. (c) Sudan Black B and carmalum.

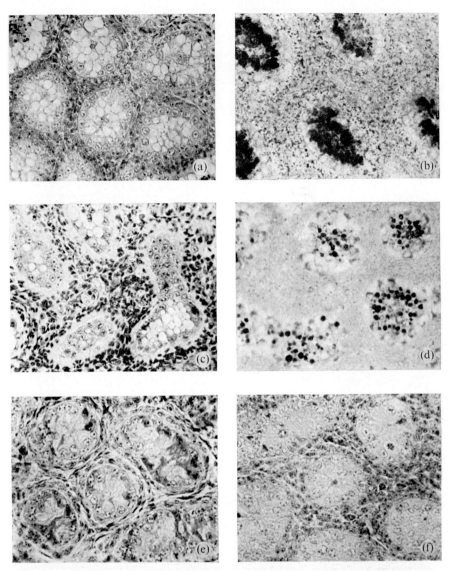

FIG. 12.8. Characteristics of the refractory and non-refractory avian testis in House Sparrows *Passer domesticus* treated with skeleton and normal photoperiods. (a) Subject held on 6L 2D 1L 15D for 35 days from late July. Seminiferous tubules are regressed and inactive with a resting ring of spermatogonia. Haematoxylin and Orange G. (b) Gelatin section of same bird as (a) showing densely lipoidal seminiferous tubules and also presence of numerous small lipoidal droplets in the juvenile interstitium. Sudan Black, B and Carmalum. (c) House Sparrow held on 6L 9D 1L 8D for 35 days (kept refractory on long-day simulation). The tubules have regressed but there has been a marked proliferation of the interstitium. Haematoxylin and Orange G. (d) Gelatin section of same bird as (c) showing lipoidal material in seminiferous tubules but absence of lipoidal droplets in the interstitium. Sudan Black B and carmalum. (e) House Sparrow held on seasonal daylengths from time of capture in early November until being killed on 15 December, refractoriness terminated. The seminiferous tubules are inactive and contain only resting spermatogonia. A generation of juvenile Leydig cells with spindle shaped nuclei has been proliferated from fibroblasts lining the seminiferous tubules. Haematoxylin and Orange G. (f) House Sparrow held on LD 16 : 8 from time of capture in late July until 8 December, that is, kept refractory. The seminiferous tubules are regressed but do show slight activity with a few primary spermatocytes. The interstitium is expanded and the Leydig nuclei are clearly stimulated having a rounded outline. They lack, however, a conspicuous nucleolus (cf. Fig. 12.5(b) but do contain coarse chromatin clumps. Haematoxylin and Orange G.

(1957), fibroblasts invade the follicle and disintegration occurs so that only scattered groups of cells remain in the stroma. These cells look identical to normal stromal interstitial cells, and, unless their origin is known, cannot be distinguished. Marshall and Coombs termed these cells 'ex-follicular gland cells' and they react positively to 3β-HSDH (Botte 1963; Chieffi and Botte 1965). Another kind of burst atresia has been reported whereby the yolk is extruded through the layer of granulosa cells into the surrounding stromal tissue where it is resorbed by phagocytes (Chieffi and Botte 1970). In yet another variation described by these authors the oocyte may be invaded by cells which result from a hyperblastic activity of the granulosa tissue. The ruptured follicles which result from ovulation also undergo a fatty atresia in like manner (see below).

Vitellogenesis

As mentioned, yolk deposition or vitellogenesis begins early, for there is a slow accumulation of material from before the time of hatching. However, the main yolk mass is added during the final growth phase, which lasts 7–14 days in the hen, and during which time the oocyte increases in weight from 0·5 g to 19 g and in diameter from 3 mm to 35 mm. In the Feral Pigeon *Columba livia* var. two follicles enlarge from about 5 mm to 20 mm within 8 days (p. 103) while in the Band-tailed Pigeon *Columba fasciata*, which lays but one egg per clutch, only one follicle develops in this way (March and Sadleir 1970). The nutrients which will be needed by the growing embryo, and also the newly hatched chick in some species, must therefore be made available from store or the daily food intake during this limited pre-ovulation growth phase. The critical timing of these events will become more apparent when the factors involved in ovulation are considered below. But first we must consider the source of the yolk nutrients.

Egg yolk comprises about 50 per cent water, while the solids are protein and lipid in the ratio 1:2, the lipid occurring as lipoprotein. There are also small quantities of inorganic compounds, free sugar as D-glucose, vitamins, and carotenoids (McIndoe 1971). Maternal antibodies which are present in yolk apparently give the newly hatched chick a degree of passive immunity to some diseases. McIndoe (1971), who has reviewed the process of yolk synthesis and structure, states that yolk consists of three main structural components. The first comprises large spherical bodies, 25–250 μm in diameter, which are distributed in the second, this being a continuous phase. The third component comprises small granules and these are distributed within the other two components. The granules contain phosvitin (a phosphoprotein) and two other high-density lipoproteins, called the lipovitellins, containing most of the calcium and iron in the yolk. These lipovitellins are composed of about 20 per cent lipid (phospholipid, cholesterol, and triglyceride) and protein. The granules account for about 23 per cent of the yolk solids and they also contain a little low-density lipoprotein. A breakdown of the yolk spheres and the continuous phase by centrifugation results in a supernatant yielding a low-density fraction, containing mostly low-density lipoprotein with little actual protein, and a water-soluble fraction. The lipoprotein fraction can be subdivided into five components which account for most of the lipid in the yolk, while the protein is made up of 18 amino acids. The water soluble fraction is composed of α-, β-, and γ-livertins.

Yolk-protein synthesis occurs in the liver and is under the control of oestrogen secretion. Following injection of oestrogen, phosvitin appears in the plasma of cockerels about 24 h later (Coolsma and Gruber 1968). Oestrone apparently causes the synthesis of RNA in the liver (Hahn, Schjeide, and Gorbman 1969). The material is transported via blood-vessels which end in the follicular walls adjacent to the vitelline membrane. The rate at which nutrients can be deposited in the growing follicle and the period over which follicles mature must affect the amount of yolk available for the embryo and influence the size of the final egg. Egg and clutch size are considered in Chapter 8 but in passing it may be noted that a large egg relative to adult body size can provide the newly hatched with a bigger food reserve or enable the chick to be hatched at a more advanced stage of development.

Ovulation

Follicle maturation either leads to ovulation or atresia. During ovulation the oocyte (sometimes termed ovum) is released from the follicle by rupture of the stigma under the influence of a surge of LH, which is sometimes referred to as ovulation inducing hormone (OIH). It seems unlikely that OIH is different from avian LH and on present evidence it is best to consider that ovulation is induced by avian LH or a substance into which it becomes altered in the blood (Ferrando and Nalbandov 1969). Mammalian pituitary extracts of LH are effective inducers of ovulation in the bird (Fraps 1965; Nelson and Nalbandov 1966) even though they are immunologically somewhat distinct from the avian gonadotrophin. Imai (1973) found that ovine FSH would not induce ovulation in the fowl and that ovine LH resulted in ovulations from 1–3 follicles. Significantly more follicles (3–5) were induced to ovulate when acetone-dried fowl pituitary was used, whereas none were ovulated using acetone-dried bovine pituitary. Partially purified fowl gonadotrophin was fractionated using (1) carboxy-methyl (CM) and (2) diethylaminoethyl cellulose (DEAE). The DEAE fraction induced about the same number of ovulations as did homogenized, dried fowl pituitary but the CM fraction was far less effective. Nevertheless, both fractions were equally good at inducing follicular growth (Imai 1972). Ovulation is not induced unless the follicle has reached a sufficient stage of development to respond and it may be that steroid hormones potentiate this capacity (see below). Prior to ovulation the granulosa cells in the region of the stigma, which is supplied by numerous capillaries, become pycnotic and they are phagocytozed. This weakens the membrane, enabling the follicle to rupture whereupon the egg is extruded (Nalbandov and James 1949; Nalbandov 1959a).

The sequence of events leading to ovulation has been much studied in the domestic fowl by Fraps (1970) in relation to the daily light cycle, although several problems remain. The domestic hen lays sequentially producing one egg per day during a cycle or sequence lasting anything from 1 day to 30 or 40 days and each cycle is then followed by one or more pause days, during which no egg is laid. Provided a photoperiod of 12–14 h light per day is given, the first egg of the sequence is laid during early or mid-morning and subsequent eggs at rather later hours on successive days, until the sequence or cycle is completed by the oviposition of the terminal egg. Thus, the interval between the laying of successive eggs is usually slightly greater than 24 h. Lag is defined as the difference in time between two successive ovulations minus 24 h. A negative lag can occur if the second ovulation occurs sooner than 24 h after the first. Nesting behaviour provides a more convenient monitor of ovarian function than does oviposition for Wood-Gush and Gilbert (1970) showed that 99 per cent of nestings are preceded by ovulation and 100 per cent of ovulations are followed by nesting, whereas only 60–95 per cent of ovulations result in oviposition. The time lag between successive ovipositions varies according to sequence being greatest between the first and second egg and then decreasing; in a long sequence it becomes negative so that the cumulative lag rarely exceeds 7 or 8 h (see Fig. 4.3).

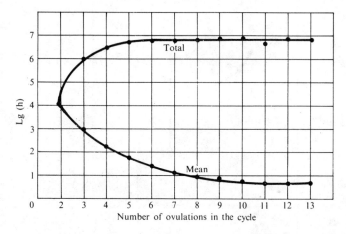

Fig. 4.3. Variation in total lag and mean lag between successive ovipositions according to the number of eggs laid in a sequence (clutch) by domestic hens. Lag represents the difference between the interval separating the laying of consecutive eggs and 24 h and so is the difference in times of day at which successive eggs are laid. From Heywang (1938) after Fraps (1970).

Fig. 4.4 is also based on Fraps (see Fraps (1970) for summary) to show how a three-egg sequence of laying occurs in an uncoupled[†] ovulation cycle with no ovulation occurring on the first day. A surge of gonadotrophin (LH) for ovulation of the first egg of the sequence is assumed to occur shortly after dawn. Actually, up to three peaks of LH have been recorded by various workers. Using the ascorbic acid depletion assay Nelson, Norton, and Nalbandov (1965) and Nelson and Nalbandov (1966) measured hourly plasma and pituitary LH titres between two successive ovulations (between the C_2 and C_3 ovulations in Fig. 4.4). Three peaks of LH were noted

[†] In an uncoupled cycle the first ovulation is separated by two or more days from any preceding cycle so that it is independent of any lag effect in the timing of the first (C_1) ovulation. In this respect an uncoupled cycle is probably more representative of the situation applying to wild bird species.

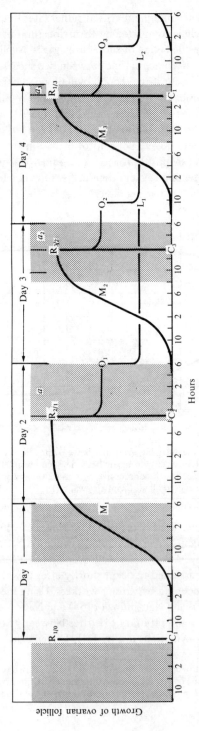

Fig. 4.4. Schematic representation of supposed relationships between ovulation and oviposition in an uncoupled 4-day maturation cycle in the domestic hen. Gonadotrophin (GTH) release ($R_{1/0}$) on day 1 causes follicle C_1 to start development (the 0 means that no follicle from any previous sequence is available for development) and its growth and maturation is shown by the curve (M_1). Gonadotrophin release ($R_{2/1}$) on the second day causes follicle 2 to start growing and provides the stimulus for ovulation of the first follicle (O_1) and maturation (M_2) of the C_2 follicle, and so on. a, a_1, and a_3 are the time differences by which gonadotrophin release is delayed relative to dawn in succeeding cycles, so that time of ovulation is pushed progressively later in the day. L is actual time of laying for eggs 1 and 2 (laying of egg 3 not depicted). From Fraps (1965).

at 8, 14, and 21 h before the C_3 ovulation with maximum titres in pituitary and plasma overlapping in time as if synthesis and release occurred together. Their results agree with measures of the pituitary content of LH made by Tanaka and Yoshioka (1967) and Tanaka (1968). Gilbert (1971b) has presented a useful diagrammatic summary of these various findings. However, in view of earlier remarks (p. 29) that argenine vasotocin can give a positive response in the OAAD assay, caution is needed in interpreting these results. Moreover, we should expect FSH to be released at some stage in the cycle, and probably at a different time from LH; it is conceivable that granulosa, theca, and interstitial components should be stimulated at discrete periods during each 24-h day, resulting in variable feed-back relationships to the hypothalamus. In fact, a peak in plasma FSH roughly corresponds with the recorded middle peak of plasma LH, if we judge from FSH assays made by Imai and Nalbandov (1971). These various measures of pituitary and plasma gonadotrophins are illustrated in Fig. 4.5 in relation to the ovulation and oviposition cycle. It is tempting to suggest that the first LH peak, which occurs about 21 h before ovulation and which approximately coincides with oviposition, is in fact caused by vasotocin. Indeed, the use of radioimmunoassay methods by Cunningham and Furr (1972) revealed only one peak of plasma LH this occurring 8 h prior to ovulation. If these authors are correct we have still to account for the middle peak of 'LH'.

Bullock and Nalbandov (1967) assayed plasma LH during the pause days which result when no ovulation occurs following the terminal (C_t) oviposition of a sequence. They found that a tonic secretion of LH did occur but there were no peaks. Heald, Rookledge, Furnival, and Watts (1968), who assayed pituitaries during the pause period, noted constant LH levels for 12–14 h post-ovulation, a small decrease 16 h before ovulation, followed by a peak at 12 h pre-ovulation and a marked decrease just before ovulation. Similarly, Tanaka and Yoshioka (1967) present data to suggest that there is a delay in release of pituitary LH which accounts for the delay in ovulation.

The manner in which successive ovulations in a sequence become delayed suggests the involvement of a circadian rhythm having a period length slightly exceeding 24 h, which is not always perfectly entrained by the environment light cycle. This topic will be more fully discussed in Chapters 11 and 12. Evidence that circadian timing mechanisms are involved in ovulation and subsequent oviposition come from scattered references to the consequences of changing the daily light cycle. When fowls are exposed to continuous light the interval of just over 24 h between successive ovipositions noted with daylengths of 12–14 h becomes increased to what is probably a 'free-running' period close to 28 h (Winget, Averkin, and Fryer 1965). The mean lag and cumulative lag are increased, the latter reaching values of 27–31 h compared with the maximum of 8 or 9 h recorded with 12–14 h daylengths (Fraps 1970). Individual hens become out of phase and may lay at any time during the 24 h if kept on continuous light (Warren and Scott 1936; McNally 1947; Morris 1961, 1962), presumably because each individual lays according to its own endogenous rhythm and not in synchronous phase with other individuals. Under such conditions entrainment of the oviposition rhythms can be effected by a noise or feeding regime (Wilson,

88

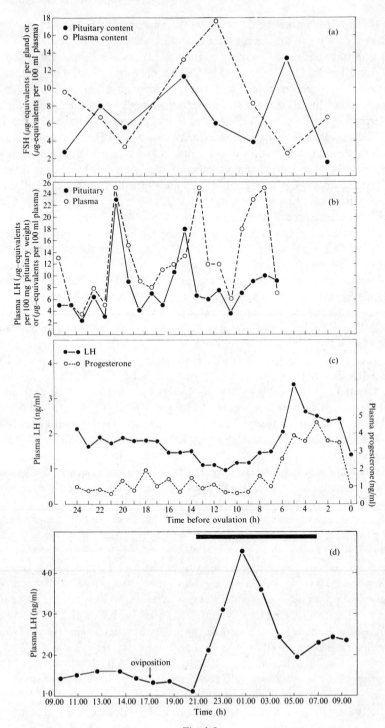

Fig. 4.5

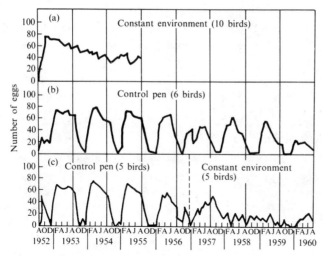

Fig. 4.6. Number of eggs laid by hens according to the light schedule. (a) Under a constant environment with LD 12:12 and ambient temperature 18 °C. (b) Held indoors but on a natural temperate seasonal photoperiod with additional light supplied to give a 12-h daylength between the autumn and spring equinoxes. (c) Held in a similar pen to (b) but then transferred to a constant environment in November 1956. From Hutchinson (1962).

Woodward, and Abplanalp 1964; McNally 1947) or a temperature cycle (Payne, Lincoln, and Charles 1965). Winget et al. (1965) used telemetry to monitor continuously the deep body temperature of an unrestrained hen; there were two temperature peaks during laying sequences spaced 15–30 min apart and corresponding with oviposition and ovulation respectively, but there was only one peak on days which were followed by no oviposition the next day (i.e. when there was no ovulation on the day measured). A detailed breakdown of the data by harmonic regression, power spectral analysis, and other sophisticated methods showed the 28-h cycle of the female to consist of two 14-h cycles repeating every 28 h whereas the cycle in the male appeared to be more stable (less affected by the constant light) and to have a period 4 h shorter than that of the female.

The involvement of an autonomous rhythm of reproduction in the domestic fowl was suspected by Hutchinson (1962) who concluded that the general pattern of photoperiodic stimuli over the year controlled the rhythm of egg laying, which depended on the history of photoperiodic stimuli rather than the length of day at the time. However, Hutchinson obtained no evidence of an autonomous periodicity in hens kept for 3 years under a constant LD 12:12, and Fig. 4.6 contrasts with Fig. 9.5 (p. 230): under a seasonal light cycle Hutchinson clearly obtained an entrained annual periodicity. Earlier, Larionov (1957) had produced a less than annual cycle of

---

Fig. 4.5. Changes in plasma and pituitary hormone concentration in domestic hens during the laying cycle. (a) FSH as measured by the HCG augmentation method. After Imai and Nalbandov (1971). (b) LH as measured by the OAAD method. From Nelson and Nalbandov (1966). (c) LH and progesterone as measured by radioimmunoassay. After Furr et al. (1973). (d) LH measured by radioimmunoassay. From Wilson and Sharp (1973).

TABLE 4.1

Relationship between frequency of the ovulatory rhythm of White Leghorn hens and that of the light–dark cycle[†] under which they were kept

Oviposition interval when held under LD 14:9	N	Length of cycle (h)[†]	Oviposition interval	Mean egg weight (g)
Less than 25 h (group A)	17	23	24·4	57·6
		25	25·2	58·5
		27	27·1	61·7
More than 25 h (group B)	31	23	26·1 ⎫ Not	59·6
		25	26·2 ⎬ significant	60·1
		27	27·3	61·9
Controls kept on 24-h days	24	24	Not measured	61·0
		24		61·5
		24		62·8

[†]For all cycles there were 14 h of light compounded with 9, 10 (controls), 11, and 13 h of dark. Differences between oviposition intervals within groups were significant except where indicated. From Rosales *et al.* (1968).

egg production by holding hens on an accelerated photoperiodic rhythm in a manner comparable with Damsté's studies of the Greenfinch *Carduelis chloris* (see Fig. 9.7, p. 233). Rosales, Biellier, and Stephenson (1968) kept 48 hens on 14 h of light and subjected tham to dark periods of 9, 11, 13 h at various times (all birds received all treatments). The average intervals between ovipositions were 25·4, 25·8, and 27·2h when t birds were held on light–dark cycles of 23, 25, and 27 h, respectively. Egg weights increased with increase in cycle length from 58·8, 59·5, to 61·8 g respectively. When the birds were initially kept on a 23-h day they segregated into groups having oviposition intervals of less than 25 h (group A, 17 birds) or more than 25 h (group B, 31 birds). The effect of different daylengths was greatest in subjects whose oviposition interval under a 23-h cycle had been less than 25 h. Anticipating the discussion of Chapters 11 and 12, we suggest that ovulation rhythms of short period or high frequency (as found in group A) are more easily entrained by the environmental oscillator[†] (length of daily light–dark cycle) than are ovulation rhythms of long period or low frequency at least within the range of light–dark periods of 23–27 h (the frequency is the reciprocal of the period, so these values represent a decrease in the frequency of the entraining cycle). Evidence for this conclusion is summarized in Table 4.1.

Fig. 4.7 shows for these same data that the standard deviation in oviposition interval was reduced under a 27-h day, presumably because a lowering of the frequency of the light–dark cycle enabled it to function as a more effective entraining oscillator. Similarly, the higher frequency of the endogenous oviposition rhythms of

[†]The ability of a controlling oscillator (in this case the length of the environmental day–night cycle) to entrain another oscillator (in this case the ovulation rhythm) depends on the relative frequencies of the two oscillators, see p. 290).

group A compared with group B subjects rendered them more amenable to entrainment by a range of environmental light–dark cycles. Similar results to those just
described were obtained by Foster (1968) who kept hens on days of 23, 24, and
25 h with 15 h of light in each cycle. Egg production was increased with increase in
cycle length the mean rates of lay, expressed as eggs per 100 birds per 24 h, being
71·6, 72·9, and 73·8 respectively. Foster's data indicate that maximum egg production
occurs when the length of the light–dark cycle is the same as the natural ovulation–
oviposition rhythm. An increase in the period of the light–dark cycle also increases
clutch length (Byerly and Moore 1941; van Albada 1958; Biellier and Ostmann 1960)
and, in addition to the interval between successive ovulations of a clutch, the interval
between successive clutches is lengthened (Lacassagne 1970). If the interval between
ovulations is increased it might be expected that egg weight should be increased.

A relative increase in the frequency of an endogenous rhythm is achieved by
lowering the frequency of the entraining oscillator, and this is what happened in the
above examples when the period of the light–dark cycle was lengthened. An alternative way of increasing the frequency of a photosensitive endogenous rhythm is by an
increase in light intensity, and this can be achieved by raising the proportion of light
in a light–dark cycle. Increasing the duration of the light and dark increases the rate
of lay (Hutchinson and Taylor 1957; Morris 1962). Studies of domestic poultry indicate that systematic relationships can be expected between clutch size and egg weight
and the photoperiodic regime. There seems no reason why similar considerations

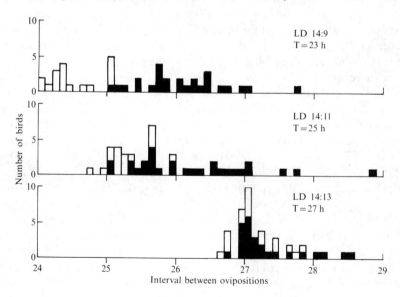

Fig. 4.7. Distribution of 17 domestic hens of group A (open histograms) and 31 hens of group
B (solid histograms) in the interval between ovipositions according to the length of the daily
light–dark cycle, giving entraining periods T of 23, 25, or 27 h. Group A subjects were birds
which when initially kept on LD 14:9 had oviposition intervals of less than 25 h whereas group
B subjects, when pre-treated under the same conditions, exhibited intervals in excess of 25 h.
From Rosales et al. (1968).

should not apply to wild species but at present no experimental details are available (see Chapter 8).

Fraps (1965) has provided evidence that the ovulatory surge of gonadotrophin for the maturation of the C_1 follicle in uncoupled cycles always occurs at about the same time of day, close to sunset, but it varies more in coupled cycles being influenced by the release of LH for ovulation of the terminal follicle of the preceding sequence. The exact relationship probably depends on the strength of the entraining light cycle as this will affect the frequency of the ovulatory rhythm and its phase of entrainment (see p. 289). Thus a period of light or dark *per se* need not be important in inhibiting or stimulating an ovulatory surge of LH, that is, light need not function inductively. There is also a possibility that some factor associated with follicle maturation and ovulation (perhaps oestrogen feed-back) influences the phasing of gonadotrophin secretion in relation to the daily photoperiod. The ability of steroids to rephase circadian rhythms will be discussed in more detail in Chapters 11 and 12.

We have described the ovulation and oviposition cycle of the laying hen in relation to gonadotrophin secretion and in a crude way this may indicate the sequence of events for a wild species producing a clutch of eggs. We must now consider the total hormonal state of the female during the egg-laying cycle and the importance of steroid feed-back mechanisms. Immature follicles can be ovulated if LH is injected into hypophysectomized hens but a combination of LH and FSH does not result in ovulation. Rupture of the follicle *in vitro* will only occur if it is removed within 2 h of the expected time of ovulation which suggests that some agent is produced endogenously that programmes the follicle for its subsequent release (see references in Gilbert (1971*b*)). This agency affects nesting behaviour for when the follicle was artificially ruptured between 3 h and 24 h before the normal time of ovulation appropriate behaviour did not occur (Gilbert and Wood-Gush 1968). Since nesting behaviour is determined by oestrogenic hormone (Wood-Gush and Gilbert 1969) this suggests that rupturing the follicle early in its cycle eliminated the production of oestrogen. With Heald *et al.* (1968) and Hawkins, Heald, and Taylor (1969), we assume that during its final growth phase the follicle produces oestrogen under the influence of FSH (or FSH + LH) secretion as already discussed. It seems unlikely that progesterone is involved during this part of the cycle as suggested by Fraps (1961); Fraps has variously implicated oestrogen, progesterone, or even androgen as the ovarian hormone which stimulates LH release. Certainly exogenous progesterone can induce ovulation but this is perhaps because it changes the balance of FSH and LH. It is effective only if administered at a critical time; if given 36 h before expected ovulation follicle atresia results, while if given 2–24 h before expected ovulation premature ovulation is induced (references in Gilbert (1967)). In view of the studies already reported detailing the effect of oestrogen and progesterone on IR-LH secretion in the male pigeon (p. 67), it might be guessed that oestrogen should similarly stimulate LH in the female and inhibit FSH release. This is, of course, the mechanism operating in the mammal. As implied already, oestrogen is probably released from the developing follicle only during a limited part of each day. Moreover, secretion perhaps does not begin, or at least reach a high rate, until the follicle enters its final growth phase, and this is suggested by behavioural evidence; the time of hypertrophy of such oestrogen-dependent structures as the oviduct is discussed below (p. 117). But it is also clear that neural changes occur in the ripe follicle which predispose it to the ovulatory action of LH. Thus the anti-adrenergic drug dibenzyline will block ovulation if injected into the follicle. Exogenous LH can only overcome the effect of the drug if administered 2 h or more before the expected time of ovulation (Ferrando and Nalbandov 1969). This and other evidence supports the scheme proposed by Fraps (see 1970 for summary).

Fraps suggests that the feed-back of stimuli from ovarian hormones is mediated via neural pathways to induce the surge of LH. Evidence for a neural involvement is that lesions in the ventro-median hypothalamus, or the fibres originating there and passing caudally to the median eminence, prevent progesterone induced ovulation (Ralph and Fraps 1959); the preoptic hypothalamus had to be kept intact for 2 h following injection of progesterone or for about 6 h before the time of expected ovulation. Electric stimulation of the preoptic area of the brain also induces prema- ture ovulation. Fraps postulates that the photoperiod times the appearance of a phase in the neural component of the LH-release mechanism which has a low threshold of response (or high sensitivity) to ovarian hormones, the precise time of LH release within the sensitive period being assumed to depend on the triggering action of ovarian hormones. However, to account for the sequential timing of ovulation, Fraps suggests that LH release also depends on a second effect of the photoperiod, this being the frequency with which follicles become mature and hence capable of ovu- lation. This hypothesis would be consistent with the involvement of a circadian rhythm mechanism which became phase-altered during the ovulation sequence. Bastian and Zarrow (1955) were probably on the right track in proposing that the ovary exhibited a cyclical variation in sensitivity to LH which was not in synchrony with LH release, causing more and more immature ova to be ovulated as the cycle progressed. However, Gilbert and Wood-Gush (1971) have summarized various pieces of evidence which refute the details of this hypothesis and it obviously has to be modified.

We have skirted the problem of why only one follicle matures at a time. Clearly the unripe follicles are not stimulated to ovulate by the endocrine balance existing when the mature follicle ruptures, though alteration of this balance by exogenous hormones can cause immature follicles to be ovulated. Gilbert and Wood-Gush (1971) have discussed the little-studied fact that an unshed follicle may remain within the ovary for 24 h longer than the others in a cycle and point out that some mechanism must then prevent the transition of the next follicle in the series from the resting to growth phase. They argue a case that the number of developing follicles remains constant so a missed ovulation should lead to a missed transition from resting to growth phase further back in the follicle hierarchy. The resultant gaps should then result in another missed ovulation. In other words, the follicle sequences seem to be determined by events occurring early in their development which indicates an ovarian control of the ovulation pattern rather than a hypothalamic–adenohypophysial one. Gilbert and Wood-Gush suggested that the extensive innervation of the follicular stalk, but not the luteal cells, might be implicated since denervated ovarian transplants show excessive growth and multiple ovulations. And they also quote Lacassagne (1957) that hens laying long egg sequences have more developing follicles than those laying short cycles. The topic is of interest in its significance for the control of clutch size in wild birds (see below).

Although the follicles of the mature, yet unmated, hen enter the final growth phase presumably when FSH and LH secretion reach a sufficiently high level under a suitable photoperiod, wild species usually need to be paired and in receipt of behavioural stimulation from the mate and nest-site. Appropriate behaviour possibly stimulates the necessary boost to FSH secretion which in turn can result in oestrogenic hormones being released from the growing follicle (see p. 114). In the mammal the ovulatory surge of LH supposedly initiates a wave of mitotic divisions in the granulosa cells, apparently programming them and thereby determining their level of secretory activity after ovulation, and also increasing their fluid content (Baker 1972; Short 1972). According to Baker, this LH surge is also reckoned to initiate the continuation of meiotic division in the oocyte, which until now has remained in the arrested stage of diakinesis. Metaphase and anaphase rapidly follow and a daughter secondary oocyte is formed plus a polar body with a small quantity of ooplasm. Immediately the secondary oocyte divides again, this time with a short prophase, and the metaphase becomes the point of arrested development. Subsequent meiotic maturation of oocyte occurs after ovulation, when the sperm enters during fertilization, and at this stage the second polar body is ejected. Presumably a similar train of events occurs in the bird for we know that the first reduction division to form a secondary oocyte and first polar body occurs 2 hours before ovulation, probably under the influence of LH (Olsen and Fraps 1950). The second maturation division occurs in the oviduct and it may depend on the penetration of the sperm (Romanoff 1960).

Luteogenesis

The ruptured follicles resulting from ovulation, or those which become atretic in the manner already described, do not constitute true corpora lutea as in mammals. In this class the cell structure of the ruptured follicle is changed following ovulation so that granulosa cells predominate. These lack the biosynthetic capacity to produce the higher steroids and so progesterone secretion begins in the developing corpus luteum. This is maintained under the influence of LH and prolactin. Progesterone feed-back on the pituitary blocks the action of oestrogen in stimulating an ovulatory surge of LH and so further ovulations are prevented. In mammals, this sequence of events has a reflex quality so that if pregnancy occurs the functional life of the corpus luteum is maintained otherwise prostaglandins from the uterus destroy the corpus luteum, progesterone secretion falls, and the animal begins a new oestrous cycle. In birds, the post-ovulatory follicle consists mainly of hypertrophied theca cells so it is histologically distinct from that of the mammal and, some authorities have claimed that it is not the source of progesterone synthesis (Deol 1955). Nevertheless, progesterone has been identified in pre- and post-ovulatory follicles (Furr and Pope 1969) while Δ^5-3β-HSDH activity has been shown to increase markedly following ovulation

(Wyburn and Baillie 1966). In fact, steroidogenesis has been noted in both the theca and granulosa cells of the atretic follicle (Woods and Domm 1966; Chieffi and Botte 1970). Electron microscopy also reveals an extensive agranular endoplasmic reticulum in the granulosa cells of the recently ovulated follicle and an increase in sudanophilia and cholesterol content (Wyburn *et al.* 1966) so the avian situation may be similar to the mammalian one. However, in *Coturnix* reared to maturity under non-stimulatory light regimes (LD 8:16) the number of atretic follicles increases and the granulosa cells only give a weak initial reaction for 3β-HSDH and then none (Sayler *et al.* 1970). This implies that steroidogenic activity is lost once atresia develops, but with present knowledge it would be unwise to exclude the atretic follicles from having a significant endocrine function. The progesterone produced at this stage of the cycle has a direct effect on behaviour (considered in the next chapter), and it is involved synergistically with oestrogen in brood patch development. Maximum plasma concentrations of progesterone in laying hens were detected 4–7 h before ovulation (Cunningham and Furr 1972) modifying Furr and Pope (1970)). As previously mentioned, progesterone given 36 h before expected ovulation causes a marked follicle atresia (Rothschild and Fraps 1949).

As in the case of the male, there is no indication of the endocrine state which promotes progesterone secretion, but it is possible that a high plasma LH concentration relative to FSH is important. It is also possible that prolactin performs some function at this stage, as in mammals. Prolactin acts synergistically with oestrogen to cause the defeathering essential for brood patch development in wild birds and it is also stimulated by the act of incubating eggs (see next two chapters).

Exogenous hormones

The detailed histological effects of exogenous steroids and prolactin on the ovary, as distinct from the general effects already mentioned, have not been well defined. By and large the action of these exogenous hormones has been monitored in terms of gross follicular morphology or ovulatory responses and there is a need for the documentation of changes detectable in sectioned material comparable with those outlined for the male. Reference has been made to androgen and it may be mentioned in passing that Chu and You (1946) reckoned that it stimulated follicular growth of hypophysectomized pigeons, as did Ringoen (1943) studying intact House Sparrows *Passer domesticus*. Testosterone has been claimed to stimulate ovulation but the time of day injections were made has not been been detailed (van Tienhoven 1961*b*). It is likely that any stimulatory effect was indirect, for example, by hypothalamic feed-back inhibition. The same arguments apply to the finding that oestrogen injections delay ovulation (Dunham and Riddle 1942; Fraps 1955). Exogenous oestrogen did not influence follicular development in hypophysectomized pigeons (Chu and You 1946). A marked increase in ovary weight was found in 6-week pullets given diethyl stilboestrol (Phillips 1959) and also domestic ducks and wild Mallard *Anas platyrhynchos* similarly treated (Phillips and van Tienhoven 1960).

Accessory organs

The right oviduct, like the right ovary, remains vestigial, except in raptors and some other taxa (see p. 48), and it is the left which becomes highly differentiated to satisfy the processes involved in producing the complex-shelled and, in most species, pigmented eggs. The developed oviduct is a tube attached to the body wall by dorsal and ventral ligaments, and also double folds of the peritoneum which cover it and in which are situated a network of blood-vessels. The whole oviduct is divisible into five regions: infundibulum, magnum, isthmus, shell gland (uterus), and vagina, a good modern account for the domestic fowl having recently been given by Aitken (1971). The walls of the oviduct embody circular and longitudinal muscle fibres variously developed in different sections and similarly there is a variable development of a lining mucous membrane; in places the membrane comprises longitudinal folds covered with a glandular and ciliated epithelium. The extruded ova are received from the ruptured follicle via an open anterior infundibulum which so encloses the ovary that the ovum does not lie free in the body cavity. The oocyte is also probably guided into this funnel by the ventral ligament. Neural factors are probably responsible for the infundibulum beginning activity rather before ovulation. Finger-like processes become engorged with blood and these are contractable by muscle fibres which grasp and engulf the egg. Eggs are not always captured and 'internal laying' frequently occurs, especially under some circumstances (Wood-Gush and Gilbert 1970).

The seasonal hypertrophy of the oviduct which occurs with sexual recrudescence in wild birds is dependent on oestrogen secretion from the ovary (Witschi and Fugo 1940; Brant and Nalbandov 1956; van Tienhoven 1961). Oestrogen also increases the size of the oviduct in immature pullets (Oades and Brown 1965) and other species, as well as an enlargement of the cloaca (Burger and Lorenz 1962). At low dose levels progesterone potentiates the action of oestrogen in stimulating the oviduct but at higher levels is antagonistic (van Tienhoven 1961; Oades and Brown 1965). Prolactin also augments the increase in weight of the oviduct following oestrogen injection in Barbary Doves (Lehrman and Brody 1957) and Canaries (Steel and Hinde 1963; Hutchison, Hinde, and Steel 1967) and a synergism between oestrogen and androgen has also been recorded (Lorenz 1954). It seems likely that the response of the oviduct to oestrogen depends on folic acid availability (Kline and Dorfman 1951) and vitamin B_{12} (Kline 1955); folic acid may reduce the metabolism of certain water-soluble proteins (Brown and Badman 1965). Oestrogen is known to stimulate the synthesis of protein (Ljungkvist 1967) and nucleic acid and increase the uptake of water and amino acids (Oka and Schimke 1969).

In the chicken, turkey, and quail the egg spends about 0·25—0·5 h in the infundibulum, 2—3 h in the magnum, 1—2 h in the isthmus and 19—24 h in the shell gland according to Gilbert (1971d). Interspecific variation may be suspected judging from the rate at which wild birds lay eggs. Fertilization of the egg occurs in the infundibulum (Olsen and Neher 1948). In domestic fowls fertile eggs can be laid 10—14 days following a single copulation, emphasizing the length of time during which sperm can remain viable, although fertility declines after about 5 days (Hammond 1952).

Sperm occur throughout the oviduct following a single copulation but then disappear from the lumen. It appears that sperm-storage glands exist in the utero-vaginal area (Bobr, Lorenz, and Ogasawara 1964), but the stimulus which leads to their release and transport, presumably by peristalsis, to the infundibulum coincident with ovulation is not known (see Gilbert (1967) for references). Wild birds normally copulate more than once before laying but during a restricted period of the courtship cycle. This is within 6 days of laying in the Pied Flycatcher *Ficedula hypoleuča* (Von Haartman 1951) and 3—4 days before laying in the Skylark *Alauda arvensis* (Delius 1965). Polyspermy occurs in that several spermatozoids enter the blastodisc (12—25 in the pigeon), although only one unites with the female pronucleus. By the time the egg is laid cell division causes the segmentation area or blastoderm to attain a diameter of nearly 5 mm in the domestic fowl.

The middle portion of the oviduct is the highly glandular magnum where the white albumen is formed; this is deposited round the ovum in four layers to make a complex structure (see Romanoff and Romanoff 1949; Gilbert 1971c,d). The rate at which the egg moves through the magnum is not really known, though it is of the order of 2—3 h. Its speed of movement influences the amount of albumen added. The structure and composition of albumen has been well reviewed by Gilbert who points out that it is mostly protein (1 part) and water (8 parts) and that of the solids (11 per cent of total) 92 per cent is protein. Of the non-protein solids about half is carbohydrate and half inorganic ions. Further reference to the chemical and physical properties of the egg contents will be deferred until Chapter 8.

The histological development of the oviduct has been studied in fowls (Richardson 1935) and in a careful investigation of the domestic Canary (Hutchison, Hinde, and Bendon 1968). While the ovaries comprise up to 1·5 per cent of body weight of the Canary there is an approximately linear relationship between oviduct and ovary weight (both represented as percentages of total body weight), as might be suspected since oviduct development depends partly on ovarian hormones. The formation of albumen coincides with the enlargement of one ovarian follicle more than the rest, and this also correlates with the completion of defeathering in the course of brood patch development and intensive nest-building behaviour. Relatively little oviduct development is apparent until nest-building activity begins in the Canary, indeed only the division of epithelial cells of the magnum was noted before the collection of the first nest-material. During the period of active nest-building the oviduct increases markedly in length—decreasing again after egg laying—and the epithelium divides and invaginates to form tubular glands. These become distended and their cytoplasm charged with albumen granules and at the same time the mucosa becomes very folded. The epithelium differentiates during this time into ciliated columnar cells and goblet cells. The histology of the magnum apparently remains uniform along its length at all stages of development. Growth and development is primarily associated with a hyperplasia, involving an increase in total DNA, and to only a lesser extent with cellular hypertrophy, that is, an increase in the ratio of dry matter and DNA (Yu and Marquart 1973). In domestic fowls oestrogen alone will cause growth of the albumen-secreting glands but progesterone or androgen are additionally necessary for the development of albumen granules (Brant and Nalbandov 1956). The relationship between oviduct development and the nesting cycle in the Canary is shown in Fig. 4.8, and will be referred to again in the next chapter.

After albumen has been deposited in the magnum the egg next passes into the less glandular and more muscular isthmus of the oviduct, where the two shell membranes, comprised of felted protein fibres cemented together with albumen, are secreted. The outermost membrane has a rough texture and so provides a key for the calcareous shell. This last, together with pigment, is added in the wider and expandable uterus or shell gland. The shell is secreted in the form of calcium salts

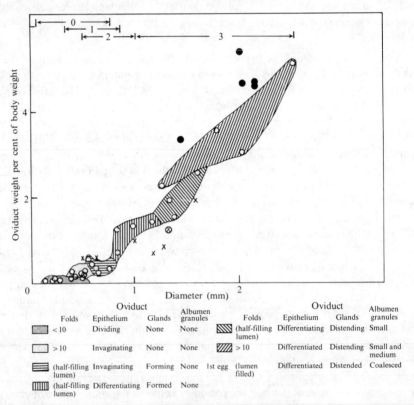

Fig. 4.8. Relation between diameter of magnum and oviduct weight, expressed as a percentage of body weight, of Canaries *Serinus canaria*. Different shadings refer to six *ad hoc* groups, the individual members of which were similar in oviduct weight, magnum diameter, and structure. The characteristic histological condition of these groups is summarized in the inset table. The lines at the top left delimit stages of nest-building reached in the breeding cycle, from 0, no nest building; 1, either removing 20 pieces of grass from hopper or placing 1–20 pieces in nest-bowl; 2, placing more then 50 pieces of grass in nest-bowl but not weaving; 3, placing more than 50 pieces of grass in nest and weaving actual nest structure. Solid circles refer to individuals which had laid their first egg, and crosses to birds which has been incubating for 7 or more days. Open circles are birds which had not yet laid and the open circle with cross an individual which had laid one egg and had the oviduct structure typical of an incubating bird. Slightly modified from Hutchison *et al.* (1968).

(about 97 per cent) supported on an organic matrix of protein fibres; Tyler (1969) discusses the snapping strength of avian egg shells. Any ground colour is deposited during the final stages of shell formation while blotches, streaks, or other surface markings are acquired after the shell has been completed. Because the egg is rotated on its long axis, surface markings are often given a spiral configuration. The rotation is achieved by a powerful sphincter muscle which when relaxed allows the egg to pass into the cloaca and to be laid pointed end first. In the domestic fowl about one third of the eggs become reversed—the vagina bulges beyond the sphincter so turning the egg—but whether this happens in other birds remains uncertain. Various environmental pollutants including DDT and its metabolites are known to cause egg shell thinning in birds (see Cooke (1973, 1975) for reviews).

Oviposition

Egg laying or oviposition involves muscular contraction of the shell gland but the integration of all the factors known to be implicated is still poorly understood. Gilbert (1971d) has reviewed the present knowledge so far as the chicken is concerned nothing being known about other bird species. He points out that no experiment to demonstrate the effectiveness of a single factor has been successful, probably because more than one is normally involved. A foreign body in the wall of the shell gland results in premature oviposition so perhaps neural stimulation from the shelled egg leads to relaxation of the sphincters. Neural mechanisms are involved, for stimulation of the telencephalon delays oviposition, while stimulation of the preoptic hypothalamus causes a premature oviposition (references given by Gilbert). Hormonal factors are also important for the posterior pituitary hormones oxytocin and argenine vasotocin are depleted before egg laying (Opel 1966). Blood levels of oxytocin reach a peak just prior to oviposition (Sturkie and Lin 1966), while a blood aminopeptidase which can deactivate oxytocin decreases about the time of egg laying (see Gilbert 1967). Injection of oxytocin or vasotocin cause muscular contractions and result in premature ovulation (Tanaka and Nakajo 1962; Gilbert and Lake 1963; Rzasa and Ewy 1970). Some priming effect of ovulation may be involved for there is a fairly constant time relationship between ovulation the passage of the egg through the oviduct and oviposition.

Removal of the post-ovulatory follicle delays egg laying (Gilbert and Wood-Gush 1965) raising the possibility that progesterone serves some function. In mammals progesterone can influence uterine prostaglandins, but although these have been identified in the fowl their involvement in egg laying has not been established.

In mammals the prostaglandins serve as local hormones and they have been extracted from a wide range of tissues (lung, brain, kidney, uterus; see Horton (1971)). They are biologically active lipids (unsaturated hydroxy-acids, with 20 carbon atoms, derived from prostanoic acid) and have been classified into A, E, and F groups. The groups differ in that one may increase while another decreases metabolic function. They are synthesized and act locally, being rapidly destroyed in the blood plasma and some tissues such as the lungs. It seems likely that they act to increase or decrease the action of other hormones on target cells, and they can be released from tissue by humoral, neural, or drug stimulation. In mammals (see review by Short (1967)) it was discovered that removal of the uterus would prolong the life of the corpora lutea and it became evident that the uterus exerted a 'lytic action'. This could be obviated by transplanting the ovaries to other parts of the body away from the site of prostaglandin activity (though to be prostaglandin $F_{2\alpha}$). Progesterone from the corpus luteum is responsible for stimulating this prostaglandin activity.

Control of clutch size

It might be suspected that circadian rhythm mechanisms could be implicated in determining clutch size. Thus, if oestrogen secretion preceding each ovulation results in the phase-shifting of gonadotrophin rhythms the system might be driven into a non-functional condition. It has long been realized that there is a general tendency for clutch size to increase with latitude (Lack 1968a), and experiments are needed to establish whether manipulation of the photoperiod can alter the size of the clutch. Whatever fundamental physiological mechanism is proved to underlie the control of

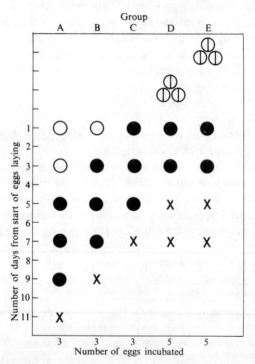

Fig. 4.9. Experiments on the addition and removal of eggs from the nests of Herring Gulls *Larus argentatus* during the laying sequence. In birds in group A the first and second eggs were taken as laid (open circles) and the subjects laid 3 more eggs (solid circles), their next follicles being suppressed (crosses); the female incubated 3 eggs. Only 1 egg was removed from the females of group B who then laid and incubated a normal clutch as typified by the control females (group C). Three eggs were added to the nests of females in groups D and E before egg laying began and this reduced the size of their own clutch and led to follicular atresia. These females incubated 3 + 2 = 5 eggs (divided circles). After Klomp (1970) based on Paludan (1952).

clutch size it seems possible that an immediate cause for the cessation of egg laying is a rise in prolactin secretion. Prolactin secretion is stimulated by the act of incubation (p. 134) and this may inhibit further ovulation. Classic experiments on the Herring Gull *Larus argentatus* by Paludan (1952) support this possibility. He showed that initially many follicles develop but that only four enter the final growth phase even though the birds normally only lay 3 eggs and have only three incubation patches. Once the first egg is laid the smallest of the remaining follicles degenerates, leaving only the second (by now already in the oviduct) and the third to mature. If the first egg laid is taken the birds often desert their nests but sometimes birds can be induced to lay another 3 eggs, showing that the fourth follicle is capable of development. It has proved feasible to get 13 eggs from one female while Salomonsen (1939a) once 'milked' a Herring Gull of 16 eggs in this way.

Paludan (1952) removed an egg the day before, or 1 egg on each of the 2 days preceding oviposition in the way shown in Fig. 4.9. With single eggs removed from the nest per day the birds laid 3 more eggs and one follicle became atretic. If 3 eggs were added at once only 2 eggs were laid and two follicles became atretic. Even if the 3 eggs were added a few days before the first egg was due, the gulls still laid 2 eggs, suggesting that the response mechanism to the stimulus of eggs in the nest becomes activated only after ovulations begin. Only then does the visual stimulus of eggs in the nest induce incubation behaviour which in turn probably induces prolactin secretion and an inhibitory feed-back of further ovulation. Similar results were obtained by Weidmann (1956) studying the Black-headed Gull *Larus ridibundus* and in this species too the fourth follicle degenerates when the bird begins to incubate the first egg. At this point the third follicle is still in the ovary but it does not degenerate. This suggests that once follicles reach some development threshold they are not affected by inhibitory feed-back induced through incubation. This could happen if prolactin is anti-FSH rather than anti-LH so that the growth phase of the follicle is selectively disturbed, but this is speculative and more study is required.

5 The endocrine basis of reproductive behaviour: 1

Summary

Appropriate timing for the fusion of male and female gametes is achieved by a complex repertoire of behaviour patterns which ensure species isolation yet synchrony between chosen partners. This synchrony results, in large measure, because the same endocrine mechanisms which regulate gametogenetic development either play a directly causal, or else permissive, rôle in mediating specific behaviour patterns. Hormonal and behavioural interactions have been particularly well studied in certain pigeon species and in the Canary *Serinus canaria*. In pigeons, the early stages of territory acquisition and mate selection are marked by sexuo-aggressive displays mediated by FSH and androgenic hormones. The early behaviour of the female is submissive and directed at reducing the males' assertiveness. Her displays somehow facilitate a reduction in the aggressive components of display—perhaps by a direct effect on FSH output—enabling courtship feeding and copulation to ensue. A second phase of courtship now develops during which the male behaves submissively and seeks to attract the female to the potential nest site where much mutual allopreening occurs. This stage appears to result from a declining androgen sensitivity, possibly resulting from a feed-back effect of androgen on hypothalamic centres as much as from a decrease in circulating androgen titres. Experimentally oestrogen injections markedly stimulate nest-demonstration, perhaps again by a feed-back inhibition of pituitary function. As the male becomes more submissive the female becomes more assertive and finally pushes the male from the nest-site stimulating him to collect nest-material. The stimulus of nest-building encourages the ovarian follicles to grow and produce oestrogen leading to oviduct development, ovulation, and egg laying; in Canaries brood patch development results from the endocrine milieu occurring at this stage of the cycle. Progesterone secretion increases in both male and female before egg laying and can be shown experimentally to stimulate incubation behaviour. The chapter describes numerous experiments performed with exogenous hormones in attempts to understand the hormonal background to pre-incubation behaviour.

Introduction

Selection for reproductive efficiency has evidently placed a high premium on appropriate timing, with regard to breeding both at the most opportune season of the year, and in a manner that ensures compatibility between the sexes for the union of the male and female gametes. In this and the next chapter the hormonal basis of the behavioural patterns which ensure that the sexes unite their reproductive capacities to maximize the prospect of leaving progeny is considered; the close physical proximity of gametogenetic and endocrine tissue in the vertebrate gonad emphasizes the evolutionary advantages of a functional integration. This section draws heavily on studies of pigeon species and the Canary *Serinus canaria,* work done particularly in the laboratories of the subdepartment of Animal Behaviour at Cambridge, and at the Institute of Animal Behavior, Rutgers University, New Jersey. Here, until his untimely death in 1973, Professor D. S. Lehrman and his colleagues began to demonstrate the interrelationships between hormones and behaviour in birds, developing a tradition started in America by such pioneers as F. A. Beach in the 1940s. The pigeons and Canary have proved particularly valuable for endocrine–ethological studies, a simple reason being that they can be induced to complete a relatively natural reproductive cycle under laboratory conditions which are convenient for observation and manipulation. The domestic hen is too 'unnatural' to appeal in this context.

When experienced and reproductively mature male and female Feral Pigeons *Columba livia* are introduced together in a cage they usually begin a sequence of courtship which, provided nesting material is available, leads after about 10 days to egg laying by the female and incubation by both sexes. Many other wild birds accomplish a comparable behaviour cycle by spending only a fraction of the day in each other's company for they must devote the rest of their time to food searching. This is true of the wild Wood Pigeon *Columba palumbus* in which courtship occurs in the nesting territory during only one or two hours in the morning and evening; although the pair may sometimes associate together at other times they are then occupied in seeking food in the company of conspecifics and virtually no effective sexual activity occurs in the feeding flock (Murton 1965). Fabricius and Jansson (1963) made the astute observation that a pair of Feral Pigeons could participate in a full behaviour cycle leading to egg laying if they were allowed access to each other, by removing an opaque partition separating adjacent cages, for only short periods each day. This experimental approach renders it easy for the observer to watch and quantify all the behaviour occurring between the birds, behaviour which might otherwise take place in his absence. In fact, these authors showed that Feral Pigeons allowed access to each other for only ½ h per day took 12–15 days to egg laying compared with 10 days for birds left permanently together. The pre-incubation behaviour cycle obtained in this way is summarized in Fig. 5.1. Essentially the same kind of cycle occurs in the Barbary Dove[†] *Streptopelia 'risoria',* the favourite study animal of Lehrman and his associates (Lovari and Hutchison 1975).

[†] American workers usually refer to *'risoria'* as the Blonde Ring Dove, Ring Dove, Domestic Pigeon, or Domestic Dove but it is quite distinct from the Feral Pigeon which is a descendent of the Rock Dove *Columba livia,* a bird which is about 33 cm long. From the Rock Dove have been bred the racing pigeons, dovecote pigeons (so they are often called domestic pigeon) and other fancy breeds. Birds of dovecote stock and from other sources have 'escaped' and established the free-living populations which inhabit many towns of Europe and, following introduction, America and other parts of the world, hence the term 'Feral' meaning wild or untamed (see Murton, Thearle, and Thomson 1972). As will be discussed (p. 276), these feral populations have been subject to selection and are variably adapted to the man-made environments in which they now live. In Europe, *Streptopelia 'risoria'* is often called the Barbary Dove, a complete misnomer as is the name White Java Dove given to a white mutant with dark eyes of the same species. In fact, *'risoria'* is a long domesticated form of the African Collared Dove *Streptopelia roseogrisea* and the specific status *'risoria'* is not justified, albeit widely used. The closely related Collared Dove *Streptopelia decaocto* will be considered in some detail in Chapter 10 (p. 267).

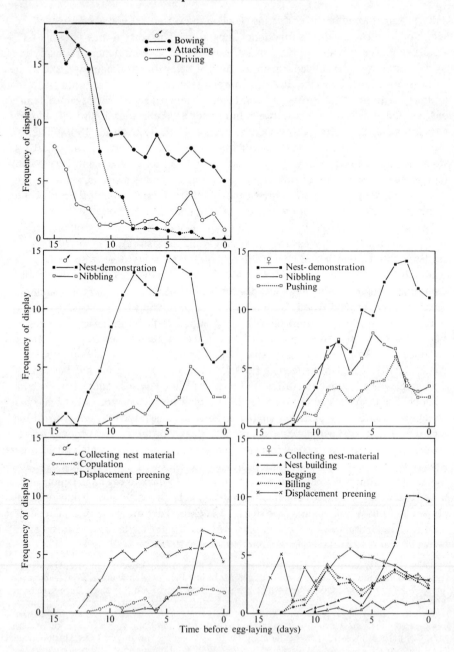

Fig. 5.1. Changes in the frequency of various display components of male and femal Feral Pigeons *Columba livia* var. during the pre-incubation courtship cycle. Males and females were allowed access to each other for 34 min per day and the scale refers to the mean number of 2-min observation periods during which the given behaviour occurred. Day 0 is the day of egg laying. From Fabricius and Jansson (1963).

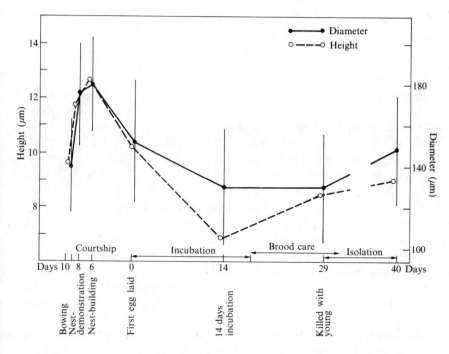

Fig. 5.2. Change in diameter of the epididymis and in the height of its columnar lining epithelium during the reproductive cycle of the Feral Pigeon. From Murton and Westwood (1975).

Assuming a compatible male and female, the following three major phases in the display cycle can be distinguished: at first, the male pursues the female and displays at her with the 'bowing display' and this causes her to behave submissively as he delivers the 'nest-demonstration display' from potential nest-sites. One of these sites is selected for nest-building and in caged subjects the male usually chooses the nest-bowl, although a few contrary individuals select another corner of the cage. During this phase of behaviour the female becomes more dominant until she eventually pushes the male off the nest-site and takes his place. This marks a critical stage in the cycle, for the male's aggression is again partly roused but becomes channelled into the collection of nesting material which he brings back to the female. The third stage of the cycle lasts 3—4 days and is marked by intensive nest-building, oviduct growth in the female, ovulation, and the start of incubation. The endocrine and behavioural event occurring during this cycle will be reviewed in the following pages. But first of all comparison of Figs. 5.1, 5.2 and 5.7 shows how the behaviour sequences are related to gross morphological changes in the accessory reproductive tracts of the male and female. Development of the epididymides and efferent ducts of the male certainly implicates androgen as was discussed in Chapter 2 (p. 76). The involvement of oestrogenic hormones in oviduct growth is a topic discussed again later in this chapter. For the present it is worth noting how hypertrophy of the male ducts precedes comparable development in the female and this reflects a general trend for endocrine dependent functions in the female to lag behind comparable ones in the male.

Pair formation

In the wild the first act of a reproductively active male pigeon is the acquisition of a territory, which he defends and proclaims by an advertisement song (perch-coo) and by a display flight. Both song and display flights serve to encourage unpaired females to approach thereby ensuring an initial meeting of the sexes.[‡] For birds in general, vocalizations are of prime importance in nocturnal species and those inhabiting places such as thick woodland where visibility is reduced, while birds of open country tend to rely to a greater extent on display flights, these often incorporating song. Advertisement songs or postures must be species-specific for hybrid matings are likely to be infertile and hence biologically disadvantageous. This is the case with the songs of different *Streptopelia* species (Lade and Thorpe 1964). However, in many pigeons the form of the bowing display is very similar and seemingly identical in related species otherwise exhibiting considerable morphological divergence. This led Goodwin (1966) to the view that selection for species recognition in pigeons had influenced display plumage pattern and colour to a greater extent than the actual movements used during display. Nevertheless, when Davies (1970) made a more detailed quantitative study of the bowing display of five closely related *Streptopelia* doves (Barbary Dove, Collared Dove *S. decaocto,* Turtle Dove *S. turtur,* Palm Dove *S. senegalensis,* and Spotted Dove *S. chinensis*) and their hybrids, he found that the form and intensity of the bowing display was characteristic for each species. For instance, the rate of bowing for each species was constant and differed from that of the other species. In some cases the F_1 hybrid proved to be intermediate between the parents, in some it resembled one of the parents but not the other while in others the characters were exaggerated beyond the range of either parent. It is worth noting that Lind and Poulsen (1963) found no correlation in duck hybrids between the pattern of inheritance of a behavioural character and the morphological structure used.

Once a female pigeon enters a male's territory he advances with the 'bowing display', this usually being the first display recorded under caged conditions. 'Bowing' is an ambivalent display, for the male is motivated by conflicting tendencies. On one hand, he is aggressive and will attack intruders into the territory; since male and female pigeons are identical in appearance a territorial male cannot recognize immediately whether an intruder is another male trying to usurp his sovereignty, although intruding males usually exhibit agonistic displays. A territorial male is also influenced by hormones which tend to make him behave sexually, and if the intruder is a female which behaves submissively, crouching and furtively avoiding close proximity, the male is motivated to mount her and copulate. And so the male as he advances at the female with the bowing display, may do one of three things, (1) grab the female's

[‡] Bird vocalizations could be important in entraining circadian rhythms of activity in the female (Gwinner 1966a) thereby ensuring synchrony in reproductive behaviour : data obtained by Catchpole (1973) illustrate this thesis.

neck feathers as he would attack another male, (2) jump on the female's back and try to copulate, or (3) turn away as if afraid. Meanwhile, the female retreats, running round and round the cage or in the wild hopping from branch to branch. In ontogeny the bowing display probably derives from the tail-fanning, head-lowering movements of alighting, these movements having become ritualized into a sexual–aggressive appeasement display. But, although all the components of the bowing display are ritualized and inherited in their fine detail, their expression requires an appropriate endocrine background so that males with regressed gonads show little or no interest when presented with a female in the same circumstances. Presumably appropriate hormones sensitize central nervous mechanisms for using permanently implanted electrodes in the brains of unanaesthetized pigeons Åkerman (1966a,b) induced specific displays by electrical stimulation; this work is considered at the end of this chapter after the various courtship displays have been discussed in detail. The bowing display is accompanied by a vocalization (the boo-coo). In *Columba palumbus* this differs from the advertising or perch-coo while in the Barbary Dove the perch-coo and the call made during bowing are indistinguishable (see Davies (1974a) for details and for changes in frequency of these calls during the reproductive cycle of the Barbary Dove).

Natural selection has probably favoured those males which fertilize a female as quickly as possible, certainly before another male gets the chance. Perhaps for this reason, it is possible to see males performing the bowing display at any other con-specific with which they cohabit in a feeding flock and such behaviour can be readily seen among street pigeons. Rarely do such advances meet with success, for most importuned individuals, particularly females, retreat, while other males will retaliate with aggressive pecking and/or wing-banging, that is, the elements of fighting. It is to the female's advantage to avoid being fertilized except by a male capable of providing good conditions for rearing young, in other words, females only become interested in a territory-holding male who performs appropriate displays.

Hormones and courtship

Two main methods can be adopted in elucidating the hormonal basis of display. Injection of exogenous hormones may induce specific behaviour patterns whereupon it may be assumed, but with reservation, that the same hormones produced endogenously underlie the display. This technique has depended on the preparation of purified hormones and even now suitably purified avian, as distinct from mammalian, extracts are not readily available. The second approach involves assaying the plasma concentration of circulating hormones at different stages of the cycle but sufficiently sensitive methods are only just becoming available for many of the hormones. The first method was adopted in some experiments in which intact Feral Pigeons were paired for ½ h per day in the manner adopted by Fabricius and Jansson (1963; see above). However, about 2 h before the birds were allowed access to each other the males (but not the females) were variously injected with hormones (Murton et al. 1969b). Males given testosterone at first showed an increase in the incidence of the bowing display and subsequently they began nest-calling allowing the cycle to progress in a

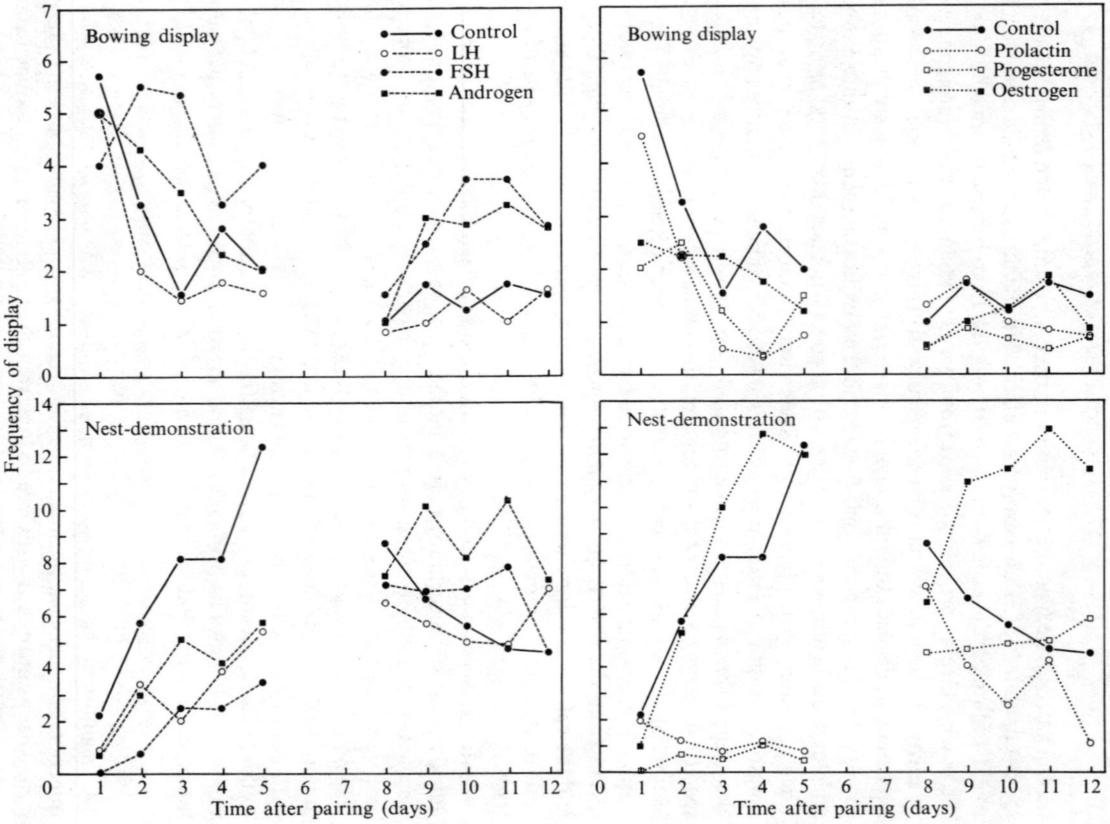

Fig. 5.3. Effect of exogenous hormones (LH, FSH and androgen in the two left-hand panels and prolactin, progesterone, and oestrogen in the two right-hand panels) on the bowing display (upper two panels) and nest demonstration display (lower two panels) of the Feral Pigeon. Hormones were administered daily for 14 days covering the period of pre-incubation courtship. Solid lines, repeated in left and right panels, depict the display sequence of untreated controls and can be compared with the data in Fig. 5.1. Injections comprised 14 × 2 mg of testosterone propionate or oestradiol benzoate, progesterone or Armour prolactin, or 14 × 1 mg N.I.H.-LH (bovine) or of N.I.H.-FSH (bovine). From Murton et al. (1969b).

manner closely resembling that seen in control pairs (Fig. 5.3). All the females laid eggs, but about 3 days later than did females paired with untreated controls. Since none of the females was injected, variations in their reproductive behaviour depended on the nature of the stimulus they received from their mates. Mammalian FSH consistently enhanced the sexuo-aggressive components of the bowing display and led to much driving, but mammalian LH was without very noticeable effect: statistically it could be shown that LH increased the probability that attacking would follow bowing (see Fig. 5.3). This was a surprise discovery for until then the gonadotrophins had not been thought of as having a direct influence on behaviour. It was not clear whether FSH elevated aggressive behaviour by a direct action or by synergizing with androgen. An additional possibility was that FSH was responsible for releasing endogenous androgen (to account for the observed differences in effect of exogenous FSH and exogenous testosterone it would have to be assumed that endogenous and exogenous androgen produced slightly different effects).

Various lines of evidence (see p.130) indicate that endogenous androgen titres are high when the male pigeon begins his bowing sequence and it is clear that androgens mediate bowing behaviour. Erickson and Lehrman (1964) found that castrated male Barbary Doves failed to bow-coo when subsequently introduced to females for a 30-min test period, although in their experiments there was 5 weeks' delay between castration and introduction to the female. Male Barbary Doves can vary in the range of displays they exhibit when first introduced to a female. In tests by Hutchison (1970a), 29 per cent of males performed the bowing display, exhibited driving (chasing as defined by Hutchison), and nest-solicited at each test (designated the PS group); 32 per cent showed pursuit but only intermittent nest-soliciting (PS1 group); 29 per cent showed pursuit but gave no nest-demonstration (P group); 10 per cent performed only the nest-demonstration display (S group). Those males which initially failed to give the nest-demonstration display gradually came to do so following successive meetings with the female. Castration does not immediately eliminate courtship in castrated Barbary Doves and display can persist for 12 days or more. Hutchison showed that if he castrated P-category males and then tested them with females they never did give the nest-demonstration display while this display declined in PS males which were similarly castrated and tested. In other words, differences between the males depended on endocrine factors which were affected by the removal of the source of the gonadal steroids. Castration of photostimulated subjects would be expected to cause an initial boost to gonadotrophin secretion by removing any inhibitory feed-back effect of androgen and indeed a 5-fold increase in plasma gonadotrophin occurred following the castration of Japanese Quail *Coturnix coturnix* (Follett *et al.* 1972a). In some further experiments, Hutchison (1970b, also 1969) quantified the elimination of courtship displays following castration. His results are summarized in Fig. 5.4, together with the consequences of subsequently injecting testosterone.

Hutchison's results are consistent with a view that both FSH and an androgen are involved in early courtship, as suggested by the results summarized in Fig. 5.3. The fact that driving was not completely eliminated by castration, whereas bowing was,

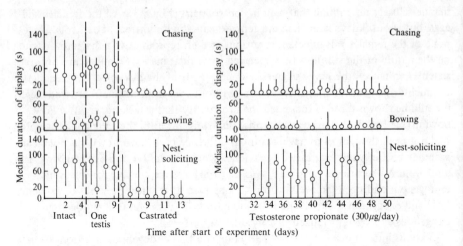

Fig. 5.4. Decline of the courtship displays of Barbary Doves *Streptopelia roseogrisea* var. following castration and the effects of testosterone propionate therapy (300 μg/day intramuscular in re-establishing courtship. Dashed lines represent 5-day recovery period. From Hutchison (1970*a*).

gives further support to the view that FSH is primarily responsible for driving (chasing and the more aggressive components of the early courtship sequence, and that androgen facilitates bowing and nest-demonstration, that is, sexual elements. Hutchison (1970*b*) performed a neat experiment in which he used hypothalamic micro-implants of testosterone proprionate, and found that bowing was stimulated by high-diffusion implants and nest-demonstration by those giving lower dose levels. It is tempting to suggest, with Hutchison, that the progression of the cycle from the bowing to nest-demonstration phase results from a diminution of the response threshold of a hormone-sensitive system in the anterior hypothalamus. In discussing his results Hutchison neglected to consider the possible feed-back effects of androgen on FSH secretion (but see Hutchison (1975)). Thus, if androgen does not inhibit nest-demonstration while FSH does, then the emergence of nest-demonstration could conceivably depend on the blocking of FSH by an androgen feed-back mechanism. Of course, we now know that inhibin feed-back is likely to be important (see p. 54). To resolve these questions it would be helpful to work with hypophysectomized castrates but the experimental difficulties are considerable. For this reason we attempted to examine the problem by studying the behaviour of oestrogenized males which were subsequently given either testosterone or FSH (Murton and Westwood 1975). As discussed in Chapter 3, oestrogen treatment appears to block FSH secretion and elevates IR-LH but it is not clear what it does to androgen titres.

Subjects for the above experiment were 12 adult male Feral Pigeons, which all had fully mature testes, as revealed by an exploratory laparotomy. The birds were withdrawn from a communal holding flight, and kept in isolation in the middle section of a battery of three cages with two sliding opaque partitions separating the three compartments. In each left-hand compartment there was a reproductively active female (12 females) which two days after laparotomy was allowed access to the adjacent male for 15 fnin. In this way it was confirmed that the males would perform the bowing display. The males were then kept in isolation while being given injections of oestradiol benzoate (10 X 2 mg) over a period of 14 days. Two rest days followed the last injection and then the males were introduced to the adjacent females by removal of the opaque partition around 10·00–11·00 h. All behaviour was recorded for 15 min, whereupon a second reproductively active male, which had been held in the third compartment, was introduced to the female and first male. The behaviour of the three birds was recorded for a further 15 min and then all subjects were sep- arated and confined in their own cages, in visual but not auditory isolation. The 12 second males were known to be capable of giving the bowing display, but they were switched from cage to cage so that a different male was presented to the 12 pairs during the experiments. In the same way, the females were moved to different cages so that in subsequent tests the males experienced a new female and then a new male. In this way we hoped to avoid pair bonds being established and the behaviour cycle progressing to nest-demonstration. On the second day all the 12 test males were lap- arotomized again and were then given 2 days rest until day 5 when they were divided into two equal groups:

Group 1. On day 5, 6 of the males were given 0·2 mg testosterone proprionate in arachis oil and after ½ h were introduced to a female. Behaviour was recorded for 15 min and then a second, but different (as mentioned above) male was introduced and behaviour recorded for another 15 min, after which the 3 subjects were isolated in their own cages. Exactly the same procedure was followed on days 6 and 7, that is, testosterone injection followed by behaviour observations each time involving different partners. On day 8 the experimental males were autopsied.

Group 2. On day 5 these 6 males were given 0·2 mg N.I.H.-FSH in saline. In exactly the same way as the androgen-treated birds, they were re-introduced to females after ½ h and behaviour was noted for 15 min before a second male was introduced for a further 15 min. These tests continued on days 6 and 7 and test sub- jects were then killed on day 8, i.e. there were three 15 min test periods per bird with the female alone, followed by three similar periods with a second male present.

The results of these experiments are detailed in Murton and Westwood (1975) and can be summarized as follows. At the termination of oestrogen treatment, the males displayed to the females with bowing and some attack movements, even though their testes were reduced in size and on histological criteria possibly inactive. The combined frequency of bowing and attacking was not statistically altered when a second male was introduced, but the males now delivered a higher proportion of attacks compared with bowing displays (40 per cent attacks) than they had when the female only was present (23 per cent attacks). The presence of a second male did not significantly alter the quantity of display (perhaps with a bigger sample some rise might have been recorded) but the quality was distinctly changed and it became mostly directed at the 'rival' male. The experimental males had had only 15 min experience of the female before the second male was introduced. That the males did not attack their females and perform more bowing to the second male appeared to depend on the behaviour of the sexes; females usually retreated and adopted submissive postures whereas the second males exhibited more aggression.

Subsequent treatment of oestrogenized subjects with testosterone resulted in a decrease in the total frequency of display. Compared with the situation immediately following oestrogen treatment, testosterone increased the proportion of bowing to attacking when only females were present (89 per cent bows against 77 per cent; $\chi^2 = 5.913; P < 0.05$) but not when another male was present (55 per cent bows against 60 per cent; $\chi^2 = 1.284$; n.s.). The total incidence of bowing and attacking combined per individual was not reduced following FSH treatment, unlike the situation pertaining when testosterone was used. In males given FSH the ratio of bowing to attacking while only a female was present was not significantly altered compared with the situation following oestrogen treatment (77 per cent bows against 69 per cent; $\chi^2 = 2.670$; n.s.). However, following FSH injection, the presence of a second male resulted in much aggression and first males now failed to discriminate their own mates from the rival. Nevertheless, many of their displays were not directed at either the female or rival male and when the total number of display bouts was considered the ratio of bowing to attack was the same as it had been just after oestrogen administration (61 per cent bows against 60 per cent; $\chi^2 = 0.098$; n.s.).

In the series of experiments just described the testes of a group of sexually mature birds (primarily of the wild-type or blue variety, see p. 276) were caused to regress by holding the subjects under a LD 8:16 in late June and July. Such males were also held in cages sandwiched between others holding a sexually active male and female and one or both were introduced as described above. The total number of bowing and attack displays given by the males with small testes did not differ significantly from the number given by oestrogen treated males. However, birds with small testes exhibited a high proportion of attack to bowing displays when tested with females only, but did not differ from oestrogen-treated males when a second male was present. Participation in courtship appeared to facilitate a recrudescence of testicular condition and after a month, during which time the testes had recovered to about half their normal size, the behaviour of the males changed so that they now delivered a higher proportion of bowing displays; the frequency of bowing was not altered when a second male was present (details in Murton and Westwood (1975)).

Testicular growth involves spermatogenesis under the influence of gonadotrophins with androgen assuming importance in the later stages of germ cell development (Lofts and Murton 1973). It is conceivable that the aggressive behaviour of the birds when they had regressed testes was influenced by gonadotrophins (FSH) and not gonadal steroids. Presumably, gonad growth indicated an increased gonadotrophin output, and was associated with interstitial cell rehabilitation and an increase in androgen synthesis and release. Injections of oxytocin and arginine vasotocin have been shown to increase the number of sexual activities in both domestic cocks *Gallus gallus* and Feral Pigeons (Kihlström and Danninge 1972). These authors question whether such augmented sexual behaviour depends on the release of gonadotrophins, especially as only one minute elapsed between injection and intensification of behaviour.

We conclude that a varying balance of FSH and androgen may account for the ambivalent sexuo-aggressive displays so typical of the early stages of pair formation. If FSH is dominant, the cycle tends to centre on bowing, attack, and driving, whereas androgen facilitates a progression towards courtship feeding and nest demonstration. During this part of the courtship cycle, the female's behaviour must be directed towards reducing the male's aggression. When she is not being pursued she crouches and generally behaves submissively and occasionally she will displacement-preen and beg food from the male. The male also does much displacement-preening at this stage and sequences of behaviour emerge involving: displacement-preening by the male or female → begging by the female → courtship feeding → displacement-preening by both sexes. Such behaviour sequences increase in frequency to reach a peak after about 3–5 days, by which time they have gradually come to be terminated by the female crouching in order to solicit the male to mount and copulate; copulation is followed by displacement-preening. Thus, courtship feeding and the first copulations are noted at least 8–9 days before ovulation. It is now realized that courtship feeding in birds serves the valuable function of providing the female with extra nourishment during the period when she is forming eggs; this topic is elaborated in more detail in Chapter 8 (p. 200).

Nest demonstration and nesting

During the period when courtship feeding and copulation first occur the male also begins to give the nest-demonstration display, indeed, the first signs of this behaviour may appear on the same day that a male and female have been introduced to each other. Nest-demonstration is eliminated in castrates (Hutchison 1970b) so gonadal hormones are required for its expression. Testosterone injections enable nest-demonstration behaviour to develop in both intact and castrated males (Murton et al. 1969b; Hutchison 1970b). Hutchison found that if testosterone propionate was given to castrated P-group male Barbary Doves, that is, those males which had not given the nest-demonstration display prior to castration (see p. 109) they did not nest-demonstrate (although they would bow) although PS-group males, that is, those which prior to castration had nest-demonstrated, would give the display following androgen therapy. Hutchison concluded from the experiments involving hypothalamic implants of testosterone, that a declining androgen sensitivity was involved in the onset of nest-demonstration. He (1970b) also implanted cholesterol, which was without effect, and oestradiol monobenzoate, which resulted in all subjects giving sustained nest-demonstration displays; castrates also began nest-demonstration when injected with oestradiol benzoate. Earlier we had demonstrated that exogenous oestrogen caused a marked stimulation of nest-demonstration behaviour in intact Feral Pigeons, as Fig. 5.3 shows, the display being continued throughout the whole pre-incubation cycle.

Does the marked ability of oestrogen to stimulate nest demonstration behaviour in the male imply that there is an endogenous release of oestrogen? Alternatively, perhaps oestrogen is effective because it is a powerful inhibitor of FSH and allows androgen behaviour to dominate. Unfortunately, the evidence remains equivocal. We know that nest-demonstration is elicited by exogenous oestrogen in birds whose testes are spermatogenetically regressed (Fig. 3.8, Fig. 5.3). The interstitium of such oestrogenized subjects becomes heavily charged with sudanophilic lipoids, but it is not clear whether the Leydig cells cease to secrete or whether the rate of synthesis exceeds the rate of hormone release. It appears that LH titres are elevated (p. 131) but exogenous LH given to intact males reduces rather than increases the incidence of nest demonstration (Murton *et al.* 1969*b*). An experiment by Erickson *et al.* (1967) is relevant here. Male Barbary Doves were castrated and tested 7 days later when both the bowing display and nest-demonstration had disappeared. Next they were injected with testosterone propionate (0·2 mg/day/bird) for 14 days, whereupon both behaviours were reinstated. The subjects were now divided into four groups which during the next 7 days were treated and responded as follows:

1. Testosterone treatment continued as did both bowing and nest-demonstration displays.
2. Testosterone treatment was continued but additionally the birds received 0·1 mg/day/bird of progesterone. The bowing display was eliminated but nest-demonstration continued unchanged.
3. Progesterone (0·1 mg/day/bird) only was given, whereupon both displays were eliminated.
4. No hormone treatment was given and both displays were eliminated.

Erickson *et al.* concluded that progesterone is selectively capable of inhibiting the reactions of some neural mechanisms to androgen while leaving others unaffected, that is, it does not act by inactivating androgen or generally antagonizing it, but possibly by specifically insulating some neural sites against it. These authors supposed that endogenous progesterone was produced during these early stages of the natural cycle (by the adrenal) but this interpretation raises difficulties, as will emerge later in this chapter.

Fig. 5.1 shows how the increasing incidence of nest-demonstration in the male is paralleled a day or two later by a similar build-up of this pattern of behaviour in the female. At this stage, the male and female spend long periods together on the potential nest-site and much caressing (nibbling) and bill-fondling occurs, which appears to function to cement the pair bond. Spells of nest-demonstration with feather-caressing are interspersed with bill-fondling, begging, and courtship feeding. The waning of the male's aggressiveness during this period appears to encourage a reciprocal increased assertiveness on the part of the female. This is seen in her determination to occupy the nest-site and she increasingly attempts to push the male from the nest. It seems very likely that the submissive behaviour of the male encourages an increase in dominance in the female which stimulates FSH secretion, but this possibility has not been investigated. It would agree with the fact that during this period oestrogen secretion becomes important resulting in growth of the oviduct as Fig. 5.7 below shows. A female paired with a passive male, who has not performed any kind of courtship, is not stimulated to become sexually aggressive. This was noted in females which were paired with prolactin- or progesterone-treated males, for they performed virtually no pushing, whereas in the presence of a nest calling male stimulated by oestrogen they exhibited an abnormally high incidence of pushing and also of pecking (Fig. 5.5). Such females also laid eggs and although the time taken appeared to be longer than with untreated controls the differences were not statistically significant (Murton *et al.* 1969*b*). No egg laying occurred in females paired with prolactin- or progesterone-treated males, while oviduct development was significantly more retarded in females associating with progesterone-treated males compared with those given prolactin. The only marked differences in the behaviour of the males given progesterone compared with those receiving prolactin was an absence of mounting and copulation.

As was mentioned in the preceding chapter, many experiments have demonstrated the dependence of oviduct development on oestrogen secretion, both in wild species and the domestic hen (p. 96). Some further experiments by Lehrman and Brody

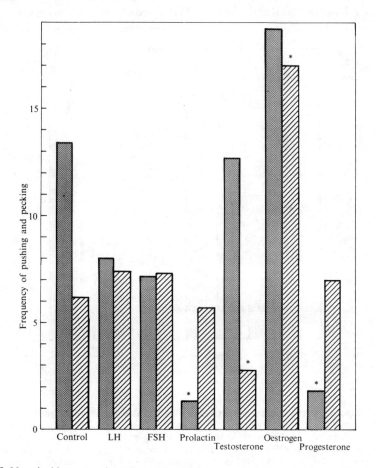

Fig. 5.5. Mean incidence (total number of 2-min periods during pre-incubation cycle) of pushing (solid) and pecking (hatched) by paired female Feral Pigeons depending on the hormone treatment of the males. Asterisks denote results which were significantly different from the control of situation. Based on data in Murton *et al.* (1969*b*).

(1957), which refer specifically to the Barbary Dove, are summarized in Fig. 5.6. Progesterone injected by itself will not stimulate oviduct growth, oestrogen alone is effective, and oestrogen plus progesterone cause maximum development. Priming the oviduct by giving the subject oestrogen (diethylstilboestrol) for 8 days followed by a mixture of progesterone and oestrogen for 2 days produced as much response as giving a mixture of oestrogen and progesterone for 10 days. When this experiment was performed the workers had suspected that the natural cycle involved a spell of oestrogenic influence followed by a short period of progesterone secretion immediately preceding the onset of incubation, a hypothesis subsequently confirmed by various experiments.

The response of the oviduct to exogenous hormones is, however, complex, as was demonstrated by the response of ovariectomized Canaries treated in winter (about 8½-h daylength) (Hinde, Steel, and Hutchison 1971). Oestrogen was found to have a dose-dependent effect on oviduct weight and when injected at low dose-

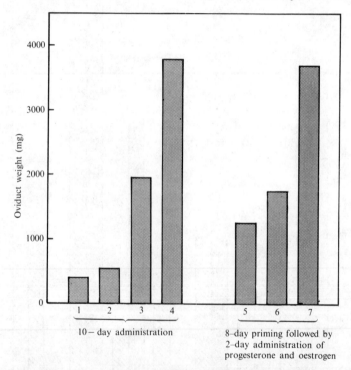

Fig. 5.6. Effect of hormone treatment on oviduct weight of Barbary Doves. Groups of 9 birds each were treated as follows: 1, Control, 0·1 ml sesame oil daily, × 10; Progesterone, 0·1 mg daily, × 10; 3, Oestrogen, 0·4 mg diethylstilboestrol daily, × 10; 4, Progesterone + oestrogen, 0·1 mg progesterone plus 0·4 mg diethylstilboestrol daily, × 10; 5, Control primed, 0·1 ml sesame oil daily, × 8; followed by 0·1 mg progesterone plus 0·4 mg diethylstilboestrol daily, × 2; 6, Progesterone primed, 0·1 mg progesterone daily, × 8, followed by 0·1 mg progesterone plus 0·4 mg diethylstilboestrol daily, × 2; 7, Oestrogen primed, 0·4 mg diethylstilboestrol daily, × 8, followed by 0·1 mg progesterone plus 0·4 mg diethylstilboestrol daily, × 2. From Lehrman and Brody (1957).

rate levels (3 × 0·05 mg/week for 3 weeks) it disrupted the normal weight to diameter relationship and produced structural abnormalities. When a high dose (9 × 0·3 mg) was given it caused only one oviduct to become secretory. When testosterone, progesterone, or prolactin were administered by themselves they resulted in less increase in oviduct weight than did 0·05 mg oestrogen but a high dose of testosterone (9 × 2·0 mg) stimulated some development of the tubular glands and the precocious appearance of goblet cells. When certain hormone combinations were tested they appeared to act synergistically judging by the total weight achieved by the oviduct, but in reality the increase resulted from an abnormal development. In particular, the magnum diameter tended to be large relative to oviduct weight. A differential effect of hormones on the magnum was suggested by the fact that its diameter was often large relative to the height of the mucosal folds, as if the processes of increase in fold height and diameter expansion are separate but concurrent events, presumably having different endocrine requirements. The addition of 2·0 mg progesterone or 0·25 mg testosterone to a dose of 0·05 mg oestrogen gave an oviduct development which most closely resembled the natural situation. It should be recalled that subjects were kept on a 8½-h photoperiods so that gonadotrophin secretion would have been inhibited and it is not certain whether these are also involved in the complete development process.

Nest-building behaviour has been shown to be closely related to oestrogen secretion in female doves, and Canaries, using oviduct growth as an indication of oestrogen secretion; access to a mate and nest material also facilitates ovulation and preparedness to begin incubation; as Fig. 5.7 shows for the Barbary Dove (see also Martinez-Vargas and Erickson 1973). More recent studies of the Barbary Dove show that treatment with a combination of oestrogen and progesterone was the most effective hormone regimen for eliciting both nest-building and incubation behaviour in ovariectomized females and that treatment with either progesterone or oestrogen alone was virtually ineffective (Cheng and Silver 1975). On the other hand, the nest-related activity of the male seemed to be mainly dependent on behaviour cues from the female. 17β-oestradiol monobenzoate induces ovariectomized Budgerigars *Melopsittacus undulatus* to enter nest boxes 8–10 days before egg laying in the manner of untreated controls (Hutchison 1975). Male Feral Pigeons begin to collect nesting material after the aggressive elements of the first phase of courtship sequence have been replaced by nest-demonstration and caressing. But there is a resurgence of bowing, courtship feeding, and copulation with the onset of twig collection (Fig. 5.1). This may be due to an increased influence of FSH consequent on waning androgen titres but this suggestion needs to be substantiated. In the experiments mentioned above, the only exogenous hormone which allowed the male Feral Pigeon to perform anything like a normal amount of twig-carrying and nest-building was FSH and all other hormones, including oestrogen, severely depressed this behaviour (Murton *et al.* 1969b). The increased aggressiveness of the male at this stage of the cycle seems to be in turn consequent on the marked change in assertiveness on the part of the female which was mentioned above, and this change is reflected in a bimodal pattern in the frequency of several of the displays detailed in Fig. 5.1. When the female assumes a more dominant part and attempts to push the male off the nest he will respond by bowing but it seems that his agonistic tendencies can be channelled into pecking at twigs and gathering nest material. In the Barbary Dove, the state of the nest has been shown to be an important determinant of the type of nest-building exhibited (White 1975*a*). Moreover, the presence of a proper nest has a profound effect on later breeding success in that it is essential for the full establishment of incubation behaviour and the participation of the sexes (White 1975*b*).

In many bird species nest-building is associated with a high level of aggressive behaviour and in several, perhaps in the primitive condition, the nest is built as a prelude to the courtship cycle. This is so in Ardeidae, for the males build a nest-platform as soon as they have acquired a territory and they then use the platform for display, at first vigorously driving off any intruder, be it male or female. Nest-building, at least in the Black-crowned Night Heron *Nycticorax nycticorax*, is apparently mediated by androgen (Noble and Wurm 1940). An important display which has been well described and discussed in the Grey Heron *Ardea cinerea* (Verwey 1930; Milstein, Prestt, and Bell 1970) is the 'bow-snap' display. This appears as a ritualized nest-building movement and sometimes the display is terminated by the

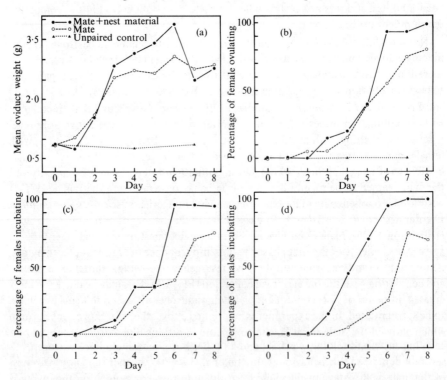

Fig. 5.7. Effect on Barbary Doves of association with mate, or mate plus nesting material on (a) oviduct weight, (b) percentage of females ovulating, (c) percentage of females incubating, and (d) percentage of males incubating. Each point is derived from tests on 20 birds with no individual bird being represented in more than one point. From Lehrman *et al.* (1961*a*).

bird actually seizing a twig; the display is strongly attractive to females. The 'bow-snap' display is very similar to the 'bill-snap' display which features prominently in threat posturing, and indeed the two displays are linked by some authors (Baerends and Van der Cingel 1962). The Brush Turkey *Alectura lathami* (Megapodiidae) builds a mound in which the eggs are incubated by heat from the sun and the decaying vegetation. During the most intensive phase of nest-building the male becomes extremely aggressive and will not allow the female near the mound (Fleay 1937).

Returning to pigeons, it is clear that behavioural stimuli from the male are important in stimulating oestrogen secretion in the female, perhaps by stimulating FSH secretion. When experienced but castrated male and intact female Barbary Doves were introduced for 7 days in the presence of a nest-bowl plus nesting-material oviduct development was significantly reduced compared with the control situation (Erickson and Lehrman 1964). The oviduct weights and ovulation frequency of females which were caged with castrates given androgen reflected the vigour of the male's resultant courtship. Correlations of male activity with female response suggested that wing-flipping was more important than bow-cooing or the nest-call. Castrated males performed no courtship displays but none the less their presence did facilitate a small degree of oviduct development. Moreover, two females caged together can stimulate each other to ovulate (Harper 1904). Matthews (1939) showed that the stimulus causing ovulation was mostly visual rather than tactile because birds separated by a sheet of glass or placed in a cage with a mirror also ovulated. Since then, Lott and Brody (1966) showed that 21 of 25 female Barbary Doves having previous breeding experience laid eggs when kept in isolation in sound-proof cubicles, which were provided with mirrors in which the subjects could see their own images and, in addition, the cubicles were equipped with microphones through which noises coming from nearby breeding birds were relayed. Noises from the colony milieu without the mirror reflection did not result in any egg laying by 12 test subjects whereas 2 out of 10 females exposed to their own mirror image, but unable to hear other doves, did lay. Lott, Scholz, and Lehrman (1967) discovered that auditory stimulation (colony sounds) would increase ovarian and oviduct development in females kept with intact males above the level attained if intact males were used without additional colony sounds. It was demonstrated by Lehrman and Friedman (1969) that the follicles of females isolated in sound-proof chambers regressed over 21 days. When some individuals were now allowed to hear the sounds relayed by a microphone from a breeding colony their follicles again increased in size by a factor of two and their oviducts increased slightly in weight. In Budgerigars a specific vocalization given during courtship has been shown to induce ovarian development and ovulation (Brockway 1965). Auditory stimuli may well be important in synchronizing egg laying in certain colonial species (p. 362).

Male Canaries do little nest-building and the size and texture of the nest-cup which the female has constructed provide the important stimuli for oviduct development and egg laying (Warren and Hinde 1961*a*). This stimulation has an influence on the frequency of certain nest-building movements and the kind of nesting material selected by the female and this in turn affects the amount of nest-building, brood patch development, and egg laying (Hinde 1965). It is doubtful whether oestrogen alone is responsible for nest-building behaviour for almost lethal doses are needed to make Canaries build in winter (Warren and Hinde 1959) and neither progesterone nor prolactin can augment its effect. Interestingly, female Canaries given pregnant mare's serum (PMS) may lay eggs without nest-building while paired females which are photostimulated in winter will nest-build (Steel and Hinde 1966a,b). As pointed out by Hinde (1967) this suggests that stimulation from an active male or secondary hormones in addition to oestrogen are implicated. If sufficient exogenous oestrogen is given the presence of the male makes no difference to the female's building behaviour (Warren and Hinde 1961b). Canaries kept on LD 20:4 while receiving PMS injections would nest-build to a much greater extent than others given similar injections on LD 9½:14½ so the gonadotrophins present in PMS do not seem to provide the essential factor (Steel and Hinde 1972a). Nevertheless, the PMS is presumably responsible for stimulating the follicles to produce oestrogen.

In some other experiments Steel and Hinde (1972b) found that nest-building was not completely eliminated in ovariectomized birds and more nest-building occurred if such subjects were kept on LD 20:4 than on short days. This nest-building could be suppressed by methallibure, a gonadotrophin inhibitor. When Canaries were exposed to various lighting schedules in January–February it was found that LD

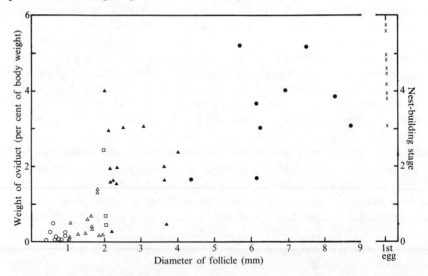

Fig. 5.8. Relation between size of largest ovarian follicle and weight of oviduct expressed as a percentage of body weight in Canaries *Serinus canaria* killed before laying their first egg. The approximate correlation between oviduct development and follicle size is indicated as follows: open circles, all follicles the same size and no obvious oviduct development; open triangles, beginnings of invagination of epithelium of the magnum; open squares, formation of tubular glands and enlargement of 5–6 follicles; solid triangles, early formation of albumen, enlargement of one follicle larger than the others, nest-orientation behaviour and brood patch formation; solid circles; granule formation in the tubule glands, one enlarged follicle, and advanced defeathering and vasularity of the brood patch with marked nest-building. Compiled from Fig. 4 and Table 6 in Hutchison *et al.* (1968).

12:12 and LD 14:10 caused substantial development of the ovaries and oviduct growth and a significant elevation of plasma and pituitary IR-LH (Follett) (Follett, Hinde, Steel, and Nicholls 1973). Oestrogen treatment of birds exposed to LD 8:16 decreased plasma and pituitary IR-LH (Follett) but some of the subjects showed an increase in gathering and placing of nest-material. Oestrogen treatment of birds exposed to LD 14:10 had no effect on plasma IR-LH, but the birds did more gathering and placing and their oviduct weights were increased. The effect of oestrogen treatment on FSH was not established. In further experiments (Hinde, Steel, and Follett 1974) nest-building was induced at various times of year in ovariectomized photosensitive Canaries with a standard dose of oestradiol. On day lengths from 6 h to 18 h light per day, building behaviour was more intense with increase in photoperiod. A similar result was obtained in photo-refractory birds, both intact and ovariectomized, but there was a marked latency of response to oestrogen in these subjects. The effect of a long photoperiod in augmenting the induction of nest-building by oestrogen could not be reproduced by injection of ovine or avian LH but the effect of FSH was not discussed. In the Bengalese Finch *Lonchura striata* egg laying has been found to be stimulated by the mate and nest-material in a complex fashion (Slater 1969, 1970).

The relationships between nest-building, ovarian growth, and oviduct development have been studied in detail in the Canary, using for the most part unpaired females. This has the advantage that the reproductive cycle becomes spread out and distinct events more readily identifiable. Moreover, Canaries, unlike pigeons (Maridon and Holcomb 1971), produce a brood patch in response to hormone changes and this provides a convenient indicator of internal events.[†] Hutchison *et al.* (1968) showed how the total weight of the ovary increased as the follicles developed, there being a sudden increase in weight when the largest follicle attained a diameter of 2 mm. In the same way, the oviduct also increased more suddenly in weight when the largest follicle reached 2 mm diameter. Fig. 5.8 shows the temporal relationship between follicle growth, oviduct development, and nest-building stage. It was noticeable that albumen formation did not occur until one follicle was enlarged more than the others and this coincided with marked defeathering and vascularization of the brood patch and also with intensive nest-building behaviour. Before this stage, which occurs at rather a constant interval before egg laying, the cycle does not necessarily progress so that follicles may become atretic, only tubular glands are formed in the magnum and little defeathering is apparent. Evidently a profound endocrine change is associated with the marked growth of one of the follicles and it seems likely that at this point progesterone secretion begins and augments oestrogen. In the chicken either progesterone or androgen are additionally necessary for the formation of albumin granules (Brant and Nalbandov 1956) and there are several other lines of evidence which implicate progesterone at this stage of the pre-incubation cycle of the Canary.

The development of the Canary brood patch progressively involves defeathering, vascularization of the ventral apterium, and an increase in tactile sensitivity. A series

[†]Jones (1971) has reviewed the hormonal control of brood patch development in 8 passerine and 3 other bird species.

TABLE 5.1

Effect of exogenous hormones on brood-patch development in intact or ovariectomized (ovar.) Canaries and in hypophysectomized White-crowned Sparrows (hypo.).

Stage of brood patch development	Oestrogen			Oestrogen plus prolactin			Oestrogen plus progesterone		Prolactin		Progesterone	
	Intact	Hypo.	Ovar.	Intact	Hypo.	Ovar.	Intact	Ovar.	Intact	Ovar.	Intact	Ovar.
Defeathering	+	0	+	+	+	+	+	+	0	0	0	0
Vascularization	+	+	+	+	+	+	+	+	0	0	0	0
Sensitivity	+	+	0	+	0	0	+	+	0	0	0	0

+ denotes a significant effect, 0 no effect, and a space means the experiment was not performed.
From Hinde (1967), summarizing data from Steel and Hinde (1963, 1964); Hinde and Steel (1964); Bailey (1952).

of experiments by Hinde and his colleagues involving intact or ovariectomized
Canaries, and other experiments on hypophysectomized White-crowned Sparrows
Zonotrichia leucophrys by Bailey (1952) have been collated by Hinde (1967), and
Table 5.1 is a summary of the results.

Neither progesterone or prolactin alone were effective in overiectomized Canaries.
Defeathering requires oestrogen plus a pituitary hormone, judging from the experiments involving hypophysectomized White-crowned Sparrows in which oestrogen
alone was inadequate, whereas oestrogen plus prolactin was effective. Vascularity
apparently depends on oestrogen only whereas the increase in tactile sensitivity
necessitates progesterone being added to oestrogen in ovariectomized subjects. Prolactin synergizes with gonadal steroids to promote development of the incubation
patch in California Quail *Lophortyx californicus* but it will stimulate hyperplasia of
the epidermis if given alone to intact birds (Jones 1969a). Gonadal steroids do not
seem to be needed for this response since the reaction can be obtained *in vitro* in
this species and in the Bobwhite Quail *Colinus virginianus* (Jones, Bern, and Arimoto,
quoted by Jones (1969b)).

In some further experiments by White and Hinde (1968), unpaired female Canaries
which laid before 6 May were distinguished from others which did not, though these
were thereafter induced to lay by being paired with males. These late layers had a
more protracted development up to the stage of brood patch defeathering and the
onset of intensive nest-building. Nevertheless, whether or not the cycle was protracted, the completion of defeathering, attainment of full vascularity, and onset of
intense nest-building occurred within a few days of each other and within a day or
two of egg laying. Some individuals showed spells of intensive nest-building without
egg laying, which alternated with periods of low-level activity. The interval between
peaks of intense nest-building tended to be about 7 days and it was suggested that
this intermittent development was associated with the maturation of successive
groups of ovarian follicles such that if one group became atretic an approximately
constant delay occurred before the next group matured to a similar stage.

Incubation behaviour

That progesterone is important at the end of the pre-incubation cycle in stimulating
both male and female doves to begin incubation has been shown by the classical
studies of Lehrman and his colleagues (Fig. 5.9). Males or females were used which
had previous breeding experience consisting of two successful breeding cycles, the
subjects then being kept in isolation for 3–5 weeks before testing. First, a series of
tests were performed which involved placing a male or female singly or paired (the
pairs were birds which had not previously met) in a cage in the presence of a nest-
bowl and nesting material. The results of these experiments showed that:
  (1)   subjects tested singly failed to incubate;
  (2)   habituation to the test cage did not affect the time taken to incubate;
  (3)   if a male and female were allowed to associate for a week before introduc-
        tion to the test situation incubation occurred after a delay of one day;
  (4)   if a male and female could associate for a week with the addition of nesting
        material then they would incubate the test egg immediately.

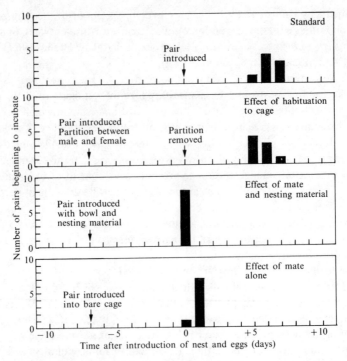

Fig. 5.9. Distribution of latencies of incubation behaviour in Barbary Doves with different conditions of association with a mate and nesting material. A test nest and eggs were introduced at 0900 h on day 0. From Lehrman (1958a).

Secondly, fresh subjects were reared in the manner described above but these were given seven daily injections of progesterone before being introduced as a pair to the test cage in the presence of a nest-bowl and nesting material. The results are summarized in Fig. 5.10 to show that progesterone injections achieved the same result as a male and female being together for a week in the presence of nesting material. Subsequently, Komisaruk (1967) induced incubation by implanting progesterone into the brain of doves. Lehrman and Wortis (1960) also demonstrated the importance of experience in progesterone-induced incubation; subjects which had never before been involved in breeding took 2–3 h longer to sit on the test eggs than did experienced birds. These experimental findings have now been essentially confirmed by the assay of plasma progesterone in doves (Silver, Reboulleau, Lehrman, and Feder 1974). This work showed that there is a rise in plasma progesterone in female doves before egg laying although no such change was detected in male doves during the cycle (but see p. 130).

These various experiments indicate that involvement in display and nest-building cause a succession of endocrine changes that culminate in progesterone secretion and this then provides the necessary stimulus for incubation. Moreover, it was noted above that the duration for which Barbary Dove pairs were kept continuously together

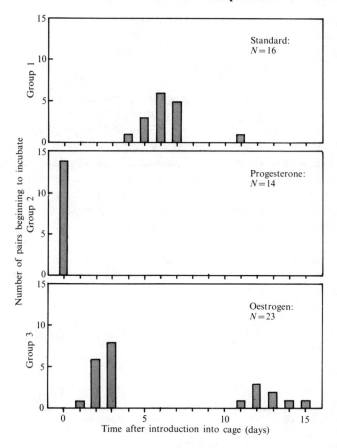

Fig. 5.10. Distribution of latencies of incubation behaviour in Barbary Doves according to hormone pre-treatment. Birds were introduced to test nest and eggs between 0900 and 0930 h on day 0. *N*, number of pairs. From Lehrman (1958*b*).

affected their readiness to incubate when they were introduced to a test nest and eggs (Fig. 5.7). Lehrman, Wortis, and Brody (1961) undertook some further experiments to find out whether a male and nesting material need to be continuously present to stimulate the female or whether they simply initiate ovarian development which then continues in the absence of the initiating stimulus. Females were housed for 7 days in cages and were then killed for autopsy. Some were kept alone for the whole 7 days but others were accompanied by a male for 1–6 days (Fig. 5.11). Comparison of Figs. 5.7 and 5.11 shows that the readiness of the female dove to incubate is closely associated with ovulation rather than with the growth of the oviduct. Again the evidence favours the interpretation that during the final maturation of the follicle progesterone is produced and that this then facilitates ovulation (p. 88) and the onset of incubation behaviour.

Progesterone induces incubation in female doves which have been secreting oestrogen. What is the situation in males in which progesterone was also shown to be effective? Stern and Lehrman (1969) castrated some Barbary Doves and then housed them in cages separated from females by an opaque partition. Intact and castrated males were given daily progesterone injections before being exposed to the females (themselves primed with progesterone to encourage incubation behaviour) through a glass plate which replaced the opaque partition; it had previously been shown that the doves incubate more readily if tested in pairs than singly (Bruder and Lehrman 1967). Stern and Lehrman now demonstrated that 13 out of 15 intact, progesterone-treated males incubated under the test conditions, compared with only 5 of 16 castrates. Intact birds given an oil placebo would not incubate, nor would castrates given oil, but the former were active in giving the bowing display when they perceived the female in the next compartment. Other birds were castrated and were then given therapy involving oil, progesterone, or testosterone propionate injections for 14 days. There then followed a treatment period of 7 days during which progesterone, oil, or testosterone injections were given. The results are summarized in Table 5.2 and they again emphasize that androgen enhances the capacity of progesterone to induce incubation. Inexperienced, but intact, males would brood squabs after injection of progesterone and they fed the squabs when both progesterone and prolactin were injected (Lott and Comerford 1968).

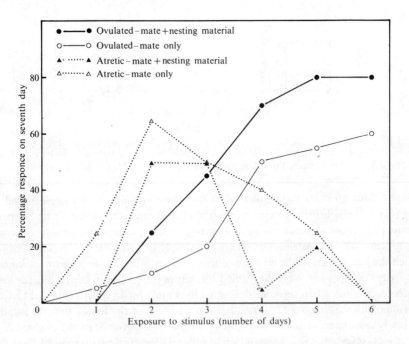

Fig. 5.11. Number of birds ovulating (light and heavy continuous lines) and developing atretic follicles (dotted lines) by the seventh day as a function of duration of exposure to a male (light lines and open symbols) or to a male plus nesting material (heavy lines and filled symbols). From Lehrman *et al.* (1961*b*).

TABLE 5.2

Effect of treatment with testosterone (TP) and progesterone on incubation behaviour of castrated male Barbary Doves

Treatment	Number of birds (out of 15) incubating within 48 h	Number of birds giving	
		bowing display	nest-demonstration
Oil → progesterone	4	0	0
Progesterone → progesterone	6	0	0
TP → TP + progesterone	11	0	10
TP → progesterone	11	0	0
TP → oil	3	0	0
TP → TP	1	8	10

From Stern and Lehrman (1969).

It is to be noted from Table 5.2, as Stern and Lehrman remark, that progesterone eliminated the bowing display in all cases, but did not cause any reduction in the incidence of the nest-demonstration display in castrates dosed with testosterone and subsequently given either testosterone only or testosterone plus progesterone. But progesterone given to intact males significantly lowered the incidence of nest-demonstration, again supporting the earlier suggestion that progesterone insulates central nervous mechanisms against the action of androgen. Thus, the bowing display requires androgen and it is inhibited by progesterone whereas nest-demonstration is induced only by androgen so that progesterone has no direct effect.

Incubation is a complex process which entails maintaining the eggs at the appropriate temperature and humidity. Furthermore, the eggs need turning at regular intervals and they have to be positioned correctly in the nest. The maintenance of these and other physical parameters within the strict tolerance limits necessary, these being in large measure species-specific and adaptive, entails a wide spectrum of stereotyped behaviour. Included in this repertoire are fixed action patterns for retrieving eggs which have rolled out of the nest and for orientating the eggs as the time of hatching approaches. It is known that light passing through the shell can stimulate photic receptors in the embryo of the domestic chick and in this way influence hatching time (Adam and Dimond 1971). Birds lay their eggs at intervals of at least a day yet they usually hatch within a short time of each other. This synchronization is partly achieved by vocal communication between the different embryos within the clutch and between the embryos and the incubating parent (reviews by Vince (1969, 1972)). The term incubation, therefore, embraces an enormous subject which we cannot discuss in detail here. For more details readers are recommended to the reviews by Landauer (1967) and Drent (1972).

The involvement of central nervous mechanisms in the reproductive cycle is emphasized by Åkerman's work, mentioned at the beginning of the chapter (p. 107). Electrical stimulation of the preoptic and anterior hypothalamus of individually caged Feral Pigeons generally induced intense bowing with all normal components

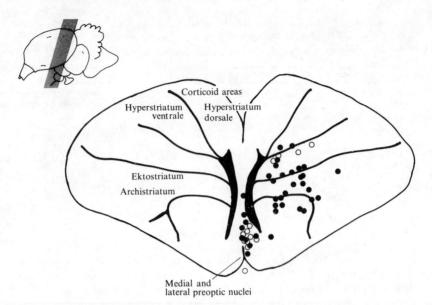

Fig. 5.12. Schematic diagram of transverse section of the pigeon brain to show sites where electrical stimulation evoked components of the bowing display (filled circles) or nest-demonstration display (open circles). In the original work several sections in the region demarcated in the inset were examined but here the results have been superimposed onto a hypothetical single section. From Akerman (1966).

in males. Confrontation of the subjects with appropriate external stimuli released complete aggressive behaviour. In females, such stimulation resulted in nest-demonstration, displacement-preening, and pecking. In males, bowing developed into nest-demonstration, especially when sites in the preoptic and anterior diencephalic paraventricular nuclear masses were stimulated. The sites at which courtship displays, and also defence and escape responses, could be elicited are summarized in Fig. 5.12. An interesting discovery was that weak stimulation, or stronger stimulation in lateral sites, induced only part of a behaviour sequence, for example, attention or feather erection. Increase in stimulation caused more components of a behaviour sequence to become manifested, and, once initiated, a sequence could continue if the electric current was switched off. This is typical of a fixed action pattern of behaviour which becomes released once the nervous mechanism is activated at a particular threshold. It may be supposed that steroid hormones act by sensitizing specific regions of the fore-brain, thereby facilitating the expression of fixed patterns of behaviour.

6 The endocrine basis of reproductive behaviour: 2

Summary
Newer techniques for assaying plasma hormones are enabling the endocrine condition of birds to be monitored throughout the breeding cycle. The *in vitro* incubation of gonad extracts has also enabled their biosynthetic capability to be monitored. Results presently available support the view that androgenic activity in the male declines from pair formation to egg laying while progesterogenic activity increases. A complex cycle of spermatogenesis and development of the epididymides occurs in the male which parallels the drastic changes in follicle size and oviduct development occurring in the female. The hormonal condition during incubation and brood care is also discussed in this chapter.

Endogenous hormones
An alternative to investigating the effect of exogenous hormones on reproductive behaviour is to measure the amount of circulating endogenous hormones during the courtship cycle. Although much has been achieved in relation to ovulation and egg laying in the domestic hen, a systematic assay of steroid and gonadotrophin concentrations has yet to be accomplished in any species performing a reasonably natural behaviour cycle. This is partly because sufficiently sensitive assay techniques are only just becoming available and those that are available have been developed by laboratories concerned with other problems. Then there is the difficulty that relatively few 'wild' species will undergo a breeding cycle in conditions of captivity at all conducive to critical assay. However, we begin by reporting on two parallel studies which help illuminate the endocrine basis of the reproductive cycle of the Feral Pigeon and then move on to consider the endocrine background to incubation, brood care, and some other behaviour.

Feral Pigeon pairs were established following an exploratory laparotomy in the manner already described in the preceding chapter (p.110). However, the male and female were allowed continuous access to each other and their progress through a complete breeding cycle was checked by periodic inspections. Subjects were killed by decapitation at well-defined stages throughout the breeding cycle. In one series of experiments individual plasma samples were collected with edetic acid as an anti-coagulant and were later assayed in the Bagshawe system already described (p. 39) for immuno-reactive luteinizing hormone (IR-LH) (see below). In another series conducted by Miss Fausta Yue (unpublished), testicular tissue was rapidly dissected out, minced with scissors, and incubated with isotopically labelled Δ^5-pregnenolone-T as a labelled precursor. The synthetic capacity of the gonad at different stages of the breeding cycle was assessed by separating the steroids produced during incubation using thin-layer chromatography and quantifying the level of radioactivity in the sample bands by a Packard liquid scintillation spectrophotometer (see Appendix 2 for summary of details). The percentage conversion indices so obtained are summarized in Fig. 6.1.

Fig. 6.1 indicates that during the first half of the pre-incubation phase of courtship the pigeon testis has a well-marked capacity to synthesize testosterone. This capacity apparently wanes after the nest-building stage has been reached, so that testes sampled at the onset of incubation converted less precursor to testosterone and proportionately more 17 α-OH-progesterone was produced. However, it must

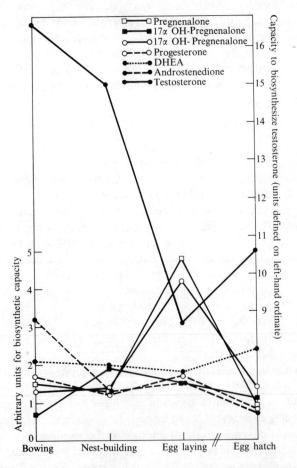

Fig. 6.1. Biosynthetic capacity of testis extracts taken from Feral Pigeons at different stages of the pre-incubation behaviour cycle and incubated *in vitro* with tritiated pregnenalone as the labelled precursor. The left-hand ordinate gives conversion indices for biosynthetic capacity (see Appendix 2), those for testosterone being plotted on the right-hand axis. Symbols refer to pregnenalone, 17a-hydroxypregnenalone, 17a-hydroxyprogesterone, progesterone, dehydro-epiandrosterone (DHEA), androstenedione, and testosterone. From Murton and Westwood (1975)

be remembered that the synthetic capacity of the testis *in vitro* may not be the same as its capacity *in vivo* in the presence of endogenous gonadotrophins.

Murton and Westwood (1975) reported on plasma levels of IR-LH for the individual males which were killed at different stages of the breeding cycle in the collateral experiment mentioned above. All subjects were killed between 10·00 h and 11·00 h. IR-LH plasma titres were low at this time of day in subjects which were performing the bowing display at initial pairing but they had become noticeably elevated by the time of nest-building and egg laying. Thus in 7 out of 12 subjects killed when they were initially paired and performing the bowing display or were involved in nest-demonstration (up to day 2) plasma LH titres were too low to measure since they fell outside the range of sensitivity of the dose—response curve. By the time of nest-building and egg laying, 8 out of 11 males had measurable quantities of plasma IR-LH and the concentration seemingly remained elevated during incubation. Some individuals progressed to the egg-laying stage more rapidly than others (see Fig. 6.8) and so gonadotrophin assays were plotted on a time scale. Apart from suggesting an increase in IR-LH between pairing and day 2 no clear trends were apparent (Murton and Westwood 1975). It follows that increases in IR-LH secretion did not automatically follow pair formation after a given time interval but were instead related to the behavioural stage which had been reached, irrespective of how long it took the subjects to reach the particular stage. This implies that involvement in some behavioural activity following initial pairing was responsible for stimulating IR-LH secretion in the males. But we must exercise caution: gonadotrophin assays were performed only on subjects killed at one time of day. If there are diurnal variations in plasma levels of LH we cannot be sure that our results do not indicate a change in phasing of the LH rhythm consequent on courtship behaviour. Thus if bowing subjects had been killed at a different time of day they may too have revealed elevated IR-LH titres. Similarly there might be a temporal variation in the biosynthetic capacity of the gonad about which we at present have no information (see p.497).

Reference to Figs. 5.2 and 6.1 shows that maximal development of the epididymides probably occurred a little after testicular extracts incubated *in vitro* exhibited their maximum capacity to synthesize testosterone and androstenedione. It will be recalled from Chapter 3 that in the mammal FSH acts on the Sertoli cell to produce a specific binding protein which draws androgen into the tubule, from whence it can pass to the epididymides; the same is likely to apply to birds. We argued in the last chapter on behavioural evidence that FSH activity was at a maximum at the start of the behaviour cycle and these results are entirely consistent with the hypothesis that the enlargement of the columnar epithelium of the epididymides followed a phase of FSH activity which allowed androgen to enter the tubules. IR-LH titres were measurable at the start of courtship, but they had risen appreciably by the time of nest-building. If androgen titres decline during the pre-incubation behaviour cycle, as was argued in the previous chapter, we might attribute the rise in circulating LH to a decline in androgen feed-back. There was a gradual decline in the capacity of testicular extracts to synthesize androgen during the pre-incubation cycle yet IR-LH levels remained high, and it will be shown below that new Leydig cells were being formed. It is possible that the old Leydigs were now spent and the young ones only had a reduced capacity to synthesize steroids and would require time to become fully active. Or, an absence of FSH might have been important. There is also the possibility the prolactin secretion was increasing and having an effect. We cannot provide answers at present and have insufficient grounds for worthwhile speculation.

Changes in testis histology

So far the pre-incubation behaviour cycle of the Feral Pigeon has been considered in relation to the endocrine state of the subject, by testing the consequence of therapy with exogenous hormones and by attempts to monitor endogenous endocrine changes. We must now examine the relationship between these changes and histological developments in the germinal epithelium and interstitial tissue of the testes; comparable data are not yet available for the female. Fig. 6.2 now details the number of different germ cells to be counted in cross-sections of the tubules from the time of initial pairing (see Murton and Westwood (1975) for raw data on germ cell counts and standard deviations, the error terms being omitted from Fig. 6.2 for the sake of clarity); Fig. 6.2 also defines the number of Leydig cells which could be counted per standard cross-section of the testis.

When males were first paired and were prepared to give the bowing display their testes already contained relatively large numbers of spermatids and few spermatozoa. Doubtless many sperm that had already been produced had migrated from the testes tubules and were already stored in the sperm sacs at the distal end of the vasa deferentia. Within a day or two from pairing a distinct wave of spermatelosis occurred so that between the beginning of nest-demonstration and intensive nest-building spermatids were rapidly metamorphosed, causing a peak in numbers of spermatozoa between nest-building and egg laying (Fig. 6.2). It may be recalled from Chapter 3 that there are grounds for implicating a part for androgen in the maturation of secondary spermatocytes into spermatozoa (p. 62) and the histological changes summarized in Fig. 6.2 are consistent with this view. There was a general increase in the number of Leydig cells coincident with the appearance of significant amounts of LH. Peak numbers of spermatozoa were noted during the period from onset of nest-building to the beginning of incubation and they were released at some stage during the incubation phase. Spermiation occurred during the period when high plasma levels of LH were recorded and it is conceivable that the endocrine mechanisms permitting sperm release show a parallel with those associated with ovulation. Very few spermatozoa remained in the tubules after the males had taken their share of incubation for 14 days and presumably they had migrated to the epididymides and vasa deferentia. No, or at most very few, spermatids appeared to be produced during the period of incubation and there was little germ cell activity until the eggs hatched; thereafter, a steady build-up of spermatid numbers was noted to the level recorded at the beginning of the cycle.

The temporal variations in spermatogenetic activity outlined above were reflected in the overall histological appearance of the testis tubules when viewed in cross-section. At pairing the tubules had a cluttered untidy appearance, because sperm which were free in the lumina were partly intermixed with other types of germ cell. Evidently, an earlier generation of germ cells had matured into spermatozoa and these had then become intermixed with subsequent generations of cells. In contrast, by egg laying the tubules presented a much more ordered appearance with the various germinal stages organized into well-defined cellular associations, displaying a clear radial co-ordination.

Changes can be seen in the appearance and number of interstitial Leydig cells and their nuclei during the reproductive cycle. The growth and maturation of the Leydig cells appears as a continuous process so that it is not possible to make definitive criteria for separating old and young cells. Nevertheless, we could readily distinguish

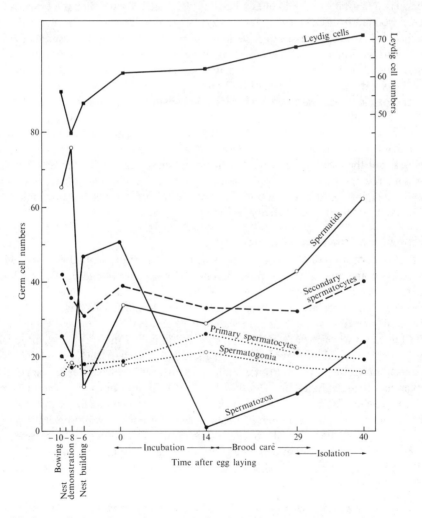

Fig. 6.2. Germ cell counts in sample segments of testis tubules at different stages of the Feral Pigeon breeding cycle. From Murton and Westwood (1975).

cells with pycnotic nuclei which had reached the end-phase of their secretory cycle. We identified mature cells, which were typically large with expanded nuclei containing diffuse chromatin, and we also distinguished small new cells with small, dark, elliptical nuclei containing dense chromatin, but intermediate categories could only arbitrarily be categorized. Mature Leydig cells were steadily depleted during the pre-incubation behaviour phase during which time the number with pycnotic nuclei increased. Although some new Leydig cells were probably formed at this time, large numbers of juvenile interstitial cells were not noted until the onset of incubation. This new generation of cells matured and probably became secretory throughout and after the period of brood care, during which time an increasing number of the nuclei became pycnotic and the total Leydig count increased. By the time the birds had been isolated from their young for 14 days large numbers of macrophages had invaded exhausted areas of interstitium. The testes of nearly all the pigeons responded positively to the Schultz test[†] for unsaturated sterols. However, the intensity of response was variable as was the number of reactive interstitial spaces per microscope field and subjects were scored on a scale of 0–3 on these criteria. At bowing five birds had scores of 1 and one was rated 0 whereas at nest-building the birds scored 3, 3, 2, 2, 1, 1, there being a significant difference in means using the Mann–Whitney U-test corrected for ties ($Z = -2.338; P = 0.018$, two-tailed). This increase in reaction to the Schultz test persisted until egg laying ($P = 0.05$ comparing bowing and egg-laying stages) but thereafter no significant differences were apparent. This means that there was a build-up of cholesterol positive material in the Leydig cells between pairing and egg laying consistent with both the recorded increase in LH and a possible reduction in biosynthetic activity, as noted in Fig. 6.1.

Hormones and parental behaviour

In vitro incubation of testes of pigeons killed at the start of incubation suggested an increase in the biosynthesis of progesterone at the expense of androgenic compounds (Fig. 6.1) and it is therefore of interest to recall from Chapter 5 that exogenous progesterone can stimulate incubation behaviour in male and female doves. Incubation has also been thought to require prolactin, the hormone discovered by Riddle and his associates when it was realized that the functioning of the crop sac gland required a then unidentified pituitary hormone (Riddle and Braucher 1931); subsequently isolated prolactin was found to resemble the substance involved in milk secretion in mammals (Riddle, Bates, and Dykshorn 1932, 1933). Injections of prolactin were shown to cause laying hens to stop laying and begin incubation, although they did not stimulate incubation in non-laying hens whose ovaries were presumably relatively inactive (Riddle, Bates, and Lahr 1935). The pituitaries of broody hens have an increased prolactin content compared with those of laying birds (Saeki and Tanabe 1955) but the critical question is whether prolactin secretion begins before the onset

---

[†]The Schultz (1952) test is based on the Lieberman–Burchard reaction. This gives a maximum response where the sterol has a double bond in position 7 and a lesser one when it is in position 5, but it does not distinguish between free and esterified sterols (Heftmann and Messettig 1960).

of incubation and so stimulates progesterone secretion. Riddle (1963b) has reaffirmed his belief that prolactin secretion is the stimulus for incubation but the experiments of Lehrman and his colleagues (see Lehrman (1963) for a review which answers Riddle's criticisms) seem to provide good evidence for the view that it is incubation behaviour which stimulates prolactin secretion.

Lehrman (1958a,b) was initially impressed by three important discoveries first made by Patel (1936). These were (1) that if incubating doves are removed from their nests during the first days of incubation the crop gland fails to grow; (2) if the doves are removed later when their crop glands have begun development, their glands regress; (3) if the male is isolated in a separate cage from which he can see his mate incubating his crop develops as if he himself were incubating. Lehrman and Brody (1961) demonstrated that prolactin given to Barbary Doves with previous breeding experience induced a much smaller proportion of birds to sit on eggs than did progesterone. Moreover, the dose levels necessary caused hypertrophy of the crop gland to an extent that never occurs at this phase of the normal cycle, and when smaller doses of prolactin were given the birds would not incubate, although they still induced increases in crop weight. In these early experiments hormones were injected into birds which had been held in isolation for several weeks. In some later experiments Lehrman and Brody (1964) allowed the birds to lay and incubate for 12 days before they were removed from their eggs. Under these conditions all subjects, both male and female, would sit on eggs presented in a test cage which was not their own, but this reaction vanished after a further 12 days of isolation. However, when these subjects were now given prolactin they at once sat on their eggs. Lehrman and Brody concluded that prolactin is important in maintaining incubation behaviour from that point in the normal cycle at.which it is first established (by progesterone), but that prolactin is relatively ineffective in inducing incubation in subjects which have not already reached the point of incubation.

This conclusion was confirmed in two further experiments. In the first it was shown that when either progesterone- or prolactin-injected doves were tested for incubation responses more would sit on eggs if tested in pairs than singly (Bruder and Lehrman 1967). In the second experiment, males and females were separated by a glass plate at various times during the pre-egg laying and incubation stages of the cycle: the males were thereafter allowed to watch the females continuously for a period ending 13 days after the laying of the second egg whereupon the males were tested for their willingness to incubate and their crop glands were examined (Friedman and Lehrman 1968). The results summarized in Fig. 6.3 show that under these conditions the males developed glandular crops, that is, secreted prolactin, in response to visual stimuli from the female only if they had previously associated freely with the female and indulged in courtship and the behaviour associated with egg laying. It may be guessed that such behaviour stimulates progesterone secretion (and perhaps LH secretion first) which primes the neuro-endocrine apparatus to respond to appropriate visual stimuli with prolactin secretion. Incidentally, Fig. 6.3 confirms that with natural levels of prolactin secretion the weight of the testes is not reduced. In one of his first experiments Lehrman (1955) showed that prolactin could also induce doves to feed squabs, provided the adults had some previous experience of rearing young. This capacity was lost if anaesthetic were locally applied to the engorged crop, but not elsewhere, suggesting that stimuli induced by the young on sensory preceptors in the crop wall could initiate feeding movements in inexperienced parents. The peripheral tension hypothesis has had to be modified since Klinghammer and Hess (1964) showed that doves will feed seeds and crop liquid to squabs that are presented to them early in the incubation period, that is, before the crop gland enlarges: this observation has been confirmed by Hansen (1966).

In pigeons the male's aggressive tendencies wane during courtship and there are indications that FSH and androgen titres have declined by the time incubation begins. How, in such circumstances, is the male able to compete with other males for the defence of his territory and nesting facilities? Conversely if FSH and androgen levels did not become reduced how could the male remain sufficiently subordinate to his female to allow her to participate in nest-construction? This apparent etho-ecological conflict partly explains a secondary function of progesterone and prolactin.

Vowles and Harwood (1966) examined the aggressive responses (these were aggressive pecking, taking the initiative in wing-slapping, and charging) and defensive

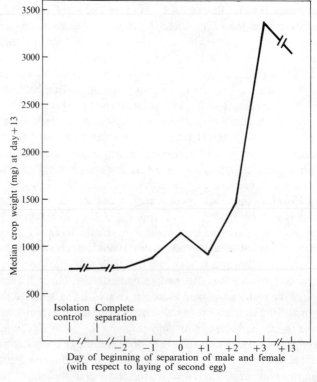

Fig. 6.3. Crop weights of male Barbary Doves at day 13 as a function of time of separation from female and nest. Visual observation of the mate and nest continued after physical separation. From Friedman and Lehrman (1968).

responses (feather-raising, wing-raising, and wing-slapping and pecking when used as a defensive movement) given by caged Barbary Doves to two test objects (see also Vowles and Prewitt 1971). These were a model black spider, or another dove which was held in the hand of the experimenter and moved up and down as if bowing. These test objects were presented at different stages in the breeding cycle with the result summarized in Fig. 6.4. Defensive responses towards the spider were given equally by both sexes and they reached a peak in incidence during incubation and brooding. Males displayed aggressively to other doves throughout the breeding cycle but there was a peak in such behaviour soon after laying and it had declined by the time of hatching.

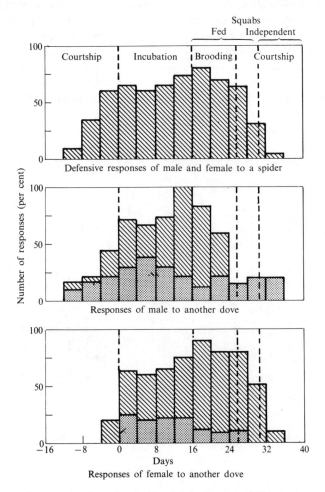

Fig. 6.4. Changes in aggressive (cross-hatched) and defensive (stippled) behaviour of male and female Barbary Doves during the breeding cycle. Day 0–day when first egg laid. Ordinate gives number of tests as a percentage of all tests in which particular behaviour was shown. From Vowles and Harwood (1966).

In a collateral series of experiments birds held in isolation were given exogenous hormones and tested with the same stimulus objects. In both sexes the defensive behaviour shown towards a model spider was stimulated by progesterone or by progesterone plus oestrogen and rather more by prolactin, but such behaviour was not enhanced by testosterone or oestrogen alone. Defensive responses towards another dove were also enhanced by progesterone or prolactin in females but in males defensive behaviour was not affected by these two hormones. However, progesterone, but not prolactin, did increase the male's aggressiveness towards other doves. As Fig. 6.4 shows, such aggressive behaviour by the male reached a high incidence at and soon after laying (*cf.* Figs. 6.1 and 6.4). It is not clear whether progesterone stimulated prolactin release or acted directly. Whereas behavioural responses to single injections of testosterone, oestrogen, oestrogen plus progesterone, or prolactin occurred after a delay of 30 min, suggesting a direct effect on the central nervous mechanisms mediating behaviour, progesterone did not result in behaviour increments until 12–18 h post-injection. This latency suggests that progesterone acted indirectly (see also p. 114).

In the experiments previously described in which exogenous hormones were injected into male Feral Pigeons (see p. 107) it was found that the incidence of driving, attacking, and pecking was maintained in progesterone-treated males compared with untreated controls (Fig. 6.5). In contrast, Fig. 6.5 also shows that these last two behaviours were markedly inhibited in the males which received prolactin.

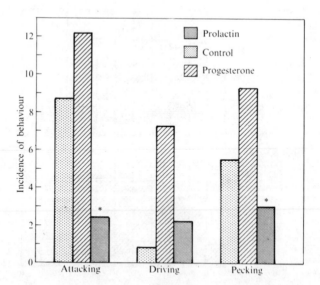

Fig. 6.5. Incidence of attacking, driving, and pecking behaviour by paired male Feral Pigeons given 10 hormone injections over a period of 14 days during which the male and female were allowed access for ½-h day. Records refer to the mean number of observation periods during which the displays were recorded and asterisks delimits results which were statistically different from the control situation at the 0·05 level. Stippled histograms, control; cross-hatched histograms, progesterone; solid histograms, prolactin. Based on data in Murton *et al.* (1969*b*).

Now, in these experiments the males were killed at the point when incubation would begin, that is, endogenous prolactin secretion would not have assumed the levels reckoned to be induced by sitting on eggs. Reference to Table 3.3 (p. 59) shows that under these circumstances the germ cell count was markedly reduced and some spermatozoa remained in the tubules. By contrast, Fig. 6.2 shows that in untreated birds at the end of incubation phase the germ cell count was noticeably reduced and very few spermatozoa remained in the tubules. This suggests that prolactin does not exert a very marked anti-gonad action if subjects are stimulated by courtship to produce other gonadotrophins and that the reduced spermatogenetic activity noted during incubation results more from a natural reduction in FSH and androgen levels. It is likely that earlier workers who noted that during incubation and prolactin secretion there was a regression of the testis attributed a causal correlation to the anti-gonad action of prolactin, when in fact the correlation was temporal. Thus Champy and Colle (1919) considered that crop gland development in the male was accompanied by a 90 per cent decrease in testis volume (see also Riddle and Bates (1933)).

Prolactin has for long been considered to inhibit FSH release (Bates, Riddle, and Lahr 1937; Schooley 1937; Payne 1942; Nalbandov 1945; Lofts and Marshall 1956; 1958). Reference again to Table 3.3 (p. 59) shows that prolactin-treated subjects differed from controls in having fewer secondary spermatocytes in contrast to oestrogen- and progesterone-treated birds. This could happen if an anti-FSH effect resulted in decreases in the synthesis of androgen-binding protein in the Sertoli cells, with a consequent failure of androgen to reach the developing germ cells (see p. 54). Prolactin did not effect feather regeneration in the Baya Weaver *Ploceus philippinus,* supposedly dependent on LH, whereas it did cause gonad regression (Thapliyal and Saxena 1964).

Hormones and other behaviour

In multi-brooded pigeon species we have recorded that appropriate environmental cues and behaviour from the mate induce a sequence of reciprocal hormone changes which serve to integrate the male and female reproductive cycles. The male initiates the process by his preliminary courtship and so is primarily responsible for timing the onset of the cycle and this is generally true in birds. Nevertheless, in pigeons and other species the females can accelerate the process. Thus, in seasonally breeding Starlings *Sturnus vulgaris,* males which were caged with a female were more advanced in spermatogenetic development than controls kept in all-male groups (Burger 1953). This observation has been assessed more critically by Schwab and Lott (1969). They kept Starlings, which initially had regressed testes, with or without females on artificial photoperiods of LD 12:12 or LD 14:10 until involution. Two groups comprised 20 males each and two comprised 15 males and 5 females. Neither the rate of testis development, nor the maximum size reached was significantly altered by the presence of females under either photoperiod. However, spermatogenesis was maintained for longer in those males which were caged with females under a photo-regime of LD 12:12 (see Fig. 9.5, p. 230) but not if the photoperiod comprised 14 h daily. It is possible, therefore, that females can stimulate additional gonadotrophin or steroid secretion in males only if they are functioning at subthreshold capacity.

It seems likely that the broad spectrum of endocrine change leading to egg laying, incubation, and brood care will be relatively stereotyped in birds since gametogenetic development must have evolved with precise control mechanisms, that is, the primary function of hormones is unlikely to vary between species. Nevertheless, avian species exhibit considerable flexibility in the uses for which these hormones can be used for secondary behavioural functions or the control of secondary-sexual nuptial plumage, and in many cases these secondary functions may be extremely specific. Moreover, endocrine-elicited behaviour may be adaptive in only a very limited context. Thus the ability of androgen to increase social status and aggressiveness in pigeons (Bennett 1940) and other species has long been recognized. Selinger and Bermant (1967) gave testosterone propionate to pairs of *male* Japanese Quail *Coturnix c. japonica* allowed contact for 5 min per day. They showed that testosterone increased the incidence of such aggressive displays as pecks and head-grabs given at conspecifics, increased the hierarchical status of the recipient, and also restored dominance relations to castrates; they were able to manipulate the level of agonistic behaviour by varying the dose of injected androgen. The effects of testosterone in the presence of a female, which might represent a more natural situation, could have been rather different (*cf.* Murton and Westwood 1975; and p. 111).

It may not be entirely accurate to use the word 'aggression' without defining more precisely the behavioural component which is affected. For example, Andrew and Rogers (1972) examined the effect of testosterone on food-searching behaviour of male domestic chicks presented with various choice situations which required different hunting strategies. They discovered that testosterone increased the persistence of the chicks enabling them to maintain 'search images' more effectively during a long series of encounters without interference from distracting stimuli. They suggested that the acquisition of a resource necessary for reproduction, for example a nest-site, mate or food for young, would perhaps be most economically achieved through an increase in persistence rather than by a ready elicitation of attack at first encounter. Under natural conditions androgen is secreted during a limited part of the year, indeed during restricted phases of the reproductive cycle, so that the biological significance of its secondary function of increasing agonistic behaviour must be related to reproductive performance. The detailed studies we consider below indicate that, in at least some species, androgen increases the status of the individual when competing for such reproductive requisites as a nest-site or building material. Undefined claims for an agonistic role of androgen need treating with caution for the reasons given by Crook and Butterfield (1968). As they point out,

hierarchies established in relation to competition for sexually relevant objectives are likely to be influenced by androgen whereas those in relation to quarrels without sexual significance will not be so affected. Many of the staged combats in laboratory situations upon which social status hierarchies are established occur in relation to space—the 'ownership' of a small cage for example or, operationally, at least the control of movement within it. Such cage space may well have territorial significance to a bird in reproductive condition and its success in controlling it thus determined by androgen levels.

Workers such as Elliott Howard (1920), who pioneered field studies of avian ethology, linked their descriptions of territorial behaviour with courtship, and, in consequence, as Davis (1963) has pointed out, the aggressive behaviour required for territory acquisition has usually been attributed to androgenic hormones. In some bird species the male and female defend separate territories in winter but share a common area for breeding, examples being the New World Mockingbird *Mimus polyglottos* (Laskey 1935) and European Robin *Erithacus rubecula* (Lack 1946) and such observations should give a clue that the hormonal control of aggression and courtship need not be the same. Davis (1957, 1959, 1963) noted that Starlings *Sturnus vulgaris* defend nest-sites in autumn, when their testes are small, and he showed experimentally that testosterone had no effect on the social rank of castrated subjects nor on their ability to defend a nest-box. It seemed that a gonadotrophin might be important, presumably LH rather than FSH since agonistic behaviour occurred with regressed testes, and Mathewson (1961) was subsequently able to demonstrate that LH would increase the social status and amount of agonistic behaviour of castrated Starlings. During Davis's studies there were good indications that castrated subjects dominated intact ones, as if a feed-back inhibition of gonadotrophin secretion had been removed.

In the Feral Pigeon, exogenous mammalian LH had relatively little effect on the pre-incubation behaviour cycle, other than to raise the probability that the male's bowing display would more often develop into an attack on the female than was the case in untreated controls; the absolute frequency of bowing, driving-attacking, and pecking did not differ from the control situation (Murton *et al.* 1969b). In the male African Weaver Finch *Quelea quelea*[†] the same mammalian LH produced substantial improvements in social status whereas testosterone did not elevate the social position of low-ranking birds in a six-bird hierarchy nor change their ability to win encounters in individual distance infringements (Crook and Butterfield 1968). Testosterone caused increments in male nest-building and increased the birds' involvement in encounters over nesting material, that is, testosterone increased aggressiveness in the specific context of collecting nest-material. It has long been appreciated that the assumption of nuptial plumage in *Quelea* depends on gonadotrophins, particularly LH, and the species was once used as a bioassay animal for LH (Witschi 1955). Butterfield and Crook (1968) kept all male groups of six of these weavers in small aviaries on seasonal daylengths at Bristol (latitude 51°N) and measured changes in plumage colour, the amount of nest-building, and the group frequency of agonistic encounters. These group frequencies of agonistic behaviour were correlated with the plumage score but not with the amount of nest-building. By the use of partial correlation techniques it was shown that if the effect of plumage (i.e. gonadotrophin output) was held constant any relationship between nest-building activity and aggression

[†]This species has been variously and confusingly called Dioch, Red-billed Dioch, Black-faced Dioch, and Weaver Finch; for this book we shall use the name *Quelea* or Red-billed Quelea.

was lost. The authors also correlated their data with an assumed knowledge of the seasonal changes in testis size based on Lofts' (1964) data as depicted in Fig. 10.2 (p. 252). However, the breeding cycle under LD 12:12 would not apply to their subjects which were kept on a seasonally changing photo-regime.

In more recent experiments Lazarus and Crook (1973) demonstrated that agonistic behaviour of females—they are normally less aggressive than males—could be increased by exogenous LH, or by ovariectomy in the breeding season, but not outside it, while oestrogen injections were shown to decrease encounter frequency. They argued that the normal low level of aggression of females and their subordination after breeding results from oestrogen inhibiting LH secretion and they quoted Eisner (1960) and Marshall (1961) for the view that oestrogen functions in this way. This interpretation conflicts with other evidence for the relationship of oestrogen to LH secretion (see p. 67) and it seems possible that any inhibition of gonadotrophin secretion implicated FSH. However, no experiments have been performed with FSH.

The contrasting importance of LH in facilitating agonistic behaviour in doves and *Quelea* can be related to their differing ecological needs. In pigeons territory acquisition occurs at the beginning of the breeding cycle, seemingly under the influence of FSH and androgen. The possession of a territory and an endocrine state which encourages sexual aggression enables the male to compete with others for a female. Intra-pair courtship can then proceed within the relative safety of a large territory which allows the male to become less aggressive towards the female, and this in turn enables her to participate in nest-building. Defence of the nest and eggs has to depend on behaviour mediated by progesterone and prolactin. In *Quelea* selection has operated to reduce the time needed to complete a breeding cycle. As Crook and Butterfield (1970) point out, it is advantageous for a male to assist a single female rearing young and not waste time competing for a territory during the restricted breeding season. Because the vegetation in the breeding areas is homogenous, and all nest-sites are equally satisfactory, selection has acted to reduce territory size to the immediate nest so these can be packed together in enormous colonies. It is important for the male to get a nest built and to attract a female as soon as conditions become favourable and this selective pressure has led the ecologically equivalent Sociable Weaver *Philetairus socius*, which inhabits the dry savannahs of Southern Africa, to build communal nests. Male *Quelea* build under the influence of androgen and then display in front of their nest-entrances. The density of nests necessitates the birds using precise routes to and from nests, but the frequency of aggressive contacts becomes greatly reduced once these routes are learned. The nest-site is crucially important for the maintenance of the short pair bond and it seems that mutual recognition of pair members is not possible without it. Unlike the pigeon, it is likely that the male *Quelea* must compete with other males for a female, after the phase of FSH and androgen secretion which lead up to nest-building. That is, *Quelea* must compete in a sexuo-aggressive context when the endocrine state which 'primitively' mediates such behaviour is probably changing in response to nest-building.

The fact that LH facilitates agonistic behaviour in the male results in them being dominant to females in the contra-nuptial feeding flocks. During the dry season the males compete with the females for declining food stocks and the sex ratio changes in favour of males, because females starve to a greater extent (Ward 1965*a*). Crook and Butterfield discussed the question of why LH-based agonistic behaviour has not evolved to a greater extent in the females to counteract their low social status, for this could conceivably increase their prospects of surviving to breed. They argued that if this happened it would endanger pair formation since it is essential for the female to show sustained submissive posturing in response to repeated aggressive approaches by the male. In consequence, the sexes experience differential success in agonistic competition for food which in turn influences the population dynamics and the psycho-endocrine mechanisms underlying reproduction. Females normally never build nests because of their subordinate position, but if given testosterone they do exhibit building behaviour, although they lack the skill to build a proper nest; perhaps this is because they do not have the opportunity during early age to develop the skills needed (see Crook and Butterfield 1970).

There is evidence that the release of some hormones is not continuous throughout the day but occurs at discrete and specific times, this appears to be so for androgen. It is possible that the diurnal variation in plasma androgen content results from a diurnal variation in thyroid secretion. It is relevant here to consider the possibility that some endocrine-based behaviour patterns will exhibit a temporal variation in their expression for this reason, especially if they are caused by the synergistic action of hormones. It has, of course, long been appreciated that birds exhibit diurnal rhythms of activity and some of these activities are known to be stimulated by hormone secretion. For example, bird song is often concentrated around dawn and dusk, this being particularly well shown by Fig. 6.6, which depicts the diurnal rhythm of 'bow-cooing' and the special vocalization given during 'nest-demonstration' in a wild population of Wood Pigeons. A similar diurnal pattern is noted in the frequency of singing by the Great Tit *Parus major* and nest-building by the Willow Warbler *Phylloscopus trochilus*; these two examples were quoted by Aschoff (1967) who pointed out that this two-peaked activity pattern is common. Unfortunately, there are as yet no studies which relate diurnal variations in endocrine state with measured aspects of dependent behaviour.

Circadian rhythms become re-phased as a result of seasonal changes in the daily ratio of light and dark (p. 290) and hormone secretion centres may be blocked or only partly stimulated or the diurnal pattern of endocrine secretion may be completely altered. Canaries which lay their first eggs late in the season take longer to develop their brood patches than do those which begin nesting early in the season, as Fig. 6.7 shows. As previously discussed, the development of vascularity probably requires only oestrogen, whereas defeathering needs a pituitary hormone, probably prolactin, as well. These seasonal variations in the speed of brood-patch development could be interpreted as reflecting a slower rate of gonadotrophin (FSH) secretion as the season advances, with a corresponding lag in oestrogen release. This could mean that the endocrine threshold for a particular behaviour to be expressed would take longer to achieve. A more precise explanation should probably implicate circadian rhythm mechanisms and the ease with which these can be phased to achieve appropriate synergisms. Fig. 6.7 emphasizes that the time interval between defeathering

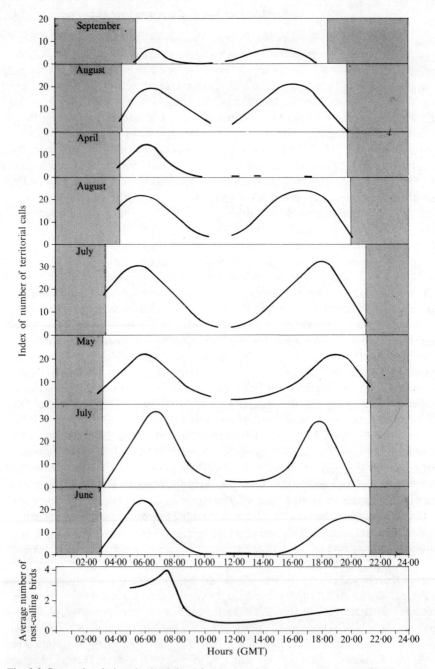

Fig. 6.6. Seasonal variations in the duirnal frequency of the bow-cooing territorial call of wild Wood Pigeons in relation to daylength and hence light intensity. Time of sunrise and sunset for the date of observation is indicated (top panels). Bottom panel refers to diurnal variations in the mean frequency of the nest-call for combined observations made between April and August. Based on Murton and Isaacson (1962).

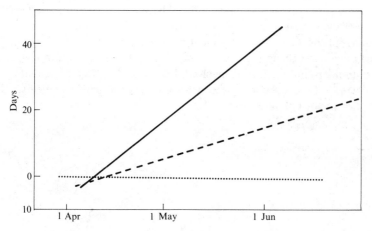

Fig. 6.7. Regression lines to show how the interval between completion of defeathering and either egg laying (solid line), achievement of advanced vascularity of the brood patch and egg laying (dashed line) or defeathering of the brood patch and assumption of vascularity (dotted line), vary with season in paired female Canaries. Based on Hinde (1962).

and the development of vascularity did not vary with season, presumably because as Hinde (1962) has suggested complete defeathering and the first occurrences of vascularity are closely correlated with and stimulated by nest-building and mark the achievement of a specific stage of development.

Male and female Feral Pigeons were taken from outside holding pens, where they had been exposed to natural photoperiods, at different times of year. Males were laparotomized and shown to have enlarged testes and were briefly paired with a female; if they did not perform the bowing display within 15 min they were discarded. All females had the largest follicle of around 5 mm diameter. Males were now caged with a strange female and the pair left to court under an artificial photoperiod of

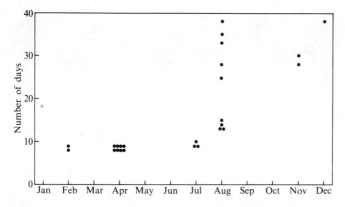

Fig. 6.8. Seasonal changes in the interval between initial pairing and egg laying in reproductively active Feral Pigeons. From Murton and Westwood (1975).

LD 17:7. As Fig. 6.8 shows, the time interval between initial pairing and egg-laying increased as the season advanced. Whether the male or female or both sexes were responsible for the delay is not at present known. Again we interpret the results in terms of progressive changes in the phasing of circadian rhythms, although at present we do not understand the factors involved. Fig. 10.11 (p. 278) shows that a progressive decline in testis size occurs in free-living birds and this perhaps reflects a changing pattern of gonadotrophin secretion.

7   Energy budgets affecting reproduction

Summary

Into the seasonal cycle birds must fit various body functions which demand energy which is surplus to the normal metabolic requirements of maintenance; these include reproduction, moult, and acquisition of diurnal or seasonal fat stores in preparation for migration or hard times. A series of peaks in energy demands therefore occur and these must never exceed the maximum available environmental energy resources. Since environmental energy supplies are variable—depending on geographical and climatic factors—the bird is faced with the need to partition conflicting demands for energy in as adaptive a manner as possible. In some regions food supplies become so plentiful at restricted seasons that the bird can afford to breed and moult simultaneously; for some species a similar end is achieved by changes in metabolic processes that reduce the need for marked peaks in energy demands. A wide range of strategies are shown by birds in adapting to a diversity of ecological conditions and species' needs. Some basic concepts in thermo-regulation and metabolic physiology are defined in this context for a range of species. The evolution of moult schedules in closely related warbler and thrush species in relation to their migratory behaviour is considered. So too are the difficulties faced by certain equatorial oceanic species living in habitats where resources change little from season to season and the population remains virtually saturated relative to resources.

The physiological processes underlying fat deposition, migratory restlessness (*Zugunruhe*) and moult are discussed and it is emphasized that migratory and non-migratory species do not differ in their actual hormone mechanisms but in the temporal phasing of the controlling hormone rhythms. A temporal association of particular functions has in the past led some authors to assume a casual relationship, for example, between gonad regression and onset of moult, but under appropriate experimental conditions in which the photo-regime is manipulated such events can be dissociated in time. The extra food needed for fat deposition is acquired as a result of changed rhythms in feeding activity.

Metabolic rates

Reproduction essentially involves the most effective budgeting of the surplus time and energy which remain after the basic daily metabolic requirements of growth and maintenance have been met. The rate at which growth and development can occur will partly determine when reproduction can begin and influence the fecundity and genotypic fitness of the species. Obviously, the reproductive process has evolved within limitations imposed by the total energy requirements of the species, for it is clear that an ostrich could not duplicate itself many times a day in the manner of a bacterial cell. Even so, the rate at which birds can divert energy towards reproductive output is impressive. According to Gilbert (1971a), a chicken daily deposits 1·8 per cent of her body weight in the egg and this is a much higher rate than the transfer of maternal body reserves into the developing foetus or placenta of the mouse (1·33 per cent), man (0·019 per cent), or cow (0·025 per cent). The need to migrate from inhospitable regions and to replace worn plumage are also competing factors in the annual energy budget. Selection will always favour the maximum effective reproductive rate and the ways in which birds can compromise their other energy demands in order to leave progeny fundamentally determines the strategies that are employed in adapting to a variable environment. As R. A. Fisher wrote in 1930: 'It would be instructive to know not only by what physiological mechanism a just apportionment is made between the nutriment devoted to the gonads and that devoted to the rest of the parental organism, but also what circumstances in the life history and environment would render profitable the diversion of a greater or lesser share of the available resources towards reproduction'.

The chemical breakdown (catabolism) of fats, carbohydrates, and protein provides heat which maintains the bird's temperature above that of the environment; most birds have a body temperature close to 41 °C when awake but this can fall at night by 4–6 °C and can rise with intense activity to more than 43 °C. The heat produced by the breakdown of body nutrients, or more conveniently the oxygen used in their oxidations, provides a measure of the metabolic rate. When at rest, at night, and starved to the point that no more nutrients are absorbed from the digestive tract, there is a range of environmental temperatures, called the zone of thermal neutrality, at which body temperature is regulated entirely by heat loss, and the metabolic rate remains constant. This is the *basal metabolic rate*. To a large extent heat regulation is achieved in consequence of the insulating properties of the plumage in some species seasonal moults result in lighter plumage in summer, while when they are cold birds fluff their feathers to provide more insulation. Birds cannot lose heat by sweating and instead rely on rapid panting to evaporate water from the respiratory passages and tongue. Some polar and marine species have insulating subcutaneous fat deposits. King and Farner (1961) discuss some of these adaptations and also mention that arterio-venous retia occur in the legs of many birds enabling the arterial blood of the upper tarsus to be transferred to the venous system without circulating to the lower tarsus and feet. This bypass operates in cold conditions, thereby preventing heat loss in the areas not protected by feathers and enabling the bird to withstand freezing conditions. With a decline in ambient temperature, body heat must be maintained by the breakdown of food reserves and if winter weather is detrimental to a bird it is usually because of difficulties in obtaining adequate food rather than low temperature *per se*.

Heat production increases as the ambient temperature decreases or increases outside the zone of thermal neutrality, as Fig. 7.1 shows. There is an upper threshold at which point further heat cannot be dissipated and the subject suffers heat stress and must soon die, but there is a wider range of temperatures below the zone of neutrality over which adjustment is possible. The resting metabolic rate which is achieved outside the zone of thermal neutrality is defined as the *standard metabolic rate*. It differs between night and day, as Fig. 7.1 demonstrates, for when the bird is not resting the zone of thermal neutrality vanishes. Kendeigh (1969) has pointed out that a more useful measure of metabolic rate is the *existence metabolism*, which is the energy balance of an active, feeding, caged bird which is maintaining a constant weight. This existence metabolism varies linearly with temperature, as Fig. 7.1 also shows. The maximum potential metabolism becomes expressed as the lowest temperature tolerance of the species, in the House Sparrow *Passer domesticus* at -35 °C, but it will be manifested at higher temperatures if the animal is performing excessive exercise. The difference between the maximum potential metabolism and the existence metabolism at any given temperature is the *productive energy*, this being the excess energy which is available for other functions. The productive energy will be modified by the food intake, since underfed birds will not achieve their maximum potential energy. Kendeigh shows how the productive energy increases with temperature until the zone of thermal neutrality is reached and how at the upper critical temperatures (see Fig. 7.1) the capacity for sustained work becomes zero, because work produces heat that cannot be dissipated. Maximum working capacity for the House Sparrow is attained at 22 °C, at the lower critical temperature. Kendeigh suggests that the decline in productive energy through the period of thermal neutrality is probably inversely proportional to the increasing rate of evaporative cooling.

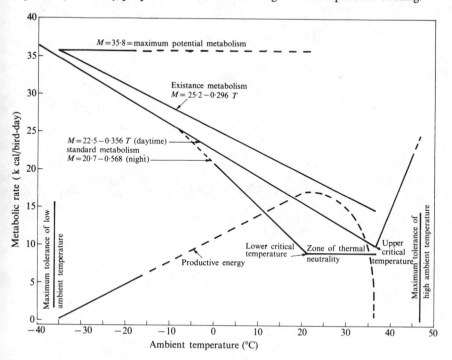

Fig. 7.1. Metabolic rate M (energy response) of House Sparrows *Passer domesticus* in relation to environmental temperature. From Kendeigh (1969*b*).

The monthly energy budget of the House Sparrow has been estimated and apportioned into various activities performed by the bird throughout the year in a helpful exercise by Kendeigh (1973), who should be consulted for the equations and criteria used in the calculations; Fig. 7.2 summarizes the information. The maximum potential metabolism varies seasonally, while the actual total metabolism can be seen from the figure to vary at a value which is approximately constantly lower. Thus, there is a fairly constant margin of safety to allow for conditions of extreme cold or food deprivation, except at the height of the moult when the birds function at almost the maximum metabolic rate possible. Energy for reproduction requires only 11 per cent of the total energy which is being used, including the basal and existence levels, but it represents 78 per cent of the productive energy which is available. Of course, the daily energy demands of moulting vary according to the speed at which feathers are replaced. In producing new feathers adult and juvenile Bullfinches *Pyrrhula pyrrhula* use a total equivalent in materials of 40 per cent and 30 per cent of their dry weights, respectively (Newton 1966). The moult takes 6–12 weeks and adults must find 40 mg of feather material per day, this being equivalent to 0·6 per cent of the dry body weight. In one season adults which moulted late were faced with

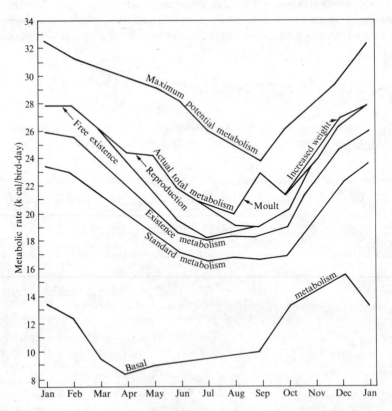

Fig. 7.2. Monthly energy budgets of the House Sparrow in central Illinois. From Kendeigh (1973

diminished food supplies and these birds survived less well than those which had moulted earlier. Newton quotes some other examples to indicate the physiological strain which can be involved in moulting. For instance, the metabolic rate increases by about 45 per cent during the moult of domestic fowls (Perek and Sulman 1945), by up to 25 per cent in the Chaffinch *Fringilla coelebs* (Koch and de Bont 1944), by 26 per cent in adult Ortolan Buntings *Emberiza hortulana*, compared with only 10 per cent in juveniles, which replace fewer feathers, and by 14 per cent in adult Yellow Buntings *E. citrinella* (Wallgren 1954). King and Farner (1961) have estimated that the House Sparrow needs 7·6 per cent more food per day for feather synthesis and that the increase in daily energy expenditure for moulting would be 1·6 kcal/bird.

Sturkie (1954) suggested that the increase in heat production (metabolism) during the moult might be compensated by heat loss resulting from the loss of plumage insulation. This was examined more critically by Blackmore (1969) in House Sparrows kept at 3, 22, and 32 °C on various photoperiodic regimes. The weight of the birds was significantly higher at 3 °C than at the other two temperatures and an inverse relationship was found between metabolized energy and temperature, the readings being 22·3, 14·4, and 11·4 kcal/bird/day at 3, 22, and 32 °C, respectively. The maximum increase in metabolized energy caused by the moult was 5·5 kcal/bird/day at 3 °C compared with 4·2 kcal/bird/day and 4·7 kcal/bird/day at 22 °C and 32 °C, respectively. The moult began later, was more intense, and was accomplished more quickly at low temperature than at high. However, the date at which peak moult and the end of moult were attained varied little with temperature and instead depended primarily on the photoperiod regime. This meant that, although the highest moulting intensities were attained at 3 °C, the total metabolic requirement was nearly the same at 32 °C because of the longer time needed to replace feathers. This is seen by summing the daily increase in metabolism per bird day during the moult to give total energy requirements calculated as 196, 165, and 193 kcal per bird at 3, 22, and 32 °C, respectively. The increased insulation provided by the growth of new feathers resulted in total savings of around 176, 96, and 70 kcal per bird at these three temperatures and almost compensated for the cost of feather growth. Thus, the actual daily cost of growing feathers worked out at about 1·5 kcal/bird/day at 22 °C and 32 °C and 2·2 kcal/bird/day at 3 °C (*cf.* results obtained by King and Farner (1961), above). Lustick (1970) has examined the effect of insulation in the Brown-headed Cowbird *Molothrus ater*, showing that in the range of thermal neutrality moulting birds increased· their oxygen consumption by 13 per cent over non-moulting controls. Below 20 °C the oxygen consumption of moulting birds was about 24 per cent above that of controls. Feather replacement accounted for about half the increase in metabolic rate and loss of insulation for the rest. Hence at low temperatures the loss of thermal regulation must modify the rate at which feathers can be discarded.

Energy thresholds

It is clear from Fig. 7.2 that House Sparrows could not find sufficient energy reserves to breed and moult simultaneously and so these events must be kept separate in the annual calendar, as was first suggested by Kendeigh (1949). The temporal ordering of conflicting requirements for energy is influenced by several environmental factors and in the next pages we shall consider some examples of the various strategies which have been adopted by birds to partition their energy resources for breeding, moulting, and migration. It will be convenient to return to a consideration of the physiological control of these functions at the end of the chapter. Discussion is aided by the schematic presentation in Fig. 7.3, to which we shall make many references. It is based on an original idea by Perrins (1970) to explain the timing of breeding seasons, the idea later being extended to include moult by Murton, Westwood, and Isaacson (1974c). For simplicity it is assumed that the existence

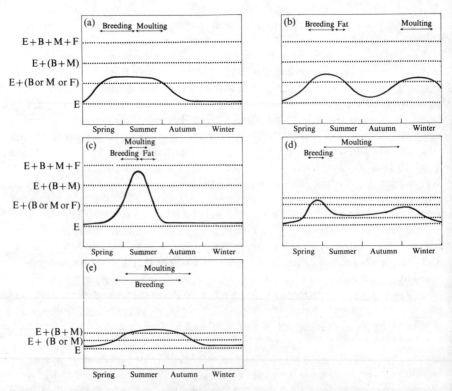

Fig. 7.3. Schematic representation of the relationship between the energy threshold required by a bird for maintenance only, indicated by the horizontal dotted base line *E*. We assume that breeding, moulting, or fat deposition require about equal amounts of extra energy and tend to be exclusive. The level of resource availability at which only breeding, moulting, or fat deposition can occur is indicated by the second dotted line, while additional thresholds which have to be reached for two or more functions to occur simultaneously are also depicted. The breeding, moult, and fattening schedule adopted by the bird will depend on seasonal changes in resource availability (solid curves) and appropriate patterns are seen in the top part of each panel. Changes in energy demands may result from a higher physiological efficiency and result in such schemes as those depicted at (d) and (e). See text for discussion.

metabolic rate remains constant, although, as Fig. 7.2 emphasizes, seasonal changes do occur as a result of ambient temperature variations and also variations in the physiological requirements of the species. The energy resources available to the bird—this incorporating the actual food supply as well as the bird's capacity to collect it—is shown in Fig. 7.3 to vary seasonally in different ways to simulate situations occurring in nature. The additional energy demands required for breeding, moult, fat deposition, and migration are depicted as thresholds in resource availability that must be met before the bird can perform the various functions, either one at a time or to some extent simultaneously. We can also think in terms of the bird achieving an appropriate body state. For the present it is assumed that the bulk of the plumage is replaced during a relatively short period as in the House Sparrow.

Energy partition

Fig. 7.3(a) represents a situation commonly experienced by the majority of the resident or partial migrant species which inhabit the temperate zones of the northern hemisphere. A period of relative food abundance occurs in spring or summer, enabling breeding to occur and to be followed by, or even partly overlapped by, the moult. In general, animal foods, particularly soil invertebrates or defoliating caterpillars, achieve peak abundance for only one or two months and many bird species depending on these foods are single- or at most double-brooded: such a cycle is illustrated in Fig. 7.4(a), referring to the breeding and moult schedule of the Blue Tit *Parus caeruleus* in north Europe and Britain. On the other hand, weed seeds are usually available over a longer period than animal foods and many temperate zone granivorous species can potentially produce three, perhaps even four, broods. Even so, the season of breeding is usually separated from that of the moult, the Bullfinch providing a fairly typical and well-studied example (Fig. 7.4(b)). It stops breeding in late July or early August whereupon the moult begins and lasts about 85 days depending on the individual (Newton 1966). The moult occurs during a period when food stocks are often abundant, although in years of particularly good supplies breeding continues later and the onset of moulting is delayed. This gives a clue that physiological factors associated with breeding can influence moult, but the situation is complicated as will emerge. Many British finches have a similar pattern of breeding and moult to the Bullfinch (Newton 1968). The Greenfinch *Carduelis chloris* takes around 85 days for feather replacement, the Goldfinch *Carduelis carduelis* about 80 days and the Linnet *A. cannabina*, Twite *A. flavirostris*, Siskin *C. spinus*, and Chaffinch about 70 days; this last has a breeding season more resembling that of the Great Tit *Parus major* (Fig. 7.4(a)) since it feeds its young on defoliating caterpillars and not seeds, and it moults in June and July rather than July and August. The Redpoll *A. flammea* begins moulting a little later than most of the finches detailed above; it does not usually begin until early August and it also takes rather less time (50 days) to complete feather replacement (Evans 1966).

Species living at lower latitudes which have the relationship with the resource level schematically shown in Fig. 7.3(a) also breed and moult in the manner just

154

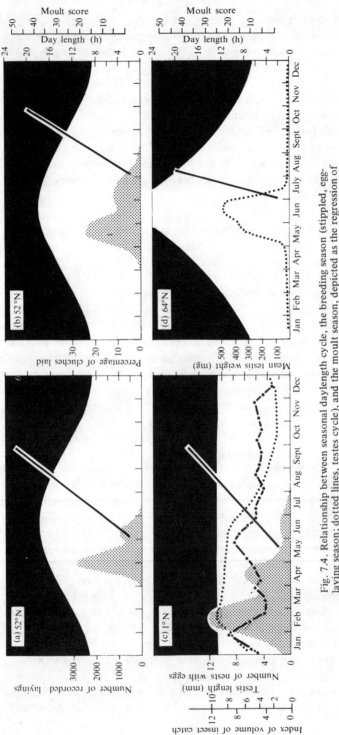

Fig. 7.4. Relationship between seasonal daylength cycle, the breeding season (stippled, egg-laying season; dotted lines, testes cycle), and the moult season, depicted as the regression of average moult score on date (see Fig. 7.6), for various bird species. (a) Blue Tit *Parus caeruleus* in Britain, based on Lack (1954) and Flegg and Cox (1969); (b) Bullfinch *Pyrrhula pyrrhula* in Britain, based on Newton (1966) and Murton (1971*b*); (c) Yellow-vented Bulbul *Pycnonotus goiavier*, with seasonal variations in food supply shown as chain line, based on Ward (1969*a*); (d) White-crowned Sparrow *Zonotrichia l. gambelii* in Alaska, based on King et al. (1966); Morton et al. (1969).

described for European finches, although the duration of the breeding and moulting seasons may be extended. For example, North American House Finch *Carpodacus mexicanus* at latitude 34 °N required 15 weeks to moult (Michener and Michener 1940) and the Boat-tailed Grackle *Cassidix mexicanus* (F. Icteridea) at latitude 31 °N takes 13–15 weeks (Selander 1958). Ward (1969*a*) whose data are summarized in Fig. 7.4(c), reported that near the equator on Singapore Island (1 °20′N) body and wing moult in the Yellow-vented Bulbul *Pycnonotus goiavier* took 119 days. A similar temporal pattern and duration of moult applies to the Black-eyed Bulbul *Pycnotus barbatus* in Africa (Moreau, Wilk and Rowan 1947) and various forest bulbuls in Sarawak (Fogden 1968).

Variations in moult pattern to suit different environmental conditions may occur between very closely related species, suggesting that the necessary physiological adjustments have been achieved quite easily during evolution. The British race of the Golden Plover *Pluvialis a. apricaria* moults immediately after breeding and moves from the uplands of England and Scotland to winter in lowland Britain. Iceland birds (*P. a. altifrons*), which winter in the British Isles and France, begin the wing moult while still incubating and finish in September before moving. But Lesser Golden Plovers *P. dominica*, which breed on the east Siberian and New World tundras, moult only the three pairs of innermost primaries before the moulting is arrested, to be completed on the wintering grounds in southern South America; the pattern is shown in Fig. 7.3(b) (see below).

Evolutionary trends are also well illustrated by the British thrushes (Turdinae), made the subject of a comparative study of Snow (1969), whose data are summarized in Fig. 7.5. Generally in these species autumn migration begins with the ending of the moult, although the Robin *Erithacus rubecula* finishes its moult somewhat before the time of migration. There is a trend in these thrushes, as well as in other taxa, for long-distance migrants to have a more rapid moult that resident species or partial migrants, presumably to ensure their readiness to make a long journey (see Fig. 7.3(b)). The Redstart *Phoenicurus phoenicurus* has a relatively short stay in Britain and it has the most rapid moult of the species studied in detail. But the Nightingale *Luscinia megarhynchos* exhibits an even more extreme adjustment, for it is able to rear a single brood only by accelerating the moult to the extent of becoming virtually flightless.[†] The allopatric Thrush Nightingale or Sprosser *Luscinia luscinia*, summering at 54 °N in Scandinavia, requires only 30–35 days to renew its plumage, and grows its primaries concurrently (Berger 1967). Since flight capability is so hindered it might be guessed that the risk of predation is increased and it may be significant that the Nightingales skulk in dense undergrowth and obtain much food on or near the ground. It is suggested below that the breeding and moulting schedule adopted by these thrushes represents a primitive condition, and that the Nightingale

---

[†] Loss of flight capacity during moult is well known in geese and ducks, especially those inhabiting high latitudes. Many inhabit desolate areas and, like the Pink-footed Goose *Anser brachyrhynchus*, can find safety from ground predators on islands or water. Some species perform a special moult migration to a safe refuge in order to moult; British Shelducks *Tadorna tadorna*, for example, move to the Heligoland Bight or Bridgwater Bay, Somerset, in autumn.

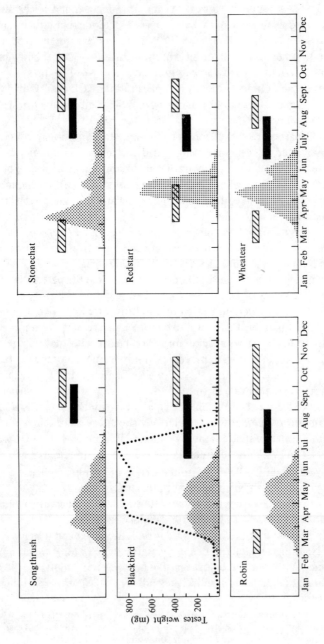

Fig. 7.5. Relationship between the breeding season (stippled, egg-laying season; dotted line, testes cycle), moulting season (solid bar), and migration season (hatched bar) of various Turdidae. From Snow (1969) with testis cycle from Murton and Westwood (unpublished).

has reached an evolutionary limit in making use of the summer habitat for both breeding and moulting. A more advanced trend would be to increase breeding capacity by deferring moult until the wintering ground is reached—a strategy illustrated by some of the warblers (see Fig. 7.3(b) and 7.6).

The type of energy cycle depicted in Fig. 7.3(c) becomes pronounced at high latitudes, where the brief summer stimulates an enormous but short-lived increase in the flora and fauna—as those who have dodged midges and mosquitoes can testify. With such conditions breeding and moulting may occur concurrently in large species like the Raven *Corvus corax* and Glaucous Gull *Larus hyperboreus* (Ingolfsson 1970), which need long incubation and brood periods (Johnston 1961). In the smaller passerines the more usual situation is for all resources to be channelled into intense breeding activity, when this ceases moulting can begin. There is survival value in breeding as early as possible, for once the young gain independence they can feed, moult, and acquire migratory fat reserves at the same time as the adults. Nevertheless, the moulting period must be short if migrants to high latitudes are to make maximum use of the period of good food supplies between arriving and departing. White-crowned Sparrows *Zonotrichia leucophrys gambelii* require only 37 days to replace their primaries and 48 days for the complete moult (Fig. 7.4(d)). In this species all nine primaries, all the retrices and about six secondaries grow concurrently (Morton, King and Farner 1969). The Lapland Bunting *Calcarius lapponicus* in Greenland manages a longer breeding season than *Zonotrichia* and, like the Thrush Nightingale in Scandinavia, must lose the power of flight when moulting (Salomonsen 1950). Its energy budget in Alaska is discussed by Custer and Pitelka (1972). Migrant species that are only marginally able to breed as well as moult in the northern summer may undergo a partial moult of body plumage, and occasionally some wing-coverts, secondaries, and tail feathers, before migrating. They then arrest moult to complete it and the replacement of primaries on the wintering ground.

The variations in moulting and breeding schedules so far discussed can be appreciated more readily if viewed in the context of the evolutionary history of a selected group of closely related species. In particular, it is necessary to realize that the centre of evolution of most of the Palaearctic passerine groups, as well as many non-passerines (Cracraft 1973) has been Eurasia, and not Africa, where many species winter. In discussing the affinities of the European and African bird faunas Moreau (1966) emphasizes the paucity of genera which are represented by breeding species in both regions; of 169 passerines on the European list only 9 breed in Africa, so that the adaptive radiation of these must be viewed in the context of a secondary adaptation to the Ethiopian region. Warm, broad-leaved, evergreen forest was continuous over North America and Eurasia in the early Tertiary and extended into North Africa. In this temperate—tropical Old World forest the muscicapid—timaliid—turdid—sylviid complex of passerine families probably evolved, and subsequently radiated following the drastic climatic changes which heralded the end of the era and which lasted throughout the Pliocene (1—11 million years ago); the climatic changes resulted in the eradication of many of the Tertiary avifaunas. The phylo-genetically older taxa contain many species which have been able to adapt to the more temperate deciduous forest conditions, maintaining a high degree of residency, and *Turdus* provides examples. The older taxa also presumably had 'first choice' in occupying the newly emerged habitats. *Sylvia*, which shows many parallels with *Turdus*, contains a larger number of resident species adapted to the more xerophytic habitats of southern Europe. Species specializing in collecting seasonally available invertebrates, particularly those from deciduous foliage, must have been forced to move further south in winter to find suitable feeding grounds. In some cases this has meant that closely related species, which share an approximately similar habitat in summer, winter in geographically distinct areas. Thus most Whitethroats *Sylvia communis* move to the thorny bush steppes in the semi-arid Sahel Zone, south-west of the African Sahara at about $12-18\ °N$, whereas Lesser Whitethroats *S. curruca* occur in a similar habitat south-east of the Sahara and they also winter in India and Ceylon. A clear difference in migration route is apparent in the recoveries of ringed birds marked in Britain, for Whitethroats pass through the maritime provinces of France, Spain, and Portugal, while British Lesser Whitethroats are recovered in south-east France, Italy, and Israel.

Most *Sylvia* warblers undergo a post nuptial moult, including replacement of the wing primaries, on the breeding ground, this being the primitive condition noted in the turdids above (Fig. 7.3(b)). The Garden Warbler *S. borin* defers[†] the post-nuptial moult until it arrives on the wintering ground in Africa and shows an advanced con-dition noted in several species of the phylogenetically younger genus *Phylloscopus* (leaf warblers) and also all species of *Locustella* (grasshopper warblers), *Acrocephalus* (reed and marsh warblers), and *Hippolais* (see Appendix 3 for details, Figs 7.3(b), and Fig. 7.6).

The *Phylloscopus* warblers are worth considering in a little more ecological detail since the photo-responses of some have been examined experimentally (see Gwinner 1969, and p. 352). Lack (1971) has provided a convenient summary of the summer and winter habitat preferences of the European species and this has been used in preparing Table 7.1. It seems reasonable to suppose that the Chiffchaff *P. collybita* is phylogenetically older than the other species in Table 7.1 and that it has managed to adapt to the desiccation of the original evergreen woodland that once covered Europe. It has a larger range than the other species and occupies a wider range of habitats, but it essentially feeds from high foliage. Where it overlaps with Bonelli's Warbler *P. bonelli* Lack suggests that ecological competition is avoided by the Chiffchaff feeding from high foliage and by Bonelli's Warbler feeding from lower vegetation. This is how the Chiffchaff and Willow Warbler *P. trochilus* appear to be separated in central Europe. Presumably the Chiffchaff had 'first choice' on wintering facilities and it is the least migratory of the European leaf warblers. The southern

---

[†]This is evidently variable, for whereas British Garden Warblers achieve a full or partial moult in Britain or during their passage through Europe, Continental birds trapped during autumn migration on the east coast of England show no evidence of a post-nuptial moult (Gladwin 1969*a*).

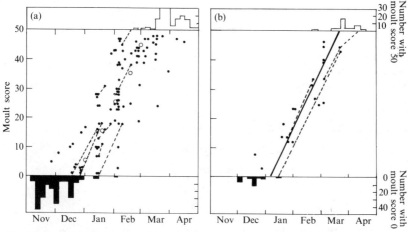

Fig. 7.6. Progression of primary moult of Garden Warbler (a) and Willow Warbler (b) at Kampala, Uganda. Individual moult scores are calculated by allocating 0 points for an old unmoulted primary, 1 point for a primary missing or in pin, 2 points for a quarter-grown primary, 3 points for a half-grown, 4 points for a three quarters-grown, and 5 points for completely new feathers, respectively. These warblers have 10 primaries on each wing so that a fully freshly moulted individual scores 50 points (moult is symmetrical). Dashed lines join records for the same bird in the same season. The solid line in (b) is the regression line for moult score on date (see also Fig. 7.4). Histograms denote number of birds having moult scores of 0 (solid) and 50 (open), that is, not begun or finished moult. Open circles are mean dates for groups of 10 moult scores. After Pearson (1973).

TABLE 7.1

Summer habitat preferences of the European Phylloscopus *warblers*

Species	Mid latitude of breeding range (°N)	Habitat preferences in		
		Northern Taiga	West-central Europe	South Europe
Arctic Warbler *P. borealis*	57	Tundra birch scrub	—	—
Greenish Warbler *P. trochiloides*	54	Spruce forest	—	—
Wood Warbler *P. sibilatrix*	52	Shady broad-leaved woodland with few shrubs		—
Willow Warbler *P. trochilus*	60	—	Broad leaved woodland with many shrubs	—
Chiffchaff *P. collybita*	54	Conifer; mostly pine	Broad leaved woodland with many shrubs	Evergreen and deciduous woodland plus bushes
Bonelli's Warbler *P. bonelli*	40	—	Low beech and oak woods with bushes	Evergreen and deciduous woodland plus bushes

Based on Lack (1971)

populations of the Mediterranean basin are mostly resident while those from further north make a comparativley short journey to winter in the Maghreb (Fig. 7.7). However, the Chiffchaff has had to modify its habitat preferences in winter and it frequents arid bush and savanna. [There has been an increasing tendency for Chiffchaffs (and Blackcaps *S. atricapilla*) to winter in the milder parts of Britain and maritime Europe in consequence of climatic amelioration (Gladwin 1969)] The advantage of not making a long migratory flight is that the Chiffchaff can

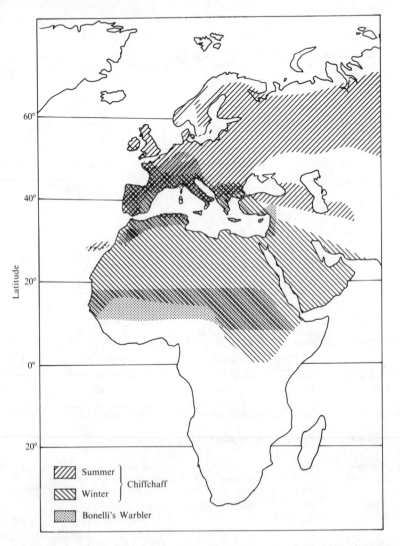

Fig. 7.7. Summer and winter distribution of the Chiffchaff *Phylloscopus collybita* (with overlapping ranges) and Bonelli's Warbler *P. bonelli* (with discrete summer and winter range). Based on Voous (1960) and Moreau (1972).

manage a longer breeding season that the other *Phylloscopus* species; it is regularly double-brooded in Britain, while the Wood Warbler *P. sibilatrix* and Willow Warbler are mostly single-brooded, although repeat laying and occasional second broods do occur. The long breeding season of the Chiffchaff may be facilitated by the fact that the species has not had to adapt to an equatorial photo-regime during the contranuptial season, as will be discussed in Chapter 13. Here more details are given of the moult (Fig. 13.6, p. 353) which is a single post-nuptial change of plumage which continues during the leisurely migration south. Fig. 7.7 shows that Bonelli's Warbler winters further south than the Chiffchaff, frequenting Acacia scrub south of the Sahara.

Desiccation of original evergreen woodland doubtless created more shrub vegetation which came to provide a feeding niche for the Willow Warbler. This species and the Wood Warbler appear to provide good examples of how the 'newer' forms have had to move further and further south in winter to avoid ecological competition with other already established congenors. Comparison of Fig. 7.7 and 7.8 demonstrates that both the Willow Warbler and Wood Warbler are geographically isolated in winter from the Chiffchaff; the Wood Warbler frequents mostly evergreen forest in winter whereas the Willow Warbler eschews such vegetation but accepts all other kinds of wooded country. These two warblers belie any contention that a quick post-nuptial moult is essential for a long migratory flight and, in fact, Willow Warblers from east Siberia must fly nearly 7500 miles to reach their wintering areas in South Africa, whereupon they moult. The British race of the Willow Warbler is unique in also having a post-nuptial moult, which is absent in the Northern race and those other *Phylloscopus* species which migrate long distances; this anomaly is discussed in Chapter 13 (pp. 354).

With the recession of the ice-sheets new habitats became available in Northern Europe and have been occupied by the Arctic Warbler *P. borealis* and Greenish Warbler *P. trochiloides* and by distinct races of the Chiffchaff *P. collybita* and Willow Warbler. In winter these northern populations had to move south and to avoid competition with congenors emanating from breeding areas further south probably had to 'leap-frog' large tracts of occupied territory in Africa and Asia north of the Equator. Hence, the apparent paradox that the higher the breeding latitude the longer the migration necessary to reach wintering grounds in the south, a trend well shown by the warblers in Fig. 7.9 (and even more by the Arctic Tern *Sterna paradisaea*, which migrates from breeding grounds within the Arctic circle to the Antarctic). Arctic Warblers winter in Indonesia and Indo-China and Greenish Warblers in India, Burma, and Indo-China and so are geographically separated from the other European leaf warblers in the contra-nuptial season.

Long-distance trans-equatorial migrants might in some circumstances experience two peaks in food availability, corresponding to the northern and southern summers. Moreau (1966) lists some birds which sporadically breed in South Africa as well as in the Palaearctic: White Storks *C. ciconia* nested in Cape Province for 7 successive years around 1940 and sporadically since; there are various scattered records of colonies of the House Martin *Delichon urbica* breeding in South Africa, this species having a primary moult in its winter quarters. Also some European Bee-eaters. *Merops apiaster* appear to breed in both hemispheres; migrants arrive in the neighbourhood of Capetown and the Orange River area in September—October and immediately begin nesting, and this continues until December—January. The birds depart again in February so that the same individuals could breed again in Europe during May—July. A double breeding season of this kind is likely to depend on unstable or transitory ecological conditions, for once a species can manage to breed and moult in its contra-nuptial quarters it would not require much ecological change

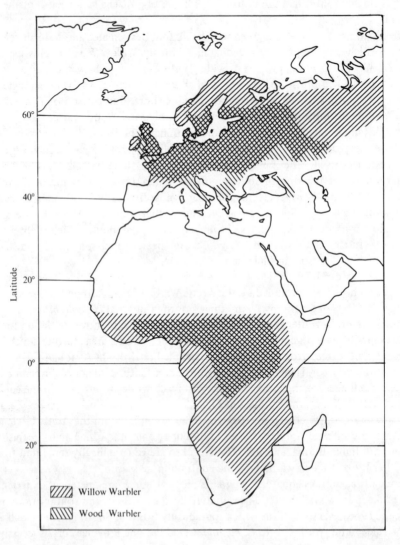

Fig. 7.8. Summer (in Europe) and winter (in Africa) distribution of the Willow Warbler *Phylloscopus trochilus* and Wood Warbler *P. sibilatrix*. Based on Voous (1960) and Moreau (1972)

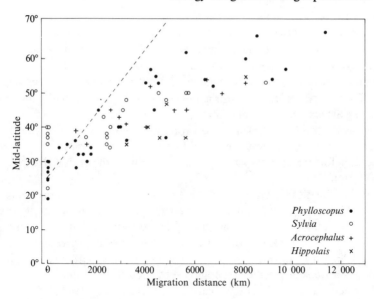

Fig. 7.9. Relationship between mid-latitude of the breeding range and distance travelled during migration to contra-nuptial quarters for various warbler genera. If *Phylloscopus* warblers simply returned to the centre of the range of the southern most species, following a north–south route the dashed line indicates the distance that they would travel. In fact, the more northern breeding species 'leap-frog' over the species living further south to wintering quarters in the southern hemisphere. Details of species contributing to graph are given in Appendix 3.

to allow a distinct resident sub-species to emerge. This may explain why a few species have resident populations at equivalent latitudes on either side of the equator, examples being Gannet *Sula bassana*,[†] Bittern *Botaurus stellaris*, Black Stork *Ciconia nigra*, Pintail *Anas acuta*, Griffon Vulture *Gyps fulvus*, Great Skua *Stercorarius skua* (see Fig. 7.10). Within Africa, for example, in Nigeria, it is becoming evident that many birds make migrations to breed in the wet or dry seasons associated with the movements of the inter-tropical weather front (Elgood, Fry and Dowsett 1973). The precise migration patterns are fixed by ecological needs and reproduction occurs when the species is stationary longest. In some, such as the Grey-headed Kingfisher *Halycon leucocephala*, breeding may occur during the period of migration resulting in a two-stage migration pattern.

The daily energy required for moulting can be reduced if feather replacement is spread over a long period; this is diagramatically illustrated in Fig. 7.3(d) for the temperate-zone Raven. This species begins the wing moult when the young hatch in March and it continues until September (Holyoak 1974). This adaptation is seen in a more extreme form in some of the temperate Columbidae; Fig. 7.3(e) gives a schematic representation. Not only is the moult extended over nine or more months in several species (Fig. 7.11), but the birds produce a small egg relative to

[†] Jarvis (1972) has made a case on behavioural and morphological criteria for treating the Cape Ganet as a distinct species *S. capensis.*

body size and compensate the chicks at hatching by feeding highly nutritive pigeon's milk (Murton *et al.* 1974c; see also p. 134). Milk production has, however, made pigeons more dependent on body-water reserves (Brisbin 1969). Since the need for energy peaks is thereby reduced breeding and moulting can occur simultaneously allowing the breeding season to be lengthened. Unlike the other European pigeons, the Turtle Dove *Streptopelia turtur* is a long-distance migrant wintering in Africa south of the Sahara. Pairs stop breeding in late summer at a time when feeding conditions are still good and their success at raising chicks achieves a seasonal maximum. In this way the birds have time to acquire lipid reserves and undergo a wing moult before departing; then moult is arrested until the birds arrive in Africa (Murton 1968).

Under the climatically stable conditions existing in some tropical oceans or equatorial rainforest, virtually predictable variations in resource availability occur. Population size may under these conditions increase to saturation so that moulting and breeding must be achieved without benefit of seasonal improvements in resources (Ashmole 1963). Two distinct strategies to cope with this situation have evolved. In some species periods of breeding are followed by moulting and the cycle repeats at intervals which may be more or less than 12 months (see Chapter 13). In the Wideawake or Sooty Tern *Sterna fuscata* of Ascension Island breeding and moulting alternate at 9½-month intervals (Fig. 7.12). Cycles of this kind seem to have arisen when synchronized colonial breeding helps to reduce egg and chick predation. If eggs were produced in small regular numbers throughout the year a predator could adapt to the food source but such predation is 'swamped' if all individuals lay their eggs at the same time. Synchronized breeding may be feasible only if the parent bird can locate food supplies within a short distance of the colony. If, on the other hand, food is not locally available individual birds may have to traverse and search large areas of ocean before returning to their colony. Under these conditions energy

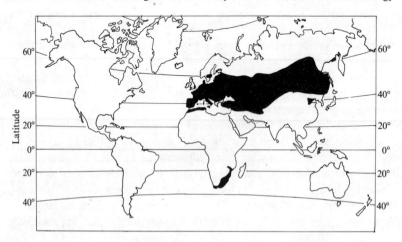

Fig. 7.10. Breeding distribution of the Bittern *Botaurus stellaris* showing separate ranges north and south of the Equator. From Voous (1960).

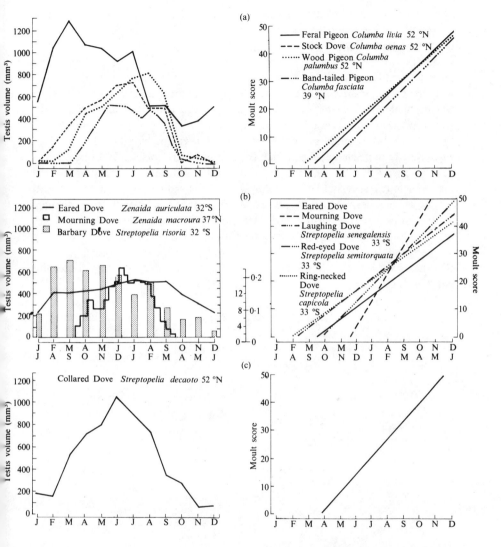

Fig. 7.11. Breeding seasons in left-hand panels (lines, testes cycles; open histogram and stippled histogram, egg-laying seasons) and moult seasons in right-hand panels (regression of moult score on date as in Fig. 7.6) of various pigeon species. The curves have been lined up to make the longest days comparable for Southern Hemisphere and Northern Hemisphere species

(a) Feral Pigeon, Stock Dove, and Wood Pigeon at latitude 52 °N (based on Lofts *et al.* 1966; Murton *et al.* 1974*b*) and Band-tailed Pigeon in U.S.A. at latitude 39 °N (breeding season based on Clait Braun *et al. in litt.* and moult on J. F. Ward *in litt.*).

(b) Eared Dove in Argentina (moult and breeding from Murton *et al.* (1974*a*)); Mourning Dove in U.S.A. (breeding from Cowan (1952); moult from Sadler *et al.* (1970); Barbary Dove introduced in Perth, Australia (from Davies (1974*b*) and moult regressions for three other *Streptopelia* spp. in Africa (from Siegfried (1971)).

(c) Collared Dove in Britain (breeding from Murton (1975), moult from Murton and Westwood (unpublished).

requirements are kept to the minimum by an extension of the moulting period, a reduction of egg size relative to adult body weight, a reduction of clutch and hence brood size, and a slowing down in the growth and development rate of the young so that the parents need provide only small quantities of food at a time. For example, the European Gannet *Sula bassana* requires 90 days to feed and fledge its chick whereas the tropical Red-footed Booby *S. sula* and White Booby *S. dactylatra*, which breed in the Galapagos Islands, take 130+ and 115 days respectively, and the Brown Booby *S. leucogaster* of Ascension Island needs 120 days (Nelson 1966). In the White Booby, although each feather is moulted annually the whole cycle, whereby all the primaries of a particular generation are replaced, takes up to three years (Dorward 1962). The shedding of the primaries begins at the innermost with the outer four feathers of larger size probably growing more slowly than the inner six. New cycles begin before the full set of primaries has been replaced so that a bird might be simultaneously growing primaries 1, 5, and 8–10. Moreover, moult ceases during some of the time when the birds are breeding, particularly when chicks are being fed. In the Great Frigate Bird *Fregata minor* and some of the large albatrosses and also the King Penguin *Aptenodytes patagonica* successful breeding plus moult requires more than a year (Nelson 1967).

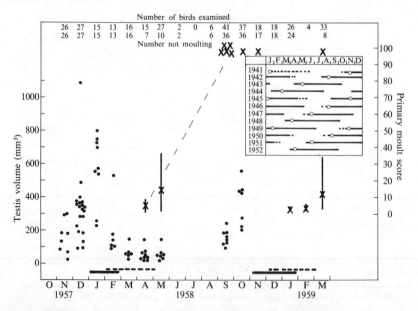

Fig. 7.12. Breeding and moult schedule of the Wideawake Tern *Sterna fuscata* on Ascension Island. Solid dots give testis volume of individual birds, heavy bars period of egg laying, and the heavy dashed bars period when chicks were in the nest. Crosses give the primary moult scores of individual birds and the fine dashed line is the estimated moult regression. Based on data in Ashmole (1963*b*).

The inset indicates the date of laying of first eggs (open circle) and length of birds stay on the island (bar) and emphasizes the 9-month periodicity of the reproductive cycle, drifting in phase relative to the calendar date. From Chapin (1954).

Migration and fat stores

Long migratory flights which enable a species to move between food sources require a surplus of energy which must be stored in readiness for the journey. This is especially so if large areas of hostile territory, such as desert or ocean, have to be traversed at a single hop. The journeys undertaken as single non-stop flights by some species are remarkable and up to half of their body weight may be stored fat: fat is a better reserve than glycogen and provides 9·5 kcal/g (King and Farner 1961). The American population of the Lesser Golden Plover travels directly across the Bering Sea from Arctic breeding grounds in order to winter in Hawaii, a non-stop flight of 2000 miles. Birds crossing the Sahara may need to fly continuously for 50—60 h (Moreau 1961) while the spring return crossing of both the desert and Mediterranean may be made in one hop, judging by radar studies (Casement 1966). Ward (1963), when he weighed passerine migrants at the southern end of Lake Chad prior to spring departure, found that 30—40 per cent of the body weight of Wheatears *O. oenanthe* and Yellow Wagtails *Motacilla flava* was made up of fat. A similar mean fat content with values of up to 49 per cent has been recorded in the case of migrants preparing to fly south across the Gulf of Mexico. Even relatively short over-land flights require the accumulation of energy in excess of normal daily metabolic requirements, and the amount deposited as fat is proportional to the length of journey expected (Odum, Connell, and Stoddards 1961; Johnston, 1966). Nisbet, Drury, and Baird (1963) calculated that the average rate of weight loss during migration corresponded to an average power consumption of about 0·076 kcal per g total weight per h, so that given 1·2 g fat and a flight speed of 32 km/h a small passerine would have a range of 195 km. This is close to a value of 145+ km calculated for the White-throated Sparrow *Zonotrichia albicollis* by Johnston (1966) who assumed, from Lasiewski (1963), that the metabolic rate in flight would be 7 or 8 times the basal rate of metabolism (see below).

Johnston pointed out that many species initiate migration with little fat reserve, but at first make short journeys and accumulate further reserves *en route*. Later migrants tend to accumulate larger reserves, presumably because they make a more rapid and direct flight to the breeding grounds, especially if these are in the far north and the breeding season is limited to a short Arctic summer. If food supplies are at all restricted it is clearly advantageous to avoid a high energy intake threshold that would be needed to accumulate large reserves in a short period. Johnston found that White-throated Sparrows beginning spring migration from wintering grounds in Florida carried only 1·2 g fat. Similarly, White-crowned Sparrows *Z. leucophrys* spending the winter in Washington State lay down about 5 g (20 per cent of body weight) before undertaking the 4000 km journey north to breeding grounds in Alaska, and this they accomplished as a series of night flights of 100—600 km each. Birds captured near Fairbanks, Alaska had accumulated only about 2·5 g in autumn in readiness for the return journey (King 1963; King, Barker, and Farner 1963; King, Farner, and Morton 1965). Usually the post-nuptial migration occurs at a more leisurely rate than the journey to the breeding grounds, when possible. Since there are disadvantages in carrying surplus weight for too long most migrants accumulate their réserves in a short space of time; White-crowned Sparrows take 6–9 days.

Similarly, in the 2 weeks or so preceding migration from Northumberland, England, in late September, Redpolls *Acanthis flammea* accumulated 2·0 g body fat in readiness for the 800–1000 km journey to winter grounds in north-west Europe (Evans 1966). Evans also noted that Redpolls which had acquired fat reserves had higher lean dry weights than individuals which had not deposited fat, so clearly some other body component had increased in weight. Available evidence indicates that birds also accumulate extra protein in the muscles which they metabolize during migration.

Dol'nik (1970) has emphasized that there must be an adaptive relationship between the weight of fat reserves and power required to carry it. He refers to Brody (1945) that the upper limit of energy that can be expanded in modèrate work is about 16 times the standard metabolic rate (or 8 times the resting energy). The standard metabolic rate increases with body weight W approximately in proportion to $W^{3/4}$[†] and so the energy expended in moderate work must bear a similar relationship

[†]An exponential relationship exists between standard energy metabolism and body weight as initially shown for mammals (see Brody 1945). Thus:

$$M = a\,W^b,$$

where a and b are constants, and M is the metabolic rate. Rewriting gives:

$$\log M = \log a + b \log W.$$

For this straight-line formula King and Farner (1961) obtained values for birds of:

$$\log M = \log 74 \cdot 3 + 0 \cdot 744 \log W \pm 0 \cdot 074.$$

This expression did not provide a good fit to the plots of small birds and they were better included if the values were altered to:

$$\log M = \log 80 \cdot 1 + 0 \cdot 659 \log W \pm 0 \cdot 076.$$

This expression has been widely used in the form

$$M_b\,(\text{kcal/day}) = 80 \cdot 1\,W^{0 \cdot 66}.$$

The reason for the difficulties experienced by King and Farner, who questioned whether the relationship might be curvilinear, was found to depend on the fact that the metabolic rate is higher in passerines than non-passerines (Lasiewski and Dawson 1967). These authors showed that the regression lines for passerines and non-passerines do not differ in slope ($b = 0 \cdot 724$ and $0 \cdot 723$, respectively) but only in their constant terms (log 129 and log 78·3). Their regression for all non-passerines included birds ranging in size from humming-bird to ostrich and it was similar to that generally accepted for mammals:

$$\log M = \log 78 \cdot 3 + 0 \cdot 723 \log W \pm 0 \cdot 068$$

Combining passerines and non-passerines gives an expression similar to that derived by King and Farner:

$$M_b\,(\text{kcal/day}) = 86 \cdot 4\,W^{0 \cdot 67},$$

but it is an artifact consequent on combining the two bird groups. Anyone using the Lasiewski and Dawson equation should take note of the fact that there is a circadian rhythm in metabolic rate and that oxygen uptake may be 25 per cent higher during activity than rest time (Aschoff and Pohl 1970).

to body weight. However, the minimum theoretical power required for flight increased with body weight in proportion to $W^{7/6}$ (Wilkie 1959) and this sets an upper limit of about 10 kg if a bird is to be able to fly. It follows that small body size is an advantage for a migrant because there is then a larger difference between available and required power. Power does not increase with the level of obesity and the theoretical maximum weight that can be lifted is close to that actually observed in several species. This emphasizes again the disadvantage of carrying surplus fat for longer than necessary.

Fat stores may also be valuable to non-migrant species in enabling them to survive seasons when it is difficult to consume sufficient food. Moreover, both temperate and tropical species exhibit a diurnal accumulation of fat to provide energy reserve for the night (Helms and Drury 1960; Helms 1968; Fig. 7.13); more fat was stored by White-crowned Sparrows during cold days than warm ones (King and Farner 1966). Of the total live weight of the Yellow-vented Bulbul in Singapore, 1·5 g is fat (equivalent to about 4—5 per cent), of which about 0·4 g is unavailable structural lipid while 0·6 g is needed for metabolism during the nearly 12-h night (Ward 1969b). During three Octobers many individuals were found to have insufficient overnight fat stores which suggests that food was hard to obtain, in spite of this being a region with apparently little seasonal change. Redpolls wintering in southern Britain and the Low Countries must store sufficient energy to last them for a 16-h night, often at temperatures close to freezing (Evans 1969). Evans calcu-

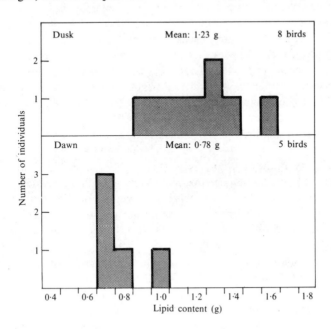

Fig. 7.13. The lipid content of Yellow-vented Bulbuls *Pycnonotus goiavier* collected at the same site in Singapore at dusk (1800—1900 h) and dawn (0600—0700 h) all subjects being killed at capture. From Ward (1969b).

lates, using data from Steen (1958), that in such conditions the birds require a total of 5—6 kcal overnight, and this compares with figures for the larger Bullfinch of 3·8 kcal in October—November and 6·4 kcal in December—January (Newton 1969). It seems likely that glycogen could be stored at a maximum rate of 22 per cent of dry liver weight and 4 per cent of pectoral muscle weight by analogy with data collected for White-crowned Sparrows (Farner et al. 1961). This means that Redpolls could only store about 0·3 kcal in this form and Bullfinches about 9·8 kcal. Finches do store some food in the expandable oesophagus but in Redpolls the usual quantity of birch seed could supply only about 2 kcal overnight, while 0·5 g of nettle seed in the Bullfinch 'crop' would yield 1·6 kcal. These species, and others like them, must build up their fat reserves in order to survive the night. Redpolls require an additional 3 kcal at least and for this 0·3 g of fat is needed. Fat reserves already present in the liver and muscle amount to only 0·05 and 0·1 g, respectively, while about 1 per cent of the total body fat is structural and cannot be metabolized. Hence, in the order of 0·15 g of subcutaneous and peritoneal fat must be acquired during the day by Redpolls if they are to survive the night.

In birds, at least in pigeons and chickens unlike laboratory rats or mice, the major site of synthesis of fatty acids is the liver (up to 96 per cent of fatty acid synthesis occurs here) and not the adipose tissue (Goodridge and Ball 1967a, b). Lipogenesis and lipophagia are usually accompanied by alterations in other body constituents. Thus the fat-free (lean) body weight of White-crowned Sparrows in spring decreased as lipogenesis occurred so that the total body weight remained unaltered (King et al. 1965). The winter increase in total weight of the Bullfinch involves an increase of 64 per cent in water content, 24 per cent of lean dry material (mainly protein) and 12 per cent of fat, consequently an increase in fat content of 0·5 g is associated with a total body weight increase of about 4·0 g (Newton 1969). This makes it misleading to express the fat content as a percentage of the total body weight since the proportion by weight of fat in the body (2—4 per cent) varies less than the actual quantity (0·2—1·1 g). In other words, in winter proportionately more of the bird is comprised of material which can be metabolized to yield a high energy output. An inverse correlation between fat and water content has also been shown in the migratory Scissors-tailed Flycatcher *Muscivora tyrannus* and Small-billed Elaenia *Elaenia parvirostris* in Venezuela; this presumably enables these species to load themselves with fuel at the beginning of migration without increasing their total body weight too much (McNeil and Itriago 1968); these Tyrannid flycatchers winter in northern South America and migrate to breeding grounds in Argentina and southern Brazil. The equatorial and non-migratory Yellow-vented Bulbul undergoes only slight seasonal changes in total body weight (2 g or 7 per cent of the total weight) although, as already mentioned (Fig. 7.13), there are marked diurnal variations in its lipid content. In addition, slight seasonal variations were found by Ward in the distribution of weight within various body tissues, particularly the flight muscles (Fig. 7.14). During periods when flight becomes least important, that is, when moulting is feasible, it appears that muscle tissue is broken down to provide amino acids for feather synthesis (Ward 1969a).

Further north in Hong Kong the closely related Chinese Bulbul *P. sinensis* exhibits a well-marked seasonal variation in body weight more typical of temperate species (Fig. 7.14). Modern research of the kind described in the next section is making it evident that total body weight is a crude, and sometimes misleading, indication of the condition of a bird.

Rhythms in energy intake and use

It is now becoming clear that many specific body constituents that may be required during particular seasons are mobilized or synthesized at appropriate times (for example, see Pohl (1971*a*); Dol'nik (1973)). This may involve the conversion of

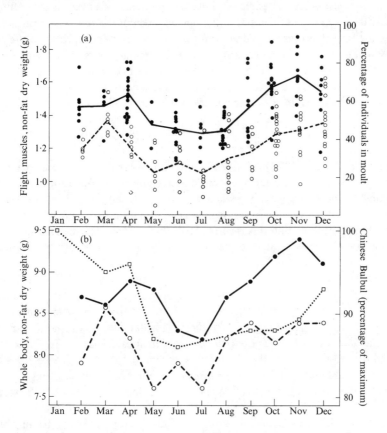

Fig. 7.14. (a) Percentage of Yellow-vented Bulbuls *Pycnonotus goiavier* in Singapore undergoing moult (tinted) in relation to seasonal changes in the non-fat, dry weight of the flight muscles of adult males (filled circles) and females (open circles); solid and dashed lines are mean values respectively. From Ward (1969*a*).

(b) Seasonal changes in the mean non-fat, dry body weight of adult male (solid circles and solid line) and female (open circles and dashed line) Yellow-vented Bulbuls in Singapore (from Ward 1969*a*). Also seasonal changes in whole body weight of the Chinese Bulbul *P. sinensis* in Hong Kong (dotted line) as a percentage of maximum body weight, i.e. 100 per cent = 36·2 g. Based on data supplied by F.O.P. Hechtel (unpublished.).

specific amino acids. Kendall, Ward, and Bacchus (1973) describe how when female *Quelea* were producing eggs the lean and dry weight of the flight muscles declined. A similar process occurs in the Yellow-vented Bulbul (1969*a*) and the House Sparrow *Passer domesticus* (Schifferli 1976, unpublished D. Phil. thesis, Oxford) and it seems that the flight muscles, or all muscles, of birds have a labile protein-rich component which can vary in quantity without affecting contractile function. Kendall *et al.* found that the quantity of saroplasm between myofibrils changed during starvation and this region appeared to provide the store of muscle protein which can be mobilized during nutrient-demanding processes. More detailed investigations by Ward and his colleague (Jones and Ward 1976) have shown how a protein store is accumulated just before breeding in both male and female *Quelea*. Its exact function in the male is still not defined but in the female much of the reserve disappears during the time that eggs are laid and almost certainly is used to provision the eggs. It may be that birds have the capacity to store limited amounts of nutrients which are particularly difficult to collect during daily feeding activities. From this it follows that concepts about food availability for birds will have to be refined towards an appreciation that certain critical resources, needed in small amounts or specific times, could be limiting even though general food supplies remain abundant. Much more attention needs to be concentrated on the nutritional status of the food of wild animals.

Kendall and Ward (1974) described how the thymus gland of *Quelea* enlarges during the breeding season and at other times and appears to function as a site of red blood cell synthesis, enlarging whenever there is a need for extra erythrocytes. Since many of these seasonally varying metabolic functions are probably controlled by circannual rhythms, and this is certainly the case for moult cycles (Ling 1972), the capacity of species to adapt to new ecological requirements must be severely restricted.

It has been demonstrated in *Zonotrichia l. gambelii* (see King (1970, 1972) for summaries) that the energy needed for fat storage results from a photoperiodically induced increase in feeding rate (hyperphagia) and not from energy-sparing alterations in metabolic rate of reduction in thermo-regulation. Birds kept on 20-h photoperiods and allowed to feed for only 9 h had a similar caloric intake to subjects allowed to feed throughout the 20 h, that is, the increased caloric intake was a true hyperphagia and did not result from a longer available feeding time (King 1961*a, b*). Morton (1967), studying free-living White-crowned Sparrows, measured the foraging index, this being defined as the product of the fraction of the flock actually engaged in feeding and a subjective estimate of the intensity at which feeding birds were procuring food. Morton's field observations showed that a change occurred between early and late April in the diurnal feeding rhythm such that the normal bimodal activity pattern was lost; a similar change was noted during hard weather (see also below) because the birds fed intently throughout the day. Morton reckoned this midday increase was the primary source of the extra energy accumulated as fat stores. Captive birds held outdoors in winter began feeding at dawn and continued steadily throughout the day, that is, they did not exhibit the bimodal pattern in feeding activity. The daily food intake of these birds was regulated at maintenance level, but increased as the photoperiod lengthened and the birds entered a migratory condition. It was noted that wild subjects continued the locomotor patterns associated with foraging even when not feeding intently, while the captive subjects displayed a bimodal rhythm of locomotor activity typical of many small birds (see below), even though this did not conform to their feeding pattern Morton remarked that this observation conformed with recognized views that feeding and foraging behaviour are motivated separately (see Lorenz 1937; Hinde 1953). Annual activity patterns in the non-migratory race Z. l. nuttalli are described by Smith, Brown and Mewaldt (1969).

It has been well established for White-crowned Sparrows that hyperphagia, fat deposition, and Zugunruhe[†] can be artificially induced by long photoperiods and they fail to develop in birds held on short days (King 1961a, b; King and Farner 1963). In the same way Morton found that in his birds the extra feeding associated with the increase in photoperiod resulted in fat deposition. Once maximum reserves had been accumulated there was a sudden change in the locomotor rhythms as migratory restlessness developed and the birds now stopped feeding 75 min instead of 30 min before the end of twilight. This early cessation of feeding was not altered if experimental subjects were deprived of food for variable periods and was also observed in free-living flocks prior to migratory departure. Morton considered this to be adaptive in that it ensures that the birds are not carrying an unnecessary burden of undigested food in the gut when they begin migration. He also questioned whether a negative feed-back stimulus of lipid reserves caused the change in feeding behaviour at dusk. Since hyperphagia results from appropriate photoperiodic induction it has been suggested that light can act directly on the hypothalamus to stimulate neural centres controlling eating (Kuenzel and Helms 1967). But there is evidence that any hypothalamic control is indirect and results from the release of hormones which themselves are directly or indirectly involved in lipogenesis (see below).

The typical feeding rhythm in which there is a peak of food intake or feeding activity soon after dawn and a bigger peak in the late afternoon can be traced in many migrant species outside the migratory season as well as in non-migrants (Baldwin and Kendeigh 1938). It is clearly adaptive for by accumulating extra food towards the latter part of the day, birds can go to roost with an overnight energy store and depleted reserves are made good next morning by a period of intensive feeding. During the post-breeding period June—September, Red-winged Blackbirds Agelaius phoeniceus studied by Hintz and Dyer (1970) preferentially selected high-energy food at the beginning of the day but the caloric intake did, however, increase at the end of the day, because more food was consumed (Fig. 7.15(a)). As mentioned,

[†]Migrants kept in cages become more active at those times before and during migration and often orientate themselves in the preferred direction of movement. This Zugunruhe, or migratory restlessness, can be quantified by fixing a micro-switch to the perch, which provides a convenient measure of the migratory urge. The duration of migratory restlessness of captive migrants reflects the time normally needed for the actual migration (Gwinner 1968a).

diurnal variations in feeding intensity become less evident during cold weather or unfavourable feeding conditions as demonstrated by Morton for the White-crowned Sparrow. Beer (1961) showed that the normal bimodal feeding pattern of House Sparrows in St. Paul, Minnesota, disappeared on some winter days when the birds had to survive 8½ h of darkness at temperatures of −21 °C to −25 °C. Under such conditions the sparrows arrived at the feeding station before dawn and fed continuously throughout the day. As noted in Fig. 7.1 the critical temperature at which the House Sparrow's metabolic rate is at a minimum is 37 °C. As Beer points out, down to about 15 °C the birds can compensate for a reduced temperature without increasing their food intake; this they do by fluffing up the feathers to increase insulation. Below 15 °C the metabolic rate must increase as temperature drops and be compensated by an increase in the energy (food) intake, the maximum rate of energy intake being achieved in the sparrow at about −20 °C (Kendeigh 1949); the gross energy intake increases as a straight line with decrease in temperature down to about −30 °C according to Kendeigh.

   As implied above, it is necessary to distinguish the separate components of feeding behaviour, that is, hunting and eating, since they need to be satiated independently. For this reason an analysis of gut remains provides limited information. For example, the Mourning Dove *Zenaida macroura* has peaks in food intake at about 10.00 and 18.00 h in August and September in North America judging by its crop contents (Fig. 7.15*b*). Fig. 7.16 demonstrates for another pigeon species, the Wood Pigeon *Columba palumbus*, that the increased food intake needed as an overnight reserve is acquired by relative changes in the rates of searching and eating. In February the birds graze on clover *Trifolium* leaves on pastures and leys. At the start of the feeding day they are highly selective and search large areas for only a few items, singling out the most nutritious and proteinaceous components of the sward. From around mid-day the pigeons gradually decrease their searching effort

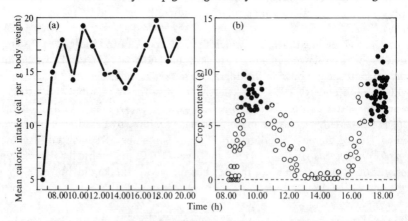

Fig. 7.15. (a) Diurnal rhythm of mean caloric intake in male Red-winged Blackbirds *Agelaius phoeniceus* between June and September. From Hintz and Dyer (1970).

   (b) Diurnal variation in weight of food contents in the crops of Mourning Doves *Zenaida macruora* either shot in fields or roosting-places (open circles) or near to water holes (solid circles). From Schmid (1965).

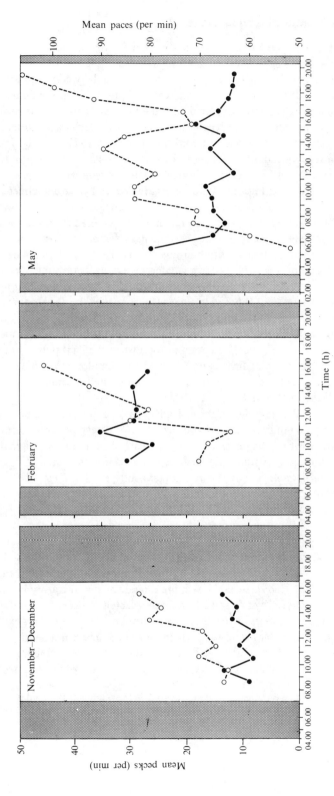

Fig. 7.16. Feeding rate (pecks per min as dashed line) and searching rate (paces per min as solid line) of Wood Pigeons *Columba palumbus* feeding on clover leaves from pastures. Food stocks decline from November to reach a minimum level in February and then recover rapidly with plant growth in April. From Murton *et al.* (1963*a*).

and increase their intake, collecting some items ignored during the morning. Food is now stored in the oesphageal diverticulum (crop) so that the birds go to roost with a food supply equivalent to 10–14 per cent of the total daily consumption. When seeking good-quality food items the area searched (measured as the number of paces per minute) was strongly correlated with the number of food items (leaf fragments) eaten per minute, but not when the searching area decreased and the birds pecked less selectively (Murton, Isaacson, and Westwood 1971). As Fig. 7.16 shows the peak in feeding activity in the morning is really a peak in hunting activity rather than in ingestion rate. Similarly, a bimodal rhythm of general feeding activity was observed in captive Feral Pigeons *Columba livia* fed *ad lib* by Ziegler, Green, and Lehrer (1971). The birds fed by a series of bursts of intake or meals which were termed bouts. When bout duration was analyzed and distinguished from the actual time spent ingesting food the early morning feeding peak was almost lost. Bout duration is presumably homologous with hunting activity. Hunting or search behaviour is related to locomotor activity and is regulated by a circadian rhythm mechanism. Morton (1967) found that the feeding rhythm would free-run in constant light in White-crowned Sparrows but it has yet to be established whether the rhythms of hunting and feeding can be phased independently under such conditions. However, it has been shown that Collared Doves *Streptopelia decaocto* held on asymmetric skeleton light schedules of LD 6½:17½ eat more food when the light pulse occurs 22 h[†] from the onset of the long light period than when the pulse occurs 8 h or 14 h afterwards, the total light time available for feeding being the same for all groups (Murton, Westwood, and Thearle 1973).

We have seen that a degree of hyperphagia leading to fat deposition can be found in both migrant and non-migrant species and that the two categories differ primarily in the timing of these events. *Zugunruhe* activity appears to be unique to migrants and could not be induced experimentally in closely related non-migrants, cf. *Zonotrichia leucophrys gambelii* and the non-migrant subspecies *Z. l. nuttalli* resident in California and Oregon (Farner 1960); similar differences between migrant and non-migrant species were demonstrated by Odum and Perkinson (1951) and Koch and de Bont (1952). This difference probably represents a difference in the phasing of locomotor activity (cf. Fig. 7.16) rather than an absolute physiological difference. We have emphasized that several functions which occur at about the same time in migrants are physiologically distinct and can be traced in resident species. In other words, the special adaptations associated with pre-migratory and migratory behaviour are that at least some physiological events have been selected to occur more or less simultaneously and this temporal association is adaptive but not causal. This explains why under experimental dietary conditions which prevented the accumulation of pre-migratory depot fat, typical nocturnal restlessness developed in the Brambling *Fringilla montifringilla*, Slate-colored Junco *Junco hyemalis*, and White-throated

---

[†] For reasons which will be explained in Chapter 11 (p. 296) a skeleton of this kind is interpreted by the bird as if the start of the shortlight pulse functions as dawn. Thus the skeleton simulates a full photoperiod of 8 hours with the 6-h light period falling at the end of the day.

Sparrow *Z. albicollis* (Lofts, Marshall and Wolfson 1963). When some species are held under constant lighting conditions for periods exceeding a year the expression of moult, gonad hypertrophy, body weight and *Zugunruhe* become out of phase again, indicating that these events are physiologically distinct (see p. 352).

Dol'nik (1970) summarized the temporal sequence of functional changes which occur in the Chaffinch *Fringilla coelebs* before migration north from Leningrad as (1) lipolysis, (2) change in activity rhythm, (3) orientation shown by caged subjects, (4) hyperphagia leading to hyper-lipogenesis, (5) onset of *Zugunruhe* He appreciated that these events were developed independently and were controlled photo-periodically from hypothalamic centres—in fact different pituitary hormone secretions and syn-ergisms can stimulate these various functions, as will be shown below. Once a full migratory state is achieved, in suitably photostimulated subjects, the various com-ponents as outlined above become co-ordinated into a single adaptive system by substrate feed-back and endocrine mechanisms now provide a metabolic regulation primarily controlled by the level of fat deposits. At this stage a difference was dis-cernible between 'lean' and 'fat' Chaffinches. The former were uniformly inactive, fed throughout the day and had an active glycolysis and glycogenesis[†] and reduced lipolysis. In contrast, fat individuals exhibited much migratory activity, especially during the second part of the day, no hyperphagia, and they fed predominantly in the second part of the day, and had an active glycolysis, lipolysis, and reduced glyco-genesis and lipogenesis. Giving lipid injections to lean birds caused their metabolism and behaviour to resemble that of fat birds, showing that the status of fat deposits directly regulated metabolism and behaviour (Dolnik and Blyumental 1967).

Effects of photoperiod

The fact that different physiological processes gradually become co-ordinated into a functional system in preparation for migration has led to some apparently conflicting experimental results. For example, it was shown some years ago that *Zugunruhe* could be developed by castrated Bramblings and Golden-crowned Sparrows *Zonotrichia atricapilla* (Lofts and Marshall 1960; Morton and Mewaldt 1962). But Weise (1967) showed that castration performed before photostimulation of *Z. albicollis* had a more inhibitory effect on *Zugunruhe* and fattening than did cas-tration given after the beginning of photostimulation, and this was confirmed by Stetson and Erickson (1972). Weise (1967) reviewed his own and other studies con-cerning the effect of castration on migratory fattening and *Zugunruhe* and showed a sequence of response depending on the time of experimentation. He castrated birds after intensive photostimulation and hardly caused any reduction in fattening and *Zugunruhe* Lofts and Marshall had castrated Bramblings *Fringilla montifringilla* in March only a few weeks before migration and caused virtually no change in *Zugunruhe*. Morton and Mewaldt castrated Golden-crowned Sparrows *Zonotrichia atricapilla* in February, 4—6 weeks before migratory behaviour, and caused very

---

[†]Glycolysis is the decomposition of glucose or glycogen by hydrolysis; glycogenesis is the transformation of glucose into glycogen; lipogenesis is the production of fat; and lipolysis the breakdown of fats to fatty acids as in digestion.

little reduction in fat deposition and *Zugunruhe*. However, Miller (1960) castrated *Z. albicollis* 2 months before *Zugunruhe* developed in controls and produced severe reductions, while Weise (1967) castrated his subjects before photostimulation and suppressed *Zugunruhe* and fat deposition completely.

The temporal association between gonad growth, migratory fattening, *Zugunruhe* and moult, and the fact that all could be induced experimentally by long photoperiods, led Meier and Farner (1964) to investigate the effect of exogenous hormones. Prolactin augmented the weight increase of White-crowned Sparrows transferred to LD 20:4 in January, whereas adrenocortical hormone ACTH (prednisone) alone or in conjunction with prolactin did not. Meier and Farner next kept this species on LD 8:16 in February and found that prolactin again increased body weight by 13 per cent. Subsequently the photoperiod was increased to 10 h (LD 10:14) and body weight increased after about 4 days by a further 13 per cent to total 126 per cent of the original weight. A similar weight increment was again induced by changing the photoperiod to LD 12:12 on day 21. It was also noted that although prolactin-treated birds weighed more than controls at the end of the period of light, they lost proportionately more weight during the night so started each day not much heavier than the controls. Possibly this represented a greater nocturnal catabolism which would be associated with migratory restlessness. The results indicate that with a lengthening of the day more gonadotrophin was produced which acted synergistically with the injected prolactin. Indeed, when exogenous hormones

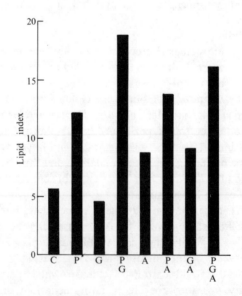

Fig. 7.17. Effect of hormone treatment on the fat content of White-crowned Sparrows *Zonotrichia l. gambelii* injected for 22 days during the refractory period. Treatments involved daily doses of prolactin (P) (300 μg); 150 μg LH + 150 μg FSH (G); 200 μg ACTH (prednisone) (A) or combinations as indicated; untreated control (C). Based on Meier and Farner (1964).

were given to photo-refractory White-crowned Sparrows a combination of prolactin and FSH/LH proved most effective in causing lipid deposition while FSH and LH given alone were ineffective (Fig. 7.17). However, gonadotrophins did cause some increment in the lean body weight. ACTH (in the form of prednisone) reduced the total body weight but raised the lipid content, because it lowered water and non-lipid fractions.

Exogenous prolactin given to White-crowned Sparrows captured from the field a month prior to vernal migration also considerably increased the amount of *Zugunruhe*, which developed only slowly in controls (Meier, Farner, and King 1965). A more critical test involved treating photo-refractory birds and using these. Neither gonadotrophins nor ACTH (prednisone) given alone were effective in stimulating *Zugenruhe*, though prolactin still proved effective and ACTH augmented the response (Fig. 7.18). The experimenters subsequently repeated their work using cortico-sterone instead of prednisone with a similar result. They also assayed pituitaries from free-living White-crowned Sparrows throughout the year and recorded signifi-cantly higher levels of prolactin during the periods of spring and autumn migration. In view of these results it remains to be shown why early castration should inhibit the development of *Zugunruhe*, unless it does so by preventing prolactin secretion. Electrolytic lesions of the ventral hypothalamus of White-crowned Sparrows elim-inated the photoperiodically induced hyperphagia and the fattening response (Stetson 1971): lesions in the posterior division of the median eminence prevent

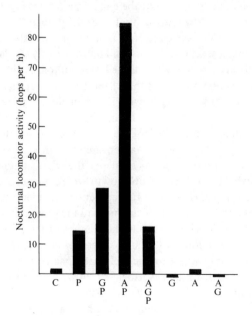

Fig. 7.18. Nocturnal locomotor activity of hormone-treated photo-refractory White-crowned Sparrows *Zonotrichia I. gambelii* held indoors on a 16-h photoperiod. Treatments involved daily injections of 300 μg prolactin (P); 150 μg LH + 150 μg FSH (G); 200 μg ACTH (prednisone) (A) or combinations as indicated; untreated controls (C). From Meier *et al.* (1965).

both the fattening response and testicular recrudescence, whereas lesions of the anterior division affected only the fattening response. On the other hand, lesions of the ventro-medial hypothalamus may cause a different kind of fattening; when Kuenzel and Helms (1970) lesioned this region in White-throated Sparrows extreme obesity followed. It is not clear whether fibre connections in this region are integral with the median eminence and a prolactin-release mechanism. More recently, Martin and Meier (1973) have shown that a temporal synergism of corticosterone and prolactin regulates migratory orientation.

Prolactin has been shown to accelerate lipogenesis by the liver of the pigeon by Goodridge and Ball (1967a, b). The rate at which labelled pyruvate could be incorporated into fatty acids was immediately increased in vitro in the presence of prolactin, while, as an apparently secondary effect, malic enzyme (coenzyme : NADP), citrate cleavage enzyme and malate dehydrogenase (coenzyme : NAD) became more active after a two-day delay. These lipogenic responses were not observed if subjects were starved, whereas the crop gland responded to prolactin irrespective of the food intake. This raised the possibility that prolactin did not have a direct action but secondarily caused the liver to become more active by stimulating a hyperphagia. These authors also discussed the possibility that the observed actions of prolactin depended on contamination with a somatotrophin.

Physiology of moult

We still lack a precise understanding of the endocrine factors regulating moult in birds. This summary can, therefore, be brief, especially since the topic has been extensively reviewed by Voitkevich (1966, as an English translation), who covers the detailed work by Russian researchers, and more recently by Palmer (1972) and Payne (1972). In listing 240 references these latter authors could between them find under 70 which were more recent than 1966 and the majority of these dealt with ecological aspects of moulting or were not immediately relevant to the topic.

Feather morphogenesis is partly determined by persistent hereditary properties vested in the individual feather follicle. Embryonic feathers develop without the evident benefit of hormones, juvenile feathers are primarily influenced by the thyroid, while additional hormonal factors influence the adult's plumage. Thyroid hormone (thyroxin), which generally affects growth, thermo-regulation, and the metabolism of electrolytes, is essential for feather development and increases oxidative processes and the balance of sulphur and nitrogen. Although feather germs can be formed by local processes in the absence of the thyroid, thyroidectomy prevents further development. The involvement of local organizational factors in the feather germ is emphasized by the fact that repeated dosing with thyroxin induces a premature moult which nevertheless follows the sequence of the normal moult. Also, according to Voitkevich, if a moult is induced by drug therapy a second application of the agent results in a reduced feather loss, those feathers that are shed coming from different follicles. Thus follicles from which feathers have just been shed cannot immediately be stimulated again, perhaps because of exhaustion of the cells of the papillae. Extirpation of the thyroid prevents moulting, while additional thyroxin stimulates an unseasonal moult and accelerates the rate of moult (the metabolic rate may increase by 25 per cent or even more during a normal moult). Thyroid hormone affects different feather tracts to a variable extent for the head and neck feathers are little influenced compared with the body plumage. The flight feathers will continue to grow following thyroidectomy provided the feather germs are formed some time before the operation. More thyroid hormone is needed to stimulate a moult in spring than in autumn. In general, the thyroid is a necessary regulator of the cellular material entering the generative zone of the feather quill at a specific stage of development and to a variable extent in the separate pteralae, and it is most important for the development of the lower part of the vanes of the contour feathers.

With the onset of sexual maturity the effects of pituitary and gonadal hormones are additionally superimposed on the background of thyroid activity. Feathers regenerate if they are artificially plucked and this emphasizes that a distinction must be made between growth and development and the organized moult. It has already been mentioned that the seasonal moult is under photoperiodic control, so it must be mediated via neurohypophysial secretions, which in turn may stimulate the thyroid and gonads. Gonadal steroids are required for the development of secondary-sexual plumage in some dimorphic species and Voitkevich distinguishes these species from those for which sex hormones are not essential for the plumage pattern. The appearance in one sex of a special display plumage dependent on gonadal or gonadotrophic hormones appears as a secondary adaptation to be discussed in Chapter 15, for it is likely that selection has operated to provide a new function for a hormone necessarily present at this stage of the cycle.

It has long been appreciated that in many species there is an inverse relationship between breeding, or more specifically the possession of a functional gonad, and moulting, as if the two events were mutually exclusive. In energy terms they often are, so that the relationship can be considered adaptive. Ward (1969a) found that the wing muscles of bulbuls on natural equatorial photoperiods declined in weight with moult, and testicular recrudescence was apparently prevented until the muscles regained their weight. Ward emphasized that there was considerable overlap in testicular and moult condition; nevertheless a check of his plotted data indicates a significant negative correlation once the primary moult begins (with $r_{34} = -0.418$; $P < 0.01$). In domestic fowls egg production and egg weight decline during moult. In view of these relationships it is not surprising that castration of birds tends to result in feather loss, as if gonadal hormones inhibit the moult. Thus, castrated fowls continuously renew their plumage instead of having a single annual moult. Moreover, in many wild species breeding may continue later in the year in favourable seasons, whereupon the onset of moulting is delayed (e.g. Bullfinch, p. 153). It seems reasonable to interpret such observations in terms of favourable environment factors stimulating gonadotrophin secretion, thereby maintaining gonadal function, conceivably with a consequent inhibitory feed-back on the hypothalamic centres controlling moult (perhaps on TSH). The administration of exogenous androgens and oestrogens to intact birds slows or arrests the normal moult in many adults (reference in Payne (1972)) including pigeons (Kobayashi 1954; and Fig. 7.19). This being the case it is not obvious why progesterone should sometimes be effective in inducing moult (Shaffner 1954; Adams 1956; Juhn and Harris 1956; Kobayashi 1958) since it would seem to introduce an inhibitory hypothalamic feed-back otherwise removed by castration. Further, in most wild birds the spring moult occurs before progesterone appears to assume a major role in the endocrine cycle. For various reasons it is thought that any permissive role of progesterone is indirect, perhaps because it inhibits gonadotrophin secretion and hence other steroid release, or because it influences thyroid activity and it is thought to be unimportant in the natural moult in birds (reference in Payne (1972)).

Photostimulation of Japanese Quail activities the delta-cells (TSH-producing cells) of both lobes of the pars distalis but at a slower pace than the beta-cells. With a return to short-day photoperiods for 18 days the delta-cells lose their chromophilic

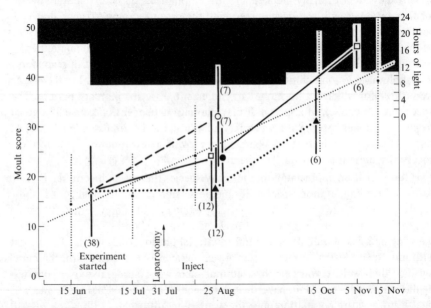

Fig. 7.19. Effect of photoperiod and hormone injection on the progression of the primary moult of Feral Pigeons. Procedure for obtaining the moult score is given in the caption to Fig. 7.6. The natural cycle of primary moult for free-living pigeons in the Manchester docks, England is shown as dotted lines which give the regression of moult score with date and the standard errors of the samples (from Murton *et al.* 1974*b*).

Subjects for the experiment were wild-caught from the Manchester population, their moult condition being defined by the cross on the Figure. Exposure to LD 8:16 caused the gonads of some birds to regress, these being mostly the 'blue' variety. The rate of moult of these individuals was increased (open circles) compared with the remaining untreated subjects whose gonads did not decrease in size (solid circles). Subjects given 15 × 0·2 mg daily injections of progesterone over the period shown did not differ from the preceding group by 25 August (open squares) but those given 15 × 0·2 mg oestradoil benzoate over the same period had a retarded moult (solid triangles). After cessation of oestrogen injections the moult proceeded at the normal rate but there is some indication that there was a delayed acceleration of the moult in subjects which had previously received progesterone (open squares). Number of subjects and standard errors of the means are indicated.

properties and the cellular volume declines to the starting level, and this loss of activity occurs more rapidly than is the case with the beta-cells (Tixier-Vidal *et al.* 1968). This activation of the delta-cells is accentuated by castration and their volume becomes even greater than following treatment with thiourea (which stimulates secretion). When Stetson and Erickson (1971) photostimulated intact and castrate White-crowned Sparrows for 120 days on LD 20:4 the thyroids of castrates on day 80 were significantly lighter than those of intact birds. (Loss of thyroid weight is assumed to indicate the mobilization and loss of colloid reserves but this was not checked by histological examination; it is known that experimentally induced exhaustion of the thyroid is associated with a total loss of reserve colloid in the vacuoles of the gland, see Voitkevich (1966, pp. 119–21).) It is of note that there were no differences in thyroid weight between castrated and intact subjects during the period when the testes of the latter birds were enlarged and that the differences between groups only became manifest with gonad regression in the intact birds. If the gonadal steroids (androgen and oestrogen) normally inhibit TSH output the effect of castration should become manifest immediately. The experiment, therefore, poses the question of whether some kind of inhibition of TSH secretion occurs at hypothalamic level, indeed perhaps the active secretion of gonadotrophin is inhibitory to TSH. This possibility is suggested by an experiment performed by Voitkevich. He induced premature moulting in Japanese Quail using three injections of a preparation isolated from the basophil zone of the adenohypophysis (the site of thyrotropic hormone formation). But six injections were needed if an extract from the whole adenohypophysis was used, that is, the action of the thyrotropic hormone was inhibited if whole extracts of the adenohypophysis containing also substances from the eosiniphil cells were used. This was confirmed in further experiments and the results do suggest that gonadotrophic hormones may inhibit TSH secretion or function.

It seems likely that moulting requires the synergism and hence appropriate phasing of several hormones. Pigeons and other birds which moult throughout the breeding season may differ from the majority of species in possessing hormone rhythms which are less easily phase shifted into non-functional relationships by changes in the photoperiod (see p. 335). Wild-caught male Feral Pigeons from an urban study population were laparotomized on 24–26 June and found to be in full breeding condition. Each was now caged singly on LD 8:16 until 31 July, when a small proportion were found to have regressed testes. These birds were predominantly of the variety called blues or wild-type, which we know to have a higher photo-response threshold than certain melanic phenotypes (see p. 277; also Murton *et al.* 1973). Moult was evidently accelerated in these individuals, and this was confirmed after they had been kept for another 25 days by which time their moult score was significantly higher than a group whose gonads did not regress in response to the reduced light regime (Fig. 7.19).

The normal moult of different pigeon phenotypes proceeds at the same rate when subjects are kept on the same photoperiod but the phasing of the moult in relation to the day-night cycle differs slightly between morphs (Fig. 10.11; p. 278). The differences noted in Fig. 7.19 could, therefore, reflect a selection for morphs which entrained in different ways to the light cycle and this, in turn, implies that their endocrine state was altered. On 31 July a proportion of the birds whose gonads showed no signs of regression were given 15 X 0·2 mg injections of oestrogen and another subgroup progesterone (15 X 0·2 mg). As Fig. 7.19 shows, oestrogen inhibited the moult which was thereafter delayed. This was very striking in the experimental rooms, for the floor in front of the cages of control birds and those given progesterone was littered with discarded contour feathers and down, while the oestrogen-treated birds were 'clean'. Progesterone did not have any immediate effect on the moult but appeared to increase the rate some while after injections were finished. The delayed effects of progesterone are reminiscent of other physiological actions of this steroid (p. 114).

8 Energy requirements for egg laying and brood care

Summary

By and large, breeding is timed to correspond with the season when food supplies for the young are optimal and when the chances of leaving the most progeny are greatest. This relationship is by no means perfect since birds must compromise. In order to have young in the nest they must first produce eggs and the production of these demands an increase in the daily energy intake which may not be met until the season is already well advanced. Earlier breeding might be possible if egg size were reduced relative to female size but this means that the chicks are born at a less advanced stage of development and this must be compensated by extra brooding and the provision of surplus food. King has produced a model to show the effect of growing several follicles simultaneously, and this emphasizes that once peak energy demands are reached further egg production does not affect the daily energy requirement. The model is extended to show how daily energy demands can be reduced by increasing the laying interval or how a bigger egg can be produced by reductions in clutch size. This model is compared with the pattern of egg size related to clutch size throughout the class Aves. The Anatidae appear to be exceptional in producing large eggs relative to female body size and yet laying at daily intervals. They also seem to be exceptional in using stored energy reserves for egg production and it is presently unclear why more species do not have this ability.

Adaptive variations in egg size and clutch size with season, year, habitat, and geographical locality are discussed. These are related to nestling growth rates and modes of development.

Clutch and egg size

Lack (1954 for summary) proposed that the clutch size of each species of nidicolous bird has been adapted by natural selection to correspond with the largest number of young for which the parents can, on the average, provide enough food. Clutches above the normal limit are at a disadvantage because the young are weakened through undernourishment and, as a result, fewer survive per brood than from clutches of normal size. This was clearly demonstrated in the Swift *Apus apus*, in which surplus young mostly died in the nest and in the Starling *Sturnus vulgaris*, in which most of the deaths occurred soon after the young fledged. Since these studies more extensive data were collected for the Great Tit *Parus major* (Kluijver 1963; Perrins 1965; Lack 1966) and it was clearly established by Perrins that nestling weight and post-fledging survival were positively correlated with brood size. Nestling weight was also shown to be dependent on brood size and post-fledging survival on nestling weight in the Pied Flycatcher *Ficedula hypoleuca* in Finland (data from von Haartman

(1954, 1957); Tompa (1967); re-analysed by Klomp (1970)), though clear-cut differences in survival related to brood size were not evident for British Pied Flycatchers, in data collected by Campbell and analysed by Lack (1966). Cody (1971) has given a useful summary of the circumstances under which clutch size shows a systematic variation: (1) age of parent (Coulson and White (1961) for Kittiwake *Rissa tridactyla*); (2) time of breeding or order of clutch within season (Kluijver (1951) for Great Tit, Snow (1958) for Blackbird *Turdus merula*); (3) food availability (Perrins 1965); (4) population density (Kluijver 1951; Lack 1966); (5) latitude (Lack 1948; Cody 1966); (6) longtitude (Lack 1968*a*); (7) altitude (with increase in Song Sparrow *Melospiza melodia* (Johnson 1960) and decrease in corvids (Holyoak 1967) and Meadow Pipit *Anthus pratensis* (Coulson 1956)); (8) habitat; (9) nest-site. As might be predicted, clutch size usually changes when a species is introduced to a new range, for example, 6 of 12 British species introduced into New Zealand lay smaller clutches than they do in the British Isles and 5 others show a similar trend but less clearly (Niethammer 1970). The physiological mechanisms by which many of these phenotypic adjustments are achieved remain to be critically defined; photoperiodic factors could be important in some cases (e.g. p. 89 and the New Zealand example just mentioned) but in others a regulation through the availability of food resources is to be expected.

The most frequent clutch size should, as a result of natural selection, be that which on average gives most survivors per brood. This was so in the Swift and Starling, but not in the Great Tit (Perrins and Moss 1974) and many other species for which the commonest clutch proved to be smaller than the most productive. There have been other various and often ingenious explanations (see for example Mountford (1968)) to account for the disparity but it seems that the most likely one is that the effect on adult survival of rearing young must be considered in any complete assessment of fitness (Kluijver 1971). This has been suggested in a model produced by Charnov and Krebs (1974) which demonstrates that if adult mortality increases with brood size, and begins to do so before the value at which the most productive brood size is reached, then the optimal clutch size will always be smaller than the most productive brood size.

For a long time the reasons limiting clutch size in nidifugous species (in which the young feed themselves from hatching), notably ducks and gallinaceous species, remained obscure, but the factors have been clarified as a result of considering the breeding seasons of birds. Lack applied a similar reasoning to the selective factors regulating breeding as he had to clutch size, that is, most birds should breed when the chances of leaving most progeny are greatest. Thus, in single-brooded species laying tends to be so timed that the young are raised when food is most plentiful for them; those raising more than one brood may start sufficiently before the most suitable time for two or more broods to be fitted in (see also Thomson 1950). It gradually became apparent in such species as the Wood Pigeon *Columba palumbus* and Bullfinch *Pyrrhula pyrrhula* (see Lack 1966 p. 272–3) that the females were physiologically capable of reproduction long before the first eggs were laid and that egg laying depends on the female obtaining sufficient energy reserves (Murton and Isaacson 1962; Murton, Isaacson, and Westwood 1963*a, b*). Fig. 8.1 shows that

young Wood Pigeons fledged at the very beginning of the breeding season in April
or May have a higher chance of surviving to become adults than those reared later
in the season, yet the peak of breeding by the population is not reached until July.
Early-hatched young can complete their moult before winter (Murton, Westwood,
and Isaacson 1974b). The same discrepancy but less marked was noted in the Stock
Dove *Columba oenas* (Murton 1966). Again advantages for early breeders have been
found in the Great Tit because young hatched early in the season have a much greater
chance of survival than those hatched later, even though the peak of egg laying
occurs after this optimum season (Perrins 1963; Kluijver 1971).

Perrins (1966) has demonstrated the advantages of early breeding particularly
clearly in the Manx Shearwater *Puffinus puffinus* (Fig. 8.1). Manx Shearwaters must

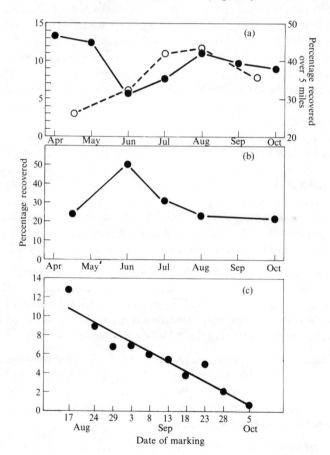

Fig. 8.1. The percentage of ringed nestling birds that were recovered after marking and were
known to survive at least two months in the case of (a) Wood Pigeon *Columba palumbus* and
(b) Stock Dove *C. oenas,* or until the following year in the case of (c) Manx Shearwater
Puffinus pufinus. The dashed line and open circles in (a) refer to the percentage of birds which
were recovered more than 5 miles from the place of marking. (a) and (b) from Murton (1961)
and (1966b) (c) from Perrins (1966).

leave the waters of western Britain and Biscay after breeding and make the long migration to the South Atlantic, where it is presumed that the adults moult for moulting birds are not found in northern seas or at the breeding stations. It seems likely that food stocks, particularly the Sardine *Sardina pilchardus*, decline in late summer and that adult Shearwaters migrate to better feeding grounds in the South Atlantic as soon as possible. They desert their young 60 days after hatching in order to prepare for the journey. However, by this time the young should have been supplied with a good reserve of fat, laid down during the first part of the breeding season when food supplies are more easily obtained; 40–60 days from hatching the young weigh 59 per cent more than the adult female. The chicks complete their nestling development and leave the nest on average 8–10 days after being deserted, having lost 27 per cent their maximum weight but still weighing 15 per cent more than the adult (Harris 1966). Now they must prepare for and make the journey to South America and it is at this stage that the disadvantages of late hatching become apparent, for such chicks have not acquired sufficient reserves and many starve. Yet, although early-fledged young survive better than the others, as Fig. 8.1 shows, only 20–30 per cent of the adults are able to lay eggs in time for the young to fledge in August and the remainder appear to be unable to gather the necessary food reserves sufficiently quickly.

If the ability of the female to acquire sufficient food reserves delays the seasonal onset of egg laying it might also influence egg size in those nidifugous species which lay particularly large eggs relative to body size. Furthermore, if environmental conditions are subject to rapid change it seems unlikely that the whole breeding cycle from courtship through egg laying and incubation to brood care could be perfectly adapted to resource availability. With these ideas in mind Lack (1968a) reappraised the subject of egg and clutch size in birds in an extremely valuable review book, and his conclusions may be summarized as follows:

1. The same quantity of food attainable per day can be used to raise a small brood rapidly or a large one slowly so that clutch size is reduced in circumstances where predation risks are high, or where rapid development is needed to avoid a bad season. So it is that hole-nesting passerines have larger clutches than similar sized open-nesting species because the young can afford to grow more slowly.

2. In those nidifugous species where the parents do not feed the young, clutch size has probably been primarily evolved in relation to the ability of the laying female to accumulate reserves in relation to the size of the egg. The size of the egg proportionate to female body weight is characteristic for each family and it varies inversely with body weight (Fig. 8.2).

3. A proportionately large egg is probably advantageous in providing the newly hatched chick with a large food reserve, and this relieves the needs of food-finding by the parents. Thus the energy which the adults can obtain at different stages in the breeding cycle has to be budgeted to best advantage; it is obviously wasteful to delay egg laying in order to produce a large egg if it is very easy to feed the chicks once they hatch.

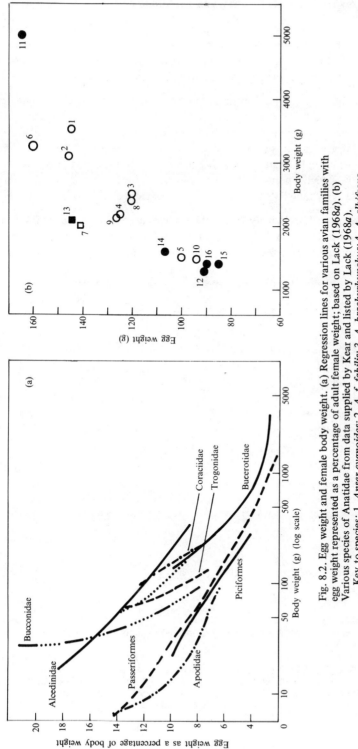

Fig. 8.2. Egg weight and female body weight. (a) Regression lines for various avian families with egg weight represented as a percentage of adult female weight; based on Lack (1968a). (b) Various species of Anatidae from data supplied by Kear and listed by Lack (1968a).

Key to species: 1, *Anser cygnoides*; 2, *A. f. fabilis*; 3, *A. brachyrhynchus*; 4, *A. albifrons albifrons*; 5, *A. erythropus*; 6, *A. a. anser*; 7, *A. indicus*; 8, *A. canagicus*; 9, *A. caerulescens*; 10, *A. rossii*, 11, *Branta canadensis canadensis*; 12, *B. c. minima*; 13, *B. sandvicensis*; 14, *B. leucopsis*; 15, *B. bernicla orientalis*; 16, *B. ruficollis*.

4. The rate of growth of the young is relatively constant, doubtless being determined genotypically. It probably depends on the average availability of food for the young, modified by brood size, and, in the young of some species, the time needed to form fat reserves as a barrier against environmental vicissitudes.

5. The incubation period of birds is positively correlated with the fledging period (Fig. 8.3) presumably because, as Lack suggests, the only way is to modify the rate of development of the embryo as well; in a sense the egg shell can be ignored for growth and development is a continuous process from the formation of the embryo to the fledged juvenile or later adult. The incubation period tends to be short relative to the fledging period in some species with proportionately small eggs, probably because the young hatch at a relatively undeveloped stage.

It follows from Lack's analysis that the reserves necessary for egg laying must mostly be drawn from the daily ration of the bird, otherwise it is not clear why the female should not take slightly longer to form her clutch and produce bigger eggs

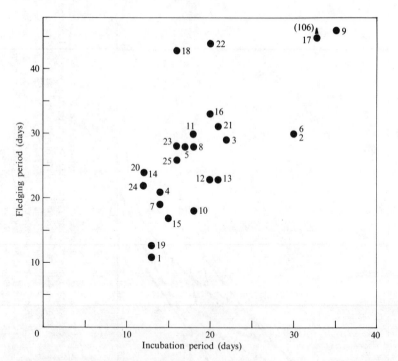

Fig. 8.3. Correlation between usual fledging period and usual incubation period for various nidicolous land bird taxa.

Key to families and orders: 1, *Crotophaginae*; 2, *Falconiformes*; 3, *Alcedinidae*; 4, *Coccyzinae*; 5, *Musophagidae*; 6, *Strigiformes*; 7, *Centropodinae*; 8, *Coraciidae*; 9, *Bucerotidae*; 10, *Trogonidae*; 11, *Caprimulgidae*; 12, *Meropidae*; 13, *Galbulidae*; 14, *Capitonidae*; 15, *Columbidae*; 16, *Psittacidae*; 17, *Steatornothidae* (fledging period = 106 days); 18, *Rhamphastidae*; 19, *Passeriformes*; 20, *Picidae*; 21, *Momotidae*; 22, *Apodidae*; 23, *Upupidae*; 24, *Cuculidae*; 25, *Trochilidae*. Based on data given in Lack (1968a).

in consequence. The clutch of 8—10 eggs of the Great Tit represents 90 per cent of the female's body weight and it is produced at daily intervals; in the Blue Tit *Parus caeruleus* the clutch is about 150 per cent of the female's weight (Perrins 1970). Even if the female could store all the egg material in a dry condition (about 25 per cent of wet weight) marked changes in the weight of the female before and after clutch completion should be observed and this is not usually the case (Perrins 1970). However, in Ross's Goose *Anser rossii*, and probably other geese, which copulate before reaching the breeding ground, the females arrive in the Canadian Arctic weighing 2400 g, lay their eggs, and then weigh only 1500 g two weeks later (Ryder 1967): clutch size is apparently limited by the total increase in body weight the female can achieve (Ryder 1970). MacInnes *et al.* (1974) found that the heaviest Snow Geese *Anser caerulescens* had the potential to lay the most eggs. At the end of the laying period the spread in adult weights was less than at the start, presumably because the heaviest females had laid the most eggs. Birds with a capacity to lay most eggs also begin laying earlier in the season, so that mean clutch weight declines rapidly over a period of only about a week, this species tending to have a fairly synchronous laying season. The eggs weigh 94 g and the mean clutch is 3·5 eggs. Gilbert (1971*a*) gives some valuable data for the domestic hen, which lays a proportionately small egg relative to her body size, equivalent to 2 per cent of the adult body weight. An egg weighing 58 g contains 7·0 g protein, 6·2 g fat, 0·3 g carbohydrate, 2·0 g calcium, 0·5 g minerals, 3·0 g other non-metallic elements, and 39·0 g water. Most of the yolk is formed during the 7—8 days period to ovulation and weighs 19 g, but the rest of the protein in the form of albumen, together with most of the water content and finally the shell must be produced in about 24 h once the egg enters the oviduct. The calcium in the shell represents 10 per cent of the body's total. In view of what was written on p. 172 it does seem that certain critical nutrient requirements for the egg may be accumulated and stored in advance of egg formation, but the capacity to do this is probably limited.

It is clear that the amount of yolk which can be deposited round the oocyte must depend on its rate of production and the period taken for the follicle to mature to ovulation, for the small quantity added before the final growth phase of the follicle can be ignored. The total amount of yolk deposited must also be a function of the number of follicles which develop simultaneously as a batch during the period of pair courtship. This is a time when oestrogen secretion is important in the female and, indeed, the synthesis of yolk proteins occurs in the liver under oestrogenic stimulation (see Lorenz 1969; McIndoe 1971). The process whereby albumen is added to each egg does not occur concurrently to affect several eggs at a time but instead each egg is treated individually during its stay in the oviduct. In considering the energetics of reproduction in birds, King (1973) has pointed out that the growth rate of the follicle is approximately sigmoid so that the energy requirement is bell-shaped in time. The separate energy needed for albumen, shell secretion, and transport through the oviduct constitutes a second peak of unknown form. However, to construct a simplified model of egg formation King assumed that the energy required for forming a single egg would follow a sine wave, so that the area beneath one cycle would be proportional to the total energy costs. Because a sine wave is used it is easy to demonstrate[†] that the peak daily energy cost E of producing one egg is given by $2A/p$ where A = the total energy in kcal to make one egg and p is the number of days over which the egg is formed.

King has plotted E against p and shows that E, the peak energy cost of producing one egg, increases sharply with decreasing period of egg formation. In seems that many species function at about the point where the curve of energy requirement begins to rise more sharply with decrease in p. He gives a series of examples showing that the period of follicular maturation varies from about 4—5 days in such passerines at the Great Tit, White-crowned Sparrow, and Jackdaw *Corvus monedula*, to 6—8 days in ducks and pigeons and 9—10 days in the Herring Gull *Larus argentatus*. Many middle- and large-sized birds, including such waders as the Stone Curlew *Burhinus oedicnemus*, take about 2 weeks to repeat if a fresh clutch is taken (personal observation). Allowing time for the initiation of courtship, this suggests that the phase of rapid follicle growth should occupy about 7—10 days. Nice (1937) noted that Song Sparrows laid the first egg of a repeat set 5 days after the loss of the clutch. This applied whether the eggs of the previous clutch had been just laid, were half-incubated, or ready to hatch. However, in many species clutches destroyed when half-incubated are replaced later than those destroyed when just laid; the interval is obviously related to the physiological condition of the pair at the time of loss. King demonstrated that the peak energy expenditure in producing a whole clutch depends on the amount of overlap among the growth cycles for each individual egg and the number of follicles growing (see Fig. 8.4(a)). If the sine model is followed, it is possible to show that the peak daily energy cost of growing several ova concurrently K reaches a peak, equivalent to the total energy cost of a single egg, on $p-1$ days after the onset of rapid yolk deposition in the first egg (p is defined above). Hence, once the peak is reached the number of eggs produced does not affect the daily energy requirement. King was primarily concerned to define the caloric cost of egg production in relation to other energy demands in reproduction, but it is instructive to extend his model to determine the circumstances in which p and clutch size become limiting.

Fig. 8.4(a) is the same as King's illustration. If the total energy A needed to form one egg over 5 days is 100 kcal, then the peak energy E required for this is equal to $2 \times 100/5 = 40$ kcal. The daily peak energy K needed to form a clutch is reached on day $p - 1 = 4$ when, as Fig. 8.4(a) shows, four follicles are growing simultaneously and require a peak daily energy of $K = 100$ kcal. Given that environmental energy resources are capable of providing a daily energy intake of 100 kcal or more per day, then the number of follicles growing simultaneously, which presumably is related to the clutch size, is not limiting once this peak is reached. In this example energy is conserved only if the total clutch is of 4 or fewer eggs. It should be noted, as King

[†]For this purpose the whole sine wave is plotted so that the amplitude is represented as the y-axis and the period as the x-axis, making $y = \sin x$. The scale of the x-axis is represented in terms of π to give $Y - 1 = (X - \frac{\pi}{2})$ so the model used is:

$$Y = \sin\left(X - \frac{\pi}{2}\right) + 1$$

also emphasized, that the peak energy requirement for the clutch is determined by, and is equivalent to, the energy content of an individual egg, and that it is independent of the period of formation of the individual egg or the energy peak needed for its formation. It is probable that many nidicolous, and particularly passerine, birds function well inside the potential energy threshold limit, so that their clutch size will be primarily geared to the number of chicks that can be effectively reared. It

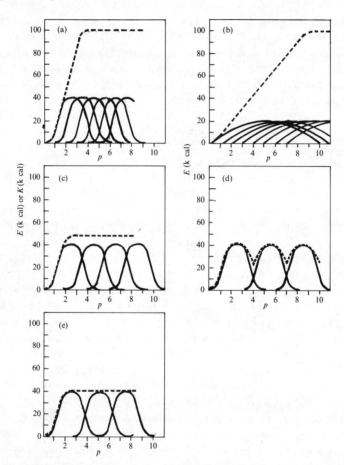

Fig. 8.4. Simplified models to show relationship between peak daily energy cost E in kilocalories of producing one egg containing 100 kcal, according to the number of days over which the egg is formed p and the interval between ovipositions R (solid curves). The model is based on King (1973a), who assumes that the energy required to produce one egg can be approximated by a sine curve which he uses in the form $y = \sin\{x - (\pi/2)\} + 1$. From this it can be shown that $E = 2A/p$, where A = the total energy needed to make one egg. In the figure A has arbitrarily been set at 100 kcal. The peak daily energy (K) required to produce a clutch is shown by the dashed line. (a) One-day interval between egg laying and each egg formed over 5 days; (b) one-day interval between egg laying and each egg formed over 10 days; (c) 2-day interval between egg laying and each egg formed over 5 days; (d) interval between egg laying is 3 days and p = 5; (e) maximum energy utilization for (c) and (d) occurs with a laying interval of 2·5 days with p = 5, given the constraints of this simplified model.

is also probable that the majority of such birds will be so-called indeterminate layers in that if eggs are removed as laid further eggs can be produced; clutch completion in such birds presumably depends on stimuli from the eggs in the nest initiating a feed-back inhibition of further laying. In some birds, including the Song Sparrow *Melospiza melodia* and House Wren *Troglodytes aedon*, the last-laid egg (clutch size varies from 4 to 9) is usually the heaviest (Nice 1937; Kendeigh, Kramer, and Hamerstrom 1956), perhaps because the last follicle does not have to share the available energy. In captive Bengalese Finches *Lonchura striata* the mean egg weight was found to increase systematically from the first to fifth egg of the clutch (Jefferies 1969). Whether or not there is any adaptive value in compensating later-hatched chicks with extra reserves is not known. It might be advantageous in the face of a high predation risk to reduce p if the rate of egg laying could thereby be increased and the clutch completed sooner. This would only apply if clutch size were smaller than $p - 1$ so that the clutch could be completed before peak was demanded, otherwise a reduction in p must lead to an increase in peak energy requirement for both egg and clutch. Therefore, in species not limited by the potential availability of energy resources p would most economically be close to the clutch size. No species lay more than one egg per day so possibly the physiological timing of ovulation in relation to the phasing of photoperiodic sensitivity, as discussed on p. 87, could set an upper limitation to the rate of egg production.

A bigger egg relative to the available energy or the maintenance of egg size with a reduction in the energy resource could be achieved by an increase in p (Fig. 8.4(b)). In the example given for a bird laying eggs needing 100 kcal, the peak energy for clutch completion is not reached until $p - 1$ days so that if p is increased to 10 days a clutch of less than $p - 1$ could be produced before peak energy resources were reached. This could provide a mechanism whereby clutch size might be adjusted to resources. Thus a species might form 4 eggs simultaneously but only begin a fifth if the energy available were particularly good, otherwise the fifth and subsequent follicles could become atretic. But to quote King again, 'as the period for producing eggs lengthens the total efficiency diminishes and is diluted by the continuing expenses of maintenance.' An increase in p also necessitates an increase in the period of courtship and a greater problem in adjusting the season of egg laying to the period of brood care when favourable environmental resources are required.

An alternative strategy for a species which finds it very difficult to reach the threshold necessary for egg production (or needing to maximize egg size) might be to produce eggs at intervals exceeding one day. Fig 8.4(c) still assumes that the total egg requires 100 kcal and shows that if the laying interval is increased to 1 egg every 2 days then the peak energy threshold for clutch formation is halved. If the interval between laying increases to 3 days the total energy required for growing a clutch drops between peaks, that is, the utilization of the energy resource becomes inefficient and the bird effectively grows a sequence of separate 1-egg clutches (Fig. 8.4(d)). The lower limit, when a clutch of continuously growing follicles changes into a series of separate eggs, occurs when the total energy curve just touches the peaks of the energy curve for each individual egg, neither dipping nor rising between the individual egg peaks; if we adopt a symmetrical sine-wave model this occurs half-way along one cycle so that a follicle should start growing each time the previous one reaches the halfway point in its formation at $p/2$ (Fig. 8.4(e)). In the example in Fig. 8.4(e) this is at intervals of 2·5 days[†] but in practice, for the reasons already given, there is probably a limitation on the fractional division of the endogenous ovulation cycle of approximately 24 h. The Shag *Phalacrocorax aristotelis* studied by Snow (1960) is possibly an example, for in this species the eggs are laid every third day although sometimes a longer interval is noted; a 4-day interval was recorded between eggs 2 and 3 and between 3 and 4 in some clutches started very early in the season—presumably a time when food supplies were most critical. It is likely that the female of this species produces eggs which are close to the size limit set by the available resources and there is some indication that egg laying ceases when the female is no longer able to form a good egg. Snow noted in 1957 that in 3-egg clutches the mean weight of second eggs was 2·5 g greater than the first and 1·3 g greater than the third; in 4-egg clutches the third egg averaged 1·2 g heavier than the second but 3·3 g more than the fourth (see also Coulson, Potts, and Horobin, 1969). The egg weight of females over 8 years in age was 13 per cent heavier than that of inexperienced 2 year old females breeding for the first time (Coulson, Potts, and Horobin 1969). The third egg of the Kittiwake (Coulson 1963) and of other gull species averages 4·4 per cent smaller in volume than the other two (see Väisänen Hildén, Soikkeli, and Vuolanto (1972) for details) and in the Herring Gull *Larus argentatus* the chick hatching from this egg has a much higher mortality than the other two, as Fig. 8.11 below from Parsons (1970) shows.

If environmental energy resources are limited, egg size can be maintained only by increasing the time taken for formation—which rapidly becomes inefficient—or by decreasing the frequency of laying. These factors set a limit to clutch size for if an increase in p leads to a wastage of energy, the daily rate of egg laying $R = p/2$ must be similarly affected. Moreover, it is likely that most species laying relatively large eggs are functioning near to the limits of food availability and must wait for a threshold in supplies to be attained. If so any increase in R or p would increase the risk of the resource level falling below the critical threshold before the bird could respond. Further, any increase in R increases the time during which the eggs are left unguarded and at risk of predation; if the female does not leave her eggs then her food-collecting ability must be impaired which by definition in the present example would be detrimental. Several discontinuities can be identified which suggest that there will be rather an abrupt transition between the conditions favouring the production of several eggs or 1 large egg. For instance, we have mentioned that there are probably limits to the fractional value of p, and there is a

[†]The greatest number of days R that can elapse between successive ovipositions before the bird effectively starts to lay separate clutches can be derived from King's formula, as:

$$p = 2A/K,$$

and since we need ½ of p this becomes:

$$R = A/K.$$

limit to the relationship between R and p unless the bird functions inefficiently. The sine-wave model has been extended (Murton and Lakhani, in preparation) to examine the way in which variations in R and p will affect clutch size. For this purpose it is assumed that the maximum available threshold of energy is fixed (at 100 kcal), that in all cases egg size is maximized, and that p is 8 days. The results are depicted in Fig. 8.5.

The actual caloric content per gram of birds' eggs varies from 1·02 kcal in the Tree Sparrow *Passer montanus* to 1·6 kcal in Galliformes and 2·02 kcal in Anseriformes, these last having a high fat content, and King assumes a net efficiency of energy transfer in egg production of about 70 per cent. He presents data for a range of species to suggest that the daily peak energy requirement for egg production is 0·7–2·4 times the basal metabolic rate. King further estimates that the daily maximum cost of egg production is equivalent to 21–30 per cent of the daily energy

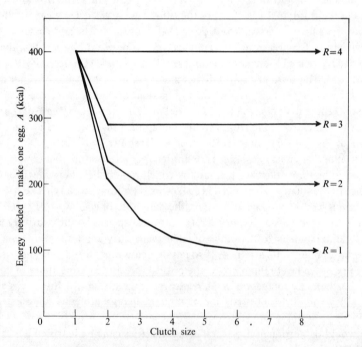

Fig. 8.5. Extension of sine-wave model as used in Fig. 8.4. to show the way in which variations in daily rate of egg-laying R and period over which eggs are formed in days p are related to the energy needed to make one egg A and to potential clutch size (number of eggs produced). The model assumes that the maximum available peak in environmental energy supplies per day E is fixed at 100 kcal, that egg size is maximized within this constraint, and that $p = 5$ days. If p is divisible by R so that $p/R = I$, where I is an integer, it can be calculated that

$$E^* = \sum_{R=0}^{I} \left\{ \sin \left(2\pi \times \frac{R}{I} - \frac{\pi}{2} \right) + 1 \right\},$$

E^* being a constant which can be calculated if p and R are given and hence I is given. Then $A = p \times E/E^*$. Based on Murton and Lakhani (in preparation).

intake at constant body weight in Galliformes, 52–70 per cent in Anseriformes, and 13–16 per cent of Passeriformes. The relative caloric increment for egg production is independent of body weight but the actual increment increases because as mentioned the B.M.R. increases with body weight as does the maintenance metabolic rate. It seems reasonable to assume that if egg weight is proportional to adult body weight W, we have an index of the energy needed for egg formation. Accordingly, Fig. 8.6 plots the egg weight as a percentage of adult body weight for different avian taxa, drawing on data from the various sources summarized by Lack (1968), against the characteristic clutch size of the group. As mentioned, proportionate egg weight decreases with increase in lg. W (Fig. 8.2) and so it is necessary to compare

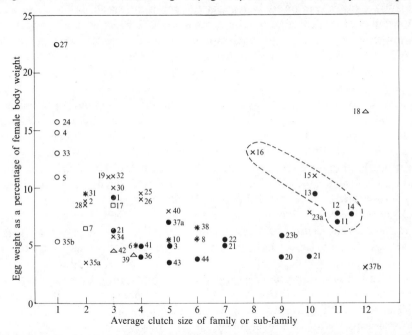

Fig. 8.6. Relationship between egg weight as a proportion of female body weight and mean clutch size for various avian families and sub-families or tribes. The percentage egg weight has been calculated for a bird weighing 500 g (see text) using data given by Lack (1968a). Percentage egg weight may be assumed to bear a relationship to E, the peak daily energy needed to form an egg, so that this figure can be compared with Fig. 8.5. Symbols depict the interval between laying of eggs: open circles, birds which lay only one egg; solid circles, eggs laid at 1-day intervals; crosses, 2 days; stars, 2–3 days; open squares, 3–5 days; open triangles, more than 5 days. The results for the Anatidae are somewhat anomolous and these are ringed.

Key to taxa: 1, *Tinamidae*; 2, *Gaviidae*; 3, *Podicipitidae*; 4, *Procellariiformes*; 5, *Phaethontidae*; 6, *Phalacrocoracidae*; 7, *Sulidae*; 8, *Ardeidae*; 10, *Threskiornithidae*; 11, *Dendrocygnini*; 12, *Anatini*; 13, *Aythyini*; 14, *Cairinini*; 15, *Mergini & Somateriini*; 16, *Oxyuri*; 17, *Falconiformes*; 18, *Megapodidae*; 19, *Cracidae*; 20, *Tetraonidae*; 21, *Phasianidae*; 22, *Numididae*; 23a, *Rallinae*; 23b, *Fulicinae*; 24, *Rhynochelidae*; 25, *Charadriidae*; 26, *Scolopacidae*; 27, *Dromadidae*; 28, *Burhinidae*; 30, *Chionidae*; 31, *Stercorariidae*; 32, *Lari-Sterni*; 33, *Alcidae*; 34, *Pteroclididae*; 35a, *Columbidae*; 35b, *Columbidae*; 36, *Psittaciformes*; 37a, *Centropodinae*; 37b, *Cuculinae*; 38, *Strigiformes*; 39, *Steatornithes*; 40, *Alcedinidae*; 41, *Coraciidae*; 42, *Bucerotidae*; 43, *Piciformes*; 44, *Passers*.

birds of similar weight. For this reason a figure of 500 g was arbitrarily chosen (to cover as many groups as possible) and the proportionate egg weight for different groups (variably the order, family, or genus) obtained from the regression lines relating body weight and percentage egg weight. Some groups had to be omitted, for example, the Anserini, because all species in the taxa either had weights well above or well below 500 g and it seemed unwise to extrapolate the regression.[*]

Fig. 8.6 demarcates a clear upper limit of egg size related to clutch size (*cf.* Fig. 8.5) although it can be seen that the Anatidae, compared with all other groups, produce eggs which are heavy relative to the number produced and frequency of laying; this is particularly noticeable when they are compared with the game birds.[†] The reason is unknown. It might relate to the high fat content of the egg alluded to above and to the ability of the female to store fat prior to egg laying, but, if this is the case, it is not clear why more groups should not have evolved a similar adaptation. If the Anatidae are omitted from consideration for the time being, it is evident that a consistent pattern is exhibited by the majority of avian species. Those birds which produce eggs weighing more than 8 per cent of the female body weight have clutches of 4 or fewer eggs and lay eggs at intervals of 2 or more days. The tinamous appear just to break the rule but it is possible that better data would indicate a modified laying frequency. In all known cases the cock incubates and rears the brood unaided by the hen. Successive polyandry and harem polygyny are common (see p. 435) and in some species different females lay in the nest of one male, who then incubates alone. Clearly, if females move round to different males depositing eggs in successive nests their laying rate is difficult to define.

Of the groups depicted in Fig. 8.6 as producing eggs at daily intervals, it might be guessed that the Centropodinae, Numidiidae, and Fulicinae are laying eggs close to the maximum size for the laying rate, that is, their plots fall on an energy threshold curve similar to the one depicted in Fig. 8.5. Groups such as the Piciformes and Passeriformes are perhaps producing eggs which are below the critical size limit. But it is impossible to prove this from Fig. 8.6 for in a group such as the Psittaciformes, which also lay small eggs relative to body size: their diet—largely fruit and buds—limit their capacity to produce larger eggs. It does, however, seem reasonable to suppose that species which produce small eggs relative to body size and also lay at less than daily intervals experience difficulty in reaching the energy threshold necessary for egg production. Examples are some Columbidae, Bucerotidae, Phalacrocoracidae, and Steatornithidae. Lack (1966) has quoted the studies of Snow (1961, 1962) to suggest that this explanation does apply to the fruit-eating Oilbird *Steatornis caripensis*, which lays eggs at variable intervals, sometimes of a week, and in which the clutch

---

[*]For this reason the Kiwi *Apteryx oweni* had to be omitted: Adult females weigh 1·2 kg and the egg comprises 25 per cent of the body weight. Moreover, the egg contains proportionately more yolk and less water than that of any other species (Reid 1971*a*, *b*).

[†]The yolk forms about 40 per cent of the weight of the egg in waterfowl and this proportion does not differ between species whose proportionate egg size to body weight varies (Lack 1968*b*; Siegfried 1969). Egg length in ducks is positively correlated with clutch size but egg breadth is not (Bezzel and Schwazenbach 1968). No explanation is currently available.

size varies with the season of laying (the incubation period is long at 33–34 days and the nestling period unusually extended and variable at 90–125 days). Hornbills are also frugivorous, with aberrant nesting habits, for the female is walled into the nesting-hole by the male, who plasters the nest-entrance with mud, leaving only sufficient space to pass food to the female (Moreau 1937).

Other groups in Fig. 8.6 that take more than 2 days to lay eggs are Falconiformes, Strigiformes, Stercorariidae, Ciconiidae, Ardeidae, Threskiornithidae, Sulidae, and Megapodidae. In the first two taxa the young are hatched at intervals and are of different ages. Lack (1954) considered asynchronous hatching an adaptation whereby if the food supply is poor the youngest and weakest chick receives no food and dies, so that brood size is rapidly adjusted to the resources available. Very often the larger chicks eat the smaller ones, and Wynne-Edwards (1962) quoted such cases of fraticide as mechanisms for the self-regulation of population size. Thus, the second chick of the Lesser Spotted Eagle *Aquila pomarina* invariably dies (Wendland 1958) and the younger chick of Verreaux's Eagle *A. verreauxi* and of other species is usually eaten by the older young (Siegfried 1968). Lack (1966) was perhaps too anxious to discredit this hypothesis of Wynne-Edwards—we think rightly—that he rather dismissed some pertinent field observations made by Ingram (1959) which were cited by Wynne-Edwards. These were that in many cases the larger nestling actually attacks and kills its sibling without waiting for it to die and it may or may not then eat its victim. We suggest that the reason for asynchronous hatching is not quite the same as Lack suggests but instead that it is a mechanism whereby a larder can be provided for the chicks. By laying a large clutch at intervals of 2 or more days the female can store energy for the chicks and reduce the need for sharp peaks in energy requirements (see also p. 172). This would explain why the chicks have a positive behavioural mechanism which induces them to kill their siblings. A similar adaptation may apply to the skuas (Stercorariidae)—the Long-tailed Skua *Stercorarius longicaudus* and Pomarine Skua *S. pomarinus* are convergently similar to the Short-eared Owl *Asio flammeus* in preying on Lemmings *Lemmus trimucronatus* and voles (*Microtus agrestis* and *M. arvalis*) and in laying larger clutches during plague years—and to the herons and other Circoniforms. In the falcons, owls, skuas, and storks, the egg weight is high relative to adult body weight so the species concerned could potentially reduce egg weight or clutch size. It is possible that the females are actually laying separate eggs and that these should not be considered as clutches in the sense discussed above. Finally, in both the Phalacrocoracidae and Sulidae incubation begins with the laying of the first egg and the young hatch asynchronously so, as with the falcons and other groups just discussed, the later-hatched and weaker chicks can be sacrificed if food is scarce. Moreover, in the tropical *Sula* species studies by Dorward (1962) the stronger chick usually fights and kills the second.

Since the Phalacrocoracidae have relatively the smallest eggs and largest clutches in marine birds the question arises of when a large egg or a large clutch will confer the greater advantage. Lack's view is that if food supplies make it difficult to rear young, brood size will be reduced in favour of a large egg, for this will in turn enable the chick to be relatively well developed and provisioned at hatching and so less at risk of starvation. Alternatively, if it is easy to feed a chick, the brood size can be increased at the expense of egg quality, for the adults can compensate when the eggs hatch. In this way clutch size and egg size depend on a balance of selective factors. When young are hatched it is possible for both parents to share in the task of providing food, at least once constant brooding is no longer necessary. In contrast the formation of the egg primarily implicates the female for the mobilization of energy cannot be partitioned equally between the sexes. Nevertheless, courtship feeding, which is widespread in birds (Lack 1940; Andrew 1961), does provide a mechanism whereby the male can contribute to the formation of the egg; a suggestion first made by Cullen and Ashmole (1963) during studies of the Black Noddy Tern *Anous tenuirostris*, in which the period of courtship feeding coincides almost exactly with the time elapsing between the loss of one egg and its replacement by a repeat. Courtship feeding performance in the Common Tern *Sterna hirundo* is correlated with the total weight of the clutch and the third egg; fledging success of the third chick is in turn dependent on the size of the third egg (Nisbet 1973). Thus, courtship feeding by the male can in favourable years increase chick production. The male Great Tit provides the female with about a third of her daily total intake, enabling her to form eggs without exhausting her own reserves (Royama 1966a; Krebs 1970). If, in such species, the female undertakes the bulk of early incubation duties the male is left free to recoup any weight loss or, if this is not necessary, he can continue to feed the sitting female. This adaptation reaches full development in the Bucerotidae (hornbills). In harsh environments which become suitable as breeding places late in the season and at relatively unpredictable times, as occurs with the melting of snow on mountain tops or its disappearance from tundra flows, it is advantageous for some female bird species to lay their eggs as soon as conditions become amenable. This may be done at the expense of the female's own body condition so that immediately after laying they must recoup the lost energy reserves. This means that the male must undertake the task of incubation and brood care and in some species there has been a reversal of sex roles, as in the Dotterel *Eudromia morinellus* and phalaropes *Phalaropus* and *Steganopus*, in which the females are larger and more brightly coloured than the male (see p. 439).

Seasonal changes in egg size

The relationship between egg weight and clutch size can be studied in those species where both vary during the course of the breeding season. In the Great Tit the weight of the eggs increases during the season whereas clutch size decreases (Perrins 1970; and Fig. 8.7). As mentioned, the first eggs must be laid as soon as the female is able so that the chicks are born in time to utilize good supplies of defoliating caterpillars. Even so, many individuals cannot lay early enough and they leave fewer survivors than those that are the first to lay eggs. Second and late repeat clutches are laid when the immediate food supply is reasonably good, that is, just after the first young are fledged, but when a decline in supplies is imminent. Under these conditions it is advantageous to lay a larger egg, thereby ensuring a better reserve for the newly hatched chick which must be reared during a period of reduced re-

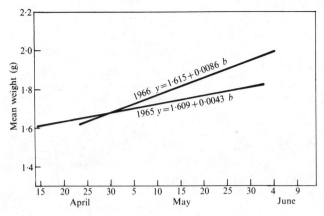

Fig. 8.7. Regression lines showing change in mean weight of Great Tit *Parus major* eggs with date of laying. Based on Perrins (1970).

sources, while a reduction in clutch size reduces the number of young needing to be fed (Fig. 8.7). An effect that could be important has been noted by O'Connor (1975*c*). He found that a large brood of chicks was a greater stimulus to the parent during the changeover from incubation to feeding behaviour and so was fed more intensively than a small brood. As a consequence the weight of chicks in large broods was found to increase more rapidly than was the case among the young of small broods. In the Kittiwake *Rissa tridactyla*, females laying late in the season produce lighter eggs than those breeding early (Coulson 1963). The same is true of the Gannet *Sula bassana* but in this case it is because older females are the first to breed, and egg size increases with age (Nelson 1966). In the Herring Gull *Larus argentatus* egg volume declines during the season (Parsons 1975). In the Wood Pigeon eggs laid at the start of the breeding season are the heaviest and egg weight declines steadily until the end, while clutch size remains constant at 2 eggs throughout (Fig. 8.8; see between pp. 82–3). Food supplies gradually improve from spring until late summer for birds breeding early depend on tree buds and only later do weed and grass seeds become available. Large eggs are laid at those seasons when the parents will have greatest difficulty in feeding their chicks. Feeding difficulties are likely to occur early in the chicks' lives when they must be continuously brooded and the adults are prevented from feeding as freely as otherwise, and this is when the squabs rely entirely on crop milk for nourishment. We suppose that a balance has evolved between the quantity of food that can be accumulated in the egg and the amount that can be supplied in compensation when the squabs hatch.

During incubation eggs lose weight† at a constant rate unless they are addled so weight at hatching bears a direct relationship to weight at laying (Figs 8.9, 8.10):

†The air space of the egg has a low specific heat and it cools faster than the rest of the egg. This rapid cooling reduces water loss when the parent leaves the nest by 50 per cent compared with that from equivalent areas of the rest of the shell. The total effect of the air space is probably to reduce water losses during incubation by 10–15 per cent (Simkiss 1974).

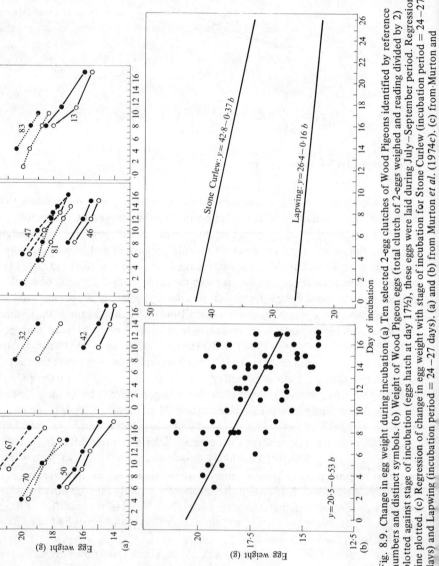

Fig. 8.9. Change in egg weight during incubation (a) Ten selected 2-egg clutches of Wood Pigeons identified by reference numbers and distinct symbols. (b) Weight of Wood Pigeon eggs (total clutch of 2-eggs weighed and reading divided by 2) plotted against stage of incubation (eggs hatch at day 17½), these eggs were laid during July–September period. Regression line plotted. (c) Regression of change in egg weight with stage of incubation for Stone Curlew (incubation period = 24–27 days) and Lapwing (incubation period = 24–27 days). (a) and (b) from Murton *et al.* (1974*c*). (c) from Murton and

in the Gannet there is a weight loss of 9—13 per cent of the initial weight (Nelson 1966), 20 per cent in the Wood Pigeon (Murton *et al.* 1974*a*). Heinroth (1922) reckoned for many different bird species that the chick at hatching weighed about two-thirds the weight of the fresh egg so, allowing 10 per cent for shell weight, his data imply a 25 per cent weight loss. The average hatching weight of Great Tits was 72·6 per cent of fresh egg weight (Schifferli 1973). Excluding predation losses, the hatching success of Wood Pigeon eggs varied according to their weight at laying: 76 per cent of eggs weighing over 20 g hatched, 42 per cent of those weighing 16—20 g did so, but no egg weighing less than 16 g was successful. This implies that there must be a critical weight below which it would not be worthwhile laying eggs. Hatching success was not related to egg weight in *incubator*-hatched Great Tit eggs, but the hatching success was much lower than in the wild (Schifferli 1973). Post-hatching survival has been related to egg volume in the Herring Gull (Fig. 8.11).

The weight of Wood Pigeon chicks at hatching was found to be uncorrelated with their weight 6 days later, during which time crop milk forms a major part of the diet, but their weight on day 6 was correlated with that a few days prior to fledging on day 16 or 17 (Murton *et al.* 1974*a*). This means that the adults were able to compensate light-weight chicks by giving them more food during the first few days of life. However, this part of the study was undertaken during the July—September part of the breeding season, when environmental food supplies are good, and we suspect that food supplies during the early few days of nestling life could be more critical at the start of the breeding season. Schifferli found that egg weight had a significant effect on Great Tit nestling weight up to the fourteenth day, for the young from lighter eggs grew more slowly. But these chicks subsequently recovered and mortality after fledging appeared not to be influenced by egg weight. Again,

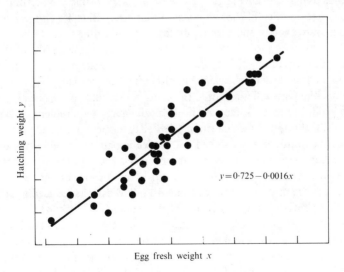

$$y = 0·725 - 0·0016x$$

Fig. 8.10. Hatching weight of nestling Great Tits *Parus major* in relation to the fresh weight of the egg. From Schifferli (1973).

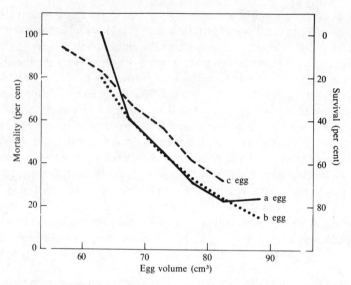

Fig. 8.11. The relationship between egg volume and post-hatching chick mortality/survival in the Herring Gull *Larus argentatus* for each egg of the laying sequence. From Parsons (1970).

most of the experiments were performed at a time of year when natural food supplies were abundant. Fortunately, with the help of data collected by Perrins, Schifferli could show that, whereas fledging success was independent of egg weight in early broods, it was positively correlated with egg weight in late broods. Clearly, it is advantageous to hatch from a large egg if food is scarce during the nestling period. Schifferli quoted Lachlan (1968) who found that heavy Blue Tit *Parus caeruleus* eggs contain more protein and carbohydrates than do lighter ones.

In the Wood Pigeon there is extremely little variation in the weight of the two eggs within any one clutch, but considerable variation exists between different clutches laid at any particular season. The first effect presumably depends on the fact that once the female has acquired the energy threshold to lay one egg, a second egg can be produced without increase in the peak energy requirement. It is possible that an additional internal mechanism may exist to ensure a matched egg weight for this would help the newly hatched chicks to compete on equal terms. By artificially manipulating brood size at hatching it has been shown that chicks from broods of two survive as successfully as do those from broods of one so that two-egg clutches are the most productive (Table 8.1). Success was reduced in artificial broods of three but nevertheless a brood of three could evidently be more proproductive than one of two, especially as there appeared to be no significant differences in post-fledging survival related to brood size (Murton *et al.* 1974a). An incapacity of the female to lay more than two eggs without reducing their viability is more likely to be the limiting factor than the adult's ability to feed the chicks. Those pigeon species which feed predominantly on fruit or buds tend to lay one-

TABLE 8.1

Weight at fledging and subsequent survival rate of Wood Pigeon squabs in different sized broods created at hatching

Brood size	Mean weight on day 16 (number weighed)	Number chicks	Percentage fledged	Number fledged per brood
1	299 ± 34 (24)	45	98	1·0
2	274 ± 38 (44)	86	98	1·7
3	246 ± 55 (44)	99	84	2·5

Based on data in Murton *et al.* 1974a.

egg clutches, egg weight representing a higher proportion of the adult's body weight (Fig. 8.6). The production of a proportionately larger egg is seen as a means of ensuring that the development rate is reduced so that the chick is hatched at a more advanced stage of development to counter the adult's difficulty of converting low-energy foods. The Band-tailed Pigeon *Columba fasciata* which in North America occupies a niche which is similar to that taken by the Old World Wood Pigeon, does appear to depend on buds to a greater extent. It lays 1 egg, which takes 19 days to hatch compared with 17½ days for the egg of the Wood Pigeon.

Variability in weight between clutches is presumably a reflection of the capacity of different females to produce good eggs. Egg weight is positively correlated with body weight in female Wood Pigeons weighing less than 480 g but not in heavier birds (Fig. 8.12). Most females (or males) in their first year of life do not attain adult weight until July or August and it may be inferred that before this time they would lay lighter eggs than adults. This cannot be proved because first-year birds are not reproductively mature until July (see p. 272) and it is possible that the capacity to lay viable eggs has been one of the selective factors involved in delaying their assumption of breeding status. Young Great Tits lay larger eggs than old birds (Winkel 1970), but it is not clear whether this depends on them laying later than old birds or whether this compensates their ability to feed chicks (see above). Limited data for the Shag suggests that females also lay eggs in proportion to their body weight (2·8–3·2 per cent) so heavy females lay heavier eggs. Snow does not specifically state whether the heavy eggs hatch more successfully as in pigeons, but this might be inferred since she established that the last egg laid had a markedly lower hatching success than the second or third egg while the first one laid had an intermediate success; these variations parallel the differences in egg weight. Egg size and female body size have also been shown to be positively correlated in the Ringed Plover *Charadrius hiaticula*, Redshank *Tringa totanus*, and Dunlin *Calidris alpina* (Väisänen 1969; Väisänen *et al.* 1972). Variations in egg weight related to the habitat have been described in the Lapwing *Vanellus vanellus* (Murton and Westwood 1974a). They are thought to reflect differences in the quality and quantity of invertebrate food available to the laying female according to the soil conditions. The factors affecting egg shape and interspecific and intra-clutch variability remain to be defined; Preston (1969) has analysed elongation, asymmetry, and bicone in more than 500 species. Egg length, but not width, in the Song Thrush *Turdus philomelos* is correlated with age and nesting habitat, for unknown reasons (Gromadzki 1966).

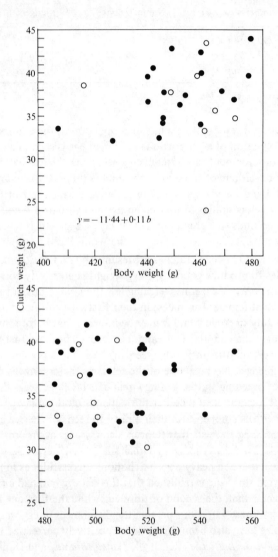

Fig. 8.12. Relationship between the body weight of female Wood Pigeons *Columba palumbus* and the weight of their clutches (2 eggs). There is no significant relationship in females weighing more than 480 g but there is for adult females weighing less than 480 g ($r_{15} = 0.562$; $P = 0.02 - 0.01$). Adults more than 12 months old, solid circles; first-summer females entering their first breeding season, open circles. From Murton *et al.* (1974c).

Nestling growth

Growth curves are available for the nestlings of several species (Fig. 8.13). Fig. 8.14 depicts a theoretically efficient growth curve in which the energy requirements for maintenance and growth equal the feeding capacity of the adult, but in practice such curves are never realized. Instead, as Ricklefs (1968, 1969a) emphasizes, growth curves are usually sigmoid in shape resulting in peaks in energy requirements and the full food gathering power of the adults cannot be achieved at first. This may be because the adults of altricial young must initially find time to brood so cannot at the same time collect food. Hence the maximum absolute rate of growth usually occurs about the mid-point of the growth period. In altricial nestlings different organs grow at different rates and maximum use is made of the food supply to grow those most needed at any time. For example, flight feathers and the pectoral musculature are not needed until the young are nearly ready to leave the nest and these do not begin growing at first. Instead, the newly hatched chick has a dispro-portionately large head and gape and digestive system so that the parents can deliver, and the chick process, as much food as possible. O'Connor (1975a) has studied some of these relationships in the Blue Tit. During the first half of the nestling period one parent feeds the chicks and the other broods and so at this stage the chicks have virtually no thermo-regulatory capacity: the chick's respiration rate increases with ambient temperature, that is, they are poikilothermic. The chicks develop partial homoiothermy half way through the nestling period, with growth of the feathers, but even so cannot maintain their body temperature with very low ambient tem-peratures. Now both adults can collect food and complete homoiothermy develops. As the chicks grow both the total dry weight and fat content increase linearly but their water content at first increases and then levels off. It is possible, therefore, that the assumption of thermo-regulation by the chick is achieved by losing water.

Royama (1966b) found that food consumption by nestling Great Tits was inversely proportional to brood size.[†] Subsequently Mertens (1969) confirmed Royama's suspicion that this depended on a saving in heat production in large broods as a result of a more favourable surface—volume ratio. Evidently, this variation in metabolic rate with brood size must influence the amount of food finding required of the adults. O'Connor (1975a) has measured the oxygen consumption of young Blue Tits in broods varying from 1 to 11 and has found that energy demands are lowest in broods of medium size; his results differ a little from those of Mertens, but the latter was studying a larger species. It follows that below a critical brood size the adults can effect no saving in the amount of food they need to supply to their chicks, presumably because extra energy is needed by the young to keep warm. Presumably too, large broods expend extra energy in keeping cool (see also Van Balen and Cavé, 1970).

In more detailed studies of the Blue Tit, O'Connor (1975b) found that at an ambient temperature of 15 °C the metabolic rate decreased with brood size in both

[†] Royama (1969) later suggested that increase in clutch size with 'latitude' could be explained by a model incorporating the energy needs of chicks as a function of ambient temperature, and the hunting time available to adults.

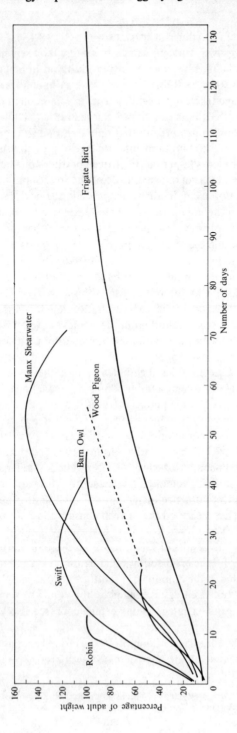

Fig. 8.13. Growth curves of birds presented in terms of percentage adult weight. In all cases solid lines indicate period in the nest and dependence on adults. Wood Pigeons leave the nest almost 50 per cent below adult weight and put on the remaining weight by feeding themselves independently; this phase of development is indicated by the dashed line. Sources of data are: Robin *Erithacus rubecula*, Swift *Apus apus* and Manx Shearwater *Puffinus puffinus*, Lack (1968a), Barn Owl *Tyto alba*, Ricklefs (1968); Wood Pigeon *Columba palumbus*, Murton et al. 1974c; Ascension Frigate Bird *Fregata aquila*, Stonehouse and Stonehouse (1963).

6-day (poikilothermic) and 12-day (homoiothermic) broods but it rose again with very large broods in 6-day chicks. This could be attributed to the lower body temperature maintained by 6-day-old nestlings in which hyperthermia was countered by evaporative cooling. At 20 °C no surface area effects were found because evaporative cooling was needed in most broods at this temperature. In contrast, in the House Sparrow *Passer domesticus* heat retention by the nest was of major importance for all age groups at the two temperatures tested and surface area effects were negligible. The normal size of the Blue Tit brood is 9–11 young and the nest is a simple cup of moss and hair placed in a tree hole. As O'Connor points out, the young of altricial birds are at first kept at a constant temperature by the brooding adult so that all the food they receive can be devoted to growth. As the chicks grow a greater feeding effort is required by the adult and selection should favour improvements in nest insulation which minimize heat loss at this stage. It seems probable that selection for a bulky, domed nest has arisen in these circumstances in such species as the House Sparrow, which has small broods (3–4 young) in which homoiothermy develops slowly. In contrast in species such as the Blue Tit, which have a large brood, this problem is less acute and there is instead a high risk of hyperthermia developing as the young approach fledging (Van Balen and Cavé 1970). In the Blue Tit the nest becomes flattened into a flat platform as the chicks grow and they can space themselves to increase heat loss when ambient temperatures are high. In some altricial species with exposed nests the parents may need to shade the chicks from the sun at times (see Fig. 15.10).

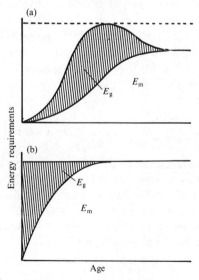

Fig. 8.14. Schematic representation of energy requirements of young birds with E_m being that required for maintenance and E_g that for growth. The dotted line is the maximum level at which adults can supply energy. (a) Inefficient because young have sigmoid growth (the usual pattern) as a function of age and the feeding potential of the adults is not fully utilized early and late in the growth period. (b) Theoretically the most efficient growth form possible. From Ricklefs (1969*b*).

As shown in Fig. 8.13, the daily growth rate of most nestling passerines does not decrease until over half the total growth is accomplished. Conceivably, in these species selection has been towards maximizing the growth rate to reduce predation risks—albeit acting at the egg stage—so that brood size is frequently well within the feeding capacity of the adults. For example, in the Blue Tit there is very little difference in the growth—weight curve for broods of 6 compared with broods of 12—what differences do exist being most apparent in the mid-phase of nestling life (Schifferli 1973). If the rearing of young is difficult the first strategy that can be adopted is to reduce clutch and hence brood size. If the adults are unable to rear even a single chick at the normal growth rate, adjustments must be made to reduce the daily energy requirements of the growing chick and this may be achieved by slowing down the rate of growth (Fig. 8.13). In the Frigate Bird *Fregata aquila* of Ascension Island the point of inflexion of the sigmoid growth curve occurs when less than one third of the total growth has been achieved (see Stonehouse and Stonehouse 1963).

The adaptations in egg size and nestling growth in face of feeding difficulties are well exemplified by the Manx Shearwater, already alluded to at the start of this chapter. Migrants which have returned from the South Atlantic begin to come ashore on dark nights in February to visit their nesting burrows, where the pair spend increasing periods so that by April up to 6 days at a time are spent together in the burrows in courtship. This phase of mutual courtship is probably important in synchronizing the physiological cycles of the pair, for they must now desert their burrow and spend all their time at sea. This enables the female to acquire the food reserves needed to form the large single egg, which represents nearly 15 per cent of her body weight (see Fig. 8.6), while the male accumulates food stocks which allow him to take the first stint of incubation during which he does not feed. Similar behaviour occurs in the Fulmar Petrel *Fulmarus glacialis* (Dunnet, Anderson, and Cormack 1963) and other sea-birds. The so-called 'honeymoon' period spent at sea immediately prior to egg laying lasts for about 35 days in the Manx Shearwater but, because there is much individual variation in the onset of egg laying, it is not particularly evident if the colony is watched for there are always some birds present (Harris 1966). However, in the Short-tailed Shearwater *Puffinus tenuirostris*, which also leaves the colony for about 35 days, the exodus, like the onset of egg laying, is more or less synchronous throughout the colony and so provides a considerable spectacle (Marshall and Serventy 1956). Egg laying in the Manx Shearwater usually extends from late April until mid-May, the mean occurring around 6—10 May (Harris 1966; Matthews 1954) whereas the laying period in the Short-tailed Shearwater (Slender-billed Shearwater) is confined to the period 20 November—3 December, with 85 per cent of the eggs being laid within 3 days of the mean (Serventy 1963). The incubation period is long and in the Manx Shearwater lasts 51 days while the fledging period involves a further 70 days (53 and 94 days respectively in *P. tenuirostris*). The protraction of growth and development over 121 days contrasts with the 21 days typically needed in many passerines.

The chick of the Manx Shearwater, unlike that of the Frigate Bird, is for a while heavier than the adult in consequence of a store of fat which serves to keep the young going during times of feeding difficulty, when the parents may go for long periods without returning with supplies. Such adaptations are presumably needed in situations where the fledging period is long and feeding conditions for the adult are at risk of fluctuating, and the same applies in the case of the aerial-feeding Swift *Apus apus* (Lack 1956), the frugivorous Oilbird (Snow 1961, 1962*a*) and North Atlantic Gannet *Sula bassana*. In equatorial oceanic regions feeding conditions may be more stable and explain the lack of fat reserves in *Fregetta* and also in the tropical species of *Sula*. It is also possible that in these hotter regions there is a risk of the young becoming hyperthermic. It is known that young Manx Shearwaters find it very difficult to lose heat at high temperature (30 °C) because of the insulating properties of their subcutaneous fat (Perrins 1973).

It was suggested above (p. 188) that it would be advantageous for the incubation period to be short in any situation where the risk of predation is high and Lack (1968*a*) argued that this accounted for the longer incubation period of species nesting in the safety of holes, compared with closely related species nesting in open sites. Lack also suggested that any increase in embryonic development rate to reduce the length of the incubation period would also affect the rate of development of the nestling (see Fig. 8.3). It is, therefore, of interest that Ricklefs (1969*b*) discovered that the rate of body growth in birds was not correlated with the nestling mortality (Fig. 8.15(a)) but was instead a function of the adult body size and the mode of development of the young (Fig. 8.15(b)). Of course, it has long been assumed that the nestlings of large birds suffer less mortality than do those of small birds, but there are many exceptions to this generalization (which in any case has not been statistically confirmed) and it is greatly influenced by nest location and habitat.

In spite of the preceding remarks, it remains evident that adaptations which reduce the vulnerable period of nestling growth and thereby reduce any predation risk must have survival value. Ricklefs (1969*b*) argued mathematically on theoretical

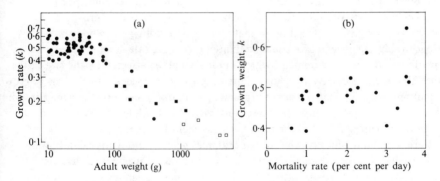

Fig. 8.15. Relationship between overall growth rate (expressed as the rate constant *k* of the logistic equation) and: (a) adult body weight of temperate-zone passerines (filled circles) and raptors (squares); open squares are raptors whose growth curves were fitted by the Gompertz equation (see original); (b) mortality rates of various temperate zone passerines. The correlation coefficient *r* = 0·445 is not significantly different from 0. From Ricklefs (1969*b*).

grounds that the most efficient conversion of energy into progeny is realized when the growth rate is maximized at the expense of brood size. Given a constant resource it is best to produce a large number of one-egg clutches and fast-growing young than a brood of several young with a slower rate of development. As Ricklefs states, this is because the number of broods raised increases relatively faster than the energy requirement of the young as growth quickens; he should be consulted for the derivation of the equations which justify this conclusion. Ricklefs advanced several reasons to account for the failure of most species to perform in this way, in fact, those having single young tend also to be the species having slow growth rates. For instance, he suggested that in environments which are variable over short periods a longer growth period would be more likely to incorporate average conditions than would a very brief growth period. It is also possible that selection for the adaptation of the egg stage of development has been stronger than for the nestling stage. It is generally true that nestlings suffer a lower predation rate than eggs (see Tables 14 and 15 in Lack (1954) and numerous subsequent studies). But it is by no means clear why eggs are at greater risk of predation, and it might be thought that a noisy and plump brood of nestlings would provide a better and more easily found meal than a nest of eggs. In the Wood Pigeon the risk of predation depends on the ability of the parents to manage proper incubation routines, their capacity to do so being directly influenced by the food supply (Murton and Isaacson 1962). For birds in general food supplies are better when young hatch than when eggs are laid so parental attentiveness is likely to increase as the season progresses. However, once the period of maximum available resources is past young will be hatched with poorer food stocks than were available to the birds during incubation. In many bird species, hatching success tends to increase at the end of the season, even exceeding nestling success, while the nestling success decreases. Examples are provided by Swallow *Hirundo rustica* (Adams 1957), Willow Warbler *Phylloscopus trochilus* (Cramp 1955), Chaffinch *Fringilla coelebs* (Newton 1964), and House Sparrow *Passer domesticus* (Summers-Smith 1963). Exceptions are the Blackbird *Turdus merula* and Song Thrush *T. philomelos* (Snow 1955).

Ricklefs draws attention to the slope of the regression relating growth rate to adult body weight, for it is close to that relating metabolic rate to body weight. Conceivably the growth rate could be limited by (a) the rate at which energy is distributed during development, which in turn would be affected by the size of the digestive and processing organs or (b) the proportion of tissues actually able to grow simultaneously. Ricklefs favours the second possibility but admits the need for more experimental data. An important discovery made by Ricklefs (1973) is that precocity of development of function (he used the attainment of flight capacity as the main criterion of adult function but quoted examples where a more critical definition was needed) in altricial, semi-altricial, and semi-precocial species was inversely correlated with the growth rate (Fig. 8.16) and directly related to future survival probability (see p. 13). Thus, a slowed growth rate is associated with the attainment of flight capability at a *relatively* early age. At present, it is impossible to decide which are the essential causal correlations. For instance, it could be reasoned that the mobility of precocial young confers an evolutionary advantage by allowing them greater flexibility in facing environmental vicissitudes. Precocial development carries the disadvantage that the egg must be well provisioned but a concomitant advantage in that it enables the growth rate to be reduced. By having a slow growth rate the embryo can spend a longer time in the egg so that food resources can be partitioned into a proportional growth that enables the chick at hatching to be relatively mature. The structure at hatching in turn limits the growth rate for these precocial species need to have well-developed legs and muscles at birth, as well as an insulation of down feathers. In consequence, more energy has to be diverted to maintenance and thermo-regulation. In contrast, the rapid growth rate of altricial species demands a high supply of energy, so that the embryo must rapidly exhaust its egg supply and the chick hatches at an immature stage of development. So much so, in fact, that its food-processing organs must be of a disproportionate size to facilitate a relatively enormous input of energy by the parents. Given a trend for a reduction in the growth rate, the adult body weight can increase and the metabolic rate decrease, presumably resulting in an increased efficiency. Perhaps this trend has enabled such species as

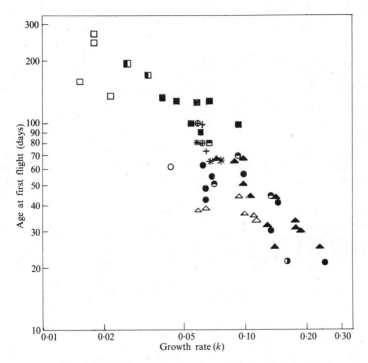

Fig. 8.16. Relationship between achievement of mature body function (judged by age at which first flight is achieved) and the growth rate expressed as a constant. Key to familes: Stars, Spheniscidae, Procellariidae, Hydrobatidae; circles containing crosses, Phaethontidae; solid squares, Sulidae; squares divided vertically, Fregatidae; open squares, Diomedeidae; Horizontally divided squares, Pelecanidae; horizontally divided circles, Phalacrocoracidae; vertically divided circles, Ardeidae; open circles, Ciconiidae; solid circles, Laridae; open triangles, Alcidae; solid triangles, raptors and Corvidae. From Ricklefs (1973).

the Manx Shearwater to occupy oceanic habitats. An argument which has been frequently expressed is that because the birds live close to the limit of their food resources the growth rate has become slowed and the reproductive rate reduced. It is difficult to avoid a circular argument of the chicken-and-egg type. Nevertheless, it is to be expected that any surviving phylogenetically long evolved species and groups are the ones most likely to have acquired a close relationship with environmental resources.

9　Some north-temperate breeding cycles

Summary

In this chapter the integration of proximate and ultimate factors (*sensu* Baker) in regulating breeding periodicity is considered for some selected temperate species, which have been adequately studies physiologically and ecologically; particular attention is given to the House Sparrow *Passer domesticus* and some closely related *Zonotrichia* species. Numerous experiments, involving a wide range of bird species and the manipulation of the photoperiod, have emphasized how the duration of the light period given to photosensitive subjects affects: (1) the rate of gameto-genetic development; (2) the time during which active gametogenesis continues before the subject enters a spontaneous state of photorefractoriness; (3) the duration of photo-induced refractoriness. In the majority of experiments the subject has been exposed to a constant photoperiod whether it be a 'short' (e.g. LD 8:16) or 'long' day (e.g. LD 16:8). When receptive subjects are exposed to stimulatory daylengths gonad development is a log-linear function of the duration of exposure within part of the range of photoperiods employed. Since any particular species exhibits responses which vary with the nature of the photo-regime, it would be expected that parallel variations would be noted in the same species living at different latitudes and in some cases the consequent alterations in response might be sufficiently adaptable. Thus, the gonad cycle of the Goldfinch *Carduelis carduelis* in Australia, where it has been introduced by man, is a little longer than in Britain. A similar effect to that obtained by altering the photoperiod cycle can be obtained by changes in 'photosensitivity threshold'. This is illustrated for two races of the Starling *Sturnus vulgaris*, for some closely related *Zonotrichia* species, for certain icterids and finches of the genus *Leucosticte*, and for various waders and other birds.

Proximate and ultimate factors

In the preceding two chapters the energetics of breeding, moulting, preparation for migration, and other, often conflicting functions, were considered and the need for them to be partitioned at the most appropriate season was emphasized. It was clear that there could be little temporal latitude between appropriate and inappropriate responses. A bird which delayed too long in beginning its moult might be faced with trying to grow new feathers when feeding conditions had become critical. In order to manage a precise temporal ordering of seasonal functions in competition with conspecifics the bird is helped if it can be guided by some environmental signal. This applies just as much in tropical forest or ocean as in markedly seasonal temperate zones. However, conditions which are suitable for breeding or moulting are not always heralded by an environmental signal which is sufficiently reliable to be used by the bird as a predictor. This is true in some desert areas where rain arrives at unpredictable times yet the birds must be able to respond as quickly as possible. A different problem is posed in almost non-seasonal equatorial habitats and consideration of these is deferred until Chapters 13 and 14.

In temperate regions, particularly the north-temperate hemisphere, the unique value of seasonal daylength variations as a primary means of timing periodic functions is evident, especially compared with such environmental cues as temperature or rainfall, which can be hopelessly variable and uncertain. Nevertheless, equatorial and tropical birds are also photoperiodic; their breeding schedules can be manipulated by experimentally imposed lighting regimes and in the wild their diurnal rhythms are regulated by light-entrained circadian oscillators as are those of all birds (Chapter 11).[†] This suggests that we should seek to understand the evolution and adaptation of photoperiodic mechanisms in terms of the responses exhibited by tropical or at least subtropical species. As we seek a coherent theory for avian photoperiodic phenomena, caution is needed, for there are tropical species which almost certainly evolved in more temperate areas. It has also been suggested by some authors that photoperiodic timing mechanisms have been evolved many times and independently in different birds groups, although it will be argued in Chapter 14 that this is not so and that many constraints have been placed on the basic photo-response mechanism of birds.

At this point, it is convenient to remind ourselves of the two separate influences involved in the regulation of breeding seasons, these being the so-called ultimate and proximate factors of Baker (1938), which are discussed in Chapter 1. Ultimate factors are those which affect the survival of progeny to a reproducible age and thereby determine the optimum period during which young should be produced. For example, if temperate birds attempted to breed in winter it is likely that, for various reasons, none of their eggs would result in the production of reproductively viable descendants; Fig. 8.1 (p. 187) provides three realistic examples. Natural selection must operate against any tendency for birds to breed at non-productive seasons, and act particularly intensely if the survival prospects of the adults are reduced in consequence of such breeding attempts. The immediate inhibition of non-seasonal or inappropriate breeding depends on the involvement of a proximate factor(s). Usually the response mechanism of the bird to environmental signals, as discussed above, provides the proximate control. For example, if a bird attained a reproductive condition only when it experienced relatively long photoperiods winter breeding would be impossible and the photoperiodic response mechanism would thereby provide proximate control. Obviously, natural selection will ensure that responses only develop to the most appropriate environmental signals so that proximate timing factors are reliably linked with those ultimate factors which ensure optimal survival prospects for any offspring. It is important to distinguish the two factors for in man-altered habitats the proximate regulation of annual events may no

[†]Phillips (1971) clearly failed to appreciate the functioning of and entrainment of endogenous rhythms when he argued that the rôle of photoperiodism had been over-emphasized (see p. 285).

longer be appropriate to an ultimate survival function. In Chapter 13 it will be shown that Feral Pigeons *Columba livia* var. living in urban areas could nowadays breed through-out the year as a result of the feeding opportunities provided by man but they are prevented from taking full advantage of the changed circumstances because of proximate responses to photo-regimes. Similar factors apply to some of the avian species which inhabit arable farmland (Murton and Westwood 1974*a*).

Various kinds of breeding schedule adopted by birds were considered in Chapter 7, both schematically (Fig. 7.3, p. 152) and with examples (Fig. 7.4, p. 154) and emphasis was placed on the way that environmental energy supplies could be partitioned to maximize survival prospects, including progeny production. In this and the next chapter we will examine how proximate factors are integrated with ultimate factors in some selected species, paying particular attention to the rôle of the photoperiod. In birds an appropriate daylength cycle allows the sexual apparatus to recrudesce so that breeding is feasible. Whether or not breeding will actually take place depends on various modifying factors, the most obvious of which being the availability of a mate and a nesting location. It is these modifying factors which Marshall (1959) classified as accelerators or inhibitors (see Chapter 1). In no temperate-zone photoperiodic species have any of these other factors been shown to cause gonad recrudescence in the absence of appropriate photostimulation. Temperature is probably the most important modifier of the gonad cycle, its effect having been demonstrated experimentally in a number of species: Starling *Sturnus vulgaris* (Burger 1948), Gambel's White-crowned Sparrow *Zonotrichia leucophrys gambelii* (Farner and Mewaldt 1952), Slate-colored Junco *Junco hyemalis* (Engels and Jenner 1956). Threshold effects may be important, for Delius (1965) found that Skylarks *Alauda arvensis* laid a few days after the mean spring air temperature rose by 10 °C. Moreover, individual females appeared to need different levels of stimulation and were more constant in the laying order in different years irrespec-tive of the starting date of egg laying in the population as a whole.

In birds the male is responsible for the initial timing of the breeding season, for it is usually he who alone acquires and advertises a territory and then courts the female, as was discussed in Chapters 5 and 6. It might be expected that the male rather than female should be most sensitive to modifying factors. Recently, Lewis and Farner (1973) have shown experimentally that temperature differences in the range 5·2–34·1 °C have a small positive effect on the rate of photoperiodically induced vernal development of the testes of Gambel's White-crowned Sparrow, but the effect was insufficient to serve a 'predictive' function. In the races *gambelii* and *pugetensis* of *Zonotrichia leucophrys* environmental temperature had no effect on the rate of development of the ovary up to the inception of vitellogenesis. Develop-ment of *Zugunruhe* and fattening appeared to be favoured by high ambient tempera-ture and the onset of pre-nuptial moult by low temperature. Lofts and Murton (1966) found that the spring temperature was correlated with the testis develop-ment of Wood Pigeons *Columba palumbus* in eastern England in March, at a time when food supplies were good, but they found it difficult to separate the effects of sunshine and photoperiod from those of temperature *per se*. Temperature did not exert any appreciable influence on spermatogenetic development in January and February but at this time the food supply appeared to assume some importance, for the testes were more advanced in their vernal recrudescence in seasons when clover food supplies were good. Experimental starvation (Bissonnette 1931) or shortage of environmental food supplies (Marshall 1951) apparently have an inhibitory effect on male birds. On the other hand, giving domestic fowl extra feed over 40 weeks to induce obesity had no effect on testis weight, sperm density in the semen, fertilizing capacity or hatchability of eggs (Parker and Arscott 1972).

The effect of temperature on the timing of egg laying by the Great Tit *Parus major* has been well studied under natural conditions, for it begins laying earlier when the weather in March and early April is warm (Fig. 9.1). Since a cumulative period of fine weather is important, the 'warmth sum', representing the total of the daily mean temperature, is used in Fig. 9.1. The relative position of the points plotted in the figure, but not the overall correlation, would vary depending on the period chosen for the warmth sum (Perrins 1973). Some of the correlation established by Perrins between temperature and laying date probably depends on food availability for the laying female, and it is known that high spring temperatures cause the caterpillars and other insects on which the birds feed to hatch earlier. In fact there was no difference in the amount of testicular growth in photostimulated groups of Great Tit that were held at a mean of −6·3 °C or 15 °C in January−February

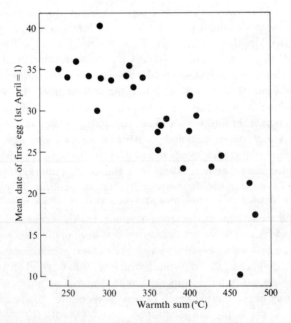

Fig. 9.1. Relationship between data at which egg-laying commences in the Great Tit *Parus major* and the spring temperature depicted as the 'warmth sum' (the sum of the mean minimum and maximum temperatures for each day during the period 1 March−20 April). Each point represents a different year. From Perrins (1973).

(Suomalainen 1938). Perrins quoted Löhrl (1957) who showed that the time of egg laying by the Pied Flycatcher *Ficedula hypoleuca* and Collared Flycatcher *F. albicollis* was correlated with temperature in March and April, yet these birds are migrants which do not return to Europe from Africa until mid-April. Evidently, the effect of temperature was indirect, probably affecting the food supply. Cavé (1968) showed how ovarian development in the Kestrel *Falco tinnunculus* occurred throughout the winter and was related to temperature so that in a mild season ovarian development was advanced. He attributed this effect to the improvements in food supply that resulted in mild seasons and similar conclusions were drawn by Southern (1970) for his Tawny Owl *Strix aluco* data.

House Sparrow

The House Sparrow *Passer domesticus* is a relatively unspecialized Old World temperate-zone passerine, which following introduction by man has successfully become established in many parts of the world. Its photo-responses have been well studied and it provides a convenient starting-point for our species review. The House Sparrow is classified in the sub-family Passerinae (sparrows) which comprise 3 genera: the snow finches *Montifringilla* with 7 species, 6 of which are confined to Africa and 1 extends to Europe; the rock sparrows *Petronia* with 5 species all occurring in Africa, although 3 extend to Asia or Europe; and the true sparrows *Passer* comprising 15 species, half of which are confined to Africa. The adaptive radiation of the European sparrows seems to have occurred in North Africa and the Mediterranean (Summers-Smith 1963). Furthermore, it seems likely that the House Sparrow speciated and extended its range northwards in close association with Paleolithic farmers during the Pleistocene so that its present commensal relationship with man is not a secondary adaptation. For most of the year it feeds on cereal grain gleaned from storage bins and the barn floor, flighting to the standing corn and stubbles in season and supplementing this cereal diet with weed seeds.

For breeding the House Sparrow must collect insects for 84 per cent or more of the chick's diet during the first 3 days of life is composed of invertebrates, this proportion decreasing to a quarter by the date of fledging (Summers-Smith 1963). Aphids are favoured for very small chicks and larger, tougher foods, such as caterpillars and various beetles, are provided as the young grow. In temperate Europe caterpillars and aphids are most readily available in May and June, when plant tissues are young and succulent, whereas seeds and cultivated cereals become more abundant from mid-summer onwards.

A relatively wide spectrum of food choice allows the House Sparrow to attempt breeding from April until August, after which the moult occurs from July—mid-October (Fig. 9.2). When records of the number of clutches laid in different localities and years are combined a smooth curve results which indicates that egg laying is at a peak in May and declines gradually until early August (Cramp quoted by Summers-Smith (1963)). At any one locality there is a tendency for egg laying to be synchronized in the population and for there to be 3, sometimes 4, distinct peaks of egg laying throughout the season.

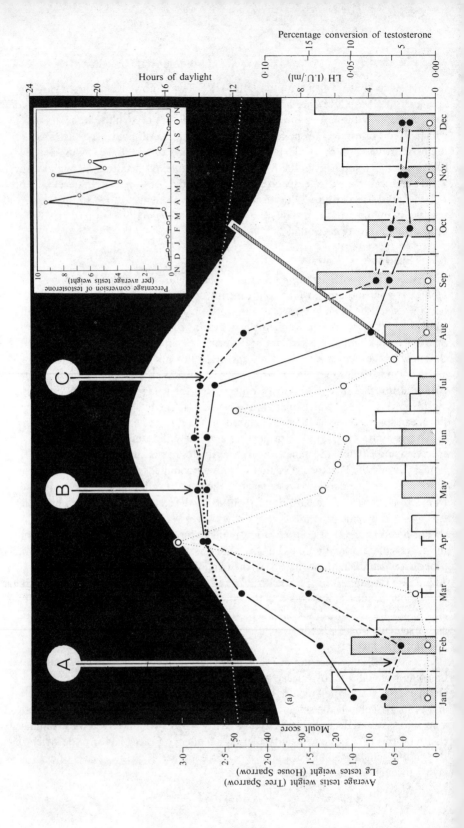

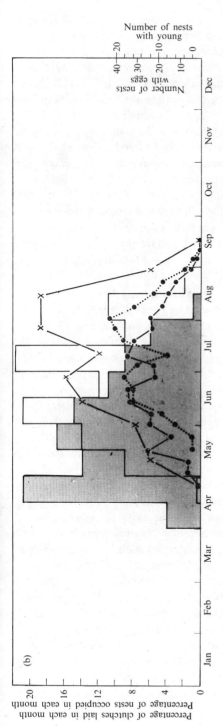

Fig. 9.2. Correlates of reproductive functions in House Sparrows *Passer domesticus* in Britain and Tree Sparrows *Passer domesticus* in Britain and Tree Sparrows *P. montanus* in Hong Kong.

(a) Mean lg. testes weight of House Sparrows collected in Cambridgeshire (dashed line) or Surrey (solid line) (from Murton and Westwood 1974b). The seasonal daylength cycle at 52 ° N is shown and the letters refer to stages in the cycle when subjects were caught and used for experiments mentioned later in the text (p. 315). The testes cycle of weight change for the Tree Sparrow in Hong Kong is shown by dotted lines and open circles and the limits of the daylength cycle in Hong Kong are indicated by heavy dots (from Lofts and Lam 1973). Stippled histograms refer to the mean plasma concentrations of immunoreactive luteinizing hormone (Bagshawe) in those subjects collected in Surrey; these were caught between 1100–1400 h (from Murton 1975). Open histograms represent the seasonal variation in the capacity of testicular tissue from Tree Sparrows (Hong Kong) to biosynthesize testosterone *in vitro*, measured by the percentage conversion from Δ^5 –pregnenolone–16α–H^3. (from Chan and Lofts 1974).

Inset: Depicts seasonal variations in the capacity of testicular tissue from Tree Sparrows to biosynthesize testosterone but in this case the results are expressed as the percentage conversion per mean testis weight. (From Chan and Lofts 1974). The diagonal line is the regression of primary moult score on date for House Sparrows in Britain. (Based on Bibby 1970).

(b) Percentage of clutches laid (open histogram) and of nests occupied with eggs or young (solid line and crosses) at a study site in Cambridgeshire (unpublished data) and of clutches laid at a site in Oxfordshire (solid histogram) (from Seel 1968a). The number of nests with eggs (solid line and filled circles) or with chicks (dotted line and solid circles) at the Cambridgeshire site is also shown. The Cambridgeshire gonad data depicted in (a) are from birds from the same site.

This pattern is masked by the fact that females breeding for the first time begin laying 15–25 days later than older females; sometimes their clutches coincide with the appearance of second clutches from old females, sometimes they are laid between the first and second clutches of the older birds (Summers-Smith 1963; Seel 1968a). Seel studied House Sparrows inhabiting urban and suburban sites at Oxford and here the laying date of first eggs varied between 4 April in 1961 to 12 April in 1964. We have studied House Sparrows at a small village in Cambridgeshire, where the birds were mostly concentrated round farm buildings and a complex of gun-dog kennels. Here first clutches generally appeared in early May and exceptionally during the last week of April. Seel established a correlation between the mean air temperature (8·3–10·6 °C at Oxford in early April) and the onset of laying. Since incubation requires extra energy, a threshold is reached with a fall in ambient temperature at which all the available surplus is required. Thus, at very low temperatures incubation must be compensated for by increases in energy intake (Kendeigh 1961, 1963; see also King 1973). The mean daily air temperature at Cambridge during the first week of April has been 7·2 °C rising to 12·8 °C at the time egg laying begins in early May. The laying season at Oxford and Cambridge is detailed in Fig. 9.2.

The ultimate control of the sparrow's breeding season must depend on the female's ability to lay eggs, the success of the eggs and chicks while in the nest, and the post-fledging survival of young to an age at which they too can reproduce. The number of eggs produced per clutch varies according to locality and season of the year. The commonest clutch for the whole season at both Oxford (Seel 1968b) and Cambridge (Table 9.1) was of 4 eggs, but the mean clutch size at Cambridge was slightly lower because a majority of females laid only 2 or 3 eggs during the first half of the season. The mean size for all clutches laid in the first half of the season at Oxford was 4·11 ± 0·76, whereas at Cambridge the maximum mean monthly clutch ever recorded was 3·59 ± 1·01 (32 clutches) in June 1968. Clutches of 3 slightly outnumber clutches of 4 in April, May, and June so that the mean clutch size is low at the beginning of the breeding season, increases during June and July, and then decline again at the end of the season in August; Cambridge data are given in Table 9.1 but a similar pattern has been noted at Oxford (Seel 1968b).

Clutch size has to be considered in relation to the success of the eggs and any resulting chicks. A proportion (33 per cent at Cambridge) of all clutches produced vanish completely. Sometimes such predators as Grey Squirrels *Sciurus carolinensis* are involved, or Jays *Garrulus glandarius* or Magpies *Pica pica* can be the culprits. Our records suggest that only 1·2 per cent of eggs are lost in this way. In most cases where the whole or part of a clutch vanishes it is the parent bird which ejects the eggs from the nest, probably because they fail to hatch or else the parent is unable to incubate properly. Table 9.1 shows that the rate of disappearance of whole clutches was related to the size of the clutches (in contrast to Seel (1968b) as was the rate of partial loss. It seems possible that females which laid clutches of suboptimal size were in some way inferior—perhaps they were bad mothers or laid infertile or thin-shelled eggs—and this was reflected in the high failure rate of their eggs.

TABLE 9.1

Clutch size and hatching success of the House Sparrow at Carlton, Cambridgeshire

Clutch size	Number of clutches laid (per cent)	Percentage of clutches failed completely	Percentage of eggs laid which hatched	Number of eggs hatched per clutch	Number of chicks subsequently fledged per clutch
Early clutches 1967–9 (April–June)					
1	16 (5)	88	13	0·1	0·1
2	33 (10)	42	45	0·9	0·8
3	124 (39)	24	55	1·4	1·1
4	119 (37)	24	59	2·4	1·6
5	25 (8)	52	36	1·8	1·2
6	–	–	–	–	–
7	1 (1)	0	[43]	[3·0]	[2·0]
Late clutches 1967–9 (July–August)					
1	17 (6)	88	12	0·1	0·1
2	28 (10)	46	36	0·7	0·6
3	54 (19)	31	55	1·6	1·0
4	146 (51)	32	51	2·0	1·4
5	40 (14)	20	58	2·9	1·7

For all years 1966–70 the mean monthly clutch size (number in bracket) was:
May 3·18 ± 1·05 (160); June 3·53 ± 1·02 (239); July 3·61 ± 1·03 (226); August 3·34 ± 1·07 (91).

Brood size at hatching can vary in consequence of the loss of some eggs from a clutch. Thus, a clutch of four eggs could give rise to a brood of 1, 2, 3, or 4 chicks. Table 9.2 shows that the adults found it easier to raise broods of 1, 2, or 3 chicks than broods of 4 or 5 in the April–June period, whereas in July and August broods

TABLE 9.2

Brood size and the fledging and post-fledging success of the House Sparrow at Carlton, Cambridgeshire

Brood size at hatching	Number of broods (per cent)	Percentage of chicks which fledged	Number chicks fledged per brood	Number nestlings ringed	Percentage recovered over one month old	Number recovered per brood
Early broods 1967–9 (April–June)				*Early and late broods*		
1	31 (15)	77	0·8	41	7·3	0·07
2	57 (28)	77	1·5	118	14·4	0·26
3	76 (37)	80	2·4	252	9·1	0·23
4	35 (17)	58	2·3	168	11·9	0·38
5	4 (2)	60	(3·0)	23	17·4	0·57
Late broods 1967–9 (July–August)						
1	23 (12)	65	0·7			
2	48 (26)	60	1·2			
3	59 (32)	68	2·1			
4	51 (27)	68	2·7			
5	5 (3)	[44]	[2·2]			

of 4 were no more difficult to rear than smaller-sized broods (Table 9.2). It should, however, be noted that overall success was lower during the latter half of the season. It is not immediately clear why the birds should be more adept at rearing broods of 4 later in the season but have a generally poorer success than earlier. Seasonal changes in the composition of the nestlings' diet might be involved; it is conceivable that a bird might sometimes be able to collect a few large food items but at other times many small items. Egg weight might increase during the season and thereby reduce the parental burden in feeding chicks.

Some of the chicks which fail to fledge are taken by predators, but the proportion lost in this way is small (1 per cent or less). Many chicks vanish and in our studies the whole brood vanished in 22 per cent of cases. The loss of chicks in this way appeared to result from starvation, and this could be proved in some cases, whereupon the dead chicks were removed from the nest by the parents. Seel (1970) noted two kinds of nutritional deficiency. Too little food usually led to the death of only a few members of the brood whereas the wrong kind of food affected the whole brood. Young either grew too slowly and died or they lost weight and succumbed. Partial incubation occurs during laying and while clutches of 2–3 eggs hatch synchronously the last egg of larger clutches is usually delayed by a day; if food shortage occurs the youngest and weakest nestling goes without while the rest of the brood is well fed (Seel 1968b). Seel (1970) weighed nestlings when near to fledging (on day 13½) and found that the mean weight decreased as the number of survivors from each initial brood size decreased. Thus, the chicks in a brood of 4 at fledging which had survived from a brood of 6 at hatching weighed more than chicks in a brood of 3 at fledging if these too originated from a brood of 6 at hatching, and similarly for all other combinations. This suggests that if a brood suffers from food shortage all chicks suffer to some extent as a result of sibling competition. Nestling weight at fledging also declined in relation to an increase in initial brood size, that is, the chick surviving from a brood of 1 weighed more than the chicks surviving from a brood of 6. This would be expected on the assumption that chicks in small broods receive proportionately more food. A similar decrease in success with increase in brood size has also been recorded in the Spanish Sparrow *P. hispaniolensis* in Kazakhstan, in Southern Russia (Gavrilov 1963).

Variations in the success of different sized clutches and broods contribute to the seasonal variations summarized in Table 9.3. If 100 pairs of sparrows lay 100 clutches of eggs in the period April—June then according to Table 9.1 5 pairs will have 1-egg clutches and produce 0·5 chicks, 10 pairs with 2 eggs will have 8 chicks and 100 pairs *in toto* will produce 122·2 chicks (1·2 chicks per clutch). Performing a similar calculation for 100 clutches in July and August shows that 100 clutches would give rise to 120·8 chicks. Evidently, the combined probability for all pairs of producing chicks which will leave the nest is similar for the two parts of the breeding season because clutch size increases in summer and compensates for a poorer breeding success. It appears that at the beginning of the season the females have some difficulty in laying eggs but can rear the chicks relatively easily but later this situation is reversed. This could happen if different kinds of food were suitable for

TABLE 9.3

Seasonal variations in the breeding success of the House Sparrow at Carlton, Cambridgeshire

Month	Number eggs laid	Percentage hatched	Percentage of chicks fledged of: eggs hatched (nestling success)	eggs laid (breeding success)
April	15	33	20	7
May	362	55	74	41
June	625	52	73	38
July	743	56	66	37
August	266	39	62	24
Total:	2011	52	69	36

the adult than for the young chick. It can be seen in Table 9.1 that clutches of 3 or 4 were commonest between April and June, those of 4 being the most productive, while in July and August clutches of 5 were the most productive but clutches of 4 the most frequent (see p. 186). The limited records detailed in Table 9.2 suggest that post-fledging survival was unrelated to brood size for there are no statistically significant differences in the results. More data are required to resolve this point.

Natural selection must favour House Sparrows attempting to breed from April until early August. To facilitate this the neuro-endocrine system is photostimulated by the short days of January, allowing gonadotrophin secretion to initiate the mitotic division of spermatogonia (Table 9.4). At first the testes increase slowly in weight and the growth rate seems to increase in March and April but if weight

TABLE 9.4

Seasonal changes in spermatogenetic development of House Sparrows in Surrey and Cambridgeshire 1968–9

		Number of individuals with:			
	Spermatogonia	spermatocytes Primary	Secondary	spermatids	spermatozoa
Jan	3	9	1	–	–
Feb	9	7	1	–	–
Mar	2	5	2	–	3
Apr	–	–	–	–	11
May	–	–	–	–	9
Jun	–	–	–	–	22
Jul	–	–	–	2†	10
Aug	3	7†	–	2†	1
Sep	17	1†	–	–	–
Oct	14	–	–	–	–
Nov	19	–	–	–	–
Dec	9	–	–	–	–

†Degenerating germ cells

increase is plotted on a logarithmic scale development is seen to occur at a constant rate (Fig. 9.2). House Sparrows collected in Surrey were more advanced in their vernal gonad recrudescence than comparable subjects from Cambridgeshire (Fig. 9.2; see Murton and Westwood (1974*b*) for statistical details).

With the assumption of reproductive capacity the bill of the male House Sparrow changes from a pale yellowish horn colour to black, a secondary sexual change absent in the female. Keck (1934) considered that the bill was pigmented during the period of greatest testicular development, although he appreciated that some birds acquired black bills by January. In our studies a fully black or blackish bill was occasionally noted in subjects with the paired testes weighing as little as 2·7 mg, while a black bill had often developed by the time the paired testes weighed about 5 mg and contained primary spermatocytes. The black pigmentation of the beak of the House Sparrow has long been regarded as an androgen-dependent secondary sexual characteristic, and has even been quoted as a reliable bioassay technique for the estimation of androgenic hormones (Keck 1932, 1933, 1934; Witschi 1955, 1961). However, purified testosterone propionate injected into House Sparrows during the post-refractory period in November failed to cause changes in beak pigmentation. Neither exogenous FSH or LH alone produced bill-darkening but given together they resulted in a good response; rather less bill-darkening occurred with a combination of FSH and testosterone (Lofts, Murton, and Thearle 1973). Bill colour scores for hormone-treated subjects are detailed in Table 3.1 (p. 53) using a scale ranging from 1 = unpigmented non-breeding condition to 4 = fully pigmented black bill.[†]

The mandibles of the males remain intensely black between February and April, but then some individuals begin to show signs of pigment loss from mid-May onwards (Murton & Westwood 1974*b*). Hence, although all the males are reproductive active and produce spermatozoa until mid-July some waning in the intensity of bill pigmentation becomes apparent 2 months earlier. There is a marked loss of bill colour beginning in mid-July as the gonads regress so that between August and December all birds have pale yellowish-horn coloured beaks.

Blood samples were collected with edetic acid from the sparrows caught in Surrey, these samples always being taken at the same time of day, between 13.00 h and 15.00 h. Pooled monthly plasma extracts assayed in the Bagshawe radioimmunoassay system, already described (p. 39), indicated seasonal changes in IR–LH content, as depicted in Fig. 9.2. Maximum plasma concentrations of IR–LH were noted in subjects killed during or just after regression in September. Thereafter, IR–LH titres fluctuated but remained relatively high throughout the period until January, increasing somewhat in February before falling to a level too low to measure in March and April, coincident with the re-establishment of spermatogenesis, rapid change in bill colour, and the presumed secretion of androgen. There was some elevation in the concentration of circulating IR-LH in May and June, a reduction

---

[†]Lofts *et al.* (1973) originally divided category 4 into grades 4 and 5 to distinguish a grey-black from jet-black bill, but this refinement is ignored here for it was not followed when field samples were graded.

in July, and then the concentration rose in August as the testes regressed. It is not yet certain that the *total* daily secretion of LH varied in the way described for it remains possible that the secretion of gonadotrophins occurs during a limited part of each day; their half-life in the plasma is short (but see p. 495). Consequently, subjects could have produced as much LH in the summer months as during the winter, but this might not be detected if secretion occurred in the early morning, well before the time when the birds were killed and autopsied.† However, data for Japanese Quail and domesticated ducks suggest that this will not be the case (see Chapter 12). It will be valuable to repeat these experiments using the more sensitive assay methods, now available. Work is in progress and rapid developments in the field can be expected. The high titres of IR-LH which were recorded in September and early winter temporally correlated with an interstitium containing large numbers of juvenile Leydig cells, which were accumulating sudanophilic lipoidal material, and with tubules which were also charged with cholesterol-positive lipids. These tubule lipids became dispersed in late January or February with the beginning of spermatogenetic development.

Photo-responses and photo-refractoriness

A striking feature of the sparrow cycle is the spontaneous gonad regression seen to occur in late July or early August (Fig. 9.2) at a time when the photoperiod is above the stimulatory threshold for a response; it is at this stage that the bird has entered its refractory phase (already alluded to in Chapter 2) during which gametogenesis cannot be elicited by photostimulation, although as soon as the phase is ended the subject is once more photo-responsive. Morphologically the refractory period is marked by the rapid collapse of the gonads, and histologically in the male by the accumulation of cholesterol-rich lipoidal material in the seminiferous tubules (Fig. 3.4). The phenomenon was first discovered by Riley (1936) when studying the photo-responses of House Sparrows. Birds experimentally photostimulated by 'long' days at this stage of the cycle do not respond and the refractory period can be dispelled only by returning the subject to a period of 'short' days as was later demonstrated by Vaugien (1955) in the case of the House Sparrow. Vaugien also showed that exogenous treatment with gonadotrophin would overcome refractoriness and enable a degree of gametogenesis to be restored in this species. But it was Burger (1949), working with Starlings *Sturnus vulgaris*, who did most during these early years to define the characteristics of refractoriness. He prematurely terminated refractoriness in Starlings by exposing them to short photoperiods and thus showed that the duration of refractoriness is determined by the light treatment and that it can be shortened by exposure to a period of short days (Burger 1947, 1949, 1953).

In 1949 Burger wrote 'refractoriness is perhaps the greatest relatively unsolved problem in reproduction' a statement that has remained true until very recently. Indeed, we still do not have a complete answer, and will defer further consideration until new data are presented in Chapter 12. For the moment it is sufficient to rely

†In captive Herring Gulls *Larus argentatus* on natural photoperiods a peak of IR-LH was noted in spring but an even higher peak occurred in November as photo-refractoriness ended (Scanes *et al.* 1974).

on the empirical observations which served until the late 1960s. These had already begun to make it clear that the photoperiodic response was not dependent on the absolute duration of daylight. For example, Fig. 9.2 shows that there were about 9·6 h of daylight (including 80 min of civil twilight) when gametogenesis started in House Sparrows in mid-January in England, increasing to 13·9 h in March when the testes were well advanced spermatogenetically. Yet Threadgold (1958, 1960a) had shown experimentally that House Sparrows could be stimulated into full breeding condition by much shorter photoperiods and successfully used cycles of LD 1:23, LD 4:20, and LD 7:17, his results being summarized in Fig. 9.3. It is seen that under short photoperiods the growth rate of the testes was reduced but active spermatogenesis persisted for longer. A period of refractoriness hardly occurred with cycles of LD 1:23 or LD 7:17 at 75 lx of illumination nor with LD 4:20 at under 10 lx, for testicular involution was immediately followed by recrudescence. Threadgold's experiments were repeated and his results subsequently confirmed by Middleton (1965). The main value of her contribution was to show that Gambel's White-crowned Sparrow would not respond to short photoperiods of 2, 4, or 7 h of light per day and even under a 20-h photoperiod it responded more slowly than the House Sparrow (Fig. 9.4). The difference is discussed below.

A proliferation of photo-experiments, using all kinds of species, captured at different seasons, and treated in multifarious ways, led during the 1950s to a rather confused picture of the photoperiodic phenomenon in birds. As early as 1932 Bissonnette and Wadlund recognized that the intensity and amount of light affected the relative length of the phases of the sexual cycle. But it was Wolfson who probably did most to collate and co-ordinate the subject, basing his interpretations on his own experiments with Slate-colored Juncos *Junco hyemalis.* in 1959 he published a valuable review which effectively demonstrated that the duration of the daily photoperiod determines:

(a)   the rate at which gametogenesis occurs in non-refractory subjects and hence the time when breeding can begin;

(b)   the amount of response, often conveniently measured in terms of absolute testis size;

(c)   the time for which a response persists before involution and regression spontaneously occur;

(d)   the length of time necessary to break refractoriness with 'short' day treatments. If subjects are kept on 'long' days without any intervening period of 'short' day treatment, gonad development can apparently be prevented indefinitely; in Golden-crowned Sparrows *Zonotrichia atricapilla* kept by Miller (1954) this was for 310 days, at which point the experiment was terminated.

Under natural conditions the duration of refractoriness varies from species to species and its adaptive value is to prevent breeding at seasons when stimulatory daylengths occur but when it is disadvantageous to attempt reproduction (Lofts

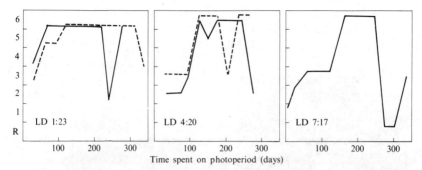

Fig. 9.3. Stage of spermatogenesis achieved (R; resting spermatogonia to 6, spermatozoa) by photosensitive House Sparrows subjected to various photoperiod regimes over the time span indicated. The light intensity was either 75 lx (solid lines) or less than 10 lx (dashed lines). Based on Threadgold (1960*a*).

and Murton 1968). For the majority of north-temperate species breeding is feasible sometime during the spring and early summer, rather than in late summer and autumn which is when refractoriness occurs.

The way in which different experimental light schedules affect the rate of testicular development and subsequent involution is illustrated by interesting data for the Starling from Hamner (1971) and Schwab (1971) as summarized in Fig. 9.5. The results generally confirm the conclusions already attributed to Wolfson, but they have the advantage that the subjects were kept for long periods on the different

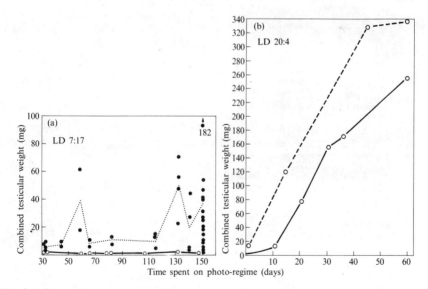

Fig. 9.4. Changes in mean testes weight of photosensitive House Sparrows *Passer domesticus* (broken lines) or White-crowned Sparrows *Zonotrichia leucophrys gambelii* (solid lines) held on photo-regimes of a (a) LD 7:17 or (b) LD 20:4. Individual readings for House Sparrows only are given in (a) to indicate the variability in response. Based on Middleton (1965).

light schedules and this raised new issues discussed below (p. 341). For the present we are interested in the behaviour of a single species when exposed to different experimental light schedules. We are led to suspect that variable responses will be

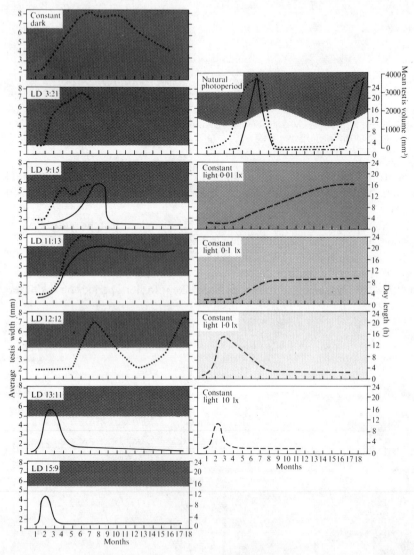

Fig. 9.5. Effect on testis size of holding photosensitive Starlings *Sturnus vulgaris* on the light regimes depicted. The testicular cycle under natural photoperiods either refers to wild birds in Britain (52 °N), as studied by Bullough (1942) and shown as dot-dashed line, or to free-living birds in California (38 °N) studied by Schwab (1971). Data in the left-hand panels refer to a light intensity of about 300 lx when lights were on and are either from Hamner (1971), shown by solid lines, or Schwab (1971) or Rutledge and Schwab (1974) (dotted lines). Data in right-hand panels for constant light of different intensities are from Hamner (1971).

manifested by any particular species residing at different latitudes in consequence of differences in the natural light cycles. A variable response might also result if changes in the sensitivity of receptor organs bring the same consequences as alterations in the intensity of the photostimulus as in Fig. 9.3. To an extent these suppositions are confirmed in House Sparrows living at different latitudes (Table 9.5). At latitude 34 °N the maximum photoperiod in mid-summer is of about 15 h duration compared with 17 h or more at 54°, and Table 9.5 shows that the breeding season is longer and refractoriness persists for a shorter time at Pasadena, U.S.A. compared with Cambridge.

The maximum size attained by the testes becomes greater in subjects of the same species exposed to longer photoperiods this being demonstrated for the House Sparrow in Fig. 9.6 at different latitudes. In a comparable manner, the absolute maximum size attained by the testes as a ratio of adult body weight in closely related species is greatest in the most photosensitive species. This is shown in Fig. 9.6 for *Columba palumbus, C. oenas, C. livia,* and its feral variety. *C. livia* can assume breeding condition under much shorter photoperiods than *C. palumbus* (see Fig. 7.11). The egg-laying season of the Tree Sparrow *Passer montanus* is a little shorter than that of the closely related House Sparrow in Britain (see p. 314) and its gonad cycle presumably very similar. Yet in Hong Kong there are two marked peaks of spermatogenetic activity in the testes cycle of the Tree Sparrow, in April and July, according to Chan and Lofts (1974) (see also Fig. 9.2). Whether this is a consequence of the less extreme seasonal light-dark cycle or of synchrony of the stage of reproductive activity from first to second broods is not yet clearly established. The House Swift *Apus affinis* has two breeding seasons at Baroda, India, and a double cycle of gonad regression and recrudescence in both breeding and non-breeding birds (Naik and Razeck 1967). Whether this represents a development of the autumn breeding of species like the Tricolored Blackbird, discussed below (p. 240), and Rook (p. 256) is not yet clear. It is interesting to consider the Black Swan in this context (p. 388).

TABLE 9.5

Duration of spermatogenesis and refractoriness in different populations of the House Sparrow

| Locality | Latitude in degrees north | Approximate duration in days of: | |
		active production of spermatozoa	refractoriness
Pasadena, U.S.A.	34	138	13
Norman, U.S.A.	36	135	31
London, Ontario	43	118	64
Cambridge, England	52	106	100

Based on Threadgold (1960*b*) plus personal data

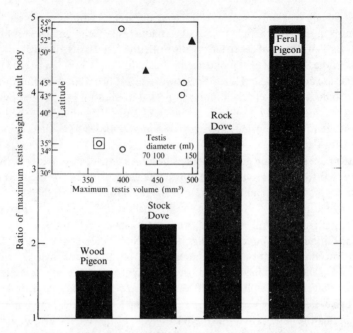

Fig. 9.6. Ratio of maximum testis weight to adult body weight in the Wood Pigeon *Columba palumbus*, Stock Dove *C. oenas*, Rock Dove *C. livia*, and Feral Pigeon *C. livia* var., species which show an increasing photosensitivity judged by their capacity to respond to shorter daily photoperiods, respectively. Inset: Relationship between maximum testis size and latitude in House Sparrows *Passer domesticus* (open circles; reading in square of doubtful validity) and Mallard *Anas platyrhynchos* (triangles, upper abscissa). Based on Lofts and Murton (1968).

Effect of altering daylight cycle

The proximate control of breeding seasons by photo-responses may be altered at different latitudes in the same, or closely related species, by virtue of differences in the natural light cycle rather than any change in the physiological apparatus of the species. There are reasons for thinking that this will not be a common phenomenon (see p. 384) and that the House Sparrow is somewhat exceptional because its range has been drastically affected by man. However, in Malaya the Great Tit *Parus major* has an extended breeding season and produces 3 broods a year rather than the 2 typical of northern Europe or one in Britain (Cairns 1956). Presumably the lessened photostimulation in Malaya delays the speed with which the cycle is forced through into a refractory phase (see also Fig. 1.1, p. 3). The same arguments are applicable when a species is introduced by man to a new range. The gonadal cycle of the introduced Goldfinch[†] *Carduelis carduelis* near Melbourne, Australia (latitude 38 °S) is a little longer than in Britain (52 °N) and the cycle is phased slightly

[†]This species should not be confused with the American Goldfinch *Spinus tristis* which is a late nesting single-brooded species whose gonads do not recrudesce into a fully active state until near the summer solstice after which refractoriness and regression follow in August–September (Mundinger 1972).

earlier judging from Middleton's (1971) data (see p. 350 for an explanation of the phase relationship of the gonad cycle in relation to the seasonal day–night cycle). This means that breeding begins slightly earlier than in Britain. The breeding biology appears to be the same as in the ancestral population (Middleton 1970*a*) and the birds depend on the introduction weed seeds man-altered arable farmland. The breeding seasons of the introduced Song Thrush *Turdus philomelos* and Blackbird *T. merula* in New Zealand (40 °S) appear also to be slightly more extended than in Britain, judging from egg-laying records presented by Bull (1943).

Experiments by Damsté (1947), using Greenfinches caught in Holland, illustrate how the photo-response of a species can become manipulated by the environmental light cycle. In this study Damsté kept his subjects in an indoor room where they

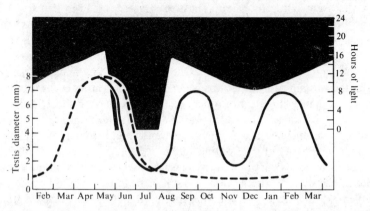

Fig. 9.7. Effect of manipulating the photo-regime in the way shown on the cycle of testis growth and involution in the Greenfinch *Carduelis chloris*. The natural seasonal testis cycle is shown by the dashed line. Based on Damsté (1947).

experienced natural photoperiods via the windows. Some birds were transferred to light-proof cupboards for just over two months between 25 May and 30 July, after which they were exposed to natural photoperiods again. Then from December to February a few of the birds were given extra artificial illumination. In this way three periods of gonad recrudescence and two of regression could be induced in one year (Fig. 9.7). Wolfson (1954) contrived a similar experiment using Slate-colored Juncos which were exposed to four periods of 9-h days and five periods of 20-h days in one year. Five periods of gonadal activity, five of fat deposition, and two moults were obtained as a result.

Changes in photo-response threshold

Many closely related species have come to have different breeding seasons in consequence of 'simple' changes in photo-response threshold and alterations of photosensitivity between species have apparently followed the same rules which apply to a single species when it is subjected to different light regimes as in Fig. 9.5. Also, when closely related species are compared it is usually found that a decrease in sensitivity to light (that is, the bird requires a longer daily photoperiod to respond) is accompanied by a more rapid gonad recrudescence when a response does begin and a slightly longer duration of refractoriness. Differences in the reproductive cycle of the resident British population of the Starling and Continental immigrants, which spend the winter in Britain, probably result from an adjustment of this kind (Bullough 1942). Spermatogonial division begins in the British birds in late September, but spermatogenesis does not normally progress beyond this point until February, when there is a burst of activity to the secondary spermatocyte stage and sometimes beyond. Continental birds show no spermatogonial division from September onwards and the first mitoses are not seen until late December or early January and then only become common in early February. Primary spermatocytes are first formed in early March. Thus, both races respond to January and February photoperiods, though this marks a beginning of spermatogenesis for Continental birds but the resumption of a process already started in the British population. During February and March the gonads of British birds grow more rapidly Unlike Continental birds, the British race shows earlier and more interstitial cell activity in the autumn and winter, this being manifest by earlier changes in bill coloration and the development of other accessory sexual organs (rete testis and *vasa deferens*), known to depend on androgen secretion. Androgen production in British birds results in autumnal sexual displays and even winter breeding, events virtually absent in the life history of the Continental birds.

The same process may apply to many passerines which migrate to the Arctic to breed during the short northern summer and is best illustrated by reference to the well-studied North American *Zonotrichia* species (family Emberizidae), a group already much alluded to, which were first made the subject of photo-experimentation by Blanchard (1941, 1942). Four races of the White-crowned Sparrow *Zonotrochia leucophrys* are of interest, for the race *nuttalli* is resident in the California area of North America, and has a reproductive cycle similar to that of the House Sparrow. *Z. l. pugetensis* replaces it to the north in Oregon and extends north to British Columbia. It is a partial migrant south to California. In contrast *Z. l. gambelii* is only a winter visitor to California and the southern States for it breeds in the east Canadian Arctic tundra, being replaced in the west by *Z. l. leucophrys*. The Golden-crowned Sparrow *Z. atricapilla* has as its nuptial quarters the sub-alpine spruce thickets of the mountains of British Columbia and Alaska, that is, its summer range is to the south of the bare tundras accepted by the White-crowned Sparrow. The Golden-crowned Sparrow is replaced to the east of the Rocky Mountains by the White-throated Sparrow *Z. albicollis*, as inhabitant of alder and willow thickets. The northern species are single-brooded and do not begin egg laying until late May and early June, whereas the southern forms produce several broods between March and July. The gonad cycles of some of these species are summarized in Fig. 9.8 to show how the duration of gametogenesis is shortened and refractoriness possibly slightly lengthened in those species which migrate to high latitudes, where they experience longer and more intense photoperiods. It has been proved experimentally that the species breeding in the Arctic require longer daylengths for gonad growth (Fig.9.8, also

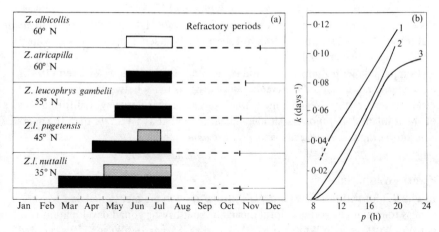

Fig. 9.8. (a) Natural breeding seasons of various *Zonotrichia* species in North America (bars). Data for solid bars are from Miller (1960) and refer to the period during which males are normally in active gametogenesis. Dotted bars refer to experimental subjects which were kept throughout on daylengths equivalent to those at the winter solstice in California and under these conditions *atricapilla* and *gambelii* did not achieve breeding condition. Open bar refers to data collected by Shank (1959) and re-analysed by Lofts and Murton (1968). Broken lines represent the duration of post-breeding photo-refractoriness under natural conditions, experimentally verified in most cases, details being given in Lofts and Murton (1968). Approximate mid-latitude of breeding range indicated.

(b) Rate of testicular growth k as a function of the duration of the daily photoperiod p for (1) *Zonotrichia leucophyrys pugetensis*, (2) *Z. atricapilla*, and (3) *Z. I. gambelii*. From Farner and Lewis (1971).

Lewis 1975). Of course, in the wild a species undertaking a post-nuptial migration from high latitudes would eventually encounter daylengths which were below the stimulatory threshold so a resurgence of gametogenesis would be prevented. For this reason it seems unlikely that the slightly longer refractory period of the high-altitude species of *Zonotrichia*[†] is ecologically adaptive but rather a consequence of a different photostimulus acting on the response mechanism. It is possible that a rapid and intense gonad recrudescence results in a greater degree of gonad exhaustion so that slightly more time is required for rehabilitation (but see p. 319).

The *Juncos* are in the same family as the crowned sparrows and the Slate-colored Junco *Junco hyemalis* has been a favourite experimental subject for North American investigators dating from the early experiments of Rowan (1926). For many years much of the photo-experimentation on this species was done by Wolfson (1959). The Slate-colored Junco breeds in the boreal forest of North America from about 45 °N to the north coast of Alaska. Wintering grounds extend from south Canada to the Gulf of Mexico, most birds occurring east of the Rockies. Wolfson's studies have been based on birds captured on autumn and spring migration at Evanston, Illinois. The refractory period has been carefully defined in this species by Shank (1959)

[†]There is evidently less interspecific difference in the time of onset of refractoriness compared with beginning of breeding activity. Compare Fig. 9.8 with Fig. 14.4 (p. 370).

following earlier work on *J. hyemalis* by Wolfson (1952) and it is slightly longer than in the *Zonotrichia* species, excepting *Z. albicollis*. Like *Z. albicollis*, *J. hyemalis* will respond, albeit slowly, to 9-h photoperiods (Wolfson and Winn 1948; Winn 1950; Engels and Jenner 1956).

Comparable with *Zonotrichia*, and some other finches below, Wolfson (1942) has demonstrated marked differences in photosensitivity between migrant and resident races of the Oregon Junco *Junco oreganus* near Berkeley, California. Testes of the resident *J. o. pinosus* recrudesced earlier and at a faster rate than the migrants *J. o. shufeldti*, *J. o. montanus*, and *J. o. oreganus*.

Farner–Wilson equation

In Fig. 9.7 logarithmic constants of testicular growth are mentioned and we should digress to explain how this helpful means of quantifying gonad development is derived. Working with White-crowned Sparrows, Farner and Wilson (1957) noted that when receptive subjects were exposed to stimulatory daylengths the growth of the testes from a resting state to approximately $\frac{1}{2}$–$\frac{3}{4}$ maximum weight was a log–linear function of the duration of photostimulation (Fig. 9.9) and they derived the formula:

$$k_p = (\lg W_t - \lg W_0)/t,$$

where k_p is the logarithmic growth constant for a specific photoperiod, W_0 is the resting testicular weight, W_t the testicular weight after t days of photostimulation.[†] Since k is a quantitative measure of the rate of gonad growth, and indirectly of gonadotrophin secretion, it enables comparative measurements of the effectiveness of different photoperiodic treatments in stimulating the pituitary–gonadal axis to be made. A value of k_p of 0·01 is equivalent to a doubling of testis weight in 30 days and a $k_p = 0\cdot057$ represents a 50-fold increase over the same period. It is seen from Fig. 9.9 that no response occurs below a threshold photoperiod (l) and that a maximum point is also reached beyond which there is no increase in response; in some species there can be a decrease. Farner and Wilson also established that k is a function of light intensity I, wavelength L, environmental temperature T, and the time elapsed following the termination of photo-refractoriness r. If all these variables are maximized k and p are directly proportional between the limits of l and the photoperiod beyond which no further increase in response occurs so that:

$$k_g = a(p - p_0),$$

where k_g is the logarithmic growth constant for the given photoperiod p, p_0 is the maximum photoperiod for which the constant is zero, and a is a constant. From this equation values of k for different temperatures and photoperiods can be cal-

---

[†]This relationship also applies to the ovarian growth resulting from photostimulation, at least in Tree Sparrows *Spizella arborea* (Morrison and Wilson 1972) and testicular growth in Harris' Sparrow *Z. querula* (Wilson 1968).

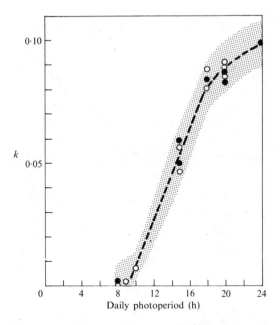

Fig. 9.9. Rate of testicular development as a function k of the duration of the daily photoperiod p in hours. Shaded area encloses the upper and lower 95 per cent fiducial limits for all points. Open circles represent samples of first-year birds; solid circles, adjusted means for samples of adults. From Farner and Wilson (1957).

culated and used to predict gonad growth in wild populations of the species using:

$$\lg W_n = \lg W_0 + \sum_{i=1}^{n} k_i,$$

where n is the number of days after the beginning of testicular development. In nature daylengths increase each day but, if this increase is linear,

$$\lg W_t = \lg W_0 + \tfrac{k}{2} t.$$

This formula enabled Dol'nik (1963) to estimate the testicular weight that would be obtained by wild Greenfinches *Carduelis chloris* and Chaffinches *Fringilla coelebs* in Baltic Russia, and Farner and Wilson (1957) to predict the date when wild White-crowned Sparrows would attain gonads of a given size. Expected and observed results were in very good agreement. Dol'nik (1963) similarly showed intraspecific variations in the photo-response of Chaffinches *F. coelebs* collected from Kaliningrad on Kurskaya Spit, Baltic Russia, (latitude 55 °N) and from near Leningrad (latitude 60 °N), when he exposed groups of these birds to the same range of photoperiods in the laboratory. Birds from near Leningrad exhibited no testicular growth when given photoperiods which were stimulatory to those from further south, although with longer photoperiods the growth rate of the Leningrad birds became more rapid

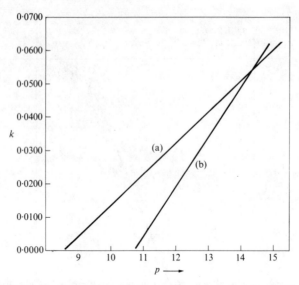

Fig. 9.10. Experimentally obtained logarithmic growth rate constants k of testicular growth from photosensitive Chaffinches *Fringilla coelebs* collected near (a) Kaliningrad (55 °N) and (b) Leningrad (60 °N) in Baltic Russia. From Dol'nik (1963).

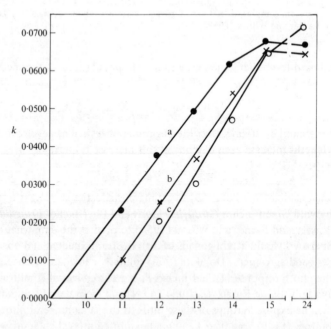

Fig. 9.11. Logarithmic constants of testis growth of photosensitive Greenfinches *Carduelis chloris* (filled circles), (b) Chaffinches *Fringilla coelebs* (crosses), and (c) Bramblings *F. montifringilla* open circles. From Dol'nik (1963).

than those from the south (Fig. 9.10). As Dol'nik explains, the response is adaptive in that birds living near Leningrad must breed later than those living in more equable climes further south, but after the spring solstice daylengths are longer further north. Some kind of increase in response threshold initially appears to inhibit the more northern population (*cf. Zonotrichia*). This was not the case in some studies made by Berthold (1969) when he compared the photo-responses of representatives of various bird species collected in Finland and Germany. He found that when kept in captivity under identical conditions Starlings and Chaffinches from Germany had similar gonad development to others from Finland so that differences in the field were phenotypic and not genotypic. Gonad development was thought to be related to migratory behaviour because captive Starlings prevented from migrating in autumn and spring had a premature gonad cycle compared with migrating birds (Berthold 1967). This experiment needs to be repeated for several criticisms are possible of the present conclusion.

Dol'nik (1963) also compared the photo-responses of Greenfinches, Chaffinches and Bramblings *Fringilla montifringilla*, captured in the autumn near Kaliningrad, by exposing them to artificial photoperiods in December. The Greenfinches could respond with gonad growth at photoperiods that failed to affect Chaffinches and Bramblings (Fig. 9.11). All species responded when given 11-h daily photoperiods but gonad growth was most rapid in the Greenfinch. All these species winter to the south of Kaliningrad and the daily photoperiod seems to be the only external factor initially responsible for gonad development, for temperature and other modifying factors can have little or no biological relevance until the species reach their breeding areas.

Photosensitivity variation in closely related species

Differences in photosensitivity have been noted in two closely related sympatric icterids in central California (Fig. 9.12). The Tricolored Blackbird *Agelaius tricolor* and Red-winged Blackbird *A. phoeniceus* are ecologically similar for they overlap in food preference, live in the same habitat, and often nest in the same marsh. In neither species do the males breed until their second year,[†] although females nest in their first year. Nesting occurs in large colonies—sometimes as many as 10 000 Tricoloreds may congregate together— it having been shown by Orians (1961) that the size of the colony depends on the local food supply. Large invertebrates such as grasshoppers and locusts are important food items for the young but when not breeding both species eat grass seeds, such as *Echinochloa*, and are pests of rise and grain crops. In its feeding and breeding ecology the Tricolored Blackbird is convergently similar to the African *Quelea* (see below). Both Tricolor and Red-wing are polygynous. The latter holds a permanent territory, mates with a succession of four or more females at intervals throughout the breeding season, and takes no share in the rearing

---

[†]Orians showed that the removal of established adult males brought young males into the breeding population at an earlier age than normal and in sub-adult plumage. This suggests that some of the restriction on breeding in the first year is environmentally induced rather than a direct physiological effect.

of young. In contrast, Tricolored Blackbirds have only a temporary territory, one, two, or rarely more females, but at the same time and they can mate with them all in one day. They move to new territories for repeat nestings and the male helps to feed the chicks. Unlike the Red-winged Blackbird, the Tricolored nests both in spring and to some extent in autumn, when daylengths are declining. There is little or no difference in the date on which first eggs are laid in the two species but Red-winged Blackbirds begin singing in December or January whereas according to Payne (1969), the Tricoloreds do not begin until a month or more later.

The gonad cycle of the Red-winged Blackbird was described many years ago (Wright and Wright 1944) but more recently Payne showed that the Red-winged Blackbird has a higher testicular growth rate constant k than the Tricolored for the range of photoperiods 9·5–14·5 h (Fig. 9.12). The functional significance of this difference is not immediately obvious for in the wild in California daylength increases are such that both species are in reproductive condition by mid-March, almost a month before eggs are laid. It may be that there is some advantage to the Red-winged in being able to sing and establish a territory earlier than the Tricolored and at lower latitudes the difference in response rate would be greater and perhaps of more significance. However, in Costa Rica (latitude 10 °N) near the southern limits of their range, Red-winged Blackbirds breed from late May until early September, coincident with the rainy season (Orians 1973).

We note that the Red-winged is more sensitive to short photoperiods than the Tricolored Blackbird and has a slightly longer refractory period (Fig. 9.12). This means that under natural photoperiods in California Red-winged Blackbirds do not become responsive to photostimulation until late October or November, whereas Tricoloreds end refractoriness in late July and can exhibit a degree of autumn gonad development. In fact, a proportion of the population achieves full spermatogenetic development by late September or early October and opportunistic nesting occurs if suitable food supplies are available. Spring breeding is followed by gonad regression refractoriness and moult, and then a resurgence of activity before a winter regression.

In California today autumn nesting is possible in the proximity of rice fields although the success rate is very low and few young are reared; these are given spiders snails, and a few insects but evidently, although the adults can acquire the energy reserves for egg laying, invertebrate food for the young is scarce. Accordingly, Payne questioned the value of autumn breeding since the success rate is so low that it might be offset by the extra energy expenditure required. He argued that selection for autumn breeding during the 50 years since the advent of rice farming must have been very weak and likely to have caused a change in the proportion of autumn breeders of less than 0·1 per cent.[†] Instead, the occurrence of autumn breeding is better explained by assuming it to have been more prevalent in the past. Payne

---

[†]Payne allowed a selective factor i of 0·001 (because only 20 chicks were fledged from 1900 nests) and a heritability coefficient h of 0·1 and then used the expression ih^2, after Lerner (1958), to estimate the change in proportion of autumn breeders in the population in a single generation. According to experiments by Lerner, the heritability, or proportion of variance in phenotypic characters accountable for by genetic differences, is in the order of 0·1.

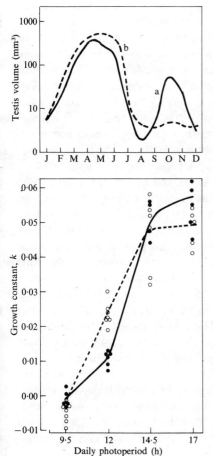

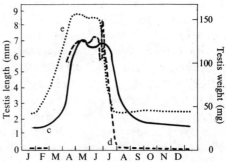

Fig. 9.12. Upper panels: seasonal testicular cycles in California (39°N) of (a) Tricolored Blackbird *Agelaius tricolor* and (b) Red-winged Blackbird *A. phoeniceus* (from Payne 1969); of (c) Brown-headed Cowbird *Moluthrus ater* in California (from Payne 1973*a*) and (d) in Ontario, 43°N (from Scott and Middleton 1968) and of (e) Great-tailed Grackle *Quiscalus mexicanus* in Texas 30°N (from Selander and Hauser 1965).

Species (a), (b) and (c) all end refractoriness at about the same time, this being in October (Payne 1967).

Lower panel: Logarithmic constants of testis growth according to duration of photoperiod for Tricolor Blackbird (solid line and symbols) and Red-winged Blackbird (dashed line and open circles). From Payne (1969).

points out that in prehistoric times plagues of grasshoppers occurred in the autumn in California and birds might have exploited these. Several outbreaks have been reported in more recent times and it appears that Indians in the Sacramento Valley had widespread cultural techniques for catching and preparing grasshoppers as food (Kroeber 1925; Holt 1946). Hence, autumn breeding in relation to a present-day agricultural food supply appears to be a secondary consequence of behaviour that once had natural survival value.

Two other icterids may be considered at this juncture. The first is the parasitic Brown-headed Cowbird *Molothrus ater* with a gonad cycle essentially similar to the blackbirds just described (Fig. 9.12); gonad recrudescence occurs in March, refractoriness develops in late July and ends in October (Payne 1967, 1973*a*). Experimentally, the log of the testicular growth rate constant under a LD 17:7 has been calculated and found not to differ from the non-parasitic species. Further discussion of this species is deferred until p. 449. The other icterid to be mentioned is a trans-equatorial migrant, the Bobolink *Dolichonyx oryzivorus*, which migrates from temperate North to subtropical South America after breeding and which, in consequence does not experience 'short' winter daylengths. It must, therefore, be able to break refractoriness on relatively long days. Birds leave latitude 40 °N–50 °N by mid-October on 15-h days, experience 12-h days briefly as they cross the equator and then 15-h days on their wintering grounds at latitude 15 °N–25 °S. Daylengths then decrease yet again to 12-h by the March equinox. When they migrate north (probably in March) these birds experience rapidly increasing photoperiods, especially as they get near to their northern breeding grounds (Engels 1959)

Captive Bobolinks held in the north throughout the summer on a 16-h photoperiod moult and acquire henny plumage in August concurrent with a loss of bill coloration. Engels (1959, 1961, 1962) was able to show that birds given 14-h photoperiods starting in September remained refractory until June, but exposure to a relatively long daylength of 12-h until December was sufficient to break their photo-refractoriness, and a return to a 14-h photoperiod produced gonadal recrudescence to full breeding condition by April. Under natural daylengths in the Northern Hemisphere birds end their refractory period by 1 November, a situation comparable with most north-temperate species. During migration the average Bobolink experience only a few weeks in the autumn of daylengths of less than about 12 h 45 min. If captive birds in October (daylength 12 h 41 min) are exposed either to photoperiods of 12·5 or 12·75 h for 4–6 weeks, and are then given a 14-h photoperiod, testicular recrudescence occurs in both groups. Hence, in this species refractoriness can be overcome by a few weeks' exposure to photoperiods comparable with those experienced during migration. This is in contrast to Juncos and White-throated Sparrows which, given 12-h photoperiods for eight weeks in October and November, do not respond to 14-h photoperiods, though they do if pretreated with 10-h photoperiods (Engels 1961). Thus, as Lofts and Murton (1968) showed in summarizing Engels' studies as above, the special adaptation of the Bobolink is the raising of the photosensitivity threshold so that the response to 14- or 15-h photoperiods is slower than in species such as *Zonotrichia*. Also the amount of light which can allow the ending of refractoriness is much higher than in most north-temperate residents. Similar findings pertain to the tropical wintering Dickcissel *Spiza americana* (Zimmerman and Morrison 1972). A difference in photosensitivity threshold also appears to account for the timing of migration and gonad development in the north-temperate wintering Hermit Thrush *Hylocichla guttata* compared with the trans-equatorial migrating Swainson's Thrush *H. ustulata* (Annan 1963).

The above adaptations may apply to some migrant waders wintering near the equator. Whimbrel *Numenius phaeopus* collected on their winter quarters in Gambia (13 °N) still had their testis tubules occluded by dense sudanophilic lipids in January and these were only just clearing in April in birds collected in England (Lofts 1962*b*). When these and other waders—those examined were Spotted Redshank *Tringa erythropus*, Sanderling *Calidris alba*, and Turnstone *Arenaria interpres*—leave their wintering grounds in late April–May, the gonads are starting their usual recovery but rarely contain germ cells in advance of primary spermatocytes, though spermatozoa are being produced by the time they arrive in England in mid-May–June. The indications are that these species respond slowly to the relatively short photoperiods of the tropics.

It is possible that the closely related Stonechat *Saxicola torquata* and Whinchat *S. rubetra* (fam. Turdidae) have the same basic photo-response mechanism and that differences in their breeding seasons partly depend on small differences in the response threshold compounded by their geographical distribution (Fig. 9.13) and the light regimes to which each is exposed. These chats presumably speciated in the early Pleistocene[†] and by virtue of their preference for open scrub-covered country may well have been able to penetrate further north during the glaciations than woodland species. The Stonechat must be the older species, for it has a very wide breeding range from Europe across Russia to Japan, extending from latitude 65 °N to 20 °N. It is the only Palaearctic passerine which also breeds, and is widely distributed, in Africa. In Britain its first eggs are laid in March followed by second and third broods until early August (Davis 1972). Indeed in its breeding, and presumably gonad, cycle it resembles the Blackbird *Turdus merula* (*cf.* Fig. 7.5, p. 156). In Africa the breeding season embraces all months of the year, according to locality, and it may be surmised that the gonads remain functional over a long period because the seasonal daylength cycles are less extreme than in Europe. The Whinchat has presumably had to avoid competition with the Stonechat, and is only a summer migrant to the Palaearctic from wintering areas on the African savannahs south of the Sahara and north of the equator. In areas of range overlap it occupies a different habitat from the Stonechat, being at home in marshy meadows and hayfields and in grassy clearings in forest, and it penetrates further north than the Stonechat, as Fig. 9.13 shows, and has possibly prevented the Stonechat from so doing. In fact, the Whinchat could be considered a derivative of the Stonechat stock which has become better adapted at higher latitudes from which it must migrate in winter. Again these ecological adaptations appear to have involved only a slight modification in threshold of photoperiodic control mechanisms.

The American Rosy Finches of the genus *Leucosticte* are of interest, for they illustrate how photo-responses applicable to nesting in the Arctic can be modified to suit lower latitudes if the species is an altitudinal migrant to the alpine or tundra zone of high mountains. King and Wales (1965) have demonstrated differences in photosensitivity in three species of Rosy Finch wintering near Salt Lake City, Utah. The length of the migratory flight is very different in each species since they occupy different geographical breeding grounds. Nevertheless, they all lay some time in June, when the habitat becomes free of snow. The Grey-crowned Rosy Finch *L. t. tephrocotis* breeds in the alpine zone of the mountains of north Montana, the Canadian Rockies, and the Brooks Range of Alaska. Hepburn's race *L. t. littoralis* occupies a range to the east, on the mountain-tops of the Casades to the Alaska

[†]The relationship of the three Old World chats with numerous African species is not yet clear and so Voous (1960) considers that the Stonechat may have had either an African or Palaearctic origin.

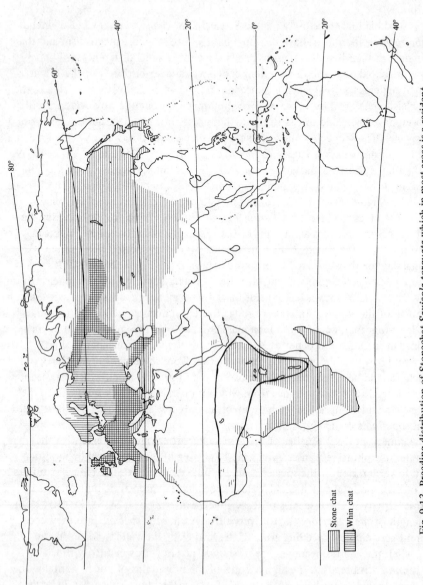

Fig. 9.13. Breeding distribution of Stonechat *Saxicola torquata* which in most areas is a resident, and Whinchat *S. rubetra* which winters in Africa, mostly in the area enclosed by the heavy line. Based on Voous (1960) and Moreau (1972).

range. The Black Rosy Finch *L. atrata* breeds in the alpine zone of the western ranges of the Rocky Mountains from south-west Montana, Idaho to Utah to north-east Nevada and so performs a short migratory flight from low to high grounds. Birds leave their wintering areas some three months before the breeding season and all arrive at the nuptial grounds well before June. Here they loiter until the snow melts and the breeding sites become tenable. In so doing, the most southerly placed species, *L. atrata*, does not experience maximum daylength conditions until the beginning of June, whereas the north-westward migration of *tephrocotis* brings it into long photoperiods at a much earlier data. In spite of this earlier exposure to longer photoperiods, the gonads remain unstimulated until nearer June, so that both species breed nearly at the same time.

Appropriate photostimulatory experiments have demonstrated that *L. t. tephrocotis* has a much higher threshold of photosensitivity than *L. atrata* and it requires much longer daylengths to initiate gonadal recrudescence. The Hepburn Rosy Finch, whose breeding grounds are somewhat intermediate in distance between the other two, is also intermediate in photosensitivity. Given a short duration (15 days) of exposure to a highly stimulatory daylength (LD 20:4) *L. t. tephrocotis* responds most, *L. t. littoralis* next, and *L. atrata* least but the degree of response is reversed with a longer duration (53 days) of exposure. In other words the Black Rosy Finch differed from the other two species in showing an initial low response to a 20-h photoperiod followed later by a higher response. These differences parallel those already noted between races of Chaffinch (Fig. 9.9). All the *Leucosticte* species show similar pre-migratory fat deposition in response to photostimulation, in spite of the fact that the altitudinal migrants move very short distances whereas the northern populations may move 2000 miles (3218 km). As King and Wales (1965) point out, this metabolic adaptation may exist for subsidiary reasons (see also King *et al.* 1963) and be maintained in response to a diversity of selection pressures. In general, non-migrant species usually show little vernal fat deposition.

Evolutionary aspects

In this chapter we have examined some fairly typical breeding cycles which are applicable to the ecological conditions experienced by the majority of temperate passerine species in the northern hemisphere, and for that matter by many non-passerines as well. In general, energy supplies encourage a breeding season beginning in spring and lasting until around, or sometimes after, the summer solstice, but ending by about July or August. Such a season suits many omnivorous or graminivorous resident species and is well illustrated by the House Sparrow *Passer domesticus*. Proximate control of a breeding season of this general nature is achieved by a photoperiodic response, which allows the gonads to begin recrudescence between winter and spring and for refractoriness to develop with the longer days of mid-summer. Variations on this basic theme, which could well be achieved by alterations in response threshold, can account for a large number of temperate breeding cycles. This appears to represent a relatively unspecialized condition and it is of interest to consider the time span which has been available for adaptive changes to this basic mechanism in passerines.

Climatic amelioration from the last glaciation did not begin until 18 000 years ago. Until this time more Palaearctic[†] species, and particularly those inhabiting deciduous woodland, would have ranged little further north than latitude 40 °N, if we can accept the vegetation map of Frenzel and Troll (1952). In the west of the Palaearctic typical present-day passerines would be found in southern Iberia, Italy, and the Aegean. In addition, during the late Pleistocene the Sahara desert supported woodland with plant species typical of the Palaearctic today and desiccation set in only about 5000 years ago. The radiation of passerines following the last glaciation to occupy northern climes and the evolution of the associated migration system must have occurred in the 10 000 years up till about 5000 years ago. Indeed, the majority of passerines probably speciated following the penultimate glaciation so that of about 8600 present-day birds, 5100 (59 per cent) are passerines. This total includes 1100 of the more primitive sub-oscines which must have radiated earlier, probably during the mid-Tertiary (Darlington 1957). The rate at which selective changes have been found in House Sparrows *Passer domesticus* introduced to the U.S.A. in 1850s (Johnston and Selander 1964), and also the fact that some Egyptian larks have evolved subspecific differences in cryptic colouring to suit local soil conditions over a period that can be dated on geological evidence, led Moreau (1966) to suggest that subspeciation at least can occur in 4000 years. Brodkorb (1971a) has taken the median longevity of species as the time necessary for half a fauna to be replaced by new species, and reckons on a species longevity in the order of ½ a million years in the Pleistocene, but more like 3 million years during the Tertiary when stable conditions persisted far longer. Moreau regards the first as a pessimistic estimate and gave good reasons for assuming that full species could emerge in 10 000–15 000 years.

It seems likely that the majority of present-day Palaearctic passerines will have radiated from species having photoperiodic responses applicable to latitudes 20–40 °N. At 30 °N the longest daylength, including civil twilight, is 15·0 h and the shortest 11·1 h. A migrant moving to the far north would experience continuous daylength in summer and a maximum of 14-h during the contra-nuptial season at the tip of South Africa; obviously migrants wintering nearer the equator would experience daylengths closer to 12-h. On the other hand resident colonists of Britain and northern Europe would experience much shorter photoperiods in winter and con-trasting long ones in summer.

As mentioned above, many species appear to have adapted to the changing con-ditions of the post-Pleistocene Palaearctic by some kind of increase or decrease in photosensitivity threshold, or by occupying new ranges where the new seasonal photoperiods result in slightly modified physiological responses. The kind of modification envisaged can be induced in a single species by artificial manipulation of the photoperiod. An increase in response threshold, such as is typically found in those passerines which migrate to high latitudes to breed, appears to be associated

[†]The Palaearctic region is so named because it is that part of the Old World with an arctic climate during the Quaternary Ice Ages.

with a more rapid growth of the gonad once recrudescence begins, a shorter period of active spermatogenesis, and a slightly lengthened refractoriness. Changes of this nature are generally adaptive in that food at high latitudes is restricted to the short summer. How well adapted is a question which is difficult to answer in the absence of other species which make better use of resources. Since we are dealing with the most recently evolved and most rapidly radiating avian group, that is, the passerines, comparison is possible only with phylogenetically older species, many of which can only have colonized high latitudes since the glaciation. However, the corollary of the thesis advanced here is that species which have evolved markedly different photo-responses for the proximate control of annual functions will be taxonomically distinct. A brief survey confirms this expectation. For example, within the Fringillidae all the European *Carduelis* finches are multi-brooded with long breeding seasons, but a distinct modification is noted in *Fringilla* with a single-brooded spring-breeding season, while the winter-breeding Crossbills *Loxia* sp. are also well differentiated generically.

We shall pursue this topic in more detail in the next chapter, in reviewing some more kinds of tropical and temperate breeding cycles. Before doing so it is worth acknowledging that this chapter has been primarily concerned with the male, and this is becuase it is he who initially determines the start of the breeding season and then stimulates the female by providing a nesting territory and appropriate displays. Males can complete their spermatogenetic development in captivity, given appropriate light regimes. Females are partly stimulated photoperiodically and develop refractoriness to long days, as shown for the Canary *Serinus canaria* (Kobayashi 1957), but their follicles do not attain full ovulatory size given light alone.

10

More tropical and temperate breeding cycles

Summary

The theme of Chapter 9 is extended in this chapter to include tropical birds and the way in which photo-responses have been modified to suit more specialized ecological needs. Attention is first focused on the Red-billed Quelea *Quelea quelea*–which belongs to a phylogenetic line that has probably had a long evolutionary history in Africa– and the north-temperate Rook *Corvus frugilegus*; for the latter it is demonstrated how the gonad cycle and photo-responses are adapted to allow breeding to occur during the weeks when peak supplies of earthworms are available. *Zonotrichia capensis* of tropical and temperate S. America derives from north-temperate *Zonotrichia* stock and, unlike *Quelea*, is secondarily adapted to the tropics–its gonad cycle at different latitudes is described. The temperate-zone pigeons are rather unusual in having extended breeding seasons, and some species do not enter a refractory phase under normal seasonal photo-cycles; however, refractoriness can be induced experimentally; this is probably a primitive condition. The cycles of temperate forms are compared with those of related species in sub-tropical regions. A polymorphism is described in *Columba livia* in which certain melanic morphs have an apparently reduced photo-threshold which allows them to have a longer breeding season, and in some cases an increased fertility. The birds show negative assortative mating preferences. The Crossbill *Loxia curvirostra* is unusual in having a long breeding season and being in reproductive condition over the short days of winter.

Introduction

In this chapter we shall examine the way in which avian photo-responses have been modified to suit more specialized ecological needs. The suggestion was made in the previous chapter that the House Sparrow type cycle is fairly generalized one, and we shall now consider how more complex responses might have been derived. The majority of present-day equatorial and tropical species probably evolved in these regions and they are not considered until Chapter 13. However, there is some affinity in terms of photoperiodic mechanisms between the avifaunas of low temperate latitudes and the tropics which we shall begin this chapter by discussing. In this context the African Red-billed Quelea *Quelea quelea*, already mentioned on p. 141, is of interest for it shares the same highly specialized family as the House Sparrow (Ploceidae[†]) being placed in the sub-family Ploceinae. This comprises 90 species confined to Africa south of the Sahara, except for 5 in India and Malaysia (Moreau 1966). The

[†]The family Ploceidea is conveniently divided into the following subfamilies: Bubalornithinae (buffalo weavers), Passerinae (including the sparrows [p. 219], sparrow weavers and scaly weavers), and Ploceinae (true weavers).

photo-responses of *Quelea* have been well studied under controlled conditions, while extensive ecological studies by Ward (1965*a*, *b*, 1966, 1971) have been stimulated by its pestiferous habits.

Red-billed Quelea

Throughout its range in Africa (see Crook and Ward 1968; Magor and Ward 1973) *Quelea* is restricted in its breeding to the wet seasons and first few weeks of the dry season (see Fig. 10.1). Early in the dry season (January–June near Lake Chad, Nigeria, where Ward made his studies) the birds feed on small grass seeds such as *Chloris, Dactyloctenium, Digitaria, Panicum* spp., and *Echinochloa colonum,* and as these are exhausted they transfer to the larger seeds of wild sorghum species and wild rice *Oryza barthii* (a diet that much resembles that of the Eared Dove *Zenaida auriculata* in Argentina to be discussed later in this chapter). If the switch to large seeds has to occur early it is likely to coincide with the ripening of cultivated cereals, such as rice and guinea-corn (a cultivated sorghum locally called Mazakwa) and the birds then cause extensive economic damage (see Crook and Ward 1968). During the dry season the males compete with the females for food and so the proportion of females declines because they are more at risk of starvation. This leads to an unequal sex ratio when breeding begins and contributes to the monogamous breeding habits, which in this species are secondarily derived from the polygamy characteristic of the majority of savanna weavers, which live in relatively stable environments compared with *Quelea* (see Crook and Butterfield 1970). With the onset of rains grass seeds germinate and cease to be available to the birds (late June–July) and so *Quelea* lay down fat reserves in preparation for this season of poor feeding prospects. They migrate several hundred kilometres south, to areas where rain has been falling (Ward 1971), but return after a month, because the plants which were previously germinating and growing are now beginning to produce seeds and these the birds collect

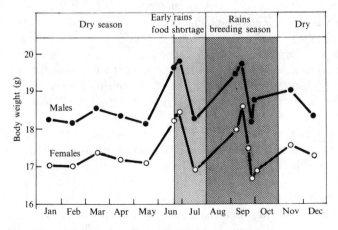

Fig. 10.1. Diagramatic representation of annual changes in body weight of a population of *Quelea quelea* in relation to the breeding season. Based on Ward (1965*b*).

from the standing vegetation. It is now (August–October) that the vast breeding colonies become established, though at all times *Quelea* live gregariously, as discussed on p. 142.

Ward (1965*b*, *c*) pointed out that large communal roosts could serve as information centres to facilitate food finding. Thus birds which had not found feeding grounds could wait about and then follow any departing flock which left the roost in the morning with purposeful flight; such behaviour would be indicative that the individuals concerned had located worthwhile food sites and were flying directly to them. This interpretation was subsequently supported by studies of Pied and White Wagtail *Motacilla alba* sub-sp. feeding on sewage farms (Zahavi 1971*a*, *b*). The flock habit also confers advantages once birds reach the feeding site, for by social facilitation individuals can copy those individuals which have found the most appropriate food items or feeding situation (Murton 1971*a*). Because gregarious behaviour in effect enables the birds to exchange information about the whereabouts of food in what is otherwise a vast and patchy environment it has strong survival value and imposes a social order which must persist during breeding.

It has been suggested that the onset of rains or an associated factor causes *Quelea* in East Africa to assume a breeding condition, and they certainly cannot respond directly to the small increments of daylength (Disney, Lofts, and Marshall 1959). In East Africa the rainfall is irregular and the start of breeding is erratic (Haylock 1959), whereas there is a regular start to the breeding season in West Africa, where the rainfall is regular and predictable and food supplies become available at about the same time each year (Ward 1965*c*) On the shores of Lake Chad breeding starts when supplies or ripe seed of *Echinochloa pyramidalis* are most abundant at the beginning of September. Insect food is also important for breeding, nestlings being fed on grasshopper nymphs and the larvae of lepidoptera. Disney and Marshall (1956) found that the inclusion of insects in the diet of captive *Quelea* did not facilitate gonad recrudescence. They supposedly demonstrated experimentally that the stimulus of green grass caused young birds, caught while being fed by their parents, to begin immediate nest-building and to moult into full breeding plumage (Marshall and Disney 1957). In reality, and as discussed on p. 142, there is probably an innate urge to nest-build, given the stimulus of green grass (or artificial substitute), and, according to Ward (1965*c*), the habit is prevalent throughout the dry season in Nigeria. Crook (1960) also noted much building activity outside the breeding season. Once adult, the birds may respond to rainfall itself and begin building behaviour in anticipation of the appearance of green grass.

Disney *et al.* (1959, 1961) considered the external timing to be achieved in conjunction with an internal sexual rhythm of the gonad–pituitary axis so that the presence of green grass stimulates only those birds that have spontaneously begun spermatogenesis. However, it would probably be truer to say that *Quelea* are in a state of readiness to breed–or near to such a state– for the greater part of the year. Given suitable releasers such as the availability of nesting material and, presumably, a satisfactory food supply they can respond by breeding behaviour. Neither rain nor green grass can provide the final stimulus for, as noted, the first rains in the Lake Chad area are associated with food shortage and a migration south. Nevertheless, this is when the birds assume a breeding dress, a preparation that continues during the period of migration south. Ward has shown that only individuals able to obtain sufficient food manage the pre-nuptial moult in which all the contour feathers of the male are replaced so that he acquires a buff crown and breast, often with a suffusion of pink. *Quelea quelea* has a dimorphic sexual plumage[†], in that about three-quarters of the males gain a black facial mask and in the remaining segment of the population it is buff or white. Females have a sparrow-like plumage at all seasons but they acquire fresh feathers before breeding and their bill changes from yellow to red. Ward noted that during the early part of the rains the synchronization of gonad enlargement and plumage changes in the population was not very close but when breeding began there was extreme synchrony between members of a breeding colony. Possibly those individuals at a similar stage of development separate from the main flock and heighten the impression of a highly synchronous attainment of reproductive capacity by the population as a whole. Nevertheless, among birds which are breeding together nest-building and egg laying is very synchronized, so that 91 per cent of all chicks are hatched within three days of each other—a remarkable co-ordination considering that in most nests the period over which the whole clutch of three eggs hatches is 30 h. Nest-building takes only about 4 days, incubation 10–12 days, and the chicks are fledged after 11–13 days though they cannot fly for a few days more. Social stimulation is probably important, as in the Black-headed Weaver *Ploceus cucullatus* and Vieillot's Black Weaver *Melanopteryx nigerrimus* (Hall 1970).

For some time it has been appreciated that *Quelea* is responsive to photostimulation, this having been demonstrated for both East African (Marshall and Disney 1956) and West African (Morel and Bourlière 1956; Morel, Morel and Bourlière 1957) populations. When Lofts (1964) kept males with an *ad lib* food supply on a photo-regime of LD 12:12 for 2½ years the testes were enlarged for about 7 months. At the end of this time spontaneous regression occurred and the ensuing refractoriness lasted for about 42 days and could not be shortened by exposing a sample of the subjects to LD 11·25:12·75 for 3 weeks. Refractoriness under LD 12:12 ended with a spontaneous recrudescence of the testes (Fig. 10.2). Kept on a photoperiod of LD 12:12 the gonad cycle became out of phase with the plumage one and many individuals moulted from one nuptial dress to another without assuming the intermediate eclipse plumage. This implies that gonadotrophin secretion was maintained, and also that the normal sequence of plumage change is not controlled by light, nor a simple endogenous alternation of eclipse and nuptial dress. If an environmental factor has to be implicated perhaps we should wonder about the nutritive value of

[†]It will be recalled (p. 141) that the attainment of breeding plumage depends on LH secretion (Witschi 1961), at least extracts of mammalian LH restore the sexual plumage. It has not been proved that avian extracts would be equally efficacious and so some caution is needed in assuming that birds with sexual plumage are secreting endogenous LH (see also p. 226).

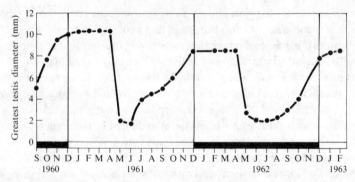

Fig. 10.2. Seasonal variations in testis size of *Quelea quelea* on a constant LD 12:12 photoperiod showing an autonomous periodicity with a constant 42-day refractory period. From Lofts (1964).

the diet. Lofts argued that in this equatorial species a semi-autonomous[†] internal rhythm of reproduction operated, so that the refractory period was independent of photoperiodic fluctuations. He considered that the shortened refractory period compared with that of temperate species was adaptive, enabling *Quelea* to respond rapidly when rain created conditions suitable for breeding, although it was later questioned why in such circumstances the refractory period was needed at all (Lofts and Murton 1968). It is apparent that in Africa *Quelea* should be able to breed at virtually any time of the year except during refractoriness; refractoriness presumably develops immediately after breeding in free-living birds. *Quelea* is by no means unique among equatorial species in being responsive to light; another weaver *Euplectes orix,* from latitude 3–14 °N, responds to experimental alterations in the photoperiod (Rollo and Domme 1943) as doubtless would all equatorial birds (see p. 370).

Quelea kept in captivity in London were transferred to LD 17:7 after being maintained on LD 12:12 (Fig. 10.3). Initially there was an increase in testis size, although no change in spermatogenetic development, and then spontaneous regression to a point where the gonads contained only resting spermatogonia and dense deposits of sudanophilic tubule lipid (Lofts 1962c). Subjects now kept on LD 8:16 remained regressed, as would typical temperate-zone passerines. Others kept for 3 weeks on LD 8:16 and then stimulated by LD 17:7 returned to breeding condition at the same time as subjects held on LD 17:7 throughout, that is, the duration of refractoriness could not be shortened by exposure to 8-h days. It is also to be noted that gonad recovery was possible under a long northern 17-h photoperiod, whereas in the House Sparrow refractoriness continues under such a regime. Nevertheless, as Lofts and Murton (1968) showed, the gonad cycle of *Quelea* could come to resemble that of the House Sparrow given appropriate photoperiod manipulation. Conversely, if sparrows were kept on only weak stimulatory photoperiods

---

[†]Hamner (1971) criticized the view that the cycle was an autonomous one and argued for the reasons given on p. 345 that a truly autonomous periodicity should be manifest only with continuous light or dark.

(*cf.* Threadgold 1960*a*) their gonad cycle could be made to resemble that of *Quelea*. Adaptation of the House Sparrow to temperate latitudes has involved a drop in photo-sensitivity threshold so that apparently unlike *Quelea* gonad recrudescence can occur on short days (LD 9:15). However, it needs to be checked whether *Quelea* would exhibit a slow rate of gonad recovery if kept for a long time, say 12 months, on similar short days, for Lofts (1962*c*) kept his subjects on LD 8:16 for only about 55 days. Additionally, photo-refractoriness is maintained in the House Sparrow by long days (LD 16:8 or LD 17:7), whereas *Quelea* can regain photosensitivity under such daylengths, for it has not had to evolve any 'protective mechanism' to prevent unseasonal breeding under long days. Are a drop in photo-threshold and modifi-cation of refractoriness consequent on the same alteration in the photo-mechanism or do they represent distinct processes? We cannot discuss this further with the evidence we have gathered so far and must defer further analysis until Chapter 12. It should be mentioned that another weaver, Baya Weaver *Ploceus philippinus*, is claimed to lack a refractory phase for subjects kept for 12 months or more on constant 15-h photoperiods failed to exhibit testicular involution (Thapliyal and Saxena 1964). Although in general the subjects were laparotomized at monthly

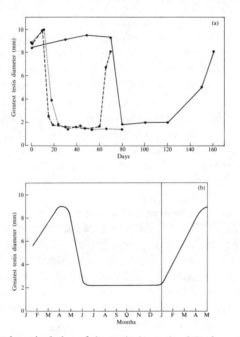

Fig. 10.3. Experimental manipulation of the testicular cycle of *Quelea quelea*. (a) Solid lines indicate the change in testis volume of birds held on LD 12:12 as in Fig. 10.2. In subjects exposed to LD 17:7 (dashed line) the period during which peak breeding condition was main-tained was shortened and the birds entered a photo-refractory state earlier. If subjects were transferred from LD 17:7 to LD 8:16 (dotted line) recrudescence did not occur. Based on data in Lofts (1962*c*). (b) By holding *Quelea quelea* on a north-temperate photoperiod the testis cycle approximates to one which is typical of a species such as *Passer domesticus* (see Fig. 9.2.). Based on Lofts and Murton (1968).

intervals, there were some months when no check was made. Accordingly, it is not possible to be sure that regression and recrudescence did not occur and it would be worthwhile making further studies.

Some spring breeders

We return now to consider some north-temperate species that have a short breeding season early in the year and which must, therefore, be able to regain gametogenetic activity on daylengths of 10 h or less, and yet cease to breed well before the summer solstice. The Rook *Corvus frugilegus* is a good example. It is a crow (Corvidae[†]) with a bill adapted for probing for invertebrate food in soft soil so that earthworms, larvae or Coleoptera, and Diptera comprise important components of its diet (Fisher, unpublished; Lockie 1955; Murton 1971*b*; Holyoak 1972). In most of Europe it does not breed further south than latitude 46—48 °N, doubtless because the drier ground is not conducive to probing, although it does range further south to 30° latitude in the moist riverine pastures of the River Tigris on the Iraq—Iran border. The northern limit is mostly at 60 °N and all the central Palearctic birds are migratory for cold winters and extensive frost make soil feeding impossible in winter and early spring. Rooks from Poland and west Russia migrate west in autumn to winter in the milder maritime countries of the Atlantic sea-board and Britain. Thus the Rook must have evolved under a north-temperate photoperiod, perhaps in the last interglacial period, for it is a bird of open steppes and grassland as distinct from the wood and parkland of the Carrion Crow *Corvus corone*. It has prospered in consequence of man's agriculture, first through de-afforestation and the creation of a pastoral economy and subsequently the expansion of arable farming: cereal seed is nowadays an important constituent of the diet being gleaned in season from sowing, stubble, or stack yard.

The breeding season is ultimately adjusted so that most of the nestlings are being fed in April and early May to coincide with a peak in earthworm availability (Lockie 1955, 1959). To achieve this, egg laying normally takes place in late March and early April and is preceded by a rapid gonadal recrudescence in late February and March (Marshall and Coombs 1957) stimulated by the relatively short February—March photoperiod. Rooks have not been subjected to artificial light regimes but pioneer experiments by Rowan (1932) demonstrated that the closely related American Crow *Corvus brachyrhynchos* is photo-responsive. The relationship between earthworm availability (based on sampling in a Cambridgeshire study area), the gonad cycle (based on Marshall and Coombs), the season of egg laying (based on records collected in Oxfordshire by Owen (1959) and our own Cambridgeshire data), are set out in Fig. 10.4. It is seen that at the end of June and throughout July earthworm supplies

---

[†]The Corvidae comprise a long-evolved passerine family, which, judged by size, social behaviour, and capacity to perform mental tasks, represents one of the further stages reached in avian evolution (Thomson 1964). The present distribution is cosmopolitan but most species occur in the temperate parts of the Northern Hemisphere. A centre of evolution in Eurasia in the late Cretaceuous—early Tertiary is most likely, with a radiation in the late Tertiary (Cracraft 1973). North America has apparently been colonized from Eurasia and South America from North America. However, the typical crows *Corvus* are absent from South America.

become restricted because during dry weather they move deeper into the soil where they are beyond probing (and sampling) range. Furthermore, in these months cereal seed is not yet available and the Rooks are hard pressed to find sufficient food (Lockie 1956; Feare, Dunnet and Patterson 1974). It is reasonable to suppose, therefore, that any eggs laid after early April would not produce many surviving young, because these young would be born too late in relation to the critical period for food. Hence, selection should operate strongly against late-breeding birds. Indeed, Table 10.1 shows that eggs laid late in the season (after mid-April) have a poorer success than those laid in March and early April. The table also suggests that for eggs laid in March the most frequent clutch size was the most productive and it shows that clutch size declines with season.

The success of chicks after leaving the nest in relation to the date of fledging has not been examined but it might be suspected that late-fledged chicks would fare

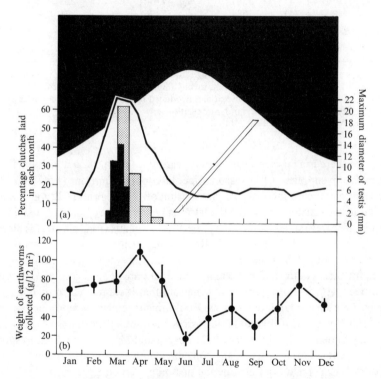

Fig. 10.4. (a) Seasonal changes in testis diameter of Rooks *Corvus frugilegus* in Britain as solid line (from Marshall and Coombs 1957) plotted against the seasonal light cycle at 52 °N. Percentage of clutches laid in each month at Oxford shown by dark histogram (data from Owen (1959) and in a Cambridgeshire study area by stippled histogram (Murton and Westwood, unpublished). Diagonal line is the regression of primary moult score on date (based on Holyoak (1974)).

(b) Seasonal variations ± standard errors in the biomass (g per 12 m²) of earthworms collected in the Cambridgeshire study area by watering sample plots on pasture and arable fields with a dilute solution of formalin in a standard manner (Murton and Westwood, unpublished).

TABLE 10.1

Breeding success and productivity of Rook in relation to clutch size and to season

	Number clutches laid (no. eggs)	Percentage of eggs laid which hatched	Percentage of eggs hatched which gave fledged young	Percentage of young fledged from eggs laid	Number chicks fledged per brood
Total for season					
March	138 (505)	66	84	56	2·0
1–15 April	58 (197)	73	85	62	2·1
16–30 April	20 (55)	51	93	47	1·3
May	7 (19)	(58)	(27)	(16)	0·4
Clutch size in March					
1	4 (4)	(25)	(100)	(25)	(0·3)
2	17 (34)	68	96	65	1·3
3	37 (111)	73	85	62	1·9
4	50 (200)	72	84	60	2·4
5–7	30 (156)	56	79	44	2·3

There was a significant decline in clutch size for the periods defined in the top half of the Table from $3·7 \pm 1·1$ for eggs laid in March, through $3·4 \pm 0·8$ in early April, $2·8 \pm 1·7$ in the second half of April, and $2·7 \pm 1·1$ for clutches produced in May. Number of clutches as given in the main Table. Error terms are standard deviations.
(Murton and Westwood, unpublished data.)

badly since there is a marked peak in first-year mortality in June and July (Fig. 10.5), and earthworm supplies at this season affect later population size (Fig. 10.6). It is interesting to compare the seasonal distribution of first-year mortality in the Rook with that in the Carrion Crow, for the latter is relatively independent of soil invertebrates and, therefore, less affected by the June droughts (Fig. 10.5). It is also seen from Fig. 10.5 that adult mortality in Rooks and Carrion Crows is highest during the breeding season, when they are competing for nest-sites and additional food supplies for their young, and not when food stocks are at an absolute minimum.

Equipped with a photo-response mechanism that is stimulated to breeding condition by February–March daylengths, Rooks might remain in reproductive condition until November if they did not possess a refractory mechanism which causes the gonads to regress spontaneously in late April and May and remain unresponsive to environmental stimuli, particularly long daylengths, until early August (Marshall and Coombs 1957). During this time the post-nuptial moult occurs. Judged by gonad histology, many birds probably do not end their refractory period until September, and some even in October. When refractoriness does end the low threshold of photo-sensitivity enables them to begin gonad recrudescence in the autumn and nest-building and sexual behaviour occur, and some females even lay eggs. Generally though, gametogenesis does not advance beyond the production of primary spermatocytes in the male, by which time the daily photoperiod and mean temperature have fallen below the stimulatory level. Autumn sexual behaviour may well

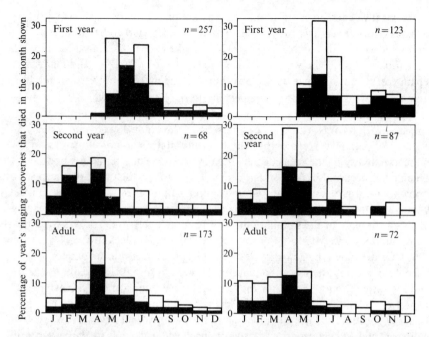

Fig. 10.5. Seasonal variations in the mortality of Rooks *Corvus frugilegus* (left panels) and Carrion Crows *C. corone* (right panels) in Britain. Histograms give the percentage of the year's total of ringing recoveries which dies in the month shown, with unshaded portions detailing the percentage of birds shot and black portions the percentage dying from other causes or reported as 'found dead'. From Holyoak (1971).

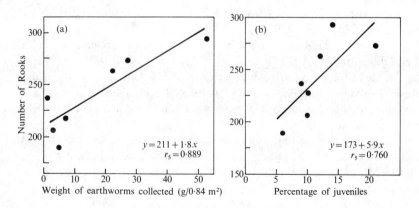

Fig. 10.6. Maximum number of Rooks present in a Cambridgeshire study area in March–April, 1965–71 in relation to: (a) The minimum biomass (g per 0·84 m²) of earthworms recorded in the previous June–July. (b) Percentage of juveniles (lacking white face) in the population in the same March–April. From Murton and Westwood (1974*a*).

have survival value in enabling adults to complete pair formation and establish
nest-sites and breeding facilities in readiness for a quick start to breeding as soon as
conditions become favourable in spring. They make regular visits at dawn and dusk
to the rookery during the winter months (Coombs 1961). The jays and magpies
(separated as the Garrulinae by some authorities) are more primitive than the true
crows and there is a group of genera peculiar to the New World which probably
represent an early radiation to North America. In general, they too nest early
compared with most North American passerines, the Piñon Jay *Gymnorhinus
cyanocephalus* in New Mexico normally breeding from February—June, depending
on the availability of the seeds of the Piñon *Pinus edulis* or alternative foods.
Breeding has occurred in August and September, during which time the normal post-
nuptial moult was arrested, when a bumper crop of pine seed was available. Ligon
(1971), who reported the incident, guessed the birds to have remained in breeding
condition throughout the summer but it seems more likely that this was the equivalent
of autumn sexual activity in the Rook (*cf.* the Tricolored Blackbird, p. 240). Erpino
(1969) has described the gonad cycle of the Magpie *Pica pica* in Wyoming.

Rooks compete intensely for a place in the nesting colony, especially the centre,
and those attempting to breed for the first time have some difficulty in establishing
themselves and are usually relegated to the edge. It is possible that social breeding,
like social roosting, enables an individual to follow birds that have discovered good
feeding grounds. Local variations in soil drainage and dry winds in March and April
cause the numbers of earthworms and other invertebrates in the surface layers of
the soil to fluctuate drastically and the best feeding sites can change from day to day.
The speed with which marked individuals locate and exploit profitable fields suggests
that they cannot search the whole area at random but instead rely on the sampling
experience of the other colony members. Evidence accrues that within gull colonies
adjacent individuals tend to collect the same food objects as if they learn from each
other (Davis 1975). Perhaps this is why birds which are synchronized in their breeding stage have a higher reproductive success than other individuals (Parsons 1975;
Feare 1976). Usually in colonial species those birds nesting in the colony centre
have a higher breeding success and suffer less from predation than those on the periphery; this is true in the Kittiwake *Rissa tridactyla* (Coulson 1971, 1972) and the
Black-headed Gull *Larus ridibundus* (Patterson 1965). This need not mean that
colonial behaviour evolved as an anti-predator device and the effect could be consequent on central birds being more efficient at finding food and therefore making
better parents.

Rooks were introduced by man to New Zealand between 1862 and 1873. According to Bull (1957) the present thriving population lays in the southern spring
in the middle of September so that most eggs hatch in October, a breeding season
comparable with that found in the northern hemisphere and open to the same
photoperiodic control.[†] Other examples in which species which have been introduced
from one hemisphere to another have simply adapted to the re-phased daylength

---

[†] It is possible that the cycle is phased to begin very slightly earlier (by a week or two) as if
the lower-amplitude day—night cycle in New Zealand was allowing a more negative phase-
difference during entrainment (see p. 289).

cycle, so that the breeding season remains essentially unaltered allowing for any changes consequent on the latitude being higher or lower, but 6 months out of phase, were given in the preceding chapter.

Another temperate-zone species with a low photo-threshold is the common wild Mallard *Anas platyrhynchos*. This holarctic duck resembles the Rook in having a limited spring-breeding season and migrating within the confines of the northern latitudes, in Europe most movements being from east to west in winter. Seasonal changes in daylength in January, February, and March provide the main proximate factor bringing the gonads into reproductive condition, somewhat earlier than in the Rook (Höhn 1947; Johnson 1961). Raitasuo (1964) has comprehensively shown how various social and reproductive behaviours are integrated into the timing process in Finland, although Desforges (1972) suggests that these do not affect ovarian development. There is some geographical variation in laying dates but most eggs are produced in March in Britain and in April in Finland. Some clutches may be found in February and in fact the egg-laying season can be more extended, with autumn clutches in semi-domesticated populations living close to man and artificial food supplies (see p. 389). The refractory period is longer than that of the Rook, extending from late June until around the last week of October (Höhn 1947; Johnson 1961). Lofts and Coombs (1965) have shown that Mallards kept under a 16·5 h photoperiod from late August or early September, when they were in a refractory state, showed no spermatogenetic recovery when subsequently exposed to stimulatory photo-periods given from late October or early November, whereas subjects held on 8-h days throughout September and October did so. The highest level of photo-stimulation that can be tolerated in dispersing the refractory state is not known. The interstitium of birds held on summer daylengths was more developed than those subjected to 8-h days.

The Starling *Sturnus vulgaris*, like the Rook, is a prober which specializes particularly on Tipulid and similar larvae in grassland. It begins breeding in late March and early April and may manage two broods during the season that soil larvae are readily available (Dunnet 1955) and before obligatory gonad regression occurs in late May or June (Bullough 1942). We saw in Fig. 9.5 (p. 230) how it was affected by different photoperiods and are now interested in the unique response which emerged under LD 12:12. Unlike *Quelea*, which underwent an approximately annual cycle of gonad recrudescence and involution when kept on an artificial equatorial photoperiod (Fig. 10.2), the Starling manifested peaks in testes size at intervals of about 9½ months (Fig. 9.5). The reason for the interspecific difference is not established and we speculate whether or not it is consequent on the photo-threshold being set lower in *Sturnus* than *Quelea*. Apparently the same process appears to have occurred in the wild in *Zonotrichia capensis* (sub-family Emberizinae[†]) of South America, which we shall now consider.

[†]The family Emberizidae is represented by three neotropical sub-families, the tanagers, swallow-tanagers, and honeycreepers, by the primarily North America sub-families the cardinals and grosbeaks and the buntings (Emberizinae). This last is here regarded as of New World origin with a radiation of two genera (the monotypic *Melophus* and *Emberiza* with 37 species) in the Old World, and, depending which authority is followed, at least 52 genera and 157 species in the New World.

Rufous-collared Sparrow

Hellmayr (1938) gives four good species of *Zonotrichia* in North America, one of which has four, perhaps five, sub-species. But in Central and South America there is only one species, with numerous sub-species, the Andean Sparrow or Rufous-collared Sparrow *Zonotrichia capensis* (King 1974). This is good evidence for Chapman's (1940) view that *capensis* has arisen from a northern stock and is secondarily adapted to the New World tropics (see also Lofts and Murton (1968)), where it now has an enormous distribution from Mexico to Cape Horn and from sea-level to 3500 m in the Andes. It is very much a bird of secondary vegetation and has prospered through man's clearance of native vegetation. As with most of the buntings, a seed diet and occasional green food suffices the adults for most of the year (seeds of *Solanum nigrum, Amaranthus* sp., and *Portulaca oleracea* were reported by Davis 1971), but insects are needed for breeding.

In Colombia, at 3°30′N, where the longest and shortest day differ by only 12 min, Miller (1959, 1965) found that there were two main periods of reproductive activity each year, although some birds could be found with active nests in all months of the year (Fig. 10.7). The bimodal pattern of breeding activity followed a similar pattern in the annual rainfall. Miller considered that the gonad cycle also went through two peaks of activity and he based his view on either direct laparotomie or else he measured the length of the cloacal protuberance of wild-caught subjects; a technique which is justified by the correlation in size between the gland and the testes (Sachs 1967; Siopes and Wilson 1975). If birds were found with partially regressed testes he extrapolated to get a date when he imagined that the testes would be fully regressed. Evidence from other tropical species (see p. 356) suggests that full regression might not have occurred and that Miller's assumption in this respect need not be justified. In fact, in Peru at 12°30′S Davis (1971) found that males rarely had completely regressed testes, but he did show the length of the cloacal protuberance to be closely correlated with testis size. He also showed that even during the main moult the testes contained at least actively dividing spermatogonia (Fig. 10.7). So too in Panama (8°8′N), where Kalma (1970) studied the species. As previously noted, the amplitude of the cycle of gonad enlargement and regression in equatorial regions is less marked than in temperate species (p. 356). In Colombia, two complete moults, including the primaries, occurred each year at 6 month intervals, alternating with the two main periods of nesting (Fig. 10.7). The moult lasted about 2 months in individuals but there was a considerable spread in the moulting season within the population. It was generally timed to occur at the end of the rainy season. Miller noted some competition between the energy demands of moulting and the state of the gonads. Although birds could moult with a functional gonad the frequency with which moult coincided with a regressed gonad was twice the rate expected by chance. The conflict between moulting and breeding was greater in females. Miller claimed that two full moults per year were essential for Andean Sparrows because the birds foraged primarily in wet grassland and scrub and the plumage became extremely worn. This seems an unlikely explanation and instead we suspect that the cycle is a consequence of exposing a higher latitude

species to LD12:12 (*cf.* Starling in Fig. 9.5). Miller also made the subjective judge-
ment that food resources did not vary seasonally, but this view might be unfounded
and other studies of equatorial species in seemingly unvarying environments have
shown that seasonal changes in food availability do occur (*cf.* Yellow-vented Bulbul
(Fig. 7.4, p. 154)).

It is of interest to compare the photo-responses of the Rufous-collared Sparrow
with those of *Quelea* (*cf.* Figs. 10.3 and 10.7). An immediate point of difference is
that the Rufous-collared Sparrow shows less intolerance of long daylengths, for when
transported to and confined at Berkeley, California (latitude $38°N$) its breeding
season was longer than under 12-h days (Miller 1965), and it also showed some capacity
for gametogenetic recovery from involution on LD 8:16 (Epple, Orians, Farner, and
Lewis 1972). When *Quelea* was held on LD 17:7 it could not be prevented from
entering a refractory phase. Miller (1965) was of the opinion that *capensis* lacks a
refractory period altogether and it has subsequently been shown to maintain
spermatogenesis continuously when kept for 17 months on LD 20:4 (Epple *et al.*
1972; Lewis, King, and Farner 1974).[†] Testicular regression did occur in captives when
the light was reduced in steps from LD 14·4:9·6 to LD 11·2:13·8, simulating a natural
photoperiod in Argentina (see below). But in contrast again, regression did not occur
in birds kept on constant LD 10:14 (no regression following 8 months at peak size)
so shorter photoperiods than normally encountered can in this species maintain an
active gonad. It has already been explained that the temperate *Zonotrichia* species
have rather a high threshold of photosensitivity, as Fig. 9.7 showed. Accordingly, if
these species are transferred to an equatorial light cycle the natural photoperiod
remains below the stimulatory threshold. Hence for *Zonotrichia* to invade the tropics
there must have been a marked drop in photosensitivity. This perhaps helps to
explain why it responds differently from *Quelea* when returned to a temperate light
cycle, for *Quelea* ancestors have never had to evolve any adaptive resistance to long
temperate daylengths.

Rufous-collared Sparrows have also been studied at other latitudes (Fig. 10.7).
The importance of ecological factors is indicated by a marked difference in the
breeding season in different parts of Panama (8 °N). Populations inhabiting lowland
pasture and a coffee plantation nested in the main dry season, whereas 3 km away
on grassy volcanic slopes nesting occurred only at the peak of the main rainy season;
the differences reflected the condition of the grass (and presumably the insects it
supported) and whether it dried out or not (Kalma 1970). It is also instructive to
consider information collected by Wolf (1969) at Costa Rica at latitude $10°10'N$,
where the longest and shortest day differ by only 76 min. Here there is a generally
high rainfall but peaks occur in June and July and again in September and November,
that is, there is a very short second 'dry' season in late July and August (Fig. 10.7).
Wolf found there to be a distinctly bimodal pattern in testis development with some

---

[†]The Baya Weaver *Ploceus philippinus* may be in the same category, for irrespective of the
stage at capture peak breeding condition could be maintained continuously under 15-h photoperiods (Thapliyal and Saxena 1964). The cycle in nature probably proceeds to regression
because of seasonal decreases in daylength.

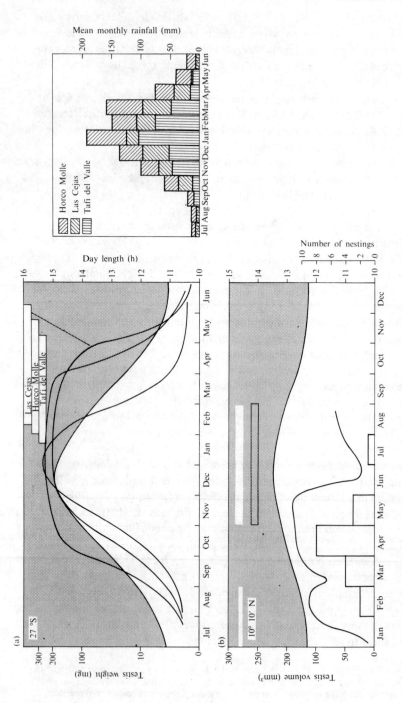

Fig. 10.7. Breeding and moult seasons of the Rufous-collared Sparrow *Zonotrichia capensis* when exposed to different day-night schedules at different latitudes, based on data given by (a) King (1973b) (for three sites), (b) by Wolfe (1969); (c) by Davis (1971); and (d) by Miller (1961, 1962, 1965). In the left-hand panels the duration of the primary moult is indicated by white bars, the testicular cycle based on testis weight or volume by solid lines and the egg-laying season by

open histograms. In addition, seasonal changes in the mean length of the cloacal protuberance of monthly samples is shown for (c) as a dashed line while seasonal changes in insect abundance are indicated by the dotted line. The three shaded right-hand panels show the seasonal incidence of rainfall for the localities against which they are placed and are due to the same authors.

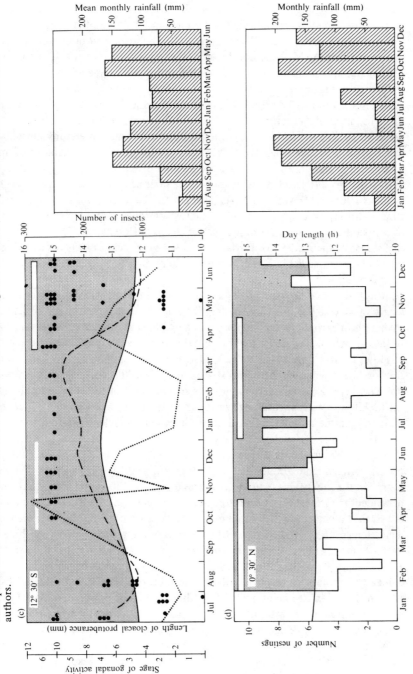

regression in June and partial recrudescence in August followed by a degree of regression lasting until about December. The main breeding period lasted from January until May, being preceded by an incomplete pre-nuptial moult of contour feathers lasting about 2 months in individuals. The main post-nuptial moult began in some individuals in May and continued until August and September, occupying 2 months per individual and it was associated with some degree of gonad regression. Reference to Fig. 10.7 suggests that the terms pre- and post-nuptial are arbitary and represent an intermediate condition between the double moult in Colombia and single moult at higher latitudes (see below). At Chilca, near Lima, in the coastal desert of Peru (latitude 12°30'S) *capensis* also has an incomplete pre-nuptial moult, which is often arrested (Davis 1971). Seasonal changes in the availability of insect food occur at this station judged by the number of insects caught by Davis in sweep nets (Fig. 10.7). In this dry area very little rainfall occurs (15·2 mm between 10 October 1968 and 9 August 1969) and the natural and crop vegetation depends almost entirely on irrigation; shrubs steadily lose foliage from November until June or July and the grass dries. Conditions are clearly different only 70 km away on the east slopes of the Andes, for Davis quoted Blancas Sánchez (1959) as showing that here the sparrows nest in March, April, and May, that is, they start breeding as those at Chilca finish.

Rufous-collared Sparrows living further south in Argentina experience a day-length cycle which is of sufficient amplitude to regulate the pituitary–gonad axis so that photoperiodic induction results in a breeding season which is reasonably close to the one which is ecologically appropriate. This was demonstrated at three localities in Argentina at latitude 29°S (King 1973b) embracing a lowland station in the arid chaco plain, a subtropical station in the foothills and a semi-desert montane site. The region experiences summer rains and winter drought but the total rain at the middle site was approaching twice that of the chaco and between a half to third more than in the mountains. At all sites the gonads of *capensis* started to recrudesce at the same time, when daylengths reached a stimulatory threshold in August, and it is reasonably certain that regression subsequently followed when the daily photo-period declined from about 14 to 11 h (Fig. 10.7). But regression began 74 days sooner in the mountains than in the lowland plain and King suggested that this depended on differences in threshold sensitivity between the three populations, although it seems more likely that ecological conditions facilitated or retarded regression (p. 153). In spite of slight variations in the gonad cycle at the three localities in Argentina, the moult sequence was virtually identical at the three stations. There was no pre-nuptial moult while the post-nuptial moult begain in late January and early February and ended in May. It lasted for 80 days in individual birds (longer than in Columbia, see below) but 110–120 days for the population. An arrested moult occurred in a quarter of the males and females at the two lowland sites, coincident with continued breeding activity, but not in the montane site.

Near to the equator it is unlikely that the seasonal changes in daylength could provide for a reliable proximate regulation of the breeding season, but the daily light-dark cycle could entrain and drive a circadian rhythm of photosensitivity as discussed in the next chapter. Indeed, perusal of Fig. 10.7 suggests the photoperiod was affecting the moult cycle so that as a LD 12:12 was approached two annual moults were possible instead of the single one typical of latitudes with more seasonal change. Similarly, the gonad cycle under an equatorial photoperiod bears a similarity to that of *Sturnus vulgaris* kept on an artificial 12-h daylength (see Fig. 9.5). This suggests that some kind of endogenous periodicity becomes manifest in individuals living near the equator but that the birds are only approximately in phase with one another. Nevertheless, a degree of temporal synchronization in the moult, and to a lesser extent the gonad cycle, may be imposed as a response to seasonal variations in food resources. Miller established that the moult was more synchronized in the population than the breeding season and concluded that its timing determined the timing of the breeding season. He was probably correct in this view but it begs the question of whether the moult was timed by an environmental signal or whether it occurred with an endogenously controlled periodicity. There is also a problem in resolving the mechanism which results in an arrested moult in some breeding individuals for this suggests that in some situations reproduction can have precedence over moulting. At higher latitudes there are seasons when the daylengths fall below the stimulatory threshold and no breeding is possible. This proximate response to light provides an approximate means of timing the breeding season though even in Argentina ecological factors must still be of paramount importance, judging by the variations in the actual egg-laying season noted in populations living comparatively close to each other.

Adaptation of photo-refractory phase

In species which require to breed around the summer solstice at latitudes above about 50 °N the development of refractoriness is generally advantageous, as in the northern *Zonotrichia* species already mentioned. In some species in which the photo-threshold is set at a high level refractoriness may hardly be expressed, for as it is about to begin natural daylengths may fall below the critical level causing gonad regression to occur without true refractoriness developing (see p. 390). It will be shown (p. 392), that the acquisition of a photo-refractory phase is an advanced condition which would be a disadvantage for species which require to breed early in the spring (say in late March) yet continue until September. However, the majority of temperate passerines have cycles of the House Sparrow type in which obligatory gonad regression occurs in late July or August. Some of the longest cycles of this kind are noted in the finches. The testes of the American Goldfinch *Spinus tristis* slowly recrudesce from March/April and spontaneous regression occurs in late August (Mundinger 1972), while occupied nests can be found until September (Stokes 1950). The European Goldfinch *Carduelis carduelis* has a longer gonad cycle for egg laying begins in April but is even more frequent in July and August (Conder 1948). In Australia, where it has been introduced, the breeding season is also extended (see p. 232). In these passerines the moult follows the breeding season and occurs during the refractory phase. A limit must be imposed on the length of the breeding season in order that moulting can be accomplished before unfavourable conditions

develop in winter. This problem does not arise in such groups as the raptors in which
the moulting season is extended and coincides with nesting. This may have helped
Eleonora's Falcon *Falco eleonorae* and the Sooty Falcon *F. concolor* to breed in the
autumn rather than spring, unique breeding seasons for temperate species. Both these
falcons depend on catching migrating passerines to feed their young and the availability of these is greatest when juveniles swell the numbers returning to Africa
(Vaughan 1961; Clapham 1964). It is not clear how the breeding season of these
species is controlled. They could be derivatives of the Peregrine Falcon *F. peregrinus*
that have evolved a resistance to summer photoperiods thereby gaining a long
potential breeding season, but an actual one which is determined by food supplies.
Alternatively, autumn breeding may represent a post-refractory recovery of gametogenesis as discussed above for the Rook (however, see p. 256).

Absence of photo-refractory phase in some pigeons

An effective absence of a refractory period is advantageous to many pigeon
(Columbidea) species, for they can breed over a very long season and in many cases
moult at the same time (Fig. 7.11, p. 165). The Columbidae are a phylogenetically
old family which probably emerged and evolved in Gondwanaland at the end of the
Cretaceous Period (Cracraft 1973).[†] Considerable speciation has occurred in
Australasia but we are concerned here with only three genera. *Columba* is the
biggest pigeon genus and almost certainly evolved in sub-tropical woodlands in the
Southern Hemisphere (see Appendix 4). The genus is represented by three Palaearctic
species, one of which, the Wood Pigeon *Columba palumbus,* shows considerable
morphological and ecological convergence with the New World Band-tailed Pigeon
Columba fasciata. Streptopelia is an Old World genus, very well represented in Africa
and Asia, which is close to *Columba* and probably emerged from the same ancestral
stock. Its members are small- to medium-sized seed-eaters whose ecological niche in
the New World is occupied by *Zenaida* and to which they bear a strong convergent
similarity. However, Goodwin (1967) considers that *Zenaida* is not closely related
to *Streptopelia* but has arisen from the New World ground doves (*Columbina* and
allies) which in turn derive from Old World bronze-wings, emerald doves, and
related groups (*Chalcophaps, Phaps, Petrophassa, Geopelia,* etc.) (but see Appendix 4)
    The Turtle Dove *Streptopelia turtur* and Collared Dove *S. decaocto* probably had
a distributional range around latitude 20–40 °N in the Old World during and just

[†]Cracraft questions Mayr's (1946) statement 'the rich development of the family in the
Australian region, where the most aberrant members of the family occur (e.g. *Caloenas, Goura,
Otidiphaps* and *Didunculus*), and the fact that most American species belong to just a few
phyletic lines, prove an Old World origin.' Cracraft argues that it is not proven yet that the
American lineages are in fact the most primitive within the family and there is evidence that
some of the Australasian genera like *Otidiphaps* are very derived (Glenny and Amadon 1955).
But if the pigeons and parrots *are* closely related Cracraft suggests that their common ancestor
was probably distributed in the Southern Hemisphere during the Cretaceous period and that
late Cretaceous and early Cenozoic palaeogeography subsequently influenced the dispersal of
the group. At the beginning of the Tertiary continuous land connections existed between South
America, Antarctica, and Australia, Australia probably drifting north from Antarctica in the
Eocene.

after the glaciations and are of interest for they have both reached more temperate regions in relatively recent times. The Turtle Dove has evolved a photo-refractory phase of the kind discussed in the previous section and has, therefore, come to have a breeding cycle which is approximately similar to that found in many north temperate species. For this reason it is considered first and its breeding cycle compared with that of its congenor the Collared Dove before we turn to discuss the cycles of the more primitive *Columba* pigeons, which effectively lack refractory phases. The Turtle Dove is a summer visitor to Europe, north to about $60°$, and winters in the African savannas at $10-15$ °N. It favours farmland hedges and the woodland edge and is essentially a seed-eating dove of secondary vegetation and cultivation; it has doubtless prospered by the argicultural activities of man (Murton, Westwood, and Isaacson 1964*a*; Murton 1968*b*). It possibly did not colonize Britain and the more northern parts of its range until relatively recent times; in Britain one of its major food plants, Fumitory *Fumaria officinalis*, a characteristically Mediterranean genus, is not represented in interglacial pollen deposits and first appears in the Romano-British pollen record (Bunting 1960). Fumitory has become established only in the artificial and disturbed conditions associated with cultivation (Warburg 1960) and it favours the drier, lighter, and often calcareous soils. It may be that the Turtle Dove spread north in Europe with the advance of the Roman legionnaires who certainly introduced many agricultural crops and their associated weeds to Britain.

The Collared Dove's ancestral home is northern India, where it lives in fairly close association with agricultural man, obtaining millet, sorghums, and other cereal and weed seeds from the fields and farmyard.[†] At the end of the sixteenth century its eastern outpost was Constantinople and the Levant from where it spread slowly to reach Belgrade in 1912. It was stabilized in the Balkans until 1930 and then suddenly exploded north-west across Europe to reach Britain in 1952 (Fisher 1953). Here its numbers have increased geometrically from 4 in 1955 to nearly 19 000 in 1964 and around 30 000–50 000 in 1970 (Fig. 1.3, p. 8). For such an expansion to have occurred a vacant ecological niche must have existed for it to occupy. Throughout Europe the Collared Dove is closely associated with man-altered habitats, such as villages and smallholdings (Keve-Kleiner 1944), and Gladkov (1938) reports that it is rigidly confined to cultivated areas in Turkestan. So too in Britain where it was initially a bird of parks, gardens, and cultivated land, favouring such sites as chicken-runs and other places where grain is available for animal stock. More recently it has become established in certain urban areas alongside the Feral Pigeon, avoiding competition by taking smaller food items. In such places selection for a lengthening of the breeding season is probably occurring (see below).

The breeding and moulting seasons of these two *Streptopelia* doves are depicted in Fig. 7.11 (p. 165) and have been shown experimentally to be under photoperiodic control. The Turtle Dove resembles the House Sparrow in becoming refractory in July, but a refractory stage does not have time to develop in the Collared Dove in

[†]The Little Brown Dove or Palm Dove *S. senegalensis* which is more closely related to the Turtle Dove is even more commensal on man in India for it takes grain from hen runs and grain stores and like the House Sparrow nests in crevices and rafters of his buildings.

Britain before declining daylengths in September result in gonad regression (see below). The chicks of Turtle Doves reared at the beginning of the breeding season in May and June have more than twice the probability of surviving than those reared in August or September (Murton 1968*b*). Because seed supplies increase during the summer, as crops and their associated seed floras ripen, the ability of Turtle Doves to hatch their eggs and rear their chicks to fledging improves between May and July; the percentage of chicks fledged of eggs laid increased from 34 per cent in May to 48 per cent in July. Under these circumstances breeding might continue throughout August and September but is prevented by the onset of photo-refractoriness. Experimentally, Turtle Doves caught wild in July were exposed to artificial LD 17:7 and LD 8:16 photoperiods (Lofts, Murton, and Westwood 1967*c*). The testes of birds held on 17-h daylengths spontaneously regressed and remained inactive until the experiment ended in mid-winter, but in subjects allowed a spell on 8-h days they could be re-stimulated when the birds were returned to LD 17:7. Young fledged late in the season in the wild would be unlikely to achieve the necessary plumage, body weight and fat reserves necessary to begin the long migration south to Africa. Murton (1968*b*) also suggested that the adults must stop breeding in order to find time and resources for a moult of the primaries before departure. Since other species can defer the primary moult until they reach Africa this view is probably mistaken.

Lofts *et al.* showed that the post-nuptial moult,[†] which had already begun when the subjects were caught in July, continued to completion irrespective of whether the birds were held on short or long daylengths. However, the next moult, under 17-h photoperiods, could be initiated only in subjects first exposed to short days for sufficiently long to break refractoriness. In the wild, refractoriness is presumably broken during the sojourn in Africa, where the doves experience an approximately 11—12-h daily photoperiod which prepares them to respond to long photoperiods when they migrate north again. The gonads must be stimulated by shorter photoperiods than are needed to initiate a moult in receptive subjects, for the gonads of birds reaching England in April are fully recrudesced whereas moulting does not begin until July. It was also demonstrated that migratory fat is deposited on long or short days as subjects enter a refractory phase and it persists until refractoriness is broken by short day schedules.

The gonad cycle of the Collared Dove in India has not been defined but the breeding season is obviously extended as egg laying has been recorded in all months of the year. In England the gonads are regressed during the short winter days so that for most individuals reproduction is only possible between March and September—early October, when the daily photoperiod exceeds about 12-h, and this is when eggs are produced (Fig. 10.8). Many of the subjects whose gonad records are included in Fig. 10.8 were collected at Ellesmere Port, where the birds were dependent for food on a flour mill. Quite a lot of individual variation in gonad size is apparent in Fig.

[†]Stresemann (1967) has shown the Turtle Dove to be peculiar in its moult, for the moult of the primaries is begun on the breeding grounds at the end of the nesting season, and it is then arrested until the birds reach their contra-nuptial quarters in Africa, whereupon body and tail moult occur and the primary moult is finished.

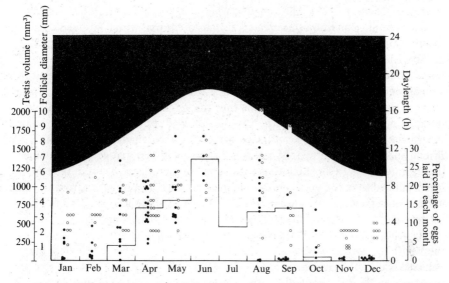

Fig. 10.8. The breeding season of the Collared Dove *Streptopelia decaocto* in England. Testicular volume (solid circle) or diameter of largest follicle (open circle) based on birds collected at Ellesmere Port in 1970–1 and percentage distribution by month of egg laying from a free-living population in Cambridgeshire 1971–2 (187 eggs laid). From Murton and Westwood (1975).

10.8, especially in spring and autumn when the birds were coming in or going out of breeding condition; for example, in September 6 males had completely regressed testes while 3 others were in breeding condition.[†] Variability might be expected in a species that has been expanding its range and also occupying new habitats. In fact, variations in photoperiodic response have been demonstrated experimentally in subjects caught wild at Ellesmere Port which were confined under artificial LD 16:8 or LD 8:16 photoperiods (Murton 1975; Murton and Westwood 1975). The testes of some males collected from the field in July—that is, about 2 months before regression occurred in free-living birds—and held on LD 8:16 collapsed while others remained unchanged. When the photoperiod was increased to 11-h on 25 September the testes of 4 birds (including 2 that were newly caught from the wild) increased in volume whereas 2 decreased. In early March, 5 out of 7 wild-caught subjects responded to 8-h photoperiods while the other 2 were unstimulated. Six out of 7 responded to a 16-h photoperiod, the rate of gonad recrudescence not being noticeably different from the recovery under 8-h photoperiods. Selection could, therefore, be favouring a reduction in the photosensitivity threshold so that gametogenesis can be initiated under 8-h days, this being consistent with the fact that ecological conditions are conducive to all-the-year breeding in some urban habitats.

Doves which were caught in breeding condition in July and then held on LD 16:8 until December exhibited spontaneous gonad regression in late November and early

---

[†] This is to be explained in terms of the gonad cycles of individual birds not being closely in phase with each other (see p. 391).

December (Murton 1975). Allowing that when caught these subjects had already been in breeding condition since March, it is apparent that with continued exposure to stimulatory daylengths refractoriness developed after about 9–10 months. That is, the Collared Dove has a breeding cycle which last about 9½ months before refractoriness develops. It will be shown in Chapter 12 (Fig. 12.17, p. 338) that there is an autonomous periodicity of this duration which can be expressed under a wide range of constant photoperiods. By the time the 9½-month cycle should end and refractoriness develop in Britain (in November) gonad regression has already occurred in response to declining daylengths (see Fig. 10.8). Accordingly, in the wild in Britain and northern Europe the seasonal light cycle is such that gonad regression occurs in response to short photoperiods before refractoriness. Thus, in the experiments alluded to in the previous paragraph, when subjects were taken from the field in November they had already experienced sufficient 'short' autumn days to break refractoriness and so could immediately respond to artificial 'long' days.

The whole of the New World is occupied by a *Zenaida* super-species comprising the Mourning Dove *Zenaida macroura* in North America to Florida, Mexico, the Bahamas, Cuba, and Hispaniola and the Eared Dove *Z. auriculata* from the southern Carribean south throughout all South America.[†] In many parts of South America it has become a pest of cultivation thriving on the sorghum, millet, and cereal crops; this is true in Argentina where in its general ecology and feeding habits it has been shown to be very similar to the North American Mourning Dove and to the Old World Collared Doves (Murton, Bucher, Nores, Gómez, and Reartes 1974a). The Mourning Dove was shown to be photoperiodic many years ago (Cole 1933) and it has a well-defined, long breeding season in North America (see Fig. 7.11, p. 165). The breeding season and gonad cycle of the Eared Dove near Córdoba (latitude 31 °S) were also depicted in Fig. 7.11. We suspect that the cycle of the Collared Dove in India might be similar but have yet to establish this. Under the photo-regime experienced in Argentina mean testis weight did not vary much between the fully recrudesced and fully regressed stage, compared with the degree of seasonal change noted in the size of the testes of Collared Doves in Britain. There was even more individual variation in the testis size of Eared Doves during any particular month than was noted in the Collared Doves (see Murton *et al.* 1974a). This suggests that the gonad cycles of individuals could not have been well synchronized and that any proximate control of breeding periodicity by light or temperature was imprecise: the same was found to be true for pigeon species of the genera *Ducula, Ptilinopus,* and *Geopelia* which were examined by Frith, Braithwaite, and Wolfe (1974) in New Guinea (latitude 9°29′S). Eared Doves roost and nest in enormous gatherings and in one site studied near Córdoba we estimated that in the order of 1 million birds were involved. When visited in March some segments of the colony had fresh eggs all the birds having

---

[†]There are three other *Zenaida* species. The White-winged Dove *Z. asiatica* is distributed in Central America, preferring arid desert areas, where its range overlaps the Mourning Dove. The allopatric Zenaida Dove *Z. aurita* is distributed in the Carribean Islands, West Indies, Bahamas and has 12 instead of 14 tail feathers typical of *Z. macroura* and *Z. auriculata*; it probably is a connecting link between the latter and the Galapagos Dove *Z. galapagoensis* (Goodwin 1967).

clearly laid within a few days of each other, other segments were feeding small young while in other sections birds were all singing or nest-building or were not breeding at all. Evidently, subsections of the colony were closely synchronized in their breeding activities and we believe that these were birds which had located good feeding grounds. In many subtropical and tropical species it seems likely that good food sources allow the birds to participate in reproductive behaviour which in turn stimulates gonadotrophin secretion and gonad growth. This may apply especially where the gonads only partially regress so that a degree of spermatogenesis is maintained throughout the year. We have found that the testes of some isolated Feral Pigeons *Columba livia* caused to regress under LD 8:16 recrudesced when the males were introduced to females, without alteration in the photoperiod (Murton and Westwood 1975).

The *Columba* pigeons of Europe do not exhibit an obligatory photo-refractory phase under natural temperate photoperiods and they can have breeding seasons which extend symmetrically on either side of the summer solstice. The Wood Pigeon *Columba palumbus* evolved as a tree-feeding, bud-eating species of temperate deciduous forest in the western Palaearctic, which can also subsist on weed leaves and seeds and the fallen fruits of such trees as Beech *Fagus sylvatica* and Oak *Quercus robur*. Cereal grains and clover leaves provided by the arable farmer are superior to many natural foods and the Wood Pigeon has expanded its range and increased in numbers in the niche created by cultivation (Murton, Westwood, and Isaacson 1964a; Murton 1965). This has been possible only because active gametogenesis occurs over a long season, from March until September, and within this long genotypically determined season considerable phenotypic variation is possible, as Fig. 10.9 demonstrates.

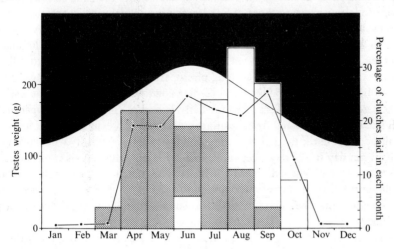

Fig. 10.9. Phenotypic variation in the egg-laying season of the Wood Pigeon *Columba palumbus* in England. Seasonal changes in testis weight based on a sample of 79 birds collected in Eastern England (lines and solid dots). Hatched histograms give the percentage number of clutches laid per month in urban London (from Cramp 1972) while open histograms give the percentage number of clutches laid in a rural locality in Cambridgeshire. From Murton (1975).

In suburban areas the Wood Pigeon breeds in relation to the development of tree foods. Cramp (1972) demonstrated that, in London, tree buds, leaves, and weed seeds increased in the diet of Wood Pigeons between January and March, remained prominent in April and May, and then declined to a low level between June and September. For the rest of the year the diet was mostly grass, and bread supplied by an indulgent public. In arable farmland tree foods are less readily available, especially as over large areas many woods and hedgerows have been destroyed. In contrast cereal grain is abundantly available from late July until the stubbles are ploughed in late summer, and in arable areas of Britain most breeding occurs in relation to this food supply. In the wetter west of England, pastoral farming predominates, woodland is more extensive, and proportionately more of the population depends on this more natural food source, especially since there may be little or no cereal production; the breeding season is generally earlier than in the east. Breeding success is higher, and first-year birds can breed relatively successfully in the east of England compared with the west. As a result the productivity rate is higher in the east and is compensated for by an increased mortality rate during the early years of life (Murton and Westwood 1974a). These spatial variations are probably also indicative of changes that have occurred with time, following a transition from a mixed arable—pastoral agriculture of the past to the intensive arable systems practised today. The Wood Pigeon has a suffciently flexible physiological mechanism to adjust its breeding season to suit farming change, but this is not true of many other species.

The season when tree buds are available, which are rich in proteins and sugars destined for the flowers and pollen, is reliably heralded by the lengthening daylengths in early spring, and Wood Pigeons become stimulated into breeding condition when photoperiods equivalent to those of late March are experienced. The rôle of daylength in stimulating gametogenesis has been demonstrated experimentally (Lofts et al. 1967a). Adult Wood Pigeons collected from the field in January or February, when their testes were still regressed, could be prematurely stimulated into full breeding condition by exposure to long photoperiods. But, while boosting the daily ration of light could bring adults into breeding condition prematurely, this was not the case for first-year birds caught in early February. Similarly, first-year males which had been hand-reared from nestlings collected the previous summer and kept throughout in captivity, failed to respond spermatogenetically when given 17-h light per day for 21 days, beginning in early March. Individuals from the same stock could respond when given extra light from 23 March until 14 April. Evidently, juveniles at first lack the capacity to respond to photostimulation, which accounts for the observation that they do not naturally assume a reproductively active condition until June or July.

Daylengths of around 12—13 h (sunrise to sunset plus civil twilight) occur in eastern England (latitude 52—53 °N) when Wood Pigeons naturally begin gametogenetic development in the spring. Assuming that this is the level necessary for a response we note that regression naturally occurs in September when daylengths become shorter than this supposed threshold. Such regression could be prevented if in September subjects were exposed to artificial 16-h daylengths but not if they experienced LD 8:16. When pigeons whose testes had just regressed under 8-h days were exposed to 18-h photoperiods their gonads were immediately stimulated back into reproductive condition until 21 December, when the experiment finished. Whether with further exposure to long days they would eventually develop refractoriness remains to be elucidated. For practical purposes the Wood Pigeon lacks a refractory period in the range of photoperiods that are experienced in the wild. It has a seasonal breeding cycle in Britain because between September and March daylengths fall below the stimulatory level. Actually, there is no latitude at which it would not have a seasonal cycle though exactly how it would behave on an equatorial LD 12:12 remains to be determined. We may remind ourselves from Fig. 7.11 that the seasonal moult occurs throughout the period when the gonads are active, this being generally true of those pigeon species which do not manifest photo-refractoriness under natural seasonal daylength. In contrast, the Turtle Dove has a moult pattern more typical of those avian species which exhibit a post-nuptial gonad refractoriness.

Female Wood Pigeons also lack a refractory period as the experiments summarized in Fig. 10.10 demonstrate. In contrast, Lofts *et al.* (1967c) showed that the female Turtle Dove, like the male, does enter a post-nuptial photo-refractory phase. It was also evident in their experiment that females, like males, are responsive to photo-stimulation. However, as already mentioned, males could reach full spermatogenetic development under experimental photoperiods whereas the maximum follicle diameter attained by the females was around 5 mm. This compares with a diameter exceeding 20 mm at ovulation.

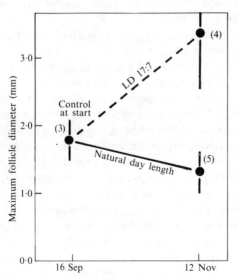

Fig. 10.10. Effect on maximum follicle size of keeping female Wood Pigeons on long or short photoperiods at the end of the natural breeding season. Birds kept on LD 17:7 showed no follicle regression so that females, like males, do not enter a photo-refractory phase at the end of the natural breeding season in Europe. Based on data in Lofts *et al.* (1967a).

The gonad cycle of the Band-tailed Pigeon *Columba fasciata* of North America appears to resemble that of the Wood Pigeon, judging by data collected by Braun (1973) and March and Sadleir (1970) (see Fig. 7.11, p. 165). The testes of birds wild caught in Colorado in May were already enlarged and regression began in the wild in early September; birds kept all the year round in outdoor aviaries assumed breeding condition in March. Birds with regressing testes in September which were placed on artificial LD 12·5:11·5 immediately responded with testicular growth. Birds with regressed gonads in November and December can respond to 10-h photoperiods and this means that Band-tails may sometimes breed in December or January in southern parts of the range (e.g. Mexico) if good stocks of Piñon seed or acorns (*Quercus*) are available. In some of these southern regions the birds wander nomadically in search of local food supplies while more northern populations comprise regular migrants with relatively restricted summer and winter ranges.

In Britain, Stock Doves are more efficient at early breeding than are Wood Pigeons, probably helped by their hole-nesting habits. Thus 15 per cent of eggs laid by Wood Pigeons in March and April gave rise to flying young compared with 38 per cent of those laid by Stock Doves; comparable figures during the period July—October were 37 per cent and 47 per cent, respectively (Murton 1966*b*). The post-fledgling survival of young Stock Doves reared early in the year is also better than that of Wood Pigeon fledglings (see Fig. 8.1, p. 187). This difference in prospects for successful early breeding may explain why juvenile Stock Doves entering their first breeding season assume breeding condition at the same time as the adults, in contrast to first-year Wood Pigeons; the few wild-caught first-year Stock Doves we have examined in January, February, and March have been at exactly the same stage of gonad development as the adults.

Fig. 7.11 (p. 165) showed that in England the Stock Dove *Columba oenas* has a potential breeding season, as judged by the activity of the testis, which is about 3 months longer than that of the Wood Pigeon. This depends on the ability of the Stock Dove to respond to a shorter daily photoperiod than the Wood Pigeon, as was shown experimentally by the experiments conducted by Lofts *et al.* (1967*b*). Thus the testes of Stock Doves kept under a natural light schedule from 22 November until 20 December were found to contain only resting spermatogonia at the end of the treatment. Other individuals given an artificial light schedule increasing from 9·1 to 10·9 h over 28 days responded with the production of secondary spermatocytes. Yet the testes of Wood Pigeons given this same treatment remained inactive with resting spermatogonia and a light regime increasing from 10·9 to 12·5 h over 28 days was needed for the production of secondary spermatocytes and some spermatids. There must be adaptive advantages for the Stock Dove in having a potentially longer breeding season than the Wood Pigeon. Unlike the Wood Pigeon it is not a tree feeder and instead favours the open fields, where it can specialize on finding weed seeds. For this reason it is much more frequently located in dry semi-desert areas and dunelands, while on farmland it resorts to the fallow fields, old stubbles, and cereal sowings more frequently than its congenor (Murton *et al.* 1964*a*).

A third British Columbid, the Rock Dove *Columba livia* is confined to coastal cliffs. About 150 years ago the species inhabited much of the rocky coastland of the British Isles. It subsequently declined in numbers coincident with the spread of the Stock Dove and Wood Pigeon, ecological competition consequent on the changes initiated by an expanding agriculture being deemed responsible (Murton *et al.* 1964*a*). Today only a few pure-line Rock Dove populations remain in the northern and western coasts and the islands of Scotland and coastal Ireland. Other coast dwelling populations of *Columba livia* do exist in south-west and north-east England, in Wales and the Isle of Man, and south-west and east Scotland and these, though intermixed with birds of domesticated origin, live ecologically like wild-type populations. At Flamborough Head, Yorkshire, 70 per cent of the population has plumage characteristics which are close to those of the pure-line wild-type (Lofts *et al.* 1966).

The physiological breeding season of the Rock Dove is even longer than that of the Stock Dove (Fig. 7.11, p.165). Gonad regression begins in September or October in northern Britain but by December spermatogenetic activity starts again and some birds achieve full spermatogenesis at this time, one to two months sooner than do Stock Doves.[†] Ecologically, the Rock Dove is separated from the Stock Dove and Wood Pigeon by habitat but in its food preferences it combines the dietary spectrum of the other two species (Murton and Clarke 1968). It is not clear whether its breeding season is potentially longer because it can exploit a wider range of food sources or because in milder maritime habitats weeds flower and seed for a longer period. It is also possible that its evolutionary history may explain the difference, for the Rock Dove essentially belongs to the Turkestanian—Mediterranean fauna of the warmer and drier regions of southern Europe and south-west Asia, whereas the other pigeons belong to the more temperate European-Turkestanian fauna (Voous 1960). The true distribution today is difficult to assess for throughout its range the Rock Dove has been domesticated and escapes and releases have re-mixed with wild populations or established new colonies; the occurrence of *Columba livia* in the New World and Australia depends on introductions by man. The mid-latitude of the present range is close to 30 °N (with a photoperiod range from 11·1 h to 15·0 h).

Town pigeons

The capacity of the Rock Dove to breed during a large segment of the year, to use ledges and holes in buildings or the girders of bridges *in lieu* of natural rock cliffs, and its ability to forage for food supplies incidentally or purposely provided by man, have enabled it to adapt to man-made habitats in urban centres. The present-day flocks of town pigeons derive from domesticated stock. In Britain until the eighteenth century every large farmhouse maintained a loft in which squabs could be raised to provide a protein food source for the squirearchy, a manorial tradition beginning with the Norman lords. Before then natural breeding caves were farmed by the local populace, this practice going back to antiquity. Indeed, there is evidence from Iraq that man

---

[†]The population is not in complete synchrony in the gonad cycle, so a few pairs can be in breeding condition when the others are inactive. Accordingly, eggs can be found in all months (Lees 1946; Murton and Clarke 1968).

was bringing wild doves into domestication as long ago as 4500 B.C. As Pliny could comment, 'pigeon-fancying is carried to insane lengths by some people'. In Europe during the Middle Ages it is evident that the lofts of free-flying doves were to a large extent dependent on the corn and other crops grown by the peasants who must have at times revolted against such exploitation. A law of 1424 stated that 'breakers of mennes Orchardes, stealers of frute, destroyers of Cunningaires [rabbit warrens] and Daw-catles [*sic*] . . . sall paie fourtie shillings to the King'. By 1579 under James VI of Scotland the penalty was £10 for a first offence, hanging to death for a third. By 1617 the Scottish county of Fife alone had 360 dove-cotes containing about 36 000 pairs of doves, and a rough extrapolation indicates a national population not much different from that of the wild Wood Pigeon, which has been estimated to be in the order of 5 million prior to breeding (Murton 1965). The decline of the dove-cote in the eighteenth century coincided with the improved farming methods of the Agrarian revolution and enclosure and was linked to an appreciation that there were more profitable means of producing protein. Releases and escapes from dove-cote sources, and to a lesser extent from the lofts of pigeon-racing enthusiasts and other fanciers, have provided the free-living (hence feral or wild) populations seen today and which, as mentioned, have in places intermingled again with wild lines.

In such places as the Salford Docks, Manchester, spillage of waste associated with the transport of materials to and from storage and provender mills provides a food supply which, because industrial processes are geared to function as much as possible at a constant rate, does not vary in quantity throughout the year. The main seasonal factor, apart from climate, is that in winter short daylengths restrict the time available to the birds for feeding, after workmen go home at the end of the day. As with many equatorial bird species, the population is saturated relative to food resources throughout the year so individual pairs must find the extra energy needed to breed in a competitive situation. In effect, the seasonally breeding photoperiodic Rock Dove has invaded a new habitat where selection favours an extension of the physiological breeding season to allow all-year-round breeding, and this has required that the photosensitivity threshold be lowered. Eggs may be laid at any time of the year and various authors have commented on winter breeding in the species (Lees 1946; Dunmore and Davis 1963; Häkkinen, Jokinen, and Tast 1973). In fact, individual birds undergo a gonad cycle in which regression occurs and lasts about a month and this tends to synchronous in the population during September (Murton *et al.* 1972; Murton, Thearle, and Coombs 1974*b*). Nevertheless, there is usually a proportion of the birds in any population which is not in phase with the majority and these may produce eggs when the rest of the pairs are incapable. The same probably applies to the Speckled Pigeon *Columba guinea* (Skead 1971).

One of the interesting features of the Feral Pigeon is the fact that selection for a change in photosensitivity is somehow associated with a plumage polymorphism involving melanism. At first impression, these town populations exhibit a bewildering array of colours and patterns but in reality certain categories predominate. Most (90 per cent and over) birds are of a blue-grey colour though a small proportion can be red as the result of a recessive gene; other colours, including albino varieties, occur very rarely in free-living populations and can be ignored for present purposes.

A *pattern* of melanic markings is caused by the spreading of pigment situated in the chromatophores of the black wing-bars which are characteristic of the wild-type (+) Rock Dove. Distribution of this pigment is controlled by a series of multiple alleles at an autosomal locus. In the recessive 'barless' condition (c) the black wing bars are lost. We have never seen this variety living free and know of only two records of it in captivity. Many free-living Feral Pigeons resemble the wild-type Rock Dove in plumage characters, although they may have grey instead of white rumps, heavier bills, and more pronounced ceres; they are sometimes called 'blues' by pigeon fanciers. Checker (C) is dominant to wild-type and it produces black flecks over the wings, while the extreme dominant allele (C^T) produces even more dark flecks which partly coalesce on the back and wings so that when the bird is held up with wings outstretched a 'T-pattern' is seen. A different kind of melanism is caused by the spreading of the kind of melanin deposit found in the tail and it results when another autosomal locus carries the dominant allele (S). When expressed the resulting 'spread-pattern' is epistatic to the checker series just described. The frequency distribution of morphs carrying the alleles described above is detailed in Table 10.2 for various populations; details of the possible genotypic composition of phenotypes are given. There is some evidence that London's Feral Pigeon population has changed from predominantly pale to dark morphs over the last 100 years (Ingram 1971).

Blue-coloured morphs belonging to the checker series, and captured at Salford in May and June, were taken to the laboratory where a laparotomy showed them to be in breeding condition. All were kept on artificial LD 8:18 for 36 days and this caused the gonads of 26 per cent of the wild-type to regress, this not being significantly different from the proportion of blue-checks so responding (23 per cent). But T-pattern subjects were much less influenced by the reduction in daylength and only 4 per cent of them exhibited testicular regression. This need not imply that the birds would stay in permanent breeding condition for, like the Collared Doves already discussed, regression might eventually occur according to an endogenous periodicity. Fig. 10.11 shows how the annual gonad cycle differs slightly in phase between the different phenotypes. It would be instructive to perform a more criticial analysis with birds of known genotype and experiments are in progress.

A simple increase in the potential length of the breeding season cannot be the main reason for the success of certain melanic forms at Salford. Indeed, spread-pattern birds should be more common were this the case. Recent studies have shown that only about a third of the total population obtains the necessary food resources to breed, and the morph distribution of the breeding population is the same as the non-breeding component (Murton *et al.* 1974*b*). There is a high degree of mate fidelity, and the pairs occupy their nesting sites throughout the year. This is in spite of the fact that the gonads of breeding birds regress in the same manner as those of non-breeders. Thus, during September the males have small testes but they nevertheless manage to defend their territories and keep their females. Although clear differences in productivity related to the length of the breeding season are not evident between morphs it has been shown that pairs in which the male is a T-pattern leave more progeny than blues (Murton *et al.* 1974*b*). Furthermore, T-pattern males contain more germ cells per segment of the testis tubules than do blue checks or wild-type blues (Murton *et al.* 1973). Since the ratio of different morphs differs between males and females (see Table 10.2), it is likely that sexual differences in selective advantage exist. One of the problems in resolving this topic is that two of the phenotypes can conceal two or three genotypes (which cannot be distinguished without breeding experiments). We suspect that certain genotypes have an advantage since the adults do not mate at random, but, as Fig. 15.4 (see between pp. 434–5) shows, prefer to mate with a partner different in plumage characters from themselves (it is reasonably certain that it is the female who makes the choice). Negative-assortative mating of this kind increases the proportion of heterozygotes and leads us to suspect that factors associated with melanism in the homozygous condition are harmful. It could account for the maintenance of the polymorphism (see p. 429). It ought to be mentioned that there is some evidence that melanic morphs are at a selective disadvantage in wild Rock Dove populations. They occur infrequently but were particularly prevalent in the Faeroe Islands after one particular breeding

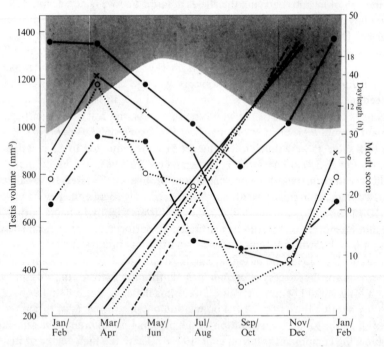

Fig. 10.11. Seasonal changes in mean testicular volume of various Feral Pigeon *Columba livia* var. phenotypes at Salford, Manchester. Solid line and solid circles, spread-pattern birds; solid line and crosses, T-pattern birds; dotted line and open circles, blue-checkers; chain line and solid circles, wild-type blues (see text for explanation of terms.) The diagonal lines are regression lines for increase in moult score with date with spread pattern shown as dashed line, T-pattern as solid and other conventions as above. Breeding cycles based on Murton *et al.* (1973).

TABLE 10.2

Proportion of different pattern phenotypes in populations of the Feral Pigeon Columba livia *var.*

	Percentage of phenotypes[†] which were:					
Source	wild-type (+)	blue check (C)	T-pattern (C^T)	spread (S)	other	sample size
Flamborough, Yorkshire[‡]	70	16	8		6	102
Leeds[‡]	23	25	33		19	265
Liverpool[‡]	23	37	27		13	305
Salford, Manchester						
Breeding adults[§]	26	33	29	4	8	1239
Nestlings[§]	21	31	37	2	9	1292
Juveniles[§]	21	36	30	4	9	1291
Males[§]	17	23	41	7	12	657
Females[§]	22	31	29	8	10	544
New York[¶]	29	28	21	13	9	647

[†]Phenotypes can be of the following genotypes, excluding the rare barless recessive gene from consideration:

Wild-type ++
Blue check C+ or CC
T-pattern $C^T C^T$ or $C^T C$ or C^T+

The spread-pattern locus is different from the checker series above.
[‡]From Lofts *et al.* (1966); [§]from Murton *et al.* (1973);
[¶]from Dunmore (1968).

season though they suffered a much higher mortality than the wild-type during the ensuing winter (Petersen and Williamson 1949).

Other species

Before ending this chapter mention should be made of some other taxa which effectively display no refractory phase in their breeding cycle. The Quail *Coturnix coturnix* has a trans-Palaearctic—Oriental summer breeding distribution at 25—62 °N and migrates to winter north of the equator in South Asia and Africa. In Britain, breeding usually begins in June but fresh eggs have also been produced in August and early September, so genuine second broods may occasionally be laid. Moreau (1951) has raised the possibility that Quail may also breed outside the Palaearctic, for the species nests in the Maghreb in early spring and then in June both parents and young cross the Mediterranean northwards and could well breed again. *Coturnix coturnix* also has a population in Africa distributed from south of the equator and including Madagascar, but confined to high altitudes. This population is presumably a glacial relict. The domesticated race of Quail originating from Japan *Coturnix c. japonica*[†] has been a favourite experimental subject and does not show indications of spontaneous gonad regression when exposed to constant long photoperiods.

[†]The exact status of Japanese birds and European and Asiatic Quail is still in debate and it is likely that the species has reached an intermediate stage of differentiation (Moreau and Wayre 1968).

Gametogenesis can be initiated with a 12-h daylength (see Fig. 12.1). The Bobwhite Quail *Colinus virginianus* occupies in North America the ecological niche of the Old World partridges and Quail and it too has been shown to lack a refractory period (Kirkpatrick 1959), presumably a convergent adaptation. So too may the Californian Quail *Lophortyx californicus* (Lewin 1963). Anthony (1970) has documented the testicular cycle of free-living birds and, as Fig. 10.12 shows, the season of gonad enlargement extends symmetrically on either side of the summer solstice.

Judging from Fig. 10.12 the photo-response threshold of the Californian Quail is set relatively high whereas the opposite must apply in the crossbills *Loxia* sp., for breeding may occur in any month of the year, including mid-winter, in some part of the range. Newton (1972) has shown how the breeding season of the Red Crossbill *Loxia curvirostra* varies depending on whether spruce or pine seeds or a combination form the main food supply. In European spruce forests breeding can begin in August soon after the cones have formed and continue until April or May. In English pine forests nesting does not normally begin until November or December and finishes when the pine seeds drop from the cone in April. In European mixed conifer woods breeding may occur from August to May and then from May to early July in pine areas (Haapanen 1966). According to Newton, in parts of the Tien Shan mountains of central Asia, where the spruce *Picea shrenkiana* keeps its cones for only 4 months, the Crossbill nests only in late summer and early autumn. In the Rocky Mountains

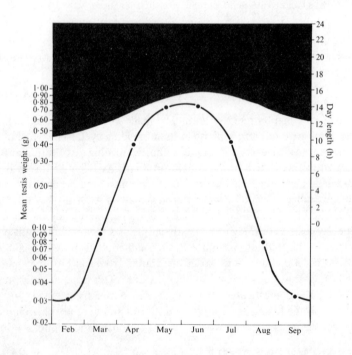

Fig. 10.12. Mean monthly testes weights of California Quail *Lophortyx californicus* and the natural photoperiod in Washington, U.S.A. Based on Anthony (1970).

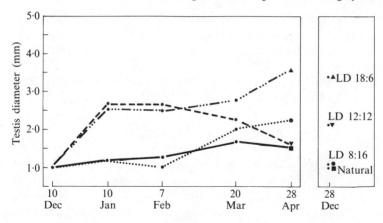

Fig. 10.13. Changes in mean testis diameter of four groups of Red Crossbills *Loxia curvirostra* exposed to various photo-regimes for 12 months. Based on Tordoff and Dawson (1965).

the Red Crossbill breeds from December to June in the low hills but continues until September on the high tops (Bailey, Niedrach, and Bailey 1953). In this species moulting can occur during periods of breeding. Seasonal changes in gonad size are far less marked than in other temperate finches but the testes do remain partly enlarged for most of the year and contain spermatozoa. Experimentally, birds placed on schedules of LD 16:8; LD 12:12; LD 8:16 for just over 12 months did not differ very markedly for much of the time and it was evident that long days were not essential for breeding (Tordoff and Dawson 1965). It is a pity that the experimental subjects were not laparotomized more frequently between April and December, for, as Fig. 10.13 shows, it is likely that there were differences in the phasing of the gonad cycle depending on the light schedule. In Chapter 13 it will be seen that the gonad cycle of the Crossbill is very reminiscent of many tropical species in that the amplitude of change in gonad size is reduced and the period of active gametogenesis is extended.

Continuous breeding has also been recorded for Moorhens *Gallinula chloropus* in South Africa, and two pairs of marked birds were seen to produce eggs in all months of the year (Siegfried and Frost 1975). It has also been shown how the food supply can modify gonad development in experimental subjects (Huxley 1976).

Finally, although experimental details are lacking it is of interest to consider the nocturnal owls. The Barn Owl *Tyto alba* is presumably uninfluenced by the day–night ratio for it is multi-brooded and egg laying may occur in any month (e.g. Smith, Wilson and Frost 1970). This is in keeping with its cosmopolitan range which includes equatorial, tropical, and temperate regions of both hemispheres and the Old and New Worlds. In contrast, the Old World temperate Tawny Owl *Strix aluco* is single-brooded and markedly seasonal in breeding and lays in the early spring. The diurnal Short-eared Owl *Asio flammeus* has a wide distribution in the temperate holarctic and temperate South America. It is mostly absent from the tropics except for a high-altitude population in the Andes at the level of the equator, and isolated island populations in the Galapagos and Caroline Islands and the West Indies. In the north temperates it is usually a seasonal breeder laying in relation to the availability of small rodents but winter breeding has been recorded when food conditions have been favourable.

11   Circadian rhythms and avian photoperiodism

Summary

Photoperiodism resolves into a question of time measurement, that is, how is a 'long' day recognized as such. It is now appreciated that this depends on a circadian rhythm mechanism, and so is one manifestation of the so-called biological clock. In this chapter the Bünning hypothesis of photo-induction, as explicitly expounded by Pittendrigh and his colleagues, is described. These authors stressed that the whole light cycle could act to phase and induct a light sensitive portion of a rhythm of photosensitivity as well as entraining this to a 24-h periodicity. Before discussing this hypothesis we describe the terminology and ideas pertinent to understanding the nature of circadian rhythms, and their properties as biological oscillations capable of entrainment by controlling environmental oscillators (*Zeitgebers*).

The Pittendrigh model of the photoperiodic clock was developed from invertebrate examples and it does not explain all observed photoperiodic phenomena. Tests with House Sparrows have, nevertheless, provided evidence in favour of the hypothesis. Complete photoperiods can be simulated by pulses of light given in a 24-h cycle which define the start and end of the normal light period: in symmetric skeleton schedules the pulses are of equal length in contrast to those used in asymmetric schedules. Such experiments have been held to support the Bünning hypothesis in that only with particular combinations of light pulses does gonad growth occur in photosensitive subjects, as if one of the pulses engaged a light-sensitive phase for photo-induction. But it is also possible, especially when the response of target tissues that respond to several hormones is being measured, that the light pulses by defining the cycle length serve to phase two or more hormone rhythms into an appropriate temporal synergism. This interpretation is less likely in experiments that measure specific plasma hormone titres following light stimulation. Better proof for the involvement of circadian rhythms is provided by light cycles in which an initial pulse of light is given to entrain rhythmic activity and then different groups of subjects are allowed to free-run in constant dark for variable periods (several days) before a second pulse is given—such resonance experiments depend on ahemeral light cycles which do not have periodicities of 24 h. Techniques of this kind were used with House Finches *Carpodacus mexicanus* to show that photo-induced testicular involution apparently depended on a circadian rhythm of photosensitivity. In this chapter other techniques for distinguishing between induction and phasing effects of light are described and the relationship between diurnal activity rhythms and photo-induced gonad development explored. The site of photosensitive clock and the part played by the pineal gland in regulating avian activity rhythms is also considered.

Introductory comments

Birds exhibit a wide spectrum of adaptive photo-responses which phase the hypothalamic neurosecretions which eventually regulate migratory fattening, moult,
gonad development, and other periodic functions. Photoperiodic phenomena are by
no means unique to birds, but have many diverse manifestations, such as the vertical
migrations of plankton or the diurnal leaf movements of plants. Could there exist
some fundamental mechanism common to the rhythms expressed by all these forms
of life which could also explain the seasonal rhythm of hormone secretion in birds? In
Fig. 2.4 (p. 40) it was shown that the stimulation of specific gonadotrophin secretion
in Greenfinches *Carduelis chloris* under long-day light schedules did not depend on
the length of the day *per se* but rather on the subject receiving light stimulation at
a certain phase during its daily rhythm of activity. The essence of this and other
related problems resolves into one of time measurement, for what mechanism within
the bird enables it to distinguish and respond with appropriate gonadotrophin
secretion to a light pulse occurring 12 h from the subjective dawn but not to react
to a pulse occurring 8 h from dawn? This leads us to consider the workings of the
biological clock and to explore whether the clock concept can also account for
refractoriness and the observed species-specific differences in breeding response.
Photoperiodic phenomena have been well studied in invertebrates (see Lees (1972)
for review) and current hypotheses that are applicable to birds depend considerably
on these taxa. Several hypotheses have been propounded to explain avian photoperiodicity but, while they all have historical interest, only two deserve mention
here.

The first hypothesis imagines time measurement to be based on the hour-glass
principle whereby some active chemical (e.g. an enzyme or hormone) accumulates
during the dark period in the sites of manufacture and is then dispersed to target
organs during the day; the length of the night determines how much material is
accumulated while light, so to speak, turns the 'hour-glass' over. There is strong evidence that a sophisticated mechanism of this kind is implicated in the clock of the
Vetch Aphid *Megoura viciae* and that it controls morph determination (Lees 1965,
1971). But in birds an oscillator model seems more appropriate.

The 'Bünning hypothesis' of photoperiodicity was first propounded in 1936 to
explain the diurnal rhythms of plant function. Bunning recognized the existence of
rhythmic cellular functions, present even in unicellular organisms, having a period
of about 24 h and suggested a daily cycle of cell metabolism comprising two halves,
one requiring light (the photophil) and the other dark (the scotophil). Processes
requiring long days were imagined to occur only if the natural daylight extended
into the dark half of the cycle, allowing the coincidence of light and a light-sensitive
phase (Bünning 1960). In this form the hypothesis does not explain how the subject
recognizes dawn, but there is a more explicit version, due to Pittendrigh (1960, 1965,
1966) and Pittendrigh and Minis (1964). These authors stressed that the whole light
cycle could act as the photoperiodic inducer of the sensitive phase in addition to
entraining the 24-h rhythm. Thus the animal is not dependent on a particular dawn,
but the characteristics of the whole light–dark cycle determine the phasing of the

light-sensitive phase. Pittendrigh proposed the term 'photoperiodically inducible phase' to signify that portion of the daily oscillation which, if illuminated, leads to photoperiodic induction. Before we pursue this hypothesis we ought to digress a little to explain the nature of daily rhythms.

The existence of endogenous 24-h rhythms has long been appreciated from such observations as the leaf movements of *Canavalia* in LD 12:12 and in constant dark (Kleinhoonte 1929), the pigment dispersion in the isopod *Ligia baudiniana* kept in constant dark (Kleitman 1940), or of locomotor activity in the mouse *Peromyscus* (Hemmingsen and Krarup 1937; Johnson 1939). The fact that under constant conditions (that is, LL or DD) the rhythms persist and free-run with a period close to 24 h, in fact deviating by only ± 10 per cent, is strong evidence that they must be endogenous and, moreover, self-sustained oscillations, for which Halberg coined the term circadian rhythm (*circa* = about, *dies* = a day). Further progress has depended on improvements in the mathematical treatment of periodicities, aided by the availability of computers. The physiology of insect circadian rhythms has been comprehensively reviewed by Brady (1974).

Characteristics of free-running circadian rhythms

During the 1950s Aschoff (1952, 1958, 1959, 1960) demonstrated how under conditions of constant light (LL) or dark (DD) various animals exhibited rhythms of activity and rest having a period of about 24 h; the fact that the period length (see Fig. 11.1 for definition of period) was not exactly 24 h can be taken as good evidence that such rhythms must be endogenous. Aschoff (1960) established an important general relationship, which Pittendrigh (1960) proposed should be designated 'Aschoff's rule', between the lengths of such free-running periods (τ_{FR}) in nocturnal and diurnal organisms. This rule states that light-active organisms exhibit a shorter free-running circadian period τ in constant light than in constant dark whereas the opposite applies to dark active organisms, that is, $\tau_{LL} < \tau_{DD}$ in nocturnal animals and $\tau_{LL} > \tau_{DD}$ in diurnal animals. Similarly, in constant light the circadian frequency (reciprocal of period) increases with increased intensity of illumination in diurnal and decreases in nocturnal species. Aschoff's rule as stated is also known as the 'circadian rule' and while exceptions to it do exist (Hoffmann 1967) it is a frequently valid statement. In fact, the relationship between circadian period and light intensity is probably best described by a parabola if a sufficiently wide range of light intensities is studied. For day active animals the circadian period probably increases from very low light intensities to low intensities $(10^{-3}-1$ lx) and then decreases for the range low to high $(1-10^3$ lx) and examples are given by Gwinner (1975). The converse appears to apply to dark-active animals. The total circadian period can be divided into one phase of activity a and one of rest ρ and it was early discovered that under conditions of high light intensity the amount and duration of activity was increased.

To account for these various empirical observations Aschoff and Wever (1962*a, b*) proposed a model in which activity was reckoned to be initiated when the level of some controlling circadian oscillator passed above a threshold, as depicted schematically in Fig. 11.1. In preparing this model Aschoff and Wever assumed that they were dealing with self-sustained oscillations which continue indefinitely, that is, there is some continuing energy input, rather than damped oscillations which decrease in amplitude and stop if energy is not fed into the system. It is worth emphasizing at this point that we are discussing rhythms which are expressed under constant conditions, that is not subject to external forces, so that they oscillate with a frequency which is characteristic of the system. The model shown in Fig. 11.1 shows how the ratio between activity and rest time (a:ρ ratio), and the amount of activity should increase in bright compared with dim light in day-adapted animals (a converse model would apply to nocturnal species). Since the amount of activity relative to rest increases as light intensity increases it is evident that the position of the level of the oscillation relative to some threshold must increase (see Fig. 11.1). Light must either influence both level and frequency independently or as Wever (1964, 1965) suggests level and frequency may be internally coupled in self-sustained oscillations so that the higher the level the higher its frequency. Thus an increase in light intensity in day-active animals may only increase the frequency through its action on level. Hence the circadian rule states that level is positively correlated with light intensity in diurnal species and negatively correlated in nocturnal ones.

Lately, Aschoff and several collaborators (Aschoff, Gerecke, Kurek, Pohl, Rieger, Saint Paul, and Wever 1971) have examined in some detail the variability of a and ρ and considered whether they summated to give the total variability of τ. They discovered that in Chaffinches *Fringilla coelebs* the activity time was inversely related to rest time, but that a short a and long ρ totalled about 26 h whereas a long a and short ρ totalled about 22 h. In other words the a:ρ ratio was an inverse function of τ. These and other observations led Aschoff and his co-workers to consider various models of a circadian oscillation into which the concept of a threshold was introduced, and Fig. 11.2 depicts two such oscillations which apparently satisfied their experimental results. Given the existence of a variation in threshold or of oscillation level (the equivalent of 'noise' in electronics) it can be seen that with a skewed oscillation quite different standard errors would be obtained for the time of activity onset compared with activity end. It is significant that a computer simulation by Wever (see below) had earlier demanded a trend to negative skewness with a long oscillation period and positive skewness (to the left) with a short period. Aschoff and his colleagues argued that these new results added strength to a model which assumed that the circadian oscillator twice passes through a threshold and that the animal is only active when the oscillation is above the threshold. Moreover, they showed that the shape, amplitude, and level of the oscillation must depend on its period. The activity records of caged Chaffinches detailed in Fig. 11.3 give some idea of the arguments used by Aschoff et al., for it can be seen that peak activity occurred at the beginning of each daily record and then became sporadic, as would be predicted if an underlying oscillation was positively skewed in form. It was found that with increase in the period of the activity rhythm peak activity occurred progressively later, consistent with an oscillation becoming negatively skewed. The diurnal range in volume of oxygen used by Chaffinches per hour was considered a measure

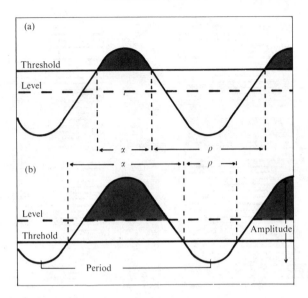

Fig. 11.1. Diagrammatic representation of two circadian oscillators to illustrate the concept of threshold and level. Activity or some other biological response occurs when the oscillation is above a threshold—conceivably a biochemical threshold—and the ratio of activity (α) to rest (ρ) is seen to vary accordingly. Modified slightly from Aschoff (1967).

of the amplitude of the underlying circadian oscillator (Aschoff *et al.* 1971) and for 7 birds with free-running circadian periods in constant light ranging from 21 h to 25 h, amplitude decreased as τ increased.

So far we have assumed that activity rhythms depend on a single oscillator. Actually, there is much evidence to suggest that activity rhythms depend on two coupled oscillators which are normally phase locked and some of this evidence is reviewed below. It has not been presented at this stage since it would confuse our

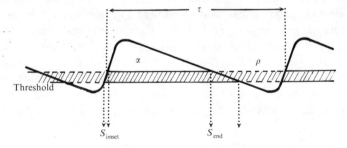

Fig. 11.2. Skewed oscillation subject to noise to illustrate how the standard error or onset of activity S_{onset} will differ from standard error of end of activity S_{end}. Based on Aschoff *et al.* (1971). Different standard errors for onset and end of activity could also be recorded if activity cycles were controlled by a system of coupled oscillators which responded slightly differently to changes in light intensity.

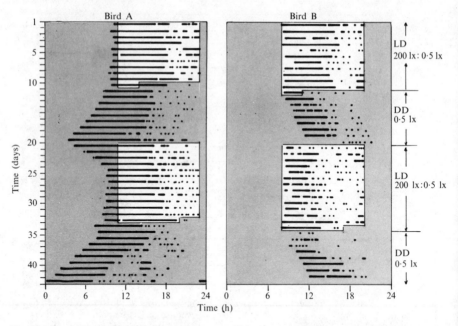

Fig. 11.3. Activity rhythms as measured by on–off perch recorders for two Chaffinches *Fringilla coelebs* kept on either a light–dark cycle (with 200 lx in light and 0·5 lx in dark) or under a constant dim light (DD of 0·5 lx). In bird A the circadian period is about 22·5 h and under constant conditions its activity rhythm free-runs to begin sooner each day, whereas bird B has a free-running circadian period of about 24·5 h. In both cases exposure to a light–dark cycle entrains the rhythm to a period of 24-h. From Aschoff (1967) quoting Aschoff and Wever (1966).

attempt to clarify the terminology adopted in this subject. However, it is important to appreciate that the results detailed in Fig. 11.2 and discussed above could well be explained by a coupled oscillator system, which could nevertheless be skewed in form as oscillation theory predicts.

The period length of a circadian oscillation can vary spontaneously for reasons that at present are poorly understood. Conditions applying before free-run influence the period length. For example, working with House Sparrows *Passer domesticus*, Eskin (1971) showed that after exposure to LD 20:4 the period length under subsequent free-run was shorter than if previous exposure had been to LD 6:18. Given sufficient time (around 50–60 days) such after-effects vanish. Eskin also showed that previous exposure of subjects to light dark cycles of short period T (less than 24 h) gave shorter τ values when subjects subsequently free-ran in total dark than did cycles of long T.

Self-sustained oscillators obviously derive energy from some source; the energy for the swing of a pendulum clock is gained from the main-spring from which energy is delivered once each cycle to restore that lost during the swing, and biological oscillators presumably depend on biochemical processes. Mechanisms involving messenger RNA, and other molecular processes, have been implicated and so too have membrane electro-permeabilities, because heavy water can alter the frequency of circadian oscillators (Enright 1971). But this topic, which remains unresolved, is outside the scope of this treatise and interested readers are referred to reviews by Hastings (1970), Schweiger (1972), and Pavlidis (1973).

Entrainment

In nature the approximate 24-h periodicity of the organism τ is brought into strict synchronization* with the exact 24-h period of the earth's rotation T by the daily cycle of night and day (Aschoff and Wever 1962c). This means that the endogenous oscillation of the organism is entrained by the 24-h oscillation of the environmental light cycle which synchronizes the endogenous rhythm and which in consequence Aschoff termed a *Zeitgeber* (synchronizer). This is seen in Fig. 11.3, where the activity rhythms of caged Chaffinches can be seen to free-run in constant dim light but to become entrained when the birds were exposed to a light–dark cycle. It is a fact, still unexplained in physiological terms, that the control of a circadian (endogenous) oscillation by an exogenous driving oscillator has been shown to obey all the laws of oscillation theory of physics (Aschoff and Wever 1962a, c).[†]

---

*When an oscillation is modified by an outside force so that it assumes the period of the driver it is synchronized. If both oscillations are self-sustained the term entrainment can be applied and in biological parlance the driving oscillation is called a *Zeitgeber.*

[†]Wever (1960, 1965) was able to construct an electronic model, based on a pendulum oscillator rather than a relaxation oscillator, and derive mathematical expressions to describe the entrainment of animal circadian rhythms. To do so it was necessary to assume only that (1) circadian rhythms are indeed self-sustaining and (2) the existence of a threshold to allow for the observation of active and rest periods and the empirical carcadian rule. Wever was able to control the length of the oscillation period in his electronic oscillator by either varying the (1) inductance or (2) the capacity. In addition, a damping resistance (3) affected the amplitude of oscillation and a grid bias (4) the average level of the oscillation. These four elements in the circuit were coupled to a light-dependent resistance in such a way that each could be varied independently. Control of inductance by increase in light intensity led to a decrease in the period of the rhythm, as in nature, whereas, in contradistinction to the natural situation, control of capacity caused an increase. The $a:\rho$ ratio was affected by changes in grid bias, which affected level, in a way typically found in real systems, by changes in threshold. In oscillation systems that are not capable of self-oscillation only the frequency is a property of the system for the amplitude (and phase angle) can be altered by manipulation. In auto-oscillating systems both frequency and amplitude are properties of the system, a wide range of such oscillations being expressed by the Van der Pol differential equation:

$$\ddot{y} - \Sigma(1 - a^2 y^2)y + \omega^2 y = 0 \,.$$

Wever simply considered light–dark cycles as the controlling oscillation and then studied the effect of single, short perturbations of light given as a pulse in constant dark or dark as a pulse in constant light. In all cases the same two types of electronic control mentioned above (that is, of inductance or grid bias) gave results identical in all respects with the observed situation. A change in frequency of the controlled oscillation or of the L:D ratio of the *Zeitgeber* affected the phase relation of the oscillations and the range of entrainment. But, if only the amplitude of an oscillation was controlled (using the damping resistance), no synchronization and no phase control was possible, for a change of amplitude causes no phase shifting. A more extensive mathematical analysis of biological oscillators has become possible with the knowledge accumulated since Wever's pioneer studies. Pavlidis (1973) has reviewed this topic and is recommended to the reader with a grasp of calculus and oscillation theory.

Synchronization of one oscillation by another must involve some form of phase control whereby there exists a definitive angular relationship between the phases of the driving and driven oscillations. This phase relationship is determined by the ratio of the frequencies of the two oscillations, but it is impossible for one oscillation to control another if their frequencies are too dissimilar. The maximum permissible phase shift to affect entrainment is 180°, and there is a 180° forbidden zone when a driven oscillator would transfer energy to, instead of receive energy from, the driver (this is an approximation for sinusoidal oscillations and in special cases deviations occur). The process of entrainment is illustrated in Fig. 11.4, which incorporates a schematic representation of the movement of the sun above the horizon changing in light intensity and duration with season to provide a *Zeitgeber* of variable strength but an invariable period of 24 h. An endogenous entrained oscillation is also depicted to show how, in this case, as the *Zeitgeber* increases in strength the controlled oscillation is advanced, that is, the phase angle measured from the mid-points of the two oscillations develops a positive shift; the total period of the rhythm remains constant. These changes depend on the increase in the level of the oscillation with increase in illumination. In the same manner the higher the natural frequency of a circadian rhythm (the spontaneous frequency assumed in the absence of entrainment) the more it leads the entraining oscillation, so that the phase angle becomes more positive and vice versa. This is shown in Fig. 11.5 from Aschoff and Wever (1962c) for Chaffinches *Fringilla coelebs* entrained on a cycle of LD 12:12 before release into LL at 0·5 lx. It follows from Aschoff's rule that in day-active animals (and the reverse applies to nocturnal species) activity begins later in a LD 11:11 and earlier in a LD 13:13 than in a LD 12:12.

Fig. 11.4 also illustrates an important practical point discussed by Aschoff (1965a)

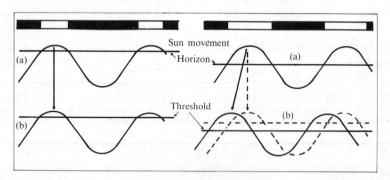

Fig. 11.4. Diagrams to show how movement of the sun above the horizon serves as a controlling oscillator or *Zeitgeber* (a) to entrain a circadian oscillator (b). In the left-hand panel winter days are short and light intensity low so that the level and frequency of the controlled oscillator are low and it is almost in phase with the *Zeitgeber*. The increased light intensity of summer (right-hand panel) causes the level and frequency of the entrained oscillator to rise and the phase angle during entrainment becomes more positive, that is the entrained oscillator (b) leads the controlling oscillator (a). If it was adaptive for the animal to have its circadian oscillators in phase with the light cycle under these conditions (dashed lines) some endogenous adjustment would be needed to reduce the frequency of the entrained oscillator. Derived from Aschoff (1967).

which is pertinent to the measurement of phase-angle differences. Because of variations in the $a:\rho$ ratio with level it is only the mid-point of activity and mid-point of the light cycle that give unambiguous phase reference points. Now a prediction from the circadian rule is that the phase angle measured between the mid-point of an activity rhythm and the mid-point of the light–dark cycle should decrease (become more negative) with a reduction in light intensity. Daan (1976) kept Chaffinches and Greenfinches in a room where they experienced natural seasonal daylengths through a window. He alternated such periods, lasting about 6 weeks, with others when the intensity of the daylight was reduced by pulling down a curtain. His results were the opposite of those predicted by the circadian rule and he therefore questioned whether (a) the mid-point was really a satisfactory reference point and (b) whether the effect of light during free-run does predict the effect of light during entrainment. Daan found that the standard deviations for the half-monthly mean phase-angle differences measured to the onset of activity were larger the longer after sunrise these means applied, whereas variations in the phase-angle differences measured to the end of daily activity increased the earlier activity ended before sunset. He next used the reciprocal of the mean of these standard deviations as a measure of the precision of activity onset and activity end, plotting these against a computed rate of change in light intensity at sunrise and sunset. The results confirmed a conclusion also reached in studies of rodents (Daan and Aschoff 1975) that the precision of the timing of onset and end of daily activity is positively correlated with the relative rate of change of the light intensity at that time of the day. Moreover, the correlation proves to be most significant for the end of activity. For reasons best understood by consulting the original publications, the results indicate that the

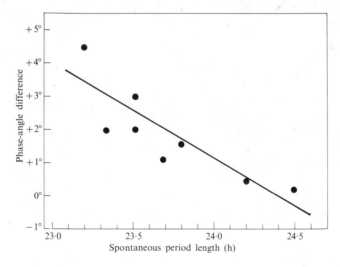

Fig. 11.5. Phase-angle difference between the activity rhythm of 8 Chaffinches *Fringilla coelebs* and an entraining light–dark *Zeitgeber* according to the spontaneous period length of each Chaffinch when measured under constant, free-run conditions. Based on Aschoff and Wever (1962c).

end of activity is more subject to a direct influence of light than the onset, whereas the onset is a better indicator of the phase of the driving oscillator.

The above results are rather difficult to understand without reference to detailed results. But they are important in demonstrating that the phases of onset and end of activity relative to the light–dark cycle show opposite changes in response to variations in daylength or light intensity, as Daan emphasizes. He draws attention to the widespread bimodal pattern of circadian activities (*cf.* Figs. 6.6 (p. 144), 7.15 (p. 174), and 7.16 (p. 175)), and mentions the phenomenon of rhythm splitting[†] (see below) to support a suggestion that there are two mutually coupled major components involved in the oscillator controlling circadian activity rhythms (*cf.* Pittendrigh 1974). These appear to have their frequencies affected by light in opposite ways so that the 'morning oscillator' responsible for activity onset is accelerated by light and the 'evening' oscillator is slowed. This could explain the delay in onset and advance in end of activity with a reduction in light intensity. It is thought that two such coupled oscillators would mostly behave as a single system with a frequency intermediate between the two natural frequencies. Hoffmann (1971) has found that under some circumstances, such as a low light intensity, the overt locomotor activity rhythm of the Tree Shrew *Tupaia belangeri* can split into two components 180° out of phase with each other. This is very suggestive evidence that the rhythm depends on at least two oscillators, which are normally phase-locked. Activity rhythms often become arrhythmic under some light conditions, depending on the species; constant bright light can often cause arrhythmicity rather than the free-running of activity rhythms. This may be explained if the coupling of two oscillators becomes dissociated so that the morning activity extends earlier in one direction and the afternoon activity extends into the evening and joins with the next days onset.

Any change in phase relationship between a controlled oscillation and its *Zeitgeber* can result from an alteration in the cycle frequency of either oscillation. An entrained animal will, however, exhibit the same overt rhythm frequency as the *Zeitgeber* frequency used in entrainment. Therefore, under natural conditions (that is, not free-running under LL or DD) we cannot talk in terms of species-specific circadian frequencies, and it is only possible to measure the frequency of a circadian oscillation in terms of its phase-angle difference to a defined *Zeitgeber*. Further caution is necessary in that the frequency of an oscillation may change with time as a consequence of physiological factors: for example, there is evidence that in birds the oscillators assume a higher frequency in spring than in autumn. Nevertheless, circadian oscillators are generally stable, which is why they free-run so precisely and remain unaffected by temperature changes. They are more resistant to alterations in their phasing at some times in their cycles than at others, just as it is easier to alter the

---

[†] Pavlidis (1973) has suggested that the term 'rhythm splitting' be reserved for phenomena such as the loss of synchrony between two rhythms involving different activities in the same organism. When a rhythm splits in a manner that results in a phase difference of about 180° a different process is involved and it is preferable to use the term 'frequency doubling', although the change in frequency will rarely be an exact doubling because of the simultaneous direct effect of light on frequency.

movement of a pendulum by applying a force in a particular direction at a particular time in the cycle. Fig. 11.3 illustrates one method of measuring the phase shifting effected during entrainment. The free-running activity rhythm in constant DD is exposed to a dark–light transition at a known time in the daily activity rhythm. Thereafter, the *Zeitgeber* is repeated as a daily stimulus. It can be seen that the activity rhythm *advances* through a series of transients before locking on to the *Zeitgeber*. If entrainment is effected by a delaying shift the new value is usually achieved immediately. Aschoff (1965*b*) has reviewed the various methods available for measuring phase shifts, including phase shifts effected by a single pulse as distinct from the continuously repeated stimulus represented in Fig. 11.3.

Phase-response curves

This last approach is more informative in understanding the nature of the driving oscillator, for masking effects of light on the overt rhythm can be avoided; for example, the direct, as distinct from phasing, effects of light noted in Daan's work above. Pittendrigh (1954), studying the rhythm of emergence of adults of *Drosophila pseudoobscura* from their pupal cases—the pupal eclosion rhythm—discovered that a single perturbation of light given as a 15-min pulse in constant DD could shift the phase of the rhythm, according to when the pulse was given. He and Bruce (Pittendrigh and Bruce 1957) scanned the whole circadian period in this way with light pulses and produced the phase-response curve depicted in Fig. 11.6. It has a configuration common to many other animal and plant rhythms. It has been found

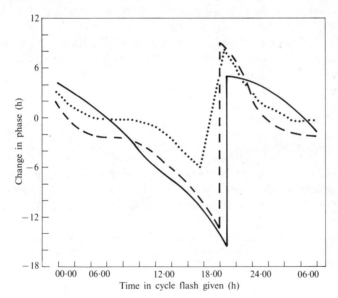

Fig. 11.6. Phase-response curves for the *Drosophila pseudoobscura* pupal eclosion rhythm obtained using single light flashes of 0·5 ms (dotted line), 4 h (dashed line), or 12 h (solid line) duration. The mid-point of the light signal was used as the reference point in calculating the phase difference to mid-point of eclosion rhythm. From Pittendrigh (1960).

that both the beginning and end of a light pulse may serve the animal as a reference point in setting its rhythm which makes it desirable to use the midpoint of a rhythm for measuring phase shifts. This was illustrated by Aschoff (1965b) when he caused phase shifts in Chaffinches by using a single transition from light to dark or from dark to light for opposite transitions gave opposite phase shifts.

Increase in signal length or intensity increases the degree of phase shifting and this means that phase-response curves must vary according to the nature of the *Zeitgeber* (Fig. 11.6). As if this did not pose sufficient difficulties to using response curves for comparative studies, they also vary if the background illumination is changed (Wever 1964). Nonetheless, the response of an organism to a defined light stimulus gives a measure of the rhythm's phase and is the only way in which the parameters of the real waveform of the circadian oscillation can be treated experimentally. It used to be a convention to consider the whole circadian period (arbitrarily reckoned as exactly 24 h and called circadian time C.T.) as the abscissa of a response graph and to mark on the ordinate delaying phase shifts as negative values above zero, and advancing shifts as positive values below the line. Some workers prefer the convention followed by physicists in which positive values are plotted above the zero mark.

From a knowledge of the phase-response curve of *Drosophila* (see Fig. 11.6) Pittendrigh (1960, 1965, 1966) Pittendrigh and Minis (1964) developed their model of the photoperiodic clock. It was suggested that in light lasting for more than 12 h the *Drosophila* oscillation is damped out being held at a fixed state, but as soon as the animal experiences dark the oscillation enters its steady-state motion. Entrainment is possible when the slope of the response curve does not exceed 2·0, depicted as the thick part of the response curve. This defines the range over which steady-state entrainment is possible, and in *Drosophila* the maximum utilizable phase advance or delay is ± 5·9 h. Thus in the entrained steady state the period τ of the free-running oscillation of the subject is transformed to that of the driving cycle T, provided this is not too different (in the range 18–30 h in *Drosophila*)* and in so doing a distinct phase shift ($\Delta\phi$) occurs which is given by

$$\tau - T = \Delta\phi . \tag{11.1}$$

In other words, the phase shift corrects the frequency of the rhythm to match the *Zeitgeber* frequency. This specific phase shift can be achieved only if the light hits the oscillation at a unique phase point which is defined by the phase-response curve.[†]

---

*Limited data for birds suggest a range in the order 20–28 h. In birds it would be expected that occasional animals with higher or lower values would occur and be eliminated by natural selection; in man it is probable that inappropriate periodicities can lead to various forms of mental illness.

[†]Pittendrigh and Minis (1964) give an example for *Drosophila* which makes this easier to understand. τ for the eclosion rhythm at 21 °C is 24 h 10 min. So in entraining to a 24-h light–dark cycle the light pulse must cause a phase change of 24 h − 24 h 10 min = − 10 min. The response curve (Fig. 11.6) shows that to do this the light must hit the oscillation around circadian time (C.T.) 18.00 h. The median of the observed eclosion rhythm peak is known to

This is a good point at which to consider whether each rhythmic function that can be observed in an animal, for example, the activity cycle, the hormone secretions leading to gonad development or moulting, and so on, are each independently photosensitive and phased by the *Zeitgeber*. Alternatively, perhaps there is only one, or relatively few, light-sensitive master clocks, which secondarily control the other rhythms. If the coupled rhythms each have different frequencies and hence different phase relationships with the master oscillation, they need not provide a direct indication of the phase of the light-controlled master clock. For various reasons the Princeton model does indeed assume a multiplicity of circadian oscillations themselves coupled to one, or perhaps more, driving or pacemaker oscillation(s). Only the pacemaker is entrained by the daily cycle of day and night. In the case of the eclosion clock a two-oscillator system is envisaged. This accounts for the fact that during entrainment to a single short pulse of light the rhythm goes through a series of transients before achieving a steady state. The model envisages that one photosensitive oscillator is immediately shifted by the light signal to the proper steady phase but that the second oscillator, which itself directly controls the actual eclosion rhythm, is less well coupled to the photosensitive oscillator and goes through a series of transients in achieving stability. Another reason for concluding that more than one oscillator was involved was that artificial selection of strains of *Drosophila* for early or late emergence affected the driven rhythms responsible for the timing of emergence, by an alteration of the phase relationship, whereas the phase of the driving oscillation relative to the light–dark cycle remained unchanged (Pittendrigh 1967). In the Pink Bollworm Moth *Pectinophora gossypiella* selection for early or late pupal eclosion gave correlated responses for the egg-hatch rhythm as if a commonly driven oscillation was involved, whereas the oviposition rhythm of the adults was apparently distinct (Pittendrigh and Minis 1971).

Fig. 11.7, from Pittendrigh and Minis (1971), helps to clarify the kind of mechanism we are discussing by showing how three driven oscillators and a light-inducible phase vary with the duration of the photoperiod. Some phase points are always in darkness in natural situations, others always in light, and others again may fall in the day or night depending on the photoperiod and the entrained steady state. The curves have the same slope, suggesting that the light-inducible phase (ϕ_i) has the same dependence on the photoperiod. Accordingly, Pittendrigh and Minis proposed that in measuring the phase of any particular rhythm the light-inducible phase of the driving oscillator should always bear a fixed relationship to it; if substantiated this would clearly provide a means of identifying the photosensitive oscillation. At present it is not certain whether the light-inducible phase is a phase-point on a driving oscillation or whether it is incorporated as part of one or more driven systems. It becomes clear that in any discussion of circadian rhythms or entrainment models that it is import-

occur at C.T. 02.00, which means that at 21 °C the observed eclosion peak will occur 03.40–02.00 = 1 h 40 min before the light pulse. Similar computations were performed for other periods T of the driving cycle, using $T - \tau = \Delta\phi_{ss}$ in which $\Delta\phi_{ss}$ is the phase shift that light must effect in each cycle of the steady state. Observed and predicted values were extremely close in the range $T = 20 \cdot 40$ h to $T = 25 \cdot 00$ h.

ant, as Pittendrigh and Minis (1971) have stressed, to select convenient phase reference points ϕ. This will enable the phase relation ψ of any rhythm R and the light
cycle L to be defined ψ_{RL}, and distinguished from the phase relationship of the
underlying circadian oscillator O and the light cycle, defined as (ψ_{OL}).

Night interruption experiments

A class of experiments which have been particularly valuable in deciphering the
nature of photoperiodic responses have relied on 'night interruptions' with short
light pulses. These simulate the action of a complete photoperiod and so have been
referred to as skeleton photoperiods. For example, it was early discovered that two
short pulses of light, say of 15 min duration each, could provide a *symmetrical* skeleton simulation (Fig. 11.8) of complete photoperiods of up to 12 h duration in
Drosophila (Pittendrigh and Minis 1964). In similar manner, *asymmetric* skeletons
consisting of one long and one short pulse simulate the action of a single pulse of
duration equal to the interval between the start of the first pulse and end of the
second, as Fig. 11.9 shows. As the interruption of the night by the short light pulse
occurs later and later the steady-state phase of the rhythm shifts to the right (ψ_{RC}
becomes more negative, for this is a night-active animal). The same would happen
if complete photoperiods were used. However, with skeletons simulating photoperiods whose duration is 15 h or longer a failure occurs for the rhythm now assumes

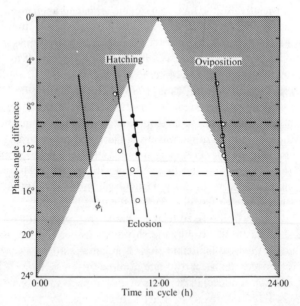

Fig. 11.7. Phase-angle difference between the hatching, eclosion, and oviposition rhythms of the
moth *Pectinophora gossypiella* and the light cycle with supposed phase of light induction ϕ_i
postulated to occur 5 h earlier than the phase-reference point of the pupal eclosion rhythm.
Dashed lines demarcate the range of natural photoperiods in Texas. From Pittendrigh and Minis
(1971).

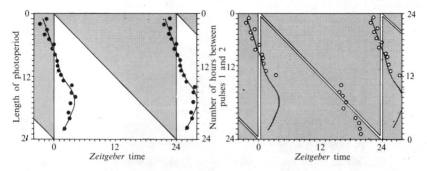

Fig. 11.8. The phase of the *Drosophila pseudoobscura* eclosion rhythm as a function of complete (*left*) and symmetrical skeleton (*right*) photoperiods. Plotted points are medians of the steady-state distributions of eclosion. The solid curve in the right hand panel is the same as that fitted to the medians for complete photoperiods in the left panel. From Pittendrigh and Minis (1964).

a phase angle relative to the light regime such that the short light pulse functions as dawn, that is, the rhythm jumps to the phase of the shorter of the alternative photoperiods. Exactly the same occurs if symmetrical pulses are used as is shown in Fig. 11.8, which refers to the *Drosophila* pupal eclosion rhythm. Pittendrigh has explained the behaviour of skeleton schedules in terms of his model of the clock by assuming that the two light pulses each generate a phase shift to equal the total phase change:

$$\tau - T = \Delta\phi_1 + \Delta\phi_2.$$
11.2

So, in order to lengthen or shorten the circadian rhythm to equal to period of the light cycle, discrete and instantaneous phase shifts are caused when the light pulse engages the circadian oscillation. In the special case when $T = \tau$ no phase shift is needed and the two shifts must equate to zero, one being positive and the other negative.

When entrainment occurs with whole or skeleton photoperiods the two oscillations do not immediately reach stability, for until the temporal coincidence of the *Zeitgeber* and circadian oscillator have reached the point where the phase shifts summate to give the stable value, the oscillation is advanced or delayed relative to the driver, and transients are observed in the overt rhythm under investigation. When skeleton schedules simulating long days are employed the oscillation interprets the *Zeitgeber* as a short day. This is because with a 'short' interval between pulses the second pulse will fall on the insensitive part of the phase-response curve and cause virtually no difference in phasing relative to that induced by one long light period of duration T_1. If the period of light exceeds the flat part of the phase-response curve the stimulus induced by one long light pulse is different from that caused by separated pulses. This shows that the organism must perceive light at different parts of the rhythm and that it is not dependent on a simple on–off switch type mechanism. A special situation arises when two short light pulses are used to simulate a photoperiod of LD 11:13, for this can be equally well read as LD 13:11. The way

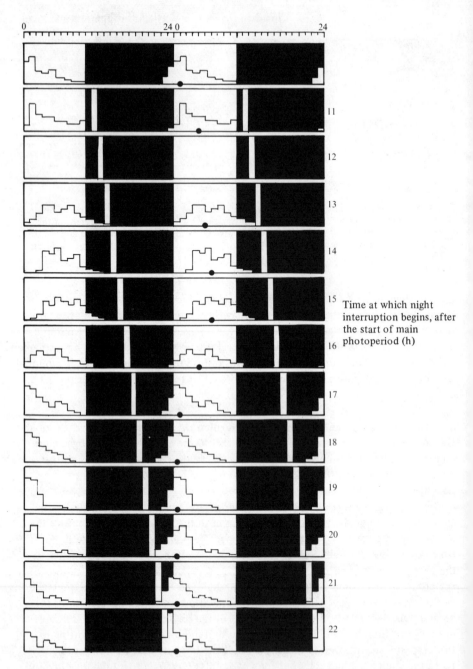

Fig. 11.9. The pupal eclosion rhythm of *Pectinophora gossypiella* with asymmetric skeleton photoperiods: main photoperiod lasts 10 h and the night interruption for 1 h. The time at which night interruption begins, after the start of the main photoperiod is indicated on the right. Solid dots represent the median of the rhythm and are used as the phase reference point for the rhythm. Top panel is control with no night break. From Pittendrigh and Minis (1971).

in which the cycle is interpreted depends on which phase of the animal's oscillation initially encounters the first pulse of the entraining cycle and the length of the first interval (11 or 13 h) in the entraining cycle. Hence, when the skeleton simulation is close to $\tau/2$ two alternative and equally stable entrained steady states can occur. This has been called the bistability phenomenon (Ottesen *et al.*, in press).

The results of experiments with skeleton photoperiods provide some evidence for the general Bünning hypothesis, which ascribes an inductive and entraining rôle to the photoperiod, and for the Princeton model of entrainment in particular. However, it is necessary to remember that the hypothesis does not account for all photoperiodic phenomena in animals, particularly all insects (Lees 1972). Even so, we have developed here some of the reasoning and results which contribute to the Princeton modification of the Bünning model because it does seem to be especially appropriate to birds and many experiments using skeleton light schedules have been performed. Before discussing these it is worthwhile quoting some studies of the perch-hopping activity rhythm of the House Sparrow *Passer domesticus* made by Eskin (1971) in an attempt to test the Princeton entrainment model.

Eskin obtained a phase-response curve for the activity rhythm by giving 6-h pulses of light at different phases in the free-running period of sparrows, which had been kept in constant DD for 5–7 months, and Fig. 11.10(a) summarizes the result. Eskin next measured the mean phase angle for groups of birds entrained to cycles of $6L + D$ to give different total periods T, that is $T = 18$, $T = 20$..., as in Fig. 11.10(b). As can be seen the phase angle of the rhythm became more positive (increased) as the length of the total period increased (*cf.* p. 290). If fewer than 75 per cent of the sparrows in a group could entrain to a particular period length this was regarded as being the point at which entrainment began to fail. Defined in this way the lower range of entrainment proved to be between $T = 15.8$ h and 17.8 h and the upper range $T = 28.0–28.7$ h. The mean free-running period (FRP) of the sparrows was 24.9 h so eq (11.1) above predicts that the phase shift between a light cycle of period $T = 28$ h and a circadian rhythm (τ) = 24.9 h (say 25 h) should be $\Delta\phi = 25 - 28 = -3.0$ h for a bird pulsed at a phase of $175°$. This last datum is obtained from the phase response curve which defines the point in the free-running cycle at which a single pulse of light will produce a phase shift equal to the difference between the free-running period and T, i.e. FRP $- T$. Eskin was in this way able to compare the phase shifts calculated from the entrainment experiment (Fig. 11.10(a)) with the phase shifts actually measured in free-running birds pulsed at different phases of their circadian periods (Fig. 11.10(b)) and obtained an excellent agreement between observed and expected results (Fig. 11.10(c)). This is good evidence that during the entrainment of the free-running circadian oscillator to the period of the light–dark cycle a phase shift occurs equal to the difference between the two frequencies. Earlier Enright (1965) had tried to obtain a response curve for the North American House Finch *Carpodacus mexicanus* but though he obtained similar results he had worked with only one bird and his results were therefore somewhat ambiguous.

If skeleton light schedules are used in isolation in the manner depicted in Figs.

11.7 and 11.8 they do not prove the involvement of a circadian rhythm in photo-periodism. For example, it could still be argued that the light period somehow embodied a rate-limiting process (if the hour-glass principle be adopted) with a carry-over effect which was removed by the provision of a pulse of light at night (Kirkpatrick and Leopold 1952). It might also be claimed that the short light pulse was interacting with the preceding or succeeding main light period. The length of the dark period had been shown to be unimportant in experiments by Wolfson (1959), using *Junco hyemalis*. In these the length of the dark period was varied being followed by long or short light periods; testis growth could be induced by long photoperiods irrespective of the length of the night. It fell to Hamner (1963, 1964, 1965) working with the House Finch to prove the involvement of a circadian rhythm mechanism in the photoperiodic control of gonad development in birds. In one of his most important experiments he employed a short non-inductive light period

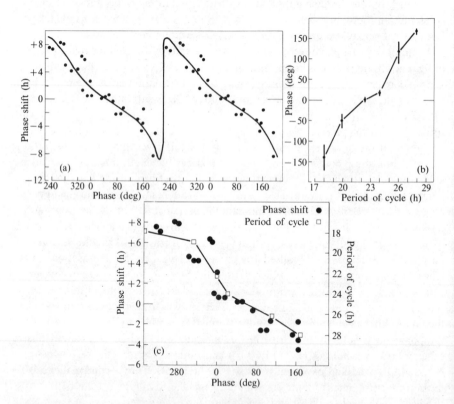

Fig. 11.10. (a) Phase-response curve of individual House Sparrows *Passer domesticus* to 6-h pulses of light. (b) Mean phase angles of groups of House Sparrows entrained by light–dark cycles of different total period length. (c) Solid circles represent phase shifts obtained by pulsing free-running sparrows with light. Open squares are phase-shift values calculated from the experimental results in (b). The right ordinate plots the periods of the cycles that correspond to the phase-shift values on the left ordinate for a free-running period of 25 h. Slightly modified from Eskin (1971).

followed by variable durations of dark during which any oscillators could free-run. Subjects with regressed winter-condition gonads were given periods of light lasting 6 hours, between which were dark intervals lasting 6, 18, 30, 42, 54, and 66 h in six different groups (Fig. 11.11). Testicular recrudescence occurred in those cycles where the light and dark phase totalled 12 (LD 6:6), 36 (LD 6:30), and 60 (LD 6:54) but not when the cycles totalled 24, 48, and 72 h. The term'resonance experiments' has been suggested for such experiments by Dr. Klaus Brinkman (quoted by Pittendrigh and Minis (1971)) and they provide good evidence in favour of some form of the Bünning hypothesis. That is, it is envisaged that a circadian oscillation embodying some kind of light-sensitive phase free-runs throughout the dark period so that it may be engaged if light is experienced at the appropriate times; the phase relationship between the light pulse and the free-running oscillation is critical to the inductive mechanism. This was essentially the interpretation of Follett and Sharp (1969), who obtained similar results with Japanese Quail *Coturnix coturnix* (Fig. 11.12).

Menaker (1965) was the first worker to employ systematically an asymmetric skeleton schedule to study the perch-hopping activity and testis stimulation of a bird, and he worked with House Sparrows (Fig. 11.13). In all cases, and sometimes by as much as 4 hours, the onset of locomotor activity preceded the dark to

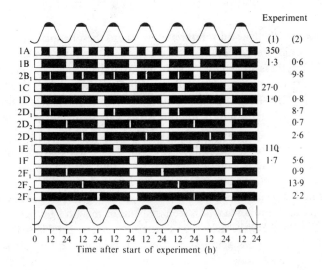

Fig. 11.11. Summary if two resonance-type experiments using ahemeral light cycles performed by Hamner (1965) in which House Finches *Carpodacus mexicanus* were pulsed with bursts of light. In experiment 1 pulses of 6 h light were combined with different intervals of dark, indicated by letters 1A–1F. For experiment 2, the light schedules 1B, 1D, and 1F, which previously failed to evoke testicular growth, were used as controls for experiments in which a further 1-h light pulse was given to interrupt the dark period ($2B_1 2D_1$, $2D_2$ ... etc.). Mean testicular weights (g) after 20 or 54 days according to the experiments are given to the right (the light schedules are only shown as extending over 7 days). An interpretation of the results is presented in terms of a light-sensitive oscillator which free-runs through the dark periods and whose sensitive phase is only engaged by certain of the light pulses. From Murton and Westwood (1975).

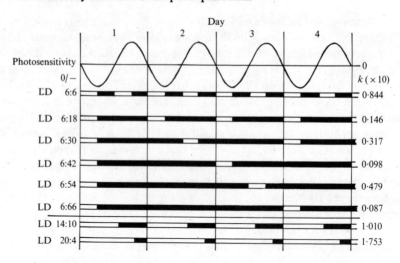

Fig. 11.12. Testicular response of Japanese Quail to ahemeral light cycles in a resonance type of experiment. Testicular response indicated as the growth rate constant k as derived in Fig. 9.9 (p. 237), with values for subjects kept on normal stimulatory 24-h schedules shown in the bottom two panels for comparison. After Follett and Sharp (1969).

light transition which the bird interpreted as dawn; the phase-angle shift being measured as the difference between the onset of activity and the subjective dawn. In each 24-h day there were two possible dawns and the one chosen by the birds depended on the relative position of the two light periods. When given a light cycle of 4L:10D:2L:8D, two experimental subjects behaved totally differently. The first bird became active just before the lights were switched on for the 4-h light period and remained active throughout the 10-h dark period and into 2 h of light; that is it treated the 8-h dark period as the night, as had all the birds tested on shorter simulations. The second bird, however, became active with the onset of the 2-h light pulse and remained active throughout the 8-h dark and 4-h light periods, that is, it treated the 10-h dark phase as the subjective night. All birds in the groups given 2-h light pulses more than 10 h from the end of the 4-h light period behaved as if the start of the 2-h were the dawn. The results are comparable with those depicted in Fig. 11.8 for the eclosion rhythm of *Pectinophora*, for with skeletons simulating a long day a phase jump occurred to the shorter of the two alternative photoperiods. To do so the birds phase-shifted from the original point of entrainment by more than 7½ h.

    Knowing that some of the subjects were treating the start of the 2-h light pulse as dawn, Menaker re-calculated the phase shifts of these birds accordingly and obtained the result depicted in Fig. 11.13. This shows that minimum phase shifts were correlated with the greatest rate of testicular involution. Thus, when the sparrows interpreted the skeleton schedule as simulating an 8-h day (either as LD 4 + 2 : 18 or LD 2 + 4 : 16 + (2)) the onset of activity occurred nearly 4 h before

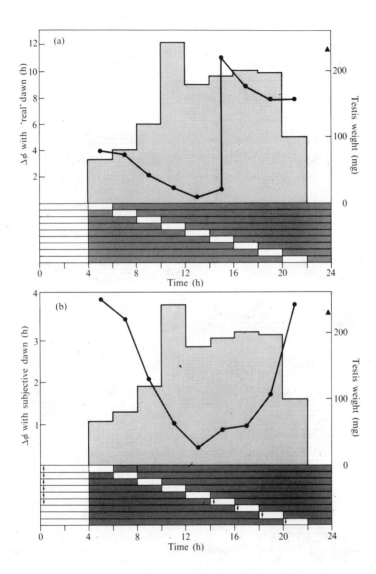

Fig. 11.13. Capacity of different skeleton light schedules to maintain the testicular weight of House Sparrows *Passer domesticus* that were already in full breeding condition. The solid lines plot the phase difference between the activity rhythm and the light cycle, that is, the number of hours by which the birds' activity preceded lights-on. The shaded histogram indicates the mean weight of the testes for groups of birds held under the various light schedules depicted at the bottom of the figure. The arrows incorporated into some of the boxes indicate the point in the light cycle to which the subjects entrained, that is, the point they treated as dawn. (a) Data plotted as if the start of the long light pulse was always treated as dawn by the birds. (b) Data plotted to allow for fact that for some birds the start of the short light pulse served as dawn. Based on Menaker (1965).

the real dawn (large positive phase shift) and in consequence the photosensitive phase for gonad maintenance was not stimulated. Involution of the gonads was most effectively prevented by a skeleton simulation of a 16-h photoperiod, whereupon the activity cycle was virtually in phase with the light cycle.

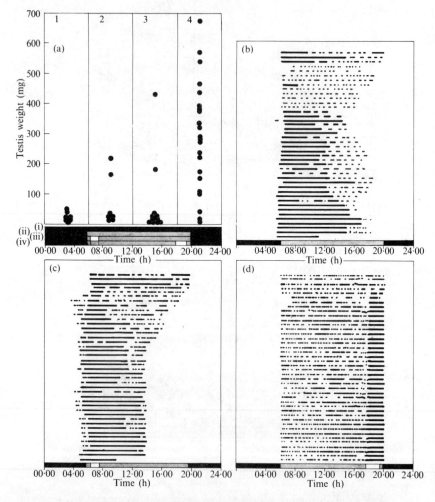

Fig. 11.14. (a) Each point is the combined testicular weight of individual House Sparrows *Passer domesticus* in groups 1–4 which were subjected to the four light schedules (i)–(iv) illustrated at the bottom of the figure. In this, black indicates no light, stippled is dull green light, and white represents a pulse of white light. (b), (c), and (d) give representative activity patterns of the birds exposed to regimes (ii), (iii), and (iv). The ordinate represents day 1 onwards and for each day the bars give the periods during which the subject was active (solid) or inactive (open). The abscissa gives the 24-h day cycle with the distribution of the light schedule during the day. Only birds in group 4, which exhibited marked activity during the pulse of white light given late in the cycle (regime (iv)), responded with testis growth. From Menaker and Eskin (1967).

Photo-induction in birds

Menaker's results suggest a close link between the rhythms controlling gonad development and locomotor activity which would be expected if one pacemaker oscillation controlled all rhythmic functions. This possibility was tested in some experiments using House Finches by Hamner and Enright (1967). They used 6-h photoperiods in cycles which were 22 h or 26 h long ($T = 22$ h or 26 h). With $T = 26$ h the 6-h light period corresponded to the end of activity but with $T = 22$ h entrainment was such that the activity rhythm was initiated by lights-on. As predicted, gonad stimulation occurred in the group in which the light pulse occurred at the end of the day but usually not when it was at the beginning. Although statistically significant correlations were obtained, there were sufficient exceptions to make the authors question whether a single clock could be involved. Their caution was well founded in view of what has since been discovered about the activity rhythm in finches (see p. 291).

Menaker and Eskin (1967) tackled the problem in another way, for they relied on the ability of dim green light to entrain the locomotor activity rhythm without causing gonad induction. Sparrows were given a cycle of LD 14(dim green):10 and this entrained their activity rhythm. Some were now pulsed with 75 min of bright light at the beginning of the green light and others received the white light pulse at the end. It happens that if such pulses are given at the beginning or end of the daily locomotor activity they have relatively little effect on the phase of the activity rhythm, as Menaker and Eskin found. The results are summarized in Fig. 11.14, and they do suggest that induction occurs when the activity rhythm is pulsed late in the subjective day. The problem of studying the inductive effects of light independently of its phasing effect would appear to have been solved except for the inconsistency of Hamner and Enright's work and the more recently acquired knowledge that the activity rhythm may not depend on a single oscillator (p. 495 below). It is possible that only one gonadotrophic effect was being monitored, and it is worth remembering that in all these experiments post-refractory subjects which had rehabilitated gonads were used. In view of the importance of these results they should be the basis for more experiments.

Extra-retinal light receptors

Individual cells in a wide range of tissues evidently have circadian clocks, and this capacity for time measurement is perhaps not surprising in view of the complex circadian periodicities exhibited by unicellular plants and animals. More interesting is the location of the master clock and the way in which it is entrained by the environmental *Zeitgeber*. In insects, the primary central nervous system clock appears to reside in the optic lobes and to be entrained via the photo-reception of the compound eyes (Brady 1974) but there is also good evidence that light can also be perceived directly via the brain cells. In birds, it is also the case that there are extra-retinal centres for light perception, a suggestion made long ago by Benoit (1961, 1964). He and his co-workers established that testis growth in blinded ducks could be induced by light impinging directly on the hypothalamic area of the brain via fine quartz rods. Benoit (1970) has since quoted Kordon as stimulating testicular development by implanting fine fibres (0·3 mm diameter) conducting white light to regions in the ventro-medial, supraoptic, and paraventricular nuclei, but not in surrounding areas. The circadian rhythm of locomotor activity of House Sparrows blinded by bilateral optic enucleation free-runs in constant dark but entrains to 24-h cycles in which visible light alternates with darkness (Menaker 1968a). Menaker showed that extra-retinal photo-receptors must be extremely sensitive, for about half of his experimentally blinded birds could entrain to cycles in which the light portion comprised green light of only 0·1 lx. The birds also obeyed Aschoff's rule in that the period length of free-running activity shortened with increase in light intensity, while the ratio of activity to rest time increased. But whereas the activity cycle of normal birds becomes arrhythmic on constant light of 500 lx and above, and they are active almost continuously, blinded subjects did not respond in this way and indeed continued to free-run when the light was increased to 2000 lx.

Just as the activity rhythm can be entrained in blinded sparrows, so can testicular recrudescence be achieved via extra-retinal photo-receptors. Bilaterally enucleated birds exhibited the same degree of testicular growth on a LD 16:8 as did intact controls, whereas a LD 6:18 was only very slightly stimulatory (Menaker and Keatts 1968, Menaker 1971). By plucking feathers from the top of the head, the amount of light reaching the brain could be increased. In contrast, Indian ink deposited under the skin of the head by injection decreased the light by a factor of 10. Now if blinded sparrows were held on a light cycle of subthreshold intensity, to which they did not entrain, entrainment could be induced by plucking feathers from the top of the birds' heads, whereas injecting Indian ink to reduce the light again resulted in the subjects free-running (Menaker 1968b; Menaker, Roberts, Elliott, and Underwood 1970). In similar manner, two groups of sparrows (not blinded) were exposed to 10 lx of light in a cycle of LD 16:8. This is about the threshold for photo-induced testis growth. In one group the subjects were injected under the skull with Indian ink and these did not respond. The second group had the head feathers plucked to increased the light intensity and they showed full testicular development (Menaker et al. 1970).

Removal of the pineal gland has been shown not to prevent blinded House Sparrow from entraining to light–dark cycles so the pineal cannot be an exclusive site of photo-reception (Menaker 1971). But pinealectomy does affect free-running circadian rhythm in constant dark, causing the subjects to become arrhythmic in their locomotor cycles (Gaston and Menaker 1968; Gaston 1971) and body temperature (Binkley, Kluth, and Menaker 1971); it may affect the coupling of oscillators. The pineal does not appear to be necessary for testicular recrudescence in the sparrow (Menaker et al. 1970) and the same apparently applies to the Japanese Quail (Homma et al. 1967; Sayler and Wolfson 1968a, b) and White-crowned Sparrow (Kobayashi 1969). Nor does ablation affect the onset of photo-refractoriness in Harris' Sparrows (Donham and Wilson 1970). In the Japanese Quail injections of melatonin were also found to have no effect by Homma and his colleagues. Pinealectomy of ducks in spring inhibits normal seasonal testicular growth (Cardinali, Cuello, Tramezzani, and

Rosner 1971) and it is possible that disruption of normal circadian rhythm patterns does occur. In all cases where testicular induction has been achieved, post-refractory subjects were used. As will emerge in Chapter 13, many such species will undergo testicular recrudescence if held on constant dark. A more critical test would involve keeping subjects for longer periods to see how the whole cycle was phased. The topic is pertinent to the question of the elements of induction and phasing in the photoperiodic response and will be raised again in the next chapter (p. 312). At this stage we note that the evidence for a neural connection between the retina and hypothalamus is conflictory. Oksche (1970) summarizes his own and other evidence against the existence of any such tracts in White-crowned Sparrows but quotes Blümcke (1961) who did find connections in the domestic fowl. However, different histological techniques have been used by the various workers which prevents direct comparison of their results.

The exact site of extra-retinal photo-reception remains to be identified. Homma and Sakakibara (1971) implanted small discs coated with radio-luminous paint at different sites in the brain and suggested that those planted against the sphenoid bone or the fissura longitudinalis cerebri, and which were effective in stimulating testicular growth, could have been acting on a single photo-receptor in the hypothalamic area; this would support the original observations of Benoit. Implants near the pineal were not effective.

12 Endocrine secretion and photoperiodism

Summary

The interpretation of results in which gonad growth is stimulated by skeleton photoperiods is complicated by the fact that birds do not necessarily entrain to the first light pulse; they may instead interpret the cycle as a 'short' rather than 'long' day. Gonad growth ultimately depends on circadian rhythms of photosensitivity so it might be assumed that the secretion of gonadotrophins would in turn be so regulated. In fact, this is not the case for once stimulatory daylengths are received gonadotrophin secretion (FSH and LH) continues throughout the 24 hours, admittedly with fluctuations in output (episodic release) that appear to fit no daily rhythm. Evidently, there is some hypothalamic mechanism which transduces the rhythmic measurement of photoperiodic inputs into a continuous stimulation of gonadotrophin releasing factor and in turn gonadotrophin production. It seems likely that FSH and LH are controlled by the same oscillator systems in birds for they are apparently stimulated by a single releasing factor. However, other hormone secretions are separately controlled so that phasing effects could be important. The differential phasing of hormone rhythms might account for avian photo-refractoriness, a feedback mechanism from thyroid hormones (first stimulated by TSH) being involved. Other feed-back effects can be discerned on a daily and seasonal basis. For example, unlike the gonadotrophins, plasma testosterone titres do exhibit a diurnal rhythm with a peak output in the morning. This possibly depends on an increase in thyroid activity during the day which inhibits androgen secretion in the presence of circulating LH. Thyroid stimulating hormone release therefore appears to be separately phased from that of the gonadotrophins. FSH induces the production of a testicular substance termed inhibin (probably a product from the germinal epithelium) which feeds back to inhibit FSH release.

Evidence which shows that the phasing of hormone rhythms is important in regulating breeding periodicity comes from experiments in which mixtures of exogenous hormones have been injected in varying temporal relationships: much attention has also been given to the effect of corticosterone and prolactin and of thyroxine and prolactin in stimulating crop gland development in the pigeon. The end of the chapter considers how circadian rhythms might be related to longer-term periodicities.

Response of birds to skeleton photoperiods

The involvement of circadian rhythms in the photoperiodic response mechanism of birds was discussed in the previous chapter in terms of the stimulatory effect of appropriate daylengths on gonad growth and spermatogenetic development. Since the gonad is a target tissue for several hormones, some of which act synergistically, we need to consider the possibility that different hormones are controlled by different oscillator systems. Indeed, it was suggested for the Greenfinch *Carduelis chloris* that immuno-reactive luteinizing hormone is secreted under a different light regime from the hormone, presumably FSH, responsible for spermatogenetic development though this interpretation may need modification (p. 39 and Fig. 2.4, p. 40). Any modification of the Bünning hypothesis of photoperiodism must also be able to account for the phemonenon of avian photo-refractoriness and be capable of explaining the wide range of inter-specific differences in photosensitivity and refractory periods examples of which were discussed in Chapters 9 and 10. These topics are pursued here.

Many workers have now employed asymmetrical skeleton photoperiods in diel (daily) cycles as distinct from the ahemeral cycles (that is, cycles in which the total of day and night is not 24-h) discussed in the previous chapter, and measured their effectiveness in inducing gonad development, and less often their effect on some other function. Various examples are given in Figs. 12.1 and 12.2 that show that gonad growth can be stimulated by some skeletons but not by others. The total amount of light remained constant during particular experiments, emphasizing that it was the distribution of the photoperiod that was important. In *Zonotrichia leucophrys* testicular induction was correlated with acetylcholinesterase (AChE) activity in the interior pituitary, this enzyme being apparently associated with the gonadotrophic cells (Haase and Farner 1969); no AChE cells appeared in photo-refractory birds (Haase and Farner 1970). The data for the Japanese Quail emphasize that both asymmetric and symmetric skeletons are effective and that female ovarian follicle development can also be stimulated by interrupted-night experiments (Follett and Sharp 1969). Light schedules which stimulate maximum testis development in male Wood Pigeons *Columba palumbus* also stimulate the most growth in the ovarian follicles (Murton 1975). Female Budgerigars *Melopsittacus undulatus* kept in separate cages with nest boxes in which they could hear male vocalizations and given a LD 6 + 2:16 laid more eggs when the light pulse ended 14 h from dawn (Shellswell, Gosney and Hinde 1975). They also laid more rapidly under a LD 14:10 than other photoperiods (Putman and Hinde 1973).[†] An apparently wide inter-specific difference in response is evident in Fig. 12.1. This poses the question to what extent different responses depend on the nature of the light stimulus and to what extent on the state of the subject, as apparently in the case of House Finches studied by Hamner in both autumn and spring (Fig. 12.1). We note that there is a trough in the response curve of the House Sparrow data collected by Menaker (1965), whereas no such trough occurs in the *Carpodacus* response curve. Yet seasonal changes in the gonad cycle of the House Finch under natural conditions, as reported by Hamner (1966, 1968), are virtually the same as those depicted for the House

[†] The relative darkness of a nest box was also important and the availability of a box accelerated egg laying. In the absence of any nest-box breeding occurs sooner on continuous dark than under LD 14:10 (Hinde and Putman 1973).

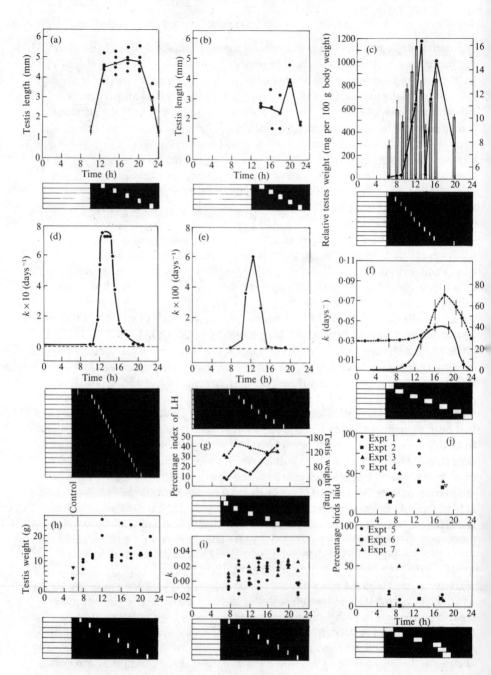

Fig. 12.1. Effect of asymmetric skeleton light schedules on reproductive function in various bird species. In most cases the effect on absolute testis size or testis growth rates k are depicted using filled circles and continuous lines to join points. Other conventions and measurements detailed below. In all cases different experimental groups of subjects were held on the schedules depicted as a vertical sequence below each figure, thus in (a) there were 5 experimental groups

Sparrow in Fig. 9.2 (p. 220). Further, the duration of the refractory period is about the same in the two species, lasting from late July until October. Do Hamner's results as presented in Fig. 12.1 differ from those of Menaker because different light schedules were used or because these two species differ in their photo-response mechanism?

Menaker's data for the activity rhythm of the House Sparrow, as plotted in Fig. 11.13 (p. 303), give the clue that entrainment to the light schedules was not necessarily as implied by Figs 12.1 and 12.2, and that the bimodal pattern of gonad induction seen in some species is an artifact of the experimental technique. Fig. 11.13 shows that when an asymmetric schedule was presented to House Sparrows to stimulate a daylength of 14 h or more the birds entrained to the start of the short light pulse and so effectively treated the photoperiod as a shorter one (p. 302). In this way a cycle of 4L 12D 2L 6D, which supposedly simulates a long day, was read by the birds as a 12 h day of 2L 6D 4L 12D. It will be remembered too that the *Pectinophora* and *Drosophila* eclosion rhythms phased-jumped to the shorter of two alternative daylength simulations when skeleton schedules were employed (see Fig 11.8 and 11.9, pp. 297 and 298). With the benefit of Menaker's studies the results in Figs. 12.1 and 12.2 are re-interpreted in Figs. 12.3 and 12.4 according to the manner in which the birds probably entrained, and it seems unlikely that any species truly has a two-phased rhythm for photo-induction.

The decisions taken in Figs 12.3 and 12.4 are arbitrary in the sense that the mode of entrainment could not be checked. Reference to the House Sparrow and Tree Sparrow data detailed in Figs 12.2 and 12.4 suggests that in spring the birds might have entrained to the start of the long light period so far as the 14-h simulation was concerned. It could be reasoned that had they entrained to the start of the short pulse, then some part of the long light pulse must then have fallen in the area of photosensitivity, at least if consistency with the results using other schedules be assumed. This rationale assumes that one of the light pulses coincided with a sensitive phase in a rhythm of photosensitivity and it is, therefore, a development of

---

and in (c) there were 11. Groups are numbered from left to right, although the numbers are omitted from the figure for clarity.

(a) House Finch *Carpodacus mexicanus* wild-caught in January. (b) House Finches as above but wild-caught in October. From Hamner (1968). (c) Tree Sparrows *Passer montanus* wild-caught in spring with regressed gonads (filled circles). Histograms depict height of columnar lining cells of the epididymides. From Lofts and Lam (1973). (d) Japanese Quail *Coturnix coturnix*. After Follett and Sharp (1969). (e) Japanese Quail *Coturnix coturnix*. From Follett (1973b). (f) White-crowned Sparrow *Zonotrichia leucophyrs gambelii* captured when photosensitivity had been regained in late winter. Testis growth rate k as solid line. Acetylcholinesterase activity of *pars distalis* as dotted line. From Russell and Farner (1968). (g) House Sparrows *Passer domesticus* taken from field in winter. Filled circles mean testis weight of group and triangles mean plasma IR-LH (Bagshawe) content. From Murton *et al.* (1970a). (h) Collared Dove *Streptopelia decaocto*: lg testis weight of individual birds initially captured in November. From Murton and Westwood (1975). (i) Wood Pigeon *Columba palumbus*. Growth rate constants of testes (filled circles) and ovarian follicles (triangles) of individual birds captured from field in late winter. From Murton (1975). (j) Percentage of Budgerigars which laid eggs by day 30 of pairing. (Some birds had already been used in other photoperiod experiments and the different experiments group subjects according to the length of time they were separated and held on natural days.) From Shellswell *et al.* (1975).

the Bünning hypothesis of photoperiodic induction. If the critical factor in stimulating photo-induced gonad growth is not induction *per se* but, instead, the correct phasing of two or more driven hormone rhythms to ensure temporal synergism, a different interpretation is possible. The light pulses might serve simply to define the length of the *Zeitgeber* and by this means determine the phase angles of any driven

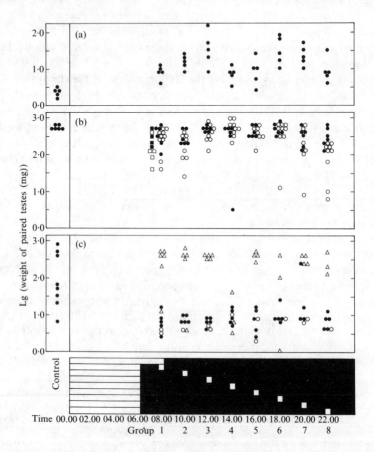

Fig. 12.2 Effect of asymmetric skeleton light schedules on lg weight of paired testes of House Sparrows collected from the wild at the three times of year indicated by the letters A, B, and C in Fig. 9.2 (p. 220). Abscissa gives the 24-h day indicating the hours from time 0 at which the various light pulses *finished* and therefore the length of full photoperiod that was simulated. Controls were killed at the start of each experiment and reflect the field condition (see Fig. 9.2). In experiments (a) and (b) some birds were held on the light schedules for 20 days (filled circles, the filled squares representing a few subjects which were held on LD 8:16) and others for 40 days (open circles). In experiment (c) some birds were held on the skeleton schedules for 30 days (solid circles) and some for a further 67 days (total 97 days) (open circles). The survivors from experiment (c), that is, those birds not already killed, were exposed to LD 16:8 for 28 days (open triangles). In all groups except 4 and perhaps 6 refractoriness was clearly broken. The solid circles at the left indicate testis' weight before the start of the experiment. From Murton and Westwood (1974b).

oscillators. This would make the length of the light pulses irrelevant and there would be no evidence for assuming induction. This is an extremely difficult point to resolve because with the Bunning hypothesis the light cycle is supposed to phase and induct simultaneously. Adopting only a phasing hypothesis we are led to the assumption that 14-h (which if a phase-jump occurred would be equivalent to 16½-h)

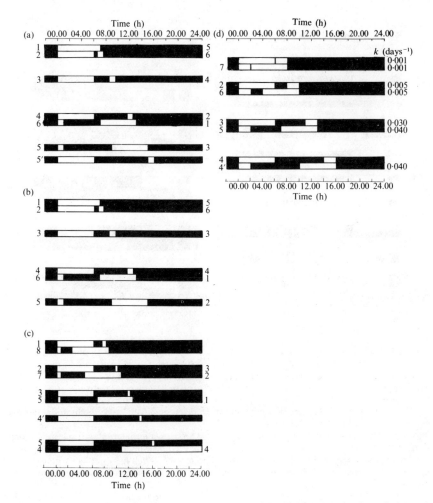

Fig. 12.3. An interpretation of some of the results in Fig. 2.4 and 12.1 showing how skeleton photoperiods, which simulate continuous photoperiods of various duration, might stimulate spermatogenesis: (a) Greenfinch, (b) House Sparrow, (c) Wood Pigeon, (d) White-crowned Sparrow. The skeletons have been plotted according to the way the subjects probably entrained, that is, by making the start of the long or short pulse the beginning of their daily activity rhythm (time 0). Some alternative arrangements for certain of the light schedules are also shown, and identified by subscripts following the group letter. For the three species in parts (a), (b), and (c), numerals at the right rank the schedules that were effective in stimulating testis development beyond secondary spermatocytes in photosensitive subjects that began with regressed organs. In the right panel growth rate constants k are given. Extended from Murton (1975).

simulations were ineffective because days of this length do not enable an appropriate hormone phasing (see Fig. 12.4). Such a view is contradicted by the observation that whole photoperiods of 14-h or more are very effective in inducing gonad growth, as too are 12 h days.

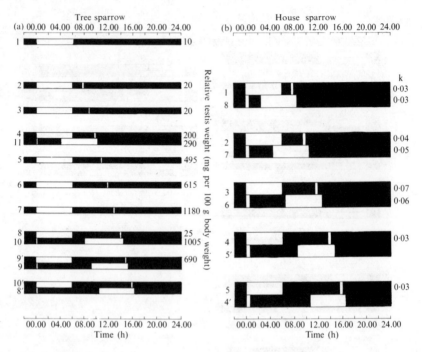

Fig. 12.4. The likely manner in which skeleton light schedules presented to photosensitive Tree Sparrows *Passer montanus,* as in Fig. 12.1, and House Sparrows *P. domesticus,* as in Fig. 12.2 (a), were interpreted by the subjects. Group numbers are given to the left of the schedules. Subjects had regressed testis at the start and the amount of growth is shown as relative testis weight (actual testis weight per 100 mg body weight) in (a) and as a growth rate constant k in (b).

   As entrainment was not necessarily to the start of the long light pulse the bimodal response depicted in Figs. 12.1 and 12.2 is an artifact.
Maximum gonad stimulation occurred in the House Sparrow on skeleton photoperiods which simulated 12-h days whereas there are indications that a longer photoperiod was needed by the Tree Sparrows. However, the House Sparrow received 6 + ½ h light per day whereas the Tree Sparrows were given 6 + ¼ h and this could have contributed to the different response, that is, the effective intensity of light would have been greater for the House Sparrows and their endogenous rhythms therefore of higher frequency. For both species it is known that long continuous photoperiods are extremely effective in stimulating gonad growth in photosensitive subjects, that is, photoperiods of LD 14:10 and longer. But in both these experiments a simulation of a 14-h photoperiod, assuming entrainment to the start of the long light pulse, was not effective. Had the birds reversed the schedule by entraining to the start of the short light pulse then the simulation would be of a 16½-h photoperiod for the House Sparrow (4′) and 16¼ h for the Tree Sparrow (8′). Either way, the skeleton should have been effective if the duration which was simulated was the important factor. The evidence does not entirely support this viewpoint and argues instead for the induction theory. In all cases the reference numbers to light schedules (experimental groups) which carry suffixes are thought to represent the inappropriate alternative methods of interpretation by the birds.

Since there appeared to be some foundation for the idea of photoperiodic induction – and it helps the discussion to assume that this occurs – a schematic band of shading to delimit this phase was incorporated on the original figures (Murton 1975). This band was thought of as that part of a circadian oscillator occurring above some threshold. The phase of photo-induction was drawn in a manner connecting the light parts of the photoperiod, which were correlated with testicular recrudescence. Although straight lines were employed, it is more realistic to think that curves would be appropriate. It may be helpful to compare Figs 12.3 and 12.4 with Fig. 11.7 (p. 296). Follett (1973*b*) was of the opinion that the use of short light pulses to illuminate the scotophase would provide a more critical indication of the photo-induction phase and he and Sharp therefore used 15-min pulses in studies of the Japanese Quail (Fig. 12.1). In fact this is not necessarily true, for it depends on whether the short pulse serves to begin or end the *Zeitgeber*. More important is the range of light schedules used, for the more combinations covered the more clearly defined must be the photo-inducible phase. Nevertheless, the re-analysis of the Quail data as in Fig. 12.3 does suggest a phase duration of about 3½ h, which is the same as that deduced by Follett. This occurs between about 12 h and 15½ h from the subjective dawn, irrespective of the total photoperiod, and would be appropriate to the supposed natural breeding season (p. 279). The photosensitive phase of the House and Tree Sparrows appears to shift relative to the subjective dawn depending on the length of the daily photoperiod. We interpret this as implying that the real oscillators also become phase-shifted with change in *Zeitgeber* strength and that the oscillators of quail are more stable to phase alteration than those of sparrows. All this needs to be verified and the techniques are now available for the systematic assay of plasma hormones under a wide array of light schedules.

Hamner (1968) was of the opinion that the phase relationships of the circadian clock were adjusted during the refractory period. He considered that it should be possible to collect birds which had just ended refractoriness in late January and demonstrate by interrupted-night techniques that the phases of the rhythm depend on season. Unfortunately, although he argued that his results supported his hypothesis they are equivocal because he used different light schedules; LD 12 + 1:11 in October and LD 10 + 1:13 in January. Thus, although Fig. 12.1 suggests big differences in the response of House Finches between autumn and winter, Fig. 12.3 demonstrates that the light schedules could be interpreted as providing illumination at nearly the same inductive phase. Now, the frequency of a circadian oscillator is expected to increase with increase in light intensity (Aschoff's rule) and an increase in frequency should result in a phase advance relative to the *Zeitgeber*. Hence, the difference in the phasing of the band of photo-induction between spring and autumn might depend entirely on the differences in total light used in the two experiments (13-h and 11-h), and so on the effective light intensity.

Some evidence for changes in entrainment with season can be seen in the House Sparrow data plotted in Fig. 12.2. Birds were taken from the field at three different times of year as indicated at A, B, and C in Fig. 9.2 (p. 220); Fig. 9.2 also shows the seasonal changes in testis size in the free-living population from which the samples were drawn. During each experiment groups of birds were kept on the same skeletons of LD 6 + ½:17½. When first captured in late January the testes of all birds contained only spermatogonia and a few prophase primary spermatocytes. All the skeleton schedules to which the birds were transferred resulted in a degree of spermatogenetic development to the point where primary spermatocytes were numerous. Only light flashes beginning 11½ h and 17½ h from the start of the long pulse caused the appearance of secondary spermatocytes and spermatozoa. Extrapolating from Fig. 9.2 suggests that in the wild the testes would have increased to about 19 mg during the 20 days beginning 3 February, natural photoperiods increasing from 10 h to 12 h over this time. The testes of birds given interrupted-light schedules for 20 days in which the ½-h light pulse began 9½ h after the start of the long light pulse reached an average weight of 17·3 mg; those birds on a 12-h skeleton simulation had testes weighing on average 50·3 mg. Whereas the sparrows tested in February apparently treated the skeletons simulating 14-h and 16-h photoperiods as if they actually were complete photoperiods of these durations, they responded differently in May. Of course, birds caught in May already had fully enlarged testes so the stimulatory capacity of the light schedules was judged in terms of preventing involution. In May the 16-h light pulse was much more effective at stimulating the testes than it had been in February, suggesting that it was interpreted differently and that the birds entrained to the start of the short light pulse. Presumably the higher light intensity in May increased the frequency of the birds' circadian oscillators resulting in a more positive phase relation to the *Zeitgeber*. At a critical range of skeletons entrainment was evidently more effectively achieved by 'reading' the *Zeitgeber* as a shorter day in May than had been the case in February.

When we caught and exposed House Sparrows to skeletons of LD 6 + 1:17 for 14 days in early January 1968 we found that a 1-h pulse ending 12½ h from 'dawn' was not very effective at stimulating testicular development (Fig. 12.1). In the 1970 series of experiments, involving subjects tested in early February, a ½-h pulse ending 12-h from 'dawn' was highly inductive (Fig. 12.2). The discrepancy is again to be explained in terms of the higher light intensity in a cycle of LD 6 + 1:17 compared with LD 6 + ½:17½ and perhaps a natural difference between subjects in January and February.

Hormone secretion under skeleton light schedules

A third experiment, performed in 1970, utilized House Sparrows caught in late July, as they were spontaneously entering a refractory state. Irrespective of the light schedule the testes continued to involute so that at autopsy the gonads of the birds contained only spermatogonia or a few primary spermatocytes. We note for later reference that involution was slower with the skeletons simulating 18-h and 20-h photoperiods (see Fig. 12.2). After being kept for 97 days all the subjects were exposed to constant 16-h photoperiods (LD 16:8) whereupon nearly all responded by rapidly achieving fully recrudesced testes. That is, all skeleton schedules had broken refractoriness except for the 14-h simulation, while one subject on an 18-h simulation failed to respond. One way to explain this result is to assume that a phasing of hormone secretion (FSH, LH and probably also TSH) that allows rehabilitation of the testes is only possible on photoperiods shorter than about 15-h. The 14-h simulation might then have been the longest photoperiod used if the birds interpreted it as a 16½-h day by entraining to the start of the short light pulse; the 16-h simulation could have been read as such, but more likely as a 14½-h day. (Some consistency would be expected in the manner by which the birds entrained to the two simulations. They might entrain to make each simulation as short as possible so that 16 h became 14½ h and 14 h remained 14 h. Alternatively, the birds could have entrained to the short light pulse in each group, making the 16-h simulation 14½ h and the 14-h simulation 16½ h. This would be consistent with the interpretation that the 14 h was effectively the longest used.) It is not clear whether refractoriness could be ended under a complete 14½-h photoperiod (equivalent to the natural photoperiod in early September when sparrows are still refractory) but the lower light intensity of a skeleton 14½-h might make the difference. Presumably the 16-h simulation could break refractoriness because it was read as a 14½-h day, but then why should not the 14-h simulation have been effective? An alternative explanation for the failure of a 14-h skeleton to break refractoriness invokes again the concept of induction and the distribution of the light pulses.

Early interrupted night experiments with Greenfinches *Carduelis chloris* showed that the interstitial Leydig cells could be markedly stimulated without concomitant spermatogenetic development (Figs 2.4 (p. 40), 12.5: see between pp. 82–3). This condition was correlated with a high plasma IR-LH (Bagshawe) titre. It might be inferred that either FSH was not secreted under these conditions or that it was not phased relative to IR-LH in a way that facilitated spermatogenesis. Another possibility is that with the longer day simulations feed-back mechanisms became operative which inhibited gonadotrophin secretion. There is now evidence from a few species that the production of thyroid hormones increases under long day regimes (p. 319). Long days stimulate gonadotrophin secretion and androgen production and androgen then feeds-back to inhibit gonadotrophin secretion (p. 321). To what extent this facilitates thyroid stimulating hormone (TSH) production is not known. We have no information regarding TSH output so far as the experiment discussed here is concerned and the topic is mentioned to indicate the complexities involved in interpreting the results. Nevertheless, the results do suggest that with some lighting schedules endocrine rhythms may become dissociated. For example, the fact that abnormal spermatogenetic development (Fig. 35 (c), see between pp. 82–3) could be achieved under some skeleton schedules also suggested that inductive effects were operating which resulted in an unnatural endocrine balance for such manifestations are not noted with complete photoperiods.

We, therefore, have some evidence (and more is given below) to suggest that photoperiodic responses depend on an *internal* coincidence system. The *external* coincidence model assumes that light has to coincide with a particular phase of a light sensitive circadian oscillator. But once we consider target tissue responses that depend on several hormone interactions it is less certain that such a model is applicable. Instead we have evidence for a multi-oscillatory circadian system which only functions properly when the various hormone rhythms are in a critical phase relationship one to the other. This can only be achieved with certain light schedules. It may be suspected that both induction and phasing effects will eventually be shown to be important.

It ought to be easy to demonstrate whether plasma titres show any evidence of phase change with photoperiod. Unfortunately'nobody has yet compared the pattern of secretion in the same species using a complete photoperiod and a skeleton simulation. There is a little evidence that the results would not be the same. Fig. 12.6 (a) gives some results for the House Sparrow which suggest that the time of IR-LH (Bagshawe) release varied with the length of the photoperiod. (In passing, it may be noted that the assay did not cross-react with exogenous mammalian FSH which had been injected just before a blood-sample was collected, whereas injected mammalian LH caused a marked elevation in recorded plasma IR-LH.) The same conclusion can be drawn from Fig. 12.6 (b) which presents data collected by Nicholls and Follett (1974) working with Japanese Quail. However, Follett and Davies (1975) point out that although the results encourage a belief in a rhythmic secretion of LH, in fact plasma levels of this hormone are elevated in all photo-stimulated quail, regardless of the time of sampling. Hormone release may be rhythmic initially but thereafter any such tendency becomes obscured by a general increase in the level of LH output. The pattern of LH secretion in intact and gonadectomized cockerels sampled at 10–30 min intervals throughout the day was in sequences of high and low output, described as an episodic release (Wilson and

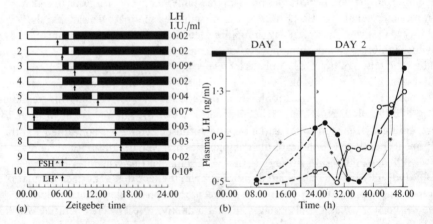

Fig. 12.6. Diurnal variations in plasma IR-LH levels in two bird species according to the light cycle.

(a) Pooled plasma titres of IR-LH (Bagshawe) in groups of 4 House Sparrows *Passer domesticus* held on the schedules indicated and killed for assay at the times indicated by the arrows. Group 9 received exogenous bovine NIH.FSH and group 10 bovine NIH.LH at the times indicated. Significant titres above base-line readings indicated by asterisks. From Murton (1975).

(b) Plasma LH levels during first 2 days of exposing young male (filled circles) or female (open circles) Japanese Quail *Coturnix coturnix* to LD 20:4 after rearing from hatch on LD 8:16. Groups of 8–10 birds killed at varying times. Most important significant differences indicated by asterisks (p < 0·05). With further exposure by long days LH secretion becomes continuous (see text). From Nicholls and Follett (1974).

Sharp 1975). During the next year or so the diurnal pattern of hormone release in a range of avian species will certainly be measured so that primary rhythms and those resulting from differential feed-back will be defined. It would be foolish to extrapolate results obtained with Japanese Quail and domestic fowl to wild species that have markedly seasonal breeding cycles.

Photostimulation also causes FSH plasma levels to rise simultaneously with those of LH in Japanese Quail, consistent with the view that a single releasing factor is involved (Follett 1975, 1976). However, with further exposure to long days and the establishment of full spermatogenesis, FSH levels decline (presumably as a consequence of inhibin feed-back) while LH titres remain relatively high: some inhibition due to androgen feed-back is apparent (Follet, *loc. cit.*). Thus, following exposure to long days there is a rise in both LH and FSH output and these elevated plasma titres eventually persist throughout the 24 hours, even though for two days there may be a rhythmic output of LH. There seems to be no evidence for a circadian rhythm of gonadotrophin output but there is clearly a diurnal rhythm in androgen secretion, with peak plasma titres detectable in the morning (work in progresss). It is difficult to explain this observation in terms of LH secretion but it could depend on the production, later in the day, of TSH and thyroid hormones, which as explained below both inhibit androgen release from the Leydig cells and also increase the rate of metabolic breakdown of circulating androgen.

Endocrine basis of photo-refractoriness

It is evident that the gonad itself does not become refractory, contrary to Vaugien's (1955) suggestion, since exogenous gonadotrophins can stimulate during the refractory period (e.g. Table 3.1 and Benoit, Mandel, Walter, and Assenmacher 1950 *b*). One of the most important recent studies is due to Wieselthier and van Tienhoven (1972). They found that thyroidectomy of Starlings *Sturnus vulgaris* prior to exposure to long photoperiods resulted in the failure of the testes to regress after an initial size increase. However, if thyroidectomy was delayed until after four weeks of exposure to long days regression did occur and it was followed by a second period of testicular size increase. Testicular size was correlated with androgen secretion as indicated by a yellow bill. The evidence indicated that thyroidectomy results in a higher gonadotrophin secretion. Thyroidectomy performed in the autumn did not affect the termination of the photo-refractory period. Supporting evidence comes from the work of Thapliyal (1969) and Chandola (1972) working with Indian finches and Jallageas and Assenmacher (1974) working with domestic strains of *Anas platyrhynchos.*

In ducks there is an elevation in thyroid activity during the long days of May and July (Astier, Halberg, and Assenmacher 1970). During this time LH plasma titres remain high (Jallageas, Follett, and Assenmacher 1974) but androgenic activity decreases because there is an increased metabolic clearance rate of testosterone (see Fig. 12.7). Assenmacher (1974) states that these effects can be simulated by thyroxine injections during the period of gonad enlargement, while an experimental thyroxine load prevents the stimulating effects of artificial long days (LD 18:6) on the gonads of ducks in winter. Moreover, the moult which usually follows sectioning of the hypothalamus in ducks is prevented by testosterone therapy or thyroid inhibition (references given by Assenmacher (1974)). These observations led Assenmacher (1974) to suggest that a seasonal increase in thyroid activity (presumably long days stimulate TSH secretion) result in an inhibition of androgen activity. While it remains likely that LH and FSH secretion rhythms depend on the same oscillator mechanism (it should be recalled that there is perhaps only one releasing factor for these two hormones, p. 28) it can be postulated that TSH (and ACTH) secretion can be independently phased and so may assume varying phase relationships according to the photoperiod, be it a full photoperiod or a skeleton simulation. Under constant illumination the rhythms might free-run until they assumed the appropriate phase relationship for gonad rehabilitation and subsequent recrudescence. With flash-lighting it is conceivable that one rhythm is immediately phase-locked

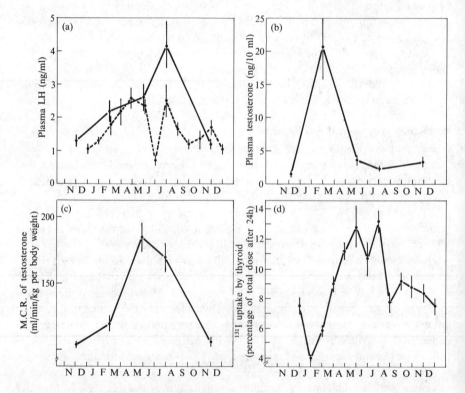

Fig. 12.7. Annual variations in plasma hormone levels in various domestic strains of *Anas platyrhynchos*.
(a) Plasma levels of IR-LH in Pekin drakes studied by Jallageas, Follett and Assenmacher (1974) (solid line) which were sampled at 2-month intervals and in certain Khaki Campbell drakes sampled monthly by Haase, Sharp, and Paulke (1975) (dashed line). (b) Plasma levels of testosterone in Pekin drakes. (c) Metabolic clearance rate (MCR) of testosterone in Pekin ducks. ((b) and (c) from Jallageas *et al.* (1974).) (d) Seasonal change in thryoid function judged in terms of rate of ^{131}1 uptake by the gland. From Astier, Halberg, and Assenmacher (1970).

and the other free-runs and eventually reaches the appropriate phase relationship.

The following sequence of events seem to occur during the reproductive cycle, the account being somewhat modified from one proposed earlier by Assenmacher (1974):

1. Given a species-appropriate photoperiod gonadotrophin secretion is initiated and causes androgen secretion by the Leydig cells, under the influence of LH, and spermatogenetic development, under the influence of FSH and androgen.

2. There is possibly a seasonal increase in the sensitivity of the hypothalamus to feed-back from testosterone and certainly testosterone affects LH release. Gwinner (1974) has discovered that the onset of a breeding condition or exogenous treatment with testosterone can cause splitting of the free-running activity rhythm of Starlings. Perhaps the secretion of testosterone during the normal breeding cycle can also cause a phase change in gonadotrophin secretion, and this could be some of the basis for its feed-back action on the hypothalamus. Testosterone propionate implanted in the basal infundibular nucleus of photostimulated American Tree Sparrows *Spizella arborea* caused regressive changes in the testis and often total collapse; cholesterol implants had no effect. When the androgen antagonist cyproterone (6-chloro-17-hydroxy-1α, 2α-methylenepregna-4, 6-diene-3, 20-dione) was implanted the onset of spontaneous regression was prevented in subjects held for 17 weeks on LD 20:8 (Cusick and Wilson 1972). When intact male and female Tree Sparrows were transferred from LD 8:16 to LD 20:4 plasma IR-LH increased 11-fold greater than in short-day controls (Wilson and Follett 1974). It remained at this level until day 56 and decreased to pre-stimulation levels between days 56 and 77 with the onset of photo-refractoriness. In castrated males the level of circulating IR-LH was higher than intact birds and it remained elevated through day 56 but eventually declined to the same level as in the intact birds. There was a 4-fold increase in pituitary LH preceding the reduction in plasma content. The cycle of plasma LH was, therefore, the same in intact and castrated males, but plasma titres were higher in the castrates. This suggests that the gonad exerts an inhibition on the LH-release mechanism, but that it is not the final determinant of the temporal pattern of circulating LH. That is, the LH-release mechanism becomes refractory in castrates just as it does in intact males; perhaps this is because the thyroid hormones are the important factor. Plasma IR-LH levels were also elevated in castrates kept on short days so that a relationship exists between the hypothalamohypophysial—gonad axis even in non-photostimulated males. Differences in gonad condition and steroid output could account for the common observation that end of season gonad regression occurs earlier in paired males, than in unmated birds (e.g. Wright and Wright 1944), for it is possible that males are stimulated to produce more testosterone. FSH stimulates the production of androgen binding protein and a substance from the germinal epithelium which becomes modified in the epididymides, referred to as inhibin, which feeds back to inhibit FSH release.

3. Further increase in the photoperiod allows TSH secretion and thyroid activity to increase. This possibly occurs independently but is perhaps also facilitated by the feed-back action of androgen and inhibin on gonadotrophin secretion, which as a result allows TSH secretion and thyroid activity to become effective; no details are available. Hyperactivity of the thyroid induces a further inhibition of testosterone secretion and an increased catabolism of the steroid. A relatively selective inhibition of FSH secretion allows LH secretion to continue at a moderate level. As mentioned above, measurement of both FSH and LH plasma levels in photostimulated Japanese Quail shows that the plasma content of both hormones increases but that after a few days plasma titres of FSH decrease while LH levels remain elevated (Follett 1975, 1976).

4. A stage is now reached at which LH titres are only partly elevated, androgenic activity is inhibited and FSH output is markedly reduced. A changed testis—thyroid axis may be one of the reasons for the failure of the gonad to respond to circulating LH as at the beginning of the cycle (see below). Once the hypothalamus—thyroid—gonad axis has reached this stage the system is effectively blocked and gonad regression results from the suppression of FSH and androgen influence; the bird has now entered a state of photo-refractoriness. There appears to be some reduction in the sensitivity of the hypothalamus to androgen feed-back in photo-refractory birds (Nicholls and Storey 1976) and it may be supposed that the inhibition of LH secretion, no longer due to androgen, is mediated via thyroid hormones. The details remain to be determined. Storey and Nicholls (1976) working with Canaries, found that although plasma LH levels rapidly increase when photosensitive subjects are switched from short to long days, the mean plasma LH levels of photo-refractory birds are only slightly higher under long than under short photoperiods. It seems that the birds can still measure the photoperiod but the response to long days is diminished. Castration of photosensitive Canaries held on short days causes an elevation of plasma LH but this does not apply to castrated photo-refractory subjects kept on short days. Storey and Nicholls imagined that this was because photo-refractory birds produce little or no androgen, a view consistent with the supposed inhibitory effects of the thyroid at this stage of the cycle. Refractoriness can only be broken by the species-specific need for exposure to short photoperiods, and the first effect of reduced daylengths is probably the inhibition of TSH output and hence thyroid activity. This probably allows the production of androgen to begin. Storey and Nicholls noted that plasma LH levels of photo-refractory castrates rise suddenly to those typical of photosensitive castrate Canaries after 21—28 days of treatment. Since no increase in LH is noted in intact photo-refractory birds under the same light regime it may be guessed that the secretion of gonadotrophin is prevented in intact birds by the start of androgen secretion.

5. Following exposure of ducks to the shorter days of autumn there is a marked elevation in circulating androgen titres consequent on the continued presence of LH and reduction in the metabolic clearance rate of testosterone (Fig. 12.7).

High androgen titres at the end of the summer can in some species result in a resurgence of sexual activity but in the absence of FSH this androgen does not enter the tubule and so spermatogenesis does not begin immediately.

6. Further exposure to short days allows the new interstitium which was partly formed under the influence of LH secretion on long days to become rehabilitated with steroid precursor material and now the whole cycle is reinstated with exposure once again to stimulatory daylengths.

Evidence that some skeleton light schedules can prevent or inhibit rehabilitation of the testes and thereby prolong refractoriness comes from an experiment performed on House Sparrows collected from the field in July 1968, when they were spontaneously entering their refractory phase and their gonads were regressing (Murton *et al.* 1970*c*). Subjects were divided into three groups and exposed to skeleton light schedules of LD 6 + 1:17 simulating 8-h, 12-h, and 16-h complete photoperiods. After 35 days half the birds in each group were killed. The interstitial cells of those given an 8-h simulation, but not those given 12-h and 16-h schedules, had acquired lipoidal droplets. Moreover, although remaining small, the Leydig cells in the birds kept on an 8-h simulation exhibited an enlarged nucleolus, and contained numerous chromatin granules. In contrast, the interstitium was more expanded in the other two groups, owing to the proliferation of more Leydig cells, most of which contained small spindle-shaped nuclei. Subsequent exposure to a straight LD 16:8 for 34 days showed that the birds held under an 8-h simulation could now respond with full spermatogenetic recovery but the other two groups remained regressed. At the half-way stage of the experiment the tubules of birds held under a 16-h simulation contained much less post-nuptial sudanophilic lipid than those under an 8-h or 12-h simulation. After further exposure to LD 16:8 there were no tubule lipids in the groups previously given 8-h and 16-h simulations, although such lipids remained in the birds previously experiencing a 12-h photoperiod. It was suggested that there had been a tonic secretion of FSH for longer in the group receiving the light pulse 16 h from dawn. This cycle may have been interpreted as a 15-h day with the long light pulse occurring at the end of the birds' activity rhythm. The variable histological manifestations that can be induced by manipulating skeleton schedules emphasizes that several factors are involved in the photo-response mechanism. Turek (1972) subsequently performed a similar experiment with Golden-crowned *Zonotrichia atricapilla* and White-crowned Sparrows *Z. leucophrys* as did Sansum and King (1975).

During the 1970 series of experiments a 12-h light skeleton given for 97 days in a LD 6 + ½:17½ schedule allowed gonad rehabilitation and recovery from refractoriness, whereas in the experiment started in July 1968, and just described, a 12-h light skeleton of LD 6 + 1:17 given for 34 days did not break refractoriness. We attribute the differences partly to the time available in the two experiments for the Leydig cells to mature and partly to slight differences in the phasing of the endocrine rhythms consequent on the differences in light intensity in the different schedules.

In the temperate-zone species that have been examined the new generation of Leydig cells appears at the end of the breeding season at a time when long days are experienced yet gonad regression begins. In several species, for example, the Mallard Duck *Anas platyrhynchos* (Lofts and Murton 1968) and Turtle Dove *Streptopelia turtur* (Lofts *et al.* 1967c) it has been demonstrated that if the birds are artificially held on long photoperiods the interstitium becomes abnormally expanded with juvenile Leydig cells, but refractoriness persists. Farner and Mewaldt (1955) noted that the Leydig cells of refractory White-crowned Sparrows were larger than in birds ending their refractory state. Hypothalamic neurosecretion judged by aldehyde—fuschin staining has been shown to continue during the refractory period in White-crowned Sparrows (Laws 1961). Fig. 12.8 (see between pp. 82–3) illustrates the marked effect of keeping birds on long days at the end of the breeding season. Evidently the factors responsible for the differentiation of Leydig cells from fibro-blasts are to be distinguished from those which make the newly formed interstitium capable of secretory activity. As mentioned, LH seems to be secreted under 'long' days but thyroid hormones may be the factor inhibiting the full secretory activity of the Leydigs from developing. With 'short' days the Leydig cell cytoplasm accumu-lates sudanophilic lipoidal meterial, presumably because a shortage of LH inhibits rapid biosynthesis of the sterol precursors. The lipid is subsequently dispersed as the cell enters a secretory phase following exposure to long days (Fig. 12.8).

In the Pekin Duck kept on natural daylength cycles, and as above (Fig. 12.7), there was a peak in plasma testosterone in February but maximum plasma IR-LH titres were not noted until July, corresponding in time to gonad regression. At this time of the year a new generation of Leydig cells appears and the interstitium becomes expanded, presumably under the influence of LH. Yet, according to Tixier-Vidal (1963), the LH cells of the pars distalis are fully active in March, whereas the FSH cells are not stimulated until later in the season. Jallageas, Assenmacher, and Follett (1974) quote unpublished data by Nicholls, Lofts and Follett which describes how in Teal Ducks *Anas crecca* which were photoperiodically manipulated to produce an expanded interstitium without spermatogenetic activity, plasma immuno-reactive LH levels were high. Thus, high plasma LH titres seem to occur after the breeding season and are associated temporally with a stimulation of the interstitial cells. On the other hand, when Pekin Ducks were exposed in December to artificial long days (LD 18:6) plasma titres of LH and testoterone rose simultaneously, and there was a concomitant increase in testis size, that is, these various functions were not dissociated in time. Therefore, just as skeleton light schedules may lead to unusual phase relationships in hormone secretions and their action on target tissues, so can artificial photostimulation at inappropriate times of the year cause an unnatural endocrine mileau. Two groups of wild Mallard *Anas platyrhynchos* were held on either LD 8:16 or LH 16:8 photoperiod between June and November at which point half were killed and their pituitaries assayed for LH using the rat ovarian ascorbic acid depletion assay (see p. 38). The remainder were all exposed to LD 16:8 until February. Previously, a similar experiment had been performed by Lofts and Coombs (1965), without the LH essay, and they had demonstrated how long days maintained refractoriness and inhibited the post-nuptial moult. Table 12.1 shows that refractoriness was associated with low pituitary

LH titres but that photostimulation of subjects in which refractoriness was broken led to a high pituitary LH level. Clearly, a long photoperiod could be associated with high or low pituitary LH concentrations; plasma levels might well have been different, a high pituitary content indicating that less was being released and *vice versa*.

TABLE 12. 1

Pituitary LH concentrations (OAAD assay) in the Mallard according to photo-regime

Treatment	Pituitary LH (μg per 100 mg)	Post-nuptial moult	Peritoneal fat
LD 8:16 June–November	101·1 ± 15·2 (7)	Nearly complete	Much fat
LD 16:8 June–November	35·2 ± 65·6 (7)	In eclipse plumage	Trace
LD 8:16 June–November then LD 16:8 November–February	293·7 ± 35·7 (6)	Complete	Little fat
LD 16:8 June–February	11·2 ± 7·8 (5)	Only partly accomplished	No fat

Murton and Westwood, unpublished

In continuously breeding species it is likely that there is a steady renewal and rehabilitation of the interstitial component. In seasonal breeding species the manner of phasing of the hormones controlling gonad function results in marked seasonal changes in testicular histology, and Leydig cell rehabilitation appears to be very dependent on the light cycle and time. The degree of rehabilitation should be reflected in the capacity of the testis to respond to long days. This is apparent in Fig. 12.8 for the House Sparrow, which uses data presented by Underwood and Menaker (1970). Earlier, Hamner (1968) had exposed House Finches captured in the wild at different times during the autumn to LD 18:6 for 25 days, and showed how the birds initially were absolutely refractory but then gradually regained the ability to respond to these long photoperiods (Fig. 12.9). (Similar experiments were done even earlier on the White-crowned Sparrow by Laws (1961), but his results do not illustrate the phenomenon quite so clearly.)

Hamner imagined that the physiological circadian rhythm has two approximately 12-h phases of differing sensitivity to light, this basic rhythm being capable of phase shifting by laboratory lighting cycles (Hamner and Enright 1967). Cycles of 24 h with photoperiods of differing duration can alter the length of these phases of differential light sensitivity so that under some conditions they may have a 9:15 relationship and under other conditions a 14:10 relationship. During the 45 days of absolute refractoriness Hamner suggested that the clock was re-set, enabling the annual cycle to begin again. He considered that a partial test of his hypothesis was the demonstration that House Finches responded differently to skeleton light schedules according to whether they were captured and tested in autumn or spring (Fig. 12.1), but we have suggested another explanation for this observation. While

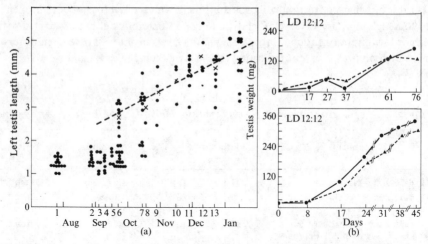

Fig. 12.9. Capacity of testicular recrudescence at the termination of photo-refractoriness in:
(a) House Finches *Carpodacus mexicanus* collected on the dates indicated and exposed to
LD 18:6 for about 22 days. Crosses are means for subjects with testes longer than 2 mm.
From Hamner (1968). (b) Testicular recrudescence in blinded House Sparrows *Passer domesticus*
(dashed line) and normal subjects wild caught and exposed to LD 12:12 beginning 15 November
(*upper*) or 7 February (*lower*). From Underwood and Menaker (1970).

Hamner clearly appreciated that refractoriness might be a phase shift phenomenon
he treated the testicular response as a unitary response. He had, nevertheless,
observed histological changes in the testis which give a clue that the cycle might
be controlled by more than one hormone rhythm. As mentioned, as many species
enter their refractory phase the testis tubules collapse and are occluded with dense
sudanophilic lipids. These gradually disperse during the autumn and the interstitial
cells acquire cholesterol positive lipid droplets. Similarly, in the House Finch the
post-nuptial tubule lipids disappear in mid-Summer and lipids then begin to accumulate in the Leydig cells (Fig. 3.4, p. 57).

   While we studied the gonad of the House Sparrow in England, Lofts and Lam
(1973) conducted a parallel study of the closely related Tree Sparrow *Passer montanus*
in Hong Kong.[†] Recrudescence of the gonads of the Tree Sparrow in Hong Kong
begins later than that of the House Sparrow in Britain. Both species respond in a
similar manner to skeleton photoperiods, as Fig. 12.1 indicates. Exact comparison
is difficult for ¼-h, and not ½-h, pulses were used in the Tree Sparrow studies and

---

[†] Both *Passer* species are sympatric in Europe and are separated ecologically by their habitat
preferences. Only the Tree Sparrow occurs in South-East Asia, where it occupies the habitat of
the House Sparrow, living as a commensal of man and using holes in buildings for nesting. The
breeding season of the Tree Sparrow in Britain begins about 15 days later and ends around 15
days earlier than that of the House Sparrow if the species are compared in the same locality
(Seel 1968a), although national records collected under the British Trust for Ornithology Scheme
do not reveal this difference (Seel 1964). If the photosensitivity threshold is set a little higher
in the Tree Sparrow this may allow it to have a shorter breeding season under a subtropical
photoperiod than could the House Sparrow.

these did not always fall at exactly comparable times as the ½-h pulses used for testing the House Sparrow.

The studies of Lofts and Lam are of interest in showing that light schedules which stimulated testicular growth also caused enlargement of the columnar cells of the epididymis and were correlated with Δ^5-3β-hydroxysteroid dehydrogenase (HSD) content of the right testis; this enzyme catalyses the oxidation of Δ^5-3β-hydroxysteroids to Δ^5-3β-ketosteroids (Baillie, Ferguson and Hart 1966) and indicates steroid biosynthesis. Light schedules which caused the most testicular growth appeared, therefore, to stimulate the maximum conversion of testosterone (Lofts 1975). During the natural breeding cycle of the Tree Sparrow HSD activity becomes noticeable with testicular enlargement (Chan and Lofts 1974). When these workers allowed for the size of the testis they estimated that maximum testosterone production would be during the breeding season, when testis size is at a peak (see inset to Fig. 9.2, p. 220). However, the conversion rate of testosterone per unit weight of testicular tissue, using tritiated pregnenalone precursor, indicated that the maximum rate of testosterone synthesis would not, in fact, be during the breeding season (see Fig. 9.2). Chan and Lofts (1974) data for the conversion rate of testosterone in testicular isolates from Tree Sparrows are approximately correlated on a seasonal basis with our measurements of plasma IR-LH levels in House Sparrows and, as Fig. 9.2 shows, these indicate considerable activity in autumn. Hence, testosterone production and testis size need always be correlated, just as we noted above that plasma LH and testosterone levels need not always vary in parallel.

Jallageas and Assenmacher (1974) found that during the first breeding season that they kept Pekin Ducks on a natural light cycle, plasma testosterone titres and testis weight were closely correlated, but in the second breeding season peak plasma testosterone titres occurred first, in February, before the testes achieved peak weight in July; the two functions became markedly dissociated. Green-winged Teal held in captivity in Hong Kong, instead of returning to Siberian breeding grounds, underwent a modified testes cycle on the local photoperiod and did not quite attain breeding condition. The peak of testosterone conversion occurred coincident with the attainment of maximum testis weight and this was even more noticeable in birds subjected to artificially long days which stimulated full gonad recrudescence (Lofts 1975).

The above results again raise the question (see p. 53) of whether the endocrine mechanisms mediating the synthesis of steroid precursors in the Leydig cells are the same as those responsible for actual androgen secretion by these cells. The histological changes in wild species at different seasons, in particular the build-up of sudanophilic lipids in autumn and winter in many species at the end of the refractory period, are impressive. This appears to be correlated in many species with high plasma levels of LH, as in the House Sparrow (see Fig. 9.2, p. 220), the Pekin Duck on a natural light cycle (see above), and probably also the Herring Gull (see p. 227).

Circadian basis of endocrine secretion

So far our evidence shows that different histological conditions can be produced by alterations in the light cycle. These seem to indicate a changing endocrine background and they correlate with plasma immuno-reactivity which it might be safest to refer to as an interstitial cell stimulating component. Because the photoperiodically induced growth of the gonads almost certainly depends on a circadian rhythm mechanism, it is logical to assume that the endocrine secretions controlling gonad development also have a circadian basis. Skeleton light schedules do not prove the existence of a circadian component but ahemeral light cycles in which any rhythm must first free-run in constant dark provide strong evidence (*cf.* Hamner's work in Fig. 11.11, p. 301). Recently, Follett, Mattocks, and Farner (1974) have measured IR–LH in white-crowned Sparrows in a similar resonance-type experiment and produced formal evidence for the involvement of a circadian oscillation and, incidently, for induction (Fig. 12.10). Similar results apply to Japanese Quail, although the pattern of IR-LH secretion in this species is evidently complex (Nicholls Scanes, and Follett 1973; Follett and Davies 1975). When young birds are held on LD 20:4 their testes gradually grow, first signs of growth being detectable after 3 days. Plasma IR-LH also increases by about 8-fold within 4 days and then stays at a plateau level for 30 days. The results confirm that LH and FSH manifestations occur

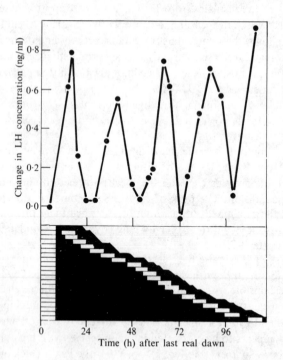

Fig. 12.10. Effect of an 8-h photoperiod given at various intervals, as in the lower panels, after entry into darkness on the plasma LH concentration of White-crowned Sparrows; subjects were previously maintained on LD 8:16. A pre-experimental blood sample was taken from all subjects early in the last 8-h period and then 7–16 h after the end of the test period and the ordinate plots the change between these two samples. From Follett *et al.* (1974).

coincidently in this species; it will be recalled that the Quail lacks a refractory period (see p. 279). In older Quail which had previously undergone a breeding cycle, exposure to a single long photoperiod was sufficient to cause an increase in gonado-trophin level on the first day. When sexually mature birds were exposed to non-stimulatory photoperiods (LD 8:16) plasma LH levels remained high for 8–16 days before falling, and testicular weight paralleled the changes in plasma gonado-trophin. As the authors' comment this result is not consistent with simple induction models of photoperiodism. Fig. 12.6 shows that on LD 20:4 secretion of IR-LH occurred 18–19 h from 'dawn' in sexually mature Quail previously held on LH 8:16 (Nicholls and Follett 1974).

Phasing of exogenous hormone treatment

Evidence that the phasing of hormone rhythms is important in controlling gonad cycles and other seasonal functions comes from experiments in which exogenous hormones have been injected at different times. When White-throated Sparrows were artificially made to go photo-refractory under 16-h photoperiods, two injections of prolactin induced fattening if given 5 h and 10 h after the start of the photoperiod, but two such injections reduced body fat levels if given at 0-h and 5-h (Meier and Davis 1967). Evidently in the first situation the exogenous prolactin could act synergistically with another hormone(s) while in the second it appeared too soon or even had some kind of inhibitory feed-back effect. Meier, Burns, and Dusseau (1969) have shown that levels of pituitary prolactin in this species are mediated by a circadian rhythm mechanism, as is also the plasma content of corticosterone (Dusseau and Meier 1971). In May a peak in pituitary prolactin content occurs around noon but in August it occurs at midnight. In May there is a 12-h interval between an increase in plasma corticosterone and release of pituitary prolactin but in August only a 6-h difference, Therefore, seasonal changes in the time relation of these two rhythms are noted between photosensitive and photo-refractory subjects. The difference in photoperiod between May and August is relatively slight and a phase change of 12 h in maximum pituitary prolactin content would hardly be expected if a simple entrained oscillator mechanism was involved. However, it is likely that adrenal corticoids are important synchronizers of daily rhythms (*cf.* Halberg 1969), and they have been shown to entrain the fattening response of the Feral Pigeon and White-throated Sparrow to prolactin (Meier, Trobec, Joseph, and John 1971*c*; Meier and Martin, 1971).

The sensitivity of the pigeon crop sac to systemic injections of prolactin given at different times of day also varies so that under a LD 12:2 midday injections were 2–5 times more effective in stimulating crop sac growth than earlier injections (Meier, Burns, Davis, and John 1971*a*). In another experiment in which injections were given for 4 days at 0, 3, 6, 9, or 12 h from the onset of light in a LD 12:12, a maximum response resulted with injections at the ninth hour and the minimum response occurred when injections were given at the onset of the light. In pigeons held on continuous light (LL) the peak of the crop sac response occurred at a different time from the peak of abdominal fattening (John, Meier, and Bryant 1972).

These authors showed that thyroid hormones alter but do not phase the rhythm of the fattening and crop sac responses; their data are summarized in Fig. 12.11. These refer to two injections of thyroxine which were given on days 1 and 3 followed by prolactin injections on days 5, 6, 7, and 8 at different times in different groups. These authors found that desynchronization of rhythms and damping out occurred in untreated subjects kept on continuous light for 15 days. In another experiment 3, 5, 3'-triiodothyronine was given daily and maintained the rhythms of fattening and crop sac responses to prolactin after they disappeared in subjects given only prolactin. Neither thyroxine nor triiodothyronine phase the response rhythms to prolactin though they may advance or delay the peaks in fattening and crop sac response rhythms so that the peaks occur at different times of day from those seen in controls. It may be noted in Fig. 12. 10 that the period lengths of the free-running circadian rhythms of responsiveness are altered by thyroxine treatment. Since it is a common observation that circadian rhythms disappear under continuous bright light, Meier *et al.* (1971*a*) suggested that continuous light may have an inhibitory effect on the thyroid. They noted the observation of Wahlstrom (1965) that triiodothyronine increased the period length of the circadian rhythm of locomotor activity in the Canary and suggested that if thyroid hormone secretion is inhibited by light it could explain the 'circadian rule' that in diurnally active organisms the

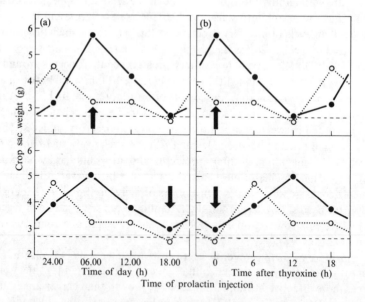

Fig. 12.11. Effect of thyroxine given at the times indicated by the arrows on the response of the pigeon crop-sac to injections of prolactin given at the times plotted (solid lines). Controls not given thyroxine shown as dotted lines. In the (a) the data are plotted chronologically but in the right-hand panels the same data are plotted with the time of thyroxine injection designated as hour 0. This emphasizes that the phasing of the rhythm is not altered compared with birds not receiving thyroxine, but the amplitude of the curve is clearly altered by thyroxine treatment. Dashed line is crop-sac weight for birds given no prolactin. Modified from John *et al.* (1972).

free-running rhythm become shorter in continuous light than continuous dark.

Although thyroxine and triiodothyronine did not phase variations in responsiveness to prolactin, corticosterone did as Fig. 12. 12, based on Meier *et al.* (1971*a*), shows for the peak in responsiveness varied according to the time of day corticosterone was administered (only the crop sac response is shown but the authors also obtained similar data for the responses in abdominal fat pad weight, liver fat, and intestine weight). Prolactin stimulated the greatest response in the crop sac when given 18 h after the injection of corticosterone, irrespective of whether corticosterone was given at 06.00 or 18.00 h (*cf.* Fig. 12. 10). As Meier and his colleagues concluded 'two or more hormones may control many gradations of several functions or conditions depending on the phase angles between them. It is not necessary that an off—on valve for hormone production be proposed to account for every effect noted involving the hormone.' For example, the timing of the fattening and crop sac responses differ although both require prolactin and it is the temporal relationship of prolactin release to that of corticosterone that determines what particular combination of responses is stimulated.

Later work by Meier and Dusseau (1973) has shown that injections of corticosterone into White-throated Sparrows held on LD 8:16 induced gonadal growth provided they were injected 18 h before the onset of the 6-h light period (Fig. 12.13). House Sparrows taken from the field in May were already in breeding

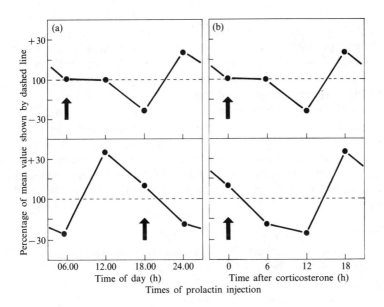

Fig. 12.12. Effect of corticosterone given at the times indicated by the arrows on the response of the pigeon crop sac to prolactin (solid lines). Data in (a) are plotted chronologically; in (b) the same data are plotted with the time of corticosterone injection designated as hour 0 to emphasize that the rhythm of response to prolactin injection is phased by corticosterone. From Meier *et al.* (1971*a*).

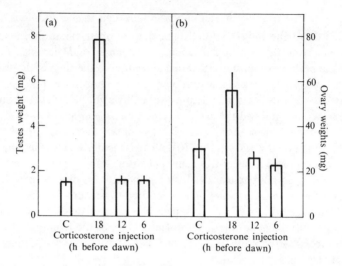

Fig. 12.13. Effect of corticosterone in setting the time of gonad ((a) testis; (b) ovary) sensitivity to light in White-throated Sparrows *Zonotrichia albicollis* kept under a non-stimulatory 6-h daily photoperiod. Injections of corticosterone were made for 13 days at 18, 12, or 6 h before the onset of light. Controls include 3 groups of saline injected birds and 1 group untreated. From Meier and Dusseau (1973).

condition and were kept on LD 16:18. The testes of controls which were not disturbed remained enlarged for the duration of the experiment whereas handling the birds (the bird was removed from the cage and its leg pricked), which might be thought to induce adrenal stress, caused gonad involution depending on when it was done (Fig. 12.14). A striking feature of the experiment was that although handling at dawn and dusk caused the same reduction in testis weight (but not handling at midday or midnight), bill colour was lost to a significantly greater extent if birds were handled at dusk. Since hormone synergisms are needed for pigmentation of the bill (see p. 226, also Lofts *et al.* 1973) it may be inferred that different hormone rhythms were being affected at the two times. Meier and Dusseau concluded that exogenous or endogenous corticosteroids altered the phasing of LH and FSH secretion.

Relation between circadian and circannual rhythms

In the next chapter we shall introduce the topic of circannual rhythms in breeding periodicity. It will help clarify the present discussion if at this point we introduce three models which Gwinner (1973) presented to suggest how circadian rhythms might be expressed as longer-term periodicities; Fig. 12.15 is a modification of his schematic presentation. The first model (Fig. 12. 15(a)) supposes that a circadian oscillator with a photosensitive phase assumes a variable phase relationship with the entraining light–dark *Zeitgeber*. This allows the light-sensitive phase to be coincident with light during summer so that a response can occur. This is essentially the

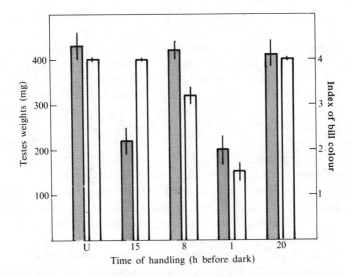

Fig. 12.14. Effect on testicular weight of handling photosensitive House Sparrows *Passer domesticus* for 6 days at 15, 8, or 1 h before the onset of dark when the subjects were kept on a LD 16:8. Unhandled birds were retained as controls. The index of bill colour (open histograms runs from 1 = grey to 4 = black. From Meier and Dusseau (1973).

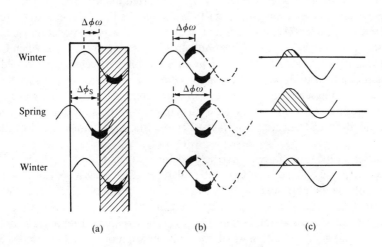

Fig. 12.15. Schematic diagram to show three hypotheses which derive circannual rhythms from circadian rhythms. In (a) the overt circannual rhythm results from changes in the phase-angle difference ($\Delta\phi$) between a circadian rhythm and its entraining day–night *Zeitgeber*. (b) from changes in the phase difference between two (coupled) circadian rhythms, perhaps consequent on a change in *Zeitgeber* and (c) from seasonal changes in the level of a circadian rhythm and hypothetical threshold; given coupled oscillators this might produce a similar result to (b). Slightly modified from Gwinner (1973).

Pittendrigh induction hypothesis except that Gwinner's model envisaged an endogenous change in oscillator frequency that causes a phase change without change in the *Zeitgeber* frequency. Like Gwinner, we reject this model because circannual rhythms persist in constant light when any supposed induction phase must be permanently illuminated. It is also difficult to account for refractoriness with this model.

The second model (Fig. 12.15(b)) assumes the existence of two oscillators which alter in phase relative to each other according to a circannual periodicity. The phasing of two oscillators might be controlled by variations in the daily light cycle. Alternatively, under constant illumination the phasing might change as a result of endogenously induced changes in oscillator frequency. If the phasing of two or more pituitary hormones were altered by feed-back from a recrudescing gonad (by the action of testosterone or oestrogen) further maintenance of active gametogenesis might be inhibited. Such an effect would be particularly pronounced if different gonad functions, for example, preparation of the interstitial component, meiotic division of primary spermatocytes, maturation of spermatids, and sperm release, necessitated a different phasing of gonadotrophic hormones relative to each other and to gonad steroids. The data presented in this chapter would favour this model.

Changes in oscillator frequency and hence phasing may result in changes in the level of an oscillator (*cf.* Fig. 11.1, p. 287). Gwinner introduced this as his third model, which we have incorporated as Fig. 12.15(c), but Gwinner considered only one oscillator to be involved. He went on to show a correlation between activity time in birds and gonad development which supported the idea that threshold changes could be important. We prefer to introduce the threshold concept into a two-oscillator model. This does not invalidate any of Gwinner's conclusions, but is in line with Daan's (1976) observations already mentioned (p. 292). When Gwinner and Turek (1971) transferred Starlings at different times of year from the wild to constant dim light or darkness the testes would either grow, regress, or stay the same depending on the season of transfer and the light conditions. But there was always a strong positive correlation between testicular size and the duration of daily activity. This reminds us of Menaker's studies (p. 301) and those of Hamner and Enright (p. 305).

Wolfson (1966) transferred Slate-colored Juncos from LD 16:8, which had stimulated gonad enlargement, to constant dark or LD 9:15 as a control. The gonads of birds on the 9-h photoperiod regressed, but those in DD remained variably enlarged. A bird whose daily activity cycle under DD remained similar to that under LD 16:8 and did not free-run before the end of the experiment retained a large testis, whereas another bird, in which the daily duration of activity shortened and the rhythm free-ran, immediately showed gonad regression. Daan's conclusion (see also p. 495) that there may be two coupled oscillations controlling the locomotor activity of finches, with opposite dependence of their frequency on light intensity, are of considerable interest in this context (see also Pittendrigh (1974)). It is likely that there are at least two factors regulating the gonad cycle and if these were shown to be coupled to the two components of the locomotor rhythm the model in Fig. 12.15

would prove to be realistic. Daan observed an increase of the morning peak of activity in spring and a forward shift in the onset of activity as if the 'morning' oscillator were specifically influenced by reproductive hormones. Daan pointed out that two mutually coupled oscillators of this kind would explain why in day-active animals activity starts later relative to sunrise, and ends earlier relative to sunset, with increase in daylength. But a complete 1:1 relationship between daylength and activity time is not found and this would follow if a coupling force between the two oscillators prevented them completely locking onto dawn and dusk respectively. Pavlidis (1973) has shown how a change in light intensity could result from a switch from first-order to second-order synchronization.

Species differences

The breeding season of the Wood Pigeon is rather different from that of the House Sparrow (*cf.* Figs.7. 11, p. 165 and 9.2, p. 220), yet casual inspection of Fig. 12.1 suggests that there is not much difference between the two species in their testicular growth response to asymmetric photoperiods.[†] Re-plotting the data as in Fig. 12.3 suggests that what we are calling the phase of photoperiodic induction occurred earlier in the subjective day and was of longer duration than in the sparrow. Two of 4 male Wood Pigeons and all females receiving a light pulse 14 h from dawn showed marked testicular development whereas others did not and it must be assumed that entrainment was different in these subjects. In fact, at the start of the breeding season most Wood Pigeons apparently responded to the 14-h simulation as if it was a 16½-h day by entraining to the short pulse, whereas all House Sparrows entrained to the long pulse. If photo-induction could be induced over a wide range of photoperiods in the pigeon it is conceivable that the level of an oscillator was higher (*cf.* more activity time) and so the frequency would be higher. Increase in frequency of an oscillator would make its phase angle to a *Zeitgeber* more positive so entrainment might be more easily achieved by a positive phase jump to the short light pulse. Fig. 12.3 also suggests that the supposed phase of photo-induction was less phase-shifted than in the House Sparrows with increase in daylength as simulated by the skeletons. This supports the interpretation that the long breeding season of Wood Pigeons, and their failure to enter a refractory stage under temperate photo-periods, results because the oscillators controlling the reproductive cycle are resistant to phase shifting.

The Feral Pigeon also has an extended breeding periodicity (p. 278) and Fig. 12.16 is of interest in demonstrating that its times of leaving and entering the roost do not vary much during the season (see also data in Miselis and Walcott (1970)) in relation to sunrise and sunset, whereas big phase changes are noted for the Jackdaw

---

[†] Gonad growth in the pigeons was recorded in terms of a growth rate constant based on the difference between log volumes (or diameter or follicle) at start and finish. It has been established that the logarithms to the base 10 of testis weight y and of testis volume x bear a straight-line relationship in the Wood Pigeon such that $y = 0.37 + 0.88x$; $r = 0.987$. This equation was a convenient means of calculating a growth-rate constant in subjects where gonad weight could not initially be obtained.

Corvus monedula: this has a gonad cycle which is very similar to that of the Rook (see Fig. 10.4, p. 255), being marked by a prominent refractory period; the same applies to the Starling (Bohnsack 1968). Since thyroid hormones are implicated in refractoriness, it may be wondered whether plasma titres of these do not reach high enough levels during the breeding season to inhibit the pituitary—gonad axis in these pigeon species. Or it is possible that it is the phasing of gonadotrophin and TSH secretion which is important. We suspect that when this question is answered it will also explain why the majority of pigeon species are exceptional in moulting throughout the breeding season. Another possibility is that stable phasing of prolactin

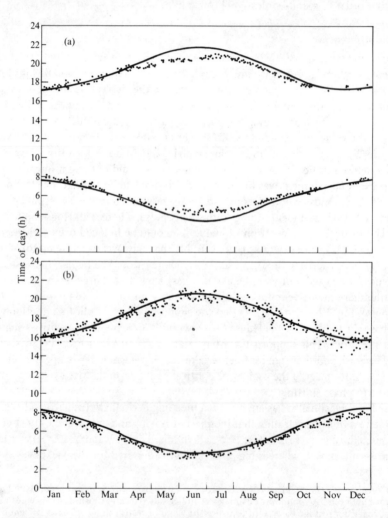

Fig. 12.16. Times of entering and leaving roost in relation to beginning of civil twilight in the morning and end of twilight in the evening for (a) Jackdaw *Corvus monedula* at 49 °N (from Aschoff and Holst 1960); (b) Feral Pigeon *Columbia livia* in Germany (from Hoffmann 1969).

secretion has enabled this hormone to become secondarily involved in the control of crop-milk production. In this context it is probably significant that in both the domestic fowl and pigeon, thyrotrophin releasing factor (TRF) causes both the release of thyroid stimulating hormone (TSH) and prolactin, but not of luteinizing hormone (LH) (Bolton, Chadwick, and Scanes 1973; Scanes 1974; Hall, Chadwick, Bolton, and Scanes 1975).

The Greenfinch has a longer breeding than the House Sparrow in northern Europe and Britain and egg laying continues until late August and September in the former compared with late July and early August for the latter. Comparison of Figs. 12.3 and 12.4 suggests that the phases of induction for testicular growth (and plasma IR-LH) do not alter so markedly with photoperiod length in the Greenfinch as they do in the House Sparrow. The Chaffinch *Fringilla coelebs* in Britain breeds from late April until mid-June and thereafter refractoriness develops. It is, therefore, of interest to compare the activity rhythms of the Chaffinch and the Greenfinch under exactly comparable conditions (Daan 1976). The Chaffinch exhibits a more pronounced phase shift in the onset and end of activity with the start of breeding in April, this phase shift persisting until the end of breeding in June so far as the activity onset is concerned, and until September with regard to the end of activity. The seasonal variation in phase angle difference between activity rhythms and light cycle is evidently much less in the Greenfinch.

It was seen in Fig. 10.9 (p. 269) that in Britain the gonads of Collared Doves *Streptopelia decaocto* assume breeding condition in March, although there can be substantial variation in the date at which individual birds achieve full gametogenetic recovery (p. 269). Experiments with skeleton photoperiods (LD 6 + ½:17½) showed that, as with House Sparrows, a 12-h simulation induced the maximum testicular growth and that a 14-h simulation was relatively ineffective, but there was much more variability in the response pattern (Fig. 12.1; Murton and Westwood 1975). It was mentioned (p. 269) that when wild-caught birds were transferred to LD 16:8 from the field in July, that is, sometime before natural gonad regression occurred in response to a fall in daylength in September, the birds could be kept in full breeding condition until December, whereupon spontaneous gonad regression occurred. Such subjects had exhibited a gonad cycle lasting about 9½ months on the assumption that they had first come into breeding condition in March. Fig. 12.17 shows what happened when Collared Doves, wild-caught in November and December, when they had regressed testes, were transferred to different unchanging light regimes and kept on these for 12 months; the size of the testes was measured at intervals by laparotomy. Of particular interest is the behaviour of the birds kept under constant dark for 12 months for they exhibited a gonad cycle which was very similar to that of birds on light—dark schedules. Clearly, the inductive action of light was not necessary for gonad growth and it seems likely that hormone rhythms free-ran to produce an autonomous periodicity lasting a little under 9 months.

In the next chapter it will emerge that birds do possess rhythms of body function having periodicities lasting many months which can be entrained by the 12-month oscillation produced through seasonal changes in daylength; such rhythms are often termed circannual rhythms and it is not yet clear how they might be compounded from circadian rhythms. In the absence of a controlling oscillation, the autonomous gonad cycle of the Collared Dove can be expressed under a wide range of constant lighting conditions, but, as Fig. 12.17 shows, the period length increases slightly with increase in light ratio; this presumably results because with an increase in light intensity the level and frequency of an oscillation increases. Not all species behave in the manner of the Collared Dove (cf. Fig. 9.5, p. 230), and the topic is discussed further in the next chapter. It may be that the response of the Collared Dove results because its oscillator system is stable to phase shifting under a wide range of *Zeitgebers*.

In the wild in Britain the approximately 9-month endogenous periodicity of gonad development in the Collared Dove is entrained by the 12-month seasonal light–dark

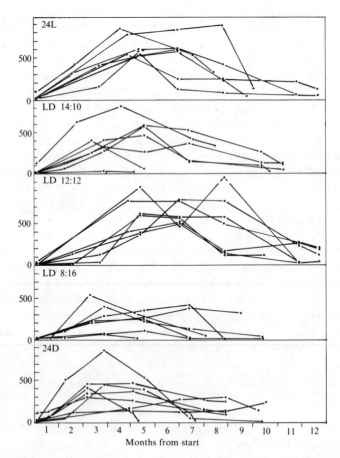

Fig. 12.17. Changes in testicular volume as measured by repeated laparotomy of individual Collared Doves *Streptopelia decaocto* held for 12 months on the constant photoperiods shown. Modified from Murton and Westwood (1975).

cycle and in the process the period of active gametogenesis is shortened to 6 months, and is almost exactly in phase with the *Zeitgeber* (Fig. 10.8, p. 269). When subjects are released from the entraining action of the *Zeitgeber* gametogenesis continues to produce a cycle lasting about 9 months (see above). For entrainment to work in this way the bird must be able to measure the daily photoperiod, and it is likely that this is the importance of induction. In the absence of induction hormone rhythms free-run to produce an engodenous periodicity of gonad function. With induction the daily rhythm can be phased to the *Zeitgeber*. It was mentioned that pigeons have relatively stable phase relationships over a wide range of light intensities. We guess that this is why they show little variation in gonad response when exposed to a wide range of skeleton simulations compared with species like the House Sparrow (see Figs 12.1 and 12.2). These last have breeding rhythms which are positively phased to the *Zeitgeber* (Fig. 9.2) so that breeding can occur in spring and early summer and cease after late July. Such birds show a wider range of response to skeleton photoperiods (see again Fig. 12.1).

It is a pity that insufficient data exist to enable the incorporation of the endocrine control of seasonal moulting schedules into this chapter. There is sufficient evidence to suspect that the phase relationships of pituitary, and the steroid hormones may alter seasonally and allow moulting at particular times but until diurnal plasma levels of all hormones are measured in a single species throughout the annual cycle it is impossible to construct an explicit hypothesis. However, it is possible to make interspecific comparisons to determine the extent to which easily measured periodic functions, such as moult schedules, gonad cycles, and seasonal fattening differ between species whose evolutionary relationships can be reasonably surmised and by such an empirical approach seek clues to the nature of the control system involved.

13

Autonomous and tropical breeding cycles

Summary

If birds are held under constant photo-regimes for more than a year they exhibit long term periodicities in gonad recrudescence and involution, moult cycles, migratory fat deposition and *Zugunruhe* behaviour. These cycles often have a period length of about 9 months and free-run. All the evidence confirms that these are endogenous periodicities which are entrained by the annual light cycle; amplitude is high in the temperates and low in equatorial zones. These so-called circannual oscillations seem to be related to circadian oscillations but the mechanisms involved are unknown. It is possible that circadian variations in photosensitivity are implicated in the process of entraining circannual rhythms and there is no justification for supposing that ultradian oscillators, as distinct from ultradian rhythms, exist. The time scales involved in experimenting with these longer-term rhythms hinder the rapid progress that could otherwise be made with simple and obvious experiments. It may be assumed that the endocrine rhythms which regulate these target tissue rhythms will themselves exhibit circannual periodicities but this has not been explicitly demonstrate although some data collected from domesticated strains of the Mallard *Anas platyrhynchos* are suggestive. Changes in the amplitude and frequency of the environmental light—dark cycle affect the entrainment of circannual rhythms. The way in which these free-run under constant conditions varies in different species with light intensity (DD, through LD 12:12, to LL) consistent with an effect of the light on the frequency and level of the circannual rhythm. Effects of this kind probably explain why the testis cycle of tropical birds is of lower amplitude than that noted in temperate regions, the gonads never achieving such a large absolute size in the tropic. Under tropical light regimes photoperiodic entrainment is less critical than in temperate zones so populations are often less well synchronized in their breeding rhythms. In *Phylloscopus* warblers there is evidence that a circannual rhythm of gonad, moult, and migratory activity is entrained under the photo-regime of the breeding ground and that it then free-runs in the relatively constant equatorial conditions experienced during the contra-nuptial season in Africa; the endogenous rhythm enables the bird to prepare for its vernal migration in the absence of marked environmental clues. Under some tropical conditions, under which ecological conditions remain fairly uniform throughout the year, free-running endogenous periodicities may be expressed; the 9-month breeding cycle of the Wideawake Tern *Sterna fuscata* being one example of several others that are described.

Autonomous cycles under artificial photoperiods

Autonomous or endogenous cycles of gonad activity have now been described for a wide range of temperate and tropical species. In temperate birds with a low photo-response threshold like the Starling *Sturnus vulgaris* an autonomous rhythm is uniquely expressed under a constant LD 12:12 and repeated cycles of testicular enlargement and regression lasting 9·5 months have been obtained (Fig. 9.5). As in other bird species, the normal testicular cycle involves a growth phase which can be induced under a fairly wide range of photo-regimes from constant light to constant dark. There follows a period during which active spermatogenesis is maintained, the length of this period depending on the amount of photostimulation already experienced. The end of this period is marked by spontaneous regression, so that stage 3 can be called the refractory period. The period following absolute refractoriness during which the capacity to respond to stimulatory light regimes is regained can be designated as stage 4, corresponding with the acceleration phase of Marshall (p. 217). As was discussed in Chapter 12 (p. 326) this period of rehabilitation is the time during which the Leydig cells develop a capacity for steroid synthesis and release.

Schwab and Rutledge (1973) have shown how in the Starling the refractory phase which develops under LD 12:12 lasts for 3 months and so is of similar length to that in subjects kept on natural daylength cycles (*cf. Quelea* below, in which the absolute refractory phase stays constant irrespective of the light cycle). A schedule of LD 12:12 appears to allow the Leydig cells to become rehabilitated whereas longer daylengths would inhibit this process. Fig. 13.1 compares the degree of recovery from refractoriness in subjects kept on natural seasonal photoperiods and LD 12:12. Recovery from refractoriness was measured by transferring subsamples to constant light (LL). Birds on natural photoperiods exhibited more testicular rehabilitation 135 days following involution (in May and June) than did birds under LD 12:12, presumably because by October the short daylengths (less than 12 h) were very favourable for interstitial rehabilitation. However, such daylengths do not stimulate much gametogenetic recovery. Under LD 12:12 rehabilitation was less complete but 12-h days did encourage more gametogenesis and so the recovery stage became advanced by 3 months to coincide with the stage that in wild subjects is manifested as the post-refractory photosensitivity gradient and that normally coincides with winter (see also Figs. 3.4, p. 57; 12.9, p. 326). The uniques feature of LD 12:12 is that it apparently allows a phasing of gonadotrophin secretion, which is conducive to both interstitial cell rehabilitation and to spermatogenesis. Periodic activity possibly results as gonadal steroids are released which at first combine with gonadotrophins to affect spermatogenesis but which with thyroid hormones also operate feed-back on hypothalamic centres to inhibit and possibly re-phase gonadotrophin secretion. Thus further interstitial cell production, or synthesis, is prevented until the system can re-cycle. Marshall was apparently correct in his view that the refractory phase times the breeding season of birds but he did not appreciate the importance of entrainment in modifying the refractory period. With photoperiods exceeding 12 h, Starlings are driven rapidly through the stage of active gametogenesis into a state of lasting refractoriness. In contrast, a photoperiod of LD 11:13 apparently allows the birds to stay in permanent breeding condition, while recovery approximating to the natural one is expressed under LD 9:15 (Fig. 9.5, p. 230). Recovery can occur in complete darkness (Rutledge and Schwab 1974).

The Slate-colored Junco *Junco hyemalis* probably has a higher photo-threshold than the Starling, judging from the timing of its natural breeding season, and under LD 12:12 it appears to remain in permanent breeding condition (Fig. 13.2); refractoriness rapidly develops with longer days and LD 9:15 causes only a slow response. *Zugunruhe* and fat deposition occur on LD 9:15, delayed by about 2 months; with continued exposure to this regime the migratory condition developed anew in the second spring of the experiment in one of the subjects (Weise 1962). We guess that there could be some regime under which an autonomous cycle is fully expressed. Wolfson (1955) found a somewhat variable response in this species

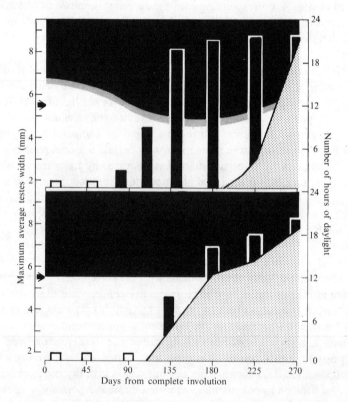

Fig. 13.1. Recovery of photosensitivity by Starlings *Sturnus vulgaris* following testicular involution into photo-refractoriness according to whether they were kept on a natural light cycle or constant LD 12:12. The x-axis gives the number of days elapsing since involution under the two light cycles. The filled columns give the mean testis size of sub-samples of birds which were removed from the two parent populations and were given a standard exposure to constant light (LL); open columns refer to subsamples that did not respond. The mean testis size of the parent populations under two light regimes is indicated by stippling. Based on Schwab and Rutledge (197

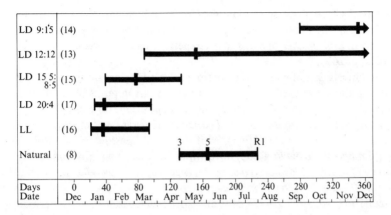

Fig. 13.2. Testicular cycles of Slate-colored Juncos *Junco hyemalis* in relation to the light cycle under which they were held. On each bar the thin and thick vertical lines indicate the points at which stage 3 and stage 5 (spermatozoa) of spermatogenesis, respectively, were attained. R1 indicates regression to stage 1 (resting spermatogonia) and arrows at the end of the bars mean that no regression occurred during the period shown. Number of birds in each group shown at left. From Wolfson (1966) based on Winn (1950).

under LD 20:4 for, although many individuals did remain refractory others displayed cycles of gonad growth and regression. Similarly, in *Zonotrichia leucophrys gambelii* held on LD 20:4 for 3 years a few individuals underwent two cycles, albeit attenuated ones, as if a periodic oscillation were damped, nevertheless the majority of the birds remained refractory (Farner and Lewis 1973). King (1968) has shown how there is also a cycling of migratory fattening and moult when this species is kept for 400 days on continuous LD 20:4, these events occurring approximately in parallel with their expression under a simulated natural photoperiod and with the cycle seen in wild subjects. Similar body-weight changes were expressed under LD 8:16 but moulting was inhibited. Only a small range of photoperiods provide conditions that allow an alternation of physiological events.

   In the equatorial *Quelea quelea* kept under constant LD 12:12 an autonomous periodicity becomes manifested (Fig. 10.2, p. 252). The testes remain fully functional for about 7 months, whereupon spontaneous refractoriness develops and persists for just over 1 month to be followed by a period of gradual recrudescence lasting about 4 months, that is, the entire cycle occupies 12 months (Lofts 1964). Some reservation is necessary since a 12-month periodicity could indicate an entrained rhythm. When exposed to north-temperate photoperiods of LD 17:7, a regime to which the species can have evolved no adaptive responses, refractoriness develops sooner than in subjects kept on LD 12:12 (Lofts 1962c). Nevertheless, refractoriness still lasts for about 1 month and spontaneous recrudescence occurs even though the daylength is held at 17 h. It seems likely that if *Quelea* were kept for long periods on LD 17:7 a periodicity of recrudescence and involution would become manifest although the frequency would differ from than seen with equal days and nights. When held under a continuous LD 16:8 the birds skip from one nuptial plumage to another without any eclipse phase being manifest (Rollo and Domm 1943). Mention should also be made of the double cycle of gonad recrudescence and regression every 12 months in *Zonotrichia capensis* when experiencing equatorial photoperiods (Fig. 10.7, p. 262).

The manifestation of endogenous periodicities under long-term constant environmental conditions has been recognized for many years in terms of moult cycles, development of migratory *Zugunruhe*, and fat deposition, for example in the Whitethroat *Sylvia communis*, Robin *Erithacus rubecula* (Merkel 1963), Dickcissel *Spiza americana* (Zimmerman 1966), White-crowned Sparrow (King 1968), Yellow Bunting *Emberiza citrinella*, Redpoll *Acanthis flammea*, Linnet *A. cannabina* (Stolt 1969), Garden Warbler *Sylvia borin*, Blackcap *S. atricapilla* (Berthold, Gwinner, and Klein 1972a), Chiffchaff *Phylloscopus collybita*, and Willow Warbler *P. trochilus* (Gwinner 1968a, b, 1971, 1972). It is of course much easier to make periodic measurements of these functions than to check gonad condition by repeated surgery and this has been done in some species, for example the Garden Warbler and Blackcap (Berthold *et al.* 1972b, Berthold 1974). It is a reasonable assumption that periodic cycles of gonad activity would also be found in the other species, and probably most

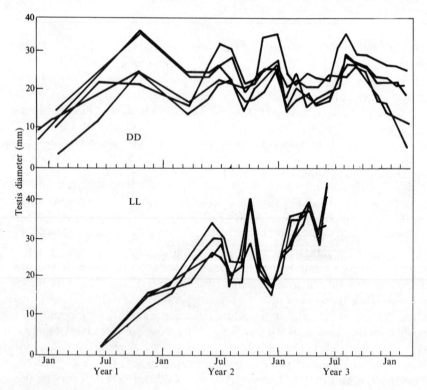

Fig. 13.3. Changes in testis diameter of 5 Pekin Ducks *Anas platyrhynchos* var. held for 3 years in total darkness (*upper*; from Benoit *et al.*, 1956b) and of 4 birds kept in constant light (*lower*; from Benoit, 1956a).

if not all bird species can exhibit apparently autonomous rhythms when kept under appropriate light regimes. Many of the periodicities approximate to a yearly cycle under constant conditions and so were called 'circannian' by Pengelley (1967). The term was subsequently changed to 'circannual' in agreement with the usage of German workers in the field (Gwinner 1968). Definitions and methodology requirements for demonstrating circannual, or other periodicities, are conveniently summarized by Reinberg (1974).

As mentioned in Chapter 9, Hamner (1971) objected to the suggestion that rhythms of the kind noted for *Quelea* or the Starling should be considered as autonomous or endogenous rhythms, for he pointed out that their expression depends on the nature of the *driving* light—dark cycle, that is, they represent responses to external light stimuli; he claimed that endogenous periodicities should be expresses as free-running rhythms only under conditions of total light or total dark. These were in fact the conditions used by Benoit, Assenmacher, and Brard (1956*a, b*) in some classic studies of Pekin Ducks, domestic forms of the Mallard *Anas platyrhynchos* (Fig. 13.3, and below). When exposed to constant light, cycles of testicular activity were expressed amongst 5 subjects but they became irregular, and the amplitude of change in testis volume was considerably less than in the normal cycle. Similarly, under constant dark, cycles of recrudescence and involution occurred and are discussed again below. Hamner's use of the word driving is the crux of the issue, for if the gonad rhythms are, in fact, entrained and driven by the light—dark cycle then his objection is justified. A significant feature of many of the rhythms which have been found is that they deviate from a 12-month periodicity and this is strong evidence that they are unentrained and hence free-running. A light cycle which lacks any amplitude, that is, all constant light—dark regimes, would not normally be expected to entrain an ultradian rhythm unless by frequency demultiplication, another point raised by Hamner.[†] Thus, just as an alternating LD 4:4 . . . can entrain a 24-h rhythm by frequency demultiplication so might an annual cycle be entrained from a 24-h light—dark regime. At present we do not know whether long-term periodicities depend on low-frequency oscillator systems, which are entrained by an annual light cycle, or whether they are compounded from higher-frequency circadian-type oscillations. If frequency demultiplication applies it could be predicted that the period length of the circannual rhythm is a function of the underlying circadian rhythm. Indeed, Gwinner (1973) has found that the circannual period of moult of 9 Starlings kept for 15 months in constant dim light was correlated with their circadian period of locomotor activity. The regression slope was not statistically different from 1 but was sufficiently large to suggest that the relationship was not proportional and this implies that a simple subharmonic entrainment is not involved. If coupled oscillators are involved in the circadian system quite complex relationships would be predicted. If the relationship noted in Starlings by Gwinner applies to other birds we should conclude that species like the Collared Dove have long breeding seasons—

[†] In oscillation physics if the frequency of an oscillator is near an integer submultiple of the frequency of the input, entrainment to the submultiple can occur this being known as subharmonic entrainment or frequency demultiplication—see Pavlidis (1973) for biological discussion.

ultradian rhythms—because they also have long circadian periods or naturally low-frequency oscillators; judging from Fig. 12.16 (p. 336) there is a less positive phase-angle difference in daily activity rhythm in pigeons, which have long breeding seasons, than in Jackdaws which have a short season.

The constant conditions necessary to allow a bird to exhibit free-running circannual rhythms of gonad or other body function vary from species to species. In some, as with the Starling (Fig. 9.5, p. 230), long days keep the subject in a permanent state of refractoriness, presumably because under long days the circadian rhythms controlling pituitary secretion remain permanently phased in a relationship which is not conducive to interstitial rehabilitation. Species whose various circadian rhythms maintain a fairly constant phase relationship under a wide range of *Zeitgeber* strengths might respond differently, as is suggested by the Collared Dove data in Fig. 12.17 (p. 338). While not conclusive, the records indicate that approximately the same, presumably endogenous, gonad periodicities are manifested under a wide range of constant light regimes, with a period length of about 9—10 months. Actually, the data suggest that the level of the oscillator system is increased with increase in light intensity, consistent with the expectation that the frequency of an oscillator should be increased with increase in light intensity (this might suggest that the circadian rule applies to circannual rhythms; see also Gwinner (1973)). The apparently endogenous gonad cycle of the Collared Dove, having an approximately 9½-month periodicity, is entrained by the amplitude of temperate seasonal photoperiodic changes to a period length of 8 months (see Fig. 10.8, p. 269). The fairly closely related Barbary Dove *Streptopelia risoria* kept captive in outdoor cages near Perth, Western Australia, over 5 years, exhibited an egg-laying cycle lasting around 9½—10 months (Davies 1974). It may be supposed that the annual light cycle at this latitude (30°S) provided a less effective entraining *Zeitgeber* than that at latitude 52°, so that the endogenous period length became manifest, although the rhythm was still held in phase with the environmental light cycle.

This is a convenient point to return to a consideration of the autonomous gonad cycle exhibited by Pekin Ducks. Fig. 13.4 extends the information in Fig. 13.3 in referring to control drakes kept for several years on natural daylight conditions (controls) and others maintained in constant dark. Also depicted is the typical monophasic annual testicular cycle of most Pekin Ducks and also a biphasic gonad cycle which is shown by a proportion of birds (D. H. Garnier quoted by Assenmacher (1974)). The two-peaked gonad cycle is reminiscent of the Tree Sparrow pattern depicated in Fig. 9.2 (p. 220) and also of the typical cycle of the Rook *Corvus frugilegus*, and those other species in which there is an autumnal recrudescence of sexual function which declines again in winter (p. 256, 240). In both the Tree Sparrow (Fig. 9.2) and the Pekin Duck (Fig. 13.4) the second peak in testis size was found to be associated with high androgen activity. In the Pekin Duck it seems that the Leydig cells in autumn have an ultrastructure typical of undifferentiated resting cells, suggesting that they do not contribute to testosterone secretion, whereas the Sertoli cells have the ultrastructural appearance of secreting cells and they could be the source of the testosterone (Garnier, Tixier-Vidal, Gourdji and Picart 1973).

This needs verifying. Assenmacher (1974) has performed a detailed analysis of the data summarized in Fig. 13.4, using the least-squares method of Halberg, Engeli, Hamburger, and Hillman (1965) and spectral analysis (Jenkins and Watts 1968). He notes that under DD there is an obvious tendency for the biphasic type of cycle to

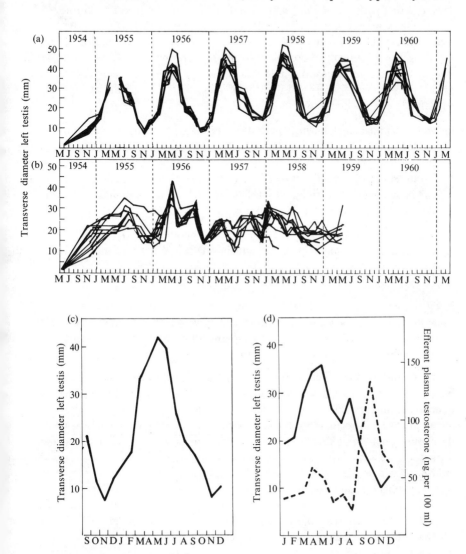

Fig. 13.4. Schematic representation of testicular cycles of Pekin Ducks *Anas platyrhynchos* var.: (a) 8 drakes held for 7 years under natural outdoor photoperiods (LD) in France. (b) 10 drakes reared in constant dark (DD) from age 18 days. (c) Classical annual monophasic type cycles of majority of Pekin ducks on a natural photo-regime. (d) Biphasic testicular cycle exhibited by a proportion of drakes on natural photo-regime (solid line) and the plasma testosterone content of such birds (dashed line). (a) and (b) are extensions of Fig. 13.3 above due to Assenmacher *et al.*, whereas (c) and (d) are new data presented by Assenmacher (1974).

emerge, such a pattern not really being evident in the control group (Fig. 13.4). The periodicity of the fluctuations in testis size seemed in consequence much shorter in DD that in the controls. Nevertheless, mathematical analysis shows that there was in reality a low-amplitude rhythm in the LD controls with a period of about 147 days, together with the 365-day annual periodicity. So even under synchronized LD conditions there is concealed a biphasic testis cycle. In DD the two periodicities are readily detected, a circasemestrial period of 156 days and a circannual period of 319 days. Interpreting these observations and the information regarding testosterone feed-back on the neurohypothalamic axis mentioned in the preceding chapter (p. 321), Assenmacher suggested that the breeding season of the ducks might be regulated in one of three ways:

1. By a circannual clock having a periodicity of 10–11 months, as seen in the free-running rhythm under DD. If such were the case it was postulated that the autumn resurgence of sexual activity, which frequently occurs in natural conditions and every second testicular cycle under free-run, results from a temporary inhibition of full testis activity by negative feed-back or other endocrine interactions which affect target tissues but not the clock itself.

2. By a circasemestrial clock with a 5–5½-month period. If such a clock was entrained to the seasonal daylength cycle an annual cycle would result of the classical monophasic type but if synchronization was incomplete a biphasic cycle could variably emerge.

3. By two endogenous rhythms fluctuating at circasemestrial and circannual periodicities. Assenmacher does emphasize, however, that his mathematical analysis did not isolate from the DD experiment a circannual periodicity that was independent of the circasemestrial period.

It is tempting to speculate that if circadian reproductive function depends on a system of coupled oscillators (p. 495) then annual rhythms might either be compounded from the same system or depend on the same principle. We notes how phase splitting of circadian rhythms can occur at low light intensities, possibly when the coupling between a driving light-sensitive pacemaker and a second pacemaker oscillator is weakened (p. 321). Now we find that at low light intensities (or perhaps in individuals with a 'high' photo-threshold) a biphasic annual cycle can become manifest Is this the phenomenon we have observed in *Zonotrochia capensis* (Fig. 10.7, p. 262) and will encounter again in the next chapter (see Fig. 14.4, p. 370). Presently we lack the information to answer these intriguing questions. Berthold (1974, for summar who has studied circannual rhythms of moult, body weight, and nocturnal restlessness on a comparative basis among *Sylvia* warblers has gonad data for Blackcaps and Garden Warblers. Under constant LD 10:14 and LD 12:12 some Blackcaps exhibited two peaks in testis size per year whereas this was never the case in the Garden Warbler (see Fig. 13.5). The Blackcap has a longer breeding season in Europe than the Garden Warbler and presumably a reduced photosensitivity threshold; it has rather a wider breeding range, tends to winter further north in Africa than the Garden

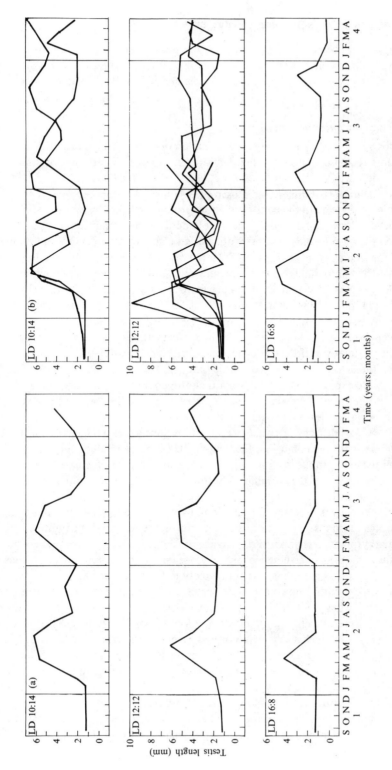

Fig. 13.5. Circannual rhythm of testis length in individual warblers kept for 3 years under the various constant photo-regimes indicated. (a) Garden Warbler *Sylvia borin* (b) Blackcap *S. atricapilla*. Based on Berthold (1974).

Warbler, and has been wintering increasingly in south-west Britain and Europe. In many ways this species pair show relationships with one another which parallel those discussed for Chiffchaff *Phylloscopus collybita* and Willow Warbler *P. trochilus* below.

Entrainment of circannual rhythms

It has not yet been shown whether manipulation of the entraining seasonal light cycle can effect circannual periodicity in birds in a manner comparable with the manipulation of circadian periodicity, but Goss (1969*a*, *b*), Goss and Rosen (1973), and Goss, Dinsmore, Grimes, and Rosen (1974) have done appropriate experiments on the photoperiodic control of antler growth in Sitka Deer *Sika nippon*. These have a rhythm of antler replacement under constant conditions of LD 24:0, LD 16:8, LD 8:16 for nearly 2 years, the period of the rhythm being about 85 per cent of the sideral year. Under LD 12:12 the rhythm disappears and antlers are not replaced. The approximately circannual and apparently endogenous periodicity is entrained to an exactly annual rhythm, given only a slight change of seasonal variation in amplitude of the annual light cycle. Thus, the duration of light and dark can vary by as little as LD 11½:12½ to LD 12½:11½ between summer and winter (equivalent to about 9° latitude.) in order to effect entrainment to a precise annual periodicity. Moreover, if the frequency of the prevailing photoperiodic cycle is increased the deer entrain and grow more than one set of antlers in a year; in fact, the deer were exposed to 0·5, 2, 3, 4, or 6 photoperiodic cycles in one year and were able to entrain to those in which the *Zeitgeber* frequency did not exceed one calendar year, nor was it less than a quarter of a year.

Goss's studies demonstrate that circannual rhythms are entrainable, according to the photoperiodic cycle, but that there are limits to the period length of the *Zeitgeber* to which the endogenous rhythm can respond. Moreover, with increase in frequency (decrease in period) of the entraining photoperiodic cycle there was found to be an increasingly negative phase-angle difference between the antler-growth rhythm and the photoperiod. Like the gonad response to light, antler growth in deer represents a target-organ response which probably depends on the temporal synergism of more than one rhythmic mechanism. For instance, it is possible to conceive that gonadotrophins and TSH can assume a phase relationship under free-running conditions which lead to gonad growth and steroid secretion. If steroid feed-back then rephases the gonadotrophin rhythm gonad maintenance could be inhibited. The cycle might continue only when the feed-back inhibition was removed and appropriate phasing of hormone secretion allowed the differentiation of new interstitial cells and the system to re-cycle. Such a scheme is covered by model (b) in Fig. 12.15 (p. 333). Circadian rhythms are universally found, from single-celled organisms to the most complex metazoa. But primitive forms like Protozoa and bacteria have life cycles that rarely last more than a day, let alone a year, and this has led various people to question how adaptive circannual rhythms could have arisen. If circannual rhythms are compounded from two or more circadian oscillators it may be unnecessary to postulate the existence of single, low-frequency pacemaker oscillators. Pohl (1971) found that regular changes occurred in the phase relationship of various circadian functions (perch-hopping activity, oxygen consumption) and the temporary dissociation of single circadian components of one function (activity) in consequence of variation in environmental conditions. He also noted that an endogenous circannual rhythm could alter the relationship between the different rhythmic components of activity. His data accord with the view that rhythmicity in vertebrates depends on a system of coupled oscillators.

Self-selection of photoperiod

Other evidence for the existence of endogenous ultradian periodicities has been obtained from subjects kept in experimental cages where they can select their own photoperiod: they may have a choice between a light and dark area (Wahlström 1964, 1965, 1971; Gwinner 1966; Aschoff, Saint Paul, and Wever 1968) or in some experiments have had to pass a simple maze before activating a light switch by settling onto a perch (Heppner and Farner 1971a, b). These last authors found that White-crowned Sparrows were cyclic in their choice of daily photoperiod, changing between sequences of long and short days, with the period length of a full cycle varying around 100–120 days. A group of photosensitive birds initially chose long photoperiods, and this induced a cycle of testicular development. However, a second cycle of gonad development was not induced during a second cycle of selection of long days. Five out of 8 photo-refractory birds also initially selected long days and this of course failed to break refractoriness and so no gonad growth occurred. But two subjects did select an initial bout of short days and became photosensitive and thereafter exhibited gonad growth. There was some evidence that the total hours of photostimulation received per circadian period S was inversely related to the circadian period τ and many cases of apparently spontaneous changes in τ were noted. Aschoff pointed out that under a self-imposed light–dark cycle the circadian period usually lengthens. In the experimental design followed by Heppner and Farner the birds had to have a long activity time to get a long photoperiod and since a long activity implies high circadian frequency the birds could find themselves in a conflict situation.

From the discussion so far we conclude that birds possess endogenous free-running long-term rhythms, but not necessarily ultradian oscillators, which are variably responsive to entrainment according to species. In most highly photoperiodic species which have short to medium breeding seasons in the temperate zone the endogenous periodicity is readily entrained by a wide range of environmental light–dark cycles and considerable synchrony is apparent in the phasing of the rhythms to the *Zeitgeber* between different individuals in a population. In tropical regions the amplitude of the seasonal light–dark cycle may be sufficient to entrain endogenous rhythms to an annual periodicity, but there may be much intraspecific variation in the phasing of the cycles. Given nearly constant equatorial photoperiods and no entrainment the rhythms of individual birds may free-run with a period length differing from 12 months. There must then exist a tendency for the cycles of individual birds to become out of phase with each other, unless some additional mechanism for ensuring synchrony arises for ecological reasons. These mechanisms will now be discussed and we begin by considering the importance of endogenous factors in the timing of annual cycles in temperate species which are migrants to equatorial regions. During the contra-nuptial season such birds receive little or no environmental cue about when they should return to the temperate zone to breed.

Autonomous periodicity in migrants

Gwinner kept Palaearctic leaf warblers under constant light conditions for long periods, in some instances nearly 3 years, during which time he regularly monitored the body weight, incidence of moulting, and development of migratory restlessness. Initially be experimented with Willow Warblers and Wood Warblers *Phylloscopus sibilatrix* (1967, 1968a) but the work considered here refers to the Willow Warbler and Chiffchaff (Gwinner 1968b, 1973). The summer and winter range of these two species was given in Figs. 7.7 (p. 160) and 7.8 (p. 162) and their origins discussed (p. 161). In Fig. 13.6 the light cycles these species experience under natural con-ditions are plotted in relation to the natural chronology of moulting and breeding. This enables comparison with some of Gwinner's results, which have been simplified by giving only the average results for groups of birds held under various artificial light regimes (see Fig. 13.6). It is immediately clear from an inspection of Fig. 13.6 that the Willow Warbler, unlike the Chiffchaff, is faced with a special problem during its winter sojourn in Africa, for it is inconceivable that changes in the daily photo-period could serve as a proximate signal to indicate when the birds should depart on the northward journey to the breeding quarters.

Willow Warblers kept on natural photoperiods in Germany, and others transferred to constant LD 12:12 in September, essentially exhibited a natural cycle of moult and *Zugunruhe* (Fig. 13.6). So too did birds which were held in Germany until September before being transported to the Congo (latitude 2°14'S), where they were kept until the following June on the local photoperiod. It appears then that the cycle of moult, fat deposition, and *Zugunruhe* (and presumably also the beginnings of gonadal recrud-escence) can continue in subjects held on medium to short days and Gwinner argued that the responses were adaptive in that they would help to compensate for a lack of reliable proximate timing devices in the quasi-stable equatorial environment frequentec during the contra-nuptial season. However, this response is evidently disturbed by a long photoperiod for when Willow Warblers were placed on LD 18:6, autumn *Zugunruhe* ended early and winter moult was also advanced by about a month, and thereafter bouts of nocturnal activity interrupted or accompanied by moulting occurred arrhythmically. Chiffchaffs behaved differently from Willow Warblers when kept on LD 12:12, for their *Zugunruhe* behaviour became irregular and the phasing of moult, body-weight change, and *Zugunruhe* became disorganized and variable between individuals. Moreover, a pre-nuptial body moult, which is normally not always expressed, became prolonged in some individuals.

Some Willow Warblers were kept for 27 months on LD 12:12 whereupon the rhythm of nocturnal activity and moult of some individuals was seen to free-run with a period length of about 10 months, strong evidence for the involvement of endogenous factors. In Chiffchaffs kept for an extended time on LD 12:12 seasonal

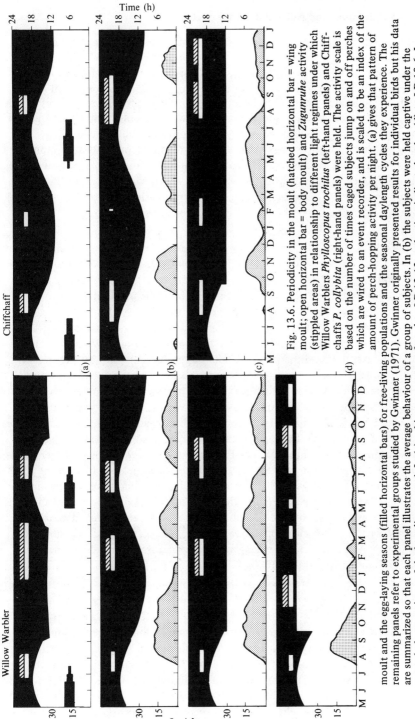

Fig. 13.6. Periodicity in the moult (hatched horizontal bar = wing moult; open horizontal bar = body moult) and *Zugunruhe* activity (stippled areas) in relationship to different light regimes under which Willow Warblers *Phylloscopus trochilus* (left-hand panels) and Chiffchaffs *P. collybita* (right-hand panels) were held. The activity scale is based on the number of times caged subjects jump on and off perches which are wired to an event recorder, and is scaled to be an index of the amount of perch-hopping activity per night. (a) gives that pattern of moult and the egg-laying seasons (filled horizontal bars) for free-living populations and the egg-laying seasons (filled horizontal bars) for free-living populations studied by Gwinner (1971). The remaining panels refer to experimental groups studied by Gwinner (1971). Gwinner originally presented results for individual birds but his data are summarized so that each panel illustrates the average behaviour of a group of subjects. In (b) the subjects were held captive under the natural photo-regimes of the breeding ground. In (c) subjects were transferred to LD 12:12 at the time indicated and at (d) to LD 18:6. In Willow Warblers the apparently entrained cycle of moult and *Zugunruhe* could be maintained under constant LD 12:12 but it was disrupted by LD 18:6. The seasonal rhythm of moult and *Zugunruhe* in the Chiffchaff was disrupted by LD 12:12. Modified from Gwinner (1971).

changes in moult, *Zugunruhe*, and body weight were lost after the first year, indicating that any endogenous component timing these events is inhibited by this photo-regime. Clearly, when Willow Warblers arrive in equatorial Africa physiological preparation for the spring return occurs at an endogenously controlled rate. Indeed, once initiated moult must require a given time for completion and Pearson's (1973) data even suggest that the ending of the winter moult could be a trigger for beginning the spring passage. The endogenous timing component had to operate over a relatively short time interval, so that there is a restricted opportunity for any individual variability in response rate to be manifested; no information is available regarding the degree of synchrony in departure date of wild Willow Warblers when leaving Africa for their nuptial quarters. However, it is clear that the birds must rely on having their 'clocks' re-set during the time they are exposed to summer photoperiods.

It is likely that the Chiffchaff has a lower photo-response threshold than the Willow Warbler, for it begins breeding earlier and is frequently double brooded, unlike the Willow Warbler. In March and April the Chiffchaff experiences rapidly increasing photoperiods exceeding 12 h which appear to inhibit the pre-nuptial or spring moult, but this becomes more clearly expressed under LD 12:12. The equivalent moult in the Willow Warbler is very developed and normally occurs as a complete moult in Africa in February. We speculate that since the Willow Warbler has a higher photo-response threshold than the Chiffchaff exposure to a LD 12:12 represents a relatively short day and allows an appropriate hormone phasing. The post-nuptial moult of the Willow Warbler which is expressed in Britain and southern Europe is lost in northern populations, perhaps because it is inhibited by exposure to long sub-Arctic summer photoperiods. So too in the European sub-species when exposed to a LD 18:6 (Fig. 13.5). The unusual double moult of some Willow Warbler populations is apparently a consequence of the phasing of the underlying hormone control mechanism. Since the Willow Warbler is a recent species, perhaps of inter- or post-glacial origin, it is conceivable that the phasing of the endocrine system regulating moult is not yet perfectly adjusted to the range of *Zeitgebers* experienced throughout its range and the double moult may not represent an ecological adaptation. Certainly the more northern populations, in which the seasonal light cycle prevents a complete post-nuptial moult, manage an even longer migratory flight without benefit of new feathers (see p. 161).

It is now established in several species kept for long periods under constant photo-regimes, that the rhythms of moult, migratory fattening, and gonad enlargement may become out of phase with each other enabling, for example, moult to occur coincidently with gonad enlargement. This emphasizes that these functions must be controlled by a distinct endocrine *milieu* and that their normal temporal associations need not imply causal interrelationships. This is further evidence for the view that rhythms of synthesis and release of the pituitary hormones at least are controlled to some extent by separate oscillator systems which can be phased independently of each other. Given natural light cycles these oscillations are phased adaptively.

Breeding cycles of tropical land birds

Many equatorial birds probably have gonad cycles of the type shown in Fig. 7.4(*c*) (p. 154) which refers to the Yellow-vented Bulbul *Pycnonotus goiavier* on Singapore Island (Ward 1969*a*). This is certainly the case in Jamaica (lat. 17 °N) (Diamond 1974) and in much of South-East Asia, where most land birds breed at the end of the heavy rains which are brought by the north-east or north-west monsoon. Hence, peak nesting by all species occurs from February to June in Malaya (Ward 1969*a*), between January and June in Sarawak at latitude 1–5 °N (Fogden 1968, 1972), between January and March in west Borneo on the equator, and between March and July in Java at latitude 6–8 °S (Voous 1950). In these birds an endogenous rhythm is presumably entrained to a 12-month periodicity by some environmental *Zeitgeber*. Although daylengths in Singapore (1°20′N) do not vary appreciably, the light intensity does change and twilight effects could be important in producing an amplitude to the succession of day and night, thereby making a weak *Zeitgeber*. A weak *Zeitgeber* implies weak entrainment, which doubtless accounts for the large measure of individual variation in gonad condition of bulbuls within the population exemplified by Ward's original data. Nevertheless, a degree of synchrony is imposed on this population, which results in a clearly defined breeding season followed by moulting. Heavy rains are brought to the area in February–June and insect food then becomes more plentiful. Breeding occurs between February and July and moulting between November and April. Ward showed that there was a decline in the fat-free dry weight of flight muscles, as well as in the total fat-free body weight, between May and August and he established that the moult of the primaries was associated with a decline in weight of the flight muscles. This led him to the view that food supplies might be critical in determining when the birds could assume a breeding condition and that the improvement in insect supplies with the monsoon provided the protein which directly led to growth of the gonads. Perhaps an improvement in food supplies facilitates courtship, which in turn stimulates the endocrine apparatus and helps synchronize different individuals. This is feasible. For example, Gwinner (1976) has kept male Starlings for long periods on constant light with and without a female in the same cage: the presence of females has been shown to alter the period of the endogenous gonad cycle, effectively advancing the onset of sexual recrudescence. To return to the bulbuls, moult occurs with a more regular seasonality than the onset of breeding and with less variability in the population. Ward's figures show a significant inverse correlation between gonad size and moult stage, as if energy cannot be used for both, but there need not be a causal relationship. However, it is possible that entrainment of an endogenous moulting cycle is more precise than that of the gonad cycle and that breeding is fitted-in between moult sequences depending on the availability of food resources.

The suggestion that the moult cycle may be timed by some factor (endogenous or exogenous) and that it in turn imposes a periodicity on the reproductive cycle was first made by Snow (1962*b*). During a 5-year study of the Black and White Manakin *Manacus manacus* in the tropical forest of Trinidad, he found that the time of onset of breeding varied by up to 5 months in different years, sometimes beginning in January and sometimes not until late May, and it probably depended on variations in the weather and food supply. In contrast, the moulting season was virtually the same in all years and for all individuals; moult mostly started in August and lasted about 80 days. Moreover, young birds, which may have left the nest in any month between January and October, all moulted into adult plumage in the months June–September of the year following their year of birth. This suggests that the moulting season is phased by an external regulator, and that breeding begins as soon as conditions become suitable, once the moult is finished. Male Manakins indulge in communal displays and hold 'courts' or 'leks'. Displays continue throughout the year and the males only devote about 10 per cent of the available time to finding food. Although they are in a state of readiness to breed throughout most of the year, the amount of display declines during the moulting period. The period of most intensive breeding activity was found to correspond with the season of maximum food supplies, the fruits of berry-bearing trees of the families Melastomaceae and Rubiaceae being important. Most other Trinidad birds, including sea, swamp, and land birds, moult at approximately the same season as the Black and White Manakin (Snow and Snow 1964). This corresponds to the onset of the wet season but it is unlikely that rain *per se* could be the proximate regulator. The Snows also thought it unlikely that the seasonal photoperiod could initiate the moult, but, in fact, it is possible that the whole annual photoperiodic cycle does drive and entrain a moulting season in the birds. Daylengths vary between 12·3 h and 13·5 h at latitude 10°, which is a sufficient amplitude judging by Goss's experiments on Sitka Deer employing a cycle ranging from 11·5 h to 12·5 h of light.

Colour-marked individuals of the hummingbird *Phaethornis superciliosus* of Costa Rica moulted at the same time in different years, usually within two weeks, but the moult cycles of individuals were up to 6 months out of phase (Stiles and Wolf 1974).

Tropical species generally have more extended seasons of gonad enlargement than related temperate forms and, as already mentioned, the moult usually takes longer (p. 155, also Fig. 13.7). An extension in the season of active gametogenesis is associated with a decrease in amplitude of the cycle and this is reflected in the difference between maximum and minimum testis size. This was first noticed by Moreau *et al.* (1947) when they studied three African species near Amani, Tanzania (latitude 5°S). They examined the Black-capped Bulbul *Pycnonotus xanthopygos micrus*, which lives in areas of bush and wooded grassland outside the forest, the Yellow-streaked Bulbul *Phyllastrephus flavostriatus tenuirostris*, living inside evergreen forest, and also the non-passerine forest-dwelling Speckled Mousebird *Colius striatus mombassicus*. In *Colius* there was an 8-fold change in total testis volume between the maximum and minimum and about 30-fold difference in *Pycnonotus*. Similarly, a 16-fold difference in the seasonal extremes of testicular weight have been recorded in the white-eye *Zosterops vatensis*, a 31-fold variation in the Golden Whistler *Pachycephala pectoralis* (Baker 1929; Baker, Marshall, and Harrison 1940) and only about 2-fold for the Eared Dove *Zenaida auriculata* (see Fig. 7.11, p. 165). These

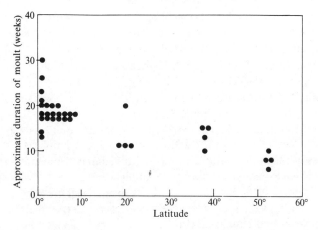

Fig. 13.7. Relationship between duration of moult in weeks and breeding latitude for 34 different passerine species (there are 35 points because in one case, *Cassidix mexicanus*, the moult of the male and female was different). From data given by Fogden (1972).

data contrast with some typical figures for temperate birds for the testes increase up to 1100-fold by weight in the House Sparrow (Fig. 9.2, p.220), 929-fold in the Red-winged Blackbird *Agelaius phoeniceus* (Wright and Wright 1944). In the bulbuls studied at Amani moult began variably in different individuals between December and April but was regular in any one individual and lasted about 5 months, but it extended over 7 months in the population as a whole. The first primaries were shed towards the end of the breeding season, when the gonads were still expanding, and moult continued as the gonads regressed to a stage where they contained only resting spermatogonia, a cycle more similar to a temperate species. In *Colius* the moult was irregular and was independent of the state of the gonads so that feather replacement continued throughout the breeding season.

Breeding cycles of tropical sea-birds
Several species of oceanic sea-bird frequenting equatorial waters experience virtually no seasonal variation in food supplies or other environmental factors and their populations remain saturated relative to resources. Moult and reproduction impose conflicting demands for energy and must be partitioned, but in the absence of seasonal changes in the environment breeding may occur at any time and individual birds alternate periods of breeding and moulting at the maximum rate of which they are physiologically capable. In species such as these, which have been evolved in a stable environment, it must be assumed that the rate of reproduction which is set endogenously is well adapted to ecological conditions.

   In these tropical oceanic birds the frequency with which the cycle of breeding and moult can be accomplished varies: 8 months in the Brown Booby *Sula leucogaster* on Ascension Island (latitude 8°S, longitude 14°W) in the Atlantic (Dorward 1962); 8 months for the Bridled Tern *Sterna anaethetus* in the Seychelles (Diamond 1976);

9 months for the Yellow-billed Tropic Bird *Phaëthon lepturus* on Ascension Island (Stonehouse 1962), Audobon's Shearwater *Puffinus lherminieri,* and the Brown Pelican *Pelecanus occidentalis* in the Galapagos Islands (latitude 0° longitude 90°W) (Snow 1965a; Harris 1969a, b, 1970); 10 months for the Swallow-tailed Gull *Creagrus furcatus* in the Galapagos (Snow and Snow 1956); 11 months for the Red-billed Tropic Bird *Phaëthon aethereus* on Ascension (Stonehouse 1962); just under 12 months for the Red-tailed Tropic Bird *Phaëthon rubricauda,* Christmas Shearwater *Puffinus nativitatus,* and Phoenix Petrel *Pterodrama alba* on Christmas Island (2°N, 157°N) in the central equatorial Pacific (Schreiber and Ashmole 1970). Actually those cycles in which the periodicity is close to 12 months may not be autonomous but they provide a comparison with those of higher frequency.

Individual pairs of those species having less than annual breeding cycles may lay during any month of the year independently of each other, or their breeding seasons may be synchronized. Thus egg laying has been recorded in all months of the year in several species including Audobon's and Christmas Shearwaters, for the Blue-faced Booby *Sula dactylatra* on Christmas Island (Schreiber and Ashmole 1970), and for the Tropic Birds of Ascension Island. Stonehouse (1962) showed that those individual Yellow-billed Tropic Birds that were successful in breeding laid again 35—47 weeks (mean 38 weeks) later, nearly one-third of the population behaving in this way. One egg is laid and incubation lasts about 41 days. Chick care requires another 75 days after which the adults can leave the breeding grounds and moult at sea, this taking about 126 days (16—20 weeks). This budget leaves about 28 days for the adults to return to the nesting station and begin courtship, egg formation, and laying. Two-thirds of the population had its breeding cycle disturbed because of the loss of an egg or chick to predators, usually within 12 weeks of initial courtship. If fresh eggs were lost a few such pairs could lay another egg about 23 days later, whereas if a 1—2 weeks old chick was lost the birds did not re-lay for about 80 days. But, only 10—12 per cent of the birds re-nested if unsuccessful and the majority left the breeding colony and completed the post-nuptial moult before returning. Hence, unsuccessful breeders may re-lay anything from 23 weeks to 39 weeks from the start of the first cycle. The highest rate of loss of eggs or young was noted at times when large numbers of the Red-billed Tropic Birds *P. aethereus* were also nesting, so that peak numbers of the two species combined were competing for nesting-sites and interfering with each other. Red-billed Tropic Birds are larger than the Yellow-billed species and while both have similar breeding cycles that of the former is more extended; 2 days extra are needed for incubation, the birds require 25—30 per cent more time to rear the chick and take 30 per cent more time over the post-nuptial moult. *In toto* the cycle is 48 weeks (20 per cent) longer in *aethereus* than in *lepturus.*

In both of the Ascension Tropic Birds egg laying became clumped at some times of year, even though any particular pair showed no evidence of a seasonal influence. Stonehouse distinguished two processes which tend to encourage a uniform pattern of egg laying. The first depends on individual variations in the length of cycle of successful and unsuccessful pairs and the second on the inverse relationship between breeding success and breeding density. Acting in the opposite direction is the tendency for successful breeders to return to the colony during the same period so that slight peaks in the annual distribution of egg laying result, the periods between peaks or troughs representing the length of the successful cycle: 8—9 months in the Yellow-billed and 11 months in the Red-billed. The fact that Red-billed Tropic Birds returned more or less synchronously to begin breeding caused peaks in the rate of disturbance and egg loss in the smaller Yellow-billed Tropic Bird and this became reflected in a degree of synchrony in their production of repeat layings. Populations of Tropic Birds which are particularly unsuccessful in their breeding attempts should, as a result of repeat layings and a uniform basic cycle, come to exhibit unsynchronized nesting and damped oscillations in breeding periodicity. In contrast, successful populations should display a marked periodicity of breeding peaks.

On Daphne Island, in the Galapagos Islands, the Red-billed Tropic Bird experiences much intraspecific competition, which causes a high rate of egg and chick loss so that the population exhibits a continuous pattern of egg laying with many repeats (Snow 1965b). Only 16 miles away, on South Plaza Island, there is little competition and a high breeding success and here there is a markedly synchronized annual periodicity in the breeding season. The precise explanation of these differences is still not forthcoming but Harris (1969c) has suggested that local breeding colonies may have recourse to discrete feeding grounds, which differ in food resources; he was able to show that apparent food shortage affected the breeding success and time of laying on Tower Island.

The Swallow-tailed Gull is an aberrant nocturnal feeding gull (Hailman 1964a) and, as mentioned, successful birds lay at 10-month intervals—more frequently if they are unsuccessful. Between breeding the adults leave the colonies for about 4—5 months to moult (Harris 1970). Although eggs may be laid at any time of the year, local colonies have their breeding synchronized, probably as a result of social facilitation (Hailman (1964b) and below). An interesting feature of this species is that it possess two incubation patches and experimentally has been shown to be capable of hatching 2 eggs and rearing both chicks. Yet only one or two individuals per thousand ever produce 2 eggs (Harris 1970). In such temperate gulls as the Great Black-back *Larus marinus* and Lesser Black-back *L. fuscus* four follicles enlarge prior to laying but only three are ovulated and the other is a reserve from which an egg can be laid in a day or two if one of those already laid is lost (p. 100). In the Swallow-tailed Gull only one follicle develops and if the egg is lost at laying, the female requires 18 days to produce a repeat (Harris 1970). The reason for the double incubation patch remains to be discovered.

The sea-birds just considered mostly re-lay and begin a new cycle as soon as they have acquired sufficient energy reserves. No proximate control of such cycles is required, nor could any suitable environmental signal be envisaged in the case of regular cycles which occupy more or less than 12 months and which shift in phase relative to calendar date. A special problem that still requires elucidation is the means by which the priorities are allocated between moulting and breeding in those species which have an arrested moult. The cycle of the Great Frigate Bird *Fregata minor* of the

Galapagos is a special case that might be mentioned at this stage for there is a 2-year periodicity which is imposed by ecological and not endogenous factors (Nelson 1957b). The adults obtain their food in the form of squids and fish by snatching it from the sea as they fly past, or else they steal it from other species by piracy. The young require a long period to learn these specialized feeding techniques before they can become independent, and while in the nest have a slow growth and the ability to fast as an adaptation to cope with periods when feeding is difficult. The breeding cycle takes a long time for these reasons and averages about 13 months; 10–20 days are needed for display and pair formation, 55 days for incubation, 130–160 days for nestling care, and a further 180 days for post-fledging care. Breeding might be repeated at 13-month intervals except that ecological conditions are such that breeding is best timed to begin at the same season each year. For this reason birds which are successful in rearing their young can breed only every 2 years. When breeding is again possible another pair may already be occupying the nest-site so that a social system based on the occupation of a territory, and waiting for the female to return can be of no value. To cope, the species has evolved a rather unusual courtship display. The males have a brilliant crimson inflatable gular sac which can be blown up like a balloon during courtship. Several males display these throat sacs as they sit around in the scrub and this combined display heightens the visual and auditory impact on any female flying overhead and she becomes attracted by and settles next to one of the males. Once pair formation has occurred one of the sexes always guards the nest-site from intruders. Obviously this method of courtship reduces the chances that a successful pair will re-mate in subsequent breeding seasons and explains why, unlike most sea-birds considered so far, the Great Frigate Bird lacks a permanent pair bond (Nelson 1967b).

Diamond (1972) has questioned whether both sexes are affected in the same way. In the Magnificent Frigate *F. magnificens* of the Lesser Antilles, the males desert the colony when the chicks are 3–4 months, leaving the female to finish rearing the brood. Diamond suggests that males could breed every year but that successful females must breed every other year. During their absence it is assumed that males undergo a full post-nuptial moult, returning to the colony after 6 months to begin a new cycle, but this needs confirmation.

Non-seasonal and less than annual breeding must depend on either an invariable food supply or a randomly fluctuating resource which cannot be predicted. In the second situation a period of good feeding prospects may induce many birds to lay and thereby impose a degree of synchrony amongst the population. Harris (1969a, 1970) could find no consistent seasonal variation in surface plankton by regular sampling round the Galapagos Islands, but famines did occur from time to time for various reasons. Similarly, Nelson (1969) showed that Red-footed Booby *Sula sula* chicks in Galapagos grew much better at some seasons than at others. Moreover, courtship and egg-laying became concentrated into those periods when chick success was high, presumably because this was correlated in time with temporary spells of good feeding. Similarly, on Ascension Island Black Noddy Terns *Anous tenuirostris* and Wideawake Terns *Sterna fuscata* had periods during which breeding success was poor and chicks died of starvation. Round Ascension Island many different sea-birds were noted feeding on the same broad spectrum of food, with squid and fish being the most important; two species of fish *Ophioblennius webbii* and the flying fish *Exocoetus volitans* predominated in the fish samples (Stonehouse 1962). Although details were lacking, it was apparent that different bird species had slight differences in food preference and perhaps more important were able to catch the various species in different ways. For instance, the White Booby *Sula dactylatra*, Brown Booby *S. leucogaster*, Black Noddy Tern, Wideawake Tern, and Fairy Tern *Gygis alba* might all congregate round the same fish shoal but each exhibited a different feeding behaviour: Fairy Terns swooped very fast from a height of about 7 m and picked objects from just above the water; Wideawakes plunge-dived into the surface water while boobies dived deeper (Dorward 1963).

In some sea-birds which have semi-annual breeding cycles there is a marked synchrony in laying date throughout the population, quite unlike the spread in laying noted in the tropic birds. A degree of synchronized egg laying was apparent amongst Brown and White Boobies on Ascension Island, both species requiring about the same period from egg-laying to the fledging of chicks (Dorward 1962). However, the 8-month cycle of the Brown Booby results in it breeding at the same time as the White Booby (which has a 12-month cycle) once every 2 years. If it is supposed that there are slight seasonal variations in the food supply which causes the White Booby consistently to favour a single annual breeding season it is difficult to understand how the Brown Booby avoids difficulties. Dorward suggested that the feeding condition might enable the White Booby to have a relatively good season once every year. Possibly some small advantage in feeding ecology may exist that enables the Brown Booby to manage some breeding during periods of suboptimal feeding conditions. Then on average the Brown Booby might experience a sequence of two 'poor' breeding seasons and one 'good' breeding season every 2 years whereas the White Booby would get two 'good' seasons. On average, this might enable the Brown Booby to raise more young than by having an annual cycle, whereas it must be presumed that the White Booby would do less well if it tried breeding more often.

More spectacular to observe, but more amenable to explanation than the example of the boobies, is the remarkable synchrony in the 9-month cycle of the Wideawake or Sooty Tern on Ascension Island as depicted in Fig. 7.12 (p. 166) (Chapin 1954; Chapin and Wing 1959). About 80 per cent of individually marked birds were found to begin a new breeding cycle 44—46 weeks after the end of the first, moulting occurring between times (Ashmole 1963*a*). Again, this is a species which appears to breed and moult at the maximum rate possible. As an adaptation against poor food supplies the chicks can grow at a variable rate so that they can survive long periods with little food and then increase rapidly in size when well fed. The marked synchrony in egg laying appears to be an adaptation to reduce predation; on Ascension Island, Frigate Birds and feral cats ensured that only 1—10 per cent of the eggs laid gave rise to flying young (Ashmore 1963). Ashmole describes how Frigate Birds would fly over the colonies and identify a brooding bird—possibly because birds brooding young fidget about more than incubating birds. The Frigate Birds would dive and snatch the brooding bird from its young and release it—adult terns are too big for Frigate Birds to swallow—away from the nest, and then quickly turn and snatch the chick before the parent could get back to provide protection. If small numbers of Wideawake Terns were always engaged in nesting activities their predators could increase in numbers—by breeding or immigration—and exploit a predictable food source. But, if all the terns breed together their numbers must swamp those of the predators, allowing each pair to have a small but significantly increased prospect of rearing progeny; the Sandwich Tern *Sterna sandvicensis* is a colonial species of temperate zones which adopts the same strategy (Cullen 1960). In the Seychelles most egg loss in Sooty Terns was due to predators but chick loss was due to pecking by other adults (Feare 1976). Birds nesting early and occupying the centre of the colony were much more successful than edge birds. Birds repeated if they lost eggs and were more successful if they laid in areas of current mass laying. These observations suggest to us that individual birds gain information from each other on feeding prospects and that synchrony in breeding and close nesting proximity are related to feeding prospects and not to predation directly (see p. 250).

The proximate regulation of synchronized egg laying is probably achieved by social stimulation. At the end of the breeding season some Wideawake Terns visit the colonies during the middle of the night, milling about and calling over favoured spots without landing. Gradually increasing numbers arrive at night and birds settle for short periods on the ground, occasionally displaying to each other. The time spent on the ground at these 'night clubs' increases as the season progresses, as does the number of individuals participating. This stage may develop over several months. There follows a brief period during which the birds remain at the colony until late in the morning and then after a few days they stay all day, whereupon the first eggs are laid. Fig. 7.12 (p. 166) shows how the testes increase rapidly in size during the 'night-club' phase. Ashmole found that birds captured when first attending the 'night-club' had small gonads but 2 months later full development had occurred. The fully recrudesced testis is about 11·5 times the size of the regressed organ, a volume increase typical of many tropical species. Involution occurs within a month of egg laying.

The exact mechanism whereby social stimulation synchronizes or phases the autonomous cycles of individual Wideawake Terns remains to be defined. There must be some quantitative effect otherwise small groups might start breeding before the rest of the colony; indeed this does seem to occur in some colonies (Feare 1976). It is conceivable that social responses are important in finding localized food sources and that only when a critical number of birds become involved can food supplies be obtained quickly enough to leave time for breeding. Vocalizations could also be of some importance. It is known that the circadian rhythms of birds can be entrained by playing appropriate songs (Gwinner 1975). Ovarian development in captive female Ring Doves *Streptopelia risoria* was stimulated more by intact than castrate males and most of all if, in addition to the intact males, sounds from the colony *milieu* were received (Lott, Scholz, and Lehrman 1967). Similarly, the follicles of females placed in auditory isolation were stimulated if the birds could hear the sounds picked up by a microphone situated in a nearby breeding room (Lehrman and Friedman 1969).

Evidently vocalizations function in part to stimulate the reproductive system of conspecifics but more information is required (Lewis and Orcutt 1971). Kunkel (1974)
provides a useful review of this little studied topic.

In the Wideawake Tern and other sea-birds, non-seasonal breeding may occur at
one locality and seasonal breeding at another. On Christmas Island the autonomous
9-month cycle of the Wideawake Tern is lost and is replaced by annual breeding
with successful adults laying around May or June, or else in late November–
December, completing one full moult and beginning another in the interval (Richardson
and Fisher 1950; Hutchinson 1950; Ashmole 1963*b*). Some or all those individuals
which are unsuccessful in rearing a chick in one breeding season attempt to breed
again 6 months later, either replacing all their primaries in the interval or else arresting
moult for breeding (Ashmole 1963*c*). A similar pattern has been noted on other
islands and, although no details are available, it perhaps depends on there being two
periods each year of peak food supplies. An annual breeding cycle is noted in the
Seychelles (Feare 1976). The Red-footed Booby breeds in all months in the Galapagos
Islands but laying is more synchronized to occur in the April–June period on
Christmas Island (Nelson 1969). The Red-tailed Tropic Bird nests seasonally in the
local summer on Kure Island (28°N, 178°W) in contrast to the situation on
Christmas Island (see p. 358).

Synchronized breeding may be advantageous if it facilitates obtaining food supplies. It must tend to result in a depletion of food reserves compared with an extended
breeding pattern and so can be advantageous only if the resource is locally distributed
in time or space or if the other individuals help in finding food. Situations could
conceivably arise in which there were conflicting selection pressures. This seemingly
applies to the Madeiran Storm Petrel *Oceanodromo castro* in the Galapagos Islands,
where it appears to have two breeding seasons, for eggs are laid in April and June and
again in December and January (Snow and Snow 1966). In reality, two entirely
separate populations are involved which use the same nest-holes, one during the
'warm' season and one during the 'cold' (Harris 1969*d*) so individual birds have an
annual cycle. In contrast, on Ascension Island egg laying occurs once a year, in
November (Allan 1962). Harris (1969*a*) emphasizes that many species can be found
laying or rearing young at all seasons in the Galapagos suggesting that in general food
supplies are not restricted seasonally (but see the discussion regarding Great Frigate
Bird above, p. 359). He suggests, therefore, that a limitation of nest-sites might make
it desirable for two populations to share the facilities, and if synchrony in breeding
was also being selected a polarization of the kind observed might occur. However,
it is also conceivable that the two populations have distinct feeding habits which lead
to differences in the best time for breeding; this possibility needs to be explored.

In most tropical terrestrial habitats there are seasonal changes, such as a wet period,
which render certain months of the year more favourable for breeding. Cycles of the
kind discussed for sea-birds are, in consequence, rarely found while species in those
habitats, such as equatorial evergreen forest, which are most likely to exhibit continuous breeding or less than annual cycles, have been poorly studied. In the Seychelles,
D. Lloyd (personal communication) has found that some of the land passerines breed at

intervals of less than 12 months. However, the Yellow Warbler *Dendroica petechia* of the Galapagos Islands breeds at regular annual intervals in association with the rains (Snow 1966).

14 Evolutionary aspects of photoperiodism

Summary

It is obviously expensive in terms of time and facilities to study comparative aspects of the circannual breeding rhythms in closely related avian species in order to seek clues to the evolution of the photoperiodic responses. If it is assumed that birds in general possess circannual rhythms in reproductive function, for which the annual day–night cycle provides the entraining *Zeitgeber*, then comparisons can be made of the reproductive behaviour of different species held under the same conditions. This is possible by considering the egg-laying seasons of a large number of species of ducks, geese, and swans (Anatidae) which are all maintained in the collection of the Wildfowl Trust at Slimbridge, England; here they all experience the same seasonal daylength cycle. There is only a small annual variance in the median date (photoperiod) at which any particular species begins egg laying and in many cases records exist for more than 20 years. In general, species originating from high latitudes need a longer daylength to initiate egg laying than those from low latitudes. However, the daylength under which refractoriness develops is more nearly the same within any particular genus; some species, including *Dendrocygna*, do not appear to enter a refractory phase under the Slimbridge *Zeitgeber*. The phase difference between the mid-point of the egg-laying season and the summer solstice is positively correlated within any particular genus with the latitude of origin and negatively with the photoperiod under which breeding begins; the data are consistent with the expectations of oscillation theory that changes in level and hence frequency of any endogenous rhythm will be reflected in phase differences during entrainment. Species of *Dendrocygna,* a primitive tropical genus, have breeding rhythms which are little altered in phase relative to the Slimbridge daylength cycle in relation to their latitude of origin while swans (*Cygnus* and *Coscoroba*) have rhythms which are markedly altered in phase according to latitude of origin; geese have egg-laying rhythms which are less changed in phase relative to the *Zeitgeber* than do swans. For some species at Slimbridge, for example, the Black Swan *Cygnus atratus*, the light cycle results in a double breeding season per annum with the two cycles about 180° out of phase. The data suggest that intrageneric variations in breeding response result from small changes of threshold in photo-response, while inter-generic differences implicate more drastic alterations which affect the duration of the breeding season and the onset refractoriness. Some limited records exist for hybrids.

Montane species have photo-responses that could hardly evolve under the local photo-regime and instead evidence shows that their responses were pre-adapted at higher latitudes from which they originate. Adaptations for desert regions are described including the unique response of the Pink-eared Duck *Malacorhynchus caryophyllacea* in which there are always small nodules of the testes which contain sperm the rest of the organ being composed of regressed tubules.

Photo-responses of waterfowl at Slimbridge

This chapter will mostly concentrate on a single family, the Anatidae, for which comparative data are available for a large number of closely related species. This is because a unique collection of captive swans, geese, and ducks is kept by the Wildfowl Trust at Slimbridge, Gloucester (latitude 51°44′N), thanks to the foresight and efforts of Sir Peter Scott. The birds are maintained in open paddocks, mostly as single species pairs or small groups, where they have a small pond and vegetation in which they can nest. All birds receive a complete diet and of course all are exposed to the same natural daylength cycle. Ducks, geese, and swans do well in captivity and most, but not all, will breed freely, indeed about 100 species do so at Slimbridge. It is also fortunate that the systematic status and taxonomy of the Anatidae have been well defined. Fig. 14.1 represents a generally well-agreed family tree of evolutionary affinities.

The keepers maintain an annual record of the date that the first eggs are laid by a particular species and in many cases such information is available over more than 20 years, although one or two species may have nested only once or twice. Once several females begin egg laying it is difficult to keep track of events. Nevertheless, annual records have been kept of the date on which the last ducklings of each species have been hatched, so that by subtracting the time needed for incubation the date of the last egg laying can be calculated. The record of last egg laying is less complete because females are generally allowed to incubate their first clutches. Only if these are taken for artificial incubation or they are lost will the females lay again. So, in order to establish the female's maximum potential for extended laying it is necessary to use the latest recorded egg-laying date rather than the median date of last eggs. Hence, the breeding season can be defined as the median date of first eggs to the latest recorded laying. Obviously the data are more reliable if breeding records are available for many years and there are several pairs of a given species in the collection. Nevertheless, the error terms are small and even records for a single pair and season can be informative. It will be recalled that male birds can be brought into full breeding condition by stimulatory photoperiods provided they are not inhibited by crowded conditions, competition from other males, or dietary deficiencies. In contrast, females can be only partly photostimulated and final maturation of the follicle leading to ovulation and oviposition requires appropriate behavioural stimulation from the male, although spontaneous but haphazard egg laying can occur. Thus, the appearance of a full clutch in a nest denotes that a male has reached the peak of spermatogenetic development and is secreting those hormones which induce display and copulatory behaviour towards a receptive female.

At Slimbridge, each species tends to lay its eggs during a clearly defined season and the order in which each closely related species begins to lay approximately parallels the sequence noted in the wild; species originating from low latitudes lay first and those from higher latitudes later. This is in spite of the fact that all birds experience the same photoperiodic cycle and species from warm countries lay eggs during the late winter and early spring, when it is cold. In initial studies of swans, geese, and shelducks, Murton and Kear (1973a) denoted the date of egg laying as the number of days succeeding the winter solstice, a measure of the increasing photoperiod. In Fig. 14.2 these same median dates are converted into the actual daylength at Slimbridge at the time of laying, and these are plotted against the mid-latitude of the natural breeding range of each species. The Whistling or Tree Ducks of the tribe *Dendrocygna* are essentially tropical and they do not breed freely at Slimbridge. It

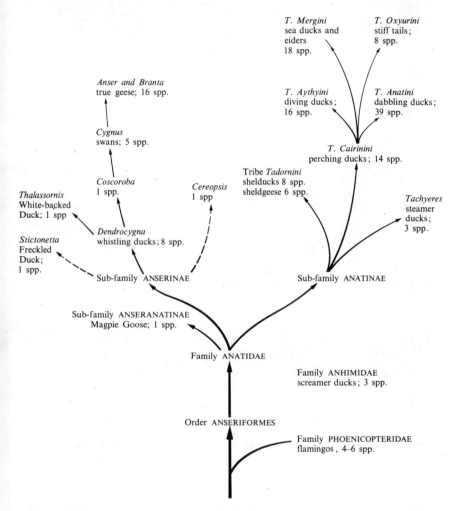

Fig. 14.1. Supposed evolutionary tree of the Anseriformes. Based on Kear (1970) in turn following the classification of Johnsgard (1965). Further evidence for placing the White-backed Duck *Thalassornis leuconotus* close to *Dendrocygna* and not the *Oxyurini* is given by Raikow (1971).

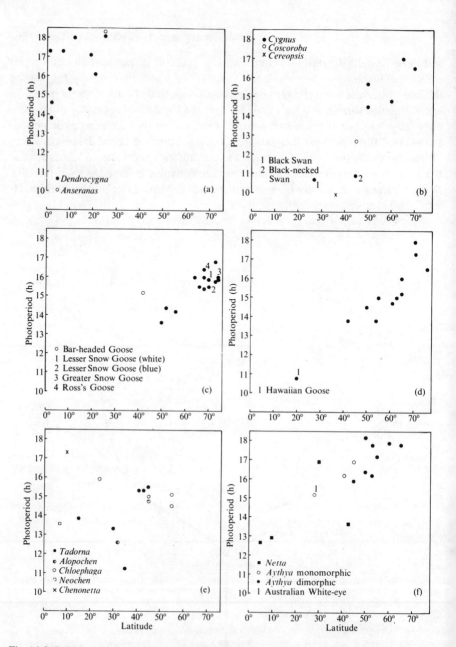

Fig. 14.2. Relationship between photoperiod (dawn–dusk plus civil twilight) under which the first eggs are laid by various species of Anatidae at Slimbridge (*y*) and the mid-latitude of their natural breeding range. Each point is a different species or sub-species but only those that are relevant to the text are named (consult sources indicated below for more detail).

(a) *Dendrocygna* and *Anseranas* (Murton and Kear 1976); (b) *Cygnus, Coscoroba,* and *Cereopsis* (Kear and Murton 1976); (c) *Anser* geese (Murton and Kear 1973*a*); (d) *Branta* geese (Murton and Kear 1973*a*); (e) Shelducks and sheldgeese (*Tadornini*) (Murton and Kear, 1973*a*). (f) *Aythyini*; 1 refers to the Australian White-eye *Aythya a. australis* (Murton and Kear 1973*b*).

appears that they are inhibited by climatic factors, for they easily get frost-bitten toes and even may die in hard winters. Accordingly, it has been necessary to use the first recorded date of egg laying for them rather than the median (Murton and Kear 1975). Fig. 14.2 emphasizes that there is a clear linear relationship between laying date and latitude of origin for most genera of Anatidae (see also Fig. 14.19); the exceptions will be discussed below.

Although some tropical species are adversely affected by the English climate (compare Cain 1973), the remarks above make it unlikely that temperature regulates the laying season at Slimbridge, otherwise the Hawaiian Goose *Branta sandvicensis* should nest after and not before the Red-breasted Goose *B. ruficollis*. It happens that a second collection of waterfowl is kept by the Wildfowl Trust at Peakirk, Northamptonshire (latitude 52°39′N; longitude 0°16′W), 100 miles to the east of Slimbridge. The daylength after the spring equinox is very slightly longer at Peakirk than at Slimbridge, but the mean temperature averages nearly a degree lower at Peakirk between December and May, but thereafter there is little difference. Fig. 14.3 depicts the presumed effect of temperature on laying date as it affects the *Branta* and *Anser* geese combined.

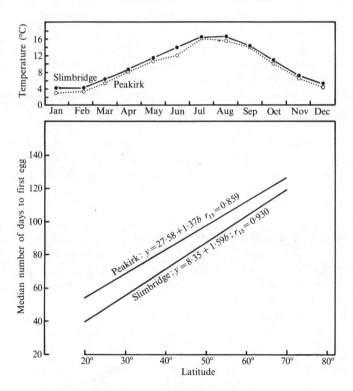

Fig. 14.3. Regression lines relating date (number of days from 1 January) of first egg laying by 17 species of *Anser* and *Branta* geese at Slimbridge, Gloucestershire or Peakirk, Northamptonshire to the mid-latitude of the natural breeding range of each species. The mean monthly temperature at the two sites is shown in the upper panel.

For those species originating from 20° latitude egg laying occurred in February at Slimbridge, 15 days earlier than at Peakirk. The mean daily temperature from 1931 to 1966 in January and February at Filton, Gloucester was 4·2 °C and 4·4 °C respectively, compared with 3·3 °C and 3·6 °C at Wittering, Northhamptonshire. Geese originating from 60° latitude mostly lay their eggs in April at Slimbridge— only 6 days earlier than at Peakirk. The mean temperature at the two locations in April averages 8·8 °C and 8·4 °C, respectively.

Species which migrate to the arctic circle cannot begin breeding until they experience long daylengths, and the serial correlation between latitude and photoperiodic response could be a reflection of the fact that tropical species have longer breeding seasons than temperate ones (cf. Baker 1938a). Inspection of Fig. 14.2 also shows what would not otherwise be obvious, namely that the photoperiod at which the onset of egg laying is initiated in closely related grey or black geese bears a straight-line relationship with the mid-latitude of origin, and that swans exhibit a significantly different regression. There would seem to be no ecological reason why some species. of geese should not closely resemble some of the swans in their photo-responses, that is, no reason why more variability should not exist in the data, unless the regressions imply some underlying physiological mechanism which is characteristic of closely related species. In fact, Murton and Kear (1973a) argued that closely related species had a photo-response mechanism which was essentially the same and that the differences between closely related swans involved only trivial changes, for example, a change in the threshold level of an oscillator which would alter its phasing relative to the Zeitgeber. A more fundamental alteration in the oscillation system was envisaged to have occurred in the evolution of the geese from swan ancestors.

Fig. 14.4 presents some of the data in a different way. The length of the breeding season (median data of first eggs to last recorded laying, except in Dendrocygna for which see the Figure caption) of some selected groups is represented as a horizontal line which is positioned against a graphical presentation of the daylength cycle at Slimbridge: the start of the breeding season is plotted against the actual date (abscissa) and the corresponding photoperiod (ordinate). With benefit of the discussion of the preceding chapter let us assume that the egg-laying seasons of the different species represent circannual rhythms of reproductive capacity which are phased to the environmental Zeitgeber. Just as with circadian rhythms, the phase difference between the Zeitgeber and the entrained rhythms can be measured and to do this the abscissae in Fig. 14.4 are converted from calendar dates to degrees. The phase differences are best measured from the mid-point of the two oscillations and these

---

Fig. 14.4. Cross-bars represent the period of each year during which different species of Anatidae lay eggs at Slimbridge. They extend from the median date on which first eggs were laid to the latest date recorded on which a clutch was produced. The bars are, therefore, a representation of the length of the breeding season and each is plotted to coincide with the daylength (y) and date (x) at which egg laying first commences. Since there are 365 days in a year the x-axis can be transformed with little loss of accuracy into 360° and this is a convenient means of recording the phase difference between the centre-point of each breeding cycle and the seasonal day-

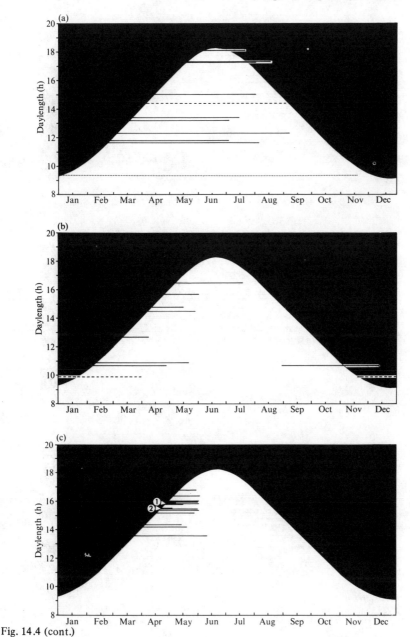

Fig. 14.4 (cont.)

length cycle, which is also depicted. (a) *Dendrocygna* (solid lines), *Anseranas semipalmata* (dashed line), and the Crested Screamer *Chauna torquata* (dotted line). (b) Swans (solid lines) and *Cereopsis* (dashed line). Swan species in order from the top are: Bewick's *Cygnus columbianus*, Trumpeter *Cygnus c. buccinater*, Whooper *C. c. cygnus*, Mute *C. olor*, Coscoroba *Coscoroba coscoroba*, Black-necked *Cygnus melanocoryphus*, and Black *C. atratus*. (c) *Anser* geese: 1 = white phase of Lesser Snow Geese *A. caerulescens* and 2 = blue phase.

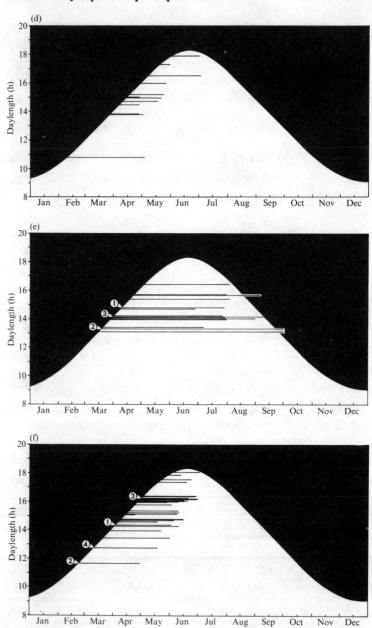

Fig. 14.4 (cont.)

(d) *Branta* geese (e) *Anas* ducks ('primitive' species). Species identified are 1. Grey Teal *A. gibberifrons*; 2. Brown Teal *A. auklandica chlorotis*; and 3. Chestnut-breasted *A. castanea*. (f) *Anas* ducks ('advanced' species). Species identified are: 1. Australian Black Duck *A. superciliosa rogersi*; 2. Australian Shoveller *A. r. rhynchotis*; 3. New Zealand Shoveller *A. r. variegata*; 4. African Black Duck *Anas sparsa*. (For details to (e) and (f) see Murton and Kear (1976).)

are plotted in Fig. 14.5 against the latitude of origin of the species concerned. The results are essentially the same as in Fig. 14.2. Variability would be reduced by working to the onset of the breeding rhythm simply because, as already explained, the

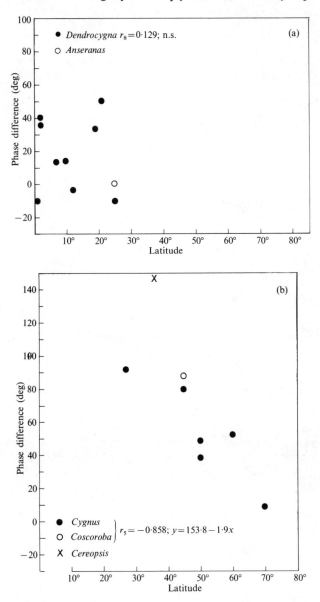

Fig. 14.5. Phase differences between the mid-point of the laying cycle of different Anatidae and the mid-point of the annual day–night *Zeitgeber* (i.e. the summer solstice), as calculated from the original drawings for Fig. 14.4, plotted against the mid-latitude of the natural breeding range of these same species: (a) *Dendrocygna* and *Anseranas*, (b) Swans and *Cereopsis*,

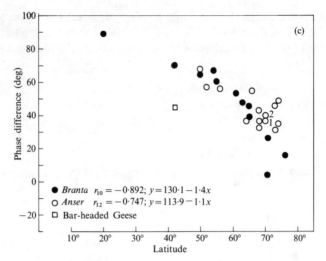

Fig. 14.5

(c) *Anser* and Branta geese: 1. Lesser Snow Goose (white phase); 2. Lesser Snow Goose (blue phase

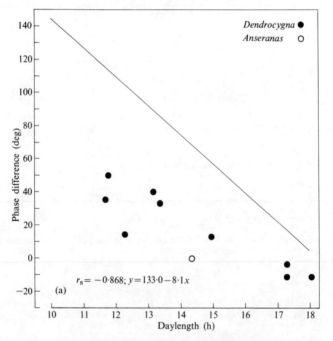

Fig. 14.6. Phase difference as calculated in Fig. 14.5, and referring to the same species, plotted against the daylength at which breeding is initiated at Slimbridge. The diagonal lines indicate the correlation that must occur when seasonal changes in daylength are plotted against date, i.e., as occurs when the y- and x-axes of Fig. 14.4 are plotted against daylength. Differences from the expected regression slope arise because the mid-points of each annual egg-laying rhythm are plotted.

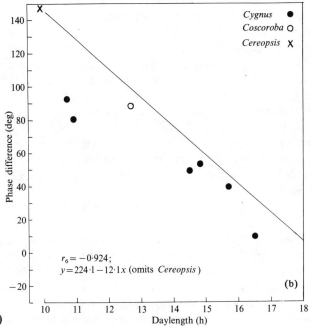

Fig. 14.6(b)

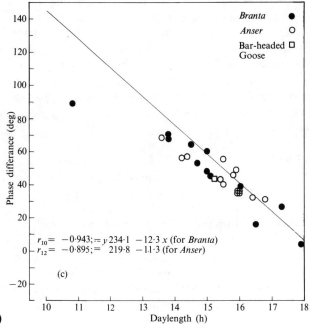

Fig. 14.6(c)

data for the median first laying dates at Slimbridge are more complete than those defining the length of the breeding season. The phase shifts can also be plotted against the photoperiod under which breeding begins and this then shows that the shorter the photoperiod under which breeding is initiated the more positively phased is the egg-laying rhythm (Fig. 14.6). Of course, this would be expected since in Fig. 14.6 the seasonal daylength cycle, represented as the number of hours of daylight and shown in Fig. 14.4 as the ordinate, is in effect being plotted against the date (abscissa) The slope of any regressions would reflect the slope of the daylength curve against date if the onset of egg laying were taken as the point for measuring phase shift and such lines are drawn on Fig. 14.6. The deviation in slope of the plotted regressions from these lines results because the mid-point of the breeding season was used to measure phase differences and the length of the season introduces an independent function (see also Fig. 14.13).

Consideration of Fig. 14.5 shows that *Dendrocygna* species differ from swans in having breeding rhythms which are little altered in phase relative to their latitude of origin, whereas geese have breeding rhythms which are less markedly altered in relation to latitude than are those of the swans. That is, geese emanating from a given range of latitudes exhibit a smaller range of phase shifts to the Slimbridge *Zeitgeber* than do swans from the same latitude range. Entrainment is not possible if the phase angle between a controlling oscillation approaches 180° and it is evident from Fig. 14.5 that this zone is approached more rapidly in swans than in geese. If swans lived at the equator they would experience a phase difference of about 160° so far as their breeding rhythms are concerned, and entrainment would be weak. The phase angle of the 'goose' breeding rhythm measured against the Slimbridge *Zeitgeber* varies relatively little with the latitude of origin. This gives the geese an advantage above about latitude 50°N (the point at which the 'swan' and 'goose' regressions cross—see Fig. 14.5), for it enables them to breed earlier than the swans but is a disadvantage below latitude 50°N where the group must begin breeding later than the swans (Fig. 14.2). The relevance of these data in helping to understand the controlling mechanisms underlying circadian and circannual periodicities will be considered again below (p. 392). First we must discuss the ecological and evolutionary implications of these various relationships.

Some primitive taxa

The screamers of the sub-order Anhimae (order Anseriformes) comprise three species (*Chauna* spp.) of goose-sized birds that live in shallow water marshes in South America They have elongated toes with only a slight development of webbing at the base, and they possess two curved, sharp pointed spurs on the front of each wing. They are held to be close to the ancestors of the Anseriformes. At Slimbridge, the Crested Screamer *Chauna torquata* has produced eggs in virtually all months of the year. In the wild, egg laying is usually between August and December at Cape San Antonio (Weller 1968) but in all months of the year at General Lavalle, Argentina (Gibson 1920). The breeding season is not critically phased relative to the environmental light—dark cyc and this appears to represent a primitive condition (Fig. 14.4(a)). The remaining swans, geese, and ducks are allocated to the sub-order Anseres which contains one family, the Anatidae, of which the Magpie Goose *Anseranas semipalmata* of tropical Australia and Southern New Guinea is the most primitive member, being allocated a sub-family (Anseranatinae) of its own.

The Magpie Goose possibly evolved from the ancestors of *Dendrocygna* and provides a link with the primitive screamers (family Anhimidae). The feet are only slightly webbed and the 14 wing feathers are moulted progressively, whereas in all other waterfowl (except some sheldgeese *Chloëphaga* spp.) the flight feathers are moulted simultaneously and the birds have a flightless period lasting 3–4 weeks. Black and white plumage and an enormous pink bill combine to make this an ugly species. The latter feature is an adaptation for digging in soft mud from which the birds obtain the tubers of the Bulkuru sedge *Eleocharis dulcis* and other roots and they also consume large quantities of seeds of water plants. The species is polygamous, an interesting feature discussed again in Chapter 15 (see also Kear (1973)). The mean date for first egg laying by the Magpie Goose at Slimbridge has been 22 June (photoperiod 18·3 h) but first eggs have appeared from as early as 26 April and as late as 15 August. Second clutches are often laid in July–August; in 1971 four clutches were produced, in May, June, August, and September. Climatic factors are probably important in modifying the laying season at Slimbridge where the species appears to have a photo-response mechanism which is similar to some of the *Dendrocygna* species (Fig. 14.4). In the Northern Territory of Australia, Frith and Davies (1961) showed that the gonads were enlarged (significant increase in size from inactive condition) at variable times in different years but during the overall period November–May. Egg laying might begin as early as February or be deferred until July depending on when rains provided a suitable degree of flooding and associated development of the swamp vegetation. In Queensland, January–April are the favoured months (Lavery 1967*a*). At northern Australian latitudes seasonal changes in daylength can be of no importance in the precise timing of reproduction, albeit they can broadly phase the breeding cycle, and it appears that, as in the case of some of the species discussed below, the gonads rapidly achieve maturity when environmental conditions are right. Whether the immediate environmental stimulus is water depth and the quality of the vegetation, as Frith and Davies suggest, has not been proved experimentally; in the wild many subtle factors, including availability of food supplies, could be correlated with what seem obvious habitat changes to the human observer and they might have more significance as proximate factors for the bird.

The whistling or tree ducks (*Dendrocygna*) are the most primitive members of the sub-family Anserinae. Primitive features are their sexual monomorphism, the involvement of both sexes in egg incubation, and their participation in mutual preening behaviour (allopreening). They, like the screamers, have a tropical distribution, as Fig. 14.7 shows. One species, the Fulvous Whistling Duck *Dendrocygna bicolor*, has an exceptionally wide range in North and South America, Africa and India. Whistling ducks frequent freshwater marshes and ponds where they feed by dabbling and by stripping seeds from the plants which emerge from shallow flood waters. Some species will graze on drier grasslands eating green vegetation and others habitu-

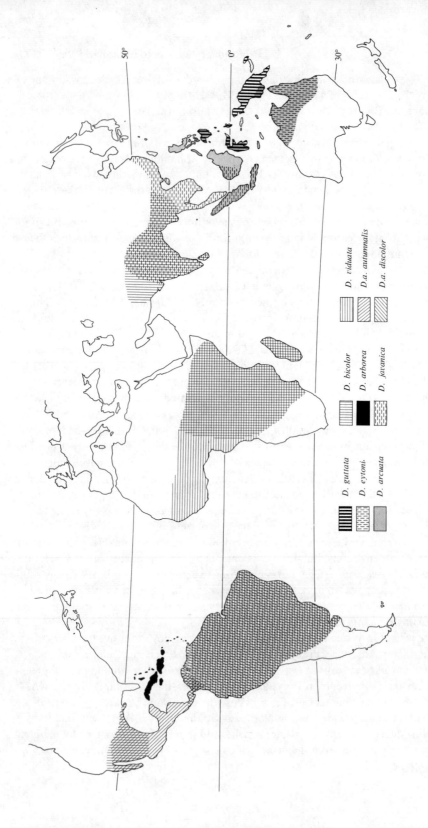

Fig. 14.7. World distribution of the genus *Dendrocygna*.

D. guttata
D. eytoni
D. arcuata

D. bicolor
D. arborea
D. javanica

D. viduata
D.a. autumnalis
D.a. discolor

ally dive down to 2–3 m and eat aquatic invertebrates. The nest-site is usually in grass or similar marsh vegetation but two species nest in tree cavities. All species presumably have photoperiodic mechanisms which are adjusted to tropical daylength cycles, and indeed several species achieve breeding capacity at Slimbridge, when daylengths reach about 14 h. Those species that have penetrated to more temperate latitudes, the Cuban Whistling Duck *D. arborea* (latitude 21°N) and the Grass or Eyton's Whistling Duck *D. eytoni* of Australia (mid-latitude of breeding range 25°S), appear to have breeding seasons that are phased more negatively to the Slimbridge *Zeitgeber* than the rest of the group (Fig. 14.4), while in Fig. 14.2 there is some suggestion that these mid-latitude forms are tending to introduce a significant correlation to the plot. This also applies to the Magpie Goose (see Fig. 14.2). However, although Eyton's Duck has laid with a photoperiod of 18 h at Slimbridge it is certain that it must respond under a shorter daylength in the wild for otherwise it would never breed in Australia (see footnote on p.380). In the wild in Queensland it has a relatively limited breeding season corresponding to the wet season, which is also the season of longest daylengths. According to Lavery (1967*b*) dispersal from drought refuges begins in late October—the birds moving to dry grasslands in the proximity of swamps and lagoons— so that egg laying takes place mostly in December and January, but sometimes from November to February, when daylengths range from about 13½ h to just over 14 h per day. Adults which have frequented inland marshes subject to seasonal drought return with their young to drought refuges from April until June. Preferred drought refuges are the deep waterlily (*Nymphaea* spp.) bilabongs which are common on the coastal plains. The Australian Wandering Whistling Duck, or Water Whistling Duck *D. arcuata australis* partly overlaps Eyton's Duck in range and is more dependent on water, that is, Eyton's Duck has evolved to occupy drier habitats when breeding. Where the two species overlap in range the Wandering Whistling Duck has the longer breeding season, by upwards of 3 months (Lavery 1967*b*).

The more tropical *Dendrocygna* species probably have the capacity to breed for most of the year in the wild depending on food availability and habitat suitability. The low amplitude of the tropical *Zeitgeber* and the lower light intensity in summer compared with the temperate daylength cycle presumably result in an even more extended breeding capacity than is noted at Slimbridge, and perhaps a less synchronized entrainment between individuals. The Fulvous Whistling Duck *D. bicolor* nests in all months of the year in Africa, depending on locality (Mackworth-Praed and Grant 1952). Egg laying is recorded from April until September in California (Bent 1925) and in the Louisiana rice fields (Bolen 1973). Similarly, egg laying from May to October has been noted in the case of the Northern Red-billed Whistling Duck *D. a. autumnalis* in Texas (Bolen 1973). The whistling ducks, therefore, manifest a characteristically tropical, and presumably primitive, photo-response mechanism and it is likely that their reproductive apparatus is active for much of the year. Adaptation to more temperate photoperiods may have involved a raising of the response threshold so that the breeding season is restricted to the longer days between spring and autumn.

All the primitive and essentially tropical waterfowl so far discussed (i.e. the screamers, Magpie Goose, and *Dendrocygna*) have rather long breeding seasons at

Slimbridge. Once reproduction begins it continues until the daylength falls below what appears to be stimulatory level, and this means that the breeding season is almost symmetrically positioned on either side of the summer solstice (Fig. 14.4). Murton and Kear (1976) considered this to be the primitive photo-response in the Anatidae. They found that the more recent genus *Anas* could be subdivided on the basis of the photo-responses of its members (Fig. 14.4). Some showed the kind of breeding response noted for *Dendrocygna* and they proved to be the most primitive species on behavioural, plumage, and morphological grounds. The advanced species of *Anas*, which have radiated in the temperate zones of the world, have breeding cycles at Slimbridge which resemble those of *Anser* and *Branta* (*cf.* Fig. 14.4). That is, a period of breeding is followed by gonad involution as a refractory phase develops following continued exposure to long daylengths (see also Fig. 13.4, p. 347).[†]

It is evident that one centre of waterfowl evolution has been the equatorial and tropical regions of the Old and New Worlds. The Magpie Goose probably colonized Australia from more tropical ancestors, just as today the Australian *Dendrocygna* species derive from more northern tropical forms. The photo-response mechanism of these Australian waterfowl is not much changed from the presumed ancestral condition except for an apparent threshold effect, that is, the breeding season as manifested under the Slimbridge *Zeitgeber* is only initiated with the longer days of late spring or summer.

Swans and geese

The swans (*Cygnus, Coscoroba*) exhibit photo-responses which are evidently adapted to a temperate seasonal daylength cycle. They all basically feed on aquatic

[†]In more recent studies Murton and Kear (in press) have distinguished those primitive species which have a high photo-threshold and so do not begin breeding until long daylengths are experienced at Slimbridge, from those with a low photo-threshold. For example, the Plumed Whistling Duck does not begin breeding at Slimbridge until early June when daylengths reach 18·2 h and it stops again in July under an 18·2-h photoperiod. Obviously, it responds to shorter photoperiod in the wild in tropical Australia (latitude 25° S). However, it was suggested that the daylength under which breeding commenced at Slimbridge was not the relevant feature in this case. Suppose long or short breeding seasons depend on the change in level of some oscillator(s) relative to a threshold as discussed in Chapter 11 (p. 287), then the short seasons of species like the Plumed Whistling Duck would indicate a low level and hence low frequency of some endogenous oscillator. A relative reduction in frequency of an oscillator might make it more easily entrainable by a low amplitude and weak *Zeitgeber*. There is some evidence that species with breeding responses at Slimbridge similar to those of the Plumed Whistling Duck have, in the wild, relatively short breeding seasons, with all the members of a population being fairly closely synchronized in their breeding stage. Moreover, the breeding seasons in the wild are approximately in phase with the *Zeitgeber* so that the peak of activity occurs around the summer solstice. In contrast, species with cycles of the kind exhibited by the Fulvous Whistling Duck at Slimbridge have much longer breeding seasons in the wild and synchrony within any population is much less noticeable; the period of peak activity is still in phase with the summer solstice. In the same way tropical species having temperate-type cycles (as in Fig. 14.4(f)) with a low photo-threshold, that is, they begin breeding under short photoperiods at Slimbridge, seem to have poorly synchronized and extended breeding seasons in their native habitats. However, the season of peak breeding in such species is not in phase with the longest day, as in the above examples, but instead occurs during the 'winter' season. The relative phasing of the breeding rhythms of different species at Slimbridge is indicative of the phasing of the breeding cycle under the natural photo-regime.

vegetation and their long necks enable them to reach down into shallow water for they do not dive. The primitive members of the group are distributed in the southern sub-tropical and temperate zone (see Fig. 14.8). The Coscoroba Swan *Coscoroba coscoroba* of South America is probably the most primitive member of the group, although its taxonomic position is rather uncertain. In anatomical characters, particularly its unconvuluted trachea and syrinx with a bulla, it resembles the Mute Swan, and its aggressive display is similar to that of both the Mute and Black Swans (Griswold 1973). Its high-pitched call, of which its name is an onomatopoeic representation, and feathering in front of the eyes, are goose-like characters, while Griswold refers to the duck-like bill, large feet, light band across the nape of the young, and lack of triumph ceremony display as reminiscent of *Dendrocygna*. Perhaps this species is close to the ancestor from which *Dendrocygna* and *Cygnus* radiated. The Black Swan *Cygnus atratus* of Australia is the next-oldest surviving member of the group, followed by the Black-necked Swan *C. melanocoryphus* of South America. The north-temperate Mute Swan *C. olor* links these primitive southern hemisphere species with the advanced, closely related, and highly migratory white-plumaged swans (Whistling, Bewick's, Whooper, and Trumpeter Swans—see index for scientific names) which breed in the sub-Arctic regions of the Holarctic. These last species probably arose during inter-glacial times to exploit the aquatic vegetation associated with the melt waters of the northern snow fields. Fossil remains of the Whistling and Bewick's Swans date from the Upper Pliocene of Idaho.

The photo-responses of the Black Swan as noted at Slimbridge show that the breeding cycle can be initiated under shorter days than occur in Australia. This led Kear and Murton (1976) to propose that the Black Swan evolved further south than present-day Australia and they argued that the centre of evolution was either Australia when it was further south (Australia was 25° of latitude further south at the end of the Cretaceous Period before moving north during the Tertiary) or the ancient land mass remaining from the break-up of Gondwanaland. A large race of the Black Swan became extinct in pre-European times in New Zealand, probably in consequence of the Polynesian invasion. Today, introduced Black Swans from Australia are thriving under the more marked seasonal photoperiod of New Zealand which causes the swans to have a regular seasonal breeding season (p. 386). Evidence accumulates that, contrary to the views of Mayr (1972), several taxa of the avifaunas of Australia are derived from a previously widespread Gondwanaland fauna (Serventy 1973; Cracraft 1973, 1976; Rich 1976) and that arrivals via the classical Indo-Malaysian route occurred towards the end of the Tertiary. This means that the Black Swan has probably been in Australia for longer than the more primitive Magpie Goose which presumably colonized Australia as the continent drifted closer to Asia during the Tertiary. The geological history of South America shows parallels with Australia and many taxa derive from a Gondwanaland element including the sub-oscine passerines, and presumably the Black-necked Swan (compare the pigeons, Appendix 5). A later invasion of oscines and other groups occurred during the Pliocene–Pleistocene Periods when the Panamanian land-bridge connection with North America was formed. Africa has probably always been in touch with the

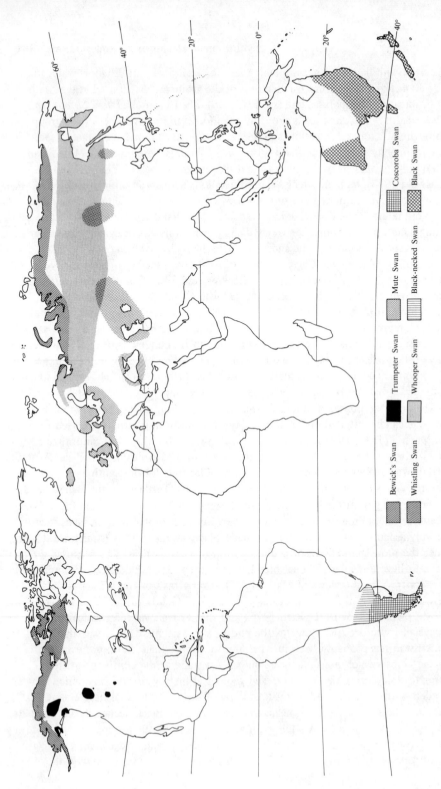

Fig. 14.8. World distribution of the Swans, *Cygnus* and *Coscoroba*.

Bewick's Swan

Whistling Swan

Trumpeter Swan

Whooper Swan

Mute Swan

Black-necked Swan

Coscoroba Swan

Black Swan

Laurasian avifauna and acted as major refuge for these elements during the late Tertiary climatic deterioration.

The Black Swan can rear its young on such aquatic weeds as Fennel Pondweed *Potamogeton pectinatus* and the floating seeds of sedges (for example *Scirpus littoralis*) but the period of cygnet care is much longer than for other members of the genus (Kear 1972; Scott and The Wildfowl Trust 1972). Vegetation that is suitable for food is often distributed in local patches, depending on variations in water depth caused by uneven rains. This perhaps explains why Black Swans breed colonially and egg laying is highly synchronized within any local population. Each family is able to congregate with others on the best feeding areas. In contrast to the Black Swan, the remaining swans feed their cygnets on aquatic invertebrates and each pair defends a nesting territory which contains adequate supplies of food. In the north-temperate tundras above latitude 60°N, the young grass which appears with the melting of the snow provides a particularly good food supply, for rich reserves of proteinaceous nutrients are concentrated into the growing parts to enable maximum plant growth and seed production during the short Arctic summer; this high protein content is soon reduced as the plants change from a vegetative to reproductive phase of development (Fridriksson 1960).[†] During glacial epochs such nutrient grass may well have occurred at lower latitudes and been a selective factor in the emergence and radiation of the geese (*Anser* and *Branta*). On morphological and behavioural criteria the geese are phylogenetically younger than the swans. Geese ancestors have been found in the Eocene of France and Utah, U.S.A., and are possibly represented in Cretaceous deposits, but true geese of the genus *Anser* do not appear in the fossil record of Europe until the Miocene. A goose-like swan is known from the Pliocene of Nebraska (Short 1969) and *Anser* and *Branta* have apparently replaced an earlier (mid-Pleistocene) radiation of sheldgeese in North America (Short 1970).

Although Fig. 14.2 demonstrates that most geese have the capacity to be stimulated into reproductive maturity by shorter photoperiods than are generally required by swans the adaptive significance is not immediately obvious. This is because above latitude 64°N stimulatory photoperiods exceeding 17 h duration occur from late April onwards so that all swan and goose species must rapidly be forced into active gametogenesis the further north they migrate. At Slimbridge the White-fronted Goose *Anser albifrons* and Bewick's Swan lay when the daylength reaches 16·7 and 16·8 h, respectively. At comparable breeding grounds in west Siberia, White-fronts lay a few days earlier than do Bewick's Swans, but both species arrive in late May, when there is already 24 h of daylight, and begin nesting activity as soon as the snow melts. On the other hand, adaptive differences are evident at lower latitudes between the Mute Swan *Cygnus olor* and Greylag Goose *Anser anser*, the centre of their ranges being latitude 50° and 56°N, respectively. At Slimbridge, the western race of the Greylag breeds on a photoperiod of 14·2 h compared with 14·5 h for the Mute Swan while the Eastern Greylag, whose mid-range is at latitude 50°N, begins breeding on a 13·6 h photoperiod. Near the Aral Sea, U.S.S.R., (40°N latitude) the Mute Swan

[†] The essential feature is that the growing cycle is telescoped at these high latitudes and a short period occurs when nutrients build up before a rapid phase of growth.

lays about the fourth week of May (cf. the Whooper Swan which is about a month earlier) whereas the Greylag begins incubation in the second half of April when freshwater is still covered with ice (Dementiev and Gladkov 1952). According to Young (1972), who studied the Greylag in Galloway, Scotland, egg laying begins from 4–10 March; the majority of birds lay in the last week of March and first 10 days of April; and 80 per cent of clutches are begun within a mean of 6·5 days of each other. Mute Swans lay about 2–3 weeks later and also over a more extended season so that repeat clutches can be found into June. It has been suggested by Haapanen, Helminen, and Suomalainen (1973) that Whooper Swans are not well adapted to breeding in the northern boreal region, where only 60 per cent of summers are long enough for a complete breeding cycle, and the southern boreal zone is apparently optimal for the species in Finland. Those races of Canada Geese that breed and winter at the same locality in North America are the first to begin breeding at Slimbridge. The Giant Canada Goose lays at Slimbridge when daylengths reach 13·8 h. It winters at latitude 44°N (see Fig. 14.9: see between pp. 434-5) and departs for its breeding grounds in Manitoba, usually between 20 March and 13 April (Gulden and Johnson 1968), that is, on a photoperiod averaging about 13·7 h. Daylengths are not very much different on arrival at the breeding grounds at 50°N. Comparable data for other races of the Canada Goose are plotted in Fig. 14.9.

Adjustment to latitude

As ancestral swan–geese radiated north they would have been exposed to a markedly increased intensity of light in summer, that is, the increased *Zeitgeber* strength should cause an increase in frequency of their circadian (circannual) oscillators and result in a more positive phasing of their breeding rhythms. That is, the birds' reproductive apparatus would be stimulated into a functional condition earlier in the season, whereas the adaptive response necessitates a delayed phasing. Figs. 14.4 and 14.5 suggest how this could be achieved by a decrease in frequency of some circadian system. If it be assumed that the swans as a group possess the same basic oscillator system which controls the onset of reproduction by allowing gonadotrophin secretion then it is conceivable that adaptation to different latitudes has involved a relatively trivial adjustment in the level of a pacemaker. This kind of change would be expressed as the interspecific differences which are observed when the species are exposed to the same standard *Zeitgeber* as in Figs. 14.2, 14.4, 14.5, and 14.6. Fig. 14.5 emphasizes that the inherent limitations of the oscillator system of the swans prevents them breeding earlier. Ecologically this may not have mattered while their young were reared on aquatic invertebrates but it presumably prevented them exploiting the grazing niche. It is therefore suggested that a more fundamental change in circadian organization accompanied the emergence of the goose ancestors (see below) and this enabled them to radiate to fill the north-temperate grazing niche, and to rear their chicks on grass.

Inspection of Fig. 14.2 shows that the regression lines relating laying date to latitude of origin in swans and *Anser* and *Branta* geese cross. 'Geese' have the capacity to respond to shorter photoperiods than 'swans' only above latitude 50°N and at

lower latitudes must be less well adapted than swans. This seemingly provides a satisfactory explanation for the observation that, with the exception of the abnormal
Bar-headed Goose *A. indicus,* which is discussed below (p.404), only one of 25 *Anser*
and *Branta* geese species breeds south of latitude 50°N. This is the Hawaiian Goose
B. sandvicensis or Ne-Ne, which certainly evolved from northern *Branta* stock to colonize
the montane slopes of Hawaii (20°N latitude). This isolated island was clearly unsaturated
by endemic species, as the success of introduced aliens has testified, and evidently
the grazing niche was vacant. In the tropics there is almost no regular periodicity in
the nutritive cycle of grass and no marked peak in protein content on which a gosling
can grow. Indeed, other wintering geese do not move further south than 30°N. The
difficulty of rearing young on grass in Hawaii is reflected in a growth rate which is
much slower than in the other *Branta* or *Anser* geese and it is not surprising that the
Ne-Ne nearly became extinct when man began altering the ecology of the islands.
Our understanding of evolutionary processes would clearly be hindered without
benefit of the Hawaiian Goose datum, so this is a good example of the scientific
value of conservation. Today the distribution of land in the southern hemisphere
does not provide the large area of high-protein grass which is so typical of the northern muskegs and tundras. But a better grazing niche must have existed in the past,
especially in the post-Gondwanaland land mass. It was probably occupied by the
ancestors of the sheldgeese *Chloëphaga* which today do feed on the grasslands of
South America.

A primitive species which is worth consideration at this point is the Cape Barren
Goose *Cereopsis novaehollandiae,* which has a relict distribution on scattered islands
off the south Australian coast. Ecologically it may always have been restricted to
the coast for it has enormously developed salt-extracting glands—a feature shared
with an extinct relative, the flightless *Cnemiornis,* of New Zealand. It leads a goose-
like existence, grazing grass on dry land, but it is only convergently similar to, and
not immediately related to, the true geese of the northern hemisphere. Actually, its
taxonomic status has long been in doubt. In several respects it resembles the South
American sheldgeese, for example, in the colour and pattern of the downy young,
the use of the 'wrists' of the wings in attack and in having an erect chest-puffing
display. In other ways, it resembles the swans, for example, the gander takes a major
share in nest-building. The goslings are occasionally oiled by the parents, a behaviour
which is otherwise only known in the screamers. Another primitive feature is that
copulation occurs on land. Kear and Murton (1973) compared the photo-response
of the species at Slimbridge with that of sheldgeese, shelducks, swans, and geese (see
Figs. 14.2, 14.4, and 14.5). They concluded that it was closer to the swans than to
any other group and close to the line from which emerged the sheldgeese and shelducks. As Fig. 14.4 shows, it breeds in winter at Slimbridge and this capacity to be
phased into reproductive condition under short days but to be inhibited by summer
photoperiods, like the Black Swan, is regarded as evidence for its evolution under a
temperate photo-regime to the south of its present-day range. In the Bass Straits,
Australia, it breeds in winter with early eggs in May, most in June and July, and a
few in August and September (Kear and Murton 1976).

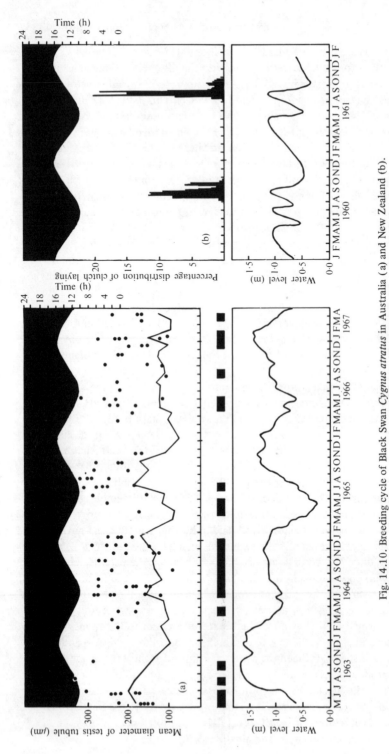

Fig. 14.10. Breeding cycle of Black Swan *Cygnus atratus* in Australia (a) and New Zealand (b).
(a) *Top*: Mean diameter of testes tubules of monthly samples from Barrenbox Swamp, New South Wales. Filled circles refer to testes containing sperm and with little sign of necrosis. Solid horizontal bars delimit the breeding season in terms of egg laying. *Bottom*: Variations in water level at Barrenbox Swamp. Based on Braithwaite and Frith (1969).
(b) *Top*: Approximate egg-laying season of the Black Swan at Lake Ellesmere, South Island, New Zealand. *Bottom*: water level at Lake Ellesmere. Derived from Miers and Williams (1969).

The swans have, in a sense, been overtaken by the geese in the north temperates. Fig. 14.2 also implies that the swans would be poorly adapted to equatorial photoperiods—though no swan species occurs in this region—because the slope of the regression relating photoperiod at egg laying to latitude of origin is steep and this indicates an imprecise timing ability at low latitudes. Perhaps this is why the more recently evolved diving ducks of the tribe *Aythyini* (see Fig. 14.2) partition the 'swan niche' in equatorial and tropical regions, aided by their capacity to dive for submerged plant material which compensates for their short necks. Murton and Kear (1973*b*) showed how, compared with swans, their photo-responses as measured at Slimbridge implied a more precise use of seasonal changes in daylength as proximate regulators of the breeding season at low latitudes. This probably explains why in Australia the breeding season of the Black Swan is not immediately influenced by the daily photoperiod whereas the Hardhead or Australian White-eye *Aythya australis* is distinctly a seasonal and photoperiodic breeder (Figs 14.10 and 14.11).

The trans-equatorial distribution of the South American and African Pochards *Netta e. erythrophthalma* and *N. e. brunnea* and of the remaining *Aythyini*, with 9 species in the northern hemisphere and 3 in the southern, suggests a centre of evolution in the tropics; the less specialized members of the group inhabit lower latitudes. *Netta* species are more primitive than the *Aythya* ducks and feed primarily on water-

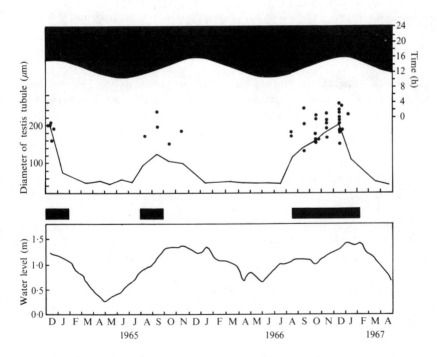

Fig. 14.11. Breeding cycle of the Australian White-eye or Hard-head *Aythya a. australis* at Barrenbox Swamp, New South Wales, Australia. Conventions as for Fig. 14.10. From Braithwaite and Frith (1969).

weeds from near the surface of lakes, although their young require this diet to be supplemented with water invertebrates. Similarly, at low latitudes the white-eyes[†] mostly take plant food, for which they often dive and they frequent pools and lakes and slow flowing rivers with dense shore vegetation and luxuriant weed growth. Although the diving ducks are potentially more efficient than swans near to the equator, and swans are absent, the situation is changed with increase in latitude (Fig. 14.2). At 40° latitude the swans require a photoperiod of about 12 h to begin breeding, whereas the *Aythya* ducks cannot respond until daylengths are nearer to 15 h. Similarly, in the wild in New Zealand, at about this latitude, eggs of the New Zealand Scaup *A. novaeseelandiae* were not laid until 4 November–22 December whereas Black Swans began egg laying in August and continued until December (Miers and Williams 1969). At this latitude the swans can make use of a given resource for feeding their young sooner than the scaup which, therefore, must be ecologically separated in their feeding preferences. The trend becomes even more pronounced at 70°N latitude, for Bewick's Swan, Whooper Swan, and the European Greater Scaup *A. m. marila* breed at about the same time on the shallow lakes formed by the melting snow. In this situation the swans can graze water weeds and the shallow waters allow the diving ducks no advantage in retrieving plant food from deeper areas. Ecological separation has been maintained because at these higher latitudes the diving ducks feed primarily on water invertebrates. The Tufted Duck *A. fuligula* feeds extensively on freshwater molluscs and the larvae of water insects (Olney 1963). In North America its place is taken by the Lesser Scaup *A. affinis* (food detailed by Dirschl (1969)) while the Pacific Greater Scaup *A. marila mariloides* favours a more marine habitat.

The Black Swan is truly photoperiodic in the sense that its breeding rhythms are phased by the seasonal day–night cycle. Only when it is exposed to the large-amplitude *Zeitgeber* experienced around latitude 50°–60° is it evident that a period of refractoriness to long summer days develops, the typically adaptive response of temperate evolved species. The phasing of the rhythms which control the gonad cycle in the Black Swan, and some other species from Australia and South America, allows a double breeding season in northern Europe (see also p. 231). Both the Black Swan and Black-necked Swans have time to lay two clutches at Slimbridge between late January and April before refractoriness to long summer days develops and inhibits further egg laying. As daylengths fall in the late summer refractoriness is ended in the Black Swan in time for the birds to respond to autumn photoperiods, so that a third clutch is sometimes laid in September. It is evident that a laying pattern involving clutch completion in January–February, March–April, and August–September cannot be adaptive in northern Europe. Coscoroba Swans lay only one clutch in spring (March–April) at Slimbridge but a pair once laid again in September. Perhaps at some time in the future Black-necked Swans will produce eggs in late summer. Two other species whose photo-responsive thresholds are set low show the

[†]Madagascar White-eye *Aythya innotata*, Ferruginous Duck *A. nyroca*, Australian White-eye *A. australis.*

beginnings of sexual behaviour in autumn, these being the Cape Barren Goose and Hawaiian Goose. The Hawaiian Goose becomes sexually active in the autumn at Slimbridge and copulation occurs. In the wild the cycle would progress but at Slimbridge the shortening of the days to winter inhibits further activity until mid-February, when egg laying occurs under a mean photoperiod of 10·8 h. There is a possibility that by now the male and female are not properly in phase and this may partly account for a high rate of infertility which has been noted (40 per cent or more eggs are infertile (Elder 1958)). In Hawaii, Elder (1958) found the breeding season to extend from late October until early February under daylengths ranging from 12·5 h to 12·0 h (the shortest day in Hawaii is around 11·8 h and the longest 14·2 h). The Radjah Shelduck *Tadorna radjah* once laid eggs at Slimbridge in August, after the post-nuptial moult, and this may have been a similar phenomenon (Murton and Kear 1976). There is a similarity between some of these low-latitude species and the Rook and other temperate-zone birds which have a low photo-response threshold and which, after having a breeding season in the spring, display a resurgence of sexual activity, with occasional egg laying, in the autumn (see p. 256).

Species having the capacity for an extended breeding season are likely to be good candidates for domestication. The Swan Goose *Anser cygnoides* begins egg laying in late March or early April at Slimbridge and remains active for almost 2 months. From it has been derived the domesticated Chinese Goose which is supposed to have two egg-laying seasons in the year, one in spring and again in autumn. In this respect it bears a degree of comparison with the Black Swan and Hawaiian Goose and it may be supposed that selection for domestication has operated to reduce the photo-response threshold.

Another domesticated species is the Muscovy Duck *Cairina moschata,* of which the wild ancestor has an equatorial and tropical distribution from Mexico south to Peru (mid-latitude of range is 7°S). The wild-type has the potential for a long breeding season at northern latitudes for at Slimbridge eggs have been laid as early as 24 February (11·7-h photoperiod) through until 20 September (daylength 13·4 h). Mention should be made of the Mallard Duck *Anas platyrhynchos,* from which has been derived the domestic ducks. The ancestral stock probably does not have a particularly long breeding season in the wild, but it is now difficult to find stocks which have not been influenced by selective breeding. At Slimbridge, Ogilvie (1964) has shown that wild Mallard Ducks may begin laying during the second week of February and continue until the last week of May. Moreover, autumn breeding may occur, for Kear (1961) found some abandoned ducklings in November 1960 (see Fig. 13.4, p. 347). Incidentally, these were reared in captivity and themselves produced a proportion of fertile eggs at the age of 7 months. Boyd (1957) recorded that a female Mallard Duck accompanied by ducklings was caught at a decoy on 6 November. The same female had been ringed the previous September as a juvenile of the year, which suggests that under some conditions birds reared in the spring can themselves breed in the autumn. Domesticated breeds are expected to produce eggs at 4–5 months.

In contrast to the Australian waterfowl discussed above, which can breed under short photoperiods, those species which require very long photoperiods before achieving full gametogenetic development may not exhibit true refractoriness under the daylength cycle experienced at Slimbridge. For example, Bewick's Swan begin egg laying on a photoperiod of 16·7 h, continues past the summer solstice, and then cease laying when the photoperiod declines again to 17·1 h. As with the Collared Dove discussed previously (see p. 268 and Fig. 10.8, p. 269) the birds probably cease breeding in consequence of a reduction in the daily photoperiod before true refractoriness becomes expressed; refractoriness should be demonstrated by artificially keeping subjects on a long (over 17·5 h) photoperiod. This may explain why the species plotted in Fig. 14.4 as requiring long days before beginning egg laying have longer egg-laying seasons than would be expected—their photo-response mechanisms appear to have become set at such a high level that they are insensitive to the inhibitory effects of long days.

Hybrids

It would be interesting to have information about the breeding season of hybrids, as this should help define the genetic basis of the differences in breeding biology of closely related waterfowl. Unfortunately, the breeding of hybrids is not encouraged at Slimbridge and data are limited. Davies, Fischer, and Gwinner (1969) studied 5 geese species held captive at Seewiesen, Austria (latitude 48°N), and noted that each nested at a characteristic time. Females paired with males of a different species laid at the same time as when paired with males of their own species, that is, the male did not modify the female's cycle. Female Greylag/Snow Goose hybrids paired to male Snow Geese started to breed later than would Greylag, but earlier than the mean time for Snow Geese. Since these hybrid females were intermediate between the parent stocks in their breeding time we have evidence that the characteristic breeding seasons of waterfowl as considered in this chapter are inherited, with perhaps several genes at different loci being involved. Mallard *Anas platyrhynchos* females paired with a conspecific drake, or with Grey Duck *A. superciliosa* males, laid at the time characteristic for the female and the same applied to Grey Duck females paired with a conspecific male or a drake Mallard (Williams and Roderick 1973). One-year-old hybrid Grey Duck/Mallards (whether G/M or M/G) began laying on a date intermediate between that of the female parent stocks, as did the geese hybrids just mentioned. Grey Ducks have a more extended egg-laying season than do Mallards (see Murton and Kear 1976) and the first-year hybrids continued laying for almost as long as the Grey Duck. Thus, while the onset of breeding was intermediate between that of the parent stocks, the length of the breeding season as determined by the onset of refractoriness was of the Grey Duck pattern. However, 2-year-old hybrids had short breeding seasons like the female Mallard and, moreover, now started laying at the same time as the Mallard. Clearly, the situation is complicated.

Black Swans in the wild

Although we have shown the Black Swan to be photoperiodic, it does not follow that the daylight cycle provides the immediate proximate factor timing its breeding season in Australia. That it does not was well shown in studies of the species near

Griffith (latitude 34°S) in New South Wales by Braithwaite and Frith (1969). Here
the rains are unpredictable in intensity rather than time, for they tend to be distributed about the winter solstice, from April to September, rather than around the
summer solstice. On average Black Swans need to be in breeding condition during
the season of short daylengths, which at this latitude reach a minimum of 10·8 h
(compared with 15·5 h in mid-December). Presumably they were pre-adapted to do
this if we are correct in our view that the species had already evolved an adaptive
response to short days at higher (more southern) latitudes. By moving to a lower
latitude (towards the equator) the species would be exposed to a *Zeitgeber* of reduced
amplitude, which would allow the cycle of active gametogenetic activity to be
extended. In fact, in New South Wales, Black Swans have partially developed gonads
throughout much of the year, as Fig. 14.10 demonstrates. There is a tendency for
synchronized involution to occur around the summer solstice, although there is much
variability in gonad condition (phasing) within the population. If it were possible to
have details for individual birds it seems likely that each would exhibit an extended
season of gonad enlargement followed by a short period of refractoriness and that
the onset of gonad involution would vary between individuals. In the autumn of
each year of study, despite dry conditions in 1965 and 1966, and severe drought in
1967, there was a fairly general resurgence of sexual activity in the population. This
kind of physiological response is imprecise and species possessing it must be at risk
of being replaced by better-adapted ones. In unpredictable environments the chances
of another species becoming better adapted are doubtless negligible so that birds
such as the Black Swan offer the best compromise in adaptation. Hence, in the Black
Swan, the breeding season is phased approximately by the seasonal photoperiodic
Zeitgeber, and the resurgence of gametogenetic activity from autumn to spring
broadly corresponds with winter rains. However, actual breeding is determined by
the condition of the habitat. It is conceivable that development of the gonads is
stimulated by nutritional factors, for breeding occurs during periods of rising or
peak water levels, and Braithwaite and Frith pointed out that this corresponded
with the availability of certain foods. They favour the view that when feeding conditions are good the birds have time to indulge in intra-pair courtship and this
provides the direct stimulus which activates the reproductive apparatus, a mechanism
first proposed by Murton and Isaacson (1962) to account for the timing of egg laying
in the Wood Pigeon.

The tendency for gonad involution to occur in Black Swans in New South Wales
in summer becomes much more marked under a north-temperate photoperiod, as
Fig. 14.4 has demonstrated. Similarly, in the more southern parts of their range the
birds tend to breed regularly in winter and early spring (Braithwaite 1970). Australian
Black Swans were introduced into New Zealand in or around 1864. At Lake Ellesmere
(latitude 43°S) in South Island, Miers and Williams (1969) have shown that the species
is markedly seasonal, egg laying usually beginning during the first week of August
but sometimes as early as mid-July (daylength about 10·5 h), to follow a winter peak
in water level (Fig. 14.10). Egg laying ceases in October as the daily photoperiod
increases to 14·5 h. It seems unlikely that the Black Swan could be successful in New

Zealand if ecological conditions made it necessary for breeding to occur round the longest day. Black Swans introduced to and captive in Japan (latitude 35°N) began laying in autumn and early winter and finished in March. Kikkawa and Yamashina (1967), who recorded this, suggested that environmental cues which trigger or inhibit breeding in Australia were not operative, but the photoperiodic response is exactly what would be predicted; Ullrich (1949) also mistakenly thought that the breeding season was not reversed in the northern hemisphere.

It is of interest to compare the testicular cycle of the Black Swan in New South Wales with that for Australian White-eye at the same locality, and data from Braithwaite and Frith are depicted in Fig. 14.11. The testes enlarge in August and September and regress in late January, whereas egg laying is at a peak in late November and early December, when the daily photoperiod is about 15 h; egg laying is noted at Slimbridge with a daylength of 15·2 h. Because this species breeds at a fixed time each year in response to photoperiodic stimuli there are marked variations in breeding success, for the birds attempt to breed irrespective of environmental conditions. It would seem that at this locality the Black Swan is better adapted to the variable environmental conditions than the Australian White-eye.

Photoperiodic mechanisms in Anatidae

Inspection of Fig. 14.4 suggests that two distinct processes control the breeding seasons of closely related waterfowl. There appears to be a level or threshold effect which is common to all taxa. In addition, there is a factor(s) which determines how long the breeding season will last once initiated, that is, under what range of photoperiods refractoriness develops and in this respect there is more similarity within than between taxa. For example, species of *Dendrocygna* differ from one another in the daily photoperiod that is required to initiate breeding and, in the same way, *Cygnus* species show such intra-generic variation. In addition, all *Cygnus* species develop refractoriness under long photoperiods, whereas the *Dendrocygna* ducks hardly seem to do so and *Anser* and *Branta* species do so more rapidly than the swans. It is noticeable from Fig. 14.4 that closely related species differ less in the daily photoperiod at which refractoriness begins than in the range of daylengths at which breeding is initiated. It will be recalled that the same conclusion applied to closely related *Zonotrichia* species (Fig. 9.8, p. 235). It is impossible to explain these differences with a one-oscillator model of photoperiodism. In fact, with our present knowledge of the nature of the oscillation systems involved in the photoperiodic response mechanism any proposed model must be speculative. But it is worthwhile trying to formulate some hypothesis to serve as a basis for future experimentation and it seems sensible to base this on the scheme outlined on p. 321.

The onset of breeding must result from gonadotrophin secretion which is stimulated by an appropriate photoregime. Differences in the photosensitive threshold presumably imply that gonadotrophin secretion is more or less easily stimulated by a given photoperiod to give the kind of target tissue response depicted in Figs. 9.11 and 9.12. This resolves into the way in which daylengths are measured by the photoperiodic clock and the sensory input is transduced via hypothalamic pathways to stimulate the secretion of gonadotrophin releasing factor(s). We might be dealing with the level and hence frequency relative to a biochemical or neural threshold of a single hypothalamic oscillator. Alternatively, regulation of the secretion of releasing factor might depend on an on—off switch provided by a double oscillator system. One oscillator might switch on the release mechanism and the second stop the system. Then the time during which releasing factor was secreted would depend on the phasing of the two oscillators and this would in turn be determined by their level and hence frequency. It will be possible to measure gonadotrophin secretion under different photoregimes in selected species to test the simple part of the hypothesis. Understanding the clock mechanism is, however, the central problem in avian photoperiodicity.

A second mechanism apparently determines the length of the breeding season, by regulating the point at which gonad regression is affected by thyroid feed-back. In 'primitive' species it is possible that there is a simple relationship between the phasing of gonadotrophin secretion and thyroid activity so that above a particular daylength threshold only gonadotrophin effects prevail. Such a scheme might explain the kind of symmetrical response whereby the gonads become stimulated under a particular daylength in 'spring' and remain active until the corresponding photoperiod is experienced in 'autumn'. Another possibility is that gonadotrophin activity and thyroid activity are simultaneously stimulated by appropriate photoperiods but that TSH and thyroid output is always below the level at which feed-back effects become operative; some pigeons seem to have this kind of cycle. The special feature of 'temperate-type' breeding cycles is that thyroid feed-back must be stimulated under a genus-specific long day regime (see Fig. 12.7). There is less variability in the photoperiod under which breeding ceases in closely related species than in the photoperiod under which breeding is initiated. This argues for a degree of independence between the initiating (gonadotrophin secretion) and terminating (thyroid feed-back) factors involved in regulating the length of the breeding season, while such an independence need not be postulated for the 'primitive-type' breeding cycles. These questions should soon be resolved by the systematic assay of plasma hormone levels in a selected range of duck species. Once this has been achieved it will be easier to formulate questions about the relationships between TSH and gonadotrophin secretion, both on a daily and on a seasonal basis.

If the photo-responses noted within a genus implicate the same oscillator system constrained by variations in threshold it would mean that the records for individual species which are plotted in Fig. 14.4 could be regarded as readings of level referring to a common system. In other words, Fig. 14.6 could be re-drawn as in Fig. 14.12, with the abscissa representing readings of threshold instead of the photoperiod at which individual species begin egg laying (the ordinate of Fig. 14.4 becomes the abscissa of Figs 14.6 and 14.12 since this must represent the independent variable or x-axis, and the phase shift must be the dependent variable). Fig. 14.12 simply shows that with increase in level or decrease in threshold the phase shift becomes more positive which would follow if the frequency is increased by increase in level. Reference again to Fig. 10.11 (p. 278) suggests that a similar process could be implicated in the breeding periodicity of different Feral Pigeon morphs. We assume that changes in

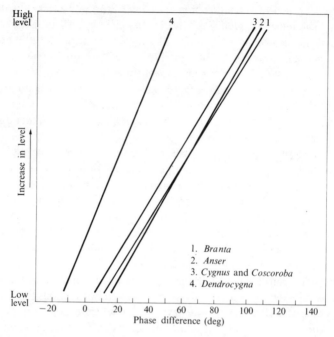

Fig. 14.12. Schematic diagram based on Fig. 14.6 in which it is assumed that a capacity to respond to a short photoperiod indicates that the level of a controlling oscillatory system is low. Lines are the regression relating photoperiod to phase difference in Fig. 14.6. See text for discussion.

testis size reflect variations in the oscillator system controlling gonadotrophin stimulation and that those morphs which achieve the biggest testes have high oscillator levels (low photosensitivity thresholds within the concept depicted in Fig. 11.1, p. 287). Fig. 10.11 shows that the gonad cycles of the birds attaining the biggest testes are more positively phased to the *Zeitgeber* than those which seemingly have a high photosensitivity threshold.[†]

Fig. 14.13 shows that there was an increase in length of the breeding season with decrease in the photoperiod needed for breeding to begin (considered equivalent to an increase in the level or frequency of the underlying oscillator system). This happened because there was more interspecific variation in the photoperiod needed to initiate breeding than in the range of photoperiods at which refractoriness developed. Because species which begin breeding under short photoperiods (have high-frequency

[†]This leads us to suspect that the series of alleles which are responsible for melanism are associated in some way with changes in frequency of the oscillation system of *Columba livia*. Melanism seems to be a feature of the morphs with a reduced threshold and hence biggest phase difference during entrainment.

Fig. 14.13. (see opposite and top p. 396). Relationship between length of egg-laying season in days for various species of wildfowl at Slimbridge and the photoperiod under which breeding begins: (a) *Dendrocygna* and *Anseranas*; (b) *Cygnus* and *Coscoroba*; (c) *Anser* and *Branta* geese.

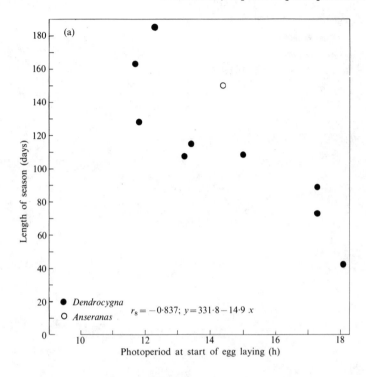

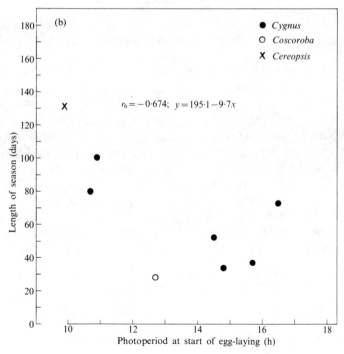

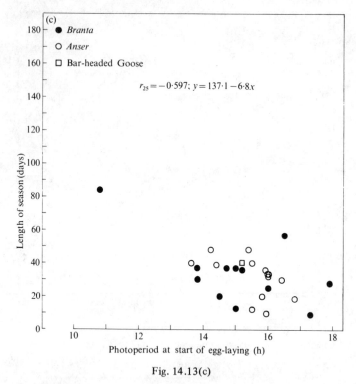

Fig. 14.13(c)

oscillator systems) tend to live at low latitudes there tends to be only a weak inverse relationship between the length of the breeding season and latitude (see Fig. 14.14).

Body and egg weight and photoresponses

In the *Anser* and *Branta* geese adult body weight is inversely correlated with the mid-latitude of the breeding range, although the Bar-headed Goose and Hawaiian Goose are clearly exceptional (Fig. 14.15(a)).[†] Body weight is, however, more strongly correlated with the daylength under which egg laying occurs at Slimbridge (Fig. 14.15(b)). Moreover, when body weight is plotted against the photoperiod at which breeding begins at Slimbridge the apparently anomalous results obtained for the Bar-headed Goose and Hawaiian Goose are corrected. This provides further evidence that these two species really 'belong' to a different latitude. The observations suggest that the correlation between body size and latitude does not depend on an ecological adaptation to latitude *per se* but reflects a metabolic adaptation which is correlated with the photo-response mechanism. This correlation is probably not directly causal but

[†] Since the species which breed the furthest north winter the furthest south the correlation becomes a direct one in winter and, therefore, the geese are not an exception to Bergmann's rule.

Fig. 14.14 (see opposite and top of p. 398). Length of egg-laying season in days for various species of wildfowl at Slimbridge and the mid-latitude of their natural breeding range.

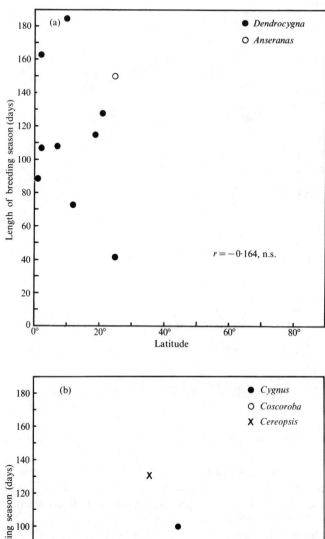

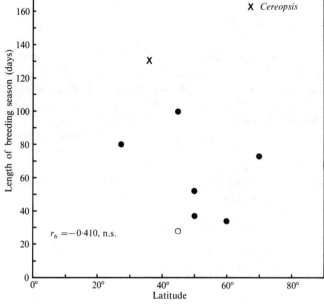

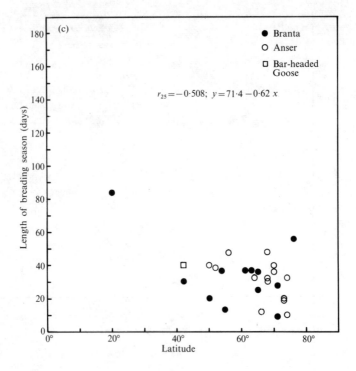

Fig. 14.14(c)

could depend on both functions, that is, metabolic rate and breeding season, being
dependent on the same basic circadian—circannual oscillator system. Since so many
of these physiological variables are interdependent it is not always easy to identify
causal mechanisms. For example, data in Scott *et al.* (1972) show that the growth
rate of swan cygnets is related to latitude and hence photo-response and that the
northern Arctic species have an increased growth rate. As mentioned, the young of
these high-latitude forms must feed on invertebrates whereas Black Swan cygnets
can be reared on a completely vegetarian diet. It is to be wondered to what extent
a change in diet was consequent on an altered circadian system. It is clearly important
to understand physiological limitations when assessing the ecological adaptations of
animals. Würdinger (1975) has shown that *Anser* and *Branta* geese exhibit species-
specific development rates when reared under identical conditions and these are
correlated with the latitude of origin.

Egg weight is correlated with adult body weight (Fig. 8.2, p. 189) and, therefore,
with latitude (Fig. 14.16(a)) and even more with photoperiod (Fig. 14.16(b)). As
in the case of body weight, plotting egg weight against the photoperiod corrects the
otherwise anomalous results for the two montane species. When egg weight was made

the dependent variable in a multiple regression analysis* which treated body weight, daylength at laying, and latitude of origin as independent variables, egg weight was found to depend primarily on adult body weight. Hence the relationship between egg weight and photoperiod depends on the correlation already established between body weight and photoperiod. Lack (1968) has shown that clutch size is highest at mid-latitudes and statistical tests confirm that there is no correlation between clutch size and latitude, nor between clutch size and the daylength at laying; it has already been emphasized that clutch size and egg weight are subject to different selection pressures (p. 192).

Other species

Instead of considering the breeding responses of different species at one latitude it is interesting to consider the responses of the same species at a range of latitudes, that is, the *Zeitgeber* is made the variable. For example, the date of first laying by captive flamingoes (family: Phoenicopteridae) in zoos in the northern hemisphere begins as early as January, February, and March at latitudes 20–30°N, is mostly in April and May between latitudes 30–40°N, and is in May and June above latitude 40°N (Hall and Kear 1975).

The laying date of the Giant Petrel *Macronectes giganteus* on islands of the Scotia Arc from South Georgia (latitude 55°S) and the Falkland Islands to the mainland of Antarctica (66°S) is shown in Fig. 14.17 and a straight-line relationship with latitude is apparent. The Giant Petrel is dimorphic with pale and dark colour phases.[†] The proportion of pale-phase birds is said to increase towards the South Pole (Murphy 1936; Bourne and Warham 1966) but this has been disputed (Hudson 1968; Conroy 1971; Shaughnessy 1971). It is not established whether the dark phase begins breeding earlier than the white morph (see p. 428). However, Carrick (quoted by Conroy (1972)) noted during the course of studies of *Macronectes* on Macquarie Island that there were two distinct populations breeding about 6 weeks out of phase with each other. Further investigations led to the recognition of a second northern sibling species designated *M. halli* (Warham 1962; Bourne and Warham 1966) which frequents the sub-Antarctic zone including Chatham, Macquarie, Kerguelen, and Gough Islands. Fig. 14.18 suggests that *M. halli* could be the original form[‡], for it has the capacity

---

*For 15 *Anser* and *Branta* geese plotted in Fig. 14.16:

$$y = 189.6 + 0.17x_b - 8.46x_d + 0.36x_l$$

where y = egg weight, x_b = adult body weight, x_d = photoperiod at egg laying, and x_l = mid-latitude of breeding range. The coefficients have the following standard errors and t-test values.

$$x_b = 0.17 \pm 0.003; \quad t = 4.972; \quad P < 0.001$$

$$x_d = 8.46 \pm 5.92; \quad t = 1.430; \quad \text{not significant}$$

$$x_l = 0.36 \pm 0.66; \quad t = 0.542; \quad \text{not significant}$$

[†]The dimorphism is controlled by two autosomal alleles with white dominant to dark and mating is at random (Shaughnessy 1970a).

[‡]Speciation can have only recently occurred, for Shaughnessy (1970b) could find no difference in the mobilities of transferrin, albumin, and haem-binding proteins in the two species using starch-gel electrophoresis.

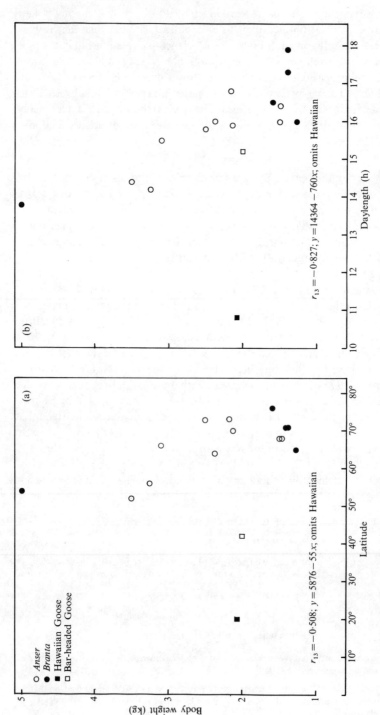

Fig. 14.15. Adult body weight of different *Anser* and *Branta* geese in relation to: (a) mid-latitude of natural breeding range; (b) photoperiod under which egg laying begins at Slimbridge.

The anomolous result for the Bar-headed Goose in (a) is discussed in the text. In (a) there is an apparent anomaly, in that the geese break Bergman's rule. This is not so for in winter the most northern breeding species tend to migrate further south so that there is a positive correlation (not shown) between body weight and mid-latitude of range during the contra-nuptial season, the Hawaiian Goose included. This species is exceptional in that its summer and winter range are effectively the same, and it is a slight altitudinal migrant.

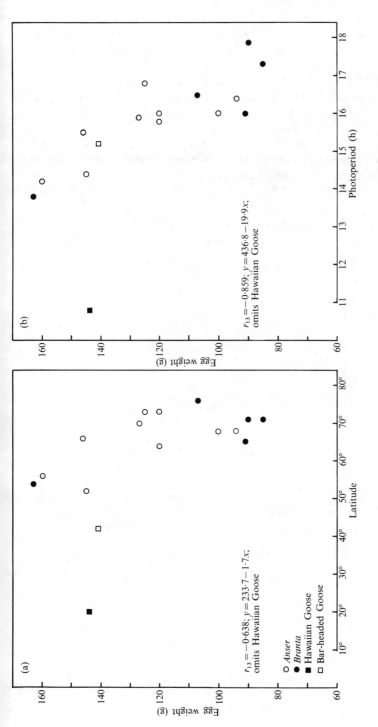

Fig. 14.16. Egg weight of different *Anser* and *Branta* geese in relation to (a) mid-latitude of natural breeding range; (b) photoperiod under which egg laying begins at Slimbridge.

to breed on shorter photoperiods than *M. giganteus*. The slope of the regression suggests that at high latitudes *M. halli* would breed much too early, before the onset of the Antarctic summer, and that its congener is equipped with a better-adapted clock.

Fig. 1.1 (p. 3) showed how the breeding season of the Great Tit *Parus major* becomes advanced and lengthened with decrease in latitude. As a result the bird is single-brooded in Britain (Lack 1966), frequently double-brooded in Holland (Kluijver 1951), and treble-brooded in Malaya (Cairns 1956). The laying seasons of the Rockhopper Penguin *Eudyptes chrysocome* at 11 stations between 36° and 55° were roughly correlated with latitude, but more exactly with the mean annual sea temperature (Warham 1972). The phasing of egg laying, moult, and migration with latitude in Chuck-will's-widow *Caprimulgus carolinensis* is illustrated by Rohwer (1971).

In order to provide data pertinent to sexual dimorphism, which is discussed in the next chapter, and to illustrate some other points information regarding the photo-

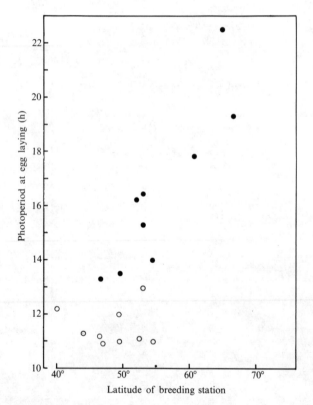

Fig. 14.17. Photoperiod at egg laying by the Giant Petrels *Macronectes giganteus* (solid circles) and *M. halli* (open circles) in relation to latitude of breeding station. There is a correlation between photoperiod at laying and latitude for *M. giganteus* with $r = 0 \cdot 888$; $y = -4 \cdot 6 + 0 \cdot 38x$, but not for *M. halli* ($r = -0 \cdot 045$, n.s.).

period at egg laying for different species of dabbling ducks (genus *Anas*) is plotted in Fig. 14.18 against the mid-latitude of the natural breeding range. The dabbling ducks are a more advanced group than the other Anatidae considered so far and they have undergone a more recent radiation. It is partly for this reason that there is much more variability in the distribution of points in Fig. 14.19 than for some of the groups depicted in Fig. 14.2.

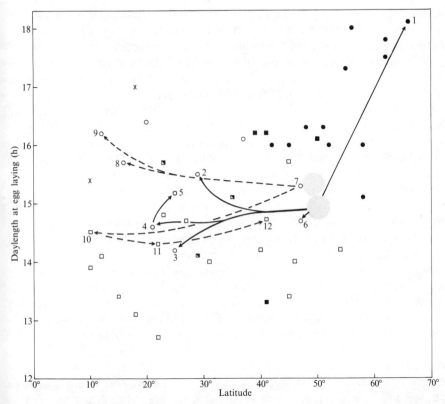

Fig. 14.18. Relation between mean date of first egg laying by different *Anas* species at Slimbridge and the mid-latitude of their natural breeding range.

Circles refer to northern hemisphere species and squares to southern hemisphere ones; open symbols are monomorphic species and solid symbols seasonally dimorphic species. Three southern hemisphere species which are permanently dimorphic are shown by half-filled squares. Crosses refer to high-altitude breeders. Satisfactory records are not available for the Mallard *Anas platyrhynchos*, for captive breeding stocks are much affected by domestication. The stippled areas indicate the presumed positions occupied by the ancestral stocks of the Mallard (lower) and Spotbill *Anas peocilorhyncha* (upper). Lines (solid or dashed) with arrows indicate the presumed radiation from these two centres; the derived forms are as follows: Mallard line: 1, Greenland Mallard *A. p. conboschas*; 2, Florida Duck *A. p. fulvigula*; 3, Mexican Duck *A. p. diazi*; 4, Hawaiian Duck *A. p. wyvilliana*; 5, Laysan Teal *A. p. laysanensis*; 6, North American Black Duck *A. rubripes*. Spotbill line: 7, Chinese Spotbill *A. poecilorhyncha zonorhyncha*; 8, Indian Spotbill *A. p. poecilorhyncha*; 9, Phillippine Duck *A. luzonica*; 10, Pelew Island Duck *A. superciliosa pelewensis*; 11, Australian Black Duck *A. s. rogersi*, 12, New Zealand Black Duck *A. s. superciliosa*. See Murton & Kear (1976) for statistical analysis ($r_{43} = 0.559; P < 0.001$, omitting two high-altitude species designated by crosses).

High-altitude breeding

Spring arrives late in the alpine zone of mountain tops relative to the surrounding lowlands, which experience the same seasonal daylength cycle. Clearly it is an adaptive advantage for species inhabiting these highland zones to require a relatively long photoperiod to initiate breeding compared with birds living in the lowlands. It will be recalled that the Bar-headed Goose *Anser indicus* was shown to be atypical, for it did not fit on the regression line relating daylength at egg laying to mid-latitude of origin of the *Anser* geese (Fig. 14.2). However, it is a high-altitude breeder which migrates from wintering grounds in northern India to nest by the small lakes and rivers which are scattered over the highlands of central Asia, from 760 m in Mongolia to 5600 m in the Pamirs. Inspection of Fig. 14.2 suggests the possibility that the Bar-headed Goose evolved from ancestors having a more northern distribution, a suspicion which is strengthened when its body weight and egg weight are also considered (see Figs 14.15 and 14.16). Johnsgard (1965) placed it close to the Greylag Goose on behavioural criteria, while Fig. 14.2 suggests an ancestry close to the line leading from *Anser* stock to the Whitefront. If a species moves to a lower latitude the new *Zeitgeber* will result in a delay in the onset of breeding activity; to breed at the same calendar date the species would require to lower its photosensitivity threshold (contrast the effect on species of radiating to higher latitudes, p. 384). It seems that the Bar-headed Goose evolved a photo-response at a 'high' latitude which without further modification pre-adapted it to montane conditions further south. In the same way the Sharp-winged Teal *Anas flavirostris oxyptera* which lives in the highland plateau of the Andes (latitude 18°S) is derived from Chilean Teal stock *A. f. flavirostris,* having a distribution further towards the South Pole and with an appropriate photoresponse (Fig. 14.19). It might be difficult for a local lowland species to evolve an adaptive oscillator phasing to suit nearby mountain-tops, for a *spontaneous* change in oscillator frequency would be needed without the stimulus of a change in the environmental *Zeitgeber*. In contrast, species from higher latitudes have experienced the selective pressure of a modified *Zeitgeber* as their stock radiated. It can be deduced from Fig. 14.2 that a reduction in the photo-threshold of the *Anser–Branta* photo-response mechanism could not have provided an appropriate phasing of the breeding season for the 'Bar-headed Goose niche'; that is, the regression line depicted in Fig. 14.2 cannot be extrapolated to pass through the required point. In contrast, a reduction of the photo-response threshold has been appropriate for the Hawaiian Goose, enabling it to adapt to the montane slopes of the Hawaiian Islands (see Fig. 14.2).

Isolated high-altitude species are often regarded as glacial relicts but this view need not always apply. Instead, it is possible that active colonization of montane niches has been most easily achieved by pre-adapted species originating from higher latitudes. The majority of bird species which breed regularly in both Europe and the Ethiopian region are typically species of high altitudes in Africa, as Moreau (1966) has remarked. Examples include the Alpine Swift *Apus melba,* the Quail *Coturnix coturnix*, the Black Stork *Ciconia nigra,* the Buzzard *Buteo buteo*—which Meinertzhagen (quoted by Voous (1960)) showed to be a southern colonist of the

Palaearctic form—and the Lammergeier *Gypaetus barbatus*; the colonizing route of this last species persists as a 'necklace' of isolated populations down the central African mountain chain. In these same mountains the Common Sandpiper *Actitis hypoleucos* and the Common Snipe *G. gallinago*[†] have been claimed to breed (Voous 1960; see also Benson and Irwin 1974). Moreau mentions two other montane birds which though given specific status are very closely related to Palaearctic species, these species pairs being the rock thrushes *Monticola saxatilis/explorator* and the wagtails *Motacilla cinerea/clara*.

When there is not a statistically significant correlation of photo-response and latitude, as with the sheldgeese *Chloëphaga* plotted in Fig. 14.2, we suspect that species have moved from their latitude of origin. This is known to have happened in several closely related *Anas* species-groups, and Fig. 14.19 indicates some of these.

Desert breeding

The major desert areas of the world mostly occur between latitudes 15° and 35° in both hemispheres (see Cloudsley-Thompson and Chadwick 1964), where they miss the influence of the westerly winds of equatorial regions and the trade-wind belts and in consequence experience anti-cyclones for most of the time. Their physical characteristics and vegetation have been usefully summarized by Serventy (1971) in relation to the special adaptations that birds require in order to live in this harsh environment. He lists the physiological adaptations of desert birds to make use of metabolic and pre-formed water, their ability to conserve water by reducing the amount wasted during excretion, and their capacities for temperature regulation (see also Dawson and Bartholomew 1968; Bartholomew 1970; Dawson 1975). Since birds must rely on evaporative cooling to control their body-temperature a premium has been placed on behaviour patterns which reduce heat stress, as well as on physiological means to prevent overheating or a degree of tolerance of hyperthermia (Dawson 1976). Adults have daily rhythms of activity which allow them to seek shade during the most oppressive times. Nests are carefully positioned to shade the incubating bird by day but allow the sun to warm the bird early in the morning. Several species, for example, the Black Wheatear *Oenanthe leucura* build a wall of stones round the nest and this probably contributes to the heat balance. Such a view is supported by temperature measurements made throughout the day at the nest of the Desert Lark *Ammomanes deserti* which uses stones in this way (Orr 1970). Some species have low metabolic levels which may also be adapted to the sporadic and often low productivity of deserts, an example being the Spotted Nightjar *Eurostopodus guttatus* of Australia (Dawson and Fisher 1969). Desert birds often have to face extreme cold and this may pose as many problems as excessive heat.

Most desert birds are able to limit renal water loss by the general avian uricotelic ability, that is, excreting uric acid as the main breakdown product of amino acids. Some have salt secreting glands, the nasal glands, which serve as extra-renal excretory

[†] The Snipe also occurs in South America, living at high altitude in the puna zone of the Andes.

pathways to maintain homeostasis, these glands being the same as those possessed by many sea-birds to enable them to ingest salt-water. Birds which possess salt glands also have larger kidneys than those without such glands (Hughes 1970) while the size of the salt gland varies with the salinity of the drinking-water to which the bird is exposed (Holmes, Butler, and Phillips 1961; Schmidt-Nielsen and Kim 1964). The ability of the nasal gland to excrete extremely concentrated solutions of Na^+, K^+, and Cl^- depends on an intact parasympathetic neural pathway from the seventh cranial nerve and also an intact pituitary–adrenal axis (Phillips and Bellamy 1962; Wright, Phillips, and Huang 1966; Holmes, Phillips, and Wright 1969; Holmes, Chan, Bradley, and Stainer 1970). The control of the nasal gland is apparently extremely complex, as the scheme depicted by Holmes *et al.* (1970), indicates.

Contrary to popular misconception, most deserts support a fair degree of plant cover. The Australian and Kalahari deserts support in places an open xerophytic woodland and even the western Sahara has grass (*Aristida*) and shrubs sufficient for nomadic pastoral tribes to use. In most desert areas avian breeding seasons do not differ much from those of the birds in adjacent non-desert habitats. In arid regions of south-west Africa the breeding season is similar to that in surrounding areas of intermediate vegetation and coincides with the flushing of vegetation, which according to latitude, and hence warmth, occurs just before or some time after rain has fallen (Moreau 1950; Immelmann 1967, 1971*a*, *b*; Immelmann and Immelmann 1968; Maclean 1971): the Immelmanns noted that exceptionally heavy rain in the Namib desert of south-west Africa was not a direct stimulus to breeding activity. Most xerophytic South American birds breed in the austral spring (from September to December in the south) as do the birds living in other habitats. The same is true of the species of the North American deserts, which are no more flexible in their breeding periodicity than species inhabiting nearby mesic areas (Dawson and Bartholomew 1968). Nevertheless, desert species are more sensitive to the ecological consequences of water shortage, even though their breeding season may be proximately phased by daylength cycles in the same manner as those of their non-desert relatives. For example, the Rufous-sided Towhee *Pipilo erythrophthalmus* of North America has a typical photoperiodically controlled gonad cycle with recrudescence occurring in March and regression in July. Further south it is likely that the cycle would be extended and that this applies to Abert's Towhee *Pipilo abertii*. In the Arizona desert this last species has been seen to nest within 10–14 days of heavy rain in March and April, continuing to breed until late summer if rainy conditions prevail. Presumably, birds in the desert have the potential to breed from February–March to July–August but the final expression of reproductive activity is only possible given appropriate ecological conditions, in particular, food supplies and nesting facilities.

Near Tucson, Arizona the Roadrunner *Geococcyx californianus* has a bimodal nesting season from mid-April to mid-June and again from late July until mid-September (Ohmart 1973). This cycle is necessitated by extreme heat and aridity in mid-summer when food supplies of lizards and insects are particularly hard to obtain, but rains in July result in improved feeding prospects. Whether the gonads involute in mid-summer so that autumn breeding is of the kind noted in the Rook (see p. 256) and Mallard (p. 346) is not known. In this area the Brown Towhee *Pipilo fuscus,* Curve-billed Thrasher *Toxostoma curvirostre,* and Rufous-winged Sparrow *Aimophila carpalis*[†] also have a bimodal nesting season, according to Ohmart, and it seems unlikely that they all have bimodal gonad cycles. Ohmart studied some of the adaptations of the Roadrunner to desert conditions. The adults must catch lizards between 07.00 h and 13.00 h and so the chicks are left unguarded at this time from the age of 4–5 days. They are featherless at this age but have black, greasy skins to aid solar brooding. The adults exercise thermo-control during incubation by positioning the nest in vegetation which casts open and shaded bands across the sitting bird. Thus, by shifting position, the incubating adult can use or avoid solar radiation according to the time of day. The birds have a gular flutter to aid in cooling in the manner of cormorants and egrets and they have nasal glands.

Non-Australian desert birds do not exhibit marked annual variations in their breeding periodicity, presumably because there is a regular occurrence of suitable ecological conditions, even if these are poor in some years. Even so, birds which have very sharply defined breeding seasons have not adapted to deserts, while, conversely, those that have tend to be species having the potential to breed over long segments of the year. Moreau (1966) lists typical species of the Saharan desert where 4 species of wheaters (*Oenanthe*) and 7 larks (Alaudidae) are particularly prominent components of the avifauna. In northern Europe, the Skylark *Alauda arvensis* lays eggs from late March until early September and it is probable that the southern larks also have this capacity. Of other species mentioned by Moreau, the Scrub Warbler *Scotocerca inquieta* probably has a long breeding season, like the closely related *Prinia* warblers; the Sand Partridge *Ammoperdix heyi* can be compared with the Red-legged Partridge *Alectoris rufa* which in northern Europe and Britain lays its first eggs in May but can lay repeat clutches through to July, unlike the Grey Partridge *Perdix perdix*, which is single-brooded. Moreau also mentioned Temminck's Horned Lark *Eremophila bilopha*, which is sometimes regarded as conspecific with the Shore or Horned Lark *E. alpestris.* This bird has a remarkably discontinuous distribution occurring above latitude 60°N in North America and the northern Palaearctic and again in the Old World desert and montane regions situated between latitude 25°N and 55°N and extending as far west as China. It frequents tundra, boreal, temperate steppe, and desert climatic zones. Voous (1960) considers that it was originally a bird of high mountains which developed during the Pleistocene Ice Ages into a bird of the tundra, when it perhaps colonized North America through the circumpolar tundra belt. It has also followed the chain of central American mountains to create an isolated population on the high plateau of Colombia at 2680 m. As there are no larks in the New World, the Shore Lark occupies a wide range of niches that are partitioned by several lark species in the Old World. It has a fairly

[†]There is circumstantial evidence that Cassin's Sparrow *Aimophila cassinii* leaves the deserts of Arizona in late May and June and breeds on the Great Plains, where the rainfall regime is different from that of the desert. The birds 'return' to Arizona after the summer solstice in early or mid-July when some breeding again occurs (Ohmart 1969).

long breeding season in northern Europe (Spjøtvoll 1970) and North America (Beason and Franks 1974), and presumably the southern forms, which are exposed to a *Zeitgeber* of lower amplitude, will have the potential for even more extended breeding. The species apparently relies on behavioural rather than physiological adaptations to cope with the desert and this may reflect a recent dispersal into this habitat (Trost 1972).

Australia is usually considered to provide the best examples of desert breeding adaptations in birds. Actually, birds in the tropical north mostly breed in the spring and with the summer rains and display a seasonal periodicity (Slater 1959; Frith and Davies 1961; Lavery, Seton, and Bravery 1968). Those inhabiting the coastal belt from about 142°W in tropical Queensland to Adelaide tend to be spring breeders, as are those in the area occupying the extreme south west corner of Australia south of Perth (Carnaby 1954; Robinson 1955; Serventy and Marshall 1957). Away from the coast towards the interior it becomes more and more likely that there will be droughts in some years (Fig. 14.19). The supposed uniqueness of some of the Australian desert birds is their capacity to breed at any time of year when conditions become favourable following rains.* It is true that drought may persist for many years, whereupon no breeding is attempted and nomadic species disappear. For example, Wedgetail Eagles *Aquila audax* in a semi-arid region of Western Australia were found to breed in 2 years out of 7 when rains provided grass for their rabbit *Oryctolagus cuniculus* food supply (M. G. Ridpath, in preparation). Seemingly more remarkable is that within days of rain falling some species can be involved with breeding duties and nomads congregate to take advantage of the local, and apparently unpredictable, flush of vegetation and invertebrates (Keast and Marshall 1954). Immelmann (1963) noted how in the Black-faced Woodswallow *Artamus cinereus* courtship behaviour began within minutes of rain which broke a drought which had lasted for many months. Copulation occurred within 2 hours and nest-building was begun the next day. Serventy (1971) has quoted similar examples for the Zebra Finch *Piephila guttata = Taeniopygia castanotis,* Budgerigar *Melopsittacus undulatus,* and Bourke's Parrot *Neophema bourkii.*[†] Unlike breeding, moulting in these arid-zone birds occurs according to a fixed schedule, often as a typical, albeit protracted, post-nuptial moult, and is apparently regulated endogenously (Keast 1968). In consequence, in species that breed erratically according to rainfall breeding may be initiated while the moult is in progress. Similarly, in species which tend to breed in spring, the post-nuptial moult occurs even if lack of rainfall inhibits breeding.

The question of whether rainfall acts as a direct stimulus of the neuro-endocrine

---

*S. S. J. Davies (personal communication) has doubts about the supposed sporadic breeding of most Australian birds and thinks that some of the claims result from biased field observations of not well-studied species. He finds that many desert plants are markedly seasonal in flowering and fruiting and that photoperiodic timing is likely to be involved. The birds experience a seasonal environment where the main variations are in the amount of resource available rather than its timing.

[†] Zebra Finches can breed at the age of 2 months if conditions are favourable (Sossinka 1975) and the Budgerigar at 3½–4 months (Pohl-Apel and Sossinka 1975).

apparatus has still to be resolved experimentally (see also Kennedy (1970)). There is no doubt that some birds can identify rain clouds or falling rain. European gardeners know that Blackbirds *Turdus merula* and Song Thrushes *T. philomeles* recognize water coming from a garden hose on a sunny day and fly to the wetted area for they can associate the falling water stimulus with a potential food source of earthworms, etc. But what still needs to be determined is whether the appearance of rain

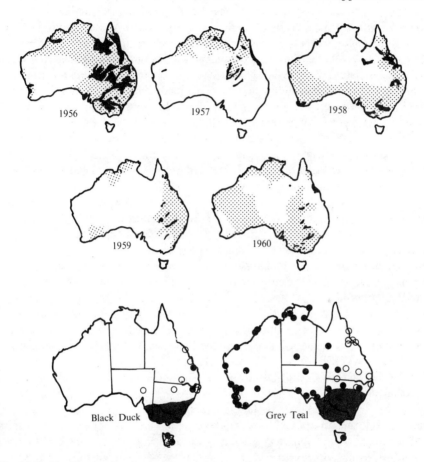

Fig. 14.19. Top five panels: places where rainfall was above average (stipple) and where flooding occurred (solid shading) in Australia in different years. From Frith 1962.

Bottom two panels: Grey Teal *Anas gibberifrons* and Black Duck *A. superciliosa* dispersal patterns in response to drought. In 1956 there was widespread flooding in Australia and associated breeding of wildfowl. Many Grey Teal and Black Duck were ringed at breeding areas in Gum Creek, New South Wales and Joanna, Southern Australia. Their initial dispersal to drought refuges in south-east Australia is indicated by open circles; dark shading encloses an area where many recoveries were made which are not itemized individually. Solid circles show subsequent dispersal from drought refuge areas. In 1957 drought occurred and as the coastal drought refuges dried there was a widespread scatter of Grey Teal over the whole continent, including the driest and most drought-affected areas. In contrast, there was little movement of Black Duck, for these remained in the few permanent swamps. Modified from Frith (1959b).

or clouds causes any of the Australian desert birds to indulge in display that in turn stimulates the hypophysial—gonad axis into activity. At present the indications are that those species which breed with the onset of rain respond to the food supplies that are made available within a day or two of rain falling.

The waterfowl of arid areas are worth consideration, for they illustrate some of the general statements made above. The African Yellowbill Duck *Anas undulata* begins breeding at Slimbridge on an average daylength of 13—14 h and it becomes refractory with summer photoperiods of 17·8 h (Fig. 14.19; Murton and Kear 1976). In Africa it would be expected to have the potential to reproduce for longer. Rowan (1963) has shown that in the winter rainfall area of the south-west Cape, 80 per cent of breeding records fall between July and October. In contrast, in summer rainfall areas in the Transvaal individuals have been recorded breeding in all months, although 70 per cent of records are for December to March. In all areas there is a time lag of from 2—3 months between rainfall, which induces dispersal from refuge areas, and the onset of breeding. Actual breeding probably depends on food supplies becoming available in the 'right' habitat conditions and the cycle is only loosely phased by the photoperiodic cycle. As with the closely related European Mallard courtship occurs about 6 months ahead of the main breeding season. There is a post-nuptial moult beginning 3—4 months after the peak of the breeding season which renders the birds flightless for a month; the peak of moulting might occur in summer in the south-west Cape or winter in the Transvaal, depending on when breeding occurred. Rowan (1963) had evidence that some birds moulted at less than annual intervals suggesting that an autonomous rhythm of moult, and perhaps gonad activity, might be manifested at these latitudes (see p. 324 regarding changed phase relation of plasma hormones in *Anas platyrhynchos*).

Of the Australian waterfowl those species which depend on shallow water habitats are seriously affected by climatic conditions and they have very variable breeding seasons in consequence. Examples are the Freckled Duck[†] *Stictonetta naevosa*, Grey Teal *Anas gibberifrons*, and Pink-eared Duck *Malacorhynchus caryophyllacea*. The Freckled Duck is an endangered species which has a southern distribution in Australia, but the other two may occur anywhere in the continent depending on suitable habitat conditions developing. The Grey Teal can breed at any time of year wherever and whenever suitable conditions occur and is usually one of the first species to arrive inland at any temporary standing water following rains (Frith 1959b). During periods of drought the birds congregate in refuge areas of permanent water from which they disperse when conditions are favourable (Fig. 14.21: see between pp. 434–5). Frith (1962) has shown that the most important refuge areas are in the south-eastern coastal districts and the most important breeding grounds are in the Murray—Darling drainage basin. With suitable conditions birds from such centres 'explode' as a random dispersal and can reach the extreme west of Australia. Although dispersal is supposedly random there is a regular return movement to the coast,

[†]The Freckled Duck is a filter-feeder which shows several affinities with the swans and geese, for example, its downy young resemble these species and on present evidence it is reckoned to belong in a separate monogeneric tribe of the Anserini (Frith 1964, 1965).

particularly the southern coast (Frith 1962). Frith (1959) demonstrated that the sexual cycle, as in so many Australian waterfowl, was initiated by a rise in water-level, which heralded an increase in food supplies.

More data from Braithwaite and Frith (1969), which are plotted in Fig. 14.20, show the testis cycle of the Grey Teal in relation to the water-level at the permanent Barrenbox swamp in inland New South Sales. The species has the capacity to achieve reproductive condition with short to medium daylengths (Murton and Kear 1976) and has occasionally laid its eggs in February at Slimbridge; that it does not do

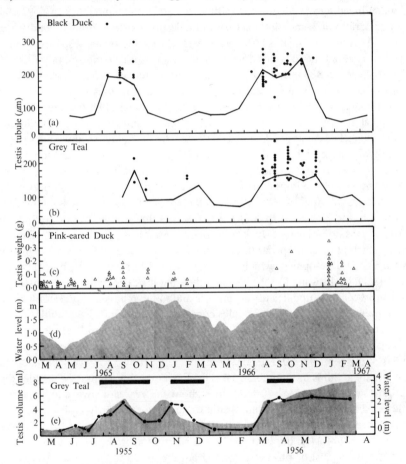

Fig. 14.20. Testis cycles of Black Duck *A. superciliosa*, Grey Teal *A. gibberifrons*, and Pink-eared Duck *Malacorhynchus membranaceus* at the Barrenbox Swamp, New South Wales, Australia. In the (a) and (b) the lines give the mean monthly testes tubule diameters and the solid circles refer to those individual birds whose testes contained spermatoza with little or no necrosis. In (c) the open triangles give the total testes weights of *all* Pink-eared Ducks examined. (d) shows the average water level at Barrenbox Swamp. Modified from Braithwaite and Frith (1969). (e) Changes in the water level at Kooba Lagoon, New South Wales and the mean monthly variation in testis volume (solid line and solid circles) and egg-laying season (horizontal bars) of Grey Teal. Based on Frith (1959*b*).

so more often is probably because it is inhibited by low temperatures and the lack of vegetation cover in winter. It becomes refractory with long summer days (18·2 h at Slimbridge) but with the shorter days of the Australian summer probably manages to remain close to reproductive condition for longer. In other words, like the Black Swan, the Grey Teal has a potentially long, albeit photoperiodically phased breeding rhythm, and is able to breed within this cycle depending on suitable ecological conditions. As Fig. 14.20 shows, the testes were not much stimulated in the spring 1965 and no breeding occurred, whereas in 1966 birds with enlarged gonads were frequent and much breeding was recorded. Braithwaite and Frith commented on the resurgence of gonad activity in some birds each autumn, despite existing drought conditions in 1967, and suggested that this reflected a residual seasonal sexual response. Although they then rejected the idea this does seem to be a reasonable interpretation (Murton and Kear 1976, p. 388). Under a more marked seasonal light cycle in New Zealand, self-introduced Grey Teal are seasonal breeders, nesting from August until November.

The Pink-eared Duck *Malacorhynchus caryophyllacea* is, like the Grey Teal, one of the most erratic of the Australian breeding species and is a nomadic inhabitant of the arid inlands of Australia. It breeds on extensive declining flood-waters for it has a remarkably shaped bill which is adapted to filter-feeding. Small plankton animals, particularly freshwater Crustacea of the sub-order Cladocera and insects including *Trichoptera* larvae, are important food items and under some circumstances plant foods, such as the floating *Azolla* and *Lemna* spp. and also seeds, may be taken (Frith, Braithwaite, and McKean 1969). Fig. 14.20 shows that irrespective of the time of year the testes of Pink-eared Ducks contained spermatozoa. In fact, the males possess gonads which at present appear unique among birds, for they display an asynchronous gametogenetic development which is clearly adapted to their erratic breeding cycle (Fig. 14.21). Braithwaite (1969) has described the testes as comprising constricted zones and expanded nodules of varying size and number which vary according to environmental conditions. The nodules comprise large seminiferous tubules of diameter 80–200 μm with advanced germ cell stages of spermatids or spermatozoa. The adjacent Leydig cells are not usually lipoidal. The constricted areas have tubules of diameter 40–60 μm, containing only one to two rows of spermatogonia and with some lipids in the interstitial cells. Evidently this adaptation enables the male to produce a small number of sperm throughout the year and always to be ready to fertilize a female; presumably androgenic steroids are constantly available in small quantities to facilitate a degree of courtship behaviour.

The endocrinological implications of this adaptation are intriguing. Presumably, gonadotrophins must be produced throughout the year and there is a possibility that the target tissues, in particular the Leydig cells, undergo cyclic changes in sensitivity to this gonadotrophin stimulation. A local availability of androgen could explain why advanced germ cells are similarly localized (see Fig. 14.21). An alternative possibility is that the peculiar condition depends on a low level of circulating gonadotrophin but this does not explain why some parts of the organ should be favoured. Nevertheless, the gonads rapidly become more generally developed given favourable ecological conditions, which suggests that a boost of gonadotrophin secretion makes the difference. This view is to some extent strengthened by the discovery that some Feral Pigeons develop asynchronous tubules, though the condition never becomes so marked as to result in nodule formation. As mentioned, a proportion of Feral Pigeons remain in partial breeding condition in autumn, whereas others in the same populations exhibit full gonad regression (see p. 277). These birds often manifest a variation in the development of the testes tubules such that in some segments these are small and contain only spermatogonia, while adjacent lengths may be expanded with spermatozoa. The condition was first noticed in birds caught in the cities of Liverpool and Leeds (Lofts *et al.* 1966). It will be recalled that some Feral Pigeons continue to hold breeding territories even though their gonads regress, and that selection seems to be favouring a lengthened breeding season. It is possible, therefore, that these asynchronously developed testes result when plasma gonadotrophin titres become reduced, but concentrations enable a degree of response in the target tissues.

The essential adaptation of opportunist breeders is a capacity to remain near to breeding condition over a long period; the ability of the Grey Teal and Pink-eared Duck to breed at any time depends on their ability to maintain active gametogenesis for a large segment of the year. Fig. 14.4, based on Slimbridge data, emphasizes that Grey Teal are not rapidly phased into a non-functional state by daylength changes. The other Austral teal, that is, the New Zealand Brown Teal *Anas aucklandica* and Chestnut-breasted Teal *A. castanea* of Australia have similar photo-response mechanisms to the Grey Teal and, as already discussed, such responses were depicted in Fig. 14.4 as being characteristic of the more primitive members of the genus *Anas* (Murton and Kear 1976). The long breeding seasons of waterfowl at Slimbridge which in the wild are opportunist breeders, contrast with the typically short seasons of more markedly photoperiodic, and often temperate forms. Two Australian species which have short breeding cycles at Slimbridge are the Australian White-eye *Aythya australis* and Australian Black Duck *Anas superciliosa rogersi* (Fig. 14.4). The former has a short and regular breeding season in Australia (Fig. 14.11) and lives in permanent swamps which are relatively little affected by climatic conditions. So too, do the Musk Duck *Biziura lobata* and Blue-billed Duck *Oxyura australis* but breeding records are not available for these species at Slimbridge; the North American Ruddy Duck *O. jamaicensis* whose range is centred on latitude 48°N (compared with 35°S for the Australian *Oxyura* species) begins breeding with a photoperiod of 15·2 h and finishes when the daylength decreases to 17·4 h, that is, breeding is concentrated from April until late July. The Black Duck has a photoperiodically regulated breeding cycle but it is more affected by immediate ecological conditions than the Australian White-eye. This is because it is not confined to the permanent swamps, although it is a much more sedentary species than the Grey Teal (Fig. 14.22). Braithwaite and Frith compare it with the Mallard (see Lofts and Coombs 1965) in that some resurgence of gonad activity occurs in autumn in Australia in some individuals and there is a general and synchronous increase of gametogenesis in spring in all birds. The Australian Shoveler *A. r. rhynchotis* probably has a gonad cycle in Australia which is similar to that of the Black Duck. At Slimbridge, its breeding season and that of the New Zealand sub-species *A. r. variegata* are of the typical temperate type (Fig. 14.4). At Slimbridge the New Zealand Shoveler requires an average photoperiod of 16·2 h to initiate breeding compared with only 14·7 h for the New Zealand Black Duck *A. s. superciliosa.* The Australian sub-species of these two ducks show a parallel difference in the wild, for the Australian Shoveler breeds about a month later than the Australian Black Duck. Ecologically this differential response is doubtless related to the fact that the shoveler's food of invertebrates, including water-snails and *Trichoptera* larvae, becomes available later than the seeds of smartweeds and sedges (Braithwaite and Frith 1969).

15

Sexual selection and the pair bond

Summary

A wide range of physiological adaptations to suit special ecological needs are reviewed in this chapter. Sexual monomorphism prevails in the tropics while marked dimorphisms become more common in temperate regions; this is well illustrated in the Anatidae, in which the sexual dimorphisms depend on endocrine factors. Sexual selection is obviously important in markedly dimorphic species, but it is becoming evident that it is important in monomorphic species, both sexes exercising choice, and that mechanisms exist which enable birds to select mating partners which are genotypically the most suitable. Examples are also accruing of physiological polymorphisms which are sometimes linked with plumage polymorphisms. Some polymorphisms are associated with a change in length of breeding season or photoperiod under which breeding is initiated. This is so in *Columba livia* (see also Chapter 11), Snow Geese *Anser caerulescens*, and the Arctic Skua *Stercorarius parasiticus*. Monogamy is the mating pattern for the majority of birds, often with a high degree of mate fidelity from season to season. Polyandry is a means of increasing the fertility of the female at the expense of the male and polygyny achieves the reverse. Ecologically such mating systems occur when one sex is differentially favoured or hampered during reproduction. In polygyny the females may have energy difficulties in producing eggs and once these are laid it is advantageous to 'force' the male to assume incubation and brood care duties. This has sometimes been achieved through an increase in female dominance, facilitated by an increase in body size relative to the male and sex reversal in plumage pattern; in this chapter we mention the consequent endocrine changes and give as examples evolutionary trends in wading birds. Promiscuity is an unusual condition but so-called co-operative breeding is known in many bird taxa. Presently sparse evidence suggests that 'co-operation' is not the basis of the habit and there is no real indication for altruistic behaviour. Instead co-operative breeding occurs in difficult habitats where ecological conditions make dispersal disadvantageous and young birds must for a time co-exist with their parents and siblings. The evolution of brood parasitism is discussed and the gonad cycles of the parasites compared with those of closely related non-parasitic forms.

Mating preferences

The striking dimorphisms, particularly of plumage, apparent in birds were attributed by Darwin to sexual selection whereby males which have gained an heritable advantage over other males are favoured by the females. Darwin was at a loss to understand how this could work unless there was an excess of males or else polygyny were practised, and he supposed that perhaps the first females to breed, probably the most vigorous, would choose the more attractive males. Fisher's (1930) theory for the evolution of sexual preferences supposed that those females which are able to mate with males favoured by natural selection would themselves derive an advantage from the advantages accruing to their male progeny, who would in turn be preferred by females. Thus male competition might enable them to breed earlier with fitter females or females might have mating preferences and select males with particular characteristics. O'Donald (1967) modelled Fisher's theory in the terms that if some females prefer to mate with males having advantageous characteristics in natural selection, then the genotypes that determine such mating preferences will also be selected so that the progeny will carry both genotypes. He showed that in a polygynous system the males have a selective advantage if they can mate with a large number of females. If a recessive gene is selected, the gene that determines the mating preference increases rapidly to reach a high frequency at equilibrium in the population, whereas if selection is for a dominant, the mating preference gene increases slowly in frequency and is at low frequency at equilibrium. When natural selection opposes sexual selection a wide range of selective coefficients can result in the establishment of a stable polymorphism.

Lack (1968a) objected to the ideas of sexual selection as advanced by Darwin and Fisher, arguing that breeding time was related not to female mating preferences but to the food supply available to her to form eggs and feed young. However, O'Donald (1972a) showed this view to be irrelevant and demonstrated by a computer model that preferred males can gain an advantage even when the earlier breeding pairs are not the fitter (see also Trivers (1972)). He showed that selection can operate on phenotypic variations in fitness of breeding time in Arctic Skuas *Stercorarius parasiticus* and was able to calculate the fitness of males which mated first as a result of sexual selection. This was extended to allow for only partial female mating preferences for the sexually advantageous males (O'Donald 1972a). Such selection is frequency-dependent: if only a few females exercise a mating preference for the advantageous males the resulting selection is negatively frequency-dependent; as the preferred males increase in frequency their selective advantage declines and selection becomes very slow; stable polymorphisms can develop when sexual selection of this kind is balanced by natural selection. If most of the females exercise a mating preference the selection of males is positively frequency-dependent so that their selective advantage increases rapidly with their increase in frequency. The discoveries of O'Donald extend Fisher's theory of sexual selection in showing that it is not necessary for the pairs breeding earlier in the season to be the fitter in order to have a selective advantage.

Mating preferences can be easily identified in species having plumage or structural polymorphisms but it is possible even in monomorphic species, which offer no obvious visual cues, that pairing preferences are commonplace. In the Kittiwake *Rissa tridactyla*, birds that retain their mates from year to year begin breeding earlier, have larger clutches, and a higher success rate than pairs which have changed mates or are breeding for the first time (Coulson 1971, 1972). Males that establish themselves in the centre of the breeding colony live longer than those at the edge so as a result more new pairings occur in the colony periphery (Coulson and Wooller 1976). In addition, edge-birds change partners more readily among those pairs in which both partners survive. A higher reproductive success and lower mortality for centre-birds means that during a lifetime a central male's production is 74 per cent greater than an average edge-living male (Coulson 1971). We suggest that the central birds have an advantage in being in the best position to derive information from the rest of the colony (see p. 250). Probably for most of the time they know where to find the best feeding grounds and it is less 'informed' edge-birds that need to learn from the central colony birds. Males fight to gain a central place in the nesting colony and it is possible that only those which are competent at finding food sources can manage to devote time to securing a favourable position. Occupying a central position in the colony could become a convention which indicates the quality of the male. Given that females might then prefer to pair with such well-endowed males it would be expected that edge-birds should exhibit a greater tendency to change mates. Once a convention becomes a reliable measure of an animal's fitness, selection can act on the convention. This may involve a plumage or structural modification, so that in mating systems which encourage intense competition for partners dimorphism may arise.

A male bird that has gained a territory, and can afford to make conspicuous displays which should seemingly make him at grave risk of being taken by a predator, demonstrates his capacity to avoid such a risk. A male bower-bird (Ptilinorhynchidae) who is able to amass an enormous collection of shells, broken glass, and other display objects outside his bower (Fig. 15.1: see between pp. 434–5) demonstrates a familiarity with his chosen terrain, and surely a bird that is able to 'waste time' collecting such objects, yet at the same time is able to feed himself, must be a potentially good father. Zahavi (1976) coined the term 'handicap principle' to account for effects of this kind. He imagined that the size of the handicap that a male bird could carry would supposedly be a measure of his quality, and in this way females could select the best husbands. Similarly, a male can choose only the most persistent females, those who in spite of his initial aggressive displays fail to be repulsed. The more time a female can devote to pursuing a male in his territory, the more she demonstrates that she has no problems in feeding herself. Birds at the bottom of the social hierarchy which have difficulty in coping with their environment can never manage the bluffs or overt signs of quality of their superiors.

This seemingly attractive theory has been criticized by Davis and O'Donald (1976a) on genetical grounds. They show both by logical argument and also be a detailed mathematical analysis that two premises of the model are false. The first is the assumption that selection continues to produce an advantage for females exercising a preference. This cannot happen because the fitness of the handicapped males cannot increase indefinitely. Secondly, selection is assumed to favour the same combination of characters in both handicapped and non-handicapped males. In fact, disruptive selection would favour a different combination of characters and those favourable to handicapped individuals would be unfavourable when passed on to non-handicapped offspring, thereby eliminating any advantage the females showing a preference might gain. Stripped of the suggestion of selection for a handicap we are left with observations that can adequately be described by Fisher's (1930) theory of sexual selection in which female mating preferences evolve through advantages gained by the females who mate with males possessing some advantageous character. The gene-determining mating preference is selected because when passed to sons it gives them an advantage by possessing the preferred and advantageous character. Thus a strong correlation develops between the gene for the preference and the gene for the preferred character. 'Runaway selection' for an exaggerated sexually advantageous character will occur but will eventually be balanced by its disadvantage in natural selection. In this context if there is a genetical capacity for some secondary-sexual character to respond quantitatively according to the individual's ability to cope with its environment, perhaps via a hormone control system, then there is a basis for sexual selection. The ability to possess a large territory may confer advantages in sexual selection in this way (see p. 429, also p. 498).

Sexual dimorphism

Many structural dimorphisms are primarily adapted to differences in feeding ecology of the sexes, as with the marked difference in bill size and shape between male and female of the extinct Huia *Heteralocha acutirostris* of New Zealand. Sexual dimorphism and differential niche utilization is well reviewed by Selander (1966). We are not concerned here with purely ecological adaptations of this kind, though it should be recognized that the different feeding requirements of the two sexes may well influence breeding periodicity. This is so in the *Accipiter* hawks, in which the normal trend in birds for the male to be bigger and stronger than the female is reversed. Cade (1960) has argued that female dominance has arisen to ensure that the male functions as the food provider, that is, differences in food habits are a consequence of the size difference. But Reynolds (1972) takes the opposite view and suggests that feeding habits have been responsible for the size dimorphism. Like Selander (1966) he appreciated that one strategy for a species to occupy a wider feeding niche, or, to reduce intraspecific competition, is for the sexes to specialize on different prey sizes. In this way three sympatric *Accipiters* of North America can utilize six different mean prey sizes and four sympatric species in Gabon feed on eight size-classes of prey (Brosset 1973). The smaller size of the male in these species allows him to feed on a wider spectrum of prey items than the female, and to function below the summit of the prey pyramid. The male, therefore, encounters more prey than the female and can be more efficient at feeding himself, the female, and the young for the first part of the breeding season. When the young are in greatest need of food the female must become involved in food capture and her size enables her to capture big as well as small prey. However, this is possible only during a limited part of the year. Hence, the breeding season may be adjusted to the food available to the young at fledging, while the actual ability of the species to lay eggs a month earlier may depend on the male's capacity to accept a more numerical prey category.

The male Capercaillie *Tetrao urogallus* is polygamous and more than twice the size of the female. In this instance size has apparently evolved in relation to fighting ability for the males compete with each other for the attention of females which visit their dispersed leks (see p. 442). This is an example of how a structural difference, which has evolved through sexual selection, may secondarily necessitate differences in feeding and other ecology between the sexes (Seiskari 1962).

When there is a balanced participation of the sexes in mate selection and courtship, plumage monomorphism prevails. Monomorphic species may be cryptically or brightly coloured. In the latter case the distinctive plumage possibly serves as an isolating mechanism to prevent attempts at interspecific hybridization with a resultant reduction in fertility. Moreover, since reduced fertility is so important at the interspecific level it may be imagined that there also exist intraspecific variations in fertility which will encourage mating preferences. When initial pair formation depends on the female choosing her mate it is only the male which needs to be distinctive and easily recognized. The extent to which males can compromise a cryptic colouration for the sake of species isolation must depend on the ecological situation and whether both sexes are required to share in parental duties. Once a male is freed from immediate parental duties, particularly incubation which may require cryptic colouring, selection can operate on his male characteristics to make them a measure of his fitness. It is not always easy to decide when male plumage has primarily evolved for species isolation or as a secondary sexual character. In mating systems in which male and female meet only briefly for copulation, or in which casual matings are frequent, there may be a need for a male to be particularly distinctive. This last function may explain the distinctive male plumage which has evolved in Paradisaeidae (birds of paradise) and Trochilidae (hummingbirds) in which group lek species are common (e.g. Snow 1974), and in the Phasianidae (pheasants) and Anatidae (swans, geese, and ducks). If the gaudy male colour in these species has evolved as a secondary sexual feature rather than to serve for species isolation a dilemma noted by Sibley (1957) and Lack (1968a) is solved, namely that hybrids are particularly frequent in these families. Moreover, the fertility of interspecific and inter-generic crosses is relatively high (Sandnes 1957). In those species which manage a long pair bond there is presumably less need for the individuals to be distinctive: 'no one knows a husband or wife better than their marriage partners.'

Plumage dimorphism may be permanent and genetically determined or seasonal and mediated by endocrine factors. Among waterfowl, plumage dimorphism has been well documented by Kear (1970). In the majority of Anserinae (swans and geese) the sexes are similar in plumage and are typically coloured brown, black, grey, or white. The tribe Tadornini (shelducks and sheldgeese) represents a transitional stage with some dimorphic and some monomorphic species, several being rather brightly coloured in both sexes (see Fig. 15.2(a): see between pp.434-5). When birds are conspicuous or possess prominent signal marks it may be suspected that they profit from group feeding or roosting and that the very visible plumage facilitates feeding by social enhancement. Magpie Geese feed on roots which are concealed under mud and, like .Rooks, must sample their feeding places before 'knowing' whether they will be rewarding (p. 462). It must be advantageous for a Magpie Goose to be able easily to recognize conspecifics that have already located worthwhile feeding sites. In the Cairinini (perching ducks), some species are dimorphic and brightly coloured with iridescent plumage, while in the remainder the male is gaudy and the female dull. Indeed, brown female plumage first appears in some of the Cairinini and not in the phylogenetically younger ground-nesting *Anas* ducks. Since the Cairinini nest in tree-holes it is unlikely that the sexual differences initially arose as an anti-predator protection for the incubating female. Perhaps a cryptic colour is generally advantageous unless there are strong selection pressures operating against. Conceivably those phylogenetically old lines which still survive and are not cryptic, for example, Magpie Geese have been successful because they have evolved group-feeding strategies. Nevertheless, Kear has shown for the Anseriformes (she includes the flamingos and *Chauna*) that monomorphism is much more common in species in which both sexes tend the young (65 cases against 9) than when only the female does so (in which category she records 18 species which are alike in plumage compared with 58 dimorphic ones). The correlation need not imply an immediate causal relationship, for monomorphic species generally have long pair bonds which must facilitate the sharing of parental duties (possible advantages of monogamy are discussed on p. 435). The swans and geese mostly have long-term pair bonds, often for life, mutual courtship displays, and preening behaviour; they copulate on land and tend to mature slowly. The shelducks are intermediate between the Anserinae and higher groups, which tend to have seasonal pair bonds: the males perform most of the courtship displays and the females choose their mates and copulation mostly occurs in water while swimming.

In her review Kear points out that the basic pattern in waterfowl evolution has been an increasing adaptation to a fully aquatic environment. The need for a waterproof plumage may have provided the stimulus for a twice-yearly moult and this in turn has enabled marked sexual dimorphism to emerge during the breeding season. Thus the primitive condition has been for both sexes to have a bright and distinctive plumage to facilitate species isolation with a secondary and advanced trend for the female to become cryptic and for the male to become the focus of intraspecific display. Kear wonders whether an initial evolutionary trend involved the young developing with their first plumage a colouring which was more cryptic than that of the adults. If so, the females might come to wear the brown juvenile dress all their lives by neotony. In the Mallard *Anas platyrhynchos* this has been achieved hormonally, and injections of oestrogenic hormones prevent the male-type plumage from being expressed. Ovariectomy causes females to don the distinctive bright male plumage at the next moult and ovarian tumours or other pathological conditions can induce a similar result. Gonadectomy of ducklings causes them to retain the normal juvenile plumage through each moult. Caridroit (1938) plucked drake Mallard and found that the feathers which regenerated between March and July were of the typical brown female type. Female feathers develop in the male during the period when plasma androgen titres are low and thyroid activity is high (Jallageas *et al.* 1974; Astier *et al.* 1970) and result in the so-called eclipse plumage. A second moult in early autumn after the refractory phase, when plasma steroid levels increase again, produces the typical male (primitive) plumage. If Mallard caught in August when in a refractory condition are kept on long (16·5 h) photoperiods they remain in eclipse, whereas controls placed on seasonal daylengths or else short artificial days (8 h) moult into the colourful cock plumage within 2 months (Lofts and Coombs 1965; Table 12.1, p. 325). Mallard drakes given extra light in February assumed the henny plumage in February instead of June (Walton 1937). Female Mallard moult into their basic plumage at the same time as the males in mid-summer but their spring moult into alternate plumage occurs as a pre-nuptial moult in spring.

A more critical and up-to-date examination of the effect of sex hormones on plumage colouration has been made for the Blue-winged Teal *Anas discors* (Greij 1973). Exogenous diethyl stilbestrol suppressed expression of a male type plumage in both sexes. It also inhibited the moult of adults but did not affect the regeneration of plucked feathers. Testosterone did not affect regenerated feathers of males but it caused immature females to develop the same colours as immature males. Castration caused males to develop the typical bright (alternate) plumage and the same bright plumage developed in females when they were ovariectomized. Greij suggested that the developing plumages of adult male teal are bright (alternate) unless the oestrogen level is high enough to induce basic feathers. Actually, it has never been shown that males produce significant amounts of oestrogen. It is conceivable that some other factor is also involved and the effect of exogenous oestrogen is indirect. Adult female plumages are always brown because feather development is very sensitive to oestrogen levels. Both basic and alternate plumages are probably dependent on the ratio of testosterone to oestrogen (female birds do produce measurable quantities of androgen even though oestrogen cannot be detected by radioimmunoassay techniques in males). Females will tend towards a male-type alternate plumage when their testosterone levels are high relative to oestrogen. In Blue-winged Teal the female alternate plumage, while dull, is certainly more patterned than the basic plumage, which develops under the influence of high oestrogen levels. A critical factor in determining plumage colour in ducks and drakes must be the time of moult in relation to plasma steroid titres.

The special adaptation of the male Mallard is in delaying the spring moult until summer, for if the moult did occur in spring in the presence of steroid hormones a female plumage would presumably develop during the season when courtship is important. Palmer (1972) has pointed out that all the Mallard group and Shoveler species have the same moults and feather generations irrespective of whether sexual dimorphism is expressed. He suggests that the feather generations are phylogenetically older than the present-day distribution or attainment of sexual dimorphism in these species. However, Kear and Murton (in press) give evidence that rhythms of gonadotrophin and hence androgen secretion have become re-phased in different *Anas* to suit the latitude at which they live (cf. Fig. 14.4e). In consequence the timing of the moult relative to plasma steroid levels has altered in different species. Many north temperate seasonally dimorphic forms have radiated back towards the equator and this has resulted in an extended season of gonadotrophin and steroid activity and the secondary acquisition of monomorphism. Examples are given below on p. 423.

Fig. 14.18 (p. 403) demonstrated for *Anas*, and Fig. 14.2 (p. 368) for *Netta* and *Aythya*, that those species which exhibit a seasonal, and presumably hormonally controlled, sexual dimorphism are those which breed at the higher latitudes above 40° and have a short, photoperiodically controlled breeding season. The Cape Shoveler *Anas smithi* of South Africa is exceptional among the *Anas* species in that it occurs below latitude 40° and does have a seasonal sexual dimorphism. It has conceivably evolved from a more northern photoperiodic line which gave rise to southern derivatives for it exhibits a temperate type photo-response at Slimbridge (Murton and Kear 1976). Certainly, the Australian and New Zealand Shovelers, which are seasonally dimorphic, are closely related to the north-temperate Common Shoveler. In the same way, the Kerguelen Pintail *Anas acuta eatoni* and the southern sub-species of the Cinnamon Teal *A. cyanoptera*, both of which show a small degree of seasonal dimorphism, are derived from northern dimorphic stock (see below). On the other hand, the Red Shoveler *Anas platalea* of South America (35°S) which is permanently dimorphic, is not immediately related to the other shovelers and it may represent a longer-evolved lineage. So too the permanently dimorphic Chestnut-breasted Teal *A. castanea*, which is probably an ancient inhabitant of Australia (Kear and Murton 1976). It is closely related to the New Zealand Brown Teal *A. aucklandica*, which variably exhibits a seasonal dimorphism. Hence, 12 of the 19 sexually dimorphic species plotted in Fig. 14.19 belong to the Holarctic and are migrants to high latitudes (Murton and Kear, in press, discuss the exceptional species in more detail than can be afforded here). Pair formation occurs in the contra-nuptial quarters where many different species which breed in separate areas or habitats come together, thereby placing a premium on isolating mechanisms. A strong and positive pair bond has to be forged which persists throughout the migratory journey so that on arrival at the breeding quarters fertilization can occur and nesting begin immediately, to take advantage of the short season available for rearing young. Movement to high latitudes for breeding brings the advantage that rich supplies of proteinaceous foods are available during the summer months, but the disadvantage of a lack of holes in which the birds could nest; the northern distribution of the Smew *Mergus albellus* in the Palaearctic and Bufflehead *Bucephala albeola* in North America (Erskine 1972) is limited by the availability of woodpecker holes. The need for *Anas* species to nest on the ground in open sites could well have contributed to the emergence of cryptic colouring in the female.

We mentioned in Chapter 1 the general trend in birds for monomorphism to prevail in the tropics and for dimorphism in plumage colour to increase towards the northern latitudes and, to a lesser extent, towards the south; in addition to the ducks just discussed the trend is well shown by the New World genera *Dendroica* (New World warblers) *Icterus* (orioles), and *Piranga* (tanagers), data being given by Skutch (1957) and Hamilton and Barth (1962). The evolutionary sequence must be viewed as a loss of monomorphism since, as in ducks, the northern forms derive from southern species (Mayr 1946) and it is necessary to distinguish between an increase in male brightness or a decrease in female brightness. Hamilton and Barth list the following theories which have been advanced to explain the emergence of dimorphism with the female being cryptic and the male more brightly coloured: (1) facilitation of mate selection (Sibley 1957); (2) rapid pair formation (Goodwin 1960; Hamilton 1961); (3) rapid sex recognition (Hamilton 1961); (4) avoidance of predation (Sibley 1957); (5) species recognition and avoidance of hybridization (Sibley 1957, 1961); (6) avoidance of competition (Mayr 1960); (7) promotion of gregariousness during the contra-nuptial season and alleviation of individual hostility at the interspecific level (Moynihan 1960); (8) co-ordination of group activity for intraspecific colonialism (Dilger 1960); (9) female chick-to-male parent imprinting if efficiency of such for females is enhanced by selection for pronounced visual plumage characters in the male (Hailman 1959).

Hamilton and Barth (1962) extended Moynihan's views and suggested that the advantage of dull females in northern regions was in reducing intraspecific, interindividual hostility, thereby facilitating rapid pair bonding. In view of what has been written about the waterfowl we modify these views slightly and conclude as follows: monomorphism prevails where species have a long pair bond and both sexes are involved in its formation and maintenance, this is the best strategy in habitats where the pair can remain together for much of the year; bright monomorphism will only prevail when species isolation or social feeding is important otherwise there will probably be a trend towards a more cryptic monomorphism; migrant species to high latitudes can only have short breeding seasons and they need to achieve rapid pair formation; mate selection depends entirely on the female so males have been subjected to increased sexual selection. In a fluctuating environment the advantages of having a long-term, stable mate relationship and the associated reduction in genetic variability are lost and there are probably advantages in changing partners in different years. Perhaps the commonness of rape in some temperate waterfowl, for example, the Mallard, is associated with an advantage for promiscuous relationships. Opportunities for migration to higher latitudes for breeding are limited in the southern hemisphere and species generally have longer pair bonds (Weller 1968a). This doubtless explains why the tropic—north temperate trend to increased dimorphism is less noticeable from the tropics to the south—temperate zone.

The breeding rhythms of species which are adapted to markedly seasonal light cycles are less affectively entrained when exposed to the reduced amplitude Zeitgebers of lower latitudes, and their gonad cycles become extended (p. 232). Hormones must be secreted over a greater part of the year and this should affect those species in which the male type plumage is suppressed by steroids. This is exactly what is found. The Mallard Anas p. platyrhynchos has a continuous Holarctic range from which colonists have established various isolated breeding populations as shown in Fig. 15.3 (see also Fig. 14.19). In all these populations the males have lost their sexually distinctive plumage and resemble the females, this applying to the Florida Duck A. p. fulvigula[†] (mid-latitude of breeding range 29°N), Mexican Duck A. p. diazi[†] (25°N), Laysan Teal A. p. laysanensis (25°N), and

---

[†]Some authors treat these as distinct species (see Aldrich and Kenard 1970; Stieglitz and Wilson 1968). Even if this is justified, they remain very closely related and can be considered a super-species.

Hawaiian Duck *A. p. wyvilliana* (21°N). Some male Hawaiian Ducks occasionally, and to a variable extent, assume some of the colour of the drake Mallard, particularly the green head plumage. Another similar example is provided by the Pintail *A. a. acuta* whose distribution in the Holarctic resembles that of the nominate form of the Mallard. The Mallard only exceptionally migrates as far south as southern Asia and the Mediterranean whereas the Pintail regularly reaches central Africa, Ceylon, the Phillipines, and Borneo. From such migrants doubtless arose the stock of Pintail which today are resident on two sub-Antarctic islands, the Kerguelen Pintail, already mentioned, on Kerguelen Island (latitude 50°S in the South Indian Ocean) and the Crozet Pintail *A. acuta drygalskii* on Crozet Island, which is 800 miles to the west of Kerguelen. In both these sub-species the males have lost their distinctive nuptial plumage, although exceptionally they may exhibit the characteristic plumage of the nominate race. Yet another example is provided by the extinct Coues's Gadwall *Anas strepera couesi,* which lived on Washington Island in the Fanning Island group (latitude·4°N) 1000 miles south of Hawaii. In this sub-species the male had come to resemble the female in plumage characteristics, whereas the nominate form is seasonally dimorphic. There is a general tendency for signal characters to be lost on remote islands, where species isolation is not important (Mayr 1942; Lack 1970) but the mechanisms involved remain to be established. The Garganey Teal *Anas querquedula* is exceptional among the north-temperate ducks in being a trans-equatorial migrant which sometimes reaches Australia. Whereas the other north-temperate *Anas* males acquire their bright plumage in autumn the Garganey does not do so until March. It is tempting to attribute this to the influence of the equatorial photo-regime it

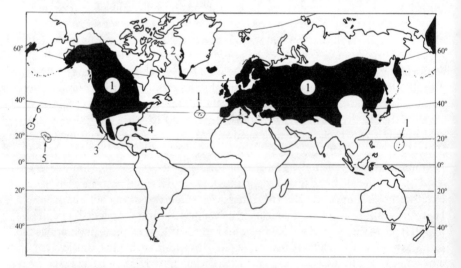

Fig. 15.3. Breeding distribution of the Mallard *Anas p. platyrhynchos* (1) and of the derived sub-species, these being (2) Greenland Mallard *A. p. conboschas*, (3) Mexican Duck *A. p. diazi*, (4) Florida Duck *A. p. fulvigula*, (5) Hawaiian Duck *A. p. wyvilliana*, (6) Laysan Teal *A. p. laysanensis*. Some workers attribute full specific status to these sub-species.

experiences during the autumn and winter months. Mallard held on long photoperiods at the end of the refractory period remain in eclipse plumage as mentioned above (p. 421 and Table 12.1).

In some ducks, steroid hormones serve to suppress the expression of the brightly coloured adult plumage thereby allowing a cryptic juvenile dress to be worn. In some other bird species the primitive condition is for the sexes to be the same and dully coloured and for the steroid dependent nuptial plumage to be bright. The showy and ornamental ruff of the male Ruff *Philomachus pugnax*—the female is called a Reeve—is acquired for the breeding season by a spring moult. Assumption of this nuptial dress is prevented by castration in early winter and it seems that it depends on androgenic steroids (van Oordt and Junge 1936). In various weaverbird species (Ploceidae), including the Red Bishop *Euplectes orix*, Golden Bishop *E. afer*, Red-billed Quelea *Quelea quelea*, Paradise Wydah *Steganura paradisea*, and the Combassou *Hypochera chalybeata*, male-type plumage depends on gonadotrophin secretion, supposedly LH. In these birds a castrated male continues to develop a nuptial plumage but ovarian hormones given to males inhibit the potentiality of LH for stimulating the nuptial plumage. Presumably this is the reason that females do not normally acquire the male-type plumage, although they can do so if ovariectomized. LH injected into *Euplectes afer* and *Steganura paradisea* several days after plucking causes the regenerated feathers to develop black bars in both intact and castrated subjects (Witschi 1961; Ralph, Grinwich, and Hall 1967a, b), while PMS (pregnant mare serum) can cause feather darkening (Witschi 1940). These hormones appear not to act locally on the feathers (Hall, Ralph, and Grinwich 1965) and it also seems unlikely that LH acts via the hypothalamus, the adrenal, thyroid, or pineal (Ralph *et al.* 1967a). Synthetic MSH does not produce black feathers, nor does extensive damage to the pituitary by electrocautery alter the response to LH (Ralph *et al.* 1967a). The half-life of exogenous ICSH in the rat is 15 min and of PMS 26 h (Parlow and Ward 1961). In response to either hormone melanocytes appear in the feathers of *Steganura* 12–24 h post-injection and not until 30–54 post-injection in *Euplectes*. The latency in response is not altered by dose rate, suggesting that an all-or-nothing response is triggered by the injection (Ralph *et al.* 1967b). The ventral feathers of *Steganura* do not normally turn black during the nuptial season and the pigment produced artificially is dark brown. In contrast, hormone treatment of *Euplectes* causes a jet-black pigmentation similar to the black pigment noted naturally during the nuptial season. A proportion of *Quelea* brought into captivity under artificial lighting moult into a melanic phase. This does not seem to be a consequence of the photoperiod and the condition could be dispersed or prevented by exposing birds to sunny natural light or ultraviolet light (Lofts 1961).

It is conceivable that plumage control by gonadotrophic hormones is a primitive, perhaps tropical, condition. An advantage in using gonadotrophic hormones to control plumage appearance may be that they are produced over a 'long' season, whereas the gonadal steroids are secreted only when the gonads become enlarged, this corresponding more precisely with the actual breeding season of birds; different selective factors will operate on other hormonally dependent secondary sexual characters, depending on the season they have to be developed. The more strictly defined reproductive seasons necessitated at temperate latitudes compared with the tropics may implicate more precise timing mechanisms. For example, development of the red bill pigmentation of the tropical weaver finches depends on LH and is inhibited by oestrogen (Witschi 1961); in the House Sparrow, gonadotrophin (FSH probably) plus androgen is needed for the black pigmentation to develop (Lofts *et al.* 1973); androgen alone is effective in causing the bill of both male and female Starlings *Sturnus vulgaris*, the most temperate of the series, to become yellow (Witschi and Miller 1938).

In some birds both sexes moult into a nuptial plumage which is mediated hormonally. This usually occurs in species in which both sexes normally carry what can be designated the male plumage, in contrast to those, such as the Song Sparrow *Melospiza melodia*, in which both sexes wear a dull female-like garb (Palmer 1972). For example, both mature male and female Herring Gulls *Larus argentatus* have a grey–white cock-type plumage whereas the immatures during their 3 years of development have a brown, henny or female plumage. Administration of androgens to immatures of either sex induces a precocious assumption of the adult's white coloration (Boss 1943). In the Black-headed Gull *Larus ridibundus* both sexes additionally acquire a dark brown head during the breeding season which in the male is prevented by castration (van Oordt and Junge 1933). Females have not been studied in this context and it remains to be shown whether in both sexes it is very high titres of gonadal steroids which cause the head colouring. If pair formation is influenced by a hormone dependent secondary sexual character, whether it be the colour of plumage or soft parts or some structural adornment, then sexual selection must operate on the amount of hormone being produced and individuals which carry too big a handicap will be at a selective disadvantage. Given that a quantitative relationship exists between the character and the amount of hormone needed to sustain it, conditions could be envisaged in which too much hormone had to be circulated. Under such circumstances one would expect there to be a readjustment of target-tissue sensitivity. Since the same hormones can be implicated in many physiological functions they are probably subject to more and different selection pressures from those acting on a single secondary sexual character. We lack information on this topic but the situation may soon be remedied now that it is becoming easier to measure titres of circulating hormones.

If hormones are to determine plumage colour only during the breeding season at least a partial pre-nuptial moult is needed. However, a device to acquire a nuptial plumage without the need for such an extra, energy-demanding moult involves feather abrasion. This occurs to some extent in *Anas* ducks but is much more developed in some examples quoted by Palmer (1972). He lists two icterids, the Orchard Oriole *Icterus spurius* and Rusty Blackbird *Euphagus carolinus*, also the Snow Bunting *Plectrophenax nivalis*, Lapland Bunting *Calcarius lapponicus*, and Rustic Bunting *Emberiza rustica*, and the neotropical Blue-black Grassquit *Volatinia jacarina*, as having coloured tips to the definitive plumage which is acquired in autumn. These coloured margins, often pale or rusty but dark in the Snow Bunting, obscure the lower part of the feather but wear off during the winter to reveal a nuptial plumage by spring. A partial moult into alternate plumage affecting the head or local body area brings the bird into full nuptial dress. The same device is used by the Bobolink *Dolichonyx oryzivorus,* another icterid, but brown margins occur on the alternate plumage acquired in spring and these wear off in time for breeding. In this species the whole plumage is changed at each moult. In the House Sparrow there is a single post-nuptial moult during which the male acquires white-tipped feathers that partly obscure the nuptial plumage, particularly the black areas of throat and breast. These white tips wear off by spring. If House Sparrows are kept on artificial 'long' photo-periods during the refractory period the post-nuptial moult is inhibited (Fig. 15.4: see between pp. 434-5).

The House Sparrow is an example of a species with permanent sexual dimorphism, the female having become dull-coloured. The plumage depends on the genetic consti-tution (XO or XX) and cannot be modified by castration or hormone therapy. The House Sparrow presumably manifests an advanced condition compared with that noted in the closely related Tree Sparrow *Passer montanus,* in which male and female are the same and brightly coloured. According to Witschi (1961), pheasants (*Phasianus* sp. and Reeve's Pheasant *Syrmaticus reevesi*) and some other gallinaceous birds may occupy an intermediate position for although their sex dimorphism is basically determined by their hereditary sex constitution, oestrogen and androgenic steroids have a partial feminizing influence (Koch 1939). Under natural conditions many old domestic hens *Gallus gallus* suffering from senile changes become mascu-linized, assuming the cock plumage and a capacity to crow, as a result of the rudi-ment of medullary (testicular) tissue in the ovary becoming functional. Such sex reversal is also relatively common, and often spectacular, in pheasant species, where, for example, a sombre-coloured female Golden Pheasant *Chrysolophus pictus* can assume the resplendent plumage of the male (see also Harrison (1932)). This suggests an evolutionary trend in which some dimorphisms which are initially of short sea-sonal duration, being mediated by hormones that are secreted during the breeding season, can be traced to dimorphisms which are controlled by steroid secretions which occur during early juvenile development, that is, the endocrine control of plumage characters is advanced during ontogeny.

Polymorphism

Examples are accruing of plumage polymorphisms that are somehow linked with physiological advantages. This promises to be an exciting field of research from which two ideas are worth pursuing. The first is that plumage colour may not always be immediately adaptive but instead be a consequence of a more important physiologi-cal adaptation. The second is that there must be countless examples of selection for physiological adaptation in monomorphic species which operate through the conven-tions of mating systems which we fail to recognize.

The Arctic Skua is polymorphic for colour, there being a dark phenotype having dark ventral feathers, a pale phase with light underparts, and an intermediate one with dark tips to the feathers of the white ventral plumage. These last are heterozygous for two alleles at a single locus, but so also are a large proportion of dark birds (45 per cent), according to O'Donald and Davis (1959). There is a variation in the proportion of colour phase with pale predominating in the north of the range and dark in the south (Fig. 15.5). On Fair Isle, Berry and Davis (1970) showed that there is an excess of pairs in which the male is darker than the female and these males mate earlier than pale males. Dark males are favoured by sexual selection because in new pairs mating with a particular female for the first time they breed about a week earlier in the breeding season than other males (O'Donald and Davis 1975). This confers a selective advantage because earlier pairs fledge more chicks than pairs breeding later in the season. Early breeding occurs when clutch size is maximal and breeding success high, and O'Donald (1972b) used this information to calculate a fitness function of breeding time, a quadratic model giving the best fit to his data

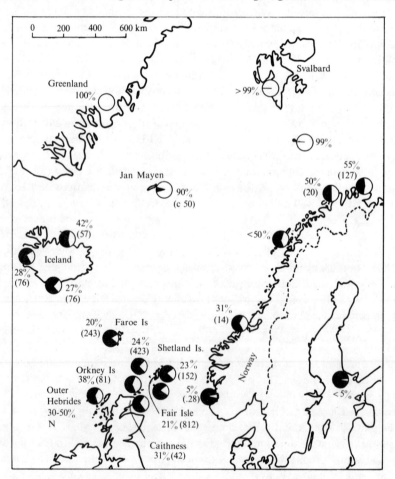

Fig. 15.5. Geographical distribution of plumage colour phase in the Arctic Skua *Stercorarius parasiticus*. Proportion of pale phase birds present in different samples shown by the white portion of the symbols (sample size indicated). From Berry and Davis (1970).

(see O'Donald, Wedd, and Davis 1974*b*). The correlation between early breeding and increased fertility, if genetically determined, must through natural selection result in more early breeding and an increase of dark-phase birds unless it is balanced by disadvantages which at present have not been identified (O'Donald 1972*c*). As O'Donald points out, just as natural selection acts on differences between species, sexual selection may be as important in determining characteristics within species and in maintaining balanced polymorphisms if countered by natural selection. Pale-phased morphs of both sexes are favoured because they start breeding at an earlier age than other phases and so they have a greater chance of surviving to breed. The component of selection in favour of pale birds is much greater than the component in favour of dark males so it is predicted that pale-phase birds will gradually replace the darks (O'Donald and Davis 1975). It has been discovered that dark-phase males hold bigger territories than pale-phase males and so on a basis of random chance it is more likely that an unmated female will encounter the territory of a dark-phase bird. This could be the basis of the female mating preferences in this species (Davis and O'Donald 1976*b*).

On Fair Isle there is a significant excess of like over unlike matings or positive assortative mating among the skuas (O'Donald 1959; O'Donald, Davis, and Broad 1974*a*). Such mating preferences will increase the frequency of homozygotes and lead to a more rapid evolution of dominance (Fisher 1930) and it is conceivable that a cline in dominance exists in the Arctic Skua for intermediate-phase birds more resemble dark than pale morphs (O'Donald and Davis 1959). Different mating preferences may apply in other parts of the Skua's range: Berry and Davis (1970) suggested that late hatching may be favoured inland in northern parts of the range, where the main food is mammals, compared with coastal and southern stations where skuas prey chiefly on birds. In Iceland unlike matings predominate (Bengtson and Owen 1973). On Foula, Shetland, 32 per cent of pale-phased birds mate assortatively whereas 47 per cent of intermediates do so on Fair Isle (O'Donald, Davis, and Broad 1974*a*).

In Chapter 10 (p. 278) it was explained that some Feral Pigeons *Columba livia* var. have mating preferences so far as the multiple alleles controlling melanic plumage pattern are concerned, and Fig. 15.6 (see between pp. 434–5) illustrates some of these preferences. From what was written in Chapter 10 it may be assumed that plumage melanism *per se* is not necessarily advantageous but rather that it is associated with other physiological advantages. These seem to be an increase in the length of the breeding season and possibly an increased fertility. Indeed, there may be a close parallel between the Feral Pigeon situation and that observed in the Arctic Skua in that two aspects of an increased reproductive potential[†] (a temporal increase in the breeding season in the Feral Pigeon and a spatial increase in the territory of the Arctic Skua) lead to female mating preferences and both are associated with melanism. Davis and O'Donald (1976*b*) have developed a model to account for female preferences in the Arctic

---

[†] A longer breeding season and an increased hormone output (gonadotrophins and androgen) could be explained by an increase in the level of the controlling oscillators as discussed in Chapter 14. It would be interesting to discover more about the apparent correlation between a reduced photo-threshold and melanism (see p. 432).

Skua. This assumes that the females at first exercise their preferences but that
as the preferred males become unavailable they are prepared to mate with other males
Interestingly the same model satisfies the situation noted in the Feral Pigeon, with
similar values being found for the fitted parameters, but in this case the females
avoid, rather than prefer, certain males (details are given in Appendix 5). The females
show no preference for wild-type but have a preference for blue checks and a rather
smaller preference for T-pattern males.

The basis for, and degree of, female preference would be expected to vary in different populations and between species. It seems likely that effects of this kind will
be found to be widespread and important but since most species show no recogniz-
able visual polymorphisms the expression of female mating preferences is likely to
go undetected. Experiments with Barbary Doves suggest that they are capable of
recognizing other individuals and prefer previous mates, the preference being maintained when physical contact is prevented or individuals are isolated for long periods
(Morris and Erickson 1971). Males were shown to display more to unfamiliar females
than to familiar ones but unfamiliar pairs were less successful in hatching eggs
(Erickson and Morris 1972). Dominant males in a penned Mourning Dove *Zenaida
macruora* population preferentially paired with dominant females, and such pairs
bred earlier and produced more surviving young than subordinate pairs (Goforth
and Baskett 1971).

Two colour phases of the Lesser Snow Goose *Anser caerulescens* are known, a
blue and a white with the latter predominating in most areas. Nevertheless, recent
warming of the eastern Canadian Arctic has been suggested as the cause of a marked
spread of blue-phase individuals since about 1930 (Cooch 1963). It has been suggested
that the two forms were once isolated, with discrete nesting, migration, and wintering areas, perhaps by the events of the Pleistocene period. Baffin Island was evidently
the distribution limit of the blue-phase birds in the immediate post-glacial period.
Blue-phase birds begin breeding earlier than do the white-phase individuals and their
photoperiodic responses are correspondingly adjusted (Figs 14.2, p. 368; 14.4, p.
371; 14.5, p. 374; 14.6, p. 375). Hence, they gained an advantage following a climatic
amelioration which possibly resulted in food supplies for chicks becoming available earlier in the season. Once ecological conditions allowed the spread and intermixing of white and blue-phase populations interbreeding was possible. Actually the
birds show a preference for mating with their own kind so that the initial spread of
the blue phase was rather slow. The polymorphism is determined by a single pair of
autosomal alleles with dominance, so that the blue-phase birds are genetically BB or
Bb and white-phase individuals are double recessives. The dominance is incomplete
so that variable amounts of white plumage occur in most heterozygotes and to some
extent in homozygotes (Cooke and Cooch 1968). Males appear to select their mates
according to the plumage of one of their parents, the choice, therefore, depending
on imprinting (Cooke, Mirsky, and Seiger 1972). Cooch has calculated that from
the time that the blue-phase phenotypes become established as 1 per cent of a population it takes 90 years for their frequency to increase to 96 per cent, a rate of
expected spread close to that observed.

Ross's Goose *Anser rossii* is a smaller, neater version of the white phase of the Lesser Snow Goose. It is confined to the Perry River region of Arctic Canada for nesting and in this area is sympatric with the Snow Goose.[†] Ross's Goose has dimorphic chicks, which can be grey or yellow, but no dimorphism is expressed in the adults. According to Cooke and Ryder (1971), the polymorphism apparently depends on a single pair of autosomal alleles with dominance so that yellow-phase chicks are double recessives. These seem to be equivalent to the yellow chicks of the Lesser Snow Goose which correspond to white adults (grey chicks later develop into blue morphs in Lesser Snow Geese). Since there is no dimorphism in adult Ross's Geese it is not surprising that mating is at random so far as the allele determining chick colour is concerned. Cooke and Ryder make the interesting suggestion that from an ancestral blue or grey goose (cf. the majority of *Anser* species) a white-phase adult (yellow chick) arose as a mutation. This gene was advantageous so that the recessive spread thereby originating a dimorphic population. Cooke and Ryder visualize that from this ancestral dimorphic stock developed the Greater and Lesser Snow Geese and Ross's Goose in the way shown in Chart 15.1.

CHART 15.1

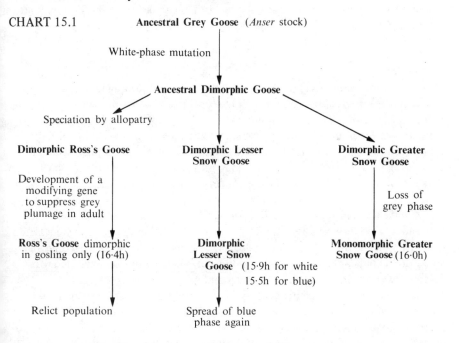

(The figures in brackets are photoperiods at which the different species or morphs begin breeding at Slimbridge—see Fig. 14.2, p. 368).

[†] Recently hybrids between Ross's and Lesser Snow Geese have been identified. Both species have extended their range and in sympatric colonies mixed clutches have been seen resulting in inappropriate imprinting and interspecific pairing. The hybrids are fully fertile with themselves and their two parent stocks and there is therefore some danger of the gene pool of the rare Ross's Goose being swamped by a hybrid swarm (Trauger, Dzubin, and Ryder 1971).

Cooke and Ryder suggest that selection for the white phase did not operate in the gosling of Ross's Goose. Expression of the dark phase in the adult might have been suppressed by modifier genes without loss of the original allele which is still manifested in the chick. Cooke and Ryder refer to a similar situation in the domesticated Emden Goose in which the gosling is pigmented but gene expression is suppressed in the adult which is white (Jerome 1959). It is interesting to speculate whether loss of melanism in these *Anser* geese is somehow associated with an increased threshold of photosensitivity and an alteration in the phase relationships of the circadian organization (see Fig. 14.5, p. 374). The Mute Swan *Cygnus olor* has a leucistic phase called the 'Polish' swan, for it was once thought to be a distinct species. The gosling has white instead of grey plumage and moults directly into the white plumage of the adult without wearing the usual grey—brown juvenile plumage. The legs and feet of the adult are pale grey or pinkish instead of black. In this case the gene responsible is sex-linked and recessive (Munro, Smith, and Kupa 1968) and it is not clear whether it has significance in the present context (see also Nelson 1976) The leucistic phase is rare in Britain but has been increasing elsewhere in Europe (Scot and the Wildfowl Trust 1972). A similar leucistic phase occurs in the Trumpeter Swan *Cygnus c. buccinator* and also exceptionally in the Black Swan *C. atratus* but the genetics have not been examined.

Melanism

It appears to be a common phenomenon that an altered breeding season among closely related species is associated with the loss or acquisition of melanin pigments. A decrease in photosensitivity threshold to acquire a shorter breeding season in polar regions seems initially to be associated with a loss of pigmentation, as in the white Arctic swans. With a readjustment of circadian organization pigmentation again becomes a feature of most of the grey geese (*Anser*), except for those with the shortest breeding seasons. The black geese (*Branta*) seem to have derived from *Anser* by further adjustments so that northern breeding forms like the Red-breasted Goose *Branta ruficollis* and Brent Goose *B. bernicla* are dark. Converse trends associated with reduced photosensitivity threshold have been noted in the Feral Pigeon. Often low-latitude derivatives of species originating at higher latitudes have become melanic, with a concomitant change in breeding periodicity. For example, Eleonora's Falcon *Falco eleonorae* breeds in late July and August on rocky islets in the Mediterranean and the Atlantic coast of Morocco and depends on catching migrant passerines during autumn migration (Vaughan 1961): so too does the Sooty Falcon *Falco concolor*, which breeds in late summer on the Dahlac Islands off Eritrea, and also in parts of the Sahara from Cairo south to the southern Red Sea (Clapham 1964). Both these species appear to be melanic derivatives of the Peregrine Falcon *F. peregrinus*, which has a photoperiodically controlled vernal breeding season. It is tempting to imagine that the melanic species are able to remain in breeding condition beyond the summer solstice because their hormone rhythms are less readily phased into a photo-refractory stage by long summer days. It is not yet clear why a relationship should exist between breeding periodicity and melanism. Perhaps the involvement of the pineal gland in regulating circadian rhythmicity and the implication of melatonin accounts for the relationship.* Alternately, ACTH activity might account for the pigmentation. The subject deserves closer study since melanin pigments evolved to protect animals from harmful light radiations.

In well-adapted species a change in plumage colour may necessitate a changed ecology, especially if plumage colour is adjusted to facilitate feeding or to escape predation. In the Snow Goose complex it seems reasonable to suppose that physiological factors have been of prime importance in the polymorphism and that this has not interfered with the ecological requirements of feeding or escaping predators. Several heron species are dimorphic with light and dark phases, particularly the Eastern Reef Egret *Egretta sacra* and Western Reef Egret *E. gularis* and Reddish Egret *Dichromanassa rufescens*. Murton (1971c) produced evidence that the different morphs adopted different hunting strategies and Holyoak (1973) then showed that some favoured particular substrates of dark rocks or sand to suit their colour. But whether these differences in feeding ecology are a secondary consequence of a dimorphism evolved for other reasons is not established; there is slight evidence that light and dark morphs of the Western Reef Egret breed at different times (Murton 1971c).

Biochemical polymorphisms

As yet we know very little about the selective advantage or disadvantage or polymorphisms which affect biochemical function (but see Bryant 1974). For example, electrophoretic and histochemical techniques have revealed that many enzymes exist in different molecular forms (isozymes) often within a particular cell. The organism is, therefore, potentially provided with a wide array of enzyme molecules to cope with the most exacting specific metabolic requirements of cells, and this is true of lactate dehydrogenase (LDH) which has long been recognized to comprise the 5 isozymes $LDH_1 \ldots LDH_5$ (Markert and Møller 1959). These LDH isozymes are composed of 4 polypeptide chains which are separable into two classes (A and B)[†] according to charge, and their synthesis depends on the activity of codominant

---

*In amphibians it is supposed that the central nervous system exerts a tonic inhibition of melanophore stimulating hormone (MSH) release so that interruption of hypothalamic–hypophysial connections (Etkin 1967), or use of agents like reserpine which depress hypothalamic activity (Iturriza 1966), cause a darkening of the skin of frogs. Similar central nervous system control is believed to operate in mammals, but the exact mechanisms remain unresolved (Taleisnik, Tomatis, and Celis 1972). Present evidence indicates that formation of a factor which inhibits release of MSH (MSH-R-IF) results from two hypothalamic enzymes acting on oxytocin. One (E_1) promotes the formation of MSH-R-IF while the other (E_2), which forms MSH-RF, competes for oxytocin thereby preventing MSH-R-IF formation. Accordingly, the amount of MSH-R-IF formed depends on the relative activity of the two enzymes and the concentration of oxytocin in the hypothalamic neurones (Celis and Taleisnik 1971a, b, 1973).

[†]'B' polypeptides are strongly inhibited by low concentrations of substrate pyruvate or lactate and this prevents accumulation of lactic acid by continued reduction of pyruvate, which is instead metabolized via the Krebs cycle. 'A' polypeptides need higher substrate concentrations for inhibition so they encourage the conversion of pyruvate to lactate when oxygen is in short supply.

alleles at two loci LDH_A and LDH_B. The polypeptide products at each locus assemble at random to form 5 tetrameric isozymes $A_4(B_0)$, A_3B_1, A_2B_2, A_1B_3, and $(A_0)B_4$, which are decreasingly inhibited by either pyruvate or lactate (see Zinkham and Isensee (1972) for summary). In 1963 a sixth lactate dehydrogenase was found in the human testis (Blanco and Zinkham 1963) and subsequently testes homogenates of other animals were found to have one or more isozymes in addition to the usual 5, including *Columba livia* (Zinkham, Blanco and Kupchyk 1964; Zinkham, Kupchyk, Blanco, and Isensee 1966). The extra isozyme(s) are localized in a particular type of mitochondria which in rats becomes the only kind present by the time the germ cells reach the stage of spermatids and spermatozoa (Domenech, Domenech, Aoki, and Blanco 1972).

These discoveries indicated that more than two genes controlled LDH synthesis. The new isozyme, originally designated LDH-X, was dissociated into another polypeptide subunit (C) which could recombine with A and B to form functional hybrid molecules of LDH. Studies of pigeons revealed that the C polypeptides are products of a third locus LDH_C, which is closely linked to, or even contiguous with the LDH_B locus. The B locus is active in both sexes and cell tissues throughout life but the C locus operates only in the testis during a specific stage of spermatogenesis. Thus, there are two alleles at the LDH_C locus and their frequency agrees with that expected by the Hardy−Weinberg law in free-living, town populations and racing stocks of *C. livia*, but not in pure-bred lines (White Carneau and Silver King) (Zinkham *et al.* 1964). There are three alleles at the B locus (Zinkham and Isensee 1972), but as yet no polymorphisms have been found at the A locus.

The iron-binding glycoproteins, transferrin from vertebrate plasma, and ovotransferrin from avian egg white are polymorphic[†] in almost all vertebrates, and this has been demonstrated for many bird species (e.g. Vohs and Carr 1969). Ferguson (1971) found two alleles (Tf^f and $TF^{s[†]}$) in *Columba livia* and *Columba palumbus* from Northern Ireland and in *Streptopelia risoria* from captive stocks. Both alleles were at intermediate frequencies in these three species, ruling out recurrent mutation as the cause of the polymorphism. Moreover, since heterozygotes did not exceed the expected proportion, a balanced polymorphism clearly did not exist. It was therefore proposed that the polymorphism was maintained by selection favouring different alleles in different populations causing frequency divergence and gene flow to equate the frequency. In Belfast pigeons, the frequency of Tf^s exceeded that of Tf^f and this was also the case in a population studied by Mueller, Smithies, and Irwin (1962) in Wisconsin, U.S.A. It has been found that ovotransferrins from the eggs of heterozygous females inhibit microbial growth more strongly than those from either homozygote and this is thought to explain why a larger proportion of eggs of heterozygous females are hatched (Frelinger 1972). This effective higher fertility of heterozygous females has been used in a genetic model that predicts a maintenance of polymorphism in the absence of a discernable excess of heterozygotes in the population (Frelinger and Crow 1973).

[†] The genetic variants show up on an electrophoretic gel as bands of altered mobility, Tf^f being fast and Tf^s slow. Frelinger (1972) has renamed these as TF^A and TF^B.

Fig. 14.9. (a) Winter and summer ranges of various races of Canada Goose *Branta canadensis* showing the main migration routes. Numbers below identify the races and the photoperiod under which breeding begins at Slimbridge is also shown. The races migrating furthest north require longer daylengths to initiate breeding. 1, Giant Canada Goose *B. c. maxima* (13·8 h); 2, Great Basin Canada Goose *B. c. moffetti* (14·5 h); 3, Dusky Canada Goose *B. c. occidentalis* (14·7 h); 4, Tavener's Canada Goose *B. c. taverneri* (15·0 h), also some Lesser Canada Goose *B. c. parvipes* (15·2 h) and Richardson's Canada Goose *B. c. hutchinsii* (?); 5, Todd's Canada Goose *B. c. interior* (15·0 h); 6, Lesser Canada Goose (15·2 h). (b) Part of a wintering flock of the Giant Canada Goose *Branta canadensis maxima*. This sub-species was thought to be extinct until this population was discovered. It breeds in parts of Manitoba and winters at around 43°N on the eastern side of the Great Lakes. Note the Mallards *Anas platyrhynchos* and the single Golden-eye *Bucephala clangula* (top right). (Near Rochester, Minnesota, U.S.A.)

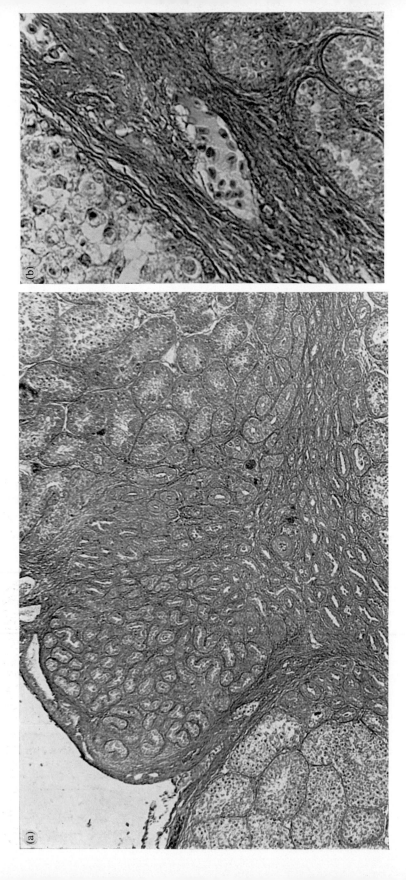

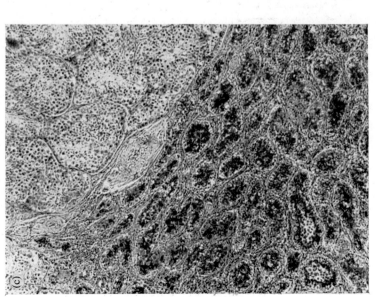

Fig. 14.21. Transverse sections of testes of Pink-eared Duck *Malacorhynchus membranaceus* showing asynchronous spermatogenetic condition of the tubules. (a) Low power wax embedded section showing small area of tissue with regressed tubules adjacent to nodules which are composed of enlarged tubules. As a result of these differences in macrostructure the whole organ has an unusual appearance bearing a superficial resemblance to an ovary; regions with enlarged tubules protrude from the matrix as 3–8 nodules some of which may reach the size of a small pea (c. 5 mm dia.). ×40. (b) High power wax section showing strand of connective tissue with contained blood vessels which separates a region of enlarged tubules, containing primary spermatocytes, from a region of small tubules, containing only spermatogonia. ×250. (c) Gelatin embedded section showing black areas of sudophilic lipoidal material in the regressed tubules whereas no such lipids are present in the recrudesced tubules. Note also black spots of lipid in the interstitial areas adjacent to the regressed tubules. ×40. The histological quality of the preparation has suffered as a result of some delay between autopsy and fixation. (a) and (b) fixed in Bouin's fluid and stained with Heidenhain's iron haematoxylin and orange G. (c) fixed in formalin and coloured with Sudan Black B. and carmalum.

Fig. 15.1. Display bower of Great Bower Bird *Chlamydera nuchalis* (near Darwin, North Australia). The bird has carefully positioned pieces of broken green beer bottles on either side of the shells to produce two bands of colour setting off the centre panel. Did the bird have an image of the end result (goal) when it first began constructing the display arena?

FIG. 15.2. (a) In the Common Shelduck *Tadorna tadorna* (tribe Tadornini) the sexes are almost identical except that the male is slightly bigger and brighter and has a knob on the base of his bill. (Suffolk, England.) (b) The Lesser Scaup *Aythya affinis* (tribe Aythyini) is markedly dimorphic. The male shown above is resplendent in iridescent purple–black, grey, and white while the female (not shown) is a drab brown bird with a white face. (California, U.S.A.)

FIG. 15.4. Effect of photoperiod on the moult of the male House Sparrow *Passer domesticus*. (a) Caught while still breeding in mid-July and held on an artificial LD 16 : 8 until October. The post-nuptial moult has been inhibited and feathers from the previous breeding season are retained. (b) Caught as (a) above but held on natural seasonal photoperiods until October. The moult has occurred resulting in numerous new white-tipped feathers. The white tips would be lost by abrasion before the next breeding season to give a black breast as at (a).

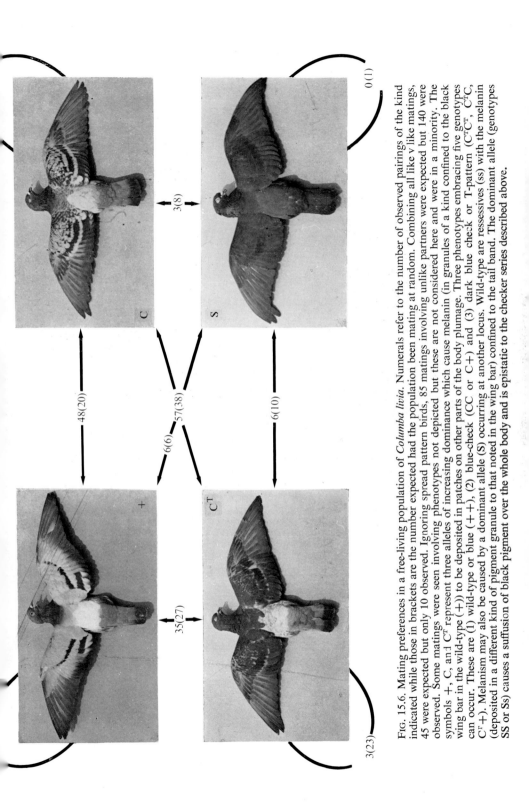

FIG. 15.6. Mating preferences in a free-living population of *Columba livia*. Numerals refer to the number of observed pairings of the kind indicated while those in brackets are the number expected had the population been mating at random. Combining all like v like matings, 45 were expected but only 10 observed. Ignoring spread pattern birds, 85 matings involving unlike partners were expected but 140 were observed. Some matings were seen involving phenotypes not depicted but these are not considered here and were in a minority. The symbols +, C, and C^T represent three alleles of increasing dominance which cause melanin (in granules of a kind confined to the black wing bar in the wild-type (+)) to be deposited in patches on other parts of the body plumage. Three phenotypes embracing five genotypes can occur. These are (1) wild-type or blue (++), (2) blue-check (CC or C+) and (3) dark blue check or T-pattern (C^TC^T, C^TC, C^T+). Melanism may also be caused by a dominant allele (S) occurring at another locus. Wild-type are ressessives (ss) with the melanin (deposited in a different kind of pigment granule to that noted in the wing bar) confined to the tail band. The dominant allele (genotypes SS or Ss) causes a suffusion of black pigment over the whole body and is epistatic to the checker series described above.

Fig. 15.8. Some of the wading birds mentioned in the text. (a) Little Stint *Calidris minuta*; (b) Sanderling *Calidris alba*;

(c) Male Dotterel *Eudromias morinellus*, incubating eggs; (d) Curlew Sandpiper *Calidris ferruginea*. (a), (b), and (d) on spring passage through Almeria, Spain; (c) Inverness-shire, Scotland.

(a)

(b)

Fig. 15.9. (a) The monomorphic, non-parasitic Bay-winged Cowbird *Molothrus badius*. This specimen, photographed near a drinking puddle in the Argentinian chaco, is in heavy post-nuptial moult and illustrates the poor plumage condition of most small birds at this season. (b) Male of the dimorphic, parasitic, Shiny Cowbird *Molothrus bonariensis*. In contrast to the iridescent black of the male, the female's plumage is a sombre uniform grey. (c) A common host of the Shiny Cowbird is the Rufous-collared Sparrow *Zonotrichia capensis* discussed in detail in Chapter 10 (see Fig. 10.7, p. 262). (Photographs taken at Córdoba Province, Argentina.)

FIG. 15.10. The nest of this Black Drongo *Dicrurus macrocercus*, built at the top of a pine t
exposed to the full glare of the tropical sun—a heat that would prove lethal for temperate
passerines. The parents occasionally shelter the chicks from the sun, as the female is doing
(Photograph: New Territories, Hong Kong.)

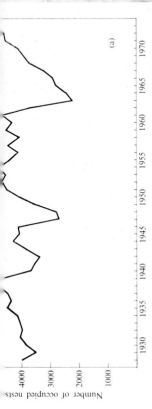

Fig. 16.1 (a). Number of occupied nests of the Heron *Ardea cinerea* in England and Wales. Arrows denote severe winters. Based on Stafford (1971) and Reynolds (1974). (b) Grey Heron *Ardea cinerea* guarding small young. (Photograph taken in Norfolk, England.)

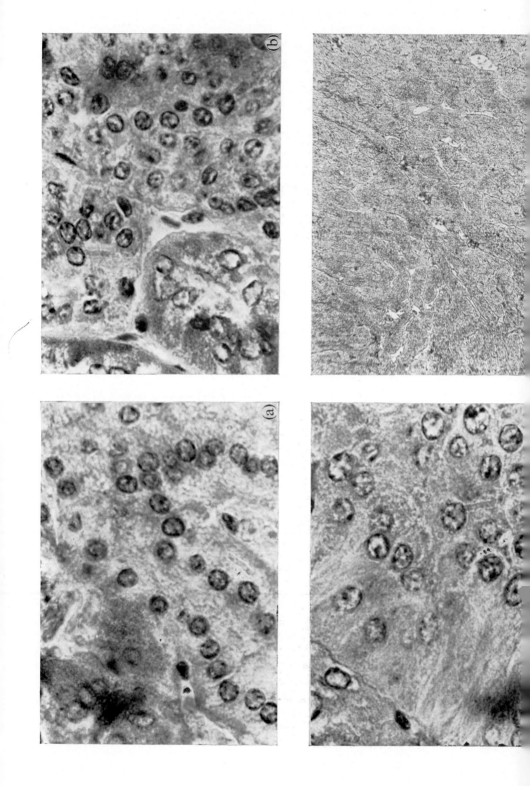

(f)

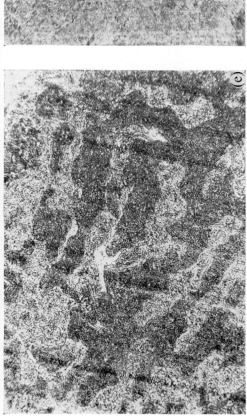

(e)

Fig. 16.7. Adrenal tissue of Wood Pigeon *Columba palumbus* showing effect of social status on the cortical (interrenal) tissue. Wax-embedded material stained with Ehrlich's haematoxylin and eosin. (a) normal adult from flock centre; (b) subordinate individual (adult) from the front of a winter feeding flock showing increase in nuclear diameter and cell hypertrophy; (c) an even more extreme example of adrenal hypertrophy in an adult which was unable to gain a feeding place in a flock.
Tissue treated with buffered methylene blue to reveal RNA (dark stain in photograph). (d) normal specimen, being the same as (a) above; (e) bird from front of flock (same as (b) above) showing marked RNA activity; (f) the same as (e) but after treatment with ribonuclease to remove RNA and to serve as a control. From Murton *et al.* (1971).

Ferguson (1971) found that the frequency of slow and fast alleles in *C. palumbus*, unlike the situation noted in *C. livia*, varied with time of year and locality and there was a deficit of heterozygotes, indicating a lowered fitness. A more detailed study of egg-white samples showed differences in the frequency of the T^f allele between Antrim and Fermanagh, Ireland, and only in the Antrim locality were phenotype numbers in agreement with Hardy—Weinberg prediction. According to Ferguson, such inter-population variation in allele frequency for polymorphic proteins is common in animals and represents an adaptation to local environmental conditions. Ferguson found only the Tf^f allele in *Streptopelia decaocto* and suggested a 'founder effect' from a presumed small population of monomorphic birds that initially colonized Britain. However, only the Tf^f allele was found in *Streptopelia risoria* from Wisconsin (Desborough and Irwin 1966). This allele also had a higher frequency in *S. 'risoria'* stock in Ireland (Ferguson 1971) and captive *S. humilis* (*S. tranquebarica*) in North America (Desborough and Irwin 1966), so it could be conferring a physiological advantage in this genus.

Polyandry and sex reversal

The majority of birds are monogamous, the pair bond lasting for the breeding season, or even for life. If it be assumed that the mechanisms of pair formation ensure that the best available mate is chosen, then the chances of changing to a better mate must be too low to make this a profitable strategy. Moreover, if two parents are needed to raise the brood and if experience improves the performance of the pair, selection will operate to cement the monogamous habit. The problems of synchronizing breeding rhythms must be lessened in long established pairs and increased in situations where casual mating is the rule. We are concerned only to define in broad terms the ecological conditions which favour a deviation from the monogamous habit, for our main interest in this respect is to examine what adjustments in reproductive physiology have been required to facilitate polygamy; this term covers a range of possible sexual relationships involving promiscuity by either sex, a male having more than one female (polygamy) or a female more than one male (polyandry). Armstrong (1964) gave a very good summary of the topic and has been only slightly modified by the more recent account of Lack (1968a). Many bird species which are otherwise monogamous may form casual pair bonds with more than one partner, for example, nesting Long-tailed Tits *Aegithalos caudatus* may form trios with two females and one male feeding the brood (Gaston 1973). Selection will favour polygamy when the reproductive rate is thereby increased.

Polyandry and sex reversal are rare and appear to have been evolved only six times in surviving bird groups, all of which are nidifugous: rheas (Rheidae), tinamous (Tinamidae), buttonquail (Turnicidae), jacanas (Jacanidae), painted snipe (Rostratulidae), the Tasmanian Native Hen *Tribonyx mortierii*, and possibly other rails (Rallidae). The Rhea *Rhea americana* is doubtfully placed in this category. The male makes a nest scrape and then calls up the females which move around in troups; he copulates with several females and each one lays an egg in the nest. When 4-5 eggs are accumulated the male incubates unaided. Meanwhile the females move to another male. In some tinamous similar behaviour occurs. The male of Bonaparte's Tinamou *Nothocercus bonapartei* apparently copulates with several females after a modified lek display and he incubates the eggs they produce for him in a common nest; it is possible that the females then favour another male (Schäfer 1954). The female Variegated Tinamou *Crypturellus variegatus* courts a male and lays an egg for him to incubate and rear the chick, and then repeats the process with other males. Similarly in the button quails, for example, the Little Button Quail *Turnix sylvatica* and both the Pheasant-tailed Jacana *Hydrophasianus chirurgus* and American Jacana *Jacana spinosa*, the female lays a series of clutches which are incubated by different males (Hoffmann 1949; Jenni and Collier 1972). Perhaps the beginnings of this trend are apparent in the Red-legged Partridge *Alectoris rufa* for the female lays a clutch which her mate incubates and rears the young while she then lays a second clutch which she herself tends (Goodwin 1953). The Red-legged Partridge is usually monogamous but has been recorded as being paired with two cocks and laying three clutches (Jenkins 1957). Compared with the monogamous and single-brooded Grey Partridge *Perdix perdix*, the Red-legged Partridge has an extended breeding season, with second broods sometimes being produced in August. A capacity to remain in breeding condition for an extended season must be a feature of polygamous species compared with closely related monogamous forms.

Polyandry is not easy to explain in ecological terms except that when it is possible it is a means of increasing the fecundity of one sex, as of course is polygyny. It could arise because of differential mortality between the sexes but is perhaps more reasonably explained in terms of the ecological problems of rearing progeny. Perhaps it is rare because it first requires an increase in female dominance thereby allowing a reversal of sexual role, other than during fertilization. Jenni and Collier (1972) suggest that polyandry allows the American Jacana to produce the maximum number of young on a limited amount of breeding habitat by reducing competition between adults and chicks. They suggest the female has become a large bird ($♀$ 161 g, $♂$ 91 g in body weight). It is also conceivable that it is the male who has become small to reduce his feeding demands and that the female has had to remain large to produce a viable egg. The essential point is the extent to which ecological conditions are selecting for a relatively large egg to compensate for feeding difficulties when the chick hatches. It is convenient to trace the trends towards sex reversal in various limicoline birds and then to return to the question of polyandry in jacanas.

A general problem faced by waders breeding in the Arctic is the shortness of the summer—although during this season food supplies may reach abundant proportions—and the variability in time of the spring snow thaw. For most species there is only time to produce one clutch of eggs and if this is lost the female is usually unable to find the reserves needed for a repeat. Even if repeats are possible the strategic timing of the season is lost. If it is possible for the female to be relieved from an energy-demanding stint of incubation and brood care, leaving this to the male, it is possible that in some situations she might accumulate reserves for further laying. A balancing process will be the male's capacity to raise the brood unaided and this may be affected by the degree of precocity of the young, in turn determined by the size of the initial egg. We begin with what seems a reasonably typical, and presumably ancestral condition of pair participation in reproduction as exemplified by the Dunlin *Calidris alpina* (Holmes 1966*a*, *b*).

Male Dunlin arrive on their tundra breeding grounds (latitude 71° near Barrow, Alaska) in late May and acquire and advertise territories. Females move into these territories and pair formation occurs. However, some individuals pair during migration, especially if bad weather delays the journey to the breeding grounds. Eggs are mostly laid in mid-June, so that the chicks hatch 3 weeks later, which is when on average peak supplies of adult chironomid and tipulid flies are available. As Holmes has shown, mid-summer is the time when the greatest number of changes occur in the insect populations, for over-wintered larvae which grew during June now pupate, imagoes emerge and lay their eggs, and new larvae begin to grow. The young Dunlin at first feed on adult insects on the surface and then later take the new generation of larvae, whereas adults need to feed mostly on larvae in mid-summer to obtain sufficient nutrients. Accordingly, adults may in bad years experience mid-summer food shortages while their young can feed well. To obviate feeding difficulties, adults drift during middle and late summer to more favourable habitats, frequenting lowland marshes which are not included in the breeding territories. It is perhaps to compensate for this that the male takes on an increasingly larger part of the task of incubation and brood care as the cycle progresses so allowing the female to compensate for weight lost in producing the eggs and during incubation (see Fig. 15.7). The

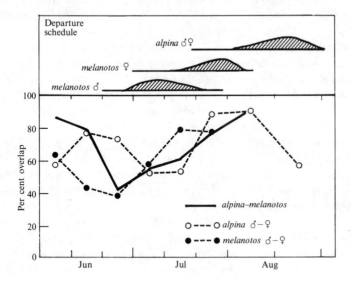

Fig. 15.7. Index of diet similarity (percentage of overlap in feeding preferences) between the sexes of Dunlin *Calidris alpina* and Pectoral Sandpiper *C. melanotus* and their departure schedule from the breeding ground. Females of *melanotus* emigrate from the breeding ground at the time when they begin to require the same foods as male and female *alpina* and so ecological competition is avoided. Similarly male *melanotus* depart when they begin to compete with their own females for food resources. From Holmes and Pitelka (1968).

pattern of weight change during the breeding season is rather different in Dunlin breeding in Finland (Soikkeli 1974). In some North American populations both sexes have a post-nuptial moult on the breeding ground but this seems not to apply to Northern Dunlin in Europe; moulting of these on the Wadden Sea, Holland, is recorded (Nieboer 1972).

The Western Sandpiper *Calidris mauri* nests near the Dunlin in Alaska but selects drier, heath tundra islands within the marshy grass-sedge tundra utilized by the Dunlin (Holmes 1973). It breeds in much denser groups in which individual territory size is small, feeding often occurs outside the territory, and there is a high intensity of interaction between territory owners. Holmes suggests that a strong pair-bond attachment keeps the parents together until the chicks hatch so that they can provide a defence against predators. The breeding behaviour of the Purple Sandpiper *C. maritima* is essentially the same except that only the male attends the young (Bengtson 1970, 1975). The Turnstone *Arenaris interpres* on Ellesmere Island (81°49'N, 71°18'W) in arctic Canada feeds on the same foods as the Dunlin during the breeding season and its breeding biology is similar to the species just discussed (Nettleship 1973).

A trend for the male to take an increasing share of incubation and brood duties is seen in the Semi-palmated Sandpiper *Calidris pusillus*, Little Stint *C. minuta* (Fig. 15.8(a) see between pp. 434-5), and Baird's Sandpiper *C. bairdii*. In *bairdii* and *pusilla* the adults leave the breeding grounds immediately the young can fly and migrate to wintering grounds in the southern hemisphere before moulting (Pitelka 1959), sometimes the female leaves before the male. It is possible that feeding conditions are more difficult for these species than for the Dunlin group just described. Another possible selective factor is the need to prepare for a long migration (the Little Stint winters in South Africa). If the female has depleted her resources in laying eggs and also needs to accumulate fat reserves for migrating there could be survival value in involving the male in more brood care. Given an evolutionary trend for male involvement in brood care, it may become possible, under some ecological conditions for the female to lay one clutch for the male to incubate and tend and immediately produce a second which she looks after. This is the strategy adopted by Temminck's Stint *Calidris temminckii*, which inhabits boggy, shrub tundra in the Old World (Hildén 1965), and by the Holarctic Sanderling *C. alba* (Fig. 15.8(b) see between pp. 434-5) (Parmelee 1970; Parmelee and Payne 1973), and it seems to have been achieved in both species because the females produce unusually small eggs, compared with other small waders, in relation to their body size (Schönwetter 1967; Lack 1966). Since the chicks, aided by only a single adult, must compensate for a relatively retarded development at hatching, it must be presumed that feeding conditions for them are comparatively good at the time of hatching. In Temminck's Stint the male waits several days before incubating his clutch, meanwhile consorting with his female to ensure that he, rather than other males, mates with her, although some females do manage to consort with a different partner (Hildén 1965), and the first stages of polyandry are evident. Polyandry seems to be much less frequent in the Sanderling but it is interesting that, unlike most waders, the female has an aerial display similar to the male and she initiates the cop lation sequence by a display at the nest-scrape (Parmelee 1970); these are signs of increasing female dominance. Parmelee and Payne suggest that in bad years the Sanderling might not raise many young by virtue of laying poorly provisioned eggs though in very bad years the females may have a reduced mortality from not taxing their reserves too heavily. They imagine that in good years the birds can compensate by the increased productivity from two clutches.

If it is relatively easy for a female to produce a clutch and for one bird to care for the brood it may be wondered why polyandry rather than polygyny has not evolved. In all the waders so far considered the female is slightly larger than the male so a trend towards female dominance might have already developed. In the Stilt Sandpiper *Micropalama himantopus* and Least Sandpiper *Calidris minutilla* (ecological equivalent in the Nearctic of Old World Little Stint), in which females are also larger than males, the larger females, smaller males, and pairs with the largest size difference between mates hatch their eggs earlier than the others and this could provide the adaptive advantage, with sexual selection the mechanism (Jehl 1970). However, in the Stilt Sandpiper the sexes take an approximately equal share in incubation and brood care duties (Jehl 1973). It is not known to what extent the sexes differ in their feeding habits. Both sexes loose weight from the time of egg laying but Jehl's data suggest that females loose proportionately more than do the males.

Far from being able to produce two clutches by involving the male in parental duties, the females of some species seem able to produce only a single—often reduced in size—clutch even if they receive all the possible help the male can give. Often this ecological need has been achieved by an increase in female assertiveness to the point at which sex reversal has occurred so that the female is bigger and more brightly coloured than the male. This is the case in the Dotterel *Eudromias marinellus* (Fig. 15.8(c) see between pp. 434-5), which has a relict distribution in the barren tundras of the Arctic and in the alpine zone of certain high mountains in Eurasia, including the Grampian and Cairngorm ranges of Scotland (Pulliainen 1970; Nethersole-Thompson 1973).[†] Small Coleoptera and their larvae and the larvae of Diptera and Tipulidae appear to be important food items, as with other plovers, and in these exposed habitats such food supplies can fluctuate violently in availability following mild rains, or drying cold winds. Female Dotterels produce 2–3 large eggs relative to their body size as soon as conditions are suitable and must now recoup their lost reserves and moult before the short summer season ends. This is made possible by relinquishing most of their parental duties to the male, although they occasionally do participate in incubation and sometimes pair with a second male. The trend towards sex reversal is seen in the New World Mountain Plover *Charadrius montanus* which is endemic to the highland short grass prairie of western North America. The fluctuating and often difficult feeding conditions experienced in spring impose a big demand on the energy budget of

---

[†] An exceptional population has been nesting in certain of the newly claimed Dutch polders since 1961. It derives from migrants which apparently found conditions sufficiently similar to their natural habitat to settle and breed. Nests have been in summer crops of flax, sugar-beet, wheat, and potatoes and Nethersole-Thompson (1973) suggests that these habitats superficially resemble the prehistoric niche which he reckons to have been the Asian steppes.

the female. Having laid eggs she must now recoup her body condition and to this end the male alone incubates. But, in the occasional good spring, the female can sometime lay two clutches, the second of which she tends (Graul 1974, 1975). Given unpredictable and variable ecological conditions it is easy to see that in some species a trend towards successive polyandry could develop rather than polygyny.

Sex reversal is even more pronounced in the phalaropes *Phalaropus* and females take virtually no part in nest duties. In Alaska, female Grey Phalaropes *P. fulicarius* vacate the tundra before the mid-summer period of food shortage, leaving the male to care for the brood. The specialized feeding actions (the birds spin round and round in shallow water to create an upward current which brings food items to the surface) and feeding site of the phalaropes may accentuate the difficulties of breeding; they frequent boggy tundra pools, are critically dependent on weather conditions, and generally arrive later at the breeding station than, for example the *Calidris* sandpipers. In both the Old World Red-necked Phalarope *P. lobatus* and the New World Wilson's Phalarope *P. tricolor* the assumption of nuptial plumage depends on androgenic steroids, for injection of testosterone propionate, but not oestradiol or prolactin, to either sex will stimulate the growth of nuptial plumage in plucked areas (Johns 1964). The female is brighter than the male because the testosterone content of phalarope ovaries exceeds that of testicular tissue from the same species, the reverse being true in species showing normal male dominance (Höhn and Cheng 1967; Höhn 1970). Incubation of testis homogenates with labelled [³H] pregnenolone showed that in Wilson's Phalarope testicular tissue generally formed more testosterone than androstenedione. The same applied to ovarian tissue from gonads weighing less than 150–200 mg but heavier ovaries formed a higher ratio of androstenedione (Fevold and Pfeiffer 1968). Thus females appear to produce maximum testosterone titres in the early part of the breeding cycle, consistent with the time nuptial plumage is acquired. Testosterone and prolactin are necessary for development of the brood patch which occurs only in the male of this species (Johns and Pfeiffer 1963). Indeed, Höhn and Cheng (1965) found that female phalaropes produced less prolactin than the males. The adenohypophyses of males captured during incubation were found by Nicoll, Pfeiffer, and Fevold (1967) to contain 3 times as much prolactin as the female gland and the authors suggested that a deficiency of prolactin might contribute to the failure of females to incubate. However, since the act of incubation supposedly stimulates prolactin secretion in doves (see p. 134) it might be argued that male phalaropes produce more prolactin than females because they alone incubate, although this would not explain the absence of the incubation patch in the female. The reduction of prolactin secretion in the female seems to have removed an anti-testosterone factor, thereby allowing testosterone dependent behaviour and plumage coloration to become dominant. It is not clear whether there has been an opposite trend in the male.

Red-necked Phalaropes may lay a second clutch and attempt to attract a second male (Tinbergen 1935) and female Wilson's Phalaropes can be promiscuous (Johns 1969, Howe 1975). In some seasons Grey Phalaropes can be promiscuous and polyandry may possibly occur, while in other years, depending on ecological conditions, females may consort with only one male (Kistchinski 1975). Therefore, given favourable environmental conditions and the availability of surplus males it is possible to visualize how successive polyandry could emerge as in the examples of the plovers and jacanas already discussed. Similarly, in the Old World Painted Snipe *Rostratula benghalensis,* which exhibits sex reversal, the female retains the territory from year to year and courts the male which enters it. She lays a clutch for this male and once incubation begins repeats the process and attracts and pairs with another male. The breeding season in this species extends throughout most of the year. Polyandry does not apparently occur in the Australian and African sub-species of the Painted Snipe (Lowe 1963), nor in the only other species in the family, the South American Painted Snipe *Nycticryhes semicollaris* in which the female incubates (Johnson and Goodall 1965).

Polygyny

This is more common than polyandry and is possible when a female can readily produce her eggs and then raise the brood unaided. In such circumstances the reproductive rate is increased if females predominate and each male has more than one female (e.g. Orians 1969). Von Haartman (1969) noted that many polygynous species have domed nests, or else nest in holes, and he argued that the increased security of such sites from predators would lessen the need for male assistance in driving off enemies. Successive polygyny occurs in some passerine birds, with the male having successive broods with different females. Its origins can be traced to species in which the males exhibit transient multiple pairing and this occurs to a variable extent in many kinds: Penduline Tit *Remiz pendulinus,* Wren *Troglodytes troglodytes,* Pied and Collared Flycatchers *Ficedula hypoleuca* and *F. albicollis,* and the Dickcissel *Spiza americana.* Successive polygyny becomes a regular practice with some of the colonial Ploceidae and those Ploceids and certain Icteridae which have grouped territories. For example, in Wagler's Oropendola *Zarhynchus wagleri* the female builds a nest and associates with a male for a few days and once fertilization has been accomplished he moves on to another female. Similar behaviour occurs in the Yellow-headed Blackbird *Xanthocephalus xanthocephalus* and the Boat-tailed Grackle *Cassidix mexicanus,* and the Red-winged and Tricolored Blackbirds *Agelaius phoeniceus* and *A. tricolor* are well-studied examples (see p. 239). In these last two species certain males (males do not mature until their second year) are able to occupy the best territories and there is a large surplus of non-breeding males, including 1-year-old males with sub-adult plumage (Orians 1961). If females are to be successful in leaving progeny it is clearly advantageous for them to prefer territory-holding males. McLaren (1972) goes so far as to postulate that territoriality has evolved in many solitary nesters for the main purposes of developing a polygynous mating system; he studied a colour-ringed population of the Ipswich Sparrow *Passerculus princeps.* In Brewer's Blackbird *Euphagus cyanocephalus* the excess of females and prevalence of polygyny varies from year to year (Williams 1952). Similarly, in the Pied Flycatcher, Dickcissel, and Long-billed Marsh Wren *Cistothorus palustris* females often pair with already mated males because, as Lack (1968a) has suggested, the other males appear to have poorer territories; this is demonstrated in the Long-billed Marsh Wren (Verner 1964; Verner and Engelsen 1970), Red-winged Blackbird (Orians 1961; Haigh 1968, quoted by Selander 1972) and Great-tailed Grackle *Quiscalus mexicanus* (Selander and Giller 1961; Kok 1970, quoted by Selander (1972)). Haigh's data showed that the number of young fledged by females mated to polygynous males was higher than for those mated to monogamous individuals. For some of the ploceid[†] weavers, nesting sites are restricted relative to food resources so that among these too some males can acquire the few territories and command the attention of the females (Crook 1962).

In simultaneous polygyny the male copulates with several females over the same period rather than successively. The females may all live in the male's large territory as in the Corn Bunting *Emberiza calandra*. The advantages of such behaviour in this species are not clear since both sexes are cryptically coloured, unlike most of the genus, and suitable habitat does not seem limited. Similar behaviour has recently been recorded for the North American Dipper *Cinclus mexicanus* (Price and Bock 1973). Male Pheasants *Phasianus colchicus* in Britain and North America hold a large territory and attract a harem of females which move off to their own areas to lay. In New Zealand the sex ratio is more even and the species is monogamous.

When the male's main function is to fertilize a female, after first acquiring a territory, conditions exist for intense sexual selection and the development of brilliant plumage and elaborate displays, as in the birds-of-paradise (Paradisaeidae), or complex display structures as in the bower-birds (Ptilinorhynchidae) (see also p. 417). Most species perform in their own individual territories but some resort to a communal display ground or lek and hold small 'courts' which they defend within the main arena. Here there can be direct competition for the female's attention. Among nidicolous birds which have leks are two species of birds-of-paradise, most manakins (Pipridae), the Cock-of-the-Rock *Rupicola rupicola* (Cotingidae) (Gilliard 1962; Snow 1971), Jackson's Widow-bird *Euplectes jacksoni* (Ploceidae) and some hummingbirds (Trochilidae). Among nidifugous species, lekking behaviour is recorded for the Great Bustard *Otis tarda* and is especially evident in the grouse (Tetronidae) and waders (Charadriiformes).

Above we discussed polyandry and sexual reversal in waders. An alternative evolutionary trend from monogamy is possible when the female can produce her eggs relatively easily and manage incubation and brood care unaided. In the Curlew Sandpiper *Calidris ferruginea* (Fig. 15.8(d)) it appears that only the female incubates (Portenko 1959; Holmes and Pitelka 1964). Males are rather bigger and more brightly coloured than the females and their courtship display bears some resemblance to that

[†]Drawing on the excellent studies of Crook and Orians, Lack (1968*a*) has been able to illustrate and emphasize the remarkable convergent evolution between the New World Icteridae (I) and African Ploceinae (P): beginning with the monogamous insectivorous Troupial *Icterus galbula* (I) and *Malimbus scutatus* (P); the polygynous grassland Redwing *Agelaius phoeniceus* (I) and *Euplectes orix* (P); the extremely colonial Tricolor Redwing *A. tricolor* (I) and *Quelea quelea* (P); the parasitic Cowbird *Molothrus ater* (I) (which parasitizes the Chestnut-sided Warbler *Dendroica pennsylvanica*) and the Cuckoo-weaver *Anomalospiza imberbis* (P) (parasite on *Cisticola* and *Prinia* warblers).

of the Pectoral Sandpiper *Calidris melanotus* mentioned below. White-rumped
Sandpipers *C. fuscicollis* nesting on the Canadian Arctic tundra also only hold territories during courtship and they abandon them once incubation begins, so the females
hatch the chicks and rear the young unaided (Parmelee, Greiner, and Graul 1968).
This allows the males to be variably promiscuous or polygynous (Parmelee and Payne
1973). Such behaviour is not recorded in the Curlew Sandpiper but it is very pronounced in the Pectoral Sandpiper (Pitelka 1959). This species arrives on the breeding ground in arctic Canada later than the Dunlin and is more opportunist in settling
to breed where it can find good food supplies. Fig. 15.7 shows how its feeding ecology
relates to that of the Dunlin. It is a useful strategy to defer breeding until the state
of the food supply can be accurately assessed and certainly this allows the females
to prepare eggs under favourable conditions. But early breeding is so generally advantageous in birds that deferment brings the risk that productivity will be low. Perhaps
this is why the male Pectorals have become polygamous. They display from a dispersed
lek, and form a brief pair bond with more than one female and these then nest within
the territory. As soon as pairing is finished the males depart for wintering grounds in
the southern hemisphere and on arrival undergo a post-nuptial body and primary moult;
a pre-nuptial body moult occurs before departure in spring. Unlike Dunlin—which
are at minimum weight in June—July and then acquire fat reserves on or near the
breeding ground in late July and August ready for autumn migration—Pectoral
Sandpipers are at their maximum weight in June. Their weight then declines in July
and fat deposition for migration is delayed until the birds reach other habitats further
south. Compared with Dunlin, Pectoral Sandpipers make more use of the Arctic
resources for breeding but they must moult in their contra-nuptial quarters in the
southern hemisphere. The Dunlin has to partition summer resources between breeding and moulting but in consequence can winter in the northern hemisphere.

The Buff-breasted Sandpiper *Tryngites subruficollis* is also polygamous with males
displaying from a dispersed lek (Holmes and Pitelka 1966). The display courts are
much closer together in the Great Snipe *Gallinago media* (Ferdinand 1966) and the
trait becomes particularly well developed and complex in the Ruff *Philomachus
pugnax* (Hogan-Warburg 1966). In this species there is much individual variation in
the colour of the wattles, which makes individual recognition easy. Resident males
have an arena which they defend from other independent males and if unsuccessful
become displaced as marginal males which may move to another lek. Satellite males
do not have arenas of their own but make use of those belonging to resident males;
they do so without displaying any aggression, although the owner may or may not
try to reject these satellites. Satellite males mostly have rather unusual coloured display plumage, a form with all white display plumage being common. The survival
value of these satellite males appears to be two-fold, (1) in attracting females to the
lek, thereby increasing the chances of copulation for all birds frequenting small leks
and (2) in promoting the establishment of new leks and the maintenance of several
leks in an area. Hogan-Warburg gives a fascinating account of the interesting balanced
polymorphism which has arisen to support this system.

The grouse (Tetronidae) also provide a good illustration of the evolution of polygyny. The Red Grouse *Lagopus l. scoticus* feeds primarily on heather which is widely and evenly distributed (see p. 471). The birds do best by dividing the feeding resources into territories; they are monogamous and essentially monomorphic, though the male does possess distinguishing red wattles above the eyes. The food of the Black Grouse *Lyrurus tetrix* is patchily distributed and the location of good feeding sites varies with the season so that a territorial system confers no advantage. Indeed, as Kruijt, de Vos, and Bossema (1972) discuss, it would be possible, following a season of limiting resources, for the males to defend small territories and for food to be available outside the territories. Or the males might try to control bigger and bigger territories and in the end be unable to prevent conspecifics from feeding. Either way it is evident that the female is not dependent on the male for a feeding territory and the males must, therefore, adopt a different strategy to attract a mate. Like the Capercaillie mentioned above (p. 418), the Black Cock is a lek species and exhibits a marked sexual dimorphism in plumage colour and, to a lesser extent, in body size.[†] Many other tetronids have lek-type displays, including the Sage Grouse *Centrocercus urophasianus,* Sharp-tailed Grouse *Pedioecetes phasianellus,* and Greater and Lesser Prairie Chickens *Tympanuchus cupide* and *T. pallidicinctus* of North America.

The Willow Grouse *Lagopus l. lagopus* and Ptarmigan *L. mutus* undergo three moults per year, in early spring, in summer, and early winter; the winter coat is white and all other plumages remarkably cryptic to suit seasonal changes in the environment. The plumage sequence is closely related to the reproductive cycle and the time of acquisition of the coats varies between individuals in parallel with the onset of gametogenetic activity (Salomonsen 1939*b*; Höst 1942; Hewson 1973; Watson 1973). The light cycle certainly regulates the plumage sequence so that long days encourage the summer plumage and short days cause a moult into white winter dress (Höst 1942), however, the photoperiodic effect is modified by temperature (Höst 1942; Hewson 1973; Watson 1973). The phasing of the moult cycle also varies between the sexes but more experimental details are required.

Promiscuity

Transient polygyny can develop into promiscuity without any sign of a pair bond developing. This represents an exceptional mating system found in a few aberrant groups. Apparently casual copulations occur in the honeyguides (Indicatoridae). These are aberrant woodpeckers (Piciformes) which are brood parasites of other birds (hole-nesting barbets Capitonidae or even open-nesting passerines such as *Zosterops* spp.). They have peculiar feeding habits, several species being specialist feeders of bee larvae, honey, or beeswax. Some have evolved the habit of leading Ratels *Mellivora capensis* (Honey Badger) and, secondarily, man to nests of honeybees and once the mammal has opened up the nest the birds are able to feed on honey

---

[†]It is not clear whether we should regard the Black Grouse or Red Grouse as most closely representing the ancestral condition. Perhaps the Black Grouse is the oldest line and the grouse descend from more brightly coloured forms (cf. brightly coloured monomorphic *Alectoris* partridges and cryptic coloured *Perdix*).

while the mammal eats the larvae. In the Black-throated Honeyguide *Indicator indicator* several males may use a call-post which is visited by females who leave following copulation. Promiscuous matings also occur in hummingbirds (Trochilidae) in most of which the sexes live independently and only come together to mate. Skutch (1951) describes the congregation of males of Longuemare's Hermit *Phaethornis longuemareus* in secondary growth, each male having a perch from which he displayed. In both these groups the feeding habits may make a degree of independence necessary so facilitating this odd mating behaviour.

Co-operative breeding

In many bird species extra individuals have been recorded assisting with feeding the young, sharing in the defence of a communal territory, or in some way apparently helping other birds to reproduce. For example, young Moorhens *Gallinula chloropus* may assist in the feeding of subsequent broods. In the Australian blue wrens, including *Malurus cyaneus,* females suffer a higher mortality than males and some of the male surplus assist their parents in raising later broods (Rowley 1965*a*). Rowley showed that the breeding success of pairs assisted by extra male helpers was much higher than those without and this was so in terms of number of young produced per adult. There are several variations in group breeding strategy. In babblers (Timaliidae), including the Arabian Babbler *Turdoides squamiceps* (Zahavi 1974, 1976; Brown 1975), there is one nest in the territory and usually only one female lays. The entire group in various combinations of adult males, females, and young from earlier broods defend the territory and care for the one communal nest and nestlings. In this case removal of the so-called helpers does not hinder the parent in rearing their brood. In the Long-tailed Tit there may be more than one nest in a territory (Gaston 1973) or, as in the Bee-eater *Merops bulocki,* an extra bird may join a breeding pair (Fry 1976).

In some species more than one individual shares in the breeding effort, for example, 1–3 or, rarely, 4 females lay in a communal nest in the Grove-billed Ani *Crotophaga sulcirostris* and all individuals contribute towards incubation and care of young: essentially the same occurs in the Smooth-billed Ani *C. ani* (Köster 1971). Similar behaviour has been noted in two Australian mudnest-builders of the sub-family Corcoracinae (Grallinidae). In the White-winged Chough *Corcorax melanorhamphos* the young stay with the parents for several years and all the group helps to build a nest into which more than one of the related females lays her eggs, and all the group help to raise the young (Rowley 1965*b*). In the group-living Bell Miner *Manorina melanophrys* (Meliphagidae) only one of the females lays eggs (Swainson 1970). Male Magpie Geese *Anseranas semipalmata* are often paired with two females (usually a mother and daughter or two sisters) and both the females lay in the same nest (Kear 1973). In the Australian Magpie *Gymnorhina hypoleuca* (family: Cracticidae), breeding flocks defend group territories and exclude the majority (73 per cent) of the population from breeding (Carrick 1972). The number of examples of co-operative breeding is already considerable, many new ones remain to be discovered, particularly in the tropics, and new variations are likely to emerge (for reviews see Rowley 1968, 1976); Grimes (1976); Woolfenden (1976); Zahavi (1976)).

In almost all cases of co-operative breeding far too little is known to justify drawing conclusions about the specific mechanisms involved. Frequently co-operative breeding involves birds which are related, for example sons and daughters, brothers and sisters. This has led some authors to account for what appears to be altruistic behaviour with no apparent survival value as being in reality kin-selection (for example, Ricklefs 1975). In kin-selection the increase in frequency of a gene depends on its effects on the survival of relatives with the same gene (Smith 1964). Consider the case of the flightless Tasmanian Gallinule or Native Hen *Tribonyx mortierii* in which females have a single male, but others pair with two or three and they all help raise the young (Ridpath 1972*a, b, c*). Groups breed more successfully than single pairs and polyandry appears to have arisen, as in *Malurus*, because there exists an excess of males. The unequal sex ratio does not appear to have followed from polyandry, for as Smith and Ridpath (1972) have pointed out selection will favour a 1:1 sex ratio even in polyandrous species (Fisher 1930). Smith and Ridpath (1972) have considered the general problem of a male who shares his female with another male, for he fathers on average only half the total young raised. Accordingly, unless the number of young raised by a trio is twice that raised by a pair a male should leave more progeny by driving out the second male. The circumstances under which the genes determining a monogamous male phenotype (D) will be replaced by genes favouring wife-sharing (d) resolves into a matter of when it is worthwhile fighting and repelling the second male. To answer this problem Smith and Ridpath invoked an unpublished analysis of ritualized intraspecific aggressive conflict due to G. R. Price.

Price demolished the widely held view that ritualized fighting is good for the species because it prevents the loser from being hurt. He pointed out instead that it is the individual which breaks the convention by 'hitting below the belt' that transmits his genes to the next generation. Accordingly, Price suggested that two levels of fighting have evolved. Ritualized fighting generally occurs but occasionally a real fight develops as a response to being hurt and this in turn damages the opponent. Hence, a selective balance should arise in which there are disadvantages in indulging in real fighting and in not fighting at all. Extending this model of intraspecific fighting, Smith and Ridpath argued that a male member of a trio might be inhibited from attempting to evict the other male if he could not predict whether or not he would win any encounter, but if he could predict the outcome he would fight. That is, a male should drive out the extra male if he knew he could win, but agree to share his wife otherwise. The necessary information might derive from previous ritualized conflicts especially if the males were brothers which had grown up together and indulged in aggressive play. Depending on whether or not males could predict the outcome of a fight, and also on whether or not they were full brothers, the conditions to be satisfied for the genes for wife-sharing (d) to spread in the population at the expense of those for monogamy (D) were given as:

	Unrelated males	Full brothers
D male knows outcome of fight	$n_2 > 2n_1$	$n_2 > 1\frac{1}{3}n_1 + \frac{2}{3}n_3$

	Unrelated males	Full brothers
D male cannot predict outcome of fight	$n_2 > n_1 + n_3$	$n_2 > n_1 + n_3$

where n_1 = number of offspring resulting from a pair; n_2 = number of offspring from a trio; n_3 = expected offspring from a male driven out of a trio. Monogamy prevails when $n_1 > \frac{1}{2}n_2$.

Smith and Ridpath concluded that once an excess of males occurs n_3 must become reduced, for the prospects of a male driven from a trio finding another mate must be small. Given $n_3 = 0$, polyandry requires $n_2 > 1\frac{1}{3}n_1$. In fact, Ridpath was able to show that trios produced an average of 1·26 young to every 1 young produced by a pair.

This example has been given in some detail because it illustrates that given the concept of kin-selection, there can be a genetical basis to account for co-operative breeding. Indeed, these conditions may apply to the Tasmanian Native Hen. However, there is a danger that workers who study group breeding and discover that relatives are involved will fall back on kin-selection as the explanation for the habit without a more critical assessment of the field situation. For example, long and detailed field observation of Arabian Babblers reveals some surprising behaviour on the part of the 'helpers'. Zahavi's patient field work indicates that they are not 'helpers' at all but 'hinderers', albeit in a very subtle fashion.* They will sometimes peck at eggs when they get the chance. They may arrive at the nest with food but instead of feeding the chicks stand back. This causes the nestlings to beg loudly and is liable to attract predators. Sometimes adults can be seen driving off so-called helpers. These and other observations indicate that so-called helpers are actually liable to indulge in infanticide at the first opportunity. If their behaviour is not altruistic no problem arises in explaining it in terms of simple natural selection and there is no need to evoke kin-selection. We suggest that the following generalization will basically account for co-operative breeding in birds. In certain habitats it may arise that the survival prospects in emigrating are virtually nil. For example, open semi-arid savannas may prove too inhospitable, and the predator risk too high, for a babbler (these birds have a weak flight) that normally lives in safety of a small clumps of thorn scrub to risk moving from cover.† Moreover, other clumps of trees, supposing they could be reached, would almost certainly be occupied. If the only prospects for surviving to leave progeny involve staying in the place of birth, adaptations that make it possible to live with the parents are needed. An obvious strategy is to become 'useful about the house', by helping the parents to feed the next gener-

---

*Babblers, and many other communal nesters such as the anis and tinamous, lay immaculate coloured eggs which are often bright and shiny. This prevents egg recognition and is thought by Zahavi to be a device to prevent females from removing or destroying each other's eggs.

†R.K.M. and A. Zahavi watched Australian White-browed Babblers *Pomotastomus superciliosus* returning to their communal nest to roost at dusk after having disturbed them from their tree cover. The birds had to fly back from some gum trees across an open space of 30 m and this they did one by one. As we watched, a Little Falcon *Falco longipennis* suddenly appeared and would probably have caught one of the babblers but for our presence.

ation. These are, of course, direct competitors and should be killed if possible and infanticide apparently does occur. The problem facing the ornithologist in the field is in distinguishing infrequent instances of 'murder' when these look like accidents.

There is no reason why species should not, so to speak, insure themselves for the future. The endocrine mechanisms contributing to co-operative breeding behaviour are probably not unique and simply require a lessening of aggressive territorial behaviour until the opportunity for an assertion of status arises—perhaps with the death of a parent. We guess that there will be a close endocrine relationship between adrenal activity and the pituitary—gonad axis so that sexual maturation is dependent on social status but no information is presently available.

Brood parasites

Once the habit of laying eggs in a conspecific's nest has emerged a further development of the trait is to lay in the nest of a different species. The parasitic habit has evolved in six families:

1. *Anatidae.* The Black-headed Duck *Heteronetta atricapilla* of South America lays in the nest of other ducks and the Red-gartered Coot *Fulica armillata* (Weller 1968*b*); this is possibly a relatively newly evolved habit. Sporadic parasitic egg laying has been also recorded in the Redhead *Aythya americana* and other North American ducks (Weller 1959).

2. *Ploceidae.* One species, the Cuckoo-weaver *Anomalospiza imberbis*, parasitizes *Cisticola* and *Prinia* warblers.

3. *Estrildidae.* Whydahs and combassous of the genera *Vidua* and *Hypochera* parasitize other weaver finches.

4. *Icteridae.* Nearly all members of the sub-family Agelainae (the cowbirds) are involved.

5. *Cuculidae.* All the sub-family Cuculinae, with 50 or so species (Europe, Africa, Asia and Australia), and some members of the sub-family Neomorphinae (mostly New World).

6. *Indicatoridae.* These aberrant woodpeckers parasitize such hole-nesting species as the barbets (Capitonidae) and sometimes open nesting passerines (e.g. white-eyes *Zosterops* spp.).

It is noticeable that aberrant mating systems including group breeding are prominent in some of these families; co-operative breeding of the Grove-billed Ani (Cuculidae) was mentioned above.

All the cowbirds are parasitic except for the monomorphic Bay-winged Cowbird *Molothrus badius* (Fig. 15.9(a) see between pp. 434–5) of South America, which, nevertheless, uses the old nests of other species in which to lay its eggs (Davis 1942). It also consorts in social groups with several females laying in the same nest (Hoy and Ottow (1964) quoting Hudson (1920)). Where their ranges overlap it is the specific host for the parasitic Screaming Cowbird *M. rufoaxillaris* and this represents the first stage in interspecific parasitism, apparently facilitated by group breeding by the host. The Shiny Cowbird *M. bonariensis* (Fig. 15.9(b) see between pp. 434–5) is also sympatric with *M. badius* and is dimorphic, the male being a resplendent iridescent black and the female a dull grey. This species victimizes a wider range of species than *M. rufoaxillaris*: one of its main hosts in South America is the Rufous-collared Sparrow *Zonotrichia capensis* (Fig. 15.8(c) see between pp. 434–5) (King 1973c), in Surinam Haverschmidt (1965) noted that the House Wren *Troglodytes aedon* was commonly used, although in marshy areas the Pied Water-tyrant *Fluvicola pica* and White-headed Marsh-tyrant *Arundinicola leucocephala* became the main hosts. In North America the commonest brood parasite is the Brown-headed Cowbird *M. ater* (see Friedman (1971) for hosts).

It might be thought that it would be advantageous for a brood parasite to have an extended breeding season to allow easy adjustment to the season of the host. This does not seem to be so, for the breeding season of the Brown-headed Cowbird is comparable with that of other icterids (Fig. 9.12, p. 241) and similarly phased by the light cycle (Middleton 1965b; Payne 1967, 1973a; McGreen and McGreen 1968). The Cuckoo *Cuculus canorus* arrives in north Europe from Africa too late to lay eggs in the first broods of the two most common hosts, the Dunnock *Prunella modularis* and Meadow Pipit *Anthus pratensis* (Lack 1963). The food supply of the parasite may limit the amount of flexibility possible in its breeding periodicity. Höhn (1972) has suggested that a lack of prolactin has led to the parasitic habit and resulted in a lack of brood patches in the parasitic species. He found, however, that the pituitaries of *Molothrus* contained as much prolactin as related non-parasitic icterids, although there was an absence of prolactin in female Koels *Eudynamys scolopacea* in breeding condition. The significance of these observations is by no means clear in view of the relationship between incubation and prolactin secretion (p. 134).

To make parasitism worthwhile a species ought to be able to produce more young by practising the habit than by raising its own brood. It might be expected that the number of eggs laid would increase, but the parasitic cowbirds lay clutches of 4–5 eggs at daily intervals, as does the non-parasitic *M. badius* (Davis 1942). The cowbirds represent a primitive stage but the more advanced Old World cuckoos (*Chrysococcyx, Clamator, Cuculus*) also lay small clutches, which are comparable in size with those of closely related non-parasitic forms, and follicular regression occurs between clutches as in other birds (Payne 1973b). However, several clutches are laid during the season so *in toto* many eggs are produced, 11–25 per female for *Cuculus canorus* (Chance 1940) and 16–26 for various African cuckoos studied by Payne (1973b). There appears to be a correlation between the availability[†] of host nests and the number of eggs laid by the parasite (Chance 1940; Payne 1973b). The laying interval is 2 or even more days in the Clamator and Cuculus cuckoos, the same frequency being noted in the closely related non-parasitic forms. The laying rate and clutch size were apparently already fixed before the evolution of the parasitic habit. But eggs can be retained in the oviduct, allowing a degree of oviducal incubation to occur (Perrins

[†] A reduced nesting density of host species was suggested as the reason for the low numbers of Cuckoos now found on English farmland (Murton 1971b, p. 126–7). Gentes cannot favour a specific host in situations where host density is low and they must parasitize more than one species. This could account for the fact that only a poor mimicry of the egg colour and pattern of the host is noted in arable areas.

1967; Payne 1973*b*). In this way, the chances are improved that the eggs of the parasite will hatch before those of the host. This adaptation, together with a fast rate of development (Perrins 1967), help ensure that eggs laid in nests where incubation has just begun may nevertheless hatch soon enough to allow a Cuckoo *Cuculus canorus* chick to evict the eggs and small young of the host. The need for the incubation period of parasite and host to match probably accounts for the fact that the two are usually reasonably closely related, the more so in the primitive examples.

The egg weight is only 2·4 per cent of the adult body weight in *Cuculus canorus,* but 7 per cent in the Great Spotted Cuckoo *Clamator glandarius,* which victimizes crows larger than itself. In the latter, the high proportionate egg weight possibly allows the chick to hatch sufficiently well developed to compete with the brood of the host, for in this species the chicks of the parasite are raised with those of the host (this is also so with the koels *Eudynamys* spp.). The incubation period for the Great Spotted Cuckoo is about 5 days less than their Magpie *Pica pica* hosts and young Magpies can survive only if they are the same age or at most 2 days younger than the cuckoo (Von Frisch 1969). Several parasite eggs may be laid in one nest, the female first removing some of the host eggs, so that intraspecific competition is possible. Many birds recognize their own eggs and would reject strange ones but the corvids appear to lack such a species-specific response to their own eggs and this may have made parasitism by the Great Spotted Cuckoo possible (Gramet 1970). The *Cuculus* species employ hosts laying smaller eggs than themselves and the reduction in egg size is an adaptation to match the egg size to that of the victim. The female Cuckoo *C. canorus* has an extensible cloaca to facilitate laying in the host's nest, and her egg has a thick shell to prevent breakage. She removes one of the host's eggs in laying her own; probably she could not remove more without causing the host to desert. Since, as mentioned, many birds are able to recognize their own eggs with considerable efficiency there has been a need for these cuckoos to produce eggs of similar colour and markings to those of their host. A degree of genetical isolation results because the cuckoos favour a specific host and this has enabled gentes laying eggs of a particular type to emerge in homogeneous habitats, that is, there is a polymorphism in egg colour and pattern (Southern 1954).

The parasite must be able to find the nests of the hosts and this is easiest when colonial species are victimized, as with the crows used by *Clamator glandarius* and the Indian Koel *Eudynamis scolopacea* (which favours the House Crow *Corvus splendens*) or *Anomalospiza* and species of *Chrysococcyx* which parasitize colonial Ploceinae weaverbirds. Most female cuckoos must find the nest of the host by watching the victim's behaviour. Hann (1941) described how the Brown-headed Cowbird first finds the nest after watching the potential host building and then visits it regularly in the absence of the hosts. Her own egg is laid 4–5 days later, which suggests that seeing nest material being manipulated provides a stimulus for oviduct development, as is the case with non-parasites (p. 120), and accounts for the close synchrony in laying time between parasite and host. The same probably applies to the European Cuckoo, judging by Chance's classic observations. He was able to stimulate Cuckoos to lay unusually large numbers of eggs by removing the foster eggs and causing the hosts to repeat. If the foster nest is destroyed at the point when the Cuckoo is ready to lay she must deposit her egg in an unusual foster parent's nest. Once oviduct development is stimulated the full complement of follicles appear to begin maturation and at least the first one or two eggs must be laid.

When cuckoos are ready to lay in their chosen hosts' nest they are liable to be attacked. This possibly explains why many species have come to resemble hawks in appearance: they lure the host away before laying to lessen the intensity of attack (see references in Lack 1968a). However, the Indian Drongo-cuckoo *Surniculus lugubris* closely resembles one of its hosts the pugnacious Black Drongo or King Crow *Dicrurus macrocercus* (Fig. 15.10: see between pp. 434–5). Lack suggests that this is the same functional adaptation as is shown by the hawk-cuckoos, that is, to mimic a frightening species, and the Drongo-cuckoo is totally unlike its more common hosts. On the other hand, the ploceid and estrildid parasites resemble their common hosts, but this can be a worthwhile strategy only when parasitizing colonial species. In some cuckoos the chicks are close mimics of the hosts' young; readers are again referred to Lack (1968a) for more details.

Southern (1964), in reviewing the topic of parasitism, poses the question of why the two most successful groups of cuckoos—the Old World Cuculinae and the New World cowbirds—should differ so much in their habits. The former tend towards strict host specificity, lay mimetic eggs, and many of them, and the cuckoo ejects its hosts' young so that it is the only nestling fed. The cowbirds are generally more plastic for they parasitize several species, their eggs are not closely mimetic with those of the host, and their chicks are reared at the expense of only one of the hosts' chicks. Southern gives tentative data to suggest that the cuculid strategy may result in a steady decline of the host population whereas with cowbirds the parasitic load can be maintained by the host more easily. More information is needed on the population dynamics of the two situations.

Zahavi (in press) has discussed the question of why, in spite of the high selection pressure resulting from the depression of reproductive potential, many hosts of brood parasites do not discriminate very markedly against the eggs or chicks of the parasite. He suggests that discrimination can be costly because of the risk of making mistakes. Imagine there are two components to the population of the parasite either temporally (European Cuckoos do not parasitise first broods of their hosts) or spatially, and the host population is composed of potentially discriminating and non-discriminating individuals. The former would be at a disadvantage in the non-parasitized segment of the population, because they might destroy their own eggs or young, whereas the latter would be at a disadvantage in the parasitized segment. Discrimination could obviously evolve only in the parasitized section of the population, but this process might be lessened or avoided if the cuckoos behaved as predators and removed eggs or nestlings, thereby reducing the breeding success of the population as a whole. It is suggested that it is advantageous for a parasite to reduce the reproductive capacity of that part of the host population which might be strongly selected for discrimination. In fact, Wyllie (1975) found that a surprisingly large number of Reed Warbler *Acrocephalus scirpaceus* nests were predated by European cuckoos in his study area. Perhaps this is the real reason why Cuculid cuckoos mimic raptorial species. In such a situation the only way to avoid being parasitized would be to abort and this commonly occurs. Zahavi suggests that only a species which can readily rebuild a new nest and produce a new clutch can evolve into a discriminating host. Non-discriminating species could, therefore, be those which are unable to avoid the predator, spatially or temporarily, and perhaps this is the situation among the cowbirds and their hosts.

16   Population regulation

Summary

The progeny resulting from reproduction could potentially lead to a geometrical increase in population size. Instead only sufficient young survive on average to make good adult mortality. The rate at which adults die depends on their survival expectation and, as discussed in Chapter 1, survival curves and rates of development are subject to natural selection. The processes that cause surplus young to die so that population size is balanced in relation to resource availability are *consequences* of reproductive rates and not *causal* mechanisms. In the past some workers have confused the evolutionary aspects of species-specific survivorship curves with the mechanical processes involved in population dynamics such as density-related mortality: natural selection does not determine the mechanisms by which animals must die but acts only to select those that may live. Animals cannot evolve mechanisms which determine an optimal allocation of resources, partly because resource availability is not really predictable, but individuals may evolve optimal strategies and life histories for coping with particular kinds of resource and their populations and disperal patterns will be affected by these adaptations. For example, some species have evolved territorial behaviour to exploit their food supply while for other kinds flock systems are more appropriate to cope with the dispersal pattern of the food source. Population-regulating processes must act via the existing social organization and the mistake made by some workers is to assume that the social organization has evolved primarily to facilitate the regulation of animal numbers. These principles and the significance of social hierarchies are discussed in detail in this chapter with reference to the Wood Pigeon *Columba palumbus*, Tawny Owl, *Strix aluco*, and Red Grouse *Lagopus lagopus*, their feeding and population ecology being dealt with in some detail, making use when possible of key-factor analysis techniques.

The rôle of natural selection

The potential of bird populations to exhibit a geometrical increase in size as a result of reproduction was mentioned in our first chapter (p. 7). Eventually some kind of regulatory brake is applied to the kind of population explosion depicted in Fig. 1.3 (p. 8) and in the majority of animal populations numbers fluctuate within narrow limits, about a mean level which stays reasonably constant (Fig. 16.1: see between pp. 434–5). Sometimes there can be a steady increase in this mean population level, as when gradual improvements in the habitat occur, or the species adapts itself to a new mode of life, and sometimes there may be declines leading to extinction for converse reasons. Stability persists for relatively long periods and the population is evidently being regulated by intrinsic or extrinsic factors. These observations conceal two quite distinct processes which affect animal populations:

1. The first process concerns the causes of adult mortality. Adult survival rates S, or conversely adult death rates $1-S$, are species specific, as was discussed in detail in Chapter 1, and there is no justification for thinking that these species-specific mortality rates are adaptations to regulate population size. Indeed, the evidence from such examples as the Pheasant *Phasianus colchicus*, already discussed, shows that adult mortality rates remain constant with wide variations in population density. This follows because, as discussed in Chapter 1, species-specific survival rates are the consequence of natural selection acting to produce the most efficient mode of development of form and function, and hence life cycle, to maximize reproductive output relative to competitors.

2. The second process is essentially a matter of population dynamics, for it is concerned with the mechanisms whereby vacancies created as a result of adult deaths are filled by new recruits so that the birth-rate and death-rate are balanced. The processes considered in this category also determine the carrying capacity of the habitat and hence absolute population density, and their extrinsic properties allow them to serve a regulatory function.

The distinction which ought to be made between (1) the factors affecting survival (mortality) rates and (2) the mechanics of population regulation has been blurred in consequence of the advocacy of two rather polarized approaches to avian population ecology. On the one hand, Lack (1954, 1966) has championed the view that the reproductive rate is set by natural selection to be the maximum possible, so that genotypes leaving the most surviving offspring are favoured. Any population excess above the carrying capacity of the environment is removed in a density-dependent manner (see below) by various interacting environmental factors, the available food supply usually being the most important, but not exclusive, factor. Lack has, therefore, concentrated his attention on population dynamics. In contrast, Wynne-Edwards (1955) has confused the processes which determine adult survival rates with those that are applicable to population dynamics in arguing that the recruitment rate, including clutch size, number of broods, and age of first breeding, have been evolved to help balance the mortality rate. He was impressed with the correlation between the low reproductive rate of Procellariiformes—and some other sea-birds—and their low mortality rates and he noted in general that larger birds tend to live longer and lay smaller clutches than smaller birds. Wynne-Edwards (1962) extended his thesis to argue that social behaviour determines the dispersion pattern of birds so as to maintain optimum numbers in relation to resources. These dispersion patterns often require that penalized individuals do not opt out of the social system and this can give the appearance (falsely) that such animals are behaving altruistically. Wynne-Edwards accepts altruism because he believes that dispersal can be maintained by group selection[†] as distinct from natural or individual selection. The idea

[†] Apart from the special case of kin-selection (involving relatives and already discussed on p. 446) no genetical basis has been established for group selection (see Haldane 1932; Smith 1964). This alone does not invalidate the concept but we, like many others, reject the idea because all the observations that supposedly depend on group selection for their explanation can be adequately accounted for by natural selection. Actually, this may also apply to some of the situations for which kin selection has been used as a 'way out' to explain apparent altruism. At the time Wynne-Edwards proposed his hypothesis accurate observational and experimental results were usually unavailable, enabling much speculation to have an inadequate factual basis.

of group selection is that whole populations can be selected for in competition with other populations as distinct from Darwinian natural selection which operates only on the individual. Wynne-Edwards writes 'the over-riding effect of group-selection, that occurs between one population or society and another, and normally results in fixing the optimum breeding rate for the population as a whole'.

Because populations are generally regulated around a stable mean (see Fig. 16.1) it is evident that either some component of the reproductive output or the factors contributing to mortality must vary in a density-dependent manner (Nicholson 1933), although not all workers accept that populations are regulated. In particular, Andrewartha and Birch (1954) and Birch (1971) reject the concepts of balance, steady density, control, and regulation and in so doing reject the need for density-

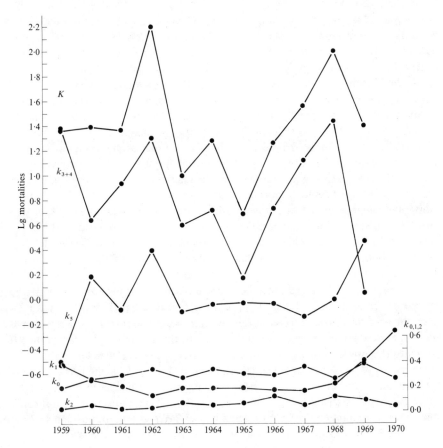

Fig. 16.2. A key-factor analysis of data from a Wood Pigeon *Columba palumbus* population study conducted at a 1072–ha site in Eastern England. Adults omitted from the analysis except in k_0, see text for explanation. k_0 represents failure to lay full potential number of eggs, k_1 represents egg mortality, k_2 represents nestling mortality, k_{3+4} represents juvenile mortality between September and March, k_5 represents juvenile mortality from March to start of breeding, K = total annual mortality. (see text for fuller explanation.)

dependent factors to be involved. In our view, these ideas have been adequately rebuked by Lack (1966), and it appears that Andrewartha and Birch were over-impressed by the harshness of the arid habitats in which they worked in the semi-deserts of Australia. In habitats that fluctuate drastically it is possible for a species to survive in one area, become extinct in another, and to recolonize and increase in the first area with a return to favourable conditions. Heterogeneity in habitate quality could well obliterate the operation of density-dependent factors, although a failure to discover density dependence would not really invalidate a logical need to evoke the concept. The factors responsible for population regulation in a few birds which have been studied in sufficient detail are now considered.

Key-factor analysis

A convenient method of examining census data involves the key-factor method devised by Varley and Gradwell (1960, 1968) for insect populations. In effect, life tables for separate generations of a population are prepared. The starting-point is the maximum potential egg production which the population could manage, assuming that all females laid the maximum number of eggs of which they are capable during one breeding season. It does not matter if this figure is an over-estimate, as it simply serves as a base line. The actual egg production must be measured, together with the number of survivors which enter the next breeding season. All records are converted to logarithms so that, in annually breeding species, the total mortality affecting a particular cohort is obtained by subtracting the number of individuals entering the next breeding generation from the maximum potential egg production initially calculated. This total mortality is K, while k_0, k_1, $k_2 \cdots k_n$ are the separate mortalities, assumed to act successively, which summate to equal K. Thus k_0 is the difference (number of eggs expected) $-$ lg (number actually produced) and is a measure of the failure to achieve maximum production; k_1 is the difference lg (number of eggs laid) $-$ lg (number that hatched) and is, therefore, the mortality due to egg failure; similar, k_2 represents chick mortality.

The number of further ks that can be calculated depends on the accuracy with which the species can be censused in the wild, and when. Fig. 16.2 depicts data of this kind which were obtained for a Wood Pigeon *Columba palumbus* population censused in a study area in Cambridgeshire, England, over 12 years. In this analysis, $k_3 + k_4$ combine the mortality k_3 between leaving the nest and the start of the winter in December and the mortality k_4 occurring between December and March;[†] k_5 is the loss occurring between March and the start of the next breeding season. Changes in the actual population converted to the lg (number birds per 100 acres) are also shown, and it can be seen that the population declined drastically after 1967 so that numbers in 1972 were only a quarter of what they had been during the late 1950s and 1960s.

In the analysis just discussed each generation of eggs was distinguished and its success followed independently of the adult component of the population, except

---

[†]Juveniles were regularly distinguished from adults in captured samples examined in the hand in late February—early March.

when k_0 was considered. Had adults not been contributing to egg production through failing to breed, it would have been recorded as part of k_0. Unlike most insects, for which the key-factor method was designed, birds in general have overlapping generations of adults and the mortality of the different age categories of adults cannot be distinguished by field observation. Moreover, in many field situations it is impossible to distinguish adults from juveniles so that population counts have to include all age groups as a single category. In some circumstances, it may be possible to count adults during the breeding season, for example, if the number of pairs inhabiting nest-boxes or holding territories is known, but it may not be possible to distinguish adults in late autumn when adults and young look alike. Some workers have overcome this problem by adding the same total of adults as measured during the initial breeding season census to subsequent stages, until the point at which the whole population can be censused. This technique was used in Fig. 16.3, for which it was assumed that the

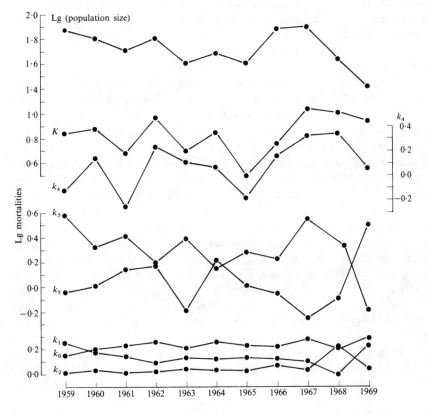

Fig. 16.3. A key-factor analysis of data from a Wood Pigeon *Columba palumbus* population study conducted at a 1072 ha site in Eastern England. Adults included in the analysis. Conventions as for Fig. 16.2. except that k_3 represents the autumn loss from September to December and k_4 the winter loss from December to February–March. The top graph depicts lg (number of pigeons in the study area). From Murton (1974).

number of adults producing eggs was the same as the number which were present when the eggs hatched and then when the young fledged. Actually, some must have died. Fortunately the adult mortality rate is nearly constant and the loss over this period small (no more than 9 per cent) so the error is of little consequence. It was corrected by the time the December census was made so far as Fig. 16.3 is concerned, and does not apply in Fig. 16.2. However, for birds with a high variable adult mortality this technique can lead to difficulties. It does not particularly affect the identification of the key factors, now to be discussed, but any bad error can affect the slope of regressions used to identify density-related factors (see below).

Figs 16.2 and 16.3 compare two different measurements of K. In the first, emphasis is placed on the first-year component of the population and adults in their second and subsequent years may enter the analysis in terms of the number of eggs actually laid and k_0. In the second analysis, the component of the population which is in its second or subsequent breeding season is included throughout and so mortality affecting the whole population is examined. During a hard winter in 1962–3 considerable emigration occurred and juvenile losses were high, causing total K in Fig. 16.2 to be high. This is less noticeable in Fig. 16.3 because many adults, but not juveniles, returned to the area after the hard spell, some of these being birds which had not been present in the previous breeding season.

Inspection of Figs 16.2 and 16.3 shows that the year-to-year changes in total K were not correlated with k_0, k_1, or k_2, and so these mortalities were not the prime contributors to the total annual mortality. This would not be an obvious conclusion from simple field observation for up to 80 per cent of all eggs laid by Wood Pigeons are eaten by corvid predators; gamekeepers, farmers, and even biologists, have falsely assumed that this scale of egg loss must keep the population in check. However, even if pigeons could lay more eggs in a season this would not lead to an increase in the numbers entering the next breeding season because of the action of k_3 and k_4. In Fig. 16.2 the autumn and winter loss of juveniles ($k_3 + k_4$) accounts for the changes occurring in K, which means that year-to-year fluctuations in the number of birds settling to breed depends on juvenile mortality occurring outside the breeding season. Considering the whole population (all adult age groups combined) in greater detail, as in Fig. 16.3, it is found that winter mortality k_4 is responsible for the bulk of the annual variation in total mortality. There is a tendency for k_4 to be inversely related to k_3. This is because much of k_3 depends on the emigration of juveniles and this can vary from year to year (Murton and Ridpath 1962; Murton 1965). If good autumn food supplies encourage the young birds to remain proportionately more die in relation to winter food shortages (see below).

Both the above analyses identified the mortality occurring in autumn and winter as the key factor that determines how population size would change from one breeding season to the next. This key factor need not be related to density so it is not necessarily the factor which *regulates* population size, even though it causes most of the change in numbers. In fact, it is possible for a very slight density-related mortality to operate on the survivors from a density-independent mortality to achieve regulation of a population size (Solomon 1964). Moreover, because population processes are cyclic, it does not matter when (before, with, or after) the density-dependent mortality operates in relation to the key factor. With Varley and Gradwell's method density dependence is tested by plotting the various ks on the lg (populations on which they act), whereupon density-related mortalities supposedly give straight-line regressions; further statistical tests can be applied to adjust for the two axes of the correlation not being truly independent variables. In the present example, neither k_3 nor k_4 alone were found to be density dependent (Murton 1974) partly because they compensate for each other. However, juvenile mortality occurring between September and February—March $(k_3 + k_4)$ is strongly related to post-breeding numbers (Fig. 16.4(a)). The points in Fig. 16.4(a) lie on a curve, the relationship between pigeons and their food supply being that of predators competing for prey. Varley and Gradwell's test for density dependence is not necessary in this example, and it would not be appropriate for it assumes a straight-line relationship between the y- and x-axes and not a competition curve. Adult mortality in autumn and winter $(k_3 + k_4)$ is also probably density dependent but in three winters exceptional mortality, caused by an experiment in which many adults were shot, masked the relationship (Fig. 16.4(b)). For the data given in Fig. 16.2, k_3 and k_4 combined also proved to be density dependent (Fig. 16.4(c)).

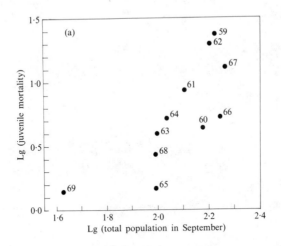

Fig. 16.4. Combined autumn and winter mortality of (a) juvenile and (b) adult Wood Pigeons as a k-value (k_{3+4}) and the initial lg (total population) in September (adults plus juveniles). (c) The autumn and winter mortality of juveniles (k_{3+4}) plotted against lg (initial number of juveniles in September). (d) Egg mortality (k_1) plotted against nest density. Open circles—years when predators were controlled by gamekeepers or nesting adults were shot. Solid circles—years when predators were not killed. Insert: the plot of x on y and y on x (two-way regression test for density-dependence for the years when predators were not killed shows that both lines lie on the same side of a slope of unity with $bxy = 1.799 \pm 0.5816$ and $byx = 0.5351 \pm 0.1687$. Partly from Murton et al. (1974d).

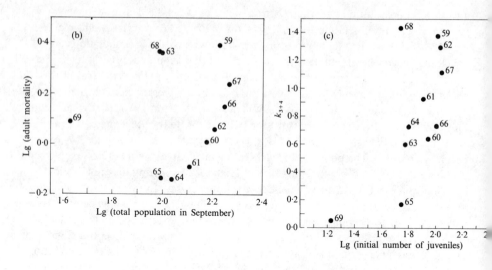

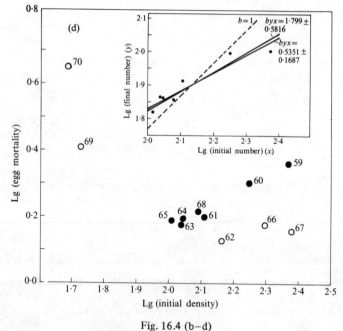

Fig. 16.4 (b–d)

Population ecology of the Wood Pigeon

The natural food of the Wood Pigeon comprises the flower and leaf buds of trees in spring supplemented by weed leaves and weed and grass seeds obtained in forest glades and adjacent grasslands (see also p. 272). In autumn, tree fruits such as those of Beech *Fagus sylvatica* or Oak *Quercus* sp. are collected from the tree or when they have fallen to the ground and this supply must last until buds again become worth

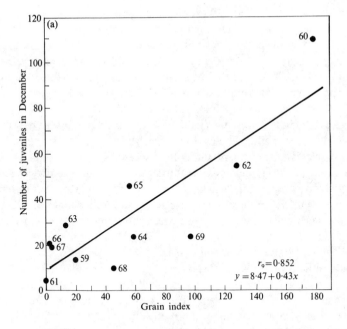

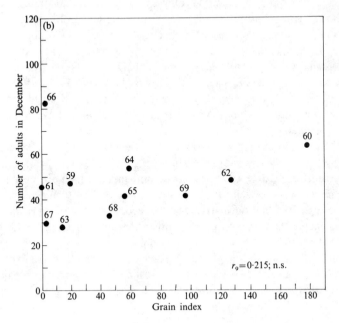

Fig. 16.5. Number of juvenile (a) or adult (b) Wood Pigeons per 100 acres of a Cambridgeshire study area in December in relation to an index of cereal grain availability in late November. From Murton *et al.* (1974*d*).

exploiting (Murton *et al.* 1964*a*). With the development of arable farming cereal grains have replaced tree food and weed seeds and clover leaves replaced weed leaves so that in some districts pigeons are almost completely dependent on these artificial sources, and, according to district, wild or cultivated foods make up a variable proportion of the diet. After the breeding season, the population detailed in Figs 16.2 and 16.3 fed on the cereal stubbles until these were exhausted of grain—usually in October or November, but sometimes not until December—whereupon the birds transferred to clover, which they obtained from leys and permanent pastures. At this season juveniles are usually still increasing in weight, and also attempting to lay down fat reserves which will serve them during the winter. As mentioned, in bad food seasons many emigrate (some die at this time) and so juvenile numbers become closely dependent on the amount of cereal grain available (Fig. 16.5(a)). Adults have already experienced this difficult transitional stage and with the advantage of a larger body size can subsist satisfactorily on clover leaves. Their numbers are not influenced at this time of year by the food supply (Fig. 16.5(b)). Reasons for these differences are examined in more detail in the next paragraph.

The natural food of the Wood Pigeon is likely to occur in localized patches, for instance, an isolated oak tree in a wood bearing a heavy crop of acorns or a patch of weed seeds in a large area of grassland. The probability of a single bird ever finding such a food supply must be low, but the probability that one bird in a big population might be successful is much higher. It follows that the optimum strategy for birds faced with such feeding problems is to respond to individuals that have successfully located good food sources (Hinde 1959; Krebs, MacRoberts and Cullen 1972). Species that rely on such feeding by 'local enhancement' (Hinde 1959), have often evolved prominent signal marks, as with the white wing bars of the Wood Pigeon, or are conspicuously coloured, as is the black Rook *Corvus frugilegus.* Ward's (1965*b*, *c*) studies of *Quelea quelea* led him to suggest that communal roosting allows birds to exchange information, enabling individuals which are unsuccessful in finding food sources to follow successful conspecifics to suitable feeding places next day (see also pp. 250 and 417). This interpretation was supported by field observations of wagtails *Motacilla* spp. by Zahavi (1971*a*) and later generalized (Ward and Zahavi 1973).

Independently, and as a result of field experiments in which different mixtures of pea, bean, or maize baits at various densities were offered to Wood Pigeons in the field, it was suggested that once arrived at the feeding ground pigeons stay together in flocks to enable them to monitor each other's feeding movements and food item selection (Murton 1971*a*). Even though a bird has arrived on a potentially worthwhile feeding station it does not 'know' what particular food item is worth exploiting. This is important since to be efficient birds must develop a search image (Tinbergen 1960) of the preferred food and an individual that specializes on a low-density or poor quality food will do badly in competition with its conspecifics.[†] The problem

---

[†] There has been some controversy about Tinbergen's hypothesis that birds develop specific search images of the food items they prefer at any particular time. Detailed discussion is rather outside the scope of this book, but the studies of Croze (1970) and M. Dawkins (1971*a*, *b*) do justify use of the concept. Royama (1970) who has criticized Tinbergen's hypothesis thinks (cont.)

is clearly increased if the food is partly hidden, as are weed seeds in grass, or soil invertebrates living beneath the surface which must be located by appropriate probing (Murton 1971b, e). If feeding in flocks is advantageous in terms of maximizing feeding opportunities there is no *a priori* need to make defence from predators the prime reason for the flock habit (Murton 1971e). Several studies have now confirmed these ideas showing that in flock-forming species those individuals which feed in flocks obtain more food per unit time than individuals which forage alone: for example, Bar-tailed Godwits *Limosa lapponica* (Smith and Evans 1973), Starlings *Sturnus vulgaris* (Powell 1974). Moreover, the idea has been extended to, and shown to be applicable to, mixed-species flocks (Krebs 1973; Croxall 1976).

It is evident that the most experienced birds will be able to manage with the minimum of information from others, whereas inexperienced individuals must seek vantage points in the flock where they can learn what to do themselves. In Wood Pigeons this is the front and front side edges of the flock, but not the rear. Since occupying good positions in the flock confers feeding advantages a hierarchy exists with dominant birds occupying the best positions. At the end of the breeding season juvenile Wood Pigeons consort in flocks which contain virtually no adults. There are plenty of foraging places available to them and they roam round the countryside from field to field. As food supplies become depleted juveniles must try to establish themselves in the adult flocks, for the adults already have the experience of where to feed in times of difficulty and have pre-empted the best sites. Accordingly, the density-dependent relationship noted in Fig. 16.5 between food supplies and juvenile numbers arises as young birds establish themselves into the adult social hierarchy or, if unsuccessful, emigrate.

If a convention confers survival advantages then selection can operate on any feature of it so long as the function is improved. When we first established a density-dependent relationship between pigeon numbers and their food supply (Murton et al. 1964b) we were puzzled as to why the processes of population regulation were mediated via the flock. It meant that subordinate flock members died of starvation when all around them there was apparently plenty of food (Murton et al. 1966).[†] Later Fretwell (1969) showed that the survival expectations of Juncos *Junco hyemalis*

---

instead that birds try to locate 'prey-niches' these being sites where prey may be found (supported by Smith and Dawkins (1971) and Smith and Sweatman (1974)). It does not follow, because they may carry an image of the food objects they expect to find, that birds normally hunt by expectation as suggested by Gibb (1962). Nevertheless, Murton *et al.* (1971) have field observations that they may sometimes hunt by carrying both an image of the place where they expect to find food and of the actual food object they expect to find. Whatever means pigeons adopt to find a feeding site (memory, presence of other birds, or direct visual stimulus) it seems that they cease their searching efforts in a given area when the food capture rate falls below an adequate food intake rate (Murton 1968a), which the birds must be able to monitor physiologically. Essentially the same concept has been extended into 'the optimal foraging model' of Charnov (1973) and Krebs, Ryan, and Charnov (1974) have experimental evidence from Willow Tits *Parus montanus* to support the model in situations where food is patchily distributed (see also Smith and Dawkins (1971) and the computer simulation of Thompson, Vertinsky, and Krebs (1974)).

[†]Watson (1971, p. 469) completely misrepresented this work and failed to appreciate that marked individuals were studied, this being clearly stated in the 1966 publication and more than adequately confirmed in subsequent papers.

at the bottom of the social hierarchy were lower than those of normal birds. These seemed to be examples which supported Wynne-Edwards (1962) thesis that birds compete for social status rather than the food *per se*, enabling dominant individuals to obtain an undisputed share of the food and causing subordinate birds to die in the face of plenty. An obvious situation to investigate was the basis of the flock habit. It was soon established that individual pigeons could obtain more food by remaining in the flock as subordinate members than if they tried to feed alone (Murton *et al.* 1966; Murton 1968*a*). The reason birds feeding alone did so badly seemed to be that they spent too much time looking around, as if frightened of being surprised by predators. We therefore rejected Wynne-Edwards' thesis on the ground that pigeons had to feed in flocks to avoid predation risks and others have followed us in applying an anti-predation explanation to the flock habit (Goss-Custard 1970; Lazarus 1972). But, this view was subsequently rejected following the discovery, detailed two paragraphs above, that flocking improves feeding prospects by allowing the birds to monitor each other's actions (see Murton 1971*a*; also p. 461). Of course, it is possible that flocking does reduce the risks of predation but this need only be a secondary function. Work in progress by R. Kenward indicates that Goshawks *Accipiter gentilis* find it easiest to catch Wood Pigeons that are feeding alone[†] and that the presence of flock-mates confers some protection. As flock size increases it becomes more difficult for the Goshawk to surprise its prey but there is no increase in the anti-predator value of the flock with more than about 30 birds present. Since the average feeding flock usually exceeds 100 birds in winter, some other factor must determine the optimum flock size.

Sometime in November or December, Wood Pigeons turn to grazing clover leaves from leys or pastures. The food searching rate is adjusted to suit the density of the preferred clover leaves and varies throughout the day (Fig. 7.16, p. 175) and from field to field, as also does the intake rate (Murton *et al.* 1971). The dominant flock members adopt the optimum feeding strategy and the subordinate animals do their best to copy, by social facilitation, these feeding actions. Subordinate birds have the same average searching rate as dominant birds but they progress by stop and starts. This is because after pausing to see what dominant animals are eating they then run ahead to avoid supplanting attacks. Hence, the pecking and pacing movements of subordinate birds are not co-ordinated and as a result they eat many inappropriate food items, such as grass and low-quality clover leaves: the nutritive value of their food is lower than that of dominant birds (Murton *et al.* 1971). Decreases in food density are countered by an increase in searching rate and this increases the interaction within the flock. Flock size is critical in this respect because any increase in interaction beginning in the flock centre is proportionately increased at the periphery. There is, therefore, an upper limit to flock size which is a function of the food density and the rate of food searching adopted by dominant birds: since the feeding rate generally decreases in the afternoon (Fig. 7.18) flock size becomes larger than in the morning. Subordinate birds unable to obtain enough food leave

[†] Since many birds that feed alone are displaced from the flock system, are underweight, and suffering marked adrenal hypertrophy their ease of capture by a Goshawk may not be surprising.

the flock and try to settle in other flocks. In this way flock size becomes strictly adjusted to food density (Murton *et al.* 1966) as does the total population (Fig. 16.6). Juveniles usually, but not always, suffer to a proportionately greater extent than adults for they generally have a subordinate status to adults. Some juveniles become established relatively easily as if they inherit some advantage. Also, juveniles hatched early in the year are more successful than later-hatched ones (p. 187).

While still in the flock subordinate Wood Pigeons begin to exhibit a noticeable hypertrophy of the cortical (inter-renal) cells (see Fig. 16.7: between pp. 434–5). These birds are thwarted from satisfying the behavioural need to correlate their feeding movements with the food distribution and it seems likely, though not yet proven, that birds have the physiological capacity to monitor their intake rate and recognize their inadequate feeding ability. Constant behavioural inhibition of this kind provides the classic condition under which neuroses develop in animals and in the Wood Pigeon it presumably causes an increase of pituitary adrenocorticotrophic hormone (ACTH) secretion and consequent stimulation of the adrenal cortex. The adaptive value must be to increase corticoid secretion (cortisol and corticosterone) (see Phillips and Chester

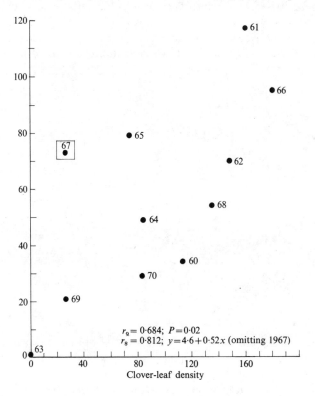

$r_9 = 0.684$; $P = 0.02$
$r_8 = 0.812$; $y = 4.6 + 0.52x$ (omitting 1967)

Clover-leaf density

Fig. 16.6. Minimum number of Wood Pigeons per 100 acres (40·5 ha) of a Cambridgeshire study area in winter (February or March) in relation to lowest clover-leaf density recorded over the same period. Omitting 1967, for reasons justified in the original, gives a higher correlation coefficient. From Murton *et al.* (1974*d*).

Jones 1957), and so increase hepatic carbohydryte, protein, and lipid metabolism to compensate for starvation. These adrenocortical steroids operate at the level of the gene by regulating the rate of DNA-dependent RNA synthesis and thereby the production of specific enzymes. They also release gluconeogenic precursors such as the amino acid from peripheral tissues. In addition, because corticosteroids have potent effects on the phasing of circadian rhythms, it may be guessed that the feeding rhythms already described (Fig. 7.16, p. 175) are influenced by feed-back from the adrenal, although this idea remains to be tested. It is not certain whether the medullary (chromaffin) tissue is stimulated in subordinate pigeons enabling the adrenaline and noradrenaline content of the sympathetic nervous system to be supplemented. Those subordinate birds which are unable to feed in the flock have markedly hypertrophied adrenal glands. Such birds need to join a flock to feed and yet once in the flock are unable to obtain sufficient food. Observation of marked birds showed this spiral of events to end in death in about a week (Murton *et al.* 1971). Superficially, the bird dies of adrenal stress in the presence of ample food and it would be easy, in the absence of detailed facts, to construct a theory that socially induced stress factors which operate at high density serve as mechanisms for population control (*cf.* Christian and Davis 1964; Christian, Lloyd, and Davis 1965; Christian 1971; Höhn 1967). Crowding *per se* is probably not important and contrived experiments with domestic fowls kept at different stocking densities have

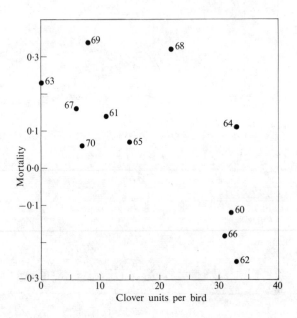

Fig. 16.8. Mortality of adult and juvenile Wood Pigeons between December and February–March (point at which minimum population recorded) in relation to the number of food units available per bird at the start of the season in December (index of clover available in December divided by number of birds). Mortality represented as a *k*-value, with negative figures denoting immigration.

given conflicting and inconclusive results (Flickinger 1961; Siegel and Siegel 1961, 1969).

In some years winter clover stocks may be good but the population already low because of a poor breeding productivity or low survival on the autumn stubbles. Or the birds may have experienced a good autumn survival rate and be faced with either good or poor feeding prospects on clover. That is, stocks of grain and clover can vary independently and because so many combinations are possible it is extremely diffi-cult to demonstrate the operation of density-dependent factors in the field. Clearly autumn and winter losses combined are related to population density (Fig. 16.4) so it would be expected that loss in relation to food stocks should also be related to density. This can be demonstrated by plotting the number of food units potentially available to each individual at the start of winter (obtained by dividing the number of food units, in this case clover leaves, by the number of birds requiring this food) against the mortality subsequently occurring (Fig. 16.8). When feeding prospects were good the mortality rate was low and birds immigrated into the area, whereas with poor feeding prospects much emigration and mortality occurred between December and March. The number of Wood Pigeons surviving the critical season of limited food resources determines the size of the next breeding population and this has gradually declined in Britain consequent on changes in arable farming which have led to a reduced acreage of clover being grown (Murton and Westwood 1974a).

Other species

Post-fledging chick mortality and the failure of young birds to establish themselves in the adult social system proved to be the important sources of density-dependent regulation in the Grey Partridge *Perdix perdix* (Blank, Southwood, and Cross 1967). So in both the partridge and pigeon, juvenile mortality has been the key factor causing fluctuations in population size between years, and also the regulatory factor. In the partridge 49 per cent of the variability in juvenile mortality in winter is unrelated to density and causes the fluctuations in population size whereas 51 per cent is depen-dent on density and is regulatory; comparable figures for the Wood Pigeon are 66 per cent and 34 per cent. Krebs (1970) used the key-factor method to analyse census data collected for the Great Tit *Parus major* by members of the Edward Grey Institute at Oxford (see Lack (1966) for details). He concluded that a density-dependent regu-lation of Great Tit numbers resulted from density-related variations in clutch size (see Fig. 16.9) and the predation rate of nestlings. Slagsvold (1975b) considered that territorial behaviour was the density-related factor. Krebs considered a weak source of density dependence in the autumn mortality rate to be spurious since a significant correlation coefficient depended on data for one year only. This population is in a sense artificial in that it depends on nesting boxes and nesting losses in these are rather high due to predation by weasels. Whether such a predation would occur under more natural nesting conditions, and what effect this might have on the post-fledging mortality remain to be defined. Both Kluijver (1951) and Lack (1964) had recognized that fecundity varied with population density but because the variations were slight, were unrelated to, and were swamped by, large post-fledgling losses, Lack felt that this

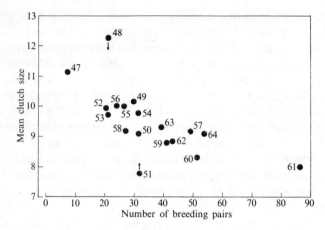

Fig. 16.9. Relation between mean clutch-size and number of breeding pairs of Great Tit *Parus major* in Marley Wood, Oxfordshire. Arrows indicate abnormally early or late seasons. From Lack (1966).

could not be the prime source of regulation. However, as mentioned, a weak source of density dependence can be regulatory if it operates in association with a strong density-independent loss.

Some idea of the stage in the life cycle when regulatory factors operate can be gleaned from examining the variability in population counts at different seasons. Ricklefs (1973) assembled some mean values and their coefficients of variation (as percentages) and noted that for the Wood Pigeon variation was greatest during the winter and declined in the breeding population. From this he rightly concluded that population changes occur in autumn without regard to density and that density-dependent processes operate during the winter to reduce the year-to-year variation. In the Red Grouse *Lagopus l. scoticus* variability in total population size decreased aft breeding (coefficient of variation of 34 per cent for the total February population, 35 per cent for the number of breeding birds per hectare, and 28 per cent for the October– January population). In the Great Tit variation was found to be the same in the breeding and winter populations, suggesting that density-dependent regulation occurs through reproduction and the survival of young, particularly in the production of fledglings per pair, as indicated by Krebs' analysis. The coefficient of variation in the productivity (number of young produced per pair) is low in the Wood Pigeon at 7 per cent because the birds compensate for egg and chick failure by re-nesting. Great Tits do this far less successfully and variability in productivity was around 25 per cent; it was as high as 56 per cent in the Tawny Owl *Strix aluco* (see below).

Ricklefs compared the coefficient of variation in size of breeding population with the variation in the number of young produced obtaining values respectively of 45 per cent and 26 per cent for Great Tit, 42 per cent and 40 per cent for Blue Tit *Parus caeruleus*, 21 per cent and 19 per cent for Pied Flycatcher *Ficedula hypoleuca*, 19 per cent and 51 per cent for Tawny Owl, and 35 per cent and 54 per cent for the Red Grouse, indicating a decreasing involvement of density-dependent factors acting at the time of reproduction. These data are of interest, for they all refer to species which defend territories and the suggestion has frequently been made that terrestrial behaviour serves to regulate the size of animal populations (see discussions and references provided by Lack 1954, 1966; Hinde 1956; Wynne-Edwards 1962; Fretwell and Lucas 1969; Watson and Moss 1970; Klomp 1972).

Territorial behaviour

In the Wood Pigeon territorial behaviour spaces out the birds which survive the winter amongst the available nesting habitat. The number of nesting pairs fluctuates more drastically in marginal habitats such as hedgerows than in mature deciduous wood-land (Murton and Westwood 1974a) so territorial behaviour apparently limits the number of birds able to settle in the best sites. If Wood Pigeons are removed from the favoured nesting habitats their place is taken by other birds, often juveniles, which move in from adjacent and presumably less preferred sites. Pairs that acquire wood-land territories have access to tree foods early in the season and they might hatch their young earlier than pairs relegated to less favourable sites. Early-hatched chicks have longer to grow and complete the juvenile moult before the onset of unfavour-able feeding conditions later in the year (Murton et al. 1974c) and this is presumably what confers on them a higher life expectancy (Fig. 8.1, p. 187). Acquisition of a good woodland territory may, therefore, confer survival advantages, but this need not mean that the territory serves for population regulation (see p. 429). In any case, so much of the birds' food is obtained outside the territory that if there is an effect on population size it is completely masked, and is unlikely to be important. The average number of young fledged per pair per season was the same in deciduous woodland, where population density was low, as it was in hedgerows and conifer woodland, where population density was much higher (Murton 1958). In the Great Tit possession of a territory could be more important since the birds mostly feed in them. Krebs (1971) removed territory owners from woodland and their place was taken by birds previously nesting in the hedgerows.

Krebs' studies support earlier discoveries of Kluijver and Tinbergen (1953) that the numbers of Great Tits vary more between years in those woods which they least prefer for nesting. Kluijver and Tinbergen called this the 'buffer effect' and supposed that as the best habitat became filled competition forced surplus birds to occupy less preferred sites, thereby allowing population size to be maintained at a less vari-able level in the preferred habitat. This observation would seem to support the con-tention that the production of young could be limited by territorial behaviour at some level below the maximal potential. Brown (1969) queried this interpretation, basing his arguments on the population data collected in Marley Wood, Oxford which were analysed by Perrins (1965). With some simplifying, but justified, assumptions, he showed that despite considerable differences in the production rates poss-ible in preferred and marginal habitats, the total production per unit area was similar in both situations (see Wood Pigeon example in preceding paragraph) and that maxi-mum production would be expected at a population density intermediate between that noted in the two sites. Unless one knows the particular distribution of birds between a good and bad habitat that would maximize total production it is not possible to estimate the amount of reduction in total productivity that supposedly results from territorial behaviour. Using a graphical method of analysis, Brown was led to the conclusion that 'the fact that many individuals do breed in poor habitats, therefore, should not necessarily be interpreted to mean that territoriality is limiting the total production of the population; it may, actually, be maximizing it!' (Watson (1973) misrepresented Brown's conclusion by quoting the paper as a supposed illustration that birds at high density in favourable habitats are driven by territorial aggression into less suitable habitats, where they breed less successfully.)

The number of breeding territories of the Skylark *Alauda arvensis* on an area of coastal sand dunes varied between 20 and 22 during the five summers from 1958 to 1962 (Delius 1965). Delius reckoned that most birds which were unable to settle formed a non-territorial, non-breeding population, estimated at 10 per cent of the total population. Following the hard winter of 1962–3 the summer population comprised only 16 pairs, territory size was bigger, and there appeared to be no non-breeding birds. Again we do not know whether failure to breed imposed the main density-dependent control of population size.

Tawny Owl

The effect of failure to breed has been fully studied in the Tawny Owl *Strix aluco* (Southern 1970). This owl depends very strictly on a woodland territory to catch its small rodent prey, principally Wood Mice *Apodemus sylvaticus* and Bank Voles *Clethrionomys glareolus*. Southern found that territory size in an area of parkland and continuous forest near Oxford declined from around 85 acres (34 ha) in 1948 to close on 30 acres (12 ha) in 1959 as the population increased from about 11 to 22 pairs. In a less preferred habitat territory size fell from around 60 acres (24 ha) to 50 acres (20 ha) and the population increased from only 7 to 10 pairs. In a good habitat territory size appears to be compressable according to the number of birds competing for space, whereas in a poorer habitat the size of territory may already be set at a minimum value which will impose a limit on population size. In the Tawny Owl it is not known what happens to birds unable to establish themselves in areas of occupied habitat but it is unlikely that marginal habitats, such as isolated spinneys and hedgerows, can absorb many birds, for these are unsuitable feeding places. Southern's key-factor analysis of his Tawny Owl population is depicted in Fig. 16.10. Having defined the maximum number of eggs which might be produced if all the pairs produced 3 eggs, Southern was able to divide k_0 (failure to achieve maximum reproductive potential) into two components: (1) failure resulting because some pairs failed to breed at all (which we here call k_{01}), defined as the difference between the total number of eggs that would be laid if *all* the females laid 3 eggs and the number that would be produced if the females which actually did lay produced 3 eggs; (2) failure to lay the maximum potential clutch, that is, the difference between number of eggs produced if the birds that did lay had each laid 3 eggs and the number of eggs they really did lay (we call this k_{02}).

Fig. 16.10 shows that k_{01} was chiefly responsible for fluctuations in total mortality and in one year (1958) when prey was scarce none of the pairs even attempted to breed. The success of Tawny Owls in laying eggs, hatching them, and fledging their young is determined by the abundance of mice and voles. There is a threshold effect at around 100 rodents per 5 ha, for beyond this density breeding success is not improved by increases in the food supply. Failure to fledge young that have been hatched is of relatively little importance (Fig. 16.10) although, as Southern points out, this is often regarded as the special adaptation of predators (see p. 199). The combined losses up to fledging, that is, $k_{01} + k_{02} + k_1$ (failure of eggs to hatch) were related in a density-dependent manner to the maximum productivity. However, the magnitude of the losses were not sufficient to adjust the number of young produced to the number needed to replace the adults lost during the year. The strong density-dependent adjustment which was needed for this occurred after fledging and during the ensuing winter, when the recently fledged young failed to secure territories in the wood and either died there or emigrated to other areas (see Fig. 16.11).

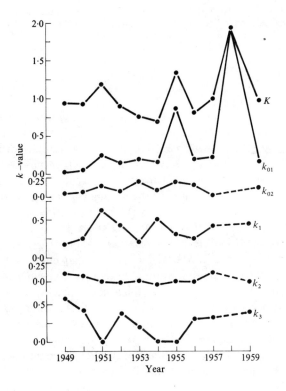

Fig. 16.10. A k-factor analysis of data from a Tawny Owl *Strix aluco* population study conducted in Oxfordshire. Adults omitted from analysis except in k_0 which is divided into two components (k_{01} represents the failure of some pairs to breed at all; k_{02} represents failure to lay the maximum potential clutch) k_1 represents egg mortality, k_2 nestling mortality, and k_3 over-winter mortality. Based on Southern (1970) as corrected by him for Dempster (1975).

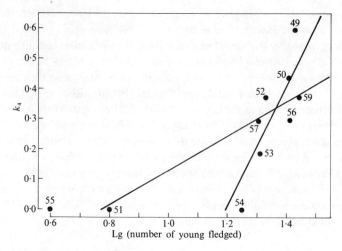

Fig. 16.11. Winter mortality (k_4) of juvenile Tawny Owls in relation to the lg (number of juveniles which fledged). As in Fig. 16.4 (a), (b), and (c), losses appear to show a non-linear relationship with density and indicate a competition curve. This invalidates the use of a two-way regression as a strict test of density-dependence. Based on Southern (1970).

Red Grouse

The Red Grouse is the best-studied example of a species in which territorial behaviour apparently determines population size. The diet comprises 48–91 per cent heather shoots from May until August and 72–100 per cent in winter (Jenkins, Watson, and Miller 1963). The food supply is evenly distributed and is best exploited by each male (joined by a female) holding a territory of sufficient size to ensure adequate opportunities for grazing. The quality of the heather varies with the substrate, fertilizer treatment, and the rotational burning which is a tradition of the grouse moors (Miller, Watson, and Jenkins 1970) and the birds select the shoots which are of highest nutritive value (Moss 1967, 1969; Moss, Miller, and Allen 1972). The parallels between the grazing habits of grouse and Wood Pigeons are striking but the same does not apply to their social organization. Doubtless this is because heather supplies are evenly dispersed whereas weed leaves, seed, and tree foods have a patchy distribution.

In July and August Red Grouse live as families, with the males paying only partial attention to their territories at dawn. Then in each year in August and September the males establish a new pattern of territories. Most of the young birds form transient flocks which range round the locality, but a proportion of young cocks remain with their parents. In October the most advanced of these cocks challenge the old males and in some cases they manage to evict the owners from their territories. The population now comprises three categories: (1) territory-owning cocks and the hens paired with them; (2) non-territorial surplus residents which failed to get a territory but which nevertheless stay on the moor; (3) the non-territorial transient birds which first formed flocks at the end of the breeding season.

A decrease in population size occurs on the moor as evicted old birds and these transient young (category 3) are displaced to marginal habitats adjacent to the moor. These birds feed in woods, scrub, stubble fields, and grassy places on the moor, whereas non-territorial residents stay on the heather moor. During the winter the transient birds do visit the moor to feed on heather but lacking a permanent home area they are at greater risk of being surprised by predators such as Hen Harrier *Circus cyaneus*, Golden Eagle *Aquila chrysaetos*, and Fox *Vulpes vulpes* (Jenkins *et al.* 1964). Non-territorial birds fare better than the transients. They must avoid the residents during the mornings when the cocks are aggressive but they manage to feed anywhere on moor in the afternoon. In January or February, territorial Grouse become more aggressive and now any surplus resident birds are displaced, no matter what the time of day, so non-territorial residents assume the status of transients and a second phase of population reduction occurs. Peak aggression and plasma LH levels are found in March (Sharp, Moss, and Watson 1975). Displaced birds suffer heavy mortality and very few exceptionally survive until the summer. It follows that the breeding population is composed of birds that essentially held territories from the previous October (Jenkins *et al.* 1963; Jenkins, Watson and Miller 1967).

The ability of a male grouse to command and hold a territory depends on its level of aggression; territory size is also related to the amount of cover, but secondarily through the effect cover exerts on aggressive behaviour (Watson 1964, 1967). Implants of testosterone increase aggressiveness and territory size whereas oestrogen lowers aggression and causes a male to give up his territory (Watson 1970). There is evidence that a high plane of nutrition during the summer predisposes birds to take smaller territories in autumn (Watson and Moss 1970) and an inverse correlation exists between territory size and the nitrogen content of the heather, as if grouse compensate for poor-quality food by taking larger territories (Lance 1971). Egg quality, clutch size, hatching date, hatching success, and the survival of chicks are all dependent on the quality of the territory (Jenkins *et al.* 1967; see also Watson and Moss (1970) for summary). Moreover, there is a carry-over effect, for it was found that young cocks reared in years of high breeding success subsequently took smaller territories and were less aggressive than those cocks from previous year classes. Conversely, in years when breeding success was low the surviving young were more aggressive and took larger territories than territorial cocks from previous year classes (Watson and Miller 1971). It is not yet clear whether this effect results from the experience of the chick in relation to the size of brood in which it is reared (Jenkins *et al.* 1967), a supposed breakdown of the stress—endocrine system in chicks reared at high population density (*cf.* Christian *et al.* 1965), or the functioning of a genetic polymorphism which favours different genotypes in a density-related manner (Chitty 1967). Watson and Moss (1972) wonder whether the plane of nutrition, known to affect the mother and her egg quality, could also predetermine the aggressive tendencies of the chick.

The proximate cause for change in population size is said to be the change in aggression between generations that leads to changes in territory size. Years of population decline follow seasons of poor breeding success and during such phases of the population cycle the young are predisposed to be more aggressive and take larger territories. The advantages of the adaptation are self-evident, but a problem is posed. If grouse that hatch from eggs produced under a poor nutritional plane are more aggressive than grouse hatched from good breeding cohorts, what is the selective factor that keeps the level of aggression low in good seasons? Manifestly a mutant with a high capacity for aggression should obtain a bigger territory, survive better, and leave more progeny, so there must be some selective disadvantage, balancing the spread of aggression. Could aggression be a consequence of an adrenal feed-back mechanism? This hypothesis imagines that poor nutritional conditions stimulate an increase in corticosteroid activity in both parent and chick as a homeostatic mechanism to mobilize body reserves. If aggressive tendencies depend directly or indirectly (via an effect of adrenal steroids on gonadal steroid production) on adrenal secretions it can be imagined that hyperactivity could be beset with disadvantages. It is still not clear what causes aggression in nature for the experiments with implants of testosterone could be misleading (see p. 113). Aggression that depends on gonadal steroids may interfere with pair formation.

On p. 467 the suggestion was made that the density-dependence which regulates Red Grouse populations is not applied during the reproductive phase of the cycle. Watson (1970) has performed a key-factor analysis on his population data but unfortunately his method was technically faulty[†] because he combined results from different study sites. Hence, final conclusions cannot be made about the regulatory factor involved, although there are good indications that mortality occurring outside the breeding season was both the key factor and that it also acted in a delayed density-dependent manner. Watson argued that winter loss was irrelevant since the real factor was territorial behaviour, resulting in the displacement of surplus individuals, which then became candidates for this post-breeding mortality. The argument becomes critical if this example is used as a basis for justifying a contention that territorial behaviour has evolved as a means of regulating population size.

Conclusion

Our thesis is that there is no difference in principal between the population dynamics of the Red Grouse, Tawny Owl, and Wood Pigeon and if the whole spectrum is considered there is no reason to attribute a primary population regulation function to the territorial habit. Density-dependent regulation of population size in the Wood Pigeon primarily results from the failure of juvenile animals to establish themselves in the social system of the adults. This occurs in winter when the adults feed in flocks. Only later and after the season of regulation do the adults establish territories (the same applies to the Grey Partridge). In the Tawny Owl the adults maintain a territory system throughout the year and so the density-dependent adjustment in recruitment operates via the territory system but outside the breeding season. In the Red Grouse there is contention about whether regulatory factors operate during or after the

[†] In a small footnote on p. 549 he writes: 'At Glen Esk from 1962 on, only part of the original area was counted. However, the number counted was multiplied by a conversion factor based on acreage so that all densities here are comparable.' This means that only population data for the seasons 1957–8 to 1960–61 can be used for a key-factor analysis.

breeding season but again they most certainly operate via the territory system. In all these three examples we believe that the social structure is determined by the feeding strategy which the bird must adopt to obtain its food. For instance, Southern and Lowe (1968) give evidence that possession of, and intimate knowledge of, a territory of sufficient size provides an insurance for the Tawny Owl that it will be able to hunt with adequate prospects of locating and catching its prey. In these three species the density-dependent regulation of population size is mostly attributable to juvenile mortality and this occurs when these birds attempt to establish themselves within the conventional feeding hierarchy system which is adopted by the adults. The number of places available for the young to occupy is, on average, in a stable environment equivalent to the loss of adults, already considered (p. 453). It is difficult to imagine how the social conventions just described could all be primarily functioning for population regulation (nor for that matter as anti-predator strategies). Watson (1971, p. 563) has commented 'The trouble about these evolutionary arguments is that they are so speculative, and thus can be used as confirmatory evidence for many possible hypotheses.' Since we must take animals as we find them, and are limited lest our experimental designs become contrived, there would seem little alternative to accepting a comparative evolutionary approach. Moreover, attempts to remove speculation from evolutionary arguments would seem to be the essence of neo-Darwinism, a philosophy we make no apology for following.

Appendix 1

Chemical composition of pigeons' milk
by percentage weight

	Needham 1942 (by wet weight)	Engelmann 1962 (by dry weight)	Harmuth 1972[†]
Water	65–81		72·4
Dry matter	19–35		27·6
Protein	13–19	58·4	
Fat	7–13	35·1	
Ash	1–2	6·5	
Carbohydrate	0	0	

[†]The dry matter contained 47·2 per cent protein with the following percentage composition of amino acids: aspartic acid 9·26 per cent, threonine 5·32 per cent, serine 5·06 per cent, glutamic acid 13·91 per cent, proline 3·04 per cent, glycine 5·36 per cent, alanine 4·90 per cent, cysteine 1·24 per cent, valine 5·29 per cent, methionine 2·91 per cent, isoleucine 4·34 per cent, leucine 8·34 per cent, tyrasine 4·23 per cent, phenylalanine 4·36 per cent, lysine 7·37 per cent, histidine 2·31 per cent, arginine 6·83 per cent, unknown 6·0 per cent.

Appendix 2

Summary flow sheet of procedure used in assaying the capacity of testicular extracts to biosynthesize steroids *in vitro* from a radioactive precursor

---

[†] Sample bands were located according to R_{DOC} value of the steroid defined for steroid A as:

$$\frac{\text{Distance travelled by A}}{\text{Distance travelled by 11–DOC in the same TLC chromatogram}}$$

[‡] Radioactivity was calculated as:

$$\text{Radioactivity } ^3H \text{ (in d.p.m.)} = \frac{(\text{Net reading from }^3H\text{ channel}) - (\text{leakage from }^{14}C\text{ channel})}{(\% \text{ recovery of }^{14}C\text{ standard}) \times (\text{efficiency of machine})}$$

From this the percentage conversion of each steroid is given by:—

$$\frac{\text{radioactivity of each steroid sample}}{\text{total radioactivity of precursors}} \times 100$$

$$= \text{conversion index} \times 100.$$

250 mg testicular extract minced with scissors

↓

Incubation flask with Δ^5-pregnenolone-T as precursor
and 10 ml Kreb–Hensleit buffer, pH = 7·4

↓

Incubation in 95% oxygen and 5% CO_2 for 4 h at 41 °C

tissue residue discarded

decant and filter through glass wool

↓

Reaction stopped by freezing pooled filtrate

Aqueous phase discarded

^{14}C-labelled testosterone, progesterone and 17 α-oestradiol added
as internal standards to check loss of steroid during processing

(1) steroids extracted with chloroform (O'Grady 1967)

(2) centrifuged

↓

Pooled organic phase dried under nitrogen

Sample aliquot tested for
estimation of recovery

Crude extract

↓

Pooled extract from 6 individual birds

↓

50 g each of testosterone, androstenedione, progesterone,
dehydroxyandrosterone (DHEA), oestriol, oestrone, pregnenolone, DOC,
17α–OH-progesterone, and 17α–OH-pregnenolone added as carriers

Thin-layer chromatography

↓

Pooled crude extract spotted on thin layer plates

developed in petroleum: benzene: ethyl acetate
in ratio 1:1:4

↓

Plates scanned under ultraviolet at 240μm for Δ^3-ketosteroids

† Sample bands and markers located

↓

Sample bands scraped off plate and eluted with
chloroform until volume about 10 ml

Eluates of sample bands

dried under stream of nitrogen in water bath at 41 °C

↓

Purified extract spotted on thin-layer plate

(1) developed in toluene and methanol in ratio 92:8

(2) eluted with chloroform

↓

Eluates of steroid samples

dissolved in chloroform/methanol (1:1) and dried
under nitrogen in a glass counting vial. 10 ml
scintillation liquid added

↓

‡ Samples counted for 10 min in Packard Tri-carb liquid
scintillation spectrophotometer Model 3003

Appendix 3
List of warbler species

Warbler species that contribute to Fig. 7.9	Breeding range		Winter range	Migratory movement (miles)	Post-nuptial wing moult in summer (S) or winter (W) quarters
	Mid-latitude	Region[†]			
Genus *Acrocephalus*					
Moustached Warbler					
A. melanopogon melanopogon	41 °N	South-western Palaearctic	North Africa	2000	W
A. melanopogon mimica	39 °N	Central Palaearctic	Iraq and north India	700	W
Sedge Warbler					
A. schoenobaenus	54 °N	Western Palaearctic	Africa	4000	W
Aquatic Warbler					
A. paludicola	53 °N	Western Palaearctic	Africa	5000	W
Schrenck's Sedge Warbler					
A. bistrigiceps	43 °N	Eastern Palaearctic	South-eastern Asia	1800	W
Reed Warbler					
A. scirpaceus	45 °N	Western and central Palaearctic	Africa	3500	W
Marsh Warbler					
A. palustris	50 °N	Western Palaearctic	Africa	4400	W

	Latitude	Breeding region	Wintering area		W/S
Blyth's Reed Warbler					
A. dumetorum	52 °N	Central Palaearctic	India	2600	W
Paddyfield Warbler					
A. agricola	45 °N	South-eastern and central Palaearctic	India	1600	W
Great Reed Warbler					
A. arundinaceus arundinaceus	45 °N	Western Palaearctic	Africa	3200	W
A. a. orientalis	40 °N	Eastern Palaearctic	South-eastern Asia	2500	S
Clamorous Reed Warbler					
A. stentoreus stentoreus	25 °N	North-eastern Africa	Resident	–	S
A. s. brunnescens	35 °N	South-central Palaearctic	Persia, India	1000	W
Genus Hippolais					
Icterine Warbler					
H. icterina	55 °N	Western Palaearctic	Africa	5000	W
Melodious Warbler					
H. polyglotta	40 °N	South-western Palaearctic	Africa	2500	W
Olivaceous Warbler					
H. pallida	35 °N	South-western Palaearctic	Africa	2000	W
Booted Warbler					
H. caligata	47 °N	Central Palaearctic	Middle East	3000	W
Olive-tree Warbler					
H. olivetorum	37 °N	Western Palaearctic	Africa	3000	W
Upcher's Warbler					
H. languida	37 °N	South-central Palaearctic	Africa	3000	W

Genus Sylvia

Blackcap					
S. *atricapilla*	50 °N	Western Palaearctic	Africa	4000	S
Garden Warbler					
S. *borin*	53 °N	Western Palaearctic	Africa	5500	W
Common Whitethroat					
S. *communis*	50 °N	Western Palaearctic	Africa	3600	S
Lesser Whitethroat					
S. *curruca curruca*	50 °N	Central-western Palaearctic	North-eastern Africa	2800	S
S. *c. blythi*	48 °N	Central Palaearctic	Middle East/India	2000	S
S. *c. minula*	45 °N	Central Palaearctic	India / Northern India	1900	S
S. *c. althaea*	34 °N	Southern Palaearctic	India	1600	S
Barred Warbler					
S. *nisoria*	48 °N	Central Palaearctic	Africa	3500	S
Orphean Warbler					
S. *hortensis hortensis*	38 °N	South-western Palaearctic	Africa	1500	S
S. *h. jerdoni*	35 °N	South-central Palaearctic	India	1500	S
Arabian Warbler					
S. *leucomelaena*	22 °N	Arabia	Resident	–	S
Desert Warbler					
S. *nana*	43 °N	South-central Palaearctic	North-eastern Africa/India	1500	S
Rüppell's Warbler					
S. *ruppelli*	37 °N	South-western Palaearctic	North-eastern Africa	1000	S
Cyprus Warbler					
S. *melanothorax*	35 °N	South-western Palaearctic	Resident	–	S
Sardinian Warbler					
S. *melanocephala*	38 °N	South-western Palaearctic	Resident	–	S
Ménétries's Warbler					
S. *mystacea*	40 °N	South-central Palaearctic	Arabia and	1600	W

Subalpine Warbler					
S. *cantillans*	37°N	South-western Palaearctic	Western Africa	1500	S
Spectacled Warbler					
S. *conspicillata*	37°N	South-western Palaearctic	Resident	–	S
Dartford Warbler					
S. *undata*	40°N	South-western Palaearctic	Resident	–	S
Marmora's Warbler					
S. *sarda*	40°N	South-western Palaearctic	Resident	–	S
Genus *Phylloscopus*					
Orange-barred Leaf Warbler					
P. *pulcher pulcher*	30°N	Oriental, South-eastern Palaearctic	Resident	Altitudinal migrant	S
Ashy-throated Leaf Warbler					
P. *m. maculipennis*	30°N	Oriental, South-eastern Palaearctic	Resident	Altitudinal migrant	S
Pallas's Leaf Warbler					
P. *proregulus*	53°N	Central Palaearctic	China	2500	S
Brooks's Leaf Warbler					
P. *subviridis*	34°N	Central Palaearctic Oriental	Northern India	300	S
Yellow-browed Warbler					
P. *inornatus inornatus*	62°N	Central Palaearctic	Malaysia	3500	S
P. *i. humei*	45°N	Oriental South-eastern Palaearctic	India	1300	S
Western Crowned Leaf Warbler					
P. *occipitalis*	32°N	Oriental	India	1100	S
Eastern Crowned Leaf Warbler					
P. *coronatus*	45°N	South-eastern Palaearctic	Indochina	2700	S

Species	°N				
Blyth's Crowned Leaf Warbler *P. reguloides*	28 °N	Oriental	Indochina	500	S
Oates's Crowned Leaf Warbler *P. davisoni*	19 °N	Oriental	Resident	Altitudinal migrant	S
Arctic Warbler *P. borealis borealis*	57 °N	Palaearctic	South-western Asia	7000–8000	W
P. b. xanthodryas	37 °N	Oriental	Indonesia	3000	?W
Greenish Warbler *P. trochiloides trochiloides*	35 °N	South-eastern Palaearctic Oriental	North-eastern India	500	W
P. t. viridances	54 °N	Eastern and central Palaearctic	South-eastern Asia	3000–8000	W
Bright Green Leaf Warbler *P. nitidus*	36 °N	South-central Palaearctic	South India	2000	W
Two-barred Greenish Warbler *P. plumbeitarsus*	53 °N	Central Palaearctic	Indochina	2800	W
Large-billed Leaf Warbler *P. magnirostris*	32 °N	South-eastern Palaearctic	Resident	800, altitudinal migrant	W
Slender-billed Leaf Warbler *P. tytleri*	36 °N	South-central Palaearctic	Resident	700, altitudinal migrant	S
Dusky Warbler *P. fuscatus*	57 °N	Central Palaearctic	India/China	2600	S
Smoky Warbler *P. fuligiventer*	29 °N	Oriental	Resident	Altitudinal migrant	S
Radde's Warbler *P. schwarzi*	55 °N	Central Palaearctic	Indochina	2700	S

	°N	Zoogeographic region	Winter area	Altitude	Status
Sulphur-bellied Leaf Warbler					
P. griseolus	40 °N	Southern and central Palaearctic	India	1800	S
Milne Edwards's Leaf Warbler					
P. armandii	34 °N	South-eastern Palaearctic	North Indochina	1100	S
Tickell's Leaf Warbler					
P. affinis affinis	28 °N	Oriental	China, Thailand	700, altitudinal migrant	S
P. a. subaffinis	24 °N	Oriental	Burma, China	Altitudinal migrant	S
Chiffchaff					
P. collybita	54 °N	Western and central Palaearctic	Central Africa	3000–4000	S
Mountain Chiffchaff					
P. sindianus	30 °N	Oriental	Northern India	1000	S
Plain Leaf Warbler					
P. neglectus	32 °N	Oriental	Resident	900	S
Willow Warbler					
P. trochilus trochilus	60 °N	Palaearctic	Western and Eastern Africa	5000	S and W
P. t. acredula	65 °N	Western Palaearctic	Eastern Africa	5300	S and W
P. t. yakutensis	68 °N	Central and eastern Palaearctic	Eastern Africa	6000–7000	?S and ?W
Wood Warbler					
P. sibilatrix	52 °N	Palaearctic	Central Africa	4200	W
Bonelli's Warbler					
P. bonelli	40 °N	South-western Palaearctic	Africa	1800	W

Appendix 4

Comments on the evolution of some pigeon taxa

To understand the nature of avian photoperiodic responses it is necessary to know the evolutionary history of the bird groups concerned. This is a complex subject which, with limited fossil evidence, must be a matter for some speculation and inspired guess-work. This short account is intended simply to highlight the kinds of problems involved, so far as a few of the Columbidae are concerned.

There is reasonable evidence to suggest that many bird orders had emerged by the late Cretaceous Period and, because the final separation of the continents did not occur until the end of the Mesozoic or early Tertiary (80 million years ago), continental drift must have been important in their radiation (Cracraft 1973). Cracraft, in an excellent review of the subject, gives reasons for believing that the parrots (Psittacidae) and pigeons (Columbidae) had their origins in Gondwanaland and so could reach Africa, South America, and Australia before the continental break-up, or at least while the sea crossings remained relatively short. There was perhaps an early radiation and dispersal of pigeons through Gondwanaland but, since the majority of present-day Austro-Asian genera are unrepresented in South America, the speciation of most genera clearly occurred after continental separation. We are primarily concerned to understand the evolutionary history of *Columba* which, in fact, is the only genus today represented in both Old and New Worlds. The simililarities between the North American and European *Columba* species are attributed by some authorities, notably Mayr (1946, 1964), to dispersal from Eurasia, possibly via the early Cenozoic North Atlantic land connection, or the later one across the Bering land bridge. Some now extinct genera, perhaps related to *Gerandia* (known from the upper Oligocene–lower Miocene of Europe according to Brodkorb (1971*b*)), may have dispersed in this way. However, we think it more likely that the present North American pigeon species are derived from South American ancestors and owe their similarity to Eurasian forms to convergence rather than an immediate relationship (they remain close enough to be in the same genus).

It is conceivable that *Columba* arose in Gondwanaland around the time of the break-up of the old land mass. There is only one *Columba* species in Australia, the White-headed Pigeon *C. leucomela,* and this is clearly closely related to several forms in the East Indian and Pacific regions, in particular the White-throated Pigeon *C. vitiensis* and the Black Wood Pigeon *C. janthina* (Goodwin 1970). The immediate origin of these two pigeons may have been New Guinea (Stresemann 1939) and this makes it probable that *C. leucomela* itself entered Australia from the north, long after the continent had drifted north during the Cenozoic. The present day distribution of *Columba*—with 16 species in Africa (several of these are island forms or else allopatric replacements), 3 in Africa and Europe, 9 in Asia, 12 in South-East Asia, 17 in America (with many good species), but only 1 in Australia—lead one to suspect that Australia was never in the centre of radiation; even allowing that the drastic change from subtropical forest to semi-arid conditions that took place during the upper Miocene (Burbridge 1960) could have exterminated many elements. Conceivably, fruit pigeons, ancestors of *Treron* and *Ducula*, initially filled the niches in Australia and prevented *Columba* from becoming established. Fig. A.1, from Cracraft (1973), in turn based on McKenzie and Sclater (1971), indicates the surmised position of the southern continents 45 million years ago, during the Eocene, to indicate the conjectural radiation pathways of *Columba* from an evolutionary centre in Gondwanaland. Essentially the same line of argument could be applied to conditions existing 36 million years ago and perhaps more recently.

Our scheme imagines that two lines of radiation into the northern hemisphere occurred from a Gondwanaland origin.

1. One line (there may have been more than one separate invasion) advanced and speciated in South America and then penetrated into North America when the Panamanian land bridge was established near the end of the Tertiary—probably in the late Pliocene. By this time climatic deterioration and cooling and a reduction in rainfall in North America created the deserts and central semi-arid plains of the subcontinent.

2. The second line(s) is imagined to have radiated north via the Pacific land masses which are now separated as Java, the Celebes and various islands to India, and thence into eastern Asia and Japan (see figure).

Africa is reckoned to have been colonized secondarily from western Asia. The above scheme is consistent with Goodwin's (1970) assessment (q.v.) of the presumed phylogenetic relationships within *Columba* which makes the *janthina* group of the East Indian, and Pacific regions primitive to the Asian and European species groups respectively. The African forms presumably reached that continent as Africa closed with Asia, that is, by an invasion from the north-east for the affinities of the olive pigeons of the *arquatrix* group are with *hogsonii* of the Himalayas. What is assumed to be a very primitive *Columba*, the Pink Pigeon *C. mayeri*, is the only *Columba* to have penetrated anywhere near to Madagascar (actually it lives on Mauritius). It conceivably reached Mauritius from primitive lines of *Columba* in the East Indies. The absence of a more recent *Columba* stock of African affinity is probably significant. There is a less striking parallel in *Streptopelia* in that the Madagascar Turtle Dove *S. picturata* seems more closely related to *S. chinensis* (from Asia) and *S. senegalensis* (India—Africa) than to the more typical doves of Africa.

The *Geopelia* doves of Australia may be related to the American Ground Doves (*Scardafella, Columbina*, and others) which in turn are not far removed from the New World genera *Leptotila* and *Zenaida*. The ancestors of these groups were possibly related to *Aplopelia* (African Lemon Doves) and *Streptopelia* (Turtle Doves), all these genera having many representatives which are adapted to arid and semi-arid habitats (see Goodwin (1970) for details). The radiation of these forms could well have occurred more recently than that of *Columba* in response to the dessication which occurred in the late Cenozoic (see below). *Streptopelia* is probably fairly close to *Columba* (see Goodwin 1970), but the inter-generic relationships of the ground feeding doves and *Columba* ancestors is presently unknown.

Some while ago Cumley and Irwin (1944) argued, on the basis of the blood antigenic relationships, that the Band-tailed Pigeon *C. fasciata* of North America has a

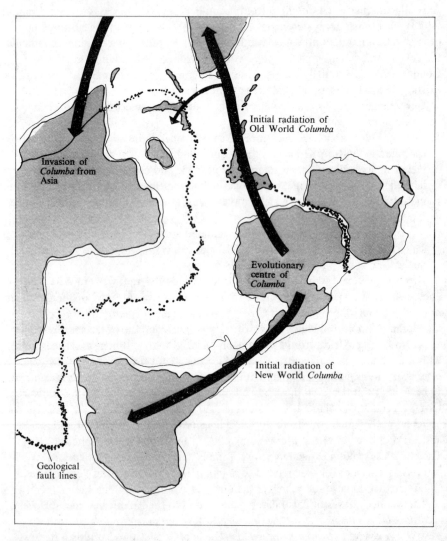

Fig. A.1. The surmised position of the continents 45 million years ago, to show the conjectural radiation pathways of *Columba* from an evolutionary centre in the land mass remaining after the break-up of Gondwanaland.

closer affinity with other New World pigeon genera than with Old World *Columba* species. More recently, Corbin (1967, 1968) has investigated the evolutionary relationships within *Columba* by an analysis of the electrophoretic chromatograms of ovalbumin tryptic peptides. He suggests a common ancestor for the American and Afro-Eurasian *Columba* species followed by speciation in separate hemispheres. Of the species examined by Corbin, the ovalbumin map of the South American Rufous Pigeon *C. cayennensis* appeared to be the least changed from the presumed ancestral condition (as Boetticher (1954) thought also). The Rufous Pigeon inhabits open forest and semi-arid regions and is partly sympatric with the Spotted Pigeon *C. maculosa*, a bird of arid and semi-arid regions, which has an ovalbumin map close to *C. cayennensis*. The White-crowned Pigeon *C. leucocephala* of the Caribbean Islands is also very close to these two species. Of the Old World species examined, the Speckled Pigeon *C. guinea* of Africa and Stock Dove *C. oenas* of Europe and Africa were evidently close. The Wood Pigeon *C. palumbus* proved equally similar in its electrophoretic pattern to the Rock Dove *C. livia*, Stock Dove, and Speckled Pigeon, but it showed the greatest similarity to *C. cayennensis*. Therefore, it is perhaps closer to the ancestral stock from which Old and New World forms emerged, while the Speckled Pigeon and Stock Dove are probably more recently derived forms. It may be significant in this respect that the Stock Dove, Rock Dove, and Speckled Pigeon nest in holes in trees and rocks, as do some other Old World doves, whereas the Wood Pigeon, like all the American species, builds its own twig nest in trees and shrubs; this presumably being the ancestral trait. The Picazuro Pigeon *C. picazuro* is a woodland species in South America, though it also frequents more open country, and is partly sympatric with *C. cayennensis*. It is rather distinct from the latter in electrophoretic pattern as if it represents a distinct wave of colonization of South America. It is possibly close to both Old and New World forms and could have spread from the south. The Band-tailed Pigeon of North America is clearly more distinct but shows close affinity with the Red-billed Pigeon *C. flavirostris* of Central America.

Corbin found the electrophoretic patterns of *Leptotila* (small to medium woodland doves of South and Central America) to be closer to *Columba* than to those of *Streptopelia* (Old World doves) and he queried whether *Streptopelia* had given rise to *Columba* and *Leptotila*, or whether *Leptotila* and *Streptopelia* derived from a *Columba*-like stock. The latter possibility seems more likely on distribution evidence. The American ground doves (*Columbina* and allies) and the *Zenaida* doves appear to be derived from *Leptotila* to occupy ground niches in arid and semi-arid regions. In like fashion *Streptopelia* comprises mostly ground feeding forms that possibly derive from a *Columba* stock, perhaps via *Aplopelia*-like (African lemon doves) ancestors; *Aplopelia* is convergently similar to *Leptotila*. This gives the presumed evolutionary scheme shown in Fig. A.2. If the above presumptions are justified it would seem that the shortened breeding season and clear refractory phase of the Turtle Dove is a secondary adaptation which has been derived from ancestors having longer breeding periodicities as typically exhibited by *Columba* and some of the other *Streptopelia* doves (see p. 266–279).

The prairies and central plains of North America have been an important isolating

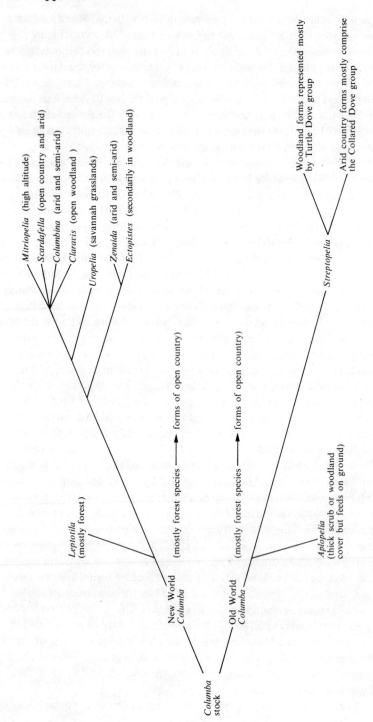

Fig. A.2. Presumed radiation from *Columba* stock.

agent in bird speciation (Mengel 1970). The woodland dwelling Band-tailed Pigeon clearly spread north from ancestors in South America (it forms a super-species with the Jamaican Band-tailed Pigeon *C. caribaea* and the Chilean Pigeon *C. araucana* (a bird of the *Nothofagus* and *Araucaria* forests of the Argentinian Andes). It has penetrated north to British Colombia along the forested Rockies and occurs throughout the west, especially where oaks, a favoured food tree, are available. However, it has never penetrated across the prairies to the forests of eastern North America, where similar suitable food supplies exist. This left a vacant niche for a tree-feeding pigeon able to exploit tree buds and fruits. This niche was filled by the Passenger Pigeon *Ectopistes migratorius,* which is clearly descended from *Zenaida* stock, and which has convergently come to have some of the features of the Band-tailed Pigeon and Wood Pigeon, particularly a plumage coloration which must be adaptive for these tree perching forms. The Mourning Dove *Zenaida macruora* forms a super-species with the Eared Dove *Z. auriculata* of South America and is evidently closely related to several other doves of Central and South America, South America must be the evolutionary centre of the genus. Because *Zenaida* ancestors were already adapted to arid zones they did not find the North American central plains an ecological and geographical barrier and could spread throughout the eastern regions and eventually give rise to *Ectopistes,* which is, therefore, secondarily adapted to a woodland existence.

Appendix 5

Model to describe negative assortative mating in the Feral Pigeon

The model, constructed by Davis and O'Donald (1976), assumes that the females have preferences α, β, and γ for matings with phenotypes A, B, and C (equivalent to wild-type, blue check, and T-pattern respectively) except when the male's phenotype is like the female's. The frequency of matings for these phenotypes is u, v, and w, and x is the frequency of mating with all other phenotypes (represented as D). For the sake of simplicity assortative matings for phenotype D are ignored and so preferential matings occur according to:

		female phenotype		
		A	B	C
		u	v	w
male phenotype	A	—	αv	αw
	B	βu	—	βw
	C	γu	γv	

The proportion of males and females left to mate at random would be:

	Males	Females
A	$u-\alpha v-\alpha w$	$u(1-\beta-\gamma)$
B	$v-\beta u-\beta w$	$v(1-\alpha-\gamma)$
C	$w-\gamma u-\gamma v$	$w(1-\alpha-\beta)$
D	x	x

The total of males and females mating at random is:

$$R = 1 - \alpha(v+w) - \beta(u+w) - \gamma(u+v).$$

The frequency of expected matings and those observed in the field by Murton, Westwood, and Thearle (1973) are shown in Table A.1.

TABLE A.1

Mating	Number pairings observed	Frequency
A × A	2	$u(1-\beta-\alpha)(u-v-\alpha w)/R$
A × B	48	$\beta u + \alpha v + u(1-\beta-\gamma)(v-\beta u-\beta w)/R + v(1-\alpha-\gamma)(u-\alpha v-\alpha w)/R$
A × C	35	$\gamma u + \alpha w + u(1-\beta-\gamma)(w-\gamma u-\gamma v)/R + w(1-\alpha-\beta)(u-\alpha v-\alpha w)/R$
A × D	12	$ux(1-\beta-\gamma)/R + x(u-\alpha v-\alpha w)/R$
B × B	5	$v(1-\alpha-\gamma)(v-\beta u-\beta w)/R$
B × C	57	$\gamma v + \beta w + v(1-\alpha-\gamma)(w-\gamma u-\gamma v)/R + w(1-\alpha-\beta)(v-\beta u-\beta w)/R$
B × D	18	$vx(1-\alpha-\gamma)/R + x(v-\beta u-\beta w)/R$
C × C	3	$w(1-\alpha-\beta)(w-\gamma u-\gamma v)/R$
C × D	13	$wx(1-\alpha-\beta)/R + x(w-\gamma u-\gamma v)/R$
D × D	2	x^2/R

Total 195 $\qquad\qquad R = 1 - \alpha(v+w) - \beta(u+w) - \gamma(u+v)$

Using a computer parameters were fitted to this model and the best fit was given by Table A.2.

TABLE A.2

Parameters fitted	Values of the fitted parameters at maximum likelihood	Log likelihood	Value of χ^2	Degrees of freedom
None	--	$-390\cdot903$	$50\cdot869$	6
α	$0\cdot340$	$-380\cdot763$	$40\cdot945$	5
β	$<0\cdot508$	$-372\cdot584$	$26\cdot856$	5
γ	$0\cdot397$	$-377\cdot682$	$35\cdot243$	5
α, β	$0\cdot327, 0\cdot520$	$-362\cdot532$	$8\cdot271$	4
α, γ	$0\cdot355, 0\cdot416$	$-364\cdot072$	$10\cdot782$	4
β, γ	$0\cdot508, 0\cdot376$	$-361\cdot747$	$6\cdot565$	4
α, β, γ	$0\cdot000, 0\cdot508, 0\cdot376$	$-361\cdot747$	$6\cdot565$	3

This shows that only the preferences for B and C, that is blue check (C) and T-pattern (C^T) are statistically significant so that if female mating preferences cause the negative assortative mating they are the result of preferences for the melanic phenotypes. The values of the parameters indicate that 0 females prefer the wild-type, that the proportion preferring blue checks is $0\cdot508$ and that $0\cdot376$ prefer the T-pattern males, the preferences being expressed only when the male is a different phenotype to the female.

Postscript

The production of a book of this scope necessitates a long gestation period and inevitably some important new discoveries have been made since we began its preparation. We have prepared a few notes to bring our text up to date, to correct the odd false impression, and to indicate some of the lines of research which are likely to influence the subject during the next decade.

It seems evident that the physiological condition of wild birds — and other animals — will prove to vary in an extremely complex manner from season to season. At present, seasonal cycles of fat deposition, of moult, or the attainment of reproductive condition are well documented. Evidence is now accruing to show that birds may store specific proteins in readiness for breeding (cf. Jones and Ward 1976); at some seasons metal-binding proteins and specific enzymes may be synthesized to store and transport those metals which are essential for vital functions, for example, zinc which is needed in large quantities for feather production (D. Osborn, in progress); the whole structure of the gut and digestive system may change to suit altered food sources and, while it is known that the nature of the food supply may induce changes in gut length (Moss 1974; Sibly and Kenward in press), it now seems possible that such morphological and physiological changes may occur in anticipation of changes in dietary intake.

It would be expected that complex programmes of this kind should be ordered by autonomous mechanisms. Our view is that the autonomous system is then entrained and phased appropriately to environmental conditions by photoperiodic mechanisms. However, on purely logical grounds, and in spite of what we have written, it cannot easily be proved that photoperiodism *per se* is important. An obvious difficulty is that birds which are exposed to long daylengths have longer in which to feed, and the quantity and quality of the food might be the major factor stimulating gonad growth, rather than the daylength. We have just performed an experiment in which Starlings *Sturnus vulgaris* were kept on LD 8:16, LD 16:8, and a skeleton of LD $6+\frac{1}{2}$:17 to simulate a 16-h day. Food was available for only six hours during the main light period. Nevertheless, subjects on a full 16-h schedule and also those on a skeleton schedule ate more food than those on the 8-h day. The testes of both long-day and simulated long-day birds were recrudescing at the end of the experiment. It could not be argued that the quantity of food or feeding time was the main factor causing gonad growth, but it was evident that the photoperiod altered the feeding behaviour of the birds, at the same time stimulating gonad growth. Clearly the photoperiod acts to stimulate a co-adapted system, although light would appear to be the primary factor timing the onset of reproductive activity.

Some sterile arguments may we hope be avoided if it is appreciated that light acts to phase as well as induct. Problems are likely to arise in experimental design, for a change in daylength will also change the phasing of feeding and hunting rhythms. For this reason, it is not really fair, and indeed naive, to standardize the time at which food is presented during experiments.

Pittendrigh and Daan (1976*a, b, c*) and Daan and Pittendrigh (1976*a, b*) have published an excellent series of papers outlining their experimental evidence for believing that the biological clock is essentially a coupled oscillator system. The bimodal pattern of locomotor or diurnal activity, which is so evident in animals (see Figs. 6.6, 7.16), is explained by assuming the morning peak to depend on one oscillator (the morning oscillator) and the afternoon peak on a second, closely coupled, oscillator. (An equivalent system is postulated for nocturnal species.)

With our colleagues Stuart and Barbara Dobson we have kept Starlings on a range of different photoperiods and have evidence to suggest that with very long daylengths (LD 22:4 and longer) a phase-jump occurs so that the bird's daily start of activity is governed by the afternoon oscillator. In other words, instead of the daily activity rhythm being controlled first by the morning (M) and then by the afternoon (A) oscillators the sequence goes AM, AM, AM. . .and the total activity time becomes increased. Our object is to discover how the oscillatory mechanisms of time measurement are translated into hormone secretion and whether both oscillators are involved in stimulating neurosecretion in the same way.

Electrical stimulation of the median eminence shows that luteinizing hormone releasing factor (LHRH) is present at all times, whether subjects are kept on long or short days (Bicknell and Follett 1975). The normal lack of secretion under short photoperiods is apparently due to an absence of neural stimulation from the photo-receptors. In passing, it is worth noting that Bolton, Chadwick, Hall, and Scanes (1976) have confirmed that the hypothalamus primarily exerts a releaser control on prolactin secretion in birds. These authors also have some evidence that a dual control system may exist, as in mammals, for rat hypothalamic extract can inhibit chicken pituitary prolactin secretion if this is first stimulated by prolactin releasing factor, that is, bird pituitaries respond to inhibitory stimulation.

It has been emphasized (pp. 28 and 318) that, although daylength is measured via a circadian rhythm of photosensitivity, the actual secretion of luteinizing hormone in gonadectomized cockerels showed no indication of a circadian rhythm, but instead occurred in apparently random sequences of high and low output termed episodic release. Gledhill and Follett (1976) have now found exactly the same situation to pertain in Japanese Quail from which blood samples were taken every 3 h for a 27-h period; subjects were either held on short (LD 8:16) or long (LD 20:4) photoperiods. The same applied to FSH. We (B. and S. Dobson, Murton, and Westwood) have taken blood samples from Feral Pigeons at 1-h intervals for 26-h and also find no evidence for a rhythm of LH secretion, although circulating levels obviously vary with photo-period length. However, there is a diurnal rhythm of androgen secretion in pigeons and this appears to result from a diurnal rhythm in the circulating levels of thyroid hormones (thyroxine and triiodothironine). A better understanding of thyroid activity may help elucidate many of our current problems, in particular the vexing question of photorefractoriness. Our research is designed to discover whether the release of thyroid hormones depends on the same oscillator system as that involved in the production of gonadotrophin releasing factor and, if so, whether there is a similar or distinct quantitative response to alterations in the daylength. Thus, we should expect the daily release of gonadotrophin with change in length of photoperiod to follow the pattern seen in Fig. 9.9. It need not follow that the slope or range of response is the same for all pituitary hormones.

From the foregoing it may be inferred that the question of whether seasonal functions depend on the appropriate phasing of hormones (an internal coincidence system) is still unresolved. Suggestive experiments are still reported. For example, Dusseau and Bosscher (1976) have found that FSH injected into immature White Leghorn cockerels at the end of a 16-h photoperiod stimulated a significantly greater weight increase of the testes than the same dose given at 0 or 8-h. This diurnal rhythm of response did not occur in chicks receiving Metopirone, an inhibitor of adrenal steroid synthesis. The authors suggest that an appropriate phase relationship between corticosterone and FSH is needed for testis growth. It will be necessary to measure circulating plasma levels of several candidate hormones in the same subject at frequent intervals before this matter can be resolved.

It would be expected that hormone secretion under a skeleton schedule should be similar to that occurring with the complete photoperiod being simulated. Storey (1976) used a 6-h primary photoperiod coupled with 2-h pulses to simulate 8, 10, 12, 14, and 16-h photoperiods. Plasma LH and testosterone levels and testicular size were significantly raised in schedules simulating 12, 14, and 16-h daylength, but not in those simulating 8 or 10-h daylength. Storey compared his results with our earlier findings for the Greenfinch (Fig. 2.4) and noted that the Canary did not respond with a night interruption occurring $6 \cdot 7 - 7 \cdot 5$-h after dawn. We suspect that our results could be mis leading in this respect. With the insensitive radioimmunoassay techniques then available we had to sacrifice our subjects after 19 days of treatment and then pool the plasma for the whole experimental group. It is conceivable that by this time there had been more feedback inhibition of LH secretion in the groups experiencing the longer day simulations than in those on short-day simulations.

It is becoming clear that daylength changes modify the sensitivity of the hypothalamus to feed-back mechanisms, but at present the interpretation of experimental results can be equivocal. For instance, Hutchison (1976) noted that the amount of courtship in Barbary Doves which could be stimulated by intra-hypothalamic implants of testosterone propionate declined with increase in time between castration and implantation. His results suggest that brain mechanisms associated with the hormonal activation of male courtship became less sensitive to androgen with time after castration. If androgen sensitizes the steroid binding properties of cellular androgen receptors in the hypothalamus this effect could depend on the long-term absence of androgen. But implants are more effective in initiating courtship in long-term (90-day) castrates kept on LD 13:11 than in those kept on LD $8 \cdot 5:15 \cdot 5$, and so a direct effect of light on hypothalamic sensitivity could be involved, or a sensitizing effect from gonadotrophin

It is also well established that male and female birds differ in their sensitivity to feedback so that following exposure to long days higher levels of LH are found in intact male quail than in intact females (Gibson, Follett, and Gledhill 1975); the same applies in chickens (Sharp 1974). Davies (1976) has results which are consistent with this observation and suggests that the disparity may result from differences in the affinity of the male and female hypothalamic and pituitary tissues for sex steroids. Testosterone appears to exercise its negative feedback on the preoptic-anterior hypo-thalamic complex, whereas oestrogenic hormones in the laying female act mainly at the level of the pituitary.

In the Japanese Quail adequate photostimulation causes plasma LH concentrations to reach their maximum levels. On the other hand, in the Feral Pigeon a photoperiod of LD 16:8 causes only about a twofold increase in plasma LH compared with one of LD 8:16. Specific courtship behaviours result in a further fivefold increase in circulating LH levels compared with those of unpaired long-day levels (Dobson, Dobson, Murton, and Westwood in preparation). This implies that courtship causes a change in hypo-thalamic sensitivity. In passing, it is of interest to note that polychlorinated biphenyls (experimentally the commercial product Aroclor was administered), which are wide-spread environmental pollutants, have been found to raise hypothalamic sensitivity causing an elevation in LH and thyroxine levels above normal during the pre-incubation courtship cycle (Dobson et al. in preparation). This work has also confirmed that following pairing LH titres immediately increase to reach a peak within 5 days (to a level five times that of unpaired long-day controls), after which they decrease markedly to the time of egg-laying, that is, between nest-building and the first egg (see also Cheng and Follett 1976). At the same time as LH titres decrease, those of thyroxine steadily increase; that is, pairing stimulates an elevation of thyroid activity which reaches a noticeable peak at the time of egg-laying.

It seems possible that many low-latitude species will prove to resemble the pigeons in that courtship will cause a marked stimulation of gonadotrophin secretion compared with the direct effect of photoperiod. In highly photoperiodic temperate species the effect of the photoperiod may be more complete, as in quail. This does not imply that species are unresponsive to immediate environmental factors. Hinde and Steele (1976) found that the tape-recorded song of the male Canary affected the amount of oestrogen-induced nest-building of ovariectomized females on three different photo-periods (LD 8:16, LD 12:12, LD 14:10). Thus, while it is known that vocalizations can influence reproductive development, by gonadotrophin or gonadal hormone secretion, it seems that they may also directly affect the behavioural response to oestrogen. Similarly, use of skeleton simulations of long days in photo-refractory Canaries suggested that there was a light-induced change in sensitivity to the effect of oestrogen which did not implicate oestrogen directly (Steel and Hinde 1976). These effects were presumably dependent on a light-induced change in hypothalamic sensitivity.

Food supplies may exert an indirect effect on the timing of reproduction by influencing the time available for courtship. In females the final growth stage of the follicle is extremely sensitive to environmental food resources (see p. vii) and probably to the existence of a store of specific nutrients, which are difficult to obtain on a daily basis (Jones and Ward 1976): the male may do his best to help the female by variously relieving her from unnecessary tasks and by courtship feeding (p. 200). If females cannot find the nutrients to grow their follicles it may be supposed that oestrogen secretion is depressed. But males too may have their endocrine state affected by nutritional factors. Scanes, Harvey, and Chadwick (1976) have found that within 12 h of the withdrawal of food, plasma levels of LH declined by 60 per cent, and they continued to fall progressively over the period of fasting. In contrast, FSH levels did not decline until after 48 h of experimental treatment. Incidentally this observation suggests that there is some separation of the control of the secretion of the two gonadotrophins.

On p. 56 we suggest that in several avian species high LH titres might occur during the photorefractory phase. Yet, on p. 322 referring to the Canary and Table 12.1 (p. 325) to the Mallard, it was mentioned that LH titres were low during photo-refractoriness. In fact, measurement of LH concentrations during the annual cycle of wild Mallard drakes by Haase, Sharp, and Paulke (1975) show that LH levels increased during the spring to a maximum in April and May. In June, when testicular regression occurred, LH levels sharply decreased to the lowest level recorded but they became partially elevated in late autumn. This is seen in Fig. 12.7 (p. 320), which emphasizes that a different pattern was found in domestic strains of the Mallard (Pekin Ducks). It is known that variations may occur in the phase relationship of androgen and LH, as was discussed on p. 324 (see also p. 410), and it seems probable that under some conditions there may be phase alterations between LH secretion and the state of photorefractoriness. This would imply that photorefractoriness does not depend on a total cessation of gonadotrophin secretion, but that it is perhaps associated with some inhibitory condition − perhaps a high thyroid activity as discussed on p. 32?

Our final comment concerns the question of sexual selection raised on p. 417. Reasons for rejecting Zahavi's 'handicap-principle' were given. The real objection was in the use of the word 'handicap'. No such objection is necessary if there is a quanti-tative relationship between a character and hormone output and if hormone output is directly influenced by the physical condition of the animal, including its nutritional status. That is, a handicap is not inherited, but is simply the capability for a character to vary as a phenotypic response so as to be a measure of potential fitness. This amounts to simple sexual selection and in these terms is not a new idea. Nevertheless, we feel that this topic has received insufficient attention and that, instead, sexual selection is usually viewed as a run-away process leading to the emergence of bizarre characters in polygamous species. We wonder whether a host of apparently unim-portant morphological characters, for example, bright plumage, colours, and patterns in waterfowl (see p. 419), will prove to be controlled by genes having important pleiotropic effects in physiological terms. If this were the case females selecting males on the basis of specific plumage traits could select for an advantageous physiology. If ecological conditions favour early breeding it will be advantageous for a female Arctic Skua to prefer a dark-phase male (see p. 428), and in this context plumage characters are an indication of physiological capability.

Bibliography

Adam, J. H. and Dimond, S. J. (1971). Influence of light on the time of hatching in the domestic chick. *Anim. Behav.* **19**, 226–9.

Adams, J. L. (1956). A comparison of different methods of progesterone administration to the fowl in affecting egg production and molt. *Poult. Sci.* **35**, 323–6.

Adams, L. E. G. (1957). Nest records of the Swallow. *Bird Study* **4**, 28–33.

Ahrén, K. and Hamberger, L. (1969). FSH and ovarian metabolism. In *Progress in endocrinology. Proceedings of the Third International Congress of Endocrinology, Mexico, 1968* (ed. C. Gual and F. J. G. Ebling), pp. 75–82. Excerpta Medica, Amsterdam.

Aitken, R. N. C. (1971). The oviduct. In *Physiology and biochemistry of the domestic fowl* (ed. D. J. Bell and B. M. Freeman), Vol. 3, pp. 1237–89. Academic Press, London and New York.

Åkerman, B. (1966a). Behavioural effects of electrical stimulation in the forebrain of the pigeon. I. Reproductive behaviour. *Behaviour* **26**, 323–38.

–– (1966b). Behavioural effects of electrical stimulation in the forebrain of the pigeon. II. Protective behaviour. *Behaviour* **26**, 339–50.

Aldrich, J. W. and Kenard, P. B. (1970). Status and speciation in the Mexican Duck (*Anas diaza*). *Wilson Bull.* **82**, 63–73.

Allan, R. G. (1962). The Madeiran Storm Petrel *Oceanodroma castro*. *Ibis* **103b**, 274–95.

Amin, S. O. and Gilbert, A. B. (1970). Cellular changes in the anterior pituitary of the domestic fowl during growth, sexual maturity and laying. *Br. Poult. Sci.* **11**, 451–8.

Andrew, R. J. (1961). The displays given by passerines in courtship and reproductive fighting: a review. *Ibis* **103a**, 315–48.

–– and Rogers, L. J. (1972). Testosterone, search behaviour and persistence. *Nature, Lond.* **237**, 343–5.

Andrewartha, H. G. and Birch, L. C. (1954). *The distribution and abundance of animals.* University of Chicago Press, Illinois, U.S.A.

Annan, O. (1963). Experiments on photoperiodic regulation of the testis cycle in two species of the thrush genus *Hylocichla*. *Auk* **80**, 166–74.

Anthony, R. (1970). Ecology and reproduction of California Quail in southeastern Washington. *Condor* **72**, 276–87.

Armstrong, E. A. (1964). Polyandry. In *A new dictionary of birds* (ed. A. Landsborough Thomson), p. 655. Nelson, London.

Aron, C., Asch, G., and Roos, J. (1966). Triggering of ovulation by coitus in the rat. *Int. Rev. Cytol.* **20**, 139.

Aschoff, J. (1952). Frequenzaenderungen der Aktivitaetsperiodik bei Maeusen in Dauerlicht und Dauerdukel. *Pflügers Arch. ges. Physiol.* **255**, 197–203.

–– (1958). Tierische Periodik unter dem Einfluss von Zeitgebern. *Z. Tierpsychol.* **15**, 1–30.

–– (1959). Periodik licht- und dunkelaktiver Tiere unter konstanten Umgebungsbedingungen. *Pflügers Arch. ges. Physiol.* **270**, 9.

— — (1960). Exogenous and endogenous components in circadian rhythms. *Cold Spring Harb. Symp. quant. Biol.* **25**, 11—28.

— — (1965*a*). The phase-angle difference in circadian periodicity. In *Circadian clocks. Proceedings of the Feldafing Summer School* (ed. J. Aschoff), pp. 262—76. North-Holland, Amsterdam.

— — (1965*b*). Response curves in circadian periodicity. In *Circadian clocks. Proceeding. of the Feldafing Summer School* (ed. J. Aschoff), pp. 95—111. North-Holland, Amsterdam.

— — (1967). Circadian rhythm in birds. *Int. orn. Congr.* **14**, 81—105.

— — and Holst, D. V. (1960). Schlafplatzflüge bei Dohlen. *Int. orn. Congr.* **12**, 35—70.

— — and Pohl, H. (1970). Der Ruheumsatz von Vögeln als Funktion der Tageszeit und Körpergrösse. *J. Orn., Lpz.* **111**, 38—47.

— — and Wever, R. (1962*a*). Aktivitätsmenge und α: ρ-Verhältnis als Messgrössen der Tagesperiodik. *Z. vergl. Physiol.* **46**, 88—101.

— — and — — (1962*b*). Beginn und Ende der Aktivität freilebender Vögel. *J. Orn., Lpz.* **103**, 1—27.

— — and — — (1962*c*). Über Phasenbeziehungen zwischen biologischer Tagesperiodik und Zeitgeberperiodik. *Z. vergl. Physiol.* **46**, 115—28.

— — and — — (1966). Circadian period and phase-angle difference in Chaffinches (*Fringilla coelebs* L.). *Comp. Biochem. Physiol.* **18**, *397—404.*

— —, Saint Paul, U. V., and Wever, R. (1968). Circadiane Periodik von Finkenvögeln unter dem Einfluss eines selbstgewählten Licht-Dunkel-Wechsels. *Z. vergl. Physiol.* **58**, 304—21.

— —, Gerecke, U., Kureck, A., Pohl, H., Rieger, P., Saint Paul, U. V., and Wever, R. (1971). Interdependent parameters of circadian activity rhythms in birds and man. In *Biochronometry* (ed. M. Menaker), pp. 3—29. National Academy of Sciences, Washington D.C.

Ashmole, N. P. (1963*a*). The regulation of numbers of tropical oceanic birds. *Ibis* **103b**, 458—73.

— — (1963*b*). The biology of the Wideawake or Sooty Tern *Sterna fuscata* on Ascension Island. *Ibis* **103b**, 297—364.

— — (1963*c*). Molt and breeding in populations of the Sooty Tern *Sterna fuscata*. *Postilla* **76**, 1—18.

Assenmacher, I. (1974). External and internal components of the mechanism controlling reproductive cycles in drakes. In *Circannual clocks: annual biological rhythms* (ed. E. T. Pengelley), pp. 197—251. Academic Press, New York and London.

— — and Tixier-Vidal, A. (1965). Hypothalamic—pituitary relations. *Proceedings of the Second International Congress of Endocrinology, August 1964*, pp. 131—45. Excerpta Medica, Amsterdam.

Astier, H., Halberg, F., and Assenmacher, I. (1970). Rythmes circanniens de l'activité thyroïdienne chez le Canard Pékin. *J. Physiol., Paris* **62**, 219—30.

Atz, J. W. (1964). Intersexuality in fishes. In *Intersexuality in vertebrates including man* (ed. C. N. Armstrong and A. J. Marshall), pp. 145—232. Academic Press, London and New York.

Axelrod, J. (1969). Control of catecholamine metabolism. In *Progress in endocrinology. Proceedings of the Third International Congress of Endocrinology, Mexico, 1968* (ed. C. Gual and F. J. G. Ebling), pp. 286—92. Excerpta Medica, Amsterdam.

Baerends, G. P. and Van der Cingel, N. A. (1962). On the phylogenetic origin of the snap display in the Common Heron (*Ardea cinerea* L.). *Symp. zool. Soc. Lond.* **8**, 7—24.

Bagshawe, K. D., Orr, A. H., and Godden, J. (1968). Cross-reaction in radioimmuno-assay between human chorionic gonadotrophin and plasma from various species. *J. Endocr.* 42, 513–8.

——, Wilde, C. E., and Orr, A. H. (1966). Radioimmunoassay for human chorionic gonadotrophin and luteinizing hormone. *Lancet* (i), 1118–21.

Bailey, A. M., Niedrach, R. J., and Bailey, A. L. (1953). The Red Crossbills of Colorado. *Mus. Pictorial, Denver* no. 9, 1–64.

Bailey, R. E. (1950). Inhibition of light-induced gonad increase in White-crowned Sparrows. *Condor* 52, 247–51.

—— (1952). The incubation patch of passerine birds. *Condor* 54, 121–36.

—— (1953). Accessory reproduction organs of male fringillid birds. Seasonal variations and response to various sex hormones. *Anat. Rec.* 115, 1–20.

Baillie, A. H., Ferguson, M. M., and Hart, D. McK. (1966). *Developments in steroid histochemistry* Academic Press, London.

—— and Mack, W. S. (1966). Hydroxysteroid dehydrogenases in normal and abnormal human testes. *J. Endocr.* 35, 239–48.

Baker, J. R. (1929). *Man and animals in the New Hebrides.* Routledge, London.

—— (1938a). The relation between latitude and breeding seasons in birds. *Proc. zool. Soc. Lond.* 108, 557–82.

—— (1938b). The evolution of breeding seasons. In *Evolution: essays on aspects of evolutionary biology* (ed. G. R. de Beer), pp. 161–77. Oxford University Press, London.

—— and Baker, I. (1936). The seasons in a tropical rain-forest (New Hebrides) 2. Botany. *J. Linn. Soc. (Zool.)* 39, 507–19.

——, Marshall, A. J., and Harrison, T. H. (1940). The seasons in a tropical rain-forest (New Hebrides) 5. Birds (*Pachycephala*). *J. Linn. Soc. (Zool.)* 41, 50–70.

Baker, T. G. (1963). A quantitative and cytological study of germ cells in human ovaries. *Proc. R. Soc. B* 158, 417–33.

—— (1971). Electron microscopy of the primary and secondary oocyte. In *Advances in the biosciences,* Vol. 6: *Schering Symposium on Intrinsic and Extrinsic Factors in Early Mammalian Development, Venice 1970* (ed. G. Raspé). Pergamon Press, New York.

—— (1972). Oogenesis and ovulation. In *Reproduction in mammals* (ed. C. R. Austin and R. V. Short), Vol. 1, pp. 14–45. Cambridge University Press, London.

Baldwin, S. P. and Kendeigh, S. C. (1938). Variations in the weight of birds. *Auk* 55, 416–67.

Barfuss, D. W. and Ellis, L. C. (1971). Seasonal cycles in melatonin synthesis by the pineal gland as related to testicular function in the House Sparrow *Passer domesticus. Gen. comp. Endocrinol.* 17, 183–93.

Barry, J., Dubois, M. P., and Poulain, P. (1973). LRF producing cells of the mammalian hypothalamus. *Cell Tissue Res.* 146, 351–66.

Bastian, J. W. and Zarrow, M. X. (1955). A new hypothesis for the asynchronous ovulatory cycle of the domestic hen (*Gallus domesticus*). *Poult. Sci.* 34, 776–88.

Bates, R. M., Riddle, O., and Lahr, E. L. (1937). The mechanism of the anti-gonad action of prolactin in adult pigeons. *Am. J. Physiol.* 119, 610–14.

Baylé, J. D. and Assenmacher, I. (1967). Contrôle hypothalamo-hypophysaire du fonctionnement thyroidien chez la Caille. *C. r. hebd. Séanc. Acad. Sci., Paris* 264, 125–8.

——, Kraus, M., and van Tienhoven, A. (1970). The effects of hypophysectomy and testosterone propionate on the testes of Japanese Quail *Coturnix coturnix japonica. J. Endocr.* 46, 403–4.

Beason, R. C. and Franks, E. C. (1974). Breeding behaviour of the Horned Lark. *Auk* 91, 65–74.

Beer, J. R. (1961). Winter feeding patterns of the House Sparrow. *Auk* **78**, 63–71.

Bengtson, S-A. (1970). Breeding behaviour of the Purple Sandpiper *Calidris maritima* in West Spitsbergen. *Ornis Scandinavica* **1**, 17–25.

—— (1975). Timing of the moult of the Purple Sandpiper *Calidris maritima* in Spitsbergen. *Ibis* **117**, 100–2.

—— and Owen, D. F. (1973). Polymorphism in the Arctic Skua *Stercorarius parasiticus* in Iceland. *Ibis* **115**, 87–92.

Bennett, M. A. (1940). The social hierarchy in Ring Doves. II. The effect of treatment with testosterone propionate. *Ecology* **21**, 148–65.

Benoit, J. (1930*a*). Hypertrophie compensatrice du testicule chez le coq et le canard après orchidectomie unilatérale. *Int. Congr. Sex Res.* **2**, 150–61.

—— (1930*b*). Castration unilatérale et hypertrophie compensatrice du testicule restant chez le coq et le canard. *C. r. Séanc. Soc. Biol.* **108**, 883–6.

—— (1936). Stimulation par la lumière de l'activité sexuelle chez le Canard et la Cane domestiques. *Bull. Biol. Fr. Belg.* **70**, 487–533.

—— (1961). Opto-sexual reflex in the duck: physiological and histological aspects. *Yale J. Biol. Med.* **34**, 97–116.

—— (1964). The role of the eye and of the hypothalamus in the photostimulation of gonads in the duck. *Ann. N.Y. Acad. Sci.* **117**, 204–16.

—— (1970). Étude de l'action des radiations visibles sur la gonadostimulation, et de leur pénétration intra-crânienne chez les oiseaux et les mammifères. In *La photorégulation de la reproduction chez les oiseaux et les mammifères* (ed. J. Benoit and I. Assenmacher), pp. 121–46. Colloques int. Cent. natn. Rech. scient. no. 172.

——, Assenmacher, I., and Walter, F. X. (1950*a*). Résponses du mécanisme gonadostimulant a l'éclairment artificiel et de la préhypophyse aux castrations bilatérale et unilatéral, chez le canard domestique mâle, au cours de la période de régression testiculaire saisonnière. *C. r. Séanc. Soc. Biol.* **144**, 573–7.

——, Mandel, P., Walter, F. X., and Assenmacher, I. (1950*b*). Sensibilité testiculaire a l'action gonadotrope de l'hypophyse chez le Canard domestique au cours de la régression testiculaire saisonnière. *C. r. Séanc. Soc. Biol.* **144**, 1400–3.

——, Assenmacher, I., and Brard, E. (1956*a*). Étude de l'evolution testiculaire du Canard domestique soumis très jeune à un éclairement artificiel permanent pendant deux ans. *C. r. hebd. Séanc. Acad. Sci., Paris* **242**, 3113–5.

——, ——, and —— (1956*b*). Apparition et maintien de cycles sexuels non saisonniers chez le Canard domestique placé pendant plus de trois ans à l'obscurité totale. *J. Physiol., Paris* **48**, 388–91.

Benson, C. W. and Irwin, M. P. S. (1974). The significance of records of the Common Sandpiper breeding in east Africa. *Bull. Br. Orn. Club* **94**, 20–1.

Bent, A. C. (1925). Life histories of North American wild fowl. 2. *Bull. U.S. natn. Mus.* **130**.

Berger, W. (1967). Die Mauser des Sprossers (*Luscinia luscinia* L.). *J. Orn., Lpz.* **108**, 320–7.

Bern, H. A. and Nicoll, C. S. (1969). The taxonomic specificity of prolactins. In *Progress in endocrinology. Proceedings of the Third International Congress of Encodrinology, Mexico 1968* (ed. C. Gual and F. J. G. Ebling), pp. 433–9. Excerpta Medica, Amsterdam.

——, Nishioka, R. S., Mewaldt, L. R., and Farner, D. S. (1966). Photoperiodic and osmotic influences on the ultrastructure of the hypothalamic neurosecretory system of the White-crowned Sparrow *Zonotrichia leucophrys gambelii*. *Z. Zellforsch. mikrosk. Anat.* **69**, 198–227.

Berry, R. J. and Davis, P. E. (1970). Polymorphisms and behaviour in the Arctic Skua (*Stercorarius parasiticus* (L.)). *Proc. R. Soc. B.* **175**, 255–67.

Berthold, P. (1967). [On the gonadal development of the Starling (*Sturnus vulgaris*) and its dependence on migration.] *Experientia* **23**, 1–6.

—— (1969). Über Populationsunterschiede im Gonadenzyklus europäischer *Sturnis vulgaris, Fringilla coelebs, Erithacus rubecula* und *Phylloscopus collybita* und deren Ursachen. *Zool. Jb. (Systematik, Ökologie und Geographie der Tiere)* **96**, 491–557.

—— (1974). Circannual rhythms in birds with different migratory habits. In *Circannual clocks: annual biological rhythms* (ed. E. T. Pengelley), pp. 55–94. Academic Press, New York and London.

——, Gwinner, E., and Klein, H. (1972*a*). Circannuale Periodik bei Grasmücken. I. Periodik des Korpergewichtes, der Mauser und der Nachtunruhe bei *Sylvia atricapilla* und *S. borin* unter vershiedenen konstanten Bedingungen. *J. Orn., Lpz.* **113**, 170–90.

——, ——, and —— (1972*b*). Circannuale Periodik bei Gräsmucken. II. Periodik der Gonadengrose bei *Sylvia atricapilla* und *S. borin* unter verschiedenen konstanten Bedingungen. *J. Orn., Lpz.* **113**, 407–17.

Bezzel, E. and Schwarzenbach, F. H. (1968). Zur Variation der Eidimensionen bei Enten und ihrer biometrischen Auswertung. I and II. *Anz. orn. Ges. Bayern* **8**, 235–54.

Bibby, C. J. (1970). Post juvenile moult of the Tree Sparrow. *Wicken Fen Group Report* **2**, 10–20.

Biellier, H. V. and Ostmann, O. W. (1960). Effect of varying day-length of oviposition in domestic fowl. *Res. Bull. Mo. agric. exp. Stn* **747**.

Binkley, S., Kluth, E., and Menaker, M. (1971). Pineal function in sparrows: circadian rhythms and body temperature. *Science, N.Y.* **174**, 311–14.

Birch, L. C. (1971). The role of environmentäl heterogeneity and genetical heterogeneity in determining distribution and abundance. In *Dynamics of populations* (ed. P. J. den Boer and G. R. Gradwell), pp. 109–28. Centre for Agricultural Publishing and Documentation, Wageningen.

Bischoff, M. B. (1969). Photoreceptoral and secretory structures in the avian pineal organs. *J. Ultrastruct. Res.* **28**, 16–26.

Bishop, D. W. (1961). Biology of spermatozoa. In *Sex and internal secretions* (3rd edn.) (ed. W. C. Young) Vol. 2, pp. 707–96. Williams and Wilkins, Baltimore.

Bissonnette, T. H. (1930). Studies on the sexual cycle in birds. III. The normal regressive changes in the testes of the European Starling *Sturnus vulgaris* from May to November. *Am. J. Anat.* **46**, 477–96.

—— (1931). Studies on the sexual cycle in birds. IV. Experimental modification of the sexual cycle in males of the European Starling (*Sturnis vulgaris*) by changes in daily period of illumination and of muscular work. *J. exp. Zool.* **58**, 281–320.

—— (1941). The 'Mule' pheasant. *Proc. Am. phil. Soc.* **85**, 49–70.

—— and Wadlund, A. P. (1932). Duration of testis activity of *Sturnus vulgaris* in relation to type of illumination. *J. exp. Biol.* **9**, 339–50.

Blackmore, F. M. (1969). The effect of the temperature, photoperiod, and molt on the energy requirements of the House Sparrow, *Passer domesticus. Comp. Biochem. Physiol.* **30**, 433–44.

Blancas Sánchez, F. (1959). Comunidades y campos de vida de Acolla y sus alrededores. *Mems Mus. Hist. nat. 'Javier Prado'* no. 7.

Blanchard, B. D. (1941). The White-crowned Sparrow *Zonotrichia leucophrys* of the Pacific seaboard: environment and annual cycle. *Univ. California Publs Zool.* **46**, 1–178.

—— (1942). Migration in Pacific coast White-crowned Sparrows. *Auk* **59**, 47–63.

Blanco, A. and Zinkham, W. H. (1963). Lactate dehydrogenases in human testes. *Science, N.Y.* **139**, 601–2.

504   Bibliography

Blank, T. H., Southwood, T. R. E., and Cross, D. J. (1967). The ecology of the partridge. I. Outline of population processes with particular reference to chick mortality and nest density. *J. Anim. Ecol.* **36**, 549–56.

Blümcke, S. (1961). Vergleichend experimentell-morphologische Untersuchungen zur Frage einer retino-hypothalamischen Bahn bei Huhn, Meerschweinchen und Zatze. *Z. mikrosk.-anat. Forsch.* **67**, 469–513.

Bobr, L. W., Lorenz, F. W., and Ogasawara, F. X. (1964). Distribution of spermatozoa and fertility in domestic birds. I. Residence sites of spermatozoa in fowl oviducts. *J. Reprod. Fert.* **8**, 39–47.

Boetticher, H. von (1954). Die Taubengattung *Columba* L. *Zool. Anz.* **153**, 3–4; 49–64.

Bohnsack, B. (1968). Uber den Tagesrhythmus des Staren (*Sturnus vulgaris*) am Schlafplatz. *Oecologia* **1**, 369–76.

Bolen, E. G. (1973). Breeding whistling ducks *Dendrocygna* spp. in captivity. *Int. Zoo Yb.* **13**, 32–8.

Bolton, N., Chadwick, A., and Scanes, C. G. (1973). The effect of thyrotrophin releasing factor on the secretion of thyroid stimulating hormone and prolactin from the chicken anterior pituitary gland. *J. Physiol., Lond.* **238**, 78–9P.

Boss, W. R. (1943). Hormonal determination of adult characters and sex behaviour in Herring Gulls (*Larus argentatus*). *J. exp. Zool.* **94**, 181–209.

— and Witschi, E. (1947). The permanent effects of early stilbestrol injections on the sex organs of the Herring Gull (*Larus argentatus*). *J. exp. Zool.* **105**, 61–77.

Botkin, D. B. and Miller, R. S. (1974). Mortality rates and survival of birds. *Am. Nat.* **108**, 181–92.

Botte, V. (1963). da localizzazione della steroide-3β-olo-deidrogenasi nell 'ovaio di pollo. *Rend. Ist. sci. Univ. libera Camerino* **4**, 205–9.

Boucek, R. J., Györi, E., and Alvarez, R. (1966). Steroid dehydrogenase reactions in developing chick adrenal and gonadal tissues. *Gen. comp. Endocrinol.* **7**, 292–303.

Bouillé, C. and Baylé, J-D. (1973). Experimental studies on the adrenocorticotropic area in the pigeon hypothalamus. *Neuroendocrinology* **11**, 73–91.

Bourne, W. R. P. and Warham, J. (1966). Geographical variation in the giant petrels of the genus *Macronectes*. *Ardea* **54**, 45–67.

Boyd, H. (1957). Early sexual maturity of a female Mallard. *Br. Birds* **50**, 302–3.

— (1962a). Mortality and fertility of European Charadrii. *Ibis* **104**, 368–87.

— (1962b). Population dynamics and the exploitation of ducks and geese. In *The exploitation of natural animal populations* (ed. E. D. Le Cren and M. W. Holdgate), pp. 85–94. British Ecological Society Symposium no. 2. Blackwell, Oxford.

Brady, J. (1974). The physiology of insect circadian rhythms. In *Advances in insect physiology* (ed. J. E. Treherne, M. J. Berridge, and V. B. Wigglesworth), pp. 1–115 Academic Press, London and New York.

Braithwaite, L. W. (1969). Testis cycles of a native duck. *J. Reprod. Fert.* **19**, 390–1.

— (1970). The Black Swan. *Aust. nat. Hist.* **16**, 375–9.

— and Frith, H. J. (1969). Waterfowl in an inland swamp in New South Wales III. Breeding. *C.S.I.R.O. Wildl. Res.* **14**, 65–109.

Brambell, F. W. R. and Marrian, G. F. (1929). Sex reversal in a pigeon (*Columba livia*). *Proc. R. Soc. B* **104**, 459–70.

Brant, J. W. A. and Nalbandov, A. V. (1956). Role of sex hormones in albumen secretion by the oviduct of chickens. *Poult. Sci.* **35**, 692–700.

Braun, C. E. (1973). Distribution and habitats of Band-tailed Pigeons in Colorado. *Proc. Western Assoc. State Game Fish Commnrs*, **53**, 336–44.

Braun, T. (1974). Evidence of multiple, cell specific, distinctive adenylate cytcase systems in rat testis. In *Hormone binding and target cell activation in the testis* (Current topics in molecular endocrinology, Vol. 1) (ed. M. L. Dufau and A. R. Means), pp. 243–64. Plenum Press, New York and London.

Breneman, W. R. (1955). Reproduction in birds: the female. *Mem. Soc. Endocr.* 4, 94–114.

—— (1956). Steroid hormones and the development of the reproductive system in the pullet. *Endocrinology* 58, 262–71.

——, Zeller, F. J., and Creek, R. O. (1962). Radioactive phosphorus uptake by chick testes as an end-point for gonadotropin assay. *Endocrinology* 71, 790–8.

Brisbin, I. L. (1969). Bioenergetics of the breeding cycle of the Ring Dove. *Auk* 86, 54–74.

Brockway, B. F. (1965). Stimulation of ovarian development and egg laying by male courtship vocalization in Budgerigars (*Melopsittacus undulatus*). *Anim. Behav.* 13, 575–8.

—— (1969). Hormonal and experiential factors influencing the nestbox oriented behaviour of Budgerigars (*Melopsittacus undulatus*). *Behaviour* 35, 1–26.

Brodkorb, P. (1971a). Origin and evolution of birds. In *Avian biology* (ed. D. S. Farner and J. R. King) Vol. 1, pp. 19–55. Academic Press, New York and London.

—— (1971b). Catalogue of fossil birds: part 4 (Columbiformes through Piciformes). *Bull. Fla St. Mus. biol. Sci.* 15, 163–266.

Brody, S. (1945). *Bioenergetics and growth.* Reinhold, New York.

Brosset, A. (1973). Evolution des *Accipiter* forestiers de l'Est du Gabon. *Alauda* 41, 185–201.

Brown, J. L. (1969). The buffer effect and productivity in tit populations. *Am. Nat.* 103, 347–54.

—— (1970). Cooperative breeding and altruistic behaviour in the Mexican Jay, *Aphelocoma ultramarina. Anim. Behav.* 18, 366–78.

—— (1975). Helpers among Arabian Babblers *Turdoides squamiceps. Ibis* 117, 243–4.

Brown, N. L., Baylé, J.-D., Scanes, C. G., and Follett, B. K. (1975). Chicken gonadotrophins: their effects on the testes of immature and hypophysectomized Japanese quail. *Cell Tissue Res.* 156, 499–520.

Brown, W. O. and Badman, H. G. (1965). The effect of gonadal hormones on the water-soluble proteins of the oviduct of the normal and folic acid-deficient chick. *Poult. Sci.* 44, 206–10.

Bruder, R. H. and Lehrman, D. S. (1967). Role of the mate in the elicitation of hormone-induced incubation behavior in the Ring Dove. *J. comp. physiol. Psychol.* 63, 382–4.

Bryant, E. H. (1974). On the adaptive significance of enzyme polymorphisms in relation to environmental variability. *Am. Nat.* 108, 1–19.

Bull, P. C. (1943). Notes on the breeding cycle of the thrush and blackbird in New Zealand. *Emu* 43, 198–208.

—— (1957). Distribution and abundance of the Rook (*Corvus frugilegus* L.) in New Zealand. *Notornis* 7, 137–61.

Bullock, D. W. and Nalbandov, A. V. (1967). Hormonal control of the hen's ovulation cycle. *J. Endocr.* 38, 407–15.

Bullough, W. S. (1942). The reproductive cycles of the British and Continental races of the Starling. *Phil. Trans. R. Soc. B* 231, 165–246.

Bünning, E. (1960). Circadian rhythms and time-measurement in photoperiodism. *Cold Spring Harb. Symp. quant. Biol.* 25, 249–56.

Bunting, A. H. (1960). Some reflections on the ecology of weeds. In *The biology of weeds* (ed. J. L. Harper), pp. 11–26. British Ecological Society Symposium no. 1 (1959). Blackwell, Oxford.

Burbridge, N. T. (1960). The phytogeography of the Australian region. *Aust. J. Bot.* **8**, 75–211.

Burger, J. W. (1944). Testicular response to androgen in the light-stimulated Starling. *Endocrinology* **35**, 182–6.

—— (1947). On the relations of day length to the phases of testicular involution and inactivity of the spermatogenetic cycle of the Starling. *J. exp. Zool.* **105**, 259–68.

—— (1948). The relation of external temperature to spermatogenesis in the male Starling. *J. exp. Zool.* **109**, 259–66.

—— (1949). A review of experimental investigations on seasonal reproduction in birds. *Wilson Bull.* **61**, 211–30.

—— (1953). The effect of photic and psychic stimuli on the reproductive cycle of the male Starling *Sturnus vulgaris. J. exp. Zool.* **124**, 227–39.

Burger, R. E. and Lorenz, F. W. (1962). Cloacal weight stimulation by estrogen in chicks. *Endocrinology* **71**, 669–70.

Burgos, M. H. and Vitale-Calpe, R. (1969). Gonadotrophic control of spermiation. In *Progress in endocrinology. Proceedings of the Third International Congress of Endocrinology, Mexico 1968* (ed. C. Gual and F. J. G. Ebling), pp. 1030–7. Excerpta Medica, Amsterdam.

Burns, J. M. (1969). Luteinizing hormone bioassay based on uptake of radio-active phosphorus by the chick testes. *Comp. Biochem. Physiol.* **34**, 727–31.

Burns, R. K. (1961). Role of hormones in the differentiation of sex. In *Sex and internal secretions* (3rd edn) (ed. W. C. Young), Vol. 1, pp. 76–158. Williams and Wilkins, Baltimore.

Butt, W. R. (1967). Chemical properties of the gonadotrophins. In *The chemistry of the gonadotrophins*, Chapter IV, pp. 58–74. Thomas, Springfield, Illinois.

Butterfield, P. A. and Crook, J. H. (1968). The annual cycle of nest building and agonistic behaviour in captive *Quelea quelea* with reference to endocrine factors. *Anim. Behav.* **16**, 308–17.

Byerly, T. C. and Moore, O. K. (1941). Clutch length in relation to period of illumination in the domestic fowl. *Poult. Sci.* **20**, 387–90.

Cade, T. J. (1960). Ecology of the Peregrine and Gyrfalcon populations in Alaska. *Univ. California Publs Zool.* **63**, 151–290.

Cain, B. W. (1973). Effect of temperature on energy requirements and northward distribution of the Black-bellied Tree Duck. *Wilson Bull.* **85**, 308–17.

Cairns, J. (1956). The Malayan Great Tit. *J. Bombay nat. Hist. Soc.* **53**, 367–73.

Cardinali, D. P., Cuello, A. E., Tramezzani, J. H., and Rosner, J. M. (1971). Effects of pinealectomy on the testicular function of the adult male duck. *Endocrinology* **89**, 1082–93.

Caridroit, F. (1938). Recherches expérimentales sur les rapports entre testicules, plumage d'éclipse et mues chez le Canard sauvage. *Trav. Stn zool. Wimereux* **13**, 47–67.

Carnaby, I. C. (1954). Nesting seasons of Western Australian birds. *West. Aust. Nat.* **4**, 149–56.

Carrick, R. (1972). Population ecology of the Australian Black-backed Magpie, Royal Penguin, and Silver Gull. *Wildl. Res. Rep. (U.S.)* **2**, 41–99.

Casement, M. B. (1966). Migration across the Mediterranean observed by radar. *Ibis* **108**, 461–91.

Catchpole, C. K. (1973). The functions of advertising song in the Sedge Warbler (*Acrocephalus schoenobaenus*) and the Reed Warbler (*A. scirpaceus*). *Behaviour* **46**, 300–20.

Catt, K. J., Tsuruhara, C., Mendelson, C., Ketelslegers, J.-M., and Dufau, M. L. (1974). Gonadotropin binding and activation of the interstitial cells of the testis. In *Hormone binding and target cell activation in the testis* (Current topics in molecular endocrinology, Vol. 1) (ed. M. L. Dufau and A. R. Means), pp. 1–30. Plenum Press, New York and London.

Cavé, A. J. (1968). The breeding of the Kestrel, *Falco tinnunculus* L. in the reclaimed area Oostelijk Flevoland. *Neth. J. Zool.* 18, 313–407.

Celis, M. E. and Taleisnik, S. (1971a). Formation of a melanocyte-stimulating hormone-release inhibiting factor by hypothalamic extracts from rats. *Int. J. Neurosci.* 1, 223–30.

—— and —— (1971b). In vitro formation of a MSH-releasing agent by hypothalamic extracts. *Experientia* 27, 1481–2.

—— and —— (1973). Interaction between paraventricular nucleus extracts and median eminence extracts on the formation of melanocyte stimulating hormone-release-inhibiting factor. *Life Sci.* 13, 493–9.

Champy, C. and Colle, P. (1919). Sur une corrélation entre la glande du jabot du pigeon et les glandes génitales. *C. r, Séanc. Soc. Biol.* 82, 818–19.

Chan, K. M. B. and Lofts,.B. (1974). The testicular cycle and androgen biosynthesis in the Tree Sparrow *Passer montanus saturatus. J. Zool.* 172, 47–66.

Chance, E. P. (1940). *The truth about the cuckoo.* Country Life, London.

Chandola, A. (1972). Thyroid in reproduction. Reproductive physiology of *Lonchura punctulata* in relation to iodine metabolism and hypothyroidism. Ph.D. Thesis no. 7774, Banaras Hindu University, Banaras (India).

Chapin, J. P. (1954). The calendar of Wideawake Fair. *Auk* 71, 1–15.

—— and Wing, L. W. (1959). The wideawake calendar 1953–1958. *Auk* 76, 153–8.

Chapman, F. A. (1940). The post-glacial history of *Zonotrichia capensis. Bull. Am. Mus. nat. Hist.* 77, 381–438.

Chapman, J. A., Henny, C. J., and Wight, H. M. (1969). The status, population dynamics, and harvest of the Dusky Canada Goose. *Wildl. Monogr., Chestertown* 18, 4–48.

Charnov, E. L. (1973). Optimal foraging: some theoretical explorations. Ph.D. thesis, University of Washington.

Charnov, E. L. and Krebs, J. R. (1974). On clutch size and fitness. *Ibis* 116, 217–19.

Chen, C. L., Bixler, E. J., Weber, A. I., and Meites, J. (1968). Hypothalamic stimulation of prolactin release from the pituitary of Turkey hens and poults. *Gen. Comp. Endocrinol.* 11, 489–94.

Cheng, M.-F. and Silver, R. (1975). Estrogen–progesterone regulation of nest-building and incubation behavior in ovariectomized Ring Doves (*Streptopelia risoria*). *J. comp. physiol. Psychol.* 88, 256–63.

Chieffi, G. and Botte, V. (1965). The distribution of some enzymes involved in the steroidogenesis of hen's ovary. *Experientia* 21, 16–20.

—— and —— (1970). The problems of 'luteogenesis' in non-mammalian vertebrates. *Boll. Zool.* 37, 85–102.

Chitty, D. (1967). The natural selection of self-regulatory behaviour in animal populations. *Proc. ecol. Soc. Aust.* 2, 51–78.

Christian, J. J. (1971). Population density and reproductive efficiency. *Biol. Reprod.* 4, 248–94.

—— and Davis, D. E. (1964). Endocrines, behaviour and population. *Science, N.Y.* 146, 1550–60.

——, Lloyd, J. A., and Davis, D. E. (1965). The role of endocrines in the self-regulation of mammalian populations. *Recent Prog. Horm. Res.* 21, 501–77.

Chu, J. P. (1940). The effects of estrone and testosterone and of pituitary extracts on the gonads of hypophysectomised pigeons. *J. Endocr.* **2**, 21–37.

— and You, S. S. (1946). Gonad stimulation by androgens in hypophysectomised pigeons. *J. Endocr.* **4**, 431–5.

Clapham, C. S. (1964). The birds of the Dahlac archipelago. *Ibis* **106**, 376–88.

Clermont, Y. (1958). Structure de l'épithélium séminal et mode de renouvellement des spermatogonies chez le canard. *Archs Anat. microsc. Morph. exp.* **47**, 47–66.

— (1967). Cinétique de la spermatogenèse chez les mammifères. *Archs Anat. microsc. Morph. exp.* **56** (3–4), Suppl.: 7–60.

Cloudsley-Thompson, J. L. and Chadwick, M. J. (1964). *Life in deserts.* Dufour Editions, Philadelphia, Pennsylvania.

Cody, M. L. (1966). A general theory of clutch size. *Evolution, Lancaster, Pa* **20**, 174–84.

— (1971). Ecological aspects of reproduction. In *Avian biology* (ed D. S. Farner and J. R. King), Vol. 1, pp. 461–512. Academic Press, New York and London.

Cole, L. J. (1933). The relation of light periodicity to the reproductive cycle, migration, and distribution of the Mourning Dove (*Zenaidura macroura carolinensis*). *Auk* **50**, 284–96.

Comfort, A. (1962). Survival curves of some birds in the London Zoo. *Ibis* **104**, 115–

Conder, P. J. (1948). The breeding biology and behaviour of the Continental Goldfinch *Carduelis carduelis carduelis. Ibis* **90**, 493–325.

Conroy, J. W. H. (1971). The white-phase Giant-petrel of the South Orkney Islands. *Bull. Br. antarct. Surv.* no. 24, 113–15.

— (1972). Ecological aspects of the biology of the Giant Petrel, *Macronectes giganteus* (Gmelin), in the maritime Antarctic. *Sci. Rep. Br. antarct. Surv.* no. 75.

Cooch, F. G. (1963). Recent changes in distribution of colour phases of *Chen c. caerulescens. Int. orn. Congr.* **13**, 1182–94.

Cooke, A. S. (1973). Shell thinning in avian eggs by environmental pollutants. *Environ. Pollut.* **4**, 85–152.

— (1975). Pesticides and egg shell formation. *Symp. zool. Soc. Lond.* **35**, 339–61.

Cooke, F. and Cooch, F. G. (1968). The genetics of polymorphism in the goose *Anser caerulescens. Evolution, Lancaster, Pa* **22**, 289–300.

— and Ryder, J. P. (1971). The genetics of polymorphism in Ross' Goose (*Anser rossii*). *Evolution, Lancaster, Pa* **25**, 483–90.

—, Mirsky, P. J., and Seiger, M. B. (1972). Colour preferences in the Lesser Snow Goose and their possible role in mate selection. *Can. J. Zool.* **50**, 529–36.

Coolsma, J. W. Th. and Gruber, M. (1968). RNA synthesis in rooster liver. *Biochim. Biophys. Acta* **169**, 306–15.

Coombs, C. J. F. (1961). Rookeries and roosts of the Rook and Jackdaw in south-west Cornwall. II. Roosting. *Bird Study* **8**, 55–70.

— and Marshall, A. J. (1956). The effects of hypophysectomy on the internal testis rhythm in birds and mammals. *J. Endocr.* **13**, 107.

Corbin, K. W. (1967). Evolutionary relationships in the avian genus *Columba* as indicated by ovalbumin tryptic peptides. *Evolution, Lancaster, Pa* **21**, 355–68.

— (1968). Taxonomic relationships of some *Columba* species. *Condor* **70**, 1–13.

Corrodi, H. and Jonsson, G. (1967). The formaldehyde fluorescence method for the histochemical demonstration of biogenic monoamines. A review on the methodology. *J. Histochem. Cytochem.* **15**, 65–78.

Coulson, J. C. (1956). Mortality and egg production of the Meadow Pipit with special reference to altitude. *Bird Study* **3**, 119–32.

— (1960). A study of the mortality of the Starling based on ringing recoveries. *J. Anim. Ecol.* **29**, 251–71.

—— (1963). Egg size and shape in the Kittiwake (*Rissa tridactyla*) and their use in estimating age composition of populations. *Proc. zool. Soc. Lond.* **140**, 211–27.

—— (1971). Competition for breeding sites causing segregation and reduced young production in colonial animals. In *Dynamics of populations,* (ed. P. J. den Boer and G. R. Gradwell) pp. 257–68. Centre for Agricultural Publishing and Documentation, Wageningen.

—— (1972). The significance of the pair-bond in the Kittiwake. *Int. orn. Congr.* **15**, 424–33.

—— and White, E. (1960). The effect of age and density of breeding birds on the time of breeding in the Kittiwake *Rissa tridactyla*. *Ibis* **102**, 71–86.

—— and —— (1961). An analysis of the factors influencing the clutch size of the Kittiwake. *Proc. zool. Soc. Lond.* **136**, 207–17.

—— and Wooller, R. D. (1976). Differential survival rates among breeding Kittiwake Gulls *Rissa tridactyla* (L.). *J. Anim. Ecol.* **45**, 205–13.

——, Potts, G. R., and Horobin, J. (1969). Variation in the eggs of the Shag *Phalacrocorax aristotelis*. *Auk* **86**, 232–45.

Cowan, J. B. (1952). Life history and productivity of a population of Western Mourning Doves in California. *Calif. Fish Game* **38**, 505–21.

Cracraft, J. (1973). Continental drift, paleoclimatology, and the evolution and biogeography of birds. *J. Zool.* **169**, 455–545.

—— (1976). Avian evolution on southern continents: influences of paleogeography and paleoclimatology. *Int. orn. Congr.* **16**. (In press)

Cramp, S. (1955). The breeding of the Willow Warbler. *Bird Study* **2**, 121–35.

—— (1972). The breeding of urban Wood Pigeons. *Ibis* **114**, 163–71.

Crook, J. H. (1960). Studies on the social behaviour of *Quelea q. quelea* (Linn.) in French West Africa. *Behaviour* **16**, 1–55.

—— (1962). The adaptive significance of pair formation types in weaver birds. *Symp. zool. Soc. Lond.* **8**, 57–70.

—— and Butterfield, P. A. (1968). Effect of testosterone propionate and luteinizing hormone on agonistic and nest building behaviour of *Quelea quelea*. *Anim. Behav.* **16**, 370–84.

—— and —— (1970). Gender role in the social system of *Quelea*. In *Social behaviour in birds and mammals: essays on the social ethology of animals and man,* (ed. J. H. Crook) pp. 211–48. Academic Press, London.

—— and Ward, P. (1968). The *Quelea* problem in Africa. In *The problems of birds as pests* (ed. R. K. Murton and E. N. Wright), pp. 211–29. Symposia of the Institute of Biology no. 17. Academic Press, London and New York.

Cross, B. A. (1972). The hypothalamus. In *Reproduction in mammals,* Vol. 3: *Hormones in reproduction* (ed. C. R. Austin and R. V. Short), pp. 29–41. Cambridge University Press, London.

Croxall, J. P. (1976). The composition and behaviour of some mixed-species bird flocks in Sarawak. *Ibis* **118**, 333–46.

Croze, H. (1970). Searching image in Carrion Crows. Hunting strategy in a predator and some antipredator devices in camouflaged prey. *Z. Tierpsychol.* **5**, 1–86.

Cullen, J. M. (1960). Some adaptations in the nesting behaviour of terns. *Int. orn. Congr.* **12**, 153–7.

—— and Ashmole, N. P. (1963). The Black Noddy *Anous tenuirostris* on Ascension Island. 2. Behaviour. *Ibis* **103b**, 423–46.

Cumley, R. W. and Irwin, M. R. (1944). The correlation between antigenic composition and geographical range in the Old and New World of some species of *Columba*. *Am. Nat.* **68**, 238–56.

Cunningham, F. J. and Furr, B. J. A. (1972). Plasma levels of luteinizing hormone and progesterone during the ovulatory cycle of the hen. In *Egg formation and production* (ed. B. M. Freeman and P. E. Lake), pp. 51–64. British Poultry Science, Edinburgh.

——, Myres, R. P., and McNeilly, J. R. (1970). Immunological studies on chicken gonadotrophins. *J. Reprod. Fert.* **23**, 538–9.

Cusick, E. K. and Wilson, F. E. (1972). On control of spontaneous testicular regression in Tree Sparrows (*Spizella arborea*). *Gen. comp. Endocrinol.* **19**, 441–56.

Custer, T. W. and Pitelka, F. A. (1972). Time-activity patterns and energy budget of nesting Lapland Longspurs near Barrow, Alaska. *Proceedings of the 1972 Tundra Biome Symp.*, pp. 160–4. (Mimeo)

Daan, S. (1976). Light intensity and the timing of daily activity of finches (Fringillidae). *Ibis* **118**, 223–36.

—— and Aschoff, J. (1975). Circadian rhythms of locomotor activity in captive birds and mammals: their variations with season and latitude. *Oecologia* **18**, 269–316.

Dahl, E. (1970*a*). Studies of the fine structure of ovarian interstitial tissue. 2. The ultrastructure of the thecal gland of the domestic fowl. *Z. Zellforsch. mikrosk. Anat.* **109**, 195–211.

—— (1970*b*). Studies of the fine structure of ovarian interstitial tissue. 3. The innervation of the thecal gland of the domestic fowl. *Z. Zellforsch. mikrosk. Anat.* **109**, 212–26.

—— (1970*c*). Studies of the fine structure of ovarian interstitial tissue. 6. Effects of clomiphene on the thecal gland of the domestic fowl. *Z. Zellforsch. mikrosk. Anat.* **109**, 227–44.

—— (1971*a*). Studies of the fine structure of ovarian interstitial tissue. 1. A comparative study of the fine structure of the ovarian interstitial tissue in the rat and the domestic fowl. *J. Anat.* **108**, 275–90.

—— (1971*b*). Studies of the fine structure of ovarian interstitial tissue. 4. Effects of steroids on the thecal gland of the domestic fowl. *Z. Zellforsch. mikrosk. Anat.* **113**, 111–32.

—— (1971*c*). Studies of the fine structure of ovarian interstitial tissue. 5. Effects of gonadotropins on the thecal gland of the domestic fowl. *Z. Zellforsch. mikrosk. Anat.* **113**, 133–56.

Dahlström, A. and Fuxe, K. (1964). Evidence for the existence of monoamine-containing neurons in the central nervous system. 1. Demonstration of monoamines in the cell bodies of brain stem neurons. *Acta physiol. scand.* **62**, suppl. 232, 1–55.

Damsté, P. H. (1947). Experimental modification of the sexual cycle of the Greenfinc *J. exp. Biol.* **24**, 20–35.

Darlington, P. J. (1957). *Zoogeography: the geographical distribution of animals.* Wiley, New York.

Das, B. C. and Nalbandov, A. V. (1955). Responses of ovaries of immature chickens to avian and mammalian gonadotrophins. *Endocrinology* **57**, 705–10.

Davidson, J. M., Feldman, S., Smith, E. R., and Weick, R. F. (1969). Localisation of steroid feed back receptors. In *Progress in endocrinology. Proceedings of the Third International Congress of Endocrinology, Mexico 1968* (ed. C. Gual and F. J. G. Ebling), pp. 542–7. Excerpta Medica, Amsterdam.

Davies, C., Fischer, H., and Gwinner, E. (1969). Die Brutzeiten einiger Gänsearten und ihrer Bastarde in identischen Bedingungen. *Oecologia* **3**, 266–76.

Davies, D. T. and Follett, B. K. (1974*a*). The effect of intraventricular administration of 6-hydroxydopamine on photo-induced testicular growth in Japanese Quail. *J. Endocr.* **60**, 277–83.

—— and —— (1974*b*). Changes in plasma luteinizing hormone in the Japanese Quail after electrical stimulation of the hypothalamus. *J. Endocr.* **63**, 31–2.

— and — (1975*a*). The neuroendocrine control of gonadotrophin release in the Japanese Quail. I. The role of the tuberal hypothalamus. *Proc. R. Soc. B.* **191**, 285–301.

— and — (1975*b*). The neuroendocrine control of gonadotrophin release in the Japanese Quail. II. The role of the anterior hypothalamus. *Proc. R. Soc. B.* **191**, 303–15.

Davies, S. J. J. F. (1970). Patterns of inheritance in the bowing display and associated behaviour of some hybrid *Streptopelia* Doves. *Behaviour* **36**, 187–214.

— (1974*a*). Studies of the three coo-calls of the male Barbary Doves. *Emu* **74**, 18–26.

— (1974*b*). The breeding season of captive Barbary Doves, *Streptopelia risoria* at Helena Valley, Western Australia. *Aust. Wildl. Res.* **1**, 85–8.

Davis, D. E. (1942). The number of eggs laid by cowbirds. *Condor* **44**, 10–12.

— (1957). Aggressive behaviour in castrated Starlings. *Science, N.Y.* **126**, 253.

— (1959). Territorial rank in Starlings. *Anim. Behav.* **7**, 214–21.

— (1963). Hormonal control of aggressive behaviour. *Int. orn. Congr.* **13**, 994–1003.

Davis, J. (1971). Breeding and molt schedules of the Rufous-collared Sparrow in coastal Perú. *Condor* **73**, 127–46.

Davis, J. C. and Schuetz, A. W. (1975). Follicle stimulating hormone enhances attachment of rat testis cells in culture. *Nature, Lond.* **254**, 611–12.

Davis, J. W. F. (1975). Specialization in feeding location by Herring Gulls. *J. Anim. Ecol.* **44**, 795–804.

— and O'Donald, P. (1976*a*). Sexual selection for a handicap; a critical analysis of Zahavi's model. *J. Theor. Biol.* **57**, 345–54.

— and — (1976*b*). Territory size, breeding time and mating preference in the Arctic Skua. *Nature, Lond.* **260**, 774–5.

Davis, P. (1972). Early, yes, but good. *BTO News* **53**, 6.

Dawkins, M. (1971*a*). Perceptual changes in chicks: another look at the 'search image' concept. *Anim. Behav.* **19**, 566–74.

— (1971*b*). Shifts of 'attention' in chicks during feeding. *Anim. Behav.* **19**, 575–82.

Dawson, W. R. (1976). Physiological and behavioural adjustments of birds to heat and aridity. *Int. orn. Congr.* **16** 455–67.

— and Bartholomew, G. A. (1968). Temperature regulation and water economy of desert birds. In *Desert biology* (ed. G. W. Brown), Vol. 1, pp. 357–94. Academic Press, New York and London.

— and Fisher, C. D. (1969). Responses to temperature by the Spotted Nightjar. *Condor* **71**, 49–53.

Delius, J. D. (1963). Das Verhalten der Feldlerche. *Z. Tierpsychol.* **20**, 297–348.

— (1965). A population study of Skylarks *Alauda arvensis. Ibis* **107**, 466–92.

Dementiev, G. P. and Gladkov, N. A. (1952). [*The birds of the Soviet Union.*] Soviet Academy of Sciences, Moscow.

Dempster, J. P. (1975). *Animal population ecology.* Academic Press, London and New York.

Deol, G. S. (1955). Studies on structure and function of the ovary of the domestic fowl. Ph.D. Thesis, Edinburgh University.

Desborough, S. and Irwin, M. R. (1966). Additional variation in serum proteins in Columbidae. *Physiol. Zoöl.* **39**, 66–9.

Desforges, M. F. (1972). Observations on the influence of social displays on ovarian development in captive Mallards *Anas platyrhynchos. Ibis* **114**, 256–7.

Desjardins, C. and Hafs, H. D. (1964). Immunochemical similarity of luteinising hormones. *J. Anim. Sci.* **23**, 903–4.

—, Zelznik, A. J., Midgley, A. R., and Reichert, L. E. (1974). *In vitro* binding and autoradiographic localization of human chorionic gonadotropin and follicle stimulating hormone in rat testes during development. In *Hormone binding and target cell activation in the testis* (Current topics in molecular endocrinology, Vol. 1) (ed. M. L. Dufau and A. R. Means), pp. 221–35. Plenum Press, New York and London.

Diamond, A. W. (1972). Sexual dimorphism in breeding cycles and unequal sex ratio in Magnificent Frigate-birds. *Ibis* **114**, 395–8.

—— (1974). Annual cycles in Jamaican forest birds. *J. Zool.* **173**, 277–301.

—— (1976). Subannual breeding and moult cycles in the Bridled Tern *Sterna anaethetus* in the Seychelles. *Ibis* **118**, 414–19.

Dilger, W. C. (1960). The comparative ethology of the African parrot genus *Agapornis*. *Z. Tierpsychol.* **17**, 649–85.

Dirschl, H. J. (1969). Foods of Lesser Scaup and Blue-winged Teal in the Saskatchewan river delta. *J. Wildl. Mgmt* **33**, 77–87.

Disney, H. J. de S. and Marshall, A. J. (1956). A contribution to the breeding biology of the Weaver Finch *Quelea quelea* (Linnaeus) in East Africa. *Proc. zool. Soc. Lond.* **127**, 379–87.

——, Lofts, B., and Marshall, A. J. (1959). Duration of the regeneration period of the internal reproductive rhythm in a xerophilous equatorial bird, *Quelea quelea*. *Nature, Lond.* **184**, 1659–60.

——, ——, and —— (1961). An experimental study of the internal rhythm of reproduction in the Red-billed Dioch *Quelea quelea* by means of photostimulation, with a note on melanism induced in captivity. *Proc. zool. Soc. Lond.* **136**, 123–9.

Dodd, J. M., Follett, B. K., and Sharp, P. J. (1971). Hypothalamic control of endocrine function in non-mammalian vertebrates. In *Advances in comparative physiology and biochemistry* (ed. O. Lowenstein), Vol. 4, pp. 114–220. Academic Press, London.

Dol'nik, T. V. (1973). [Diurnal and seasonal cycles of the blood sugar in sedentary and migrating birds.] *Zool. Zh.* **52**, 94–103.

Dol'nik, V. R. (1963). A quantitative study of vernal testicular growth in several species of finches (Fringillidae). *Dokl. Akad. Nauk SSSR* **149**, 370–2.

—— (1970). Fat metabolism and bird migration. In *La photorégulation de la reproduction chez les Oiseaux et les Mammifères* (ed. J. Benoit and I. Assenmacher) pp. 351–63. Colloques int. Cent. natn. Rech. scient. no. 172.

—— and Blyumental, T. I. (1967). Autumnal premigratory and migratory periods in the Chaffinch (*Fringilla coelebs coelebs*) and some other temperate-zone passerine birds. *Condor* **69**, 435–68.

Domenech, E. M. de., Domenech, C. E., Aoki, A., and Blanco, A. (1972). Association of the testicular lactate dehydrogenase isozyme with a special type of mitochondria *Biol. Reprod.* **6**, 136–47.

Dominic, C. J. and Singh, R. M. (1969). Anterior and posterior groups of portal vessels in the avian pituitary. *Gen. comp. Endocrinol.* **13**, 22–6.

Domm, L. V. (1939). Modifications in sex and secondary sexual characters in birds. In *Sex and internal secretions* (2nd edn) (ed. E. Allen), pp. 227–327. Williams and Wilkins, Baltimore, Maryland.

Donham, R. S. and Wilson, F. E. (1970). Photorefractoriness in pinealectomized Harris' Sparrows. *Condor* **72**, 101–2.

Dorward, D. F. (1962). Comparative biology of the White Booby and the Brown Booby *Sula* spp. at Ascension. *Ibis* **103b**, 174–220.

—— (1963). The Fairy Tern *Gygis alba* on Ascension Island. *Ibis* **103b**, 365–78.

Drent, R. (1972). Adaptive aspects of the physiology of incubation. *Int. orn. Congr.* **15**, 255–80.

Dubois, R. (1965). La lignée germinale chez les reptiles et les oiseaux. *Année biol.* **4**, 637–66.

Dunham, H. H. and Riddle, O. (1942). Effects of a series of steroids on ovulation and reproduction in pigeons. *Physiol. Zoöl.* 15, 383–94.

Dunmore, R. (1968). Plumage polymorphism in a feral population of the Rock Pigeon. *Am. Midl. Nat.* 79, 1–7.

—— and Davis, D. E. (1963). Reproductive condition of feral pigeons in winter. *Auk* 8, 374.

Dunnet, G. M. (1955). The breeding of the Starling *Sturnus vulgaris* in relation to its food supply. *Ibis* 97, 619–62.

——, Anderson, A., and Cormack, R. M. (1963). A study of survival of adult Fulmars with observations on the pre-laying exodus. *Br. Birds* 56, 2–18.

Dusseau, J. W. and Meier, A. H. (1971). Diurnal and seasonal variations of plasma adrenal steroid hormone in the White-throated Sparrow, *Zonotrichia albicollis*. *Gen. comp. Endocrinol.* 16, 399–408.

Einarsen, A. S. (1945). Some factors affecting Ring-necked Pheasant population density. *Murrelet* 26, 39–44.

Eisner, E. A. (1960). The relationship of hormones to the reproductive behaviour of birds, referring especially to parental behaviour; a review. *Anim. Behav.* 8, 155–79.

Elder, W. H. (1958). A report on the Nene. Internal report to the Board of Agriculture and Forestry.

Elgood, J. H., Fry, C. H., and Dowsett, R. J. (1973). African migrants in Nigeria. *Ibis* 115, 1–45, 375–411.

Engelmann, C. [1962]. *Ernährung und Fütterung des Gerflügels.* Neumann-Verlag, [Radebeul, Germany].

Engels, W. L. (1959). The influence of different day lengths on the testes of a trans-equatorial migrant, the Bobolink (*Dolichonyx oryzivorus*). In *Photoperiodism and related phenomena in plants and animals* (ed. R. B. Withrow), pp. 759–66. American Association for the Advancement of Science Publ. no. 55. Washington D.C.

—— (1961). Photoperiodism and the annual testicular cycle of the Bobolink (*Dolichonyx oryzivorus*), a trans-equatorial migrant, as compared with two temperate zone migrants. *Biol. Bull. mar. biol. Lab., Woods Hole* 120, 140–7.

—— (1962). Day-length and the termination of photorefractoriness in the annual testicular cycle of the trans-equatorial migrant *Dolichonyx* (The Bobolink). *Biol. Bull. mar. biol. Lab., Woods Hole* 123, 94–104.

—— and Jenner, C. E. (1956). The effect of temperature on testicular recrudescence in juncos at different photoperiods. *Biol. Bull. mar. biol. Lab., Woods Hole* 110, 129–37.

Enright, J. T. (1965). Synchronization and ranges of entrainment. In *Circadian clocks: Proceedings of the Feldafing Summer School* (ed. J. Aschoff), pp. 112–24. North-Holland, Amsterdam.

—— (1971). Heavy water slows biological timing processes. *Z. vergl. Physiol.* 72, 1–16.

Epple, A., Orians, G. H., Farner, D. S., and Lewis, R. A. (1972). The photoperiodic testicular response of a tropical finch, *Zonotrichia capensis costaricensis*. *Condor* 74, 1–4.

Erickson, C. J. (1970). Induction of ovarian activity in female Ring Doves by androgen treatment of castrated males. *J. comp. physiol. Psychol.* 71, 210–15.

—— and Lehrman, D. S. (1964). Effect of castration of male Ring Doves upon ovarian activity of females. *J. comp. physiol. Psychol.* 58, 164–6.

—— and Morris, R. L. (1972). Effects of mate familiarity on the courtship and reproductive success of the Ring Dove (*Streptopelia risoria*). *Anim. Behav.* 20, 341–4.

——, Bruder, R. H., Komisaruk, B. R., and Lehrman, D. S. (1967). Selective inhibition by progesterone of androgen-induced behaviour in male Ring Doves (*Streptopelia risoria*). *Endocrinology* **81**, 39–45.

Erpino, M. J. (1969). Seasonal cycle of reproductive physiology in the Black-billed Magpie. *Condor* **71**, 267–79.

Erskine, A. J. (1971). Growth, and annual cycle in weights, plumages and reproductive organs of Goosanders in eastern Canada. *Ibis* **113**, 42–58.

—— (1972). Buffleheads. *Can. Wildl. Serv., Monogr. Ser.* no. 4.

Eskin, A. (1971). Some properties of the system controlling the circadian activity rhythm of sparrows. In *Biochronometry* (ed. M. Menaker), pp. 55–79. National Academy of Sciences, Washington D.C.

Etkin, W. (1967). Relation of the pars intermedia to the hypothalamus. In *Neuroendocrinology* (ed. L. Martini and W. F. Ganong), Vol. 2, pp. 261–82. Academic Press, New York and London.

Evans, P. R. (1966). Autumn movements, moult and measurements of the Lesser Redpoll *Carduelis flammea cabaret*. *Ibis* **108**, 183–216.

—— (1969). Ecological aspects of migration, and pre-migratory fat deposition in the Lesser Redpoll *Carduelis flammea cabaret*. *Condor* **71**, 316–30.

Fabricius, E. and Jansson, A-M. (1963). Laboratory observations on the reproductive behaviour of the pigeon (*Columba livia*) during the pre-incubation phase of the breeding cycle. *Anim. Behav.* **11**, 534–47.

Falck, B. and Owman, C. (1965). A detailed methodological description of the fluorescence method for the cellular demonstration of biogenic monoamines. *Acta Univ. lund.* Sect. 2., **7**, 1–23.

——, Hillarp, N-Å., Thieme, G., and Torp, A. (1962). Fluorescence of catecholamines and related compounds condensed with formaldehyde. *J. Histochem. Cytochem.* **10**, 348–54.

Farner, D. S. (1959). Photoperiodic control of annual gonadal cycles in birds. In *Photoperiodism and related phenomena in plants and animals,* (ed. R. B. Withrow), pp. 717–50. American Association for the Advancement of Science, Publ. no. 55. Washington D.C.

—— (1960). Metabolic adaptations in migration. *Int. orn. Congr.* **12**, 197–208.

—— (1964). The photoperiodic control of reproductive cycles in birds. *Am. Scient.* **52**, 137–56.

—— (1965). Circadian systems in the photoperiodic responses of vertebrates. In *Circadian clocks: Proceedings of the Feldafing Summer School* (ed. J. Aschoff), pp. 357–67. North-Holland, Amsterdam.

—— and Lewis, R. A. (1971). Photoperiodism and reproductive cycles in birds. In *Photophysiology* (ed. A. C. Giese), Vol. 6, pp. 325–64. Academic Press, London.

—— and —— (1973). Field and experimental studies of the annual cycles of White-crowned Sparrows. *J. Reprod. Fert.* Suppl. 19, 35–50.

—— and Mewaldt, L. R. (1952). The relative roles of photoperiod and temperature in gonadal recrudescence in male *Zonotrichia leucophrys gambelii*. *Anat. Rec.* **113**, 612.

—— and —— (1955). The natural termination of the refractory periods in the White-crowned Sparrow. *Condor* **57**, 112–16.

—— and Wilson, A. C. (1957). A quantitative examination of testicular growth in the White-crowned Sparrow. *Biol. Bull. mar. biol. Lab., Woods Hole* **113**, 254–67.

——, Oksche, A., Kamemoto, F. I., King, J. R., and Cheyney, H. E. (1961). A comparison of the effect of long daily photoperiods on the pattern of energy storage in migratory and non-migratory finches. *Comp. Biochem. Physiol.* **2**, 125–42.

——, Kobayashi, H., Oksche, A., and Kawashima, S. (1964). Proteinase and acid-phosphatase activities in relation to the function of the hypothalamo-hypophysial neurosecretory systems of photostimulated and of dehydrated White-crowned

Sparrows. In *Progress in brain research* (ed. W. Bargmann and J. P. Schadé), Vol. 5, pp. 147–56. Elsevier, Amsterdam.

——, Morton, M. L., and Follett, B. K. (1968). The limitation of rate of photoperiodically induced testicular growth in the White-crowned Sparrow, *Zonotrichia leucophrys gambelii*. The effect of hemicastration. *Archs Anat. Histol. Embryol.* **51**, 190–6.

Fawcett, D. W. (1970). A comparative view of sperm ultrastructure. *Biol. Reprod.*, suppl. 2, 90–127.

—— and Phillips, D. M. (1970). Recent observations on the ultrastructure and development of the mammalian spermatozoon. In *Comparative spermatology* (ed. B. Baccetti), pp. 13–28. Academic Press, New York and London.

Feare, C. J. (1976). The breeding of the Sooty Tern *Sterna fuscata* in the Seychelles and the effects of experimental removal of its eggs. *J. Zool.* **179**, 317–60.

——, Dunnet, G. M., and Patterson, I. J. (1974). Ecological studies of the Rook (*Corvus frugilegus* L.) in north-east Scotland: food intake and feeding behaviour. *J. appl. Ecol.* **11**, 867–96.

Ferdinand, L. (1966). Display of the Great Snipe (*Gallinago media* Latham). *Dansk. orn. Foren. Tidsskr.* **60**, 14–34.

Ferguson, A. (1971). Geographic and species variation in transferrin and ovotransferrin polymorphism in the Columbidae. *Comp. Biochem. Physiol.* **38B**, 477–86.

Ferrando, G. and Nalbandov, A. V. (1969). Direct effect on the ovary of the adrenergic blocking drug dibenzyline. *Endocrinology* **85**, 38–42.

Fevold, H. R. and Pfeiffer, E. W. (1968). Androgen production *in vitro* by phalarope gonadal tissue homogenates. *Gen. comp. Endocrinol.* **10**, 26–33.

Fisher, J. (1953). The Collared Turtle Dove in Europe. *Br. Birds* **46**, 153–81.

Fisher, R. A. (1928). *Genetical theory of natural selection*. Oxford University Press, London and New York. (Reprinted by Dover, New York (1958)).

—— (1930). *The genetical theory of natural selection*. Clarendon Press, Oxford.

Fleay, D. H. (1937). Nesting habits of the Brush-turkey. *Emu* **36**, 153–63.

Fleet, R. (1974). The Red-tailed Tropicbird on Kure Atoll. *Orn. Monogr., Am. Orn. Un.* no. 16.

Flegg, J. J. M. and Cox, C. J. (1969). The moult of British Blue Tit and Great Tit populations. *Bird Study* **16**, 147–57.

Flickinger, G. L. (1961). Effect of grouping on adrenals and gonads of chickens. *Gen. comp. Endocrinol.* **1**, 332–40.

Fogden, M. P. L. (1968). Some aspects of the ecology of bird populations in Sarawak. D.Phil. Thesis, Oxford University.

—— (1972). The seasonality and population dynamics of equatorial forest birds in Sarawak. *Ibis* **114**, 307–43.

Follett, B. K. (1970). Gonadotrophin-releasing activity in the quail hypothalamus. *Gen. comp. Endocrinol.* **15**, 165–79.

—— (1973*a*). The neuroendocrine regulation of gonadotrophin secretion in avian reproduction. In *Breeding biology of birds* (ed. D. S. Farner), pp. 209–43. National Academy of Sciences, Washington D.C.

—— (1973*b*). Circadian rhythms and photoperiodic time measurement in birds. *J. Reprod. Fert.* Suppl. 19, 5–18.

—— (1975). Follicle-stimulating hormone in the Japanese Quail: variation in plasma levels during photoperiodically induced testicular growth and maturation. *J. Endocr.* **67**, 19P–20P.

—— (1976). Plasma follicle-stimulating hormone during photoperiodically induced sexual maturation in male Japanese Quail. *J. Endocr.* **69**, 117–26.

—— and Davies, D. T. (1975). Photoperiodicity and the neuroendocrine control of reproduction in birds. *Symp. zool. Soc. Lond.* **35**, 199–224.

516   Bibliography

—— and Sharp, P. J. (1968). Adrenergic and cholinergic systems in the hypothalamus of the Japanese Quail (*Coturnix coturnix japonica*). *Archs Anat. Histol. Embryol.* **51**, 213–22.

—— and —— (1969). Circadian rhythmicity in photoperiodically induced gonadotrophin release and gonadal growth in the quail. *Nature, Lond.* **223**, 968–71.

——, Kobayashi, H., and Farner, D. S. (1966). The distribution of monoamine oxidase and acetylcholinesterase in the hypothalamus and its relation to the hypothalamo-hypophysial neurosecretory system in the White-crowned Sparrow, *Zonotrichia leucophrys gambelii*. *Z. Zellforsch. mikrosk. Anat.* **75**, 57–65.

——, Scanes, C. G., and Cunningham, F. J. (1971). A radioimmunoassay for avian luteinizing hormone. *J. Endocr.* **51**, *v–vi*.

——, ——, and —— (1972*a*). A radioimmunoassay for avian luteinizing hormone. *J. Endocr.* **52**, 359–78.

——, ——, and Nicholls, T. J. (1972*b*). The chemistry and physiology of the avian gonadotrophins. In *Hormones glycoproteiques hypophysaires*, pp. 193–211. I.N.S.E.R.M., Paris.

——, Hinde, R. A., Steel, E., and Nicholls, T. J. (1973). The influence of photoperiod on nest building, ovarian development and luteinizing hormone in Canaries (*Serinus canarius*). *J. Endocr.* **59**, 151–62.

——, Mattocks, P. W., and Farner, D. S. (1974). Circadian function in the photoperiodic induction of gonadotropin secretion in the White-crowned Sparrow, *Zonotrichia leucophrys gambelii*. *Proc. natn. Acad. Sci. U.S.A.* **71**, 1666–9.

Foster, W. H. (1968). The effect of light-dark cycles of abnormal lengths upon egg production. *Br. Poult. Sci.* **9**, 273–84.

Franchimont, P., Chari, S., Hagelstein, M. T., and Duraiswami, S. (1975). Existence of a follicle-stimulating hormone inhibiting factor 'inhibin' in bull seminal plasma. *Nature, Lond.* **257**, 402–4.

Fraps, R. M. (1955). Egg production and fertility in poultry. In *Progress in the physiology of farm animals* (ed. J. Hammond), Vol. 2, pp. 661–710. Butterworths, London.

—— (1961). Ovulation in the domestic fowl. In *Control of ovulation* (ed. C. A. Villee), pp. 133–62. Pergamon Press, Oxford.

—— (1965). Twenty-four hour periodicity in the mechanism of pituitary gonadotrophin release for follicle maturation and ovulation in the chicken. *Endocrinology* **77**, 8–18.

—— (1970). Photoregulation in the ovulation cycle of the domestic fowl. In *La photorégulation de la reproduction chez les Oiseaux et les Mammifères* (ed. J. Benoit and I. Assenmacher), pp. 281–306. Colloques int. Cent. natn. Rech. scient. no. 172.

Fraschini, F. (1969). The pineal gland and the control of LH and FSH secretion. In *Progress in endocrinology. Proceedings of the Third International Congress of Endocrinology, Mexico 1968* (ed. C. Gual and F. J. G. Ebling), pp. 637–44. Excerpta Medica, Amsterdam.

Frelinger, J. A. (1972). The maintenance of transferrin polymorphism in pigeons. *Proc. natn. Acad. Sci. U.S.A.* **69**, 326–9.

—— and Crow, J. F. (1973). Transferrin polymorphism and Hardy-Weinberg ratios. *Am. Nat.* **107**, 314–17.

French, F. S. and Ritzén, E. M. (1973*a*). High affinity androgen binding protein in rat testis: evidence for secretion in efferent duct fluid and adsorption by epididymis. *Endocrinology* **93**, 88–95.

—— and —— (1973*b*). Androgen-binding protein in efferent duct fluid of rat testis. *J. Reprod. Fert.* **32**, 479–83.

——, McLean, W. S., Smith, A. A., Tindall, D. J., Weddington, S. C., Petrusz, P. P.,

Sar, M., Stumpf, W. E., Nayfeh, S. N., Hansson, V., Trygstad, O., and Ritzen, E. M. (1974). Androgen transport and receptor mechanism in testis and epididymis. In *Hormone binding and target cell activation in the testis* (Current topics in molecular endocrinology, Vol. 1) (ed. M. L. Dufau and A. R. Means), pp. 265–85. Plenum Press, New York and London.

Frenzel, B. and Troll, C. (1952). Die Vegetationszonen des nordlichen Eurasian während der letzten Eiszeit. *Eiszeitalter Gegenw.* 2, 154–67.

Fretwell, S. D. (1969). On territorial behaviour and other factors influencing habitat distribution in birds. III. Breeding success in a local population of Field Sparrows (*Spizella pusilla* Wils.). *Acta biotheor.* 19, 47–52.

— and Lucas, H. L. (1969). On territorial behaviour and other factors influencing habitat distribution in birds. I. Theoretical development. *Acta biotheor.* 19, 16–36.

Fridriksson, S. (1960). Eggjahvítumagn og Lostaetni Túngrasa. *Rit LandbDield, B* 12, 1–26.

Friedman, M. and Lehrman, D. S. (1968). Physiological conditions for the stimulation of prolactin secretion by external stimuli in the male Ring Dove. *Anim. Behav.* 16, 233–7.

Friedmann, H. (1971). Further information on the host relations of the parasitic cowbirds. *Auk* 88, 239–55.

Frith, H. J. (1959*a*). Ecology of wild ducks in inland Australia. *Monogr. Biol.* 8, 383–95.

— (1959*b*). The ecology of wild ducks in inland New South Wales. IV. Breeding. *C.S.I.R.O. Wildl. Res.* 4, 156–81.

— (1962). Movements of the Grey Teal, *Anas gibberifrons* Müller (Anatidae). *C.S.I.R.O. Wildl. Res.* 7, 50–70.

— (1964). The downy young of the Freckled Duck *Stictonetta naevosa. Emu* 64, 42–7.

— (1965). Ecology of the Freckled Duck, *Stictonetta naevosa* (Gould). *C.S.I.R.O. Wildl. Res.* 10, 125–39.

— and Davies, S. J. J. F. (1961). Ecology of the Magpie Goose, *Anseranas semipalmata* Latham (Anatidae). *C.S.I.R.O. Wildl. Res.* 6, 91–141.

—, Braithwaite, L. W. and McKean, J. L. (1969). Waterfowl in an inland swamp in New South Wales. II. Food. *C.S.I.R.O. Wildl. Res.* 14, 17–64.

—, —, and Wolfe, T. O. (1974). Sexual cycles of pigeons in a tropical environment. *Aust. Wildl. Res.* 1, 117–28.

Furr, B. J. A. (1969). Identification of steroids in the ovaries and plasma of laying hens and the site of production of progesterone in the ovary. *Gen. comp. Endocrinol.* 13, Abstr. 56.

— and Cunningham, F. J. (1970). The biological assay of chicken pituitary gonadotrophins. *Br. Poult. Sci.* 11, 7–13.

— and Pope, G. S. (1970). Identification of cholesterol, 7-oxocholesterol pregnenolone, progesterone, 20-hydroxypregn-4-en-3-one epimers and 5β-androstane-3, 17-dione in plasma and ovarian tissue of the domestic fowl. *Steroids* 16, 471–85.

—, Bonney, R. C., England, R. J., and Cunningham, F. J. (1973). Luteinizing hormone and progesterone in peripheral blood during the ovulatory cycle of the hen *Gallus domesticus. J. Endocr.* 57, 159–69.

Fuxe, K. and Hökfelt, T. (1967). The influence of central catecholamine neurons on the hormone secretion from the anterior and posterior pituitary. In *Neurosecretion* (ed. F. Stutinsky), pp. 165–77. Springer, Berlin.

— and Ljunggren, L. (1965). Cellular localization of monoamines in the upper brain stem of the pigeon. *J. comp. Neurol.* 125, 355–82.

Gallien, L. (1962). Comparative activity of sexual steroids and genetic constitution in sexual differentiation of amphibian embryos. *Gen. comp. Endocrinol.* Suppl. 1, 346–55.

Garnier, D. H., Tixier-Vidal, A., Gourdji, D., and Picart, R. (1973). Ultrastructure des cellules de Leydig et des cellules de Sertoli au cours du cycle testiculaire du Canard Pekin. *Z. Zellforsch. mikrosk. Anat.* **144**, 369–94.

Gaston, A. J. (1973). The ecology and behaviour of the Long-tailed Tit. *Ibis* **115**, 330–51.

Gaston, S. (1971). The influence of the pineal organ on the circadian activity rhythm in birds. In *Biochronometry* (ed. M. Menaker), pp. 541–8. National Academy of Sciences, Washington D.C.

— and Menaker, M. (1968). Pineal function: the biological clock in the sparrow? *Science, N.Y.* **160**, 1125–7.

Gavrilov, E. I. (1963). The biology of the eastern Spanish Sparrow, *Passer hispaniolensis transcaspicus* Tschusi, in Kazakhstan. *J. Bombay nat. Hist. Soc.* **60**, 301–17.

Gibb, J. A. (1962). L.Tinbergen's hypothesis of the role of specific search images. *Ibis* **104**, 106–11.

Gibson, E. (1920). Further ornithological notes from the neighbourhood of Cape San Antonio, Buenos Ayres. 3. Phoenicopteridae–Rheidae. *Ibis* 11th Ser., 2, 1–97.

Gilbert, A. B. (1967). Formation of the egg in the domestic chicken. In *Advances in reproductive physiology* (ed. A. McLaren), Vol. 2, pp. 111–79. Academic Press, London and New York.

— (1971a). The female reproductive effort. In *Physiology and biochemistry of the domestic fowl* (ed. D. J. Bell and B. M. Freeman), Vol. 3, pp. 1153–62. Academic Press, London and New York.

— (1971b). Control of ovulation. In *Physiology and biochemistry of the domestic fowl* (ed. D. J. Bell and B. M. Freeman), Vol. 3, pp. 1225–35. Academic Press, London and New York.

— (1971c). Egg albumen and its formation. In *Physiology and biochemistry of the domestic fowl* (ed. D. J. Bell and B. M. Freeman), Vol. 3, pp. 1291–1329. Academic Press, London and New York.

— (1971d). Transport of the egg through the oviduct and oviposition. In *Physiology and biochemistry of the domestic fowl* (ed. D. J. Bell and B. M. Freeman), Vol. 3, pp. 1345–52. Academic Press, London and New York.

— (1971e). The ovary. In *Physiology and biochemistry of the domestic fowl* (ed. D. J. Bell and B. M. Freeman), Vol. 3, pp. 1163–1208. Academic Press, London and New York.

— and Amin, S. O. (1969). Preliminary studies on the functional cytology of the anterior pituitary of the hen with light and electron microscope. *J. Reprod. Fert.* **20**, 363–4.

— and Lake, P. E. (1963). The effect of oxytocin and vasopressin on oviposition in the domestic hen. *J. Physiol., Lond.* **169**, 52–3P.

— and Wood-Gush, D. G. M. (1965). The control of the nesting behaviour of the domestic hen. III. The effect of cocaine in the post-ovulatory follicle. *Anim. Behav.* **13**, 284–5.

— and — (1968). Control of the nesting behaviour of the domestic hen. IV. Studies on the pre-ovulatory follicle. *Anim. Behav.* **16**, 168–70.

— and — (1971). Ovulatory and ovipository cycles. In *Physiology and biochemistry of the domestic fowl* (ed. D. J. Bell and B. M. Freeman), Vol. 3, pp. 1353–78. Academic Press, London and New York.

Gilliard, E. T. (1962). On the breeding behavior of the Cock-of-the-Rock (Aves, *Rupicola rupicola*). *Bull. Am. Mus. nat. Hist.* **124**, 31—68.

Gladkov, N. A. (1938). [Notes sur le faune ornitholigique des terrains cultivés du Turkestan.] *Byull. mosk. Obshch. Ispyt. Prir.* **47**, 360—73.

Gladwin, T. W. (1969*a*). Post-nuptial wing-moult in the Garden Warbler. *Bird Study* **16**, 131—2.

—— (1969*b*). Weights, foods and survival of Blackcaps and Chiffchaffs in the British Isles in winter. *Bird Study* **16**, 133.

Glenner, G. G., Burtner, H. J., and Brown, G. W. (1957). The histochemical demonstration of monoamine oxidase activity by tetrazolium salts. *J. Histochem. Cytochem.* **5**, 591—600.

Glenny, F. H. and Amadon, D. (1955). Remarks on the pigeon, *Otidiphaps nobilis* Gould. *Auk* **72**, 199—203.

Godden, P. M. M., Scanes, C. G., and Sharp, P. J. (1975). Variations in the levels of follicle-stimulating hormone in the circulation of birds, as determined by homologous radioimmunoassay. *J. Endocr.* **67**, 20P—21P.

Goforth, W. and Baskett, T. (1971). Social organisation of penned Mourning Doves. *Auk* **88**, 528—42.

Gogan, F., Kordon, C., and Benoit, J. (1963). Retentissement de lésions de l'eminence médiane sur la gonadostimulation du canard. *C. r. Séanc. Soc. Biol.* **157**, 2133—6.

Goodridge, A. G. and Ball, E. G. (1967*a*). The effect of prolactin on lipogenesis in the pigeon. *In vivo* studies. *Biochemistry, N.Y.* **6**, 1676—82.

—— and —— (1967*b*). The effect of prolactin on lipogenesis in the pigeon. *In vitro* studies. *Biochemistry, N.Y.* **6**, 2335—43.

Goodwin, D. (1953). Observations on voice and behaviour of the Red-legged Partridge *Alectoris rufa*. *Ibis* **95**, 581—614.

—— (1960). Sexual dimorphism in pigeons. *Bull. Br. Orn. Club* **80**, 42—55.

—— (1966). The bowing display of pigeons in reference to phylogeny. *Auk* **83**, 117—23.

—— (1967). *Pigeons and doves of the world*. British Museum (Natural History), London.

—— (1970). *Pigeons and doves of the world* (2nd edn). British Museum (Natural History), London.

Gorbman, A., Kobayashi, H., and Uemura, H. (1963). The vascularization of the hypophysial structures of the Hagfish. *Gen. comp. Endocrinol.* **3**, 505—14.

Goss, R. J. (1969*a*). Photoperiodic control of antler cycles in deer. I. Phase shift and frequency changes. *J. exp. Zool.* **170**, 311—24.

—— (1969*b*). Photoperiodic control of antler cycles in deer. II. Alterations in amplitude. *J. exp. Zool.* **171**, 233—4.

—— and Rosen, J. K. (1973). The effect of latitude and photoperiod on the growth of antlers. *J. Reprod. Fert.* Suppl. 19, 111—18.

——, Dinsmore, C. E., Grimes, L. N., and Rosen, J. K. (1974). Expression and suppression of the circannual antler growth cycle in deer. In *Circannual clocks: annual biological rhythms* (ed. E. T. Pengelley), pp. 393—422. Academic Press, New York and London.

Goss-Custard, J. D. (1970). The responses of Redshank (*Tringa totanus* L.) to spatial variations in their prey density. *J. Anim. Ecol.* **39**, 91—113.

Gourdji, D. (1965). Modifications des types cellulaires hypophysaires impliqués dans le cycle sexuel annuel chez l'Ignicolore mâle. *Gen. comp. Endocrinol.* **5**, 682.

—— (1967). Étude du déterminisme des variations du contenu hypophysaire en prolactine chez le Canard Pékin. Influence de la lumière permanente, de la castration et de la testostérone et de leurs interactions. *C. r. hebd. Séanc. Acad. Sci. Paris D* **264**, 1482—4.

── (1970). Prolactine et relations photosexuelles chez les oiseaux. *Colloques int. Cent. natn. Rech. Scient.* **172**, 233–58.

Graber, J. W. and Nalbandov, A. V. (1965). Neurosecretion in the White Leghorn cockerel. *Gen. comp. Endocrinol.* **5**, 485–92.

──, Frankel, A. I., and Nalbandov, A. V. (1967). Hypothalamic center influencing the release of LH in the cockerel. *Gen. comp. Endocrinol.* **9**, 187–92.

Gramet, P. (1970). Le parasitisme des Corvidés par le Coucou-Geai (*Clamator glandarius*). Interprétation éthologique du déterminisme des phénomènes naturels. *Rev. Comportement anim.* **4**, 17–26.

Graul, W. D. (1974). Adaptive aspects of the Mountain Plover social system. *Living Bird* **12**, 69–94.

── (1975). Breeding biology of the Mountain Plover. *Wilson Bull.* **87**, 6–31.

Greeley, F. and Meyer, R. K. (1953). Seasonal variation in testis-stimulating activity of male pheasant pituitary glands. *Auk* **70**, 350–8.

Greij, E. D. (1973). Effects of sex hormones on plumages of the Blue-winged Teal. *Auk* **90**, 533–51.

Grimes, L. W. (1976). Co-operative breeding in African birds.

Griswold, J. A. (1973). The Coscoroba *Coscoroba coscoroba*. *Int. Zoo Yb.* **13**, 38–40.

Gromadzki, M. (1966). Variability of egg-size of some species of the forest birds. *Ekol. pol.* Ser. A **14**, 101–9.

Gronau, J. and Schmidt-Koenig, K. (1970). Annual fluctuation in pigeon homing. *Nature, Lond.* **226**, 87–8.

Gulden, N. A. and Johnson, L. L. (1968). History, behaviour and management of a flock of Giant Canada Geese in southeastern Minnesota. In *A Symposium on Canada Goose management*, ed. R. L. Hine and C. Schoenfeld, pp. 59–71. Dembar Educational Services, Madison, Wisconsin.

Gwinner, E. (1966a). Entrainment of a circadian rhythm in birds by species-specific song cycles (Aves, Fringillidae: *Carduelis spinus, Serinus serinus*). *Experientia* **22**, 765.

── (1966b). Tagesperiodische Schwankungen der Vorzugshelligkeit bei Vögeln. *Z. vergl. Physiol.* **52**, 370–9.

── (1967). Circannuale Periodik der Mauser und der Zugunruhe bei einem Vogel. *Naturwissenschaften* **54**, 447.

── (1968a). Artspezifische Muster der Zugunruhe bei Laubsängern und ihre mögliche Bedeutung für die Beendigung des Zuges im Winterquartier. *Z. Tierpsychol.* **25**, 843–53.

── (1968b). Circannuale Periodik als Grundlage des jahreszeitlichen Funktionswandels bei Zugvögeln. Untersuchungen am Fitis (*Phylloscopus trochilus*) und am Waldlaubsänger (*P. sibilatrix*). *J. Orn., Lpz.* **109**, 70–95.

── (1969). Untersuchungen zur Jahresperiodik von Laubsängern. *J. Orn., Lpz.* **110**, 1–21.

── (1971). A comparative study of circannual rhythms in warblers. In *Biochronometry* (ed. M. Menaker), pp. 405–27. National Academy of Sciences, Washington D.C.

── (1972). Adaptive function of circannual rhythms in warblers. *Int. orn. Congr.* **15**, 218–36.

── (1973). Circannual rhythms in birds: their interaction with circadian rhythms and environmental photoperiod. *J. Reprod. Fert.* Suppl. 19, 51–65.

── (1974). Testosterone induces 'splitting' of circadian locomotor activity rhythms in birds. *Science, N.Y.* **185**, 72–4.

── (1975). Circadian and circannual rhythms in birds. In *Avian biology* (ed. D. S. Farner and J. R. King), Vol. 5, pp. 221–84. Academic Press, New York and London.

—— and Tureck, F. (1971). Effects of season on circadian activity rhythms of the Starling. *Naturwissenschaften* **58**, 627.

Haapanen, A. (1966). Bird fauna of the Finnish forests in relation to forest succession. *Ann. Zool. fenn.* **3**, 176–200.

——, Helminen, M., and Suomalainen, H. K. (1973). Population growth and breeding biology of the Whooper Swan, *Cygnus c. cygnus,* in Finland in 1950–70. *Finnish Game Res.* **33**, 39–60.

Haase, E. (1970). Zum Einfluss der Tageslänge auf die Gonadotropinsekretion bei Bergfinken, *Fringilla montifringilla. Verh. dt. zool. Ges.* **64**, 274–8.

—— and Farner, D. S. (1969). Acetylcholinesterase in der Pars distalis von *Zonotrichia leucophrys gambelii* (Aves). *Z. Zellforsch. mikrosk. Anat.* **93**, 356–68.

—— and —— (1970). The function of the acetylcholinesterase cells of the Pars distalis of the White-crowned Sparrow *Zonotrichia leucophrys gambelii. Acta zool., Stockh.* **51**, 99–106.

—— and —— (1971). Investigations of the butylcholinesterase-containing cells of the adenohypophysis of the White-crowned Sparrow *Zonotrichia leucophrys gambelii. Z. Zellforsch. mikrosk. Anat.* **118**, 570–8.

——, Sharp, P. J., and Paulke, E. (1975). Seasonal changes in plasma LH levels in domestic ducks. *J. Reprod. Fert.* **44**, 591–4.

Hahn, W. E., Schjeide, O. A., and Gorbman, A. (1969). Organ-specific estrogen-induced RNA synthesis resolved by DNA—RNA hybridization in the domestic fowl. *Proc. natn. Acad. Sci. U.S.A.* **62**, 112–19.

Haigh, C. R. (1968). Sexual dimorphism, sex ratios and polygyny in the Red-winged Blackbird. Ph.D. thesis, University of Washington, Seattle.

Hailman, J. P. (1959). Why is the male Wood Duck strikingly colorful? *Am. Nat.* **93**, 383–4.

—— (1964*a*). The Galápagos Swallow-tailed Gull is nocturnal. *Wilson Bull.* **76**, 347–54.

—— (1964*b*). Breeding synchrony in the equatorial Swallow-tailed Gull. *Am. Nat.* **98**, 79–83.

Häkkinen, I., Jokinen, N. and Tast, J. (1973). The winter breeding of the Feral Pigeon *Columba livia domestica* at Tampere in 1972/1973. *Ornis fenn.* **50**, 83–8.

Halberg, F. (1969). Chronobiology. *A. Rev. Physiol.* **31**, 675–725.

——, Engeli, M., Hamburger, C., and Hillman, D. (1965). Spectral resolution of low frequency small-amplitude rhythms in excreted 17-ketosteroids; probable androgen-induced circaseptan desynchronization. *Acta endocr., Copehn.,* Suppl. **103**, 1–54.

Haldane, J. B. S. (1932). *The causes of evolution,* pp. 207–10. Cornell University Press, London.

Hall, J. R. (1970). Synchrony and social stimulation in colonies of the Black-headed Weaver *Ploceus cucullatus* and Vieillot's Black Weaver *Melanopteryx nigerrimus. Ibis* **112**, 93–104.

Hall, N. Duplaix and Kear, J. (1975). Breeding requirements in captivity. In *Flamingos* (ed. J. Kear and N. Duplaix Hall). T. and A. D. Poyser, Berkhamsted.

Hall, P. F., Ralph, C. L., and Grinwich, D. L. (1965). On the locus of action of interstitial cell-stimulating hormone (ICSH or LH) on feather pigmentation of African weaver birds. *Gen. comp. Endocrinol.* **5**, 552–7.

Hall, T. R., Chadwick, A., Bolton, N. J., and Scanes, C. G. (1975). Prolactin release *in vitro* and *in vivo* in the pigeon and the domestic fowl following administration of synthetic thyrotrophin-releasing factor (TRF). *Gen. Comp. Endocrinol.* **25**, 298–306.

Hamberger, L. A. and Ahrén, K. E. B. (1967). Effects of gonadotrophins *in vitro* on glucose uptake and lactic acid production of ovaries from prepubertal and hypo-physectomised rats. *Endocrinology* **81**, 93.

Hamilton, T. H. (1961). On the functions and causes of sexual dimorphism in breeding plumage characters of North American species of warblers and orioles. *Am. Nat.* **95**, 121–3.

—— and Barth, R. H. (1962). The biological significance of season change in male plumage appearance in some New World migratory bird species. *Am. Nat.* **96**, 129–44.

Hammond, J. (1952). Fertility. In *Marshall's physiology of reproduction* (3rd edn) (ed. A. S. Parkes), Vol. 2, pp. 648–740. Longmans, Green, New York.

Hamner, W. M. (1963). Diurnal rhythm and photoperiodism in testicular recrudescence of the House Finch. *Science, N.Y.* **142**, 1294–5.

—— (1964). Circadian control of photoperiodism in the House Finch demonstrated by interrupted-night experiments. *Nature, Lond.* **203**, 1400–1.

—— (1965). Avian photoperiodic response–rhythms: evidence and inference. In *Circadian clocks: Proceedings of the Feldafing Summer School* (ed. J. Aschoff), pp. 379–84. North-Holland, Amsterdam.

—— (1966). Photoperiodic control of the annual testicular cycle in the House Finch, *Carpodacus mexicanus. Gen. comp. Endocrinol.* **7**, 224–33.

—— (1968). The photorefractory period of the House Finch. *Ecology* **49**, 211–27.

—— (1971). On seeking an alternative to the endogenous reproductive rhythm hypothesis in birds. In *Biochronometry* (ed. M. Menaker), pp. 448–61. National Academy of Sciences, Washington D.C.

—— and Enfright, J. T. (1967). Relationships between photoperiodism and circadian rhythms of activity in the House Finch. *J. exp. Biol.* **46**, 43–61.

Hann, H. W. (1941). The cowbird at the nest. *Wilson Bull.* **53**, 209–21.

Hansen, E. W. (1966). Squab-induced crop growth in Ring Dove foster parents. *J. comp. physiol. Psychol.* **62**, 120–2.

Hansson, V., Reusch, E., Trygstad, O., Torgerson, O., Ritzén, E. M., and French, F. S. (1973). FSH stimulation of testicular androgen binding protein. *Nature: New Biology* **246**, 56–8.

Harmuth, D. (1972). *American Pigeon Journal,* April 1972, p. 200.

Harper, E. H. (1904). The fertilization and early development of the pigeon's egg. *Am. J. Anat.* **3**, 349–86.

Harris, G. W. and Donovan, B. T. (eds) (1966). *The pituitary gland.* Butterworths, London.

Harris, M. P. (1966). The breeding biology of the Manx Shearwater. *Ibis* **108**, 17–33.

—— (1969a). Breeding seasons of sea-birds in the Galapagos Islands. *J. Zool.* **159**, 145–65.

—— (1969b). Food as a factor controlling the breeding of *Puffinus lherminieri. Ibis* **111**, 139–56.

—— (1969c). Factors influencing the breeding cycle of the Red-billed Tropicbird in the Galapagos Islands. *Ardea* **57**, 149–57.

—— (1969d). The biology of Storm Petrels in the Galápagos Islands. *Proc. Calif. Acad. Sci.* **37**, 95–165.

—— (1970). Breeding ecology of the Swallow-tailed Gull *Creagus furcatus. Auk* **87**, 215–43.

Harrison, J. M. (1932). A series of nineteen Pheasants (*Phasianus colchicus* L.) presenting analogous secondary sexual characteristics in association with changes in the ovaries. *Proc. zool. Soc. Lond.* **1**, 193–203.

Hastings, J. W. (1970). Cellular-biochemical clock hypothesis. In *The biological clock: two views* (ed. F. A. Brown, J. W. Hastings, and J. D. Palmer), pp. 61–91. Academic Press, New York and London.

Haverschmidt, F. (1965). *Molothrus bonariensis* parasitizing *Fluvicola pica* and *Arundinicola leucocephala* in Surinam. *Auk* **82**, 508–9.

Hawkins, R. A., Heald, P. J., and Taylor, P. (1969). The uptake of (6,7-³H) 17β-oestradiol by tissues of the domestic fowl during an ovulation cycle. *Acta endocr., Copenh.* **60**, 210–15.

Haylock, J. W. (1959). *Investigations on the habits of Quelea birds and their control.* Government Printers, Nairobi.

Heald, P. J., Furnival, B. E., and Rookledge, K. A. (1967). Changes in the levels of luteinizing hormone in the pituitary of the domestic fowl during an ovulatory cycle. *J. Endocr.* **37**, 73–81.

——, Rookledge, K. A., Furnival, B. E., and Watts, G. D. (1968). Changes in luteinizing hormone content of the anterior pituitary of the domestic fowl during the interval between clutches. *J. Endocr.* **41**, 197–201.

Hedlund, L. (1970). Sympathetic innervation of the avian pineal body. *Anat. Rec.* **166**, 406.

——, Ralph, Ch. L., Chepko, J., and Lynch, H. J. (1971). A diurnal serotonin cycle in the pineal body of the Japanese Quail: photoperiod phasing and the effect of superior cervical ganglionectomy. *Gen. comp. Endocrinol.* **16**, 52–8.

Heftmann, E. and Mesettig, E. (1960). *The biochemistry of steroids.* Reinhold, New York.

Heinroth, O. (1922). Die Beziehungen zwischen Vogelgewicht, Eigewicht, Gelegegewicht und Brutdauer. *J. Orn., Lpz.* **70**, 172–285.

Heller, C. G., Heller, G. V., and Rowley, M. J. (1969). Human spermatogenesis: an estimate of the duration of each cell association and of each cell type. In *Progress in endocrinology. Proceedings of the Third International Congress of Endocrinology, Mexico 1968* (ed. C. Gual and F. J. G. Ebling), pp. 1012–18. Excerpta Medica, Amsterdam.

Hellmayr, C. E. (1938). Catalogue of the birds of the Americas and the adjacent Islands in the Field Museum of Natural History. *Publs Field Mus. nat. Hist. (Zool.)* **13** (11), 1–662.

Helms, C. W. (1968). Food, fat, and feathers. *Am. Zool.* **8**, 151–67.

—— and Drury, W. H., Jr. (1960). Winter and migratory weight and fat: field studies on some North American buntings. *Bird-Banding* **31**, 1–40.

Hemmingsen, A. M. and Krarup, N. B. (1937). Rhythmic diurnal variations in the oestrous phenomena of the rat and their susceptibility to light and dark. *Biol. Meddr.* **13** (7), 1–61.

Heppner, F. H. and Farner, D. S. (1971a). Training White-crowned Sparrows, *Zonotrichia leucophrys gambelii*, in self-selection of photoperiod. *Z. Tierpsychol.* **28**, 62–8.

—— and —— (1971b). Periodicity in self-selection of photoperiod. In *Biochronometry* (ed. M. Menaker), pp. 463–79. National Academy of Sciences, Washington D.C.

Herlant, M. (1956). Corrélations hypophyso-génitales chez la femelle de la chauve-souris, *Myotis myotis* (Borkhausen). *Archs Biol., Liege* **67**, 89–180.

—— (1960). Étude critique de deux techniques nouvelles destinées à mettre en évidence les différentes catégories cellulaires présentes dans la glande pittuitaire. *Bull. Microsc. appl.* **10**, 37–44.

—— (1964). The cells of the adenohypophysis and their functional significance. *Int. Rev. Cytol.* **17**, 299–381.

Herrick, R. B. and Adams, J, L. (1956). The effects of progesterone and diethylstilbestrol injected singly or in combination, on sexual libido, and the weight of the testes of single comb White Leghorn cockerels. *Poult. Sci.* **35**, 1269–73.

Hewson, R. (1973). The moults of captive Scottish Ptarmigan (*Lagopus mutus*). *J. Zool.* **171**, 177–87.

Heywang, B. W. (1938). The time factor in egg production. *Poult. Sci.* **17**, 240–7.

Hildén, O. (1965). Zur Brutbiologie des Temminckstrandläufers, *Calidris temminckii* (Leisl.). *Ornis fenn.* **42**, 1–5.

Hilton, F. K. (1968). Endocrine control of seminal-glomus growth and function in Starlings of different ages. *Physiol. Zoöl.* **41**, 364–70.

Hinde, R. A. (1953). Appetitive behaviour, consummatory act, and the hierarchical organisation of behaviour–with special reference to the Great Tit (*Parus major*). *Behaviour* **5**, 189–224.

—— (1956). The biological significance of territories of birds. *Ibis* **98**, 340–69.

—— (1959). Behaviour and speciation in birds and lower vertebrates. *Biol. Rev.* **34**, 85–128.

—— (1962). Temporal relations of brood patch development in domesticated Canaries. *Ibis* **104**, 90–7.

—— (1965). Interaction of internal and external factors in integration of Canary reproduction. In *Sex and behaviour* (ed. F. A. Beach), pp. 381–415. Wiley, New York

—— (1967). Aspects of the control of avian reproductive development within the breeding season. *Int. orn. Congr.* **14**, 135–54.

—— and Putman, R. J. (1973). Why Budgerigars breed in continuous darkness. *J. Zool.* **170**, 485–91.

—— and Steel, E. (1964). Effect of exogenous hormones on the tactile sensitivity of the Canary brood patch. *J. Endocr.* **30**, 355–60.

——, ——, and Hutchison, R. (1971). Control of oviduct development in ovariectomized Canaries by exogenous hormones. *J. Zool.* **163**, 265–76.

——, ——, and Follett, B. K. (1974). Effect of photoperiod on oestrogen-induced nest-building in ovariectomized or refractory female Canaries (*Serinus canarius*). *J. Reprod. Fert.* **40**, 383–99.

Hintz, J. V. and Dyer, M. I. (1970). Daily rhythm and seasonal change in the summer diet of adult Red-winged Blackbirds. *J. Wildl. Mgmt* **34**, 789–99.

Hoffmann, A. (1949). Über die Brutpflege des polyandrischen Wasserfasans, *Hydrophasianus chirurgus* (Scop.). *Zool.Jb. (Systematik, Ökologie und Geographia der Tiere)* **78**, 367–403.

Hoffmann, K. (1967). Kritik des Erlinger Modells. *Nachr. Akad. Wiss. Göttingen* **10**, 132–3.

—— (1969). Zum Tagesrhythmus der Brutablösung beim Kaptäubchen (*Oena capensis* L.) und bei anderen Tauben. *J. Orn., Lpz.* **110**, 448–64.

—— (1971). Splitting of the circadian rhythm as a function of light intensity. In *Biochronometry* (ed. M. Menaker), pp. 134–51. National Academy of Sciences, Washington D.C.

Hogan-Warburg, A. J. (1966). Social behaviour of the Ruff *Philomachus pugnax* (L.). *Ardea* **54**, 109–229.

Hohlweg, W. (1934). Veränderungen des hypophysenvorderlappens und des ovariums nach Behandlungen mit grossen dosen von Follikelhormon. *Klin. Wschr.* **13**, 92–5.

Höhn, E. O. (1947). Sexual behaviour and seasonal changes in the gonads and adrenals of the Mallard. *Proc. zool. Soc. Lond.* **177**, 281–304.

—— (1960). Seasonal changes in the Mallard's penis and their hormonal control. *Proc. zool. Soc. Lond.* **134**, 547–58.

—— (1967). The relevance of J. Christian's theory of a density-dependent endocrine population regulating mechanism to the problem of population regulation in birds. *Ibis* **109**, 445–6.

—— (1970). Gonadal hormone concentrations in Northern Phalaropes in relation to nuptial plumage. *Can. J. Zool.* **48**, 400–1.

—— (1972). Prolactin lack in a brood parasite: a summary report and appeal for material. *Ibis* **114**, 108.

—— and Cheng, S. C. (1965). Prolactin and the incidence of brood patch formation and incubation behaviour of the two sexes in certain birds with special reference to phalaropes. *Nature, Lond.* **208**, 197—8.

—— and —— (1967). Gonadal hormones in Wilson's Phalarope (*Steganopus tricolor*) and other birds in relation to plumage and sex behaviour. *Gen. comp. Endocrinol.* **8**, 1—11.

Holmes, R. T. (1966*a*). Breeding ecology and annual cycle adaptations of the Red-backed Sandpiper (*Calidris alpina*) in northern Alaska. *Condor* **68**, 3—46.

—— (1966*b*). Feeding ecology of the Red-backed Sandpiper (*Calidris alpina*) in arctic Alaska. *Ecology* **47**, 32—45.

—— (1973). Social behaviour of breeding Western Sandpipers *Calidris mauri. Ibis* **115**, 107—23.

—— and Pitelka, F. A. (1964). Breeding behavior and taxonomic relationships of the Curlew Sandpiper. *Auk* **81**, 363—79.

—— and —— (1966). Ecology and evolution of sandpiper (Calidritinae) social systems. *Int. orn. Congr.* **14**, abstr. 70—1.

—— and —— (1968). Food overlap among coexisting sandpipers on northern Alaskan tundra. *Syst. Zool.* **17**, 305—18.

Holmes, W. N., Butler, D. G., and Phillips, J. G. (1961). Observations on the effect of maintaining Glaucous-winged Gulls (*Larus glaucescens*) on fresh and sea water for long periods. *J. Endocr.* **23**, 53—61.

——, Phillips, J. G., and Wright, A. (1969). The control of extra-renal excretion in the duck (*Anas platyrhynchos*) with special reference to the pituitary-adrenal axis. *Gen. comp. Endocrinol.* Suppl. **2**, 358—73.

——, Chan, Mo-Yin, Bradley, J. S., and Stainer, I. M. (1970). The control of some endocrine mechanisms associated with salt regulation in aquatic birds. In *Hormones and the environment* (ed. G. K. Benson and J. G. Phillips), pp. 87—108. Memoirs of the Society for Endocrinology no. 18. Cambridge University Press, London.

Holt, C. (1946). Shasta ethnography. *Anthrop. Rec. Univ. Calif.* **3**, 299—349.

Holyoak, D. (1967). Breeding biology of the Corvidae. *Bird Study* **14**, 153—68.

—— (1971). Movements and mortality of Corvidae. *Bird Study* **18**, 97—106.

—— (1972). Food of the Rook in Britain. *Bird Study* **19**, 59—68.

—— (1973). Significance of colour dimorphism in Polynesian populations of *Egretta sacra. Ibis* **115**, 419—20.

—— (1974). Moult seasons of the British Corvidae. *Bird Study* **21**, 15—20.

Homma, K. and Sakakibara, Y. (1971). Encephalic photoreceptors and their significance in photoperiodic control of sexual activity in Japanese Quail. In *Biochronometry* (ed. M. Menaker), pp. 333—41. National Academy of Sciences, Washington D.C.

——, McFarland, L. Z., and Wilson, W. D. (1967). Response of the reproductive organs of the Japanese Quail to pinealectomy and melatonin injections. *Poult. Sci.* **46**, 314—19.

Hornberger, F. (1954). Reifealter und Ansiedlung beim Weissen storch. *Vogelwarte* **17**, 114—49.

Horton, E. W. (1971). Prostaglandins. In *Physiology and biochemistry of the domestic fowl* (ed. D. J. Bell and B. M. Freeman), Vol. 1, pp. 589—601. Academic Press, London and New York.

Höst, P. (1942). Effect of light on the moults and sequences of plumage in the Willow Ptarmigan. *Auk* **59**, 388—403.

Howard, H. E. (1920). *Territory in bird life.* John Murray, London.

Howe, M. A. (1975). Behavioural aspects of the pair bond in Wilson's Phalarope. *Wilson Bull.* **87**, 248—70.

Hoy, G. and Ottow, J. (1964). Biological and oological studies of the molothrine cowbirds (Icteridae) of Argentina. *Auk* **81**, 186–203.

Hudson, R. (1965). The spread of the Collared Dove in Britain and Ireland. *Br. Birds* **58**, 105–39.

—— (1968). The white-phase Giant Petrels of the South Orkney Islands. *Ardea* **56**, 178–83.

—— (1972). Collared Doves in Britain and Ireland during 1965–70: *Br. Birds* **65**, 139–55.

Hudson, W. H. (1920). *Birds of La Plata*, Vol. 1. Dent, London.

Hughes, M. R. (1970). Relative kidney size in non passerine birds with functional salt glands. *Condor* **72**, 164–8.

Humphreys, P. N. (1972). Brief observations on the semen and spermatozoa of certain passerine and non-passerine birds. *J. Reprod. Fert.* **29**, 327–36.

Hussell, D. J. T. (1972). Factors affecting clutch size in Arctic passerines. *Ecol. Monogr.* **42**, 317–64.

Hutchinson, G. E. (1950). Marginalia: Wideawake Fair. *Am. Scient.* **38**, 613–16.

Hutchinson, J. C. D. (1962). The annual rhythm of egg production in fowls. *Wld's Poult. Congr.* **12**, 124–9.

—— and Taylor, W. W. (1957). Seasonal variation in the egg production of fowls: effect of temperature and change of day length. *J. agric. Sci., Camb.* **49**, 419–34.

Hutchison, J. B. (1969). Changes in hypothalamic responsiveness to testosterone in male Barbary Doves (*Streptopelia risoria*). *Nature, Lond.* **222**, 176–7.

—— (1970a). Influence of gonadal hormones on the hypothalamic integration of courtship behaviour in the Barbary Dove. *J. Reprod. Fert.* Suppl. 11, 15–41.

—— (1970b). Differential effects of testosterone and oestradiol on male courtship in Barbary Doves (*Streptopelia risoria*). *Anim. Behav.* **18**, 41–51.

—— (1975a). Target cells for gonadal steroids in the brain: studies on steroid-sensitive mechanisms of behaviour. In *Neural and endocrine aspects of behaviour in birds* (ed. P. Wright, P. G. Caryl, and D. M. Vowles), pp. 123–37. Elsevier, Amsterdam.

Hutchison, R. E. (1975b). Influence of oestrogen on the initiation of nesting behaviour in female Budgerigars. *J. Endocr.* **64**, 417–28.

——, Hinde, R. A., and Steel, E. A. (1967). The effects of oestrogen, progesterone and prolactin on brood patch formation in ovariectomized Canaries. *J. Endocr.* **39**, 379–85.

——, ——, and Bendon, B. (1968). Oviduct development and its relation to other aspects of reproduction in domestic Canaries. *J. Zool.* **155**, 87–102.

Hutt, F. B. (1949). *Genetics of the fowl*. McGraw-Hill, New York.

Huxley, C. R. (1976). Gonad weight and food supply in captive Moorhens *Gallinula chloropus*. *Ibis* **118**, 411–13.

Imai, K. (1972). Effects of avian and mammalian pituitary preparations on follicular growth in hens treated with Methallibure or fasting. *J. Reprod. Fert.* **31**, 387.

—— (1973). Effects of avian and mammalian pituitary preparations on induction of ovulation in the domestic fowl, *Gallus domesticus*. *J. Reprod. Fert.* **33**, 91–8.

—— and Nalbandov, A. V. (1971). Changes in FSH activity of anterior pituitary glands and of blood plasma during the laying cycle of the hen. *Endocrinology* **88**, 1465–7.

——, Tanaka, M., and Nakajo, S. (1972). Gonadotrophic activities of anterior pituitary and of blood plasma and ovarian response to exogenous gonadotrophin in moulting hens. *J. Reprod. Fert.* **30**, 433–43.

Immelmann, K. (1963). Drought adaptations in Australian desert birds. *Int. orn. Congr* **13**, 649–57.

—— (1967). Periodische Voränge in der Fortpflanzung tierischer Organismen. *Studium gen.* **20**, 15–33.

—— (1971*a*). Environmental factors controlling reproduction in African and Australian birds—a comparison. *Ostrich* Suppl. 8, 193—204.

—— (1971*b*). Ecological aspects of periodic reproduction. In *Avian biology* (ed. D. S. Farner and J. R. King), Vol. 1, pp. 341—89. Academic Press, New York and London.

—— and Immelmann, G. (1968). Zur Fortpflanzungsbiologie einiger Vögel in der Namib. *Bonn. zool. Beitr.* 19, 329—39.

Ingolfsson, A. (1970). The moult of remiges and retrices in Great Black-backed Gulls *Larus marinus* and Glaucous Gulls *L. hyperboreus* in Iceland. *Ibis* 112, 83—92.

Ingram, C. (1959). The importance of juvenile cannibalism in the breeding biology of certain birds of prey. *Auk* 76, 218—26.

—— (1971). Colour changes in London pigeons. *Country Life* 150, 1785.

Ishii, S. and Furuya, T. (1975). Effects of purified chicken gonadotropins on the chick testis. *Gen. comp. Endocrinol.* 25, 1—8.

Iturriza, F. C. (1966). Monoamines and control of the pars intermedia of the toad pituitary. *Gen. comp. Endocrinol.* 6, 19—25.

Iwamura, Y., Koshihara, H., and Noumura, T. (1975). Studies on differentiation of Müllerian ducts in the Quail, *Coturnix coturnix japonica*. II Effects of sex hormones on nucleic acid synthesis in isolated female ducts. *J. exp. Zool.* 192, 25—32.

Jackson, G. L. (1971). Avian luteinizing hormone-releasing factor. *Endocrinology* 89, 1454—9.

—— and Nalbandov, A. V. (1969*a*). Luteinizing hormone releasing activity in the chicken hypothalamus. *Endocrinology* 84, 1262—5.

—— and —— (1969*b*). A substance resembling arginine vasotocin in the anterior pituitary gland of the cockerel. *Endocrinology* 84, 1218—23.

—— and —— (1969*c*). Ovarian ascorbic acid depleting factors in the chicken hypothalamus. *Endocrinology* 85, 113—20.

Jallageas, M. and Assenmacher, I. (1972). Effets de la photoperiode et du taux d'androgene circulant sur la fonction thyroidienne du Canard. *Gen. comp. Endocrinol.* 19, 331—40.

—— and —— (1974). Thyroid-gonadal interactions in the male domestic duck in relationship with the sexual cycle. *Gen. comp. Endocrinol.* 22, 13—20.

——, ——, and Follett, B. K. (1974). Testosterone secretion and plasma luteinizing hormone concentration during a sexual cycle in the Pekin Duck, and after thyroxine treatment. *Gen. comp. Endocrinol.* 23, 472—5.

Jarvis, M. J. F. (1972). The systematic position of the South African Gannet. *Ostrich* 43, 211—16.

Jefferies, D. J. (1969). Induction of apparent hyperthyroidism in birds fed DDT. *Nature, Lond.* 222, 578—9.

Jehl, J. R. (1970). Sexual selection for size differences in two species of sandpipers. *Evolution, Lancaster, Pa* 24, 311—19.

—— (1973). Breeding biology and systematic relationships of the Stilt Sandpiper. *Wilson Bull.* 85, 115—47.

Jenkins, D. (1957). The breeding of the Red-legged Partridge. *Bird Study* 4, 97—100.

—— and Watson, A. (1970). Population control in Red Grouse and Rock Ptarmigan in Scotland. *Finnish Game Res.* 30, 121—41. (Eighth International Congress of Game Biologists, Helsinki 1967).

——, ——, and Miller, G. R. (1963). Population studies on Red Grouse, *Lagopus lagopus scoticus* (Lath.) in north-east Scotland. *J. Anim. Ecol.* 32, 317—76.

——, ——, and —— (1964). Predation and Red Grouse populations. *J. appl. Ecol.* 1, 183—95.

——, ——, and —— (1967). Population fluctuations in the Red Grouse *Lagopus lagopus scoticus. J. Anim. Ecol.* 36, 91—122.

Jenkins, G. M. and Watts, D. G. (1968). *Spectral analysis and its applications* Holden-Day, San Francisco.

Jenni, D. A. and Collier, G. (1972). Polyandry in the American Jaçana (*Jacana spinosa*). *Auk* **89**, 743–65.

Jerome, F. N. (1959). Colour inheritance in geese. *Can. J. Genet. Cytol.* **1**, 135–41.

John, T. M., Meier, A. H., and Bryant, E. E. (1972). Thyroid hormones and the circadian rhythms of fat and cropsac responses to prolactin in the pigeon. *Physiol. Zoöl.* **45**, 34–42.

Johns, J. E. (1964). Testosterone-induced nuptial feathers in phalaropes. *Condor* **66**, 449–55.

— — (1969). Field studies of Wilson's Phalarope. *Auk* **86**, 660–70.

— — and Pfeiffer, E. W. (1963). Testosterone-induced incubation patches of phalarope birds. *Science, N.Y.* **140**, 1225–6.

Johnsgard, P. A. (1965). *Handbook of waterfowl behavior.* Cornell University Press, Ithaca.

Johnson, A. W. and Goodall, J. D. (1965). *The birds of Chile,* Vol. I. Platt Establecimientos, Buenos Aires.

Johnson, M. (1939). Effect of continuous light on periodic spontaneous activity of White-footed Mice (*Peromyscus*). *J. exp. Zool.* **82**, 315–28.

Johnson, O. W. (1961). Reproductive cycle of the Mallard Duck. *Condor* **63**, 351–64.

Johnson, R. E. (1960). Variation in breeding season and clutch size in Song Sparrows of the Pacific coast. *Condor* **56**, 268–73.

Johnston, D. W. (1961). Timing of annual molt in the Glaucous Gulls of northern Alaska. *Condor* **63**, 474–8.

(1966). A review of the vernal fat deposition picture in overland migrant birds. *Bird-Banding* **37**, 172–83.

Johnston, R. F. and Selander, R. K. (1964). House Sparrows: rapid evolution of races in North America. *Science, N.Y.* **144**, 548–50.

Jones, P. J. and Ward, P. (1976). The level of reserve protein as the proximate factor controlling the timing of breeding and clutch-size in the Red-billed Quelea *Quelea quelea. Ibis* **118**, 547–74.

Jones, R. E. (1969*a*). Hormonal control of incubation patch development in the California Quail, *Lophortyx californicus. Gen. comp. Endocrinol.* **13**, 1–13.

— — (1969*b*). Epidermal hyperplasia in the incubation patch of the California Quail, *Lophortyx californicus,* in relation to pituitary prolactin content. *Gen. comp. Endocrinol.* **12**, 498–502.

— — (1969*c*). Effect of prolactin and progesterone on gonads of breeding California Quail. *Proc. Soc. exp. Biol. Med.* **131**, 172–4.

— — (1970). Effects of season and gonadotrophin on testicular interstitial cells of California Quail. *Auk* **87**, 729–37.

— — (1971). The incubation patch of birds. *Biol. Rev.* **46**, 315–39.

Juhn, M. and Harris, P. C. (1956). Responses in molt and lay of fowl to progestin and gonadotropins. *Proc. Soc. exp. Biol. Med.* **92**, 709–11.

Jungck, E. C., Thrash, A. M., Ohlmacher, A. P., Knight, A. M., and Dryenforth, L. Y. (1957). Sexual precocity due to interstitial cell tumor of the testis: report of two cases. *J. clin. Endocr. Metab.* **17**, 291–5.

Juorio, A. V. and Vogt, M. (1967). Monoamines and their metabolites in the avian brain. *J. Physiol., Lond.* **189**, 489–518.

Kalma, D. L. (1970). Some aspects of the breeding ecology and annual cycle of three populations of the Rufous-collared Sparrow (*Zonotrichia capensis*) in western Panama. Ph.D. dissertation, Yale University.

Kanematsu, S. and Mikami, S. (1970). Effects of hypothalamic lesions on protein-bound [131]iodine and thyroidal ^{131}I uptake in the chicken. *Gen. comp. Endocrinol.* **14**, 25–34.

Kannankeril, J. V. and Domm, L. V. (1968). Development of the gonads in the female Japanese Quail. *Am. J. Anat.* **123**, 131–46.

Kappers, J. A. (1969). The morphological and functional evolution of the pineal organ during its phylogenetic development. In *Progress in endocrinology. Proceedings of the Third International Congress of Endocrinology, Mexico 1968* (ed. C. Gual and F. J. G. Ebling), pp. 619–26. Excerpta Medica, Amsterdam.

Kato, Y. (1939). [Functional significance of the cell types in the anterior pituitary of the fowl with special reference to the gonadostimulating hormone.] *Jap. J. zootech. Sci.* **11**, 193–209.

— and Nishida, S. (1935). [Studies on the cytogenesis of the anterior pituitary in the fowl.] *Jap. J. zootech. Sci.* **8**, 16–44.

Kear, J. (1961). Early sexual maturity in Mallard. *Br. Birds* **54**, 427–8.

— (1970). The adaptive radiation of parental care in waterfowl. In *Social behaviour in birds and mammals: essays on the social ethology of animals and man,* (ed. J. H. Crook), pp. 357–91. Academic Press, London.

— (1972). Reproduction and family life. In *The swans,* Peter Scott and The Wildfowl Trust, p. 79. Michael Joseph, London.

— (1973). The Magpie Goose in captivity. *Int. Zoo Yb.* **13**, 28–32.

— and Murton, R. K. (1973). The systematic status of the Cape Barren Goose as judged by its photo-responses. *Wildfowl* **24**, 141–3.

— and — (1976). The origins of Australian waterfowl as indicated by their photo-responses. *Int. orn. Congr.* **16**, 83–97.

Keast, A. (1968). Moult in birds of the Australian dry country relative to rainfall and breeding. *J. Zool.* **155**, 185–200.

Keast, J. A. and Marshall, A. J. (1954). The influence of drought and rainfall on reproduction in Australian desert birds. *Proc. zool. Soc. Lond.* **124**, 493–9.

Keck, W. N. (1932). Control of the sex characters in the English Sparrow, *Passer domesticus* (Linnaeus). *Proc. Soc. exp. Biol. Med.* **30**, 158–9.

— (1933). Control of the bill color of the male English Sparrow by injection of male hormone. *Proc. Soc. exp. Biol. Med.* **30**, 1140–1.

— (1934). The control of secondary sex characters in the English Sparrow, *Passer domesticus* (Linnaeus). *J. exp. Zool.* **67**, 315–45.

Kendall, M. D. and Ward, P. (1974). Erythropoiesis in an avian thymus. *Nature, Lond.* **249**, 366–7.

—, —, and Bacchus, S. (1973). A protein reserve in the pectoralis major flight muscle of *Quelea quelea. Ibis* **115**, 600–1.

Kendeigh, S. C. (1949). Effect of temperature and season on energy resources of the English Sparrow. *Auk* **66**, 111–27.

— (1961). Energy of birds conserved by roosting in cavities. *Wilson Bull.* **73**, 140–7.

— (1963). Regulation of nesting time and distribution in the House Wren. *Wilson Bull.* **75**, 418–27.

— (1969*a*). Tolerance of cold and Bergmann's rule. *Auk* **86**, 13–25.

— (1969*b*). Energy responses of birds to their thermal environment. *Wilson Bull.* **81**, 441–9.

— (1970). Energy requirements for existence in relation to size of bird. *Condor* **72**, 60–5.

— (1972). Energy control of size limits in birds. *Am. Nat.* **106**, 79–88.

— (1973). Energetics of reproduction in birds. In *Breeding biology of birds* (ed. D. S. Farner), pp. 111–17. National Academy of Sciences, Washington D.C.

—, Kramer, T. C., and Hamerstrom, F. (1956). Variation in egg characteristics of the House Wren. *Auk* **73**, 42–65.

Kennedy, R. J. (1970). Direct effects of rain on birds: a review. *Br. Birds* **63**, 401–14.

Keve-Kleiner, A. (1944). Die Ausbreitung der orientalischen Lachtaube in Ungarn im letzten Dezennium. *Aquila* **50**, 281–98.

Kihlström, J. E. and Danninge, I. (1972). Neurohypophysial hormones and sexual behaviour in males of the domestic fowl (*Gallus domesticus* L.) and the pigeon (*Columbia livia* Gmel.). *Gen. comp. Endocrinol.* **18**, 115–20.

Kikkawa, J. and Yamashina, Y. (1967). Breeding of introduced Black Swans in Japan. *Emu* **66**, 377–81.

King, J. R. (1961a). On the regulation of premigratory fattening in the White-crowned Sparrow. *Physiol. Zoöl.* **34**, 145–57.

— (1961b). The bioenergetics of vernal premigratory fat deposition in the White-crowned Sparrow. *Condor* **63**, 128–42.

— (1963). Autumnal migratory-fat deposition in the White-crowned Sparrow. *Int. orn. Congr.* **13**, 940–9.

— (1968). Cycles of fat deposition and molt in White-crowned Sparrows in constant environmental conditions. *Comp. Biochem. Physiol.* **24**, 827–37.

— (1970). Photoregulation of food intake and fat metabolism in relation to avian sexual cycles. In *La photorégulation de la reproduction chez les oiseaux et les mammifères* (ed. J. Benoit and I. Assenmacher), pp. 365–85. Colloques int. Cent. natn. Rech. scient. no. 172.

— (1972). Adaptive periodic fat storage by birds. *Int. orn. Congr.* **15**, 200–17.

— (1973a). Energetics of reproduction in birds. In *Breeding biology of birds* (ed. D. S. Farner), pp. 78–107. National Academy of Sciences, Washington D.C.

— (1973b). The annual cycle of the Rufous-collared Sparrow (*Zonotrichia capensis*) in three biotopes in north-western Argentina. *J. Zool.* **170**, 163–88.

— (1973c). Reproductive relationships of the Rufous-collared Sparrow and the Shiny Cowbird. *Auk* **90**, 19–34.

— (1974). Notes on geographical variation and the annual cycle in Patagonian populations of the Rufous-collared Sparrow *Zonotrichia capensis*. *Ibis* **116**, 74–83.

— and Farner, D. S. (1961). Energy metabolism, thermoregulation and body temperature. In *Biology and comparative physiology of birds* (ed. A. J. Marshall), Vol. 2, pp. 215–88. Academic Press, New York and London.

— and — (1963). The relationship of fat deposition to *Zugunruhe* and migration. *Condor* **65**, 200–23.

— and — (1966). The adaptive role of winter fattening in the White-crowned Sparrow with comments on its regulation. *Am. Nat.* **100**, 403–18.

— and Wales, E. E. (1965). Photoperiodic regulation of testicular metamorphosis and fat deposition in three taxa of Rosy finches. *Physiol. Zoöl.* **38**, 49–68.

—, Barker, S., and Farner, D. S. (1963). A comparison of energy reserves during autumnal and vernal migratory periods in the White-crowned Sparrow, *Zonotrichia leucophrys gambelii*. *Ecology* **44**, 513–21.

—, Farner, D. S., and Morton, M. L. (1965). The lipid reserves of White-crowned Sparrows on the breeding ground in central Alaska. *Auk* **82**, 236–52.

—, Follett, B. K., Farner, D. S., and Morton, M. L. (1966). Annual gonadal cycles and pituitary gonadotropins in *Zonotrichia leucophrys gambelii*. *Condor* **68**, 476–87.

Kinsky, F. C. (1971). The consistent presence of paired ovaries in the Kiwi (*Apteryx*) with some discussion of this condition in other birds. *J. Orn., Lpz.* **112**, 334–57.

Kirkpatrick, C. M. (1959). Interrupted dark period: tests for refractoriness in Bobwhit Quail hens. In *Photoperiodism and related phenomena in plants and animals* (ed. R. B. Withrow), pp. 751–8. American Association for the Advancement of Science Publ. no. 55, Washington D.C.

— and Leopold, A. C. (1952). The role of darkness in sexual activity of the quail. *Science, N.Y.* **116**, 280–1.

Kistchinski, A. A. (1975). Breeding biology and behaviour of the Grey Phalarope *Phalaropus fulicarius* in east Siberia. *Ibis* 117, 285–301.

Kleinhoonte, A. (1929). Über die durch das Licht regulierten autonomen Bewegungen der Canavalia-Blätter. *Archs néerl. Sci.* Ser. 3^b, 5, 1–110.

Kleitman, N. (1940). The modifiability of the diurnal pigmentary rhythm in isopods. *Biol. Bull. mar. biol. Lab., Woods Hole* 78, 403–6.

Kline, I. T. (1955). Relationship of vitamin B_{12} to stilbestrol stimulation of the chick oviduct. *Endocrinology* 57, 120–8.

—— and Dorfman, R. I. (1951). Estrogen stimulation of the oviduct in vitamin-deficient chicks. *Endocrinology* 48, 345–57.

Klinghammer, E. and Hess, E. H. (1964). Parental feeding in Ring Doves (*Streptopelia roseogrisea*): innate or learned? *Z. Tierpsychol.* 21, 338–47.

Klomp, H. (1970). The determination of clutch-size in birds: a review. *Ardea* 58, 1–124.

—— (1972). Regulation of the size of bird populations by means of territorial behaviour. *Neth. J. Zool.* 22, 456–88.

Kluijver, H. N. (1951). The population ecology of the Great Tit, *Parus m. major* L. *Ardea* 39, 1–135.

—— (1963). The determination of reproductive rates in Paridae. *Int. orn. Congr.* 13, 706–16.

—— (1971). Regulation of numbers in populations of Great Tits (*Parus m. major*). In *Dynamics of populations* (ed. P. J. den Boer and G. R. Gradwell), pp. 507–23. Centre for Agricultural Publishing and Documentation, Wageningen.

—— and Tinbergen, L. (1953). Territory and the regulation of territory in titmice. *Archs néerl. Zool.* 10, 266–87.

Kobayashi, H. (1954). Studies on molting in the pigeon. VIII. Effects of sex steroids on molting and thyroid gland. *Annotnes zool. jap.* 27, 22–6.

—— (1957). Physiological nature of refractoriness of ovary to the stimulus of light in the Canary. *Annotnes zool. jap.* 30, 8–18.

—— (1958). On the induction of molt in birds by 17α-oxyprogesterone-17-capronate. *Endocrinology* 63, 420–30.

—— (1969). Pineal and gonadal activity in birds. In *Seminar on hypothalamic and endocrine functions in birds*, p. 72. International House of Japan, Tokyo.

—— and Farner, D. S. (1964). Cholinesterases in the hypothalamo-hypophysial neurosecretory system of the White-crowned Sparrow, *Zonotrichia leucophrys gambelii*. *Z. Zellforsch. mikrosk. Anat.* 63, 965–73.

—— and —— (1966). Evidence of a negative feedback on photoperiodically induced gonadal development in the White-crowned Sparrow, *Zonotrichia leucophrys gambelii*. *Gen. comp. Endocrinol.* 6, 443–52.

—— and Wada, M. (1973). Neuroendocrinology in birds. In *Avian biology* (ed. D. S. Farner and J. R. King), Vol. 3, pp. 287–347. Academic Press, New York and London.

Koch, E. L. (1939). Zur frage der beeinflussbarkeit der gefiederfarben der Vögel. *Z. wiss. Zool.* (*A*) 152, 27–82.

Koch, H. J. and de Bont, A. F. (1944). Influence de la mue sur l'intensité du métabolisme chez les pinsons, *Fringilla c. coelebs*, L. *Annls Soc. r. zool. Belg.* 75, 81–6.

—— and —— (1952). Standard metabolic rate, weight changes and food consumption of *Fringilla c. coelebs* L. during sexual maturation. *Annls Soc. r. zool. Belg.* 82, 1–12.

Kok, O. B. (1970). Behaviour of the Great-tailed Grackle (*Quiscalus mexicanus*). Ph.D. thesis, University of Texas, Austin.

Komisaruk, B. R. (1967). Effects of local brain implants of progesterone on reproductive behavior in Ring Doves. *J. comp. physiol. Psychol.* 64, 219–24.

Köster, F. (1971). Zum Nistverhalten des Ani, *Crotophaga ani*. *Bonn. zool. Beitr.*
22, 4–27.

Krebs, J. R. (1970). Regulation of numbers in the Great Tit (Aves : Passeriformes).
J. Zool. 162, 317–33.

— (1971). Territory and breeding density in the Great Tit, *Parus major* L. *Ecology*
52, 2–22.

— (1973). Social learning and the significance of mixed-species flocks of chickadees
(*Parus* spp.). *Can. J. Zool.* 51, 1275–88.

—, MacRoberts, M. H., and Cullen, J. M. (1972). Flocking and feeding in the Great
Tit *Parus major*: an experimental study. *Ibis* 114, 507–30.

—, Ryan, J. C., and Charnov, E. L. (1974). Hunting by expectation or optimal
foraging? A study of patch use by chickadees. *Anim. Behav.* 22, 953–64.

Kroeber, A. L. (1925). *Handbook of the Indians of California*. Smithsonian Institute,
Bureau of American Ethnology, Bulletin no. 78. Scholarly, Michigan. (2nd
printing, California Book Co., Berkeley, 1967).

Krohn, P. L. (1967). Factors influencing the number of oocytes in the ovary. *Archs
Anat. microsc. Morph. exp.* 56, 151–9.

Kruijt, J. P., de Vos, G. J., and Bossema, I. (1972). The arena system of Black Grouse.
Int. orn. Congr. 15, 399–423.

Kuenzel, W. J. and Helms, C. W. (1967). Obesity produced in a migratory bird by
hypothalamic lesions. *BioScience* 17, 395–6.

— and — (1970). Hyperphagia, polydipsia and other effects of hypothalamic
lesions in the White-throated Sparrow *Zonotrichia albicollis*. *Condor* 72, 66–75.

Kumaran, J. D. S. and Turner, C. W. (1949). The endocrinology of spermatogenesis
in birds. II. Effect of androgens. *Poult. Sci.* 28, 739–46.

Kunkel, P. (1974). Mating systems of tropical birds: the effects of weakness or
absence of external reproduction-timing factors, with special reference to
prolonged pair bonds. *Z. Tierpsychol.* 34, 265–307.

Lacassagne, L. (1957). Dynamique de l'ovogenèse. Contribution a l'étude de la phase
de grand accroissement des follicules chez la poule domestique. *Annls Zootech.*
2, 85–93.

— (1970). Analyse du déterminisme de la formation des séries chez la poule a l'aide
de cycles nycthéméraux de 21 et 26 heures. *Annls Biol. anim. Biochim. Biophys.*
10, 59–71.

Lachlan, C. (1968). Energy flow through a population of Great Tits, with notes on
Blue Tits. Ph.D. thesis, University of Durham.

Lack, D. (1940). Courtship feeding in birds. *Auk* 57, 169–78.

— (1946). *The life of the robin*. Witherby, London.

— (1948). The significance of clutch size. *Ibis* 89, 302–52; 90, 25–45.

— (1954). *The natural regulation of animal numbers*. Clarendon Press, Oxford.

— (1956). Further notes on the breeding biology of the Swift, *Apus apus*. *Ibis* 98,
606–19.

— (1963). Cuckoo hosts in England. *Bird Study* 10, 185–201.

— (1964). A long-term study of the Great Tit (*Parus major*). *J. Anim. Ecol.* 33
(Suppl.), 159–73.

— (1966). *Population studies of birds*. Clarendon Press, Oxford.

— (1968a). *Ecological adaptations for breeding in birds*. Methuen, London.

— (1968b). The proportion of yolk in eggs of waterfowl. *Wildfowl* 19, 67–9.

— (1970). The endemic ducks of remote islands. *Wildfowl* 21, 5–10.

— (1971). *Ecological isolation in birds*. Blackwell, Oxford and Edinburgh.

— and Moreau, R. E. (1965). Clutch-size in tropical birds of forest and savanna.
Oiseau Revue fr. Orn. 35 (no. spécial), 76–89.

Lacy, D. (1967). The seminiferous tubule in mammals. *Endeavour* **26**, 101–8.
—— and Lofts, B. (1965). Studies on the structure and function of the mammalian testis I. Cytological and histochemical observations after continuous treatment with oestrogenic hormone and the effects of FSH and LH. *Proc. R. Soc. B* **162**, 188–97.
—— and Pettitt, A. J. (1969). Transmission electron microscopy and the production of steroids by the Leydig and Sertoli cells of the human testis. *Micron* **1**, 15–33.
——, Vinson, G. P., Collins, P., Bell, J., Fyson, P., Pudney, J., and Pettitt, A. J. (1969). The Sertoli cell and spermatogenesis in mammals. In *Progress in endocrinology. Proceedings of the Third International Congress of Endocrinology, Mexico 1968* (ed. C. Gual and F. J. G. Ebling), pp. 1019–29. Excerpta Medica, Amsterdam.
Lade, B. I. and Thorpe, W. H. (1964). Dove songs as innately coded patterns of specific behaviour. *Nature, Lond.* **202**, 366–8.
Lahr, E. L. and Riddle, O. (1945). Intersexuality in male embryos of pigeons. *Anat. Rec.* **92**, 425–31.
Lake, P. E. (1957). The male reproductive tract of the fowl. *J. Anat.* **91**, 116–29.
—— and Furr, B. J. A. (1971). The endocrine testis in reproduction. In *Physiology and biochemistry of the domestic fowl* (ed. D. J. Bell and B. M. Freeman), Vol. 3, pp. 1469–88. Academic Press, London and New York.
Lance, A. (1970). Telemetry studies of Red Grouse. *Nature Conservancy Research in Scotland Report 1968–70*, pp. 17–18.
Landauer, W. (1967). The hatchability of chicken eggs as influenced by environment and heredity. *Monogr. Storrs agric. Exp. Stn* **1** (rev.), 1–315.
Larionov, V. F. (1957). [On the specificity of the action of light on the reproduction of birds.] *Dokl. Akad. Nauk SSSR* **112**, 779–81.
Lasiewski, R. C. (1963). Oxygen consumption of torpid, resting, active, and flying hummingbirds. *Physiol. Zoöl.* **36**, 122–40.
—— and Dawson, W. R. (1967). The re-examination of the relation between standard metabolic rate and body weight in birds. *Condor* **69**, 13–23.
Laskey, A. R. (1935). Mockingbird life history studies. *Auk* **52**, 370–82.
Lauber, J. K., Boyd, J. E., and Axelrod, J. (1968). Enzymatic synthesis of melatonin in avian pineal body: extraretinal response to light. *Science, N.Y.* **161**, 489–90.
Lavery, H. J. (1967*a*). The Magpie Goose in Queensland. *Advis. Leafl. Div. Pl. Ind. Qd* no. 901.
—— (1967*b*). Whistling-ducks in Queensland. *Advis. Leafl. Div. Pl. Ind. Qd* no. 917.
——, Seton, D., and Bravery, J. A. (1968). Breeding seasons of birds in north-eastern Australia. *Emu* **68**, 113–47.
Laws, D. F. (1961). Hypothalamic neurosecretion in the refractory and postrefractory periods and its relationship to the rate of photoperiodically induced testicular growth in *Zonotrichia leucophrys gambelii*. *Z. Zellforsch. mikrosk. Anat.* **54**, 275–306.
—— and Farner, D. S. (1960). Prolactin and the photoperiodic testicular response in White-crowned Sparrows. *Endocrinology* **67**, 279–81.
Lazarus, J. (1972). Natural selection and the functions of flocking in birds: a reply to Murton. *Ibis* **114**, 556–8.
—— and Crook, J. H. (1973). The effects of luteinizing hormone, oestrogen and ovariectomy on the agonistic behaviour of female *Quelea quelea*. *Anim. Behav.* **21**, 49–60.
Lees, A. D. (1965). Is there a circadian component in the *Megoura* photoperiodic clock? In *Circadian clocks: Proceedings of the Feldafing Summer School* (ed. J. Aschoff), pp. 351–6. North-Holland, Amsterdam.

534   Bibliography

— — (1971). The relevance of action spectra in the study of insect photoperiodism. In *Biochronometry* (ed. M. Menaker), pp. 372—80. National Academy of Sciences, Washington D.C.

— — (1972). The role of circadian rhythmicity in photoperiodic induction in animals. In *Circadian rhythmicity. Proc. int. Symp. circadian Rhythmicity, Wageningen 1971*, pp. 87—110. Pudoc, Wageningen, Holland.

Lees, J. (1946). All the year breeding of the Rock Dove. *Br. Birds* **39**, 136—41.

Lehrman, D. S. (1955). The physiological basis of parental feeding behavior in the Ring Dove (*Streptopelia risoria*). *Behaviour* **7**, 241—86.

— — (1958*a*). Induction of broodiness by participation in courtship and nest-building in the Ring Dove (*Streptopelia risoria*). *J. comp. physiol. Psychol.* **51**, 32—6.

— — (1958*b*). Effect of female sex hormones on incubation behavior in the Ring Dove (*Streptopelia risoria*). *J. comp. physiol. Psychol.* **51**, 142—5.

— — (1963). On the initiation of incubation behaviour in doves. *Anim. Behav.* **11**, 433—8.

— — and Brody, P. N. (1957). Oviduct response to estrogen and progesterone in the Ring Dove (*Streptopelia risoria*). *Proc. Soc. exp. Biol. Med.* **95**, 373—5.

— — and — — (1961). Does prolactin induce incubation behaviour in the Ring Dove? *J. Endocr.* **22**, 269—75.

— — and — — (1964). Effect of prolactin on established incubation behavior in the Ring Dove. *J. comp. physiol. Psychol.* **57**, 161—5.

— — and Friedman, M. (1969). Auditory stimulation of ovarian activity in the Ring Dove (*Streptopelia risoria*). *Anim. Behav.* **17**, 494—7.

— — and Wortis, R. P. (1960). Previous breeding experience and hormone-induced incubation behavior in the Ring Dove. *Science, N.Y.* **132**, 1667—8.

— —, Brody, P. N., and Wortis, R. P. (1961*a*). The presence of the mate and of nesting material as stimuli for the development of incubation behavior and for gonadotropin secretion in the Ring Dove (*Streptopelia risoria*). *Endocrinology* **68**, 507—16.

— —, Wortis, R. P., and Brody, P. (1961*b*). Gonadotropin secretion in response to external stimuli of varying duration in the Ring Dove (*Streptopelia risoria*). *Proc. Soc. exp. Biol. Med.* **106**, 298—300.

Leonhardt, H. (1970). Über Plasmazellen im Nervengewebe (Eminentia mediana des Kaninchens). *Acta neuropathol.* **16**, 148—53.

Lerner, I. M. (1958). *Genetic basis of selection.* Wiley, New York.

Levins, R. (1968). *Evolution in changing environments: some theoretical explanations.* Monographs in population biology no. 2. Princeton University Press, New Jersey.

Lewin, V. (1963). Reproduction and development of young in a population of Californian Quail. *Condor* **65**, 249—78.

Lewis, R. A. (1975). Reproductive biology of the White-crowned Sparrow. II Environmental control of reproductive and associated cycles. *Condor* **77**, 111—24

— — and Farner, D. S. (1973). Temperature modulation of photoperiodically induced vernal phenomena in White-crowned Sparrows (*Zonotrichia leucophrys*). *Condor* **75**, 279—86.

— — and Orcutt, F. S. Jr. (1971). Social behaviour and avian sexual cycles. *Scientia, Bologna* **65**, 447—72.

— —, King, J. R., and Farner, D. S. (1974). Photoperiodic responses of a subtropical population of the finch (*Zonotrichia capensis hypoleuca*). *Condor* **76**, 233—7.

Ligon, J. D. (1971). Late summer-autumnal breeding of the Piñon Jay in New Mexico. *Condor* **73**, 147—53.

Lind, H. and Poulsen, H. (1963). On the morphology and behaviour of a hybrid between Goosander and Shelduck (*Mergus merganser* L. X *Tadorna tadorna* L.) *Z. Tierpsychol.* **20**, 558—96.

Ling, J. K. (1972). Adaptive functions of vertebrate molting cycles. *Am. Zool.* 12, 77–93.

Ljungkvist, H. I. (1967). Light and electron microscopical study of the effect of oestrogen on the chicken oviduct. *Acta endocr., Copenh.* 56, 391–402.

Lockie, J. D. (1955). The breeding and feeding of Jackdaws and Rooks with notes on Carrion Crows and other Corvidae. *Ibis* 97, 341–69.

—— (1956). Winter fighting in feeding flocks of Rooks, Jackdaws and Carrion Crows. *Bird Study* 3, 180–90.

—— (1959). The food of nestling Rooks near Oxford. *Br. Birds* 52, 332–4.

Lofts, B. (1961). Melanism in captive Weaver-finches (*Quelea quelea*). *Nature, Lond.* 191, 993–4.

—— (1962*a*). The effects of exogenous androgen on the testicular cycle of the Weaver-finch *Quelea quelea*. *Gen. comp. Endocrinol.* 2, 394–406.

—— (1962*b*). Cyclical changes in the interstitial and spermatogenetic tissue of migratory waders 'wintering' in Africa. *Proc. zool. Soc. Lond.* 138, 405–13.

—— (1962*c*). Photoperiod and the refractory period of reproduction in an equatorial bird, *Quelea quelea*. *Ibis* 104, 407–14.

—— (1964). Evidence of an autonomous reproductive rhythm in an equatorial bird (*Quelea quelea*). *Nature, Lond.* 201, 523–4.

—— (1968). Patterns of testicular activity. In *Perspectives in endocrinology: hormones in the lives of lower vertebrates* (ed. E. J. W. Barrington and C. Barker Jørgensen), pp. 239–304. Academic Press, New York and London.

—— (1972). The Sertoli cell. *Gen. comp. Endocrinol.* Suppl. 3, 636–48.

—— (1975). Environmental control of reproduction. *Symp. zool. Soc. Lond.* 35, 177–97.

—— and Coombs, C. J. F. (1965). Photoperiodism and the testicular refractory period in the Mallard. *J. Zool.* 146, 44–54.

—— and Lam, W. L. (1973). Circadian regulation of gonadotrophin secretion. *J. Reprod. Fert. Suppl.* 19, 19–34.

—— and Marshall, A. J. (1956). The effects of prolactin administration on the internal rhythm of reproduction in male birds. *J. Endocr.* 13, 101–6.

—— and —— (1958). An investigation of the refractory period of reproduction in male birds by means of exogenous prolactin and follicle stimulating hormone. *J. Endocr.* 17, 91–8.

—— and —— (1959). The post-nuptial occurrence of progestins in the seminiferous tubules of birds. *J. Endocr.* 19, 16–21.

—— and —— (1960). The experimental regulation of *Zugunruhe* and the sexual cycle in the Brambling *Fringilla montifringilla*. *Ibis* 102, 209–14.

—— and Murton, R. K. (1966). The role of weather, food and biological factors in timing the sexual cycle of Woodpigeons. *Br. Birds* 59, 261–80.

—— and —— (1967). The effects of cadmium on the avian testis. *J. Reprod. Fert.* 13, 155–64.

—— and —— (1968). Photoperiodic and physiological adaptations regulating avian breeding cycles and their ecological significance. *J. Zool.* 155, 327–94.

—— and —— (1973). Reproduction in birds. In *Avian biology* (ed. D. S. Farner and J. R. King), Vol. 3, pp. 1–107. Academic Press, New York and London.

——, Marshall, A. J., and Wolfson, A. (1963). The experimental demonstration of pre-migration activity in the absence of fat deposition in birds. *Ibis* 105, 99–105.

——, Murton, R. K., and Westwood, N. J. (1966). Gonad cycles and the evolution of breeding seasons in British Columbidae. *J. Zool.* 150, 249–72.

——, ——, and —— (1967*a*). Photoresponses of the Woodpigeon *Columba palumbus* in relation to the breeding season. *Ibis* 109, 338–51.

——, ——, and —— (1967b). Interspecific differences in photosensitivity between three closely related species of pigeons. *J. Zool.* **151**, 17—25.

——, ——, and —— (1967c). Experimental demonstration of a post-nuptial refractory period in the Turtle Dove *Streptopelia turtur. Ibis* **109**, 352—8.

——, ——, and Thearle, R. J. P. (1973). The effects of testosterone propionate and gonadotropins on the bill pigmentation and testes of the House Sparrow (*Passer domesticus*). *Gen. comp. Endocrinol.* **21**, 202—9.

Löhrl, H. (1957). Populationsökologische Untersuchungen beim Halsbandschnäpper (*Ficedula albicollis*). *Bonn. zool. Beitr.* **8**, 130—77.

Lorenz, F. W. (1954). Effects of estrogens on domestic fowl and applications in the poultry industry. *Vitam. Horm., Lpz.* **12**, 235—75.

—— (1969). Reproduction in domestic fowl. In *Reproduction in domestic animals* (2nd edn) (ed. H. H. Cole and P. T. Cupps), pp. 569—608. Academic Press, New York and London.

Lorenz, K. (1937). Über die Bildung des Instinktbegriffes. *Naturwissenschaften* **25**, 289—331.

Lostroh, A. J. and Johnson, R. E. (1966). Amounts of interstitial cell-stimulating hormone and follicle-stimulating hormone required for follicular development, uterine growth and ovulation in the hypophysectomized rat. *Endocrinology* **79**, 991.

Lott, D. F. and Brody, P. N. (1966). Support of ovulation in the Ring Dove by auditory and visual stimuli. *J. comp. physiol. Psychol.* **62**, 311—13.

—— and Comerford, S. (1968). Hormonal initiation of parental behavior in inexperienced Ring Doves. *Z. Tierpsychol.* **25**, 71—5.

——, Scholz, D. S., and Lehrman, D. S. (1967). Exteroceptive stimulation of the reproductive system of the female Ring Dove (*Streptopelia risoria*) by the male and by the colony milieu. *Anim. Behav.* **15**, 433—7.

Lovari, S. and Hutchison, J. B. (1975). Behavioural transitions in the reproductive cycle of Barbary Doves (*Streptopelia risoria* L.). *Behaviour* **53**, 126—50.

Lowe, V. T. (1963). Observations on the Painted Snipe. *Emu* **62**, 221—37.

Lustick, S. (1970). Energy requirements of molt in cowbirds. *Auk* **87**, 742—6.

Lyons, W. R. and Dixon, J. S. (1966). The physiology and chemistry of the mammotrophic hormone. In *The pituitary gland* (ed. G. W. Harris and B. T. Donovan), Vol. 1, pp. 527—81. Butterworths, London.

Ma, R. C. S. and Nalbandov, A. V. (1963). Discussion on: Physiology of the pituitary gland as affected by transplantation or stalk transection. In *Advances in neuroendocrinology* (ed. A. V. Nalbandov), pp. 306—11. University of Illinois Press, Urbana, U.S.A.

McGeen, D. S. and McGeen, J. J. (1968). The cowbirds of Otter lake. *Wilson Bull.* **80**, 84—93.

McIndoe, W. M. (1971). Yolk synthesis. In *Physiology and biochemistry of the domestic fowl* (ed. D. J. Bell and B. M. Freeman), Vol. 3, pp. 1207—23. Academic Press, London and New York.

MacInnes, C. D., Davis, R. A., Jones, R. N., Lieff, B. C., and Pakulak, A. J. (1974). Reproductive efficiency of McConnell River small Canada Geese. *J. Wildl. Mgmt* **38**, 686—707.

McKenzie, D. and Sclater, J. G. (1971). The evolution of the Indian Ocean since the late Cretaceous. *Geophys. J. R. astr. Soc.* **25**, 437—528.

McLaren, I. A. (1972). Polygyny as the adaptive function of breeding territory in birds. *Trans. Conn. Acad. Arts Sci.* **44**, 189—210.

Mackworth-Praed, C. W. and Grant, C. H. B. (1952). *Birds of eastern and northeastern Africa.* Longmans, London.

Maclean, G. L. (1971). The breeding seasons of birds in the south-western Kalahari. *Ostrich* Suppl. **8**, 179–92.

McNally, E. H. (1947). Some factors that affect oviposition in the domestic fowl. *Poult. Sci.* **26**, 396–9.

McNeil, R. and de Itriago, M. C. (1968). Fat deposition in the Scissors-tailed Flycatcher *Muscivora t. tyrannus* and the Small-billed Elaenia *Elaenia parvirostris* during the austral migratory period in northern Venezuela. *Can. J. Zool.* **46**, 123–8.

Magor, J. I. and Ward, P.(1973). Illustrated descriptions, distribution maps and bibliography of the species of *Quelea* (Weaver-birds : Ploceidae). *Trop. Pest Bull.* **1**, 37.

March, G. L. and Sadleir, R. M. F. S. (1970). Studies on the Band-tailed Pigeon (*Columba fasciata*) in British Columbia. I. Seasonal changes in gonadal development and crop gland activity. *Can. J. Zool.* **48**, 1353–7.

Maridon, B. and Holcomb, L. C. (1971). No evidence for incubation patch changes in Mourning Dove throughout reproduction. *Condor* **73**, 374–5.

Markert, C. L. and Møller, F. (1959). Multiple forms of enzymes: tissue, ontogenetic, and species specific patterns. *Proc. natn. Acad. Sci. U.S.A.* **45**, 753–63.

Marshall, A. J. (1951). Food availability as a timing factor in the sexual cycle of birds. *Emu* **50**, 267–82.

—— (1959). Internal and environmental control of breeding. *Ibis* **101**, 456–78.

—— (1961). Reproduction. In *Biology and comparative physiology of birds* (ed. A. J. Marshall), Vol. 2, pp. 169–213. Academic Press, New York and London.

—— and Coombs, C. J. F. (1957). The interaction of environmental internal and behavioural factors in the Rook (*Corvus f. frugilegus*) Linnaeus. *Proc. zool. Soc. Lond.* **128**, 545–89.

—— and Disney, H. J. de S. (1956). Photostimulation of an equatorial bird (*Quelea quelea*, Linnaeus). *Nature, Lond.* **177**, 143–4.

—— and —— ('1957). Experimental induction of the breeding season in a xerophilous bird. *Nature, Lond.* **180**, 647–9.

—— and Serventy, D. L. (1956). The breeding cycle of the Short-tailed Shearwater, *Puffinus tenuirostris* (Temminck), in relation to trans-equatorial migration and its environment. *Proc. zool. Soc. Lond.* **127**, 489–510.

—— and —— (1957). On the post-nuptial rehabilitation of the avian testis tunic. *Emu* **57**, 59–63.

Martan, J. (1969). Epididymal histochemistry and physiology. *Biol. Reprod.* **1**, 134–54.

Martin, D. D. and Meier, A. H. (1973). Temporal synergism of corticosterone and prolactin in regulating orientation in migratory White-throated Sparrow (*Zonotrichia albicollis*). *Condor* **75**, 369–74.

Martinez-Vargas, M. C. and Erickson, C. J. (1973). Some social and hormonal determinants of nest-building behaviour in the Ring Dove (*Streptopelia risoria*). *Behaviour* **45**, 12–37.

Mathewson, S. (1961). Gonadotrophic hormones affect aggressive behaviour in Starlings. *Science, N.Y.* **134**, 1522–3.

Matsui, T. (1964). Effect of water deprivation on the hypothalamic neurosecretory system of the Tree Sparrow, *Passer montanus saturatus*. *J. Fac. Sci. Tokyo Univ.* Sect. 4, **10**, 355–68.

—— (1966). Effect of prolonged daily photoperiods on the hypothalamic neurosecretory system of the Tree Sparrow (*Passer montanus saturatus*). *Endocr. jap.* **13**, 23–38.

—— and Kobayashi, H. (1965). Histochemical demonstration of monoamine oxidase in the hypothalamo-hypophysial system of the Tree Sparrow and the rat. *Z. Zellforsch. mikrosk. Anat.* **68**, 172–82.

Matsuo, S. (1954). [Studies on the acidophilic cells of the anterior pituitary in the fowl.] *Jap. J. zootech. Sci.* **25**, 63–9.

—— and Kato, Y. (1961). [Functional cytology of the adénohypophysis in the fowl.] *Japanese Branch meeting of World Poultry Science* **12**, 1.

——, Vitums, A., King, J. R., and Farner, D. S. (1969). Light-microscope studies of the cytology of the adenohypophysis of the White-crowned Sparrow, *Zonotrichia leucophrys gambelii. Z. Zellforsch. mikrosk. Anat.* **95**, 143–76.

Matthews, G. V. T. (1954). Some aspects of incubation in the Manx Shearwater, *Procellaria puffinus,* with particular reference to chilling resistance in the embryo. *Ibis* **96**, 432–40.

Matthews, L. H. (1939). Visual stimulation and ovulation in pigeons. *Proc. R. Soc. B* **126**, 557–60.

Mayer-Gross, H. (1970). *The nest record scheme* (BTO Field Guide no. 12). British Trust for Ornithology, Tring.

Mayr, E. (1942). *Systematics and the origin of species.* Columbia University Press, New York.

—— (1946). History of the North American bird fauna. *Wilson Bull.* **58**, 3–41.

—— (1960). Isolation as an evolutionary factor. *Proc. Am. phil. Soc.* **103**, 221–30.

—— (1964). Inferences concerning the Tertiary American bird faunas. *Proc. natn. Akad. Sci. U.S.A.* **51**, 280–8.

—— (1972). Continental drift and the history of the Australian bird fauna. *Emu* **72**, 26–8.

Means, A. R. and Huckins, C. (1974). Coupled events in the early biochemical actions of FSH on the Sertoli cells of the testis. In *Hormone binding and target cell activation in the testis* (Current topics in molecular endocrinology, Vol. 1) (ed. M. L. Dufau and A. R. Means), pp. 145–65. Plenum Press, New York and London.

Meier, A. H. (1969). Antigonadal effects of prolactin in the White-throated Sparrow *Zonotrichia albicollis. Gen. comp. Endocrinol.* **13**, 222–5.

—— and Davis, K. B. (1967). Diurnal variations of the fattening response to prolactin in the White-throated Sparrow *Zonotrichia albicollis. Gen. comp. Endocrinol.* **8**, 110–14.

—— and Dusseau, J. W. (1968). Prolactin and the photoperiodic gonadal response in several avian species. *Physiol. Zoöl.* **41**, 95–103.

—— and —— (1973). Daily entrainment of the photoinducible phases for photostimulation of the reproductive system in the sparrows, *Zonotrichia albicollis* and *Passer domesticus. Biol. Reprod.* **8**, 400–10.

—— and Farner, D. S. (1964). A possible endocrine basis for premigratory fattening in the White-crowned Sparrow *Zonotrichia leucophrys gambelii* (Nuttall). *Gen. comp. Endocrinol.* **4**, 584–95.

—— and Martin, D. D. (1971). Temporal synergism of corticosterone and prolactin controlling fat storage in the White-throated Sparrow, *Zonotrichia albicollis. Gen. comp. Endocrinol.* **17**, 311–18.

——, Farner, D. S., and King, J. R. (1965). A possible endocrine basis for migratory behaviour in the White-crowned Sparrow, *Zonotrichia leucophrys gambelii. Anim. Behav.* **13**, 453–65.

——, Burns, J. T., and Dusseau, J. W. (1969). Seasonal variations in the diurnal rhythm of pituitary prolactin content in the White-throated Sparrow, *Zonotrichia albicollis. Gen. comp. Endocrinol.* **12**, 282–9.

——, ——, Davis, K. B., and John, T. M. (1971*a*). Circadian variations in sensitivity of the pigeon cropsac to prolactin. *J. Interdiscip. Cycle Res.* **2**, 161–71.

——, Martin, D. D., and MacGregor, R. (1971*b*). Temporal synergism of corticosterone and prolactin controlling gonadal growth in sparrows. *Science, N.Y.* **173**, 1240–2.

——, Trobec, T. N., Joseph, M. M., and John, T. M. (1971c). Temporal synergism of prolactin and adrenal steroids in the regulation of fat stores. *Proc. Soc. exp. Biol. Med.* 137., 408—15.

Meites, J. (ed.) (1970). *Hypophysiotropic hormones of the hypothalamus: assay and chemistry.* Williams and Wilkins, Baltimore, Maryland, U.S.A.

—— and Nicoll, C. S. (1966). Adenohypophysis : prolactin. *A. Rev. Physiol.* 28, 57—88.

—— and Turner, C. W. (1947). Effects of sex hormones on pituitary lactogen and crop glands of common pigeons. *Proc. Soc. exp. Biol. Med.* 54, 465—8.

Menaker, M. (1965). Circadian rhythms and photoperiodism in *Passer domesticus.* In *Circadian clocks: Proceedings of the Feldafing Summer School* (ed. J. Aschoff), pp. 385—95. North-Holland, Amsterdam.

—— (1968a). Extraretinal light perception in the sparrow. I. Entrainment of the biological clock. *Proc. natn. Acad. Sci. U.S.A.* 59, 414—21.

—— (1971). Rhythms, reproduction and photoreception. *Biol. Reprod.* 4, 295—308.

—— and Eskin, A. (1967). Circadian clock in photoperiodic time measurement: a test of the Bünning hypothesis. *Science, N.Y.* 157, 1182—5.

—— and Keatts, H. (1968). Extraretinal light perception in the sparrow. II. Photoperiodic stimulation of testis growth. *Proc. natn. Acad. Sci. U.S.A.* 60, 146—51.

—— and Oksche, A. (1974). The avian pineal organ. In *Avian Biology* (ed. D. S. Farner and J. R. King), Vol. 4, pp. 79—118. Academic Press, New York and London.

——, Roberts, R., Elliott, J., and Underwood, H. (1970). Extraretinal light perception in the sparrow. III. The eyes do not participate in photoperiodic photoreception. *Proc. natn. Acad. Sci. U.S.A.* 67, 320—5.

Mengel, R. M. (1970). The North American central plains as an isolating agent in bird speciation. In *Pleistocene and recent environments of the Central Great Plains* (ed. W. Dort and J. K. Jones), pp. 279—340. Department of Geology, University of Kansas Special Publication 3.

Merkel, F. W. (1963). Long-term effects of constant photoperiods on European Robins and Whitethroats. *Int. orn. Congr.* 13, 950—9.

Mertens, J. A. L. (1969). The influence of brood size on the energy metabolism and water loss of nestling Great Tits *Parus major major. Ibis* 111, 11—16.

Michener, H. and Michener, J. R. (1940). The molt of House Finches of the Pasadena region, California. *Condor* 42, 140—53.

Middleton, A. L. A. (1965a). The ecology and reproductive biology of the European Goldfinch *Carduelis carduelis* near Melbourne, Victoria. Doctoral thesis, Monash University, Australia.

—— (1965b). Testicular response to an increased photoperiod in the Brown-headed Cowbird. *Auk* 82, 504—6.

—— (1970a). The breeding biology of the Goldfinch in southeastern Australia. *Emu* 70, 159—67.

—— (1970b). Foods and feeding habits of the European Goldfinch near Melbourne. *Emu* 70, 12—16.

—— (1971). The gonadal cycle of the Goldfinch in southeastern Australia. *Emu* 71, 159—66.

—— (1972). The structure and possible function of the avian seminal sac. *Condor* 74, 185—90.

Middleton, J. (1965). Testicular responses of House Sparrows and White-crowned Sparrows to short daily photoperiods with low intensities of light. *Physiol. Zoöl.* 38, 255—66.

Midgley, A. R., Niswender, G. D., Gay, V. L., and Reichert, L. E. (1971). Use of antibodies for characterisation of gonadotrophins and steroids. *Recent Progr. Horm. Res.* 27, 235—301.

Miers, K. H. and Williams, M. (1969). Nesting of the Black Swan at Lake Ellesmere, New Zealand. *Rep. Wildfowl Trust* **20**, 23–32.

Mikami, S. (1954). [Cytochemical studies on the anterior pituitary of the fowl.] *Jap. J. zootech. Sci.* **25**, 55–63.

—— (1958). Electron microscopic study on the hypothalamo-hypophysial system. I. Fine structure of the glandular cells of fowl hypophysis. *Jap. J. vet. Sci.* **20**,131–

——, Vitums, A., and Farner, D. S. (1969). Electron microscopic studies on the adenohypophysis of the White-crowned Sparrow, *Zonotrichia leucophrys gambelii*. *Z. Zellforsch. mikrosk. Anat.* **97**, 1–29.

——, Oksche, A., Farner, D. S., and Vitums, A. (1970). Fine structure of the vessels of the hypophysial portal system of the White-crowned Sparrow *Zonotrichia leucophrys gambelii*. *Z. Zellforsch. mikrosk. Anat.* **106**, 155–74.

——, Farner, D. S., and Lewis, R. A. (1973a). The prolactin cell of the White-crowned Sparrow, *Zonotrichia leucophrys pugetensis*. *Z. Zellforsch. mikrosk. Anat.* **138**, 455–74.

——, Hashikawa, T., and Farner, D. S. (1973b). Cytodifferentiation of the adeno-hypophysis of the domestic fowl. *Z. Zellforsch. mikrosk. Anat.* **138**, 299–314.

Milhorn, H. T. (1966). *The application of control theory to physiological systems.* Saunders, Philadelphia, Pennsylvania, U.S.A.

Millar, J. B. (1960). Migratory behaviour of the White-throated Sparrow, *Zonotrichia albicollis,* at Madison, Wisconsin. Doctoral dissertation, University of Wisconsin, Madison, Wisconsin.

Miller, A. H. (1954). The occurrence and maintenance of the refractory period in crowned sparrows. *Condor* **56**, 13–20.

—— (1959). Reproductive cycles in an equatorial sparrow. *Proc. natn. Acad. Sci. U.S.A.* **45**, 1095–100.

—— (1960). Adaptation of breeding schedule to latitude. *Int. orn. Congr.* **12**, 513–22.

—— (1961). Molt cycles in equatorial Andean Sparrows. *Condor* **63**, 143–61.

—— (1962). Bimodal occurrence of breeding in an equatorial sparrow. *Proc. natn. Acad. Sci. U.S.A.* **48**, 396–400.

—— (1965). Capacity for photoperiodic response and endogenous factors in the reproductive cycles of an equatorial sparrow. *Proc. natn. Acad. Sci. U.S.A.* **54**, 97–101.

Miller, G. R., Watson, A., and Jenkins, D. (1970). Responses of Red Grouse popu-lations to experimental improvement of their food. In *Animal populations in relation to their food resources* (ed. A. Watson), pp. 323–35. Blackwell, Oxford.

Milstein, P. le S., Prestt, I., and Bell, A. A. (1970). The breeding cycle of the Grey Heron. *Ardea* **58**, 171–257.

Miselis, R. and Walcott, C. (1970). Locomotor activity rhythms in homing pigeons (*Columba livia*). *Anim. Behav.* **18**, 544–51.

Monesi, V. (1962). Autoradiographic study of DNA synthesis and the cell cycle in spermatogonia and spermatocytes of mouse testis using tritaited thymidine. *J. Cell Biol.* **14**, 1.

—— (1967). Ribonucleic acid and protein synthesis during differentiation of male germ cells in the mouse. *Archs Anat. microsc. Morph. exp.* **56**, 61–74.

—— (1971). Chromosome activities during meiosis and spermiogenesis. *J. Reprod. Fert. Suppl.* **13**, 1.

Moreau, R. E. (1937). The comparative breeding biology of the African hornbills (Bucerotidae). *Proc. zool. Soc. Lond.* **107a**, 331–46.

—— (1950). The breeding seasons of African birds. I. Land birds. *Ibis* **92**, 223–67.

—— (1951). The British status of the quail and some problems of its biology. *Br. Birds* **44**, 257–76.

—— (1961). Problems of Mediterranean—Saharan migration. *Ibis* **103a**, 373—427.
—— (1966). *The bird faunas of Africa and its islands.* Academic Press, New York and London.
—— (1972). *The Palaearctic—African bird migration systems.* Academic Press, London and New York.
—— and Wayre, P. (1968). On the Palaearctic quails. *Ardea* **56**, 209—27.
——, Wilk, A. L., and Rowan, W. (1947). The moult and gonad cycles of three species of birds at five degrees south of the Equator. *Proc. zool. Soc. Lond.* **117**, 345—64.
Morel, G. and Bourlière, F. (1956). Recherches écologiques sur les *Quelea quelea* (L.) de la basse vallée du Sénégal. II. La reproduction. *Alauda* **24**, 97—122.
——, Morel, M. Y., and Bourlière, F. (1957). The Black-faced Weaver Bird or Dioch in West Africa. An ecological study. *J. Bombay nat. Hist. Soc.* **54**, 811—25.
Morgan, W. and Kohlmeyer, W. (1957). Hens with bilateral oviducts. *Nature, Lond.* **180**, 98.
Morris, J. A. (1961). The effect of continuous light and continuous noise on pullets held in a sealed chamber. *Poult. Sci.* **40**, 995—1000.
—— (1962). The effect of changing daylengths on the reproductive responses of the pullet. *Wld's Poult. Congr.* **12**, 115—23.
Morris, R. L. and Erickson, C. J. (1971). Pair bond maintenance in the Ring Dove (*Streptopelia risoria*). *Anim. Behav.* **19**, 398—406.
Morrison, J. V. and Wilson, F. E. (1972). Ovarian growth in Tree Sparrows *Spizella arborea*. *Auk* **89**, 146—55.
Morton, M. L. (1967). Diurnal feeding patterns in White-crowned Sparrows (*Zonotrichia leucophrys gambelii*). *Condor* **69**, 491—512.
—— and Mewaldt, L. R. (1962). Some effects of castration on a migratory sparrow (*Zonotrichia atricapilla*). *Physiol. Zoöl.* **35**, 237—47.
——, King, J. R., and Farner, D. S. (1969). Postnuptial and postjuvenal molt in White-crowned Sparrows in central Alaska. *Condor* **71**, 376—85.
Moss, R. (1967). Probable limiting nutrients in the main food of Red Grouse (*Lagopus lagopus scoticus*). In *Secondary productivity of terrestrial ecosystems* (ed. K. Petrusewicz), Vol. 1, pp. 369—79. Institute of Ecology, Polish Academy of Science, Warsaw.
—— (1969). A comparison of Red Grouse (*Lagopus l. scoticus*) stocks with the production and nutritive value of heather (*Calluna vulgaris*). *J. Anim. Ecol.* **38**, 103—12.
——, Miller, G. R., and Allen, S. E. (1972). Selection of heather by captive Red Grouse in relation to age of the plant. *J. appl. Ecol.* **9**, 771—81.
Moudgal, N. R. and Li, C. H. (1961). An immunochemical study of sheep pituitary interstitial cell-stimulating hormone. *Archs Biochem. Biophys.* **95**, 93—8.
Mountford, M. D. (1973). The significance of clutch size. In *The mathematical theory of the dynamics of biological populations* (ed. M. S. Bartlett and R. W. Hiorns), pp. 315—23. Academic Press, London and New York.
Moynihan, M. (1960). Some adaptations which help to promote gregariousness. *Int. orn. Congr.* **12**, 523—41.
Mueller, J. O., Smithies, O., and Irwin, M. R. (1962). Transferrin variation in Columbidae. *Genetics, Princeton* **47**, 1385—92.
Mulder, E., Van Beurden-Lamers, W. M. O., de Boer, W., Brinkman, A. O., and van der Molen, H. J. (1974). Testicular estradiol receptors in the rat. In *Hormone binding and target cell activation in the testis* (Current topics in molecular endocrinology, Vol. 1) (ed. M. L. Dufau and A. R. Means), pp. 343—55. Plenum Press, New York and London.
Mundinger, P. C. (1972). Annual testicular cycle and bill color change in the eastern American Goldfinch. *Auk* **89**, 403—19.

Munns, T. J. (1970). Effect of different photoperiods on melatonin synthesis in the pineal gland of the Canary (*Serinus canarius*) and testicular activity. Dissertation for degree of Ḋ. Phil., St. Louis University.

Munro, R. E., Smith, L. T., and Kupa, J. J. (1968). The genetic basis of color differences observed in the Mute Swan (*Cygnus olor*). *Auk* **58**, 504–6.

Munro, S. S. (1938). Functional changes in fowl spermatozoa during their passage through the excurrent ducts of the male. *J. exp. Zool.* **79**, 71–92.

Murphy, G. I. (1968). Pattern in life history and the environment. *Am. Nat.* **102**, 391–403.

Murphy, R. C. (1936). *Oceanic birds of South America*. American Museum of Natural History, New York, U.S.A.

Murton, R. K. (1958). Breeding of Wood-pigeon populations. *Bird Study* **5**, 157–83.

—— (1961). Some survival estimates for the Wood-pigeon. *Bird Study* **8**, 165–73.

—— (1965). *The Wood-pigeon*. New Naturalist Monograph Series. Collins, London.

—— (1966*a*). A statistical evaluation of the effect of Wood-pigeon shooting as evidenced by the recoveries of ringed birds. *Statistician* **16**, 183–202.

—— (1966*b*). Natural selection and the breeding seasons of the Stock Dove and Woodpigeon. *Bird Study* **13**, 311–27.

—— (1968*a*). Some predator-prey relationships in bird damage and population control In *The problems of birds as pests* (ed. R. K. Murton and E. N. Wright), pp. 157–69. Institute of Biology Symposium no. 17. Academic Press, London.

—— (1968*b*). Breeding, migration and survival of Turtle Doves. *Br. Birds* **61**, 193–212

—— (1971*a*). The significance of a specific search image in the feeding behaviour of the Wood-pigeon. *Behaviour* **40**, 10–42.

—— (1971*b*). *Man and birds*. Collins, London.

—— (1971*c*). Polymorphism in Ardeidae. *Ibis* **113**, 97–9.

—— (1971*d*). Why do some bird species feed in flocks? *Ibis* **113**, 534–6.

—— (1972). The control of bird populations. *Gerfaut* **62**, 63–82.

—— (1974). The impact of agriculture on birds. *Ann. appl. Biol.* **76**, 358–66.

—— (1975). Ecological adaptation in avian reproductive physiology. *Symp. zool. Soc. Lond.* no. 35, 149–75.

—— and Clarke, S. P. (1968). Breeding biology of Rock Doves. *Br. Birds* **61**, 429–48.

—— and Isaacson, A. J. (1962). The functional basis of some behaviour in the Wood-pigeon *Columba palumbus*. *Ibis* **104**, 503–21.

—— and Kear, J. (1973*a*). The nature and evolution of the photoperiodic control of reproduction in certain wildfowl (Anatidae). *J. Reprod. Fert. Suppl.*, **19**, 67–84.

—— and —— (1973*b*). The influence of daylight in the breeding of diving ducks. *Int. Zoo Yb.* **13**, 19–23.

—— and —— (1976). The role of daylength in regulating the breeding seasons and distribution of wildfowl. In *Light as an ecological factor II* (ed. R. Bainbridge), pp. 337–60. Blackwell, Oxford.

—— and Ridpath, M. G. (1962). The autumn movements of the Woodpigeon. *Bird Study* **9**, 7–41.

—— and Westwood, N. J. (1974*a*). Some effects of agricultural change on the English avifauna. *Br. Birds* **67**, 41–69.

—— and —— (1974*b*). An investigation of photo-refractoriness in the House Sparrow by artificial photoperiods. *Ibis* **116**, 298–313.

—— and —— (1975). Integration of gonadotrophin and steroid secretion, spermatogenesis and behaviour in the reproductive cycle of male pigeon species. In *Neural and endocrine aspects of behaviour in birds* (ed. P. Wright, P. G. Caryl, and D. M. Vowles), pp. 51–89. Elsevier, Amsterdam.

——, Isaacson, A. J., and Westwood, N. J. (1963*a*). The feeding ecology of the Woodpigeon. *Br. Birds* **56**, 345–75.

——, ——, and —— (1963*b*). The food and growth of nestling Wood-pigeons in relation to the breeding season. *Proc. zool. Soc. Lond.* **141**, 747–82.

——, Westwood, N. J., and Isaacson, A. J. (1964*a*). The feeding habits of the Wood-pigeon *Columba palumbus,* Stock Dove *C. oenas* and Turtle Dove *Streptopelia turtur. Ibis* **106**, 174–88.

——, ——, and —— (1964*b*). A preliminary investigation of the factors regulating population size in the Wood-pigeon. *Ibis* **106**, 482–507.

——, Isaacson, A. J., and Westwood, N. J. (1966). The relationships between Wood-pigeons and their clover food supply and the mechanism of population control. *J. appl. Ecol.* **3**, 55–96.

——, Bagshawe, K. D., and Lofts, B. (1969*a*). The circadian basis of specific gonadotrophin release in relation to avian spermatogenesis. *J. Endocr.* **45**, 311–12.

——, Thearle, R. J. P., and Lofts, B. (1969*b*). The endocrine basis of breeding behaviour in the Feral Pigeon (*Columba livia*). I. Effects of exogenous hormones on the pre-incubation behaviour of intact males. *Anim. Behav.* **17**, 286–306.

——, Lofts, B., and Orr, A. H. (1970*a*). The significance of circadian based photosensitivity in the House Sparrow *Passer domesticus. Ibis* **112**, 448–56.

——, ——, and Westwood, N. J. (1970*b*). The circadian basis of photoperiodically controlled spermatogenesis in the Greenfinch *Chloris chloris. J. Zool.* **161**, 125–36.

——, ——, and —— (1970*c*). Manipulation of photo-refractoriness in the House Sparrow *Passer domesticus* by circadian light regimes. *Gen. comp. Endocrinol.* **14**, 107–13.

——, Isaacson, A. J., and Westwood, N. J. (1971). The significance of gregarious feeding behaviour and adrenal stress in a population of Wood-pigeons *Columba palumbus. J. Zool.* **165**, 53–84.

——, Thearle, R. J. P., and Thompson, J. (1972). Ecological studies of the Feral Pigeon *Columba livia* var. I. Population, breeding biology and methods of control. *J. appl. Ecol.* **9**, 835–74.

——, Westwood, N. J., and Thearle, R. J. P. (1973). Polymorphism and the evolution of a continuous breeding season in the pigeon *Columbia livia. J. Reprod. Fert. Suppl.,* **19**, 561–75.

——, Bucher, E. H., Nores, M., Gómez, E., and Reartes, J. (1974*a*). The ecology of the Eared Dove *Zenaidura auriculata* in Argentina. *Condor* **76**, 80–8.

——, Thearle, R. J. P., and Coombs, C. F. B. (1974*b*). Ecological studies of the Feral Pigeon *Columba livia* var. III. Reproduction and plumage polymorphism. *J. appl. Ecol.* **11**, 841–54.

——, Westwood, N. J., and Isaacson, A. J. (1974*c*). Factors affecting egg-weight, body-weight and moult of the Woodpigeon *Columba palumbus. Ibis* **116**, 1–22.

——, ——, and —— (1974*d*). A study of Wood-pigeon shooting: the exploitation of a natural animal population. *J. appl. Ecol.* **11**, 61–81.

Naik, R. M. and Razack, A. (1967). Studies on the House Swift *Apus affinis* (G. E. Gray). 8. Breeding seasons and their regulation. *Pavo* **5**, 75–96.

Nalbandov, A. V. (1945). A study of the effect of prolactin on broodiness and on cock testes. *Endocrinology* **36**, 251–8.

—— (1959*a*). Neuroendocrine reflex mechanisms: bird ovulation. In *Comparative endocrinology* (ed. A. Gorbman), pp. 161–73. Wiley, New York.

—— (1959*b*). Role of sex hormones in the secretory function of the avian oviduct. In *Comparative endocrinology* (ed. A. Gorbman), pp. 524–32. Wiley, New York.

—— (1966). Hormonal activity of the pars distalis in reptiles and birds. In *The pituitary gland* (ed. G. W. Harris and B. T. Donovan), Vol. 1, pp. 295–316. Butterworths, London.

—— and Card, L. E. (1946). Effect of FSH and LH upon the ovaries of immature chicks and low-producing hens. *Endocrinology* **38**, 71–8.

—— and James, M. F. (1949). The blood vascular system of the chicken ovary. *Am. J. Anat.* **85**, 347–78.

Needham, J. (1942). *Biochemistry and morphogenesis.* Cambridge University Press, London.

Nelson, C. H. (1976). The color phases of downy Mute Swans. *Wilson Bull.* **88**, 1–3.

Nelson, D. M. and Nalbandov, A. V. (1966). Hormone control of ovulation. In *Physiology of the domestic fowl* (ed. C. Horton-Smith and E. C. Amoroso), pp. 3–10. Oliver and Boyd, Edinburgh.

——, Norton, H. W., and Nalbandov, A. V. (1965). Changes in hypophysial and plasma LH levels during the laying cycle of the hen. *Endocrinology* **77**, 889–96.

Nelson, J. B. (1966). The breeding biology of the Gannet *Sula bassana* on the Bass Rock, Scotland. *Ibis* **108**, 584–626.

—— (1967*a*). The breeding behaviour of the White Booby *Sula dactylatra. Ibis* **109**, 194–231.

—— (1967*b*). Etho-ecological adaptations in the Great Frigate-bird. *Nature, Lond.* **214**, 318.

—— (1969). The breeding ecology of the Red-footed Booby in the Galapagos. *J. Anim. Ecol.* **38**, 181–98.

Nethersole-Thompson, D. (1973). *The Dotterel.* Collins, London.

Nettleship, D. N. (1973). Breeding ecology of Turnstones *Arenaria interpres* at Hazen Camp, Ellesmere Island, N. W. T. *Ibis* **115**, 202–17.

Newton, I. (1964). The breeding biology of the Chaffinch. *Bird Study* **11**, 47–68.

—— (1966). The moult of the Bullfinch *Pyrrhula pyrrhula. Ibis* **108**, 41–67.

—— (1968). The temperatures, weights and body composition of moulting Bullfinches. *Condor* **70**, 323–32.

—— (1969). Winter fattening in the Bullfinch. *Physiol. Zoöl.* **42**, 96–107.

—— (1972). *Finches.* Collins, London.

Nice, M. M. (1937). Studies in the life history of the Song Sparrow, Vol. 1. *Trans. Linn. Soc. N.Y.* **4**, 1–247.

Nicholls, T. J. and Follett, B. K. (1974). The photoperiodic control of reproduction in *Coturnix* quail. The temporal pattern of LH secretion. *J. comp. physiol. Psychol.* **93**, 301–13.

—— and Graham, G. P. (1972). Observations on the ultrastructure and differentiation of Leydig cells in the testis of the Japanese Quail (*Coturnix coturnix japonica*). *Biol. Reprod.* **6**, 179–92.

—— and Storey, C. R. (1976). The effects of castration on plasma LH levels in photosensitive and photorefractory Canaries (*Serinus canarius*). *Gen. comp. Endocrinol.* **29**, 170–74.

——, Scanes, C. G., and Follett, B. K. (1973). Plasma and pituitary luteinizing hormone in Japanese Quail during photoperiodically induced gonadal growth and regression. *Gen. comp. Endocrinol.* **21**, 84–98.

Nicholson, A. J. (1933). The balance of animal populations. *J. Anim. Ecol.* **2**, 132–78.

Nicoll, C. S. (1967). Bio-assay of prolactin. Analysis of the pigeon crop-sac response to local prolactin injection by an objective and quantitative method. *Endocrinology* **80**, 641–55.

——, Pfeiffer, E. W., and Fevold, H. R. (1967). Prolactin and nesting behaviour in phalaropes. *Gen. comp. Endocrinol.* **8**, 61–5.

——, Fiorindo, R. P., McKennee, C. T., and Parsons, J. A. (1970). Assay of hypothalamic factors which regulate prolactin secretion. In *Hypophysiotropic hormones of the hypothalamus: assay and chemistry* (ed. J. Meites), pp. 115–50. Williams and Wilkins, Baltimore.

Nieboer, E. (1972). Preliminary notes on the primary moult in Dunlins *Calidris alpina. Ardea* **60**, 112–9.

Niethammer, G. (1970). Clutch sizes of introduced European passerines in New Zealand. *Notornis* **17**, 214–22.

Nisbet, I. C. T.·(1973). Courtship feeding, egg size and breeding success in Common Terns. *Nature, Lond.* **241**, 141–2.

——, Drury, W. H., and Baird, J. (1963). Weight-loss during migration. I. Deposition and consumption of fat by the Blackpoll Warbler *Dendroica striata. Bird-banding* **34**, 107–38.

Nishiyama, H. (1955). Studies on the accessory reproductive organs in the cock. *J. Fac. Agric. Kyushu Univ.* **10**, 277–305.

Noble, G. K. and Wurm, M. (1940). The effect of testosterone propionate on the Black-crowned Night Heron. *Endocrinology* **26**, 837–50.

Oades, J. M. and Brown, W. O. (1965). A study of the water-soluble oviduct proteins of the laying hen and the female chick treated with gonadal hormones. *Comp. Biochem. Physiol.* **14**, 475–89.

O'Connor, R. J. (1975a). Growth and metabolism in nestling passerines. *Symp. zool. Soc. Lond.* **35**, 277–306.

—— (1975b). The influence of brood size upon metabolic rate and body temperature in nestling Blue Tits *Parus caeruleus* and House Sparrows *Passer domesticus. J. Zool.* **175**, 391–403.

—— (1975c). An adaptation for early growth in tits, *Parus* spp. *Ibis* **117**, 523–6.

O'Donald, P. (1959). Possibility of assortative mating in the Arctic Skua. *Nature, Lond.* **138**, 1210–11.

—— (1967). A general model of sexual and natural selection. *Heredity, Lond.* **22**, 499–518.

—— (1972a). Sexual selections by variations in fitness at breeding time. *Nature, Lond.* **237**, 349–51.

—— (1972b). Natural selection of reproductive rates and breeding times and its effects on sexual selection. *Am. Nat.* **106**, 368–79.

—— (1972c). Sexual selection for colour phases in the Arctic Skua. *Nature, Lond.* **238**, 403–4.

—— and Davis, P. E. (1959). The genetics of the colour phases of the Arctic Skua. *Heredity, Lond.* **13**, 481–6.

—— and Davis, J. W. F. (1975). Demography and selection in a population of Arctic Skuas. *Heredity, Lond.* **35**, 75–83.

——, ——, and Broad, R. A. (1974a). Variation in assortative mating in two colonies of Arctic Skuas. *Nature, Lond.* **252**, 700–1.

——, Wedd, N. S., and Davis, J. W. F. (1974b). Mating preferences and sexual selection in the Arctic Skua. *Heredity, Lond.* **33**, 1–16.

Odum, E. P., Connell, C. E., and Stoddard, H. L. (1961). Flight energy and estimated flight ranges of some migratory birds. *Auk* **78**, 515–27.

—— and Perkinson, J. D. (1951). Relation of lipid metabolism to migration in birds: seasonal variation in body lipids of the migratory White-throated Sparrow. *Physiol. Zoöl.* **24**, 216–30.

Oehmke, H.-J. (1971). Vergleichende neurohistologische Studien am Nucleus infundibularis einiger australischer Vögel. *Z. Zellforsch. mikrosk. Anat.* **122**, 122–38.

·——, Priedkalns, J., Vaupel-von Harnack, M.; and Oksche, A. (1969). Fluoreszenz-und elektronen-mikroskopische Untersuchungen am Zwischenhirn-Hypophysensystem von *Passer domesticus. Z. Zellforsch. mikrosk. Anat.* **95**, 109–33.

Ogilvie, M. A. (1964). A nesting study of Mallard in Berkeley New Decoy, Slimbridge. *Rep. Wildfowl Trust* **15**, 84–8.

O'Grady, J. E. (1967). The determination of oestradiol and oestrone in the plasma of the domestic fowl by a method involving the use of labelled derivatives. *Biochem. J.* **106**, 77–86.

Ohmart, R. D. (1969). Dual breeding ranges in Cassin Sparrow (*Aimophila cassini*). In *Physiological systems in semiarid environments* (ed. C. C. Hoff and M. L. Riedesel) Abstr., p. 105. University of New Mexico Press, Albuquerque, U.S.A.

—— (1973). Observations on the breeding adaptations of the Roadrunner. *Condor* 75, 140–9.

Oka, T. and Schimke, R. T. (1969). Progesterone antagonism of estrogen-induced cytodifferentiation in chick oviduct. *Science, N.Y.* 163, 83–5.

Oksche, A. (1970). Retino-hypothalamic pathways in mammals and birds. In *La Photorégulation de la Reproduction chez les Oiseaux et le Mammifères* (ed. J. Benoit and I. Assenmacher), pp. 151–65. Colloques int. Cent. natn. Rech. scient. no. 172.

—— and Vaupel-von Harnack, M. (1965). Elektronenmikroskopische Untersuchungen an den Nervenbahnen des Pinealkomplexes von *Rana esculenta* L. *Z. Zellforsch. mikrosk. Anat.* 68, 389–426.

—— and —— (1966). Elektronenmikroskopische Untersuchungen zur Frage der Sinneszellen im Pinealorgan der Vögel. *Z. Zellforsch. mikrosk. Anat.* 69, 41–60.

——, Laws, D. F., Kamemoto, F. I., and Farner, D. S. (1959). The hypothalamo-hypophysial neurosecretory system of the White-crowned Sparrow, *Zonotrichia leucophrys gambelii*. *Z. Zellforsch. mikrosk. Anat.* 51, 1–42.

——, Morita, Y., and Vaupel-von Harnack, M. (1969). Zur Feinstruktur und Funktion des Pinealorgans der Taube (*Columba livia*). *Z. Zellforsch. mikrosk. Anat.* 102, 1–3(

——, Oehmke, H.-J., and Farner, D. S. (1970). Weitere Befunde zur Struktur und Funktion des Zwischenhirn-Hypophysensystems der Vögel. In *Aspects of neuroendocrinology: International Symposium on Neurosecretion, Kiel 1969* (ed. W. Bargmann and B. Scharrer), pp. 261–73. Springer-Verlag, Berlin and New York.

——, ——, and —— (1971). Neuro-anatomical problems of detection and localization of neurones producing neurohormones and releasers, with special reference to the avian hypothalamo-hypophysial system. In *Subcellular organization and function in endocrine tissues* (ed. H. Heller and K. Lederis), pp. 903–8. Memoirs of the Society for Endocrinology no. 19. Cambridge University Press, London and New York.

——, Kirschstein, H., Kobayashi, H., and Farner, D. S. (1972). Electron microscopic and experimental studies of the pineal organ in the White-crowned Sparrow *Zonotrichia leucophrys gambelii*. *Z. Zellforsch. mikrosk. Anat.* 124, 247–74.

——, ——, Hartwig, H. G., Oehmke, H. J., and Farner, D. S. (1974). Secretory parvocellular neurons in the rostral hypothalamus and in the tuberal complex of *Passer domesticus*. *Cell Tissue Res.* 149, 363–70.

Olney, P. J. S. (1963). The food and feeding habits of Tufted Duck *Aythya fuligula*. *Ibis* 105, 55–62.

Olsen, M. W. and Fraps, R. M. (1950). Maturation changes in the hen's ovum. *J. exp. Zool.* 114, 475–89.

—— and Neher, B. H. (1948). The site of fertilisation in the domestic fowl. *J. exp. Zool.* 109, 355–66.

Opel, H. (1966). Release of oviposition-inducing factor from the median eminence-pituitary stalk region in neural lobectomized hens. *Anat. Rec.* 154, 396.

—— and Nalbandov, A. V. (1958). A study of hormonal growth and ovulation of follicles in hypophysectomised hens. *Poult. Sci.* 37, 1230–1.

Orians, G. H. (1961). The ecology of blackbird (*Agelaius*) social systems. *Ecol. Monogr* 31, 285–312.

—— (1969). On the evolution of mating systems in birds and mammals. *Am. Nat.* 103, 589–603.

—— (1973). The Red-winged Blackbird in tropical marshes. *Condor* 75, 28–42.

Orr, Y. (1970). Temperature measurements at the nest of the Desert Lark (*Ammomanes deserti deserti*). *Condor* 72, 476–8.

Ottesen, E., Daan, Serge, and Pittendrigh, C. S. (In press). The entrainment of a circadian system by brief light pulses. II. Criteria for stable entrainment. *Am. Nat.*

Owen, D. F. (1959). The breeding season and clutch-size of the Rook *Corvus frugilegus. Ibis* 101, 235–9.

Palmer, R. S. (1972). Patterns of molting. In *Avian biology* (ed. D. S. Farner and J. R. King), Vol. 2, pp. 65–102. Academic Press, New York and London.

Paludán, K. (1951). Contributions to the breeding biology of *Larus argentatus* and *Larus fuscus. Vidensk. Meddr dansk naturh. Foren.* 114, 1–128.

Pařízek, J. (1964). Vascular changes at sites of oestrogen production by parenteral injection of cadmium salts: the destruction of placenta by cadmium salts. *J. Reprod. Fert.* 7, 263.

Parker, J. E. and Arscott, G. M. (1972). Obesity and fertility in a light breed of domestic fowl, *Gallus domesticus. J. Reprod. Fert.* 28, 213–19.

Parlow, A. F. (1961). Bioassay of pituitary luteinizing hormone by depletion of ovarian ascorbic acid. In *Human pituitary gonadotrophins* (ed. A. Albert), pp. 300–10. Thomas, Springfield, Illinois, U.S.A.

—— and Ward, D. N. (1961). Rate of disappearance of LH, PMS, and HCG from plasma. In *Human pituitary gonadotrophins* (ed. A. Albert), pp. 204–9. Thomas, Springfield, Illinois, U.S.A.

Parmelee, D. F. (1970). Breeding behaviour of the Sanderling in the Canadian high Arctic. *Living Bird* 9, 97–146.

—— and Payne, R. B. (1973). On multiple clutches and the breeding strategy of Arctic Sanderlings. *Ibis* 115, 218–26.

——, Greiner, D. W., and Graul, W. D. (1968). Summer schedule and breeding biology of the White-rumped Sandpiper in the central Canadian Arctic. *Wilson Bull.* 80, 1–29.

Parsons, J. (1970). Relationship between egg size and post-hatching chick mortality in the Herring Gull (*Larus argentatus*). *Nature, Lond.* 228, 1221–2.

—— (1975). Seasonal variations in the breeding success of the Herring Gull: an experimental approach to pre-fledging success. *J. Anim. Ecol.* 44, 553–73.

Patel, M. D. (1936). The physiology of the formation of 'pigeon's milk'. *Physiol. Zoöl.* 9, 129–52.

Patterson, I. J. (1965). Timing and spacing of broods in the Black-headed Gull *Larus ridibundus. Ibis* 107, 433–59.

Pavlidis, T. (1973). *Biological oscillators: their mathematical analysis*. Academic Press, New York and London.

Payne, C. G., Lincoln, D. W., and Charles, D. R. (1965). The influence of constant and fluctuating environmental temperatures on time of oviposition under continuous lighting. *Br. Poult. Sci.* 6, 93–5.

Payne, F. (1942). The cytology of the anterior pituitary of the fowl. *Biol. Bull. mar. biol. Lab., Woods Hole* 82, 79–111.

—— (1961). The pituitary of the fowl: a correction and addition. *Anat. Rec.* 140, 321–7.

Payne, R. B. (1967). Gonadal responses of Brown-headed Cowbirds to long daylength. *Condor* 69, 289–97.

—— (1969). *Breeding seasons and reproductive physiology of Tricolored Blackbirds and Redwinged Blackbirds*. University of California Press, Berkeley and Los Angeles, U.S.A.

—— (1972). Mechanisms and control of molt. In *Avian biology* (ed. D. S. Farner and J. R. King), Vol. 2, pp. 103–55. Academic Press, New York and London.

—— (1973a). The breeding season of a parasitic bird, the Brown-headed Cowbird, in central California. *Condor* 75, 80–99.

—— (1973b). Individual laying histories and the clutch size and numbers of eggs of parasitic cuckoos. *Condor* **75**, 414–38.

Pearl, R. and Schoppe, W. F. (1921). Studies on the physiology of reproduction in the domestic fowl. XVIII. Further observations on the anatomical basis of fecundity. *J. exp. Zool.* **34**, 101–18.

Pearse, A. G. E. (1968). *Histochemistry*. Part 1: *Theoretical and applied*. Churchill, London.

Pearson, A. K. (1955). Natural history and breeding behaviour of the Tinamou *Nothoprocta ornata*. *Auk* **72**, 113–27.

Pearson, D. J. (1973). Moult of some Palaearctic warblers wintering in Uganda. *Bird Study* **20**, 24–36.

Pengelley, E. T. (1967). The relation of the external conditions to the onset and termination of hibernation and estivation. In *Mammalian hibernation. Proceedings of the Third International Symposium on Natural Mammalian Hibernation, Toronto 1965* (ed. K. C. Fisher, A. R. Dawe, C. P. Lyman, E. Schonbaum, and F. E. South), Vol. 3, pp. 1–29. Oliver and Boyd, Edinburgh.

Perek, M. and Sulman, F. (1945). The B.M.R. in molting and laying hens. *Endocrinology* **36**, 240–3.

Perrins, C. M. (1963). Survival in the Great Tit, *Parus major*. *Int. orn. Congr.* **13**, 717–28.

—— (1965). Population fluctuations and clutch-size in the Great Tit, *Parus major* L. *J. Anim. Ecol.* **34**, 601–47.

—— (1966). Survival of young Manx Shearwaters *Puffinus puffinus* in relation to their presumed date of hatching. *Ibis* **108**, 132–5.

—— (1967). The short apparent incubation period of the Cuckoo. *Br. Birds* **60**, 51–2.

—— (1970). The timing of birds' breeding seasons. *Ibis* **112**, 242–55.

—— (1973). Some effects of temperature on breeding in the Great Tit and Manx Shearwater. *J. Reprod. Fert. Suppl.* **19**, 163–73.

—— and Moss, D. (1974). Survival of young Great Tits in relation to age of female parent. *Ibis* **116**, 220–4.

Petersen, N. F. and Williamson, K. (1949). Polymorphism and breeding of the Rock Dove in the Faeroe Islands. *Ibis* **91**, 17–23.

Peterson, A. J. (1955). The breeding cycle in the Bank Swallow. *Wilson Bull.* **67**, 235–86.

Pfeiffer, C. A. and Kirschbaum, A. (1943). Relation of interstitial ceel hyperplasia to secretion of male hormone in the sparrow. *Anat. Rec.* **85**, 211–27.

Pfeiffer, L. A. (1947). Gonadotrophic effects of exogenous sex hormones on the testes of sparrows. *Endocrinology* **41**, 92–104.

Phillips, A. R. (1971). Avian breeding cycles: are they related to photoperiods? *An. Inst. Biol. Univ. nac. auton. Méx.* **42**, Ser. Zool. (1), 87–98.

Phillips, J. G. and Bellamy, D. (1962). Aspects of the hormonal control of the nasal gland secretion in birds. *J. Endocr.* **24**, vi–vii.

—— and Chester Jones, I. (1957). The identity of adrenocortical steroids in lower vertebrates. *J. Endocr.* **16**, iii.

Phillips, R. E. (1959). Endocrine mechanisms of the failure of Pintails (*Anas acuta*) to reproduce in captivity. Ph.D. Thesis. Cornell University.

—— and van Tienhoven, A. (1960). Endocrine factors involved in the failure of Pintail Ducks *Anas acuta* to reproduce in captivity. *J. Endocr.* **21**, 253–61.

Pierce, J. G., Liao, T. H., Howard, S. M., Shome, B., and Cornell, J. S. (1971). Studies on the structure of thyrotrophin: its relationship to luteinizing hormone. *Recent Prog. Horm. Res.* **27**, 165–206.

Pitelka, F. A. (1959). Numbers, breeding schedule, and territoriality in Pectoral Sandpipers of northern Alaska. *Condor* **61**, 233–64.

Pittendrigh, C. S. (1954). On temperature independence in a clock system controlling emergence time in *Drosophila. Proc. natn. Acad. Sci. U.S.A.* **40**, 1018–29.
—— (1960). Circadian rhythms and the circadian organization of living systems. *Cold Spring Harb. Symp. quant. Biol.* **25**, 159–84.
—— (1965). On the mechanism of the entrainment of a circadian rhythm by light cycles. In *Circadian clocks: Proceedings of the Feldafing Summer School* (ed. J. Aschoff), pp. 277–97. North-Holland, Amsterdam.
—— (1966). The circadian oscillation in *Drosophila pseudoobscura* pupae; a model for the photoperiodic clock. *Z. Pflanzenphysiol.* **54**, 275–307.
—— (1967). Circadian systems. I. The driving oscillation and its assay in *Drosophila pseudoobscura. Proc. natn. Acad. Sci. U.S.A.* **58**, 1762–7.
—— (1973). Circadian oscillations in cells and the circadian organization of multi-cellular systems. In *The neurosciences: Proceedings of the Third Intensive Study Program—Neurosciences Research Program, Boulder, Colorado 1972* (ed. F. O. Schmitt and F. G. Worden), pp. 437–58. MIT Press, Cambridge, Massachusetts, U.S.A.
—— (In press). The entrainment of a circadian system by brief light pulses. I. The empirical basis of a model. *Am. Nat.*
—— and Bruce, V. G. (1957). An oscillator model for biological clocks. In *Rhythmic and synthetic processes in growth* (ed. D. Rudnick), pp. 75–109. Princeton University Press, New Jersey, U.S.A.
—— and Minis, D. H. (1964). The entrainment of circadian oscillations by light and their role as photoperiodic clocks. *Am. Nat.* **98**, 261–94.
—— and —— (1971). The photoperiodic time measurement in *Pectinophora gossypiella* and its relation to the circadian system in that species. In *Biochronometry* (ed. M. Menaker), pp. 212–47. National Academy of Sciences, Washington D.C.
——, Ottesen, E., and Daan, Serge. (In press). The entrainment of a circadian system by brief light pulses. III. Experimental tests of the model. *Am. Nat.*
Pohl, H. (1971*a*). Seasonal variation in metabolic functions of Bramblings. *Ibis* **113**, 185–93.
—— (1971*b*). Über Beziehungen zwischen circadianen Rhythmen bei Vögeln. *J. Orn., Lpz.* **112**, 266–78.
Pohl-Apel, G. and Sossinka, R. (1975). Gonadenentwicklung beim Wellensittich *Melopsittacus undulatus* unter verschiedenen Lichbedingungen. *J. Orn. Lpz.* **116**, 207–12.
Portenko, L. A. (1959). Studien an einigen seltenen Limicolen aus den nördlichen und östlichen Sibirien II. Der Sichelstrandläufer, *Erolia ferruginea* (Pontopp.). *J. Orn., Lpz.* **100**, 141–72.
Potts, G. R. (1969). Partridge survival project. *A. Rep. Game Res. Ass.* **8**, 14–17.
Powell, G. V. N. (1974). Experimental analysis of the social value of flocking by Starlings (*Sturnus vulgaris*) in relation to predation and foraging. *Anim. Behav.* **22**, 501–5.
Preston, F. W. (1969). Shapes of birds' eggs: extant North American families. *Auk* **86**, 246–64.
Price, F. E. and Bock, C. E. (1973). Polygyny in the Dipper. *Condor* **75**, 457–86.
Priedkalns, J. and Oksche, A. (1969). Ultrastructure of synaptic terminals in the nucleus infundibularis and nucleus supraopticus of *Passer domesticus. Z. Zellforsch. mikrosk. Anat.* **98**, 135–47.
Pulliainen, E. (1970). On the breeding biology of the Dotterel *Charadrius morinellus. Ornis fenn.* **47**, 69–73.
Purves, H. D. (1966). Cytology of the adenohypophysis. In *The pituitary gland* (ed. G. W. Harris and B. T. Donovan), Vol. 1, pp. 147–232. Butterworths, London.

Putman, R. J. and Hinde, R. A. (1973). Effect of the light regime and breeding experience on Budgerigar reproduction. *J. Zool.* **170**, 475–84.

Quay, W. B. (1966). 24-hour rhythms in pineal 5-hydroxytriptamine and hydroxy indole-o-methyl transferase activity in the Macaque. *Proc. Soc. exp. Biol. Med.* **121**, 946.

— (1972). Infrequency of pineal atrophy among birds and its relation to nocturnality. *Condor* **74**, 35–45.

— and Renzoni, A. (1963). Comparative and experimental studies of pineal structure and cytology in passeriform birds. *Riv. Biol.* **56**, 363–407.

Rahn, H. (1939). The development of the chick pituitary with special reference to the cellular differentiation of the pars buccalis. *J. Morph.* **64**, 483–517.

Raikow, R. J. (1971). The osteology and taxonomic position of the White-backed Duck, *Thalassornis leuconotus. Wilson Bull.* **83**, 270–7.

Raitasuo, K. (1964). Social behaviour of the Mallard *Anas platyrhynchos,* in the course of the annual cycle. *Suom. Riista* **24**, 1–72.

Ralph, C. L. (1959). Some effects of hypothalamic lesions on gonadotrophin release in the hen. *Anat. Rec.* **134**, 411–31.

— and Fraps, R. M. (1959). Effect of hypothalamic lesions on progesterone induced ovulation in the hen. *Endocrinology* **65**, 819–24.

—, Grinwich, D. L., and Hall, P. F. (1967a). Hormonal regulation of feather pigmentation in African weaver birds: the exclusion of certain possible mechanisms. *J. exp. Zool.* **166**, 289–94.

—, —, and — (1967b). Studies of the melanogenic response of regenerating feathers in the weaver bird: comparison of two species in response to two gonadotrophins. *J. exp. Zool.* **166**, 283–7.

—, Hedlund, L., and Murphy, W. A. (1967c). Diurnal cycles of melatonin in bird pineal bodies. *Comp. Biochem. Physiol.* **22**, 591–9.

Ramírez, V. D. (1969). Positive feedback mechanisms involved in the control of LH and FSH secretion. In *Progress in endocrinology. Proceedings of the Third International Congress of Endocrinology, Mexico 1968* (ed. C. Gual and F. J. G. Ebling) pp. 532–41. Excerpta Medica, Amsterdam.

Reid, B. (1971a). The weight of the Kiwi and its egg. *Notornis* **18**, 245–9.

— (1971b). Composition of a Kiwi egg. *Notornis* **18**, 250–2.

Reinberg, A. (1974). Aspects of circannual rhythms in man. In *Circannual clocks: annual biological rhythms* (ed. E. T. Pengelley), pp. 423–505. Academic Press, New York and London.

Reiter, R. J. (1969). Pineal–gonadal relationships in male rodents. In *Progress in endocrinology. Proceedings of the Third International Congress of Endocrinology, Mexico 1968* (ed. C. Gual and F. J. G. Ebling), pp. 631–6. Excerpta Medica, Amsterdam.

Reynolds, C. M. (1974). The census of Heronries, 1969–73. *Bird Study* **21**, 129–34.

Reynolds, R. T. (1972). Sexual dimorphism in accipiter hawks: a new hypothesis. *Condor* **74**, 191–7.

Rich, P. V. (1976). The history of birds on the island continent Australia. *Int. orn. Congr.* **16**, 53–65.

Richardson, F. and Fisher, H. I. (1950). Birds of Moku Manu and Manana Islands off Oahu, Hawaii. *Auk* **67**, 285–306.

Richardson, K. C. (1935). The secretory phenomena in the oviduct of the fowl, including the process of shell formation examined by microincineration technique. *Phil. Trans. R. Soc. B* **225**, 149–95.

Richdale, L. E. (1957). *A population study of penguins.* Oxford University Press, London.

Ricklefs, R. E. (1968). Patterns of growth in birds. *Ibis* **110**, 419–51.
— (1969*a*). Natural selection and the development of mortality rates in young birds. *Nature, Lond.* **223**, 922–5.
— (1969*b*). Preliminary models for growth rates of altricial birds. *Ecology* **50**, 1031–9.
— (1973). Patterns of growth in birds. II. Growth rate and mode of development. *Ibis* **115**, 177–201.
— (1975). The evolution of co-operative breeding in birds. *Ibis* **117**, 531–4.
Riddle, O. (1925). Sex in a fraternity of pigeons obtained from an interfamily cross. *Anat. Rec.* **31**, 349–50.
— (1927). The cyclical growth of the vesicula seminalis in birds is hormone controlled. *Anat. Rec.* **37**, 1–11.
— (1963*a*). Prolactin in vertebrate function and organisation. *J. natn. Cancer Inst.* **31**, 1039–110.
— (1963*b*). Prolactin or progesterone as key to parental behaviour: a review. *Anim. Behav.* **11**, 419–32.
— and Bates, R. W. (1933). Concerning anterior pituitary hormones. *Endocrinology* **17**, 689–98.
— and Braucher, P. F. (1931). Control of the special secretion of the crop-gland in pigeons by an anterior pituitary hormone. *Am. J. Physiol.* **97**, 617–25.
— and Johnson, M. W. (1939). An undescribed type of partial sex reversal in dove hybrids from a sub-family cross. *Anat. Rec.* **75**, 509–27.
—, Bates, R., and Dykshorn, S. W. (1932). A new hormone of the anterior pituitary. *Proc. Soc. exp. Biol. Med.* **29**, 1211–2.
—, —, and — (1933). The preparation, identification and assay of prolactin—a hormone of the anterior pituitary. *Am. J. Physiol.* **105**, 191–216.
—, —, and Lahr, E. L. (1935). Prolactin induces broodiness in fowl. *Am. J. Physiol.* **111**, 352–60.
—, Hollander, W. F., and Schooley, J. P. (1945). A race of hermaphrodite-producing pigeons. *Anat. Rec.* **92**, 401–23.
Ridpath, M. G. (1972*a*). The Tasmanian native hen, *Tribonyx mortierii*. I. Patterns of behaviour. *C.S.I.R.O. Wildl. Res.* **17**, 1–51.
— (1972*b*). The Tasmanian native hen, *Tribonyx mortierii*. II. The individual, the group, and the population. *C.S.I.R.O. Wildl. Res.* **17**, 53–90.
— (1972*c*). The Tasmanian native hen, *Tribonyx mortierii*. III. Ecology. *C.S.I.R.O. Wildl. Res.* **17**, 91–118.
Riley, G. (1936). Light regulation of sexual activity in the male sparrow (*Passer domesticus*). *Proc. Soc. exp. Biol. Med.* **34**, 331–2.
Ringoen, A. R. (1943). Effects of injections of testosterone propionate on the reproductive system of the female English sparrow *Passer domesticus* (Linnaeus). *J. Morph.* **73**, 423–40.
Robinson, A. H. (1955). Nesting seasons of Western Australian birds—a further contribution. *West. Aust. Nat.* **4**, 187–92.
Rohwer, S. A. (1971). Moult and the annual cycle of the Chuck-Will's-Widow, *Caprimulgus carolinensis*. *Auk* **88**, 485–519.
Rollo, M. and Domm, L. V. (1943). Light requirements of the weaver-finch (*Pyromelana*). *Auk* **60**, 357–67.
Romanoff, A. L. (1960). *The avian embryo: structural and functional development.* Macmillan, New York.
— and Romanoff, A. J. (1949). *The avian egg.* Wiley, New York.

Romeis, B. (1940). Die Hypophyse. In *Handbuch der mikroskopischen Anatomie des Menchen* (ed. W. von Möllendorf), Vol. 6, Pt 3. Springer, Berlin.

Rosales, A. A., Biellier, H. V., and Stephenson, A. B. (1968). Effect of light cycles on oviposition and egg production. *Poult. Sci.* **47**, 586–91.

Rothschild, I. and Fraps, R. M. (1949). The interval between normal release of ovulating hormone and ovulation in the domestic hen. *Endocrinology* **44**, 134–40.

Rowan, M. K. (1963). The Yellowbill Duck *Anas undulata* Dubois in Southern Africa. *Ostrich* Suppl. (5), 1–56.

Rowan, W. (1926). On photoperiodism, reproductive periodicity and the annual migrations of birds and certain fishes. *Proc. Boston Soc. nat. Hist.* **38**, 147–89.

—— (1929). Experiments in bird-migration. *Proc. Boston Soc. nat. Hist.* **39**, 151–208.

—— (1932). Experiments in bird migration. III. The effects of artificial light, castration and certain extracts on the autumn movement of the American Crow (*Corvus brachyrhynchos*). *Proc. natn. Acad. Sci. U.S.A.* **18**, 639–54.

Rowley, I. (1965*a*). The life history of the Superb Blue Wren. *Emu* **64**, 251–97.

—— (1965*b*). White-winged Choughs. *Aust. natur. Hist.* **15**, 81–5.

—— (1968). Communal species of Australian birds. *Bonn. zool. Beitr.* **19**, 362–70.

Rowley, I. C. R. (1976). Co-operative breeding in Australian birds. *Int. orn. Congr.* **16**, 657–66.

Royama, T. (1966*a*). A re-interpretation of courtship feeding. *Bird Study* **13**, 116–29.

—— (1966*b*). Factors governing feeding rate, food requirement and brood size of nestling Great Tits *Parus major*. *Ibis* **108**, 313–47.

—— (1969). A model for the global variation of clutch size in birds. *Oikos* **20**, 562–7.

—— (1970). Factors governing the hunting behaviour and selection of food by the Great Tit (*Parus major* L.). *J. Anim. Ecol.* **39**, 619–68.

Russell, D. H. and Farner, D. S. (1968). Acetylcholinesterase and gonadotropin activity in the anterior pituitary. *Life Sci.* **7**, 1217–21.

Rutherford, W. H. (ed.) (1965). *Description of Canada Goose populations common to the central flyway*. The Central Flyway Waterfowl Council, Colorado Game, Fish, and Parks Department, U.S.A.

Rutledge, J. T. and Schwab, R. G. (1974). Testicular metamorphosis and prolongation of spermatogenesis in Starlings (*Sturnis vulgaris*) in the absence of daily photostimulation. *J. exp. Zool.* **187**, 71–6.

Ryder, J. P. (1967). The breeding biology of Ross' Geese in the Perry River region, North-west Territories. *Can. Wildl. Serv. Rep. Ser.* **3**, 1–56.

—— (1970). A possible factor in the evolution of clutch size in Ross' Goose. *Wilson Bull.* **82**, 5–13.

Rzasa, J. and Ewy, Z. (1970). Effect of vasotocin and oxytocin on oviposition in the hen. *J. Reprod. Fert.* **21**, 549–50.

Sachs, B. D. (1967). Photoperiodic control of the cloacal gland of Japanese Quail. *Science, N.Y.* **157**, 201.

Sadler, K. C., Tomlinson, R. E., and Wight, H. M. (1970). Progress of primary feather molt of adult Mourning Doves in Missouri. *J. Wildl. Mgmt* **34**, 783–8.

Saeki, Y. and Tanabe, Y. (1955). Changes in prolactin content of fowl pituitary during broody periods and some experiments on the induction of broodiness. *Poult. Sci.* **34**, 909–19.

Salem, M. H. M., Norton, H. W., and Nalbandov, A. V. (1970*a*). A study of ACTH and CRF in chickens. *Gen. comp. Endocrinol.* **14**, 270–80.

——, ——, and —— (1970*b*). The role of vasotocin and of CRF in ACTH release in the chicken. *Gen. comp. Endocrinol.* **14**, 281–9.

Salomonsen, F. (1939*a*). Oological studies in gulls. I. Egg producing power of *Larus argentatus*. *Dansk. Orn. Foren. Tidsskr.* **33**, 113–33.

—— (1939*b*). Moults and sequences of plumages in the Rock Ptarmigan (*Lagopus mutus* (Montin)). *Vidensk. Meddr dansk naturh. Foren.* 103, 1—491.

—— (1950). *The birds of Greenland.* Ejnar Munksgaard, Copenhagen.

Sandnes, G. C. (1957). Fertility and viability in intergeneric pheasant hybrids. *Evolution, Lancaster, Pa.* 11, 426—44.

Sansum, E. L. and King, J. R. (1975). Photorefractoriness in a sparrow: phase of circadian photosensitivity elucidated by skeleton photoperiods. *J. comp. Physiol.* 98, 183—8.

Sayler, A. and Wolfson, A. (1968*a*). Influence of the pineal gland on gonadal maturation in the Japanese Quail. *Endocrinology* 83, 1237—46.

—— and —— (1968*b*). Role of the eyes and superior cervical ganglia on the effects of light on the pineal and gonads of the Japanese Quail. *Archs Anat. Histol. Embryol.* 51, 213—22.

——, Dowd, A. J., and Wolfson, A. (1970). Influence of photoperiod on the localisation of Δ^5-3β-hydroxysteroid dehydrogenase in the ovaries of maturing Japanese Quail. *Gen. comp. Endocrinol.* 15, 20—30.

Scanes, C. G. (1974). Some *in vitro* effects of synthetic thyrotrophin releasing factor on the secretion of thyroid stimulating hormone from the anterior pituitary gland of the domestic fowl. *Neuroendocrinology* 15, 1—9.

—— and Follett, B. K. (1972). Fractionation and assay of chicken pituitary hormones. *Br. Poult. Sci.* 13, 603—10.

——, Follett, B. K., and Goos, H. J. Th. (1972*a*). Cross-reaction in a chicken LH radioimmunoassay with plasma and pituitary extracts from various species. *Gen. comp. Endocrinol.* 19, 596—600.

——, Nicholls, T. J., and Follett, B. K. (1972*b*). Variations in the level of plasma immunoreactive luteinizing hormone during breeding cycles in Japanese Quail. *Proc. Soc. Endocr.* 55, xv—xvi.

——, Cheeseman, P., Phillips, J. G., and Follett, B. K. (1974). Seasonal and age variation of circulating immunoreactive luteinizing hormone in captive Herring Gulls, *Larus argentatus. J. Zool.* 174, 369—75.

Schäfer, E. (1954). Zur Biologie des Steisshuhnes *Nothocercus bonapartei. J. Orn., Lpz.* 95, 219—32.

Schally, A. V., Arimura, A., Bowers, C. Y., Kastin, A. J. , Sawano, S., and Redding, T. W. (1968). Hypothalamic neurohormones regulating anterior pituitary function. *Recent Prog. Horm. Res.* 24, 497—588.

Schifferli, L. (1973). The effect of egg weight on the subsequent growth of nestling Great Tits *Parus major. Ibis* 115, 549—58.

Schmid, W. D. (1965). Energy intake of the Mourning Dove (*Zenaida macruora marginella*). *Science, N.Y.* 150, 1171—2.

Schmidt-Nielsen, K. and Kim, Y. T. (1964). The effect of salt intake on size and function of the salt gland of ducks. *Auk* 81, 160—72.

Schneider, H. P. G. and McCann, S. M. (1969). Possible role of dopamine as transmitter to promote discharge of LH-releasing factor. *Endocrinology* 85, 121—32.

—— and —— (1970). Luteinizing hormone-releasing factor discharged by dopamine in rats. *J. Endocr.* 46, 401—2.

Schönwetter, M. (1967). *Handbuch der Oologie*, pp. 3—64. Lieferung 14. Akademie Verlag, Berlin.

Schooley, J. P. (1937). Pituitary cytology in pigeons. *Cold Spring Harb. Symp. quant. Biol.* 5, 165—79.

Schreiber, R. W. and Ashmole, N. P. (1970). Sea-bird breeding seasons on Christmas Island, Pacific Ocean. *Ibis* 112, 363—94.

Schultz, A. (1925). Uber Cholesterinesterverfettung. *Verh. dt. path. Ges.* 20, 120—3.

Schüz, E. (1957). Das Verschlingen eigener Junger ('kronismus') bei Vögeln und seine Bedeutung. *Vogelwarte* **19**, 1–15.

Schwab, R. G. (1970). Light-induced prolongation of spermatogenesis in the European Starling, *Sturnus vulgaris*. *Condor* **72**, 466–70.

—— (1971). Circannian testicular periodicity in the European Starling in the absence of photoperiodic change. In *Biochronometry* (ed. M. Menaker), pp. 428–45. National Academy of Sciences, Washington D.C.

—— and Lott, D. F. (1969). Testis growth and regression in Starlings (*Sturnus vulgaris*) as a function of the presence of females. *J. exp. Zool.* **171**, 39–42.

—— and Rutledge, J. T. (1973). Effects of natural and artificial illumination on testicular metamorphosis in the European Starling (*Sturnus vulgaris*): maturation, involution, and photorefractory phases. *Int. Congr. : Le soleil au service de l'homm Paris 1973*. B.15–1 to B.15–11.

Schweiger, H. G. (1972). Circadian rhythms: subcellular and biochemical aspects. In *Circadian rhythmicity. Proceedings of the International Symposium on Circadian Rhythmicity, Wageningen 1971*, pp. 157–74. Pudoc, Wageningen.

Scott, D. M. and Middleton, A. L. A. (1968). The annual testicular cycle of the Brown-headed Cowbird (*Molothrus ater*). *Can. J. Zool.* **46**, 77–87.

Scott, P. and Wildfowl Trust (1972). *The swans*. Michael Joseph, London.

Seel, D. C. (1964). An analysis of the nest record cards of the Tree Sparrow. *Bird Study* **11**, 265–71.

—— (1968a). Breeding seasons of the House Sparrow and Tree Sparrow *Passer* spp. at Oxford. *Ibis* **110**, 129–44.

—— (1968b). Clutch-size, incubation and hatching success in the House Sparrow and Tree Sparrow *Passer* spp. at Oxford. *Ibis* **110**, 270–82.

—— (1970). Nestling survival and nestling weights in the House Sparrow and Tree Sparrow *Passer* spp. at Oxford. *Ibis* **112**, 1–14.

Seiskari, P. (1962). On the winter ecology of the Capercaille *Tetroa urogallus*, and the Black Grouse, *Lyrurus tetrix*, in Finland. *Finn. Game Res.* **22**, 1–119.

Sekeris, C. E. (1969). Increased template activity of chromatin: a first step for the expression of hormone action. In *Progress in endocrinology. Proceedings of the Third International Congress of Endocrinology, Mexico 1968* (ed. C. Gual and F. J. G. Ebling), pp. 7–16. Excerpta Medica, Amsterdam.

Selinger, H. E. and Bermont, G. (1967). Hormonal control of aggressive behaviour in Japanese Quail. *Behaviour* **28**, 255–68.

Selander, R. K. (1958). Age determination and molt in the Boat-tailed Grackle. *Condor* **60**, 355–76.

—— (1966). Sexual dimorphism and differential niche utilization in birds. *Condor* **68**, 113–51.

—— (1972). Sexual selection and dimorphism in birds. In *Sexual selection and the descent of man* (ed. B. Campbell), pp. 180–230. Heineman Educational Books, London.

—— and Giller, D. R. (1961). Analysis of sympatry of Great-tailed and Boat-tailed Grackles. *Condor* **63**, 29–86.

—— and Hauser, R. J. (1965). Gonadal and behavioural cycles in the Great-tailed Grackle. *Condor* **67**, 157–82.

Serventy, D. L. (1963). Egg-laying time table of the Slender-billed Shearwater *Puffinus tenuirostris*. *Int. orn. Congr.* **13**, 338–43.

—— (1971). Biology of desert birds. In *Avian biology* (ed. D. S. Farner and J. R. King), Vol. 1, pp. 287–339. Academic Press, New York and London.

—— and Marshall, A. J. (1957). Breeding periodicity of Western Australian birds with an account of unseasonal breeding in 1953 and 1955. *Emu* **57**, 99–126.

Shaffner, C. S. (1954). Feather papilla stimulation by progesterone. *Science, N.Y.* **120**, 345.

Shank, M. C. (1959). The natural termination of the refractory period in the Slate-coloured Junco and in the White-throated Sparrow. *Auk* **76**, 44–54.

Sharma, D. C., Racz, E. A., Dorfman, R. I., and Schoen, E. J. (1967). A comparative study of the biosynthesis of testosterone by human testes and a virilizing inter-stitial cell tumour. *Acta endocr., Copenh.* **56**, 726–36.

Sharp, P. J. and Follett, B. K. (1968). The distribution of monoamines in the hypo-thalamus of the Japanese Quail *Coturnix coturnix japonica. Z. Zellforsch. mikrosk. Anat.* **90**, 245–62.

── and ── (1969a). The blood supply to the pituitary and basal hypothalamus in the Japanese Quail (*Coturnix coturnix japonica*). *J. Anat.* **104**, 227–32.

── and ── (1969b). The effect of hypothalamic lesions on gonadotrophin release in Japanese Quail (*Coturnix coturnix japonica*). *Neuroendocrinology* **5**, 205–18.

── and ── (1969c). The effect of reserpine on the pituitary-gonadal axis in quail. In *Seminar on hypothalamic and endocrine function in birds*, p. 43. International House of Japan, Tokyo.

── and ── (1970). The adrenergic supply within the avian hypothalamus. In *Aspects of neuroendocrinology: International Symposium on Neurosecretion, Kiel 1969* (ed. W. Bargmann and B. Scharrer), pp. 95–103. Springer-Verlag, Berlin and New York.

──, Moss, R., and Watson, A. (1974). Seasonal variations in plasma luteinizing hormone levels in male Red Grouse (*Lagopus lagopus scoticus*). *J. Endocr.* **64**, 44P.

Shaughnessy, P. D. (1970a). Serum proteins of two sibling species of giant petrel (*Macronectes* spp.). *Comp. Biochem. Physiol.* **33**, 721–3.

── (1970b). The genetics of plumage phase dimorphism of the Southern Giant Petrel *Macronectes giganteus. Heredity, Lond.* **25**, 501–6.

── (1971). Frequency of the white phase of the Southern Giant Petrel, *Macronectes giganteus. Aust. J. Zool.* **19**, 77–83.

Shellswell, G. B., Gosney, S., and Hinde, R. A. (1975). Photoperiodic control of Budgerigar reproduction: circadian changes in sensitivity. *J. Zool.* **175**, 53–60.

Shirley, H. V. and Nalbandov, A. V. (1956). Effects of transecting hypophyseal stalks in laying hens. *Endocrinology* **58**, 694–700.

Short, L. L. (1969). A new genus and species of gooselike swan from the Pliocene of Nebraska. *Am. Mus. Novit.* no. 2369, 1–7.

── (1970). Mid-pleistocene birds from western Nebraska, including a new species of sheldgoose. *Condor* **72**, 147–52.

Short, R. V. (1967). Reproduction. *A. Rev. Physiol.* **29**, 373–400.

── (1972). Role of hormones in sex cycles. In *Reproduction in mammals.* Vol. 3: *Hormones in reproduction* (ed. C. R. Austin and R. V. Short), pp. 42–73. Cambridge University Press, London.

Sibley, G. G. (1957). The evolutionary and taxonomic significance of sexual dimorphism and hybridization in birds. *Condor* **59**, 166–91.

──(1961). Hybridization and isolating mechanisms. In *Vertebrate speciation, a symposium* (ed. W. F. Blair), pp. 69–95. University of Texas Press, Austin, U.S.A.

Siegel, H. S. and Siegel, P. B. (1961). The relationship of social competition with endocrine weights and activity in male chickens. *Anim. Behav.* **9**, 151–8.

Siegel, P. B. and Siegel, H. S. (1969). Endocrine responses of six stocks of chickens reared at different population densities. *Poult Sci.* **48**, 1425–33.

Siegfried, W. R. (1968). Breeding season, clutch and brood sizes in Verreaux's Eagle. *Ostrich* **39**, 139–45.

── (1969). The proportion of yolk in the egg of the Maccoa Duck. *Wildfowl* **20**, 78.

—— (1971). Moult of the primary remiges in three species of *Streptopelia* doves. *Ostrich* **42**, 161–5.

—— and Frost, P. G. H. (1975). Continuous breeding and associated behaviour in the Moorhen *Gallinula chloropus*. *Ibis* **117**, 102–9.

Siller, W. G. (1956). A Sertoli cell tumour causing feminization in a Brown Leghorn capon. *J. Endocr.* **14**, 197–203.

Silver, R., Reboulleau, C., Lehrman, D. S., and Feder, H. M. (1974). Radioimmunoassay of plasma progesterone during the reproductive cycle of male and female Ring Doves (*Streptopelia risoria*). *Endocrinology* **94**, 437–44.

Simkiss, K. (1974). The air space of an egg: an embryonic 'cold nose'? *J. Zool.* **173**, 225–32.

Singh, D. V. and Turner, C. W. (1967). Effect of melatonin upon thyroid hormone secretion rate and endocrine glands of chicks. *Proc. Soc. exp. Biol. Med.* **125**, 407–11.

Siopes, T. D. and Wilson, W. O. (1975). The cloacal gland—an external indicator of testicular development in *Coturnix*. *Poult. Sci.* **54**, 1225–9.

Skead, D. M. (1971). A study of the Rock Pigeon *Columba guinea*. *Ostrich* **42**, 65–9.

Skutch, A. F. (1951). Life history of Longuemare's Hermit Hummingbird. *Ibis* **93**, 180–95.

—— (1957). Migratory and resident warblers in Central America. In *The warblers of America*, pp. 275–85. Devin-Adair, New York.

—— (1966). A breeding census and nesting success in Central America. *Ibis* **108**, 1–16.

Slagsvold, T. (1975*a*). Breeding time of birds in relation to latitude. *Norw. J. Zool.* **23**, 213–18.

—— (1975*b*). Critical period for regulation of Great Tit (*Parus major* L.) and Blue Tit (*Parus caeruleus* L.) populations. *Norw. J. Zool.* **23**, 67–88.

Slater, P. (1959). Breeding periods of birds in the Kimberley Division, Western Australia. *West. Aust. Nat.* **7**, 35–41.

Slater, P. J. B. (1969). The stimulus to egg-laying in the Bengalese Finch. *J. Zool.* **158**, 427–40.

—— (1970). Nest building in the Bengalese Finch. II. The influence of hormonal and experiential factors on the male. *Behaviour* **37**, 24–39.

Smith, D. G., Wilson, C. R., and Frost, H. H. (1970). Fall nesting Barn Owls in Utah. *Condor* **72**, 492.

Smith, E. R. and Davidson, J. M. (1968). Role of estrogen in the cerebral control of puberty in female rats. *Endocrinology* **80**, 100–8.

Smith, J. Maynard (1964). Group selection and kin selection. *Nature, Lond.* **201**, 1145–7.

—— and Ridpath, M. G. (1972). Wife sharing in the Tasmanian Native Hen *Tribonyx mortierii*: a case of kin selection? *Am. Nat.* **106**, 447–52.

Smith, J. N. M. and Dawkins, R. (1971). The hunting behaviour of individual Great Tits in relation to spatial variations in their food density. *Anim. Behav.* **19**, 695–706.

—— and Sweatman, H. P. A. (1974). Food searching behaviour of titmice in patchy environments. *Ecology* **55**, 1216–32.

Smith, K. B. V. and Lacy, D. (1959). Residual bodies of seminiferous tubules of the rat. *Nature, Lond.* **184**, 249–51.

Smith, P. C. and Evans, P. R. (1973). Studies of shorebirds at Lindisfarne, Northumberland. I. Feeding ecology and behaviour of the Bar-tailed Godwit. *Wildfowl* **24**, 135–9.

Smith, P. M. and Follett, B. K. (1972). Luteinizing hormone releasing factor in the quail hypothalamus. *J. Endocr.* **53**, 131–8.

Smith, R. W., Brown, I. L., and Mewaldt, L. R. (1969). Annual activity patterns of caged non-migratory White-crowned Sparrows. *Wilson Bull.* **81**, 419–40.

Snow, B. K. (1960). The breeding biology of the Shag *Phalacrocorax aristotelis* on the island of Lundy, Bristol Channel. *Ibis* **102**, 554–75.

— (1974). Lek behaviour and breeding of Guy's Hermit Hummingbird *Phaethornis guy. Ibis* **116**, 278–97.

Snow, D. W. (1955). The breeding of Blackbird, Song Thrush, and Mistle Thrush in Great Britain. 3. Nesting success. *Bird Study* **2**, 169–78.

— (1958). *A study of blackbirds.* Allen and Unwin, London.

— (1961). The natural history of the Oilbird, *Steatornis caripensis,* in Trinidad, W.I. 1. General behaviour and breeding habits. *Zoologica, N.Y.* **46**, 27–48.

— (1962a). The natural history of the Oilbird, *Steatornis caripensis,* in Trinidad, W.I. 2. Population, breeding ecology and food. *Zoologica, N.Y.* **47**, 199–221.

— (1962b). A field study of the Black and White Manakin, *Manacus manacus* in Trinidad. *Zoologica, N.Y.* **47**, 65–104.

— (1965a). The breeding of Audubon's Shearwater (*Puffinus lherminieri*) in the Galapagos. *Auk* **82**, 591–7.

— (1965b). The breeding of the Red-billed Tropic Bird in the Galapagos Islands. *Condor* **67**, 210–14.

— (1966). Annual cycle of the Yellow Warbler in the Galapagos. *Bird-Banding* **37**, 44–9.

— (1969). The moult of British thrushes and chats. *Bird Study* **16**, 115–29.

— (1971). Notes on the biology of the Cock-of-the-rock (*Rupicola rupicola*). *J. Orn., Lpz.* **112**, 323–33.

— and Snow, B. K. (1964). Breeding seasons and annual cycles of Trinidad land-birds. *Zoologica, N.Y.* **49**, 1–39.

— and — (1966). The breeding season of the Madeiran Storm Petrel *Oceanodroma castro* in the Galapagos. *Ibis* **108**, 283–4.

— and — (1967). The breeding cycle of the Swallow-tailed Gull *Creagrus furcatus. Ibis* **109**, 14–24.

Soikkeli, M. (1974). Size variation of breeding Dunlins in Finland. *Bird Study* **21**, 151–4.

Solomon, M. E. (1964). Analysis of procedures involved in the natural control of insects. *Adv. ecol. Res.* **2**, 1–58.

Sossinka, R. (1975). Quantitative Untersuchungen zur sexuellen Reifung des Zebrafinken, *Taeniopygia castanotis* Gould. *Verh. dt. zool. Ges.* **1974**, 344–7.

Southern, H. N. (1954). Mimicry in Cuckoos' eggs. In *Evolution as a process* (ed. J. S. Huxley, A. C. Hardy, and E. B. Ford), pp. 209–32. Allen and Unwin, London.

— (1964). Parasitism. In *A new dictionary of birds* (ed. A. Landsborough Thomson), pp. 593–7. Nelson, London.

— (1970). The natural control of a population of Tawny Owls (*Strix aluco*). *J. Zool.* **162**, 197–285.

— and Lowe, V. P. W. (1968). The pattern of distribution of prey and predation in Tawny Owl territories. *J. Anim. Ecol.* **37**, 75–97.

Spjøtvoll, O. (1970). [On *Eremophila alpestris* in a mountain area in Norway.] *Sterna* **9**, 163–74.

Stafford, J. (1971). The Heron population of England and Wales, 1928–1970. *Bird Study* **18**, 218–21.

Stanley, A. J. and Witschi, E. (1940). Germ cell migration in relation to asymmetry in the sex glands of hawks. *Anat. Rec.* **76**, 329–42.

Steel, E. A. and Hinde, R. A. (1963). Hormonal control of brood patch and oviduct development in domesticated Canaries. *J. Endocr.* **26**, 11–24.

Steel, E. and Hinde, R. A. (1964). Effect of exogenous oestrogen on brood patch development of intact and overiectomized Canaries. *Nature, Lond.* **202**, 718—19.

—— and —— (1966*a*). Effect of artificially increased day-length in winter on female domesticated Canaries. *J. Zool.* **149**, 1—11.

—— and —— (1966*b*). Effect of exogenous serum gonadotrophin (PMS) on aspects of reproductive development in female domesticated Canaries. *J. Zool.* **149**, 12—30.

—— and —— (1972*a*). Influence of photoperiod on PMSG-induced nest-building in Canaries. *J. Reprod. Fert.* **31**, 425—31.

—— and —— (1972*b*). Influence of photoperiod on oestrogenic induction of nest-building in Canaries. *J. Endocr.* **55**, 265—78.

——, Follett, B. K., and Hinde, R. A. (1975). The role of short days in the termination of photorefractoriness in female Canaries (*Serinus canarius*). *J. Endocr.* **64**, 451—64.

Steelman, S. and Pohley, F. H. (1953). Assay of the follicle stimulating hormone based on the augmentation with human chorionic gonadotrophin. *Endocrinology* **53**, 604—16.

Steen, J. (1958). Climatic adaptation in some small northern birds. *Ecology* **39**, 625—9

Stern, J. M. and Lehrman, D. S. (1969). Role of testosterone in progesterone-induced incubation behaviour in male Ring Doves (*Streptopelia risoria*). *J. Endocr.* **44**,13—2?

Stetson, M. H. (1969*a*). The role of the median eminence in control of photoperiodically induced testicular growth in the White-crowned Sparrow *Zonotrichia leucophrys gambelii. Z. Zellforsch. mikrosk. Anat.* **93**, 369—94.

—— (1969*b*). Hypothalamic regulation of FSH and LH secretion in male and female Japanese Quail. *Am. Zool.* **7**, 1078—9.

—— (1971). Neuroendocrine control of photoperiodically induced fat deposition in White-crowned Sparrows. *J. exp. Zool.* **176**, 409—14.

—— (1972). Feedback regulation of testicular function in Japanese Quail: testosterone implants in the hypothalamus and adenohypophysis. *Gen. comp. Endocrinol.* **19**, 37—47.

—— (1973). Recovery of gonadal function following hypothalamic lesions in Japanese Quail. *Gen. comp. Endocrinol.* **20**, 76—85.

—— and Erickson, J. E. (1970). The antigonadal properties of prolactin in birds: Failure of prolactin to inhibit the uptake of ^{32}P by testes of cockerels *in vivo. Gen. comp. Endocrinol.* **15**, 484—7.

—— and —— (1971). Endocrine effects of castration in White-crowned Sparrows. *Gen. comp. Endocrinol.* **17**, 105—14.

—— and —— (1972). Hormonal control of photoperiodically induced fat deposition in White-crowned Sparrows. *Gen. comp. Endocrinol.* **19**, 355—62.

——, Lewis, R. A., and Farner, D. S. (1973). Some effects of exogenous gonadotropins and prolactin on photostimulated and photorefractory White-crowned Sparrows. *Gen. comp. Endocrinol.* **21**, 424—30.

Stieglitz, W. O. and Wilson, C. T. (1968). Breeding biology of the Florida Duck. *J. Wildl. Mgmt* **32**, 921—34.

Stiles, F. G. and Wolf, L. L. (1974). A possible circannual moult rhythm in a tropical hummingbird. *Am. Nat.* **108**, 341—54.

Stockell-Hartree, A. and Cunningham, F. J. (1969). Purification of chicken pituitary follicle-stimulating hormone and luteinizing hormone. *J. Endocr.* **43**, 609—16.

—— and —— (1971). The pituitary gland. In *Physiology and biochemistry of the domestic fowl* (ed. D. J. Bell and B. M. Freeman), Vol. 1, pp. 428—57. Academic Press, New York.

Stokes, A. W. (1950). Breeding behaviour of the Goldfinch. *Wilson Bull.* **62**, 107—27.

Stolt, B-O. (1969). Temperature and air pressure experiments on activity in passerine birds with notes on seasonal and circadian rhythms. *Zool. Bidr. Upps.* **38**, 175—231.

Stonehouse, B. (1962). The tropic birds (Genus *Phaethon*) of Ascension Island. *Ibis* **103b**, 124.–61.

—— and Stonehouse, S. (1963). The Frigate Bird *Fregata aquila* of Ascension Island. *Ibis* **103b**, 409–22.

Storey, C. R. and Nicholls, T. J. (1976). Some effects of manipulation of daily photoperiod on the onset of a photorefractory state in Canaries (*Serinus canarius*). *Gen. comp. Endocrinol.* **30**, 204–8.

Stresemann, E. (1939). Die Vögel von Celebes. *J. Orn., Lpz.* **87**, 300–424.

—— (1967). Inheritance and adaptation in moult. *Int. orn. Congr.* **14**, 75–80.

Sturkie, P. D. and Lin, Y. C. (1966). Release of vasotocin and oviposition in the hen. *J. Endocr.* **35**, 325–6.

—— and Meyer, D. (1972). Circadian rhythm in blood and pineal levels of serotonin in chickens. *Fedn Proc. Fedn Am. Soc. exp. Biol.* **31**, abstr. 629.

Summers-Smith, D. (1963). *The House Sparrow*. New Naturalist Monograph Series. Collins, London.

Suomalainen, H. (1938). The effect of temperature on the sexual activity of nonmigratory birds, stimulated by artificial lighting. *Ornis fenn.* **14**, 108–12.

Sutherland, E. W., Hardman, J. G., Butcher, R. W., and Broadus, A. E. (1969). The biological role of cyclic AMP (some areas of contrast with cyclic GMP). In *Progress in endocrinology. Proceedings of the Third International Congress of Endocrinology, Mexico 1968* (ed. C. Gual and F. J. G. Ebling), pp. 26–32. Excerpta Medica, Amsterdam.

Swainson, G. W. (1970). Co-operative rearing in the Bell Miner. *Emu* **70**, 183–8.

Szumowski, P. and Theret, M. (1965). Causes possibles de la faible fertilité des oies et des difficultés de son amélioration. *Recl Méd. vét. Éc. Alfort* **141**, 583.

Taber, E. (1951). Androgen secretion in the fowl. *Endocrinology* **48**, 6–16.

—— (1964). Intersexuality in birds. In *Intersexuality in vertebrates including man* (ed. C. N. Armstrong and A. J. Marshall), pp. 285–310. Academic Press, New York and London.

——, Clayton, M., Knight, J., Gambrell, D., Flowers, J., and Ayres, C. (1958). Ovarian stimulation in the immature fowl by desiccated pituitaries. *Endocrinology* **6284**–9.

Taleisnik, S., Caligaris, L., and Astrada, J. J. (1966). Effect of copulation on the release of pituitary gonadotrophins in male and female rats. *Endocrinology* **79**, 125.

——, Tomatis, M. E., and Celis, M. E. (1972). Role of catecholamines in the control of melanocyte-stimulating hormone secretion in rats. *Neuroendocrinology* **10**, 235–45.

Tanaka, K. (1968). [Fluctuation of pituitary LH levels during the egg-laying cycle of the hen.] *Jap. J. zootech. Sci.* **39**, 377–85.

—— and Nakajo, S. (1962). Participation of neurohypophysial hormone in oviposition in the hen. *Endocrinology* **70**, 453–8.

—— and Yoshioka, S. (1967). Luteinizing hormone activity of the hen's pituitary during the egg-laying cycle. *Gen. comp. Endocrinol.* **9**, 374–9.

Thapliyal, J. P. (1969). Thyroid in avian reproduction. *Gen. comp. Endocrinol.* Suppl., **2**, 11–122.

—— and Saxena, R. N. (1964). Absence of a refractory period in the Common Weaver Bird. *Condor* **66**, 199–208.

Thompson, W. A., Vertinsky, I., and Krebs, J. R. (1974). The survival value of flocking in birds: a simulation model. *J. Anim. Ecol.* **43**, 785–820.

Thomson, A. L. (1950). Factors determining the breeding seasons of birds: an introductory review. *Ibis* **92**, 173–84.

Thomson, A. Landsborough (1964). *A new dictionary of birds*. Nelson, London.

Threadgold, L. T. (1958). Photoperiodic response of the House Sparrow, *Passer domesticus. Nature, Lond.* **182**, 407–8.

—— (1960a). Testicular response of the House Sparrow *Passer domesticus* to short photoperiods and low intensities. *Physiol. Zoöl.* **33**, 190–205.

—— (1960b). A study of the annual cycle of the House Sparrow at various latitudes. *Condor* **62**, 190–201.

Tinbergen, L. (1960). The natural control of insects in pinewoods. I. Factors influencing the intensity of predation by songbirds. *Archs néerl. Zool.* **13**, 265–336.

Tinbergen, N. (1935). Field observations of east Greenland birds. I. The behaviour of the Red-necked Phalarope (*Phalaropus lobatus* (L.)) in spring. *Ardea* **24**, 1–42.

Tindall, D. J., Schrader, W. T., and Means, A. R. (1974). The production of androgen binding protein by Sertoli cells. In *Hormone binding and target cell activation in the testis* (Current topics in molecular endocrinology, Vol. 1) (ed. M. L. Dufau and A. R. Means), pp. 167–75. Plenum Press, New York and London.

Tixier-Vidal, A. (1963). Histophysiologie de l'adénohypophyse des oiseaux. In *Cytologie de l'adénohypophyse* (ed. J. Benoit and C. Da Lage), pp. 255–73. Editions du C.N.R.S., Paris.

—— and Assenmacher, I. (1966). Etude cytologique de la préhypophyse du Pigeon pendant la couvaison et la lactation. *Z. Zellforsch. mikrosk. Anat.* **69**, 489–519.

—— and Benoit, J. (1962). Influence de la castration sur la cytologie préhypophysaire du Canard mâle. *Archs Anat. microsc. Morph. exp.* **51**, 266–86.

—— and Follett, B. K. (1973). The adenohypophysis. In *Avian biology* (ed. D. S. Farner and J. R. King), Vol. 3, pp. 109–82. Academic Press, New York and London.

—— and Picart, R. (1971). Electron microscopic localisation of glycoproteins in pituitary cells of duck and quail. *J. Histochem. Cytochem.* **19**, 775–97.

——, Herlant, M., and Benoit, J. (1962). La préhypophyse du canard Pékin au cours du cycle annuel. *Archs Biol., Liège* **73**, 317–68.

——, Follett, B. K., and Farner, D. S. (1968). The anterior pituitary of the Japanese Quail, *Coturnix coturnix japonica.* The cytological effects of photoperiodic stimulation. *Z. Zellforsch. mikrosk. Anat.* **92**, 610–35.

Tompa, F. S. (1967). Reproductive success in relation to breeding density in Pied Flycatchers, *Ficedula hypoleuca* (Pallas). *Acta zool. fenn.* **118**, 28.

Tordoff, H. B. and Dawson, W. R. (1965). The influence of daylight on reproductive timing in the Red Crossbill. *Condor* **67**, 416–22.

Tougard, C. (1971). Recherches sur l'origine cytologique de l'hormone mélanophorotrope chez les Oiseaux. *Z. Zellforsch. mikrosk. Anat.* **116**, 375–90.

Trauger, D. L., Dzubin, A., and Ryder, J. P. (1971). White geese intermediate between Ross' Geese and Lesser Snow Geese. *Auk* **88**, 856–75.

Trivers, R. L. (1972). Parental investment and sexual selection. In *Sexual selection and the descent of man* (ed. B. Campbell), pp. 136–79. Heinemann Educational Books, London.

Trost, C. H. (1972). Adaptations of Horned Larks (*Eremophila alpestris*) to hot environments. *Auk* **89**, 506–27.

Turek, F. W. (1972). Circadian involvement in termination of the refractory period in two sparrows. *Science, N.Y.* **178**, 1112–3.

Tveter, K. J., Unhjem, O., Attramadal, A., Aakvaag, A., and Hansson, V. (1971). Androgenic receptors in rat and human prostrate. III. Progesterone receptors. In *Advances in the biosciences.* Vol. 7: *Schering Workshop on steroid hormone 'receptors', 1970* (ed. G. Raspé). Pergamon Press, Oxford.

Tyler, C. (1969). The snapping strength of the egg shells of various orders of birds. *J. Zool.* **159**, 65–77.

Ueck, M. (1970). Untersuchungen zur Feinstruktur und Innervation des Pinealorgans von *Passer domesticus* L. *Z. Zellforsch. mikrosk. Anat.* **105**, 276–302.

Uemura, H. (1964a). Cholinesterase in the hypothalamo-hypophysial neurosecretory system of the bird *Zosterops palpebrosa japonica. Zool. Mag., Tokyo* **73**, 118–26.

—— (1964*b*). Effects of gonadectomy and sex steroids on the acid phosphatase activity of the hypothalamo-hypophyseal system in the bird, *Emberiza rustica latifascia. Endocr. jap.* **11**, 185—203.

—— (1965). Histochemical studies on the distribution of cholinesterase and alkaline phosphatase in the vertebrate neurosecretory system. *Annotnes zool. jap.* **38**, 79—96.

—— and Kobayashi, H. (1963). Effects of prolonged daily photoperiods and estrogen on the hypothalamic neurosecretory system of the passerine bird, *Zosterops palpebrosa japonica. Gen. comp. Endocr.* **3**, 253—64.

Ullrich, H. (1949). Rhythmik und Periodik in individuellen und überindividuellen Leben. *Universum, Wien* **4**, 713—16.

Underwood, H. and Menaker, M. (1970). Photoperiodically significant photoreception in sparrows: is the retina involved? *Science, N.Y.* **167**, 298—301.

Väisänen, R. A. (1969). Evolution of the Ringed Plover (*Charadrius hiaticula* L.) during the last hundred years in Europe. A new computer method based on egg dimensions. *Suomal. Tiedeakat. Toim (Ser. A. Biol.)* no. 149, 1—90.

——, Hildén, O., Soikkeli, M., and Vuolanto, S. (1972). Egg dimension variation in five wader species: the role of heredity. *Ornis fenn.* **49**, 25—44.

Van Albada, M. (1958). Influencia de los periodos de iluminacion sobre la postura en ciclos en la galina domestica. *XI Congreso Mundial de Avicultura, Mexico*, pp. 275—83.

Van Balen, J. H. and Cavé, A. J. (1970). Survival and weight loss of nestling Great Tits, *Parus major,* in relation to brood-size and air temperature. *Neth. J. Zool.* **20**, 464—74.

Van Oordt, G. J. and Junge, G. C. A. (1933). Der Einfluss der kastration bei männlichen Lachmowen (*Larus ridibundus* L.). *Wilhelm Roux Arch. EntwMech. Org.* **128**, 166.

—— and —— (1936). Der Einfluss der Kastration auf männliche Kampfläufer (*Philomachus pugnax*). *Wilhelm Roux Arch. EntwMech. Org.* **134**, 112—21.

Van Tienhoven, A. (1959). Reproduction in the domestic fowl: physiology of the female. In *Reproduction in domestic animals* (1st edn) (ed. H. H. Coles and P. T. Cupps) Vol. 2, pp. 305—42. Academic Press, New York.

—— (1961*a*). The effect of massive doses of corticotrophin and corticocosterone on ovulation of the chicken (*Gallus domesticus*). *Acta Endocr., Copenh.* **38**, 407—12.

—— (1961*b*). Endocrinology of reproduction in birds. In *Sex and internal secretions* (3rd end) (ed. W. C. Young) Vol. 2, pp. 1088—169. Williams and Wilkins, Baltimore.

—— (1968). *Reproductive physiology of vertebrates.* Saunders, Philadelphia.

Varley, G. C. and Gradwell, G. R. (1960). Key factors in population studies. *J. Anim. Ecol.* **29**, 399—401.

—— and —— (1968). Population models for the winter moth. In *Insect abundance* (ed. T. R. E. Southwood), Vol. 4, pp. 132—42. Symposia of the Royal Entomological Society no. 4. Blackwell, Oxford.

Vaughan, R. (1961). *Falco eleonorae. Ibis* **103a**, 114—28.

Vaugien, L. (1955). Influence de l'obscuration temporaire sur la durée de la phase refractaire du cycle sexuel du moineau domestique. *Bull. Biol. Fr. Belg.* **89**, 294—309.

Verner, J. (1964). Evolution of polygamy in the Long-billed Marsh Wren. *Evolution, Lancaster, Pa* **18**, 252—61.

—— and Engelsen, G. H. (1970). Territories, multiple nest building, and polygyny in the Long-billed Marsh Wren. *Auk* **87**, 557—67.

Verwey, J. (1930). Die Paarungsbiologie des Fischreihers. *Zool. Jb. Abt. Allgemeine Zoologie und Physiologie der Tiere* **48**, 1—120.

Vince, M. A. (1969). Embryonic communication, respiration and the synchronisation of hatching. In *Bird vocalizations: their relation to current problems in biology and psychology* (ed. R. A. Hinde), pp. 233–60. Cambridge University Press, London.

— — (1972). Communication between quail embryos and the synchronisation of hatching. *Int. orn. Congr.* 15, 357–62.

Vitums, A., Mikami, S., Oksche, A., and Farner, D. S. (1964). Vascularization of the hypothalamo-hypophysial complex in the White-crowned Sparrow *Zonotrichia leucophrys gambelii*. *Z. Zellforsch. mikrosk. Anat.* 64, 541–69.

Vohs, P. A. and Carr, L. R. (1969). Genetic and population studies of transferrin polymorphism in Ring-necked Pheasants. *Condor* 71, 413–17.

Voitkevich, A. A. (1966). *The feathers and plumage of birds.* Sidgwick and Jackson, London.

Von Frisch, O. (1969). Die Entwicklung des Häherkuckucks (*Clamator glandarius*) im Nest der Wirtsvögel und seine Nachzucht in Gefangenschaft. *Z. Tierpsychol.* 26, 641–50.

Von Haartman, L. (1951). Der Trauerfliegenschnäpper. 2. Populationsprobleme. *Acta zool. fenn.* 56, 1–104.

— — (1954). Der Trauerfliegenschnäpper. 3. Die Nahrungsbiologie. *Acta zool. fenn.* 83, 1–96.

— — (1967). Clutch-size in the Pied Flycatcher. *Int. orn. Congr.* 14, 155–64.

— — (1969). Nest-site and evolution of polygamy in European passerine birds. *Ornis fenn.* 46, 1–12.

Voous, K. H. (1950). The breeding seasons of birds in Indonesia. *Ibis* 92, 279–87.

— — (1960). *Atlas of European birds.* Nelson, London.

Vowles, D. M. and Harwood, D. (1966). The effect of exogenous hormones on aggressive and defensive behaviour in the Ring Dove (*Streptopelia risoria*). *J. Endocr.* 36, 35–51.

— — and Prewitt, E. (1971). Stimulus and response specificity in the habituation of anti-predator behaviour in the Ring Dove (*Streptopelia risoria*). *Anim. Behav.* 19, 80–6.

Wada, M. (1972). Effect of hypothalamic implantation of testosterone on photo-stimulated testicular growth in Japanese Quail *Coturnix coturnix japonica*. *Z. Zellforsch. mikrosk. Anat.* 124, 507–19.

Wahlström, G. (1964). The circadian rhythm in the Canary studied by self-selection of photoperiod. *Acta Soc. Med. upsal.* 69, 241–71.

— — (1965). Experimental modifications of the internal clock in the Canary, studied by self-selection of light and darkness. In *Circadian clocks. Proceedings of the Feldafing Summer School* (ed. J. Aschoff), pp. 324–8. North-Holland, Amsterdam.

— — (1971). The internal clock of the Canary: experiments with self-selection of light and darkness. In *Biochronometry* (ed. M. Menaker), pp. 152–68. National Academy of Sciences, Washington D.C.

Wallgren, H. (1954). Energy metabolism of two species of the genus *Emberiza* as correlated with distribution and migration. *Acta zool. fenn.* 84, 1–110.

Walton, A. (1937). On the eclipse plumage of the Mallard (*Anas platyrhynchos platyrhynchos*). *J. exp. Biol.* 14, 440–7.

Warburg, E. F. (1960). Some taxonomic problems in weedy species. In *The biology of weeds* (ed. J. L. Harper), pp. 43–7. British Ecological Society Symposium no. 1. (1959). Blackwell, Oxford.

Ward, P. (1963). Lipid levels in birds preparing to cross the Sahara. *Ibis* 105, 109–11.

— — (1965a). Seasonal changes in the sex ratio of *Quelea quelea* (Ploceinae). *Ibis* 107, 397–9.

— — (1965b). Feeding ecology of the Black-faced Dioch *Quelea quelea* in Nigeria. *Ibis* 107, 173–214.

—— (1965c). The breeding biology of the Black-faced Dioch *Quelea quelea* in Nigeria. *Ibis* **107**, 326–49.

—— (1966). Distribution, systematics and polymorphism of the African weaver-bird *Quelea quelea*. *Ibis* **108**, 34–40.

—— (1969a). The annual cycle of the Yellow-vented Bulbul *Pycnonotus goiavier* in a humid equatorial environment. *J. Zool.* **157**, 25–45.

—— (1969b). Seasonal and diurnal changes in the fat content of an equatorial bird. *Physiol. Zoöl.* **42**, 85–95.

—— (1971). The migration patterns of *Quelea quelea* in Africa. *Ibis* **113**, 275–97.

—— and Zahavi, A. (1973). The importance of certain assemblages of birds as 'information-centres' for food-finding. *Ibis* **115**, 517 34.

Warham, J. (1962). The biology of the Giant Petrel *Macronectes giganteus*. *Auk* **79**, 139–60.

—— (1972). Breeding seasons and sexual dimorphism in Rockhopper Penguins. *Auk* **89**, 86–105.

Warren, D. C. and Scott, H. M. (1936). Influence of light on ovulation in the fowl. *J. exp. Zool.* **74**, 137–56.

Warren, R. P. and Hinde, R. A. (1959). The effect of oestrogen and progesterone on the nest-building of domesticated Canaries. *Anim. Behav.* **7**, 209–13.

—— and —— (1961a). Roles of the male and the nest-cup in controlling the reproduction of female Canaries. *Anim. Behav.* **9**, 64–7.

—— and —— (1961b). Does the male stimulate oestrogen secretion in female Canaries? *Science, N.Y.* **133**, 1354–5.

Warren, S. P. (1968). Primary catecholamine fibres in the ventral hypothalamus of the White-crowned Sparrow *Zonotrichia leucophrys gambelii*. Master's Thesis, University of Washington.

Watson, A. (1964). Aggression and population regulation in Red Grouse. *Nature, Lond.* **202**, 506–7.

—— (1967). Population control by territorial behaviour in Red Grouse. *Nature, Lond.* **215**, 1274–5.

—— (1970). Territorial and reproductive behaviour of Red Grouse. *J. Reprod. Fert. Suppl.* **11**, 3–14.

—— (1971). Animal population ecology. *Sci. Prog., Oxf.* **59**, 451–74.

—— (1973). Moults of wild Scottish Ptarmigan, *Lagopus mutus*, in relation to sex, climate and status. *J. Zool.* **171**, 207–23.

—— and Miller, G. R. (1971). Territory size and aggression in a fluctuating Red Grouse population. *J. Anim. Ecol.* **40**, 367–83.

—— and Moss, R. (1970). Dominance, spacing behaviour and aggression in relation to population limitation in vertebrates. In *Animal populations in relation to their food resources* (ed. A. Watson), pp. 167–218. British Ecological Society Symposium no. 10 (1969). Blackwell, Oxford.

—— and —— (1972). A current model of population dynamics in Red Grouse. *Int. orn. Congr.* **15**, 134–49.

Weddington, S. C., Hansson, V., Ritzén, E. M., Hagenas, L., French, F. S., and Nayfeh, S. N. (1975). Sertoli cell secretory function after hypophysectomy. *Nature, Lond.* **254**, 145–6.

Weidmann, U. (1956). Observations and experiments on egg-laying in the Black-headed Gull (*Larus ridibundus* L.). *Br. J. Anim. Behav.* **4**, 150–61.

Weiner, N. (1969). The regulation of adrenergic neurotransmitter metabolism. In *Progress in endocrinology. Proceedings of the Third International Congress of Endocrinology, Mexico 1968* (ed. C. Gual and F. J. G. Ebling), pp. 294–301. Excerpta Medica, Amsterdam.

Weise, C. M. (1962). Migratory and gonadal responses of birds on long-continued short day-lengths. *Auk* **79**, 161–72.

—— (1967). Castration and spring migration in the White-throated Sparrow. *Condor* **69**, 49–68.

Weller, M. W. (1959). Parasitic egg laying in the Redhead (*Aythya americana*) and other North American Anatidae. *Ecol. Monogr.* **29**, 333–65.

—— (1968*a*). Notes on some Argentine Anatids. *Wilson Bull.* **80**, 189–212.

—— (1968*b*). The breeding biology of the parasitic Black-headed Duck. *Living Bird* **7**, 169–207.

Wendland, V. (1958). Zum Problem des vorzeitigen Sterbens von jungen Greifvögeln und Eulen. *Vogelwarte* **19**, 186–91.

Wever, R. (1960). Possibilities of phase-control, demonstrated by an electronic model. *Cold Spring Harb. Symp. quant. Biol.* **25**, 197–206.

—— (1964). Zum Mechanismus der 24-stunden-Periodik. III. Mitteilung Anwendung der Modell Gleichung. *Kybernetik* **2**, 127–44.

—— (1965). A mathematical model for circadian rhythms. In *Circadian clocks. Proceedings of the Feldafing Summer School* (ed. J. Aschoff), pp. 47–63. North-Holland, Amsterdam.

White, S. J. (1975*a*). Effects of stimuli emanating from the nest on the reproductive cycle in the Ring Dove. II. Building during the pre-laying period. *Anim. Behav.* **23**, 869–82.

—— (1975*b*). Effects of stimuli emanating from the nest on the reproductive cycle in the Ring Dove. III. Building in the post-laying period and effects on the success of the cycle. *Anim. Behav.* **23**, 883–8.

—— and Hinde, R. A. (1968). Temporal relations of brood patch development, nest-building and egg-laying in domesticated Canaries. *J. Zool.* **155**, 145–55.

Wieselthier, A. S. and van Tienhoven, A. (1972). The effect of thyroidectomy on testicular size and on the photorefractory period in the Starling, *Sturnus vulgaris*. *J. exp. Zool.* **179**, 331–8.

Wight, P. A. L. (1971). The pineal gland. In *Physiology and biochemistry of the domestic fowl* (ed. D. J. Bell and B. M. Freeman) Vol. 1, pp. 549–73. Academic Press, New York.

—— and MacKenzie, G. M. (1970). Dual innervation of the pineal of the fowl, *Gallus domesticus. Nature, Lond.* **228**, 474–5.

—— and —— (1971). The histochemistry of the pineal gland of the domestic fowl. *J. Anat.* **108**, 261–73.

Wilkie, D. R. (1959). The work output of animals: flight by birds and by manpower. *Nature, Lond.* **183**, 1515–6.

Williams, G. C. (1966). Natural selection, the costs of reproduction, and a refinement of Lack's principle. *Am. Nat.* **100**, 687–90.

Williams, L. (1952). Breeding behaviour of the Brewer Blackbird. *Condor* **54**, 3–47.

Williams, M. and Roderick, C. (1973). The breeding performance of Grey Duck *Anas superciliosa*, Mallard *Anas platyrhynchos* and their hybrids in captivity. *Int. Zoo Yb.* **13**, 62–9.

Wilson, F. E. (1967). The tubero-infundibular neuron system: a component of the photoperiodic control mechanism of the White-crowned Sparrow, *Zonotrichia leucophrys gambelii. Z. Zellforsch. mikrosk. Anat.* **82**, 1–24.

—— (1968). Testicular growth in Harris' Sparrow (*Zonotrichia querula*). *Auk* **85**, 410–15.

—— and Follett, B. K. (1974). Plasma and pituitary luteinizing hormone in intact and castrated Tree Sparrows (*Spizella arborea*) during a photoinduced gonadal cycle. *Gen. comp. Endocrinol.* **23**, 82–93.

—— and Hands, G. R. (1968). Hypothalamic neurosecretion and photoinduced testicular growth in the Tree Sparrow, *Spizella arborea*. *Z. Zellforsch. mikrosk. Anat.* **89**, 303–19.

Wilson, S. C. and Sharp, P. J. (1973). Variations in plasma LH levels during the ovulatory cycle of the hen, *Galllus domesticus*. *J. Reprod. Fert.* **35**, 561–4.

—— and —— (1975). Episodic reelase of luteinizing hormone in the domestic fowl. *J. Endocr.* **64**, 77–86.

Wilson, W. O., Woodward, A. E., and Abplanalp, H. (1964). Exogenous regulation of oviposition in chickens. *Poult. Sci.* **43**, 1187–92.

Winget, C. M., Averkin, E. G., and Fryer, T. B. (1965). Quantitative measurement by telemetry of ovulation and oviposition in the fowl. *Am. J. Physiol.* **209**, 853–8.

Wingstrand, K. G. (1951). *Structure and development of avian pituitary*. Gleerup, Lund.

Winkel, W. (1970). Experimentelle Untersuchungen zur Brutbiologie von Kohl-und Blaumeise (*Parus major* and *P. caeruleus*). *J. Orn., Lpz.* **111**, 154–74.

Winn, H. S. (1950). Effects of different photoperiods on body weight, fat deposition, molt and male gonadal growth in the Slate-colored Junco. Ph.D. Dissertation, Northwestern University, Evanston, Illinois, U.S.A.

Witschi, E. (1935). The origin of asymmetry in the reproductive system of birds. *Am. J. Anat.* **56**, 119–41.

—— (1940). The quantitative determination of follicle stimulating and luteinizing hormones in mammalian pituitaries and a discussion of the gonadotropic quotient, F/L. *Endocrinology* **27**, 437–46.

—— (1945). Quantitative sutdies on the seasonal development of the deferent ducts in passerine birds. *J. exp. Zool.* **100**, 549–64.

—— (1955). Vertebrate gonadotrophins. *Mem. Soc. Endocr.* **4**, 149–64.

—— (1961). Sex and secondary sexual characters. In *Biology and comparative physiology of birds* (ed. A. J. Marshall), Vol. 2, pp. 115–68. Academic Press, New York.

—— and Fugo, N. W. (1940). Response of sex characters of the adult female Starling to synthetic hormones. *Proc. Soc. exp. Biol. Med.* **45**, 10–14.

—— and Miller, R. A. (1938). Ambisexuality in the female Starling. *J. exp. Zool.* **79**, 475–87.

Wolf, L. L. (1969). Breeding and molting periods in a Costa Rican population of the Andean Sparrow. *Condor* **71**, 212–19.

Wolfson, A. (1942). Regulation of spring migration in juncos. *Condor* **44**, 237–63.

—— (1952). The occurrence and regulation of the refractory period in the gonadal and fat cycles of the Junco. *J. exp. Zool.* **121**, 311–25.

—— (1954). Sperm storage at lower-than-body temperature outside the body cavity in some passerine birds. *Science, N.Y.* **120**, 68–71.

—— (1955). Absence of a refractory period in the gonadal cycle of juncos exposed to 20-hour photoperiods. *Anat. Rec.* **122**, 454–5.

—— (1959). Role of light and darkness in the regulation of spring migration and reproductive cycles in birds. In *Photoperiodism and related phemonema in plants and animals* (ed. R. B. Withrow), pp. 679–716. American Association for the Advancement of Science Publ. no. 55. Washington D.C.

—— (1966). Environmental and neuroendocrine regulation of annual gonadal cycles and migratory behaviour in birds. *Recent Prog. Horm. Res.* **22**, 177–244.

—— and Winn, H. S. (1948). Summation of day lengths as the external stimulus for photoperiodic responses in birds. *Anat. Rec.* **101**, 70–1.

Wood-Gush, D. G. M. and Gilbert, A. B. (1969). Oestrogen and the pre-laying behaviour of the domestic hen. *Anim. Behav.* **17**, 586–9.

—— and —— (1970). The rate of egg loss through internal laying. *Br. Poult. Sci.* **11**, 161–3.

Woods, J. E. and Domm, L. V. (1966). A histochemical identification of the androgen producing cells in the gonads of the domestic fowl and albino rat. *Gen. comp. Endocrinol.* **7**, 559–70.

Woolfenden, G. E. (1976). Co-operative breeding in American birds. *Int. orn. Congr* **16**, 674–84.

Wright, A., Phillips, J. G., and Huang, D. P. (1966). The effect of adenohypophysectomy on the extra-renal and renal excretion of the saline-loaded duck (*Anas platyrhynchos*). *J. Endocr.* **36**, 249–56.

Wright, P., Caryl, P. G., and Vowles, D. M. (eds) (1975). *Neural and endocrine aspects of behaviour in birds.* Elsevier, Amsterdam.

Wright, P. L. and Wright, M. H. (1944). The reproductive cycle of the male Redwinged Blackbird. *Condor* **46**, 46–59.

Wurdinger, I. (1975). Vergleichend morphologische Untersuchungen zur Jugendentwicklung von *Anser-* und *Branta-*Arten. *J. Orn., Lpz.* **116**, 65–86.

Wurtman, R. J. (1969). Control of the mammalian pineal by light and sympathetic nerves. In *Progress in endocrinology. Proceedings of the Third International Congress of Endocrinology, Mexico 1968* (ed. C. Gual and F. J. G. Ebling), pp. 627–30. Excerpta Medica, Amsterdam.

—— and Axelrod, J. (1965). The pineal gland. *Scient. Am.* **213**, 50–60.

——, ——, and Kelly, D. E. (1968). *The pineal.* Academic Press, New York.

Wyburn, G. M. and Baillie, A. H. (1966). Some observations on the fine structure and histochemistry of the ovarian follicle of the fowl. In *Physiology of the domestic fowl* (ed. C. Horton-Smith and E. C. Amoroso), pp. 30–8. Oliver and Boyd, Edinburgh.

——, Johnston, H. S., and Aitken, R. N. C. (1965). Specialised plasma membranes in the preovulatory follicle of the fowl. *Z. Zellforsch. mikrosk. Anat.* **68**, 70–9.

——, ——, and —— (1966). Fate of the granulosa cells in the hen's follicle. *Z. Zellforsch. mikrosk. Anat.* **72**, 53–65.

Wydoski, R. S. (1964). Seasonal changes in the color of Starling bills. *Auk* **81**, 542–50

Wynne-Edwards, V. C. (1955). Low reproductive rates in birds, especially sea-birds. *Int. orn. Congr.* **11**, 540–7.

—— (1962). *Animal dispersion in relation to social behaviour.* Oliver and Boyd, Edinburgh.

Yamashima, Y. (1952). Notes on experimental brooding induced by prolactin injections in the domestic cock. *Annotnes zool. jap.* **25**, 135–40.

Yates, F. E. and Brown-Grant, K. (1969). A new look at classical endocrine feedback loops. In *Progress in endocrinology. Proceedings of the Third International Congress of Endocrinology, Mexico 1968* (ed. C. Gual and F. J. G. Ebling), pp. 515–22. Excerpta Medica, Amsterdam.

Young, J. G. (1972). Breeding biology of feral Greylag Geese in south-west Scotland. *Rep. Wildfowl Trust* **23**, 83–7.

Yu, J. Y-L. and Marquardt, R. R. (1973). Development, cellular growth, and function of the avian oviduct. *Biol. Reprod.* **8**, 283–98.

Zahavi, A. (1971a). The function of pre-roost gatherings and communal roosts. *Ibis* **113**, 106–9.

—— (1971b). The social behaviour of the White Wagtail *Motacilla alba alba* wintering in Israel. *Ibis* **113**, 203–11.

—— (1974). Communal nesting by the Arabian Babbler. A case of individual selection. *Ibis* **116**, 84–7.

—— (1976). Co-operative nesting in Eurasian birds. *Int. orn. Congr.* **16**, 685–93.

Zeigler, H. P., Green, H. L., and Lehrer, R. (1971). Patterns of feeding behaviour in the pigeon. *J. comp. physiol. Psychol.* **76**, 468–77.

Zigmond, R. E. (1975). Target cells for gonadal steroids in the brain: studies on hormone binding and metabolism. In *Neural and endocrine aspects of behaviour in birds* (ed. P. Wright, P. G. Caryl, and D. M. Vowles), pp. 111–21. Elsevier, Amsterdam.

——, Stern, J. M., and McEwen, B. S. (1972). Retention of radioactivity in cell nuclei in the hypothalamus of the Ring Dove after injection of ³H-testosterone. *Gen. comp. Endocrinol.* **18**, 450–3.

Zimmerman, J. L. (1966). Effects of extended tropical photoperiod and temperature on the Dickcissel. *Condor* **68**, 377–87.

—— and Morrison, J. V. (1972). Vernal testes development in tropical-wintering Dickcissels. *Wilson Bull.* **84**, 475–81.

Zinkham, W. H., Blanco, A., and Kupchyk, L. (1964). Lactate dehydrogenase in pigeon testes: genetic control by three loci. *Science, N.Y.* **144**, 1353–4.

—— and Isensee, H. (1972). Genetic control of lactate dehydrogenase synthesis in the somatic and gametic tissue of pigeons. *Johns Hopkins med. Jnl* **130**, 11–25.

——, Kupchyk, L., Blanco, A., and Isensee, H. (1966). A variant of lactate dehydrogenase in somatic tissues of pigeons: physicochemical properties and genetic control. *J. exp. Zool.* **162**, 45–56.

Additional references

Bicknell, R. J. and Follett, B. K. (1975). A quantitative assay for luteinizing hormone releasing hormone (LHRH) using dispersed pituitary cells. *Gen. Comp. Endocrinol.* **26**, 141–52.

Bolton, N. J., Chadwick, A., Hall, T. R., and Scanes, C. G. (1976). Effect of chicken and rat hypothalamic extracts on prolactin secretion in the chicken. *IRCS med. Sci., Biochem.* **4**, 495.

Cheng, Mei-Fang and Follett, B.K. (1976). Plasma luteinizing hormone during the breeding cycle of the female Ring Dove. *Horm. & Behav.* **7**, 199–205.

Daan, S. and Pittendrigh, C. S. (1976a). A functional analysis of circadian pacemakers in nocturnal rodents. II. The variability of phase response curves. *J. comp. Physiol.* **106**, 253–66.

Daan, S. and Pittendrigh, C. S. (1976b). A functional analysis of circadian pacemakers in nocturnal rodents. III. Heavy water and constant light: homeostasis of frequency? *J. comp. Physiol.* **106**, 267–90.

Davies, D. T. (1976). Steroid feedback in the male and female Japanese Quail. *J. Endocr.* **70**, 513–4.

Dusseau, J. W. and Bosscher, J. R. (1976). Adrenal phasing of a diurnal rhythm of testicular responsiveness to FSH in chickens. *Gen. Comp. Endocrinol.* **28**, 255–63.

Gibson, W. R., Follett, B. K., and Gledhill, B. (1975). Plasma levels of luteinizing hormone in gonadectomized Japanese Quail exposed to short or to long daylengths. *J. Endocr.* **64**, 87–101.

Gledhill, B. and Follett, B. K. (1976). Diurnal variation and the episodic release of plasma gonadotrophins in Japanese Quail during a photoperiodically induced gonadal cycle. *J. Endocr.* **71**, 245–57.

Hinde, R. A. and Steel, E. (1976). The effect of male song on an estrogen-dependent behavior pattern in the female Canary (*Serinus canarius*). *Horm. & Behav.* **7**, 293–304.

Hutchison, J. B. (1976). Hormones and brain mechanisms of sexual behaviour: a possible relationship between cellular and behavioural events in doves. In *Perspectives in experimental biology*, Vol. 1: *Zoology* (ed. P. Spencer Davies). Pergamon Press, Oxford and New York.

Moss, R. (1974). Winter diets, gut lengths and interspecific competition in Alaskan Ptarmigan. *Auk* **91**, 737–46.

Pittendrigh, C. S. and Daan, S. (1976*a*). A functional analysis of circadian pacemakers in nocturnal rodents. I. The stability and lability of spontaneous frequency. *J. comp. Physiol.* **106**, 223–52.

Pittendrigh, C. S. and Daan, S. (1976*b*). A functional analysis of circadian pacemakers in nocturnal rodents. IV. Entrainment: pacemaker as clock. *J. comp. Physiol.* **106**, 291–331.

Pittendrigh, C. S. and Daan, S. (1976*c*). A functional analysis of circadian pacemakers in nocturnal rodents. V. Pacemaker structure: a clock for all seasons. *J. comp. Physiol.* **106**, 333–55.

Scanes, C. G., Harvey, S., and Chadwick, A. (1976). Plasma luteinizing hormone and follicle stimulating hormone concentration in fasted immature male chickens. *IRCS med. Sci., Biochem.* **4**, 371.

Sharp, P. J., (1974). A comparison of circulating levels of LH in intact and gonadectomized growing fowl. *J. Endocr.* **61**, viii.

Steel, E. and Hinde, R. A. (1976). Effect of a skeleton photoperiod on the daylength-dependent response to oestrogen in Canaries (*Serinus canarius*). *J. Endocr.* **70**, 247–54.

Storey, C. R. (1976). Circadian photosensitivity in the photoperiodic control of gonadotropin secretion in the Canary, *Serinus canarius*. *Gen. Comp. Endocrinol.* **30**, 198–203.

Wyllie, I. (1975). Study of Cuckoos and Reed Warblers. *Br. Birds* **68**, 369–78.

Species Index

Author Index

Subject Index